Morphogenetic Hormones of Arthropods

Recent Advances in Comparative Arthropod Morphology, Physiology, and Development

Morphogenetic Hormones of Arthropods

Roles in Histogenesis, Organogenesis, and Morphogenesis

edited by A. P. GUPTA

3

RUTGERS
UNIVERSITY PRESS
New Brunswick, New Jersey

Manufactured in the United States of America

Library of Congress Cataloging-in-Publication Data
(Revised for vol. 3)

Morphogenetic hormones of arthropods.

(Recent advances in comparative arthropod morphology, physiology, and development)
Includes bibliographies and indexes.
Contents: pt. 1. Discoveries, syntheses, metabolism, evolution, modes of action, and techniques — pt. 2. Embryonic and postembryonic sources — pt. 3. Roles in histogenesis, organogenesis, and morphogenesis.
1. Arthropoda—Morphogenesis. 2. Juvenile hormones.
3. Hormones. I. Gupta, A. P., 1928–
[DNLM: 1. Arthropods. 2. Hormones. 3. Morphogenesis.
QL 434.7 M871]
QL434.72.M67 1989 595'.2043 89-6029

ISBN 0-8135-1416-9 (pt. 3)

CONTENTS

PREFACE

This is the third and last part of Volume 1 on the morphogenetic hormones of arthropods, part of an international series entitled *Recent Advances in Comparative Arthropod Morphology, Physiology, and Development,* published by Rutgers University Press (future volumes in this series will be published by CRC Press, Boca Raton, Florida). It deals with hormonal regulation of histogenesis, organogenesis, and morphogenesis during embryonic and postembryonic stages. Unlike Parts 1 and 2, herein the authors cover more topics on insects than on other major arthropod groups, primarily because comparable studies in these groups are either unavailable or still scarce. Although the roles of morphogenetic hormones in events such as cuticle deposition, embryonic diapause, metamorphosis, color change, determination of sexual characters, regeneration, and spermatogenesis have been discussed in relation to several arthropod groups, clearly much remains to be done on the hormonal regulation of embryonic and postembryonic events in all groups, including insects.

In any multiauthored treatise some overlaps are inevitable, and this series as a whole is no exception. However, overlaps have been kept to the minimum necessary for clarity of exposition. Wherever relevant, such overlaps and divergences of opinions have been cross-referenced. Because the subject of taxonomic rankings of major arthropod groups is controversial, each contributor has used his/her preferred taxonomic categories. Furthermore, each contributor has had complete freedom to develop, interpret, and present his/her views. Each chapter attempts to present an in-depth review of the topic it covers.

The endeavor of organizing such a series could not have been successful without the cooperation and assistance of all the authors who responded to my invitation to contribute. To them, I am indebted. To numerous authors, journal publishers, and professional societies who generously and unhesitantingly allowed reproduction of published or unpublished materials in the original or modified forms, I am grateful. And I greatly appreciate the cooperation and assistance of Karen Reeds, Marilyn Campbell, Barbara Kopel, and Dina Bednarczyk at Rutgers University Press, and the meticulous copyediting by Bert N. Zelman of Publishers Workshop Inc., Brooklyn, N.Y. As always in the past, the ungrudging support of my wife and children during the preparation of the book has been enormous and has allowed me to devote to this project considerable time that rightfully belonged to them. To them, I am ever so grateful.

Ayodhya P. Gupta
New Brunswick, New Jersey
May 31, 1990

CONTRIBUTORS

Christian Bordereau
Laboratory of Zoology
CNRS 674
6 Gabriel Boulevard
21100 Dijon
France

A. Bouthier
Practical Service of Cellular Biology
1st Cycle B 1st Fl.
University of P. & M. Curie-Paris VI
9 Quay St. Bernard
75252 Paris Cedex
France

Désiré Bullière
Laboratory of Animal Biology & Physiology
University of Antilles and Guyane
97167 Point-a-Pitre
Gaudeloupe (F.W.I.)
West Indies

Françoise Bullière
Laboratory of Animal Biology
UA. 682 USTM Grenoble
BP68
38042 Saint Martin d'Herès Cedex
France

Michel Descamps
Laboratory of Invertebrate Endocrinology
Animal Biology
University of Science and Techniques of Lille
59655 Villeneuve d'Ascq Cedex
France

Thomas E. Mittler
Department of Entomology
201 Wellman Hall
University of California
Berkeley, CA 94720

Michel Mouze
Division of Animal Biology
University of Sciences and Techniques of Lille
59655 Villeneuve d'Ascq Cedex
France

Pierre Y. Noël
B.I.M.M.
Muséum National d'Histoire Naturelle
55 rue de Buffon
75005 Paris
France

Charles Noirot
Laboratory of Zoology
CNRS 674
6 Gabriel Boulevard
21100 Dijon
France

Vladimír J. A. Novák
Laboratory of Evolutionary Biology CSAV
Czechoslovak Academy of Sciences
Na Folimance 11
12000 Praha 2
Czechoslovakia

Luc Passera
Laboratory of Entomology
CNRS 333
Paul-Sabatier University
118 route de Narbonne
31062 Toulouse Cedex
France

Geneviève G. Payen
Laboratory of Reproductive Physiology
Division of Crustacean Endocrinology
University of P. & M. Curie and CNRS
4 Jussieu Place
75252 Paris Cedex 05
France

Heinz Rembold
Max-Planck Institute for Biochemistry
D-8033 Martinsried
West Germany

Franz Romer
Institute of Zoology
Johannes Gutenberg University
Saar Street 21
6500 Mainz
West Germany

Peter-Frank Röseler
Zoology Institute (II)
University of Würzburg
Röntgenring 10
8700 Würzburg
West Germany

Giovanni Sbrenna
Department of Evolutionary Biology
Via L. Borsari 45
University of Ferrara
44100 Ferrara
Italy

Constantine E. Sekeris
National Hellenic Research Foundation
Biological Research Center
48 Vassileos Constantinou Avenue
11835 Athens
Greece

P. Sivasubramanian
Department of Biology
University of New Brunswick
Fredericton, N. B.
Canada

Marinus J. Sommeijer
Laboratory of Comparative Physiology
University of Utrecht
P.O.B. 80.086
3508 TB
The Netherlands

Klaus D. Spindler
Institute for Zoology
University Street 1
University of Dusseldorf
D-4000 Dusseldorf
West Germany

Koichi Suzuki
Faculty of Agriculture
Iwate University
Morioka, Iwate 020
Japan

Jean P. Suzzoni
Laboratory of Entomology
CNRS 333
Paul-Sabatier University
118 route de Narbonne
31062 Toulouse Cedex
France

Hayo H. W. Velthuis
Laboratory of Comparative Physiology
University of Utrecht
P.O.B. 80.086
3508 TB
The Netherlands

Okitsugu Yamashita
Faculty of Agriculture
Nagoya University
Chikusa, Nagoya 464
Japan

Jan Žďárek
Institute of Organic Chemistry & Biochemistry
Insect Chemical Ecology Unit
U Salamounky 41
15800 Praha 5
Czechoslovakia

Morphogenic Hormones of Arthropods

Hormonal Regulation of Embryogenetic Events

Role of the Gradient Factor in Arthropod Morphogenesis

1

VLADIMÍR J. A. NOVÁK

1.1. Introduction

The development from egg to adult animal can be viewed as a close causal interaction of three different processes: chemical differentiation, or *chemogenesis;* differentiation of form, or *morphogenesis;* and gradual origination and accomplishment of individual functions, or *functionogenesis.* Morphogenesis is subdivided into the following gradual stages: egg segmentation, embryogenesis, postembryonic development, gamogenesis, and metagenesis, i.e., adult involution until death of the individual.

Growth of the body as a whole may be either *isometric* (proportionate), i.e., without change in form, or *anisometric* (disproportionate), when different parts of the body have unequal growth. The latter may be either *allometric,* when different parts of the body grow at different rates, or it may be of the *gradient* type, when some parts of the body do not grow at all, with growth elsewhere remaining limited to only certain parts of the body, the so-called growth gradients. Gradient growth is the main basis of morphogenesis. The term has been suggested on the basis of the findings and conclusions that are the subject of this chapter. (See fig. 1.1, p. 8)

The three insect metamorphosis hormones form a perfect tool for the study of insect metamorphosis and morphogenesis in general. This is especially true of juvenile hormone (JH) or any of its analogues. Normal morphogenesis occurs in its absence both during embryogenesis and in the postembryonic (larval) period. In its presence at an effective concentration, any morphogenesis is stopped and isometric (proportionate) growth takes place without any change in form and without any differentiation. This is the postembryonic period of larval growth in holometabolous insects, and it can also be induced by JH in any stage of metamorphosis as well as of embryogenesis since its beginning (Novák, 1969). This activity of the hormone is limited, however, to a certain period for each instar, i.e., before the start of histological decomposition of the corresponding tissue, as was clearly shown by Wigglesworth (1936); he called it the *critical period* in his first papers on JH. Thus JH can stop morphogenesis at any moment during its course. That is why Williams (1952) called JH the "status quo hormone."

If the growth of any insect tissue can be preserved and the degeneration of any structure prevented by a single chemical such as JH, then it may be assumed that the loss of a similar substance in the corresponding tissue has been the cause of its loss of ability for further growth and thus of the corresponding change in form. Morphogenesis can thus be viewed as an inherent consequence of loss of the ability for further growth of the corresponding parts of the body in a species-specific way. From this deduction arose the concept that the gradual loss of growth ability is caused by the regular inactivation of an indispensable growth factor contained in the whole insect body and originating in the zygote, i.e., at the very start of ontogenesis. Because the gradual inactivation of this factor is presumed to be the cause of the gradual change of form that we call morphogenesis, hence of the gradually appearing growth gradients in the body, it has been called gradient factor (GF) (Novák, 1951a,b, 1975a, etc.).

The hypothetical GF and its gradual inactivation can thus be viewed as the main mechanism of morphogenesis. This idea has proved to be very fruitful, first, for a full understanding of the principle of insect morphogenesis, as shown by the GF theory of insect metamorphosis (Novák, 1951b, 1956, 1975a), later, it led to a series of generalizations and conclusions that provided a better understanding and further analysis of morphogenesis, in general, and contributed to the explanation of this basic biological phenomenon. In this way, the general GF theory of morphogenesis (Novák, 1955, 1959, 1975a) was formulated. On this basis, some pathological deviations of form, such as tumor growth (blastomogenesis) and its most often pernicious case, the malignoses (oncogenesis) (the GF theory of blastomogenesis, Novák, 1959), can be better understood. The GF theory also enables one to understand the many specific deviations of structure that are especially complicated in the ontogeny of many arthropods.

This will be the subject of the present chapter. But before we discuss them, let us consider first the experimental evidence in favor of the GF theory of insect metamorphosis.

1.1.1. *Experimental Evidence Supporting the GF Hypothesis*

The first conclusion regarding the existence of the gradient factor and its role in the insect morphogenesis (Novák, 1951a,b, 1975a) was based on the experimental work on insect growth hormones that I conducted during my 10-month stay in the laboratory of Prof. V. B. Wigglesworth at the University of Cambridge, England, in 1949. It was corroborated by my attempt to synthesize the various data on this subject, which seemed rather contradictory at that time, and was stimulated by my experimental evidence of the nonspecificity of the hormones in insects (Novák, 1951a,b, 1975a). The three metamorphosis hormones proved to be excellent tools for experimental analysis of insect morphogenesis in that they allowed experimental interference with any morphogenesis phenomenon during any stage of ontogenesis.

As already mentioned, all three metamorphosis hormones play an important role in insect morphogenesis either by directly or indirectly promoting or suppressing it.

1.1.2. *Activation Hormone (AH)*

The primary role in this respect belongs to the activation hormone (AH), or brain hormone (Kopeć, 1917, 1922; Wigglesworth, 1934, 1940a,b), a neurohormone produced by the median neurosecretory cells of the pars cerebralis protocerebri. Its action is necessary to induce or reinduce development after its interruption, e.g., at the beginning of each larval instar. Its temporary absence is most often the cause of diapause, which is an adaptation to avoid various adverse conditions during development. Diapause takes place during various periods of insect development, since early embryogenesis till the end of metamorphosis, as well as during various stages of gametogenesis. The action of the AH is necessary to start the secretion of the other two metamorphosis hormones.

1.1.3. *Molting Hormone (MH)*

The molting hormone (MH) (Wigglesworth, 1934; Piepho, 1938), or ecdysone (Butenandt and Karlson, 1954), is equally necessary for growth process in all arthropods. The hormonal effect of each molting process is indispensable, because it ensures the synchronization of molting on the entire body surface, including the cuticular coverings of all internal surfaces of ectodermal origin such as those of the stomodeum and proctodeum or the tracheae. And without the temporary shedding of the cuticle, no more extensive growth—and first of all no morphogenesis—is possible. On the basis of individual molting processes, arthropod postembryonic morphogenesis is subdivided into a series of more or less independent growth processes, each of which has its own hormonal activation system. This is indeed helpful for detailed analysis.

The molting process in arthropods is so closely connected with morphogenesis and growth, especially of the epidermal cells, that it may be viewed as an inevitable component thereof. Each molting process of postembryonic development in arthropods coincides and is closely interconnected with a corresponding section of development. In each instar, all the main phases of the developmental processes repeat themselves in the same order, including all their hormonal stimulations (see Section 1.2.6, below).

1.1.4. *Juvenile Hormone (JH)*

The most important and evocative experimental evidence has been produced with JH, or neotenine, produced by the corpora allata (CA) and the various juvenoids. As shown in the first series of papers by Wigglesworth (1935, 1936, 1940a,b), JH is a true morphogenetic hormone, interfering with the very principle of morphogenesis. As mentioned above, its effective concentration prevents any morphogenesis by converting anisometric (allometric or gradient) growth into isometric (proportionate) growth, thus preventing morphogenesis without stopping growth.

The first experiments by Wigglesworth (1936, 1940a,b) with parabiosis and CA transplantations in *Rhodnius prolixus* clearly showed the true character of JH and the principle of its mode of action. In his classical parabiosis experiments, he demonstrated conclusively that joining a nymph of the penultimate or earlier instar with a nymph at the beginning of the last larval instar (up to a certain time after bloodsucking) resulted in a partial or total suppression of metamorphosis without preventing an increase in size. In this way, supernumerary sixth-instar giant nymphs were obtained. Conversely, when the CA of a penultimate (fourth-) or younger instar nymph was removed by decapitation, a small precocious adult resulted. The hormonal character of this CA action was convincingly shown by another experiment in which the same result was obtained when the corresponding specimens were connected by a glass tube so that only the hemolymph communicated.

In the same papers, Wigglesworth (1935, 1936) also demonstrated the effect of the CA hormone on ovarian development. The ovaries of adult females whose CA was removed at a certain time after bloodsucking did not produce mature eggs. In later experiments, Wigglesworth (1940b, 1952) showed that the same parabiosis-like effect with a younger nymph was obtained by implantation of its CA. Another important finding was that the occurrence of mitoses in the epidermis of the dorsal surface of abdomen in *Rhodnius* was different in the presence of the active JH concentration—the mitoses occurring regularly on the whole surface of the abdomen—whereas in its absence, or also before its active concentration had been reached, and during metamorphosis, it occurred only in the areas characteristic for the future adult structure (Fig. 1.1).

1.1.5. *Morphogenetic Analysis*

A detailed analysis of the morphogenetic effects of JH was performed in the lygaeid bug *Oncopeltus fasciatus* (Novák, 1951b, 1956, 1975a). This species has proved very suitable for the study of morphogenesis because of its prominent black pattern, which in the last-instar nymph is very different from the adult pattern: a cuticular pigment of melanin type occurs exclusively in the imaginal parts of the nymphal body, which can be traced with the pigment so perfectly that the melanin can be viewed as a kind of chemical indicator of the imaginal parts of the cuticle (Fig. 1.2). It therefore permits a detailed analysis of what happens to both the nymphal and imaginal structures during the metamorphosis period of morphogenesis. This species is more suitable for this study because metamorphosis in Hemiptera is, in many respects, of a transitional type between the incomplete metamorphosis in Exopterygota

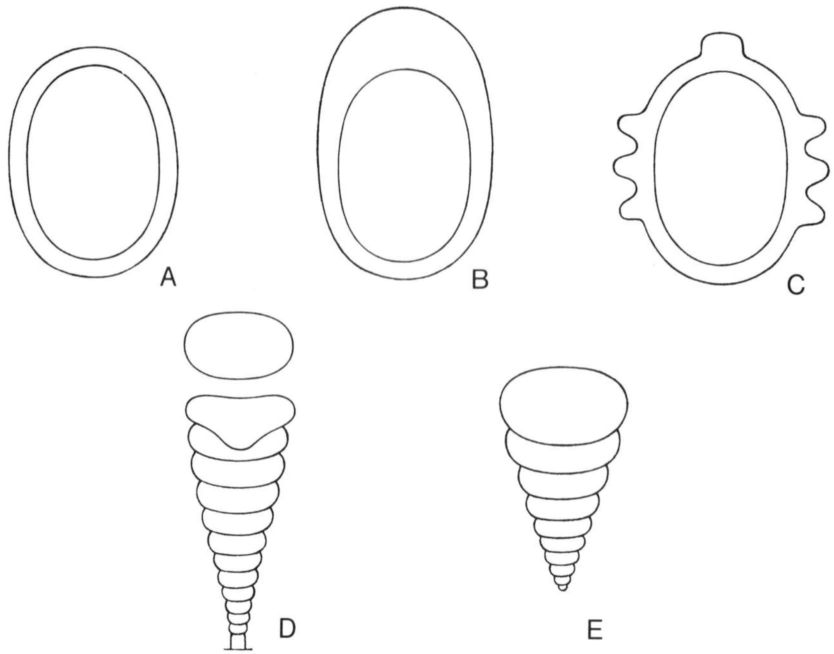

FIGURE 1.1. The main types of growth: (A) isometric (harmonic) growth; (B,C) anisometric growth—B = allometric growth, C = gradient growth; (D) vegetative reproduction of the type of strobilation (Scyphomedusae); (E) metameric growth on the basis of strobilation.

(Hemimetabola) and the complete metamorphosis in Endopterygota (Holometabola), the difference between the two being mainly of degree and not of quality, at least from the physiological point of view, as suggested by Wigglesworth (1939). The results obtained allow the generalization valid for both incomplete and complete metamorphosis. The basis for this conclusion was a complete series of transitions between the last-instar nymph and adult obtained after implanting the active CA into the fifth instar in various periods after molting. In terms of the GF concept it may be summarized as follows:

In the presence of JH in an active concentration, the entire body grows at the same rate without any change in form. This happens during the first four nymphal instars. The slight increase in the imaginal parts of the body (such as the nymphal wing pads distinguished by the black pigment) is fully understood as due to the short periods in the beginning of each instar before the active concentration of JH has been reached. This was clearly shown by Wigglesworth's (1940a) histological experiments and also by his parabiosis experiments in which joining a

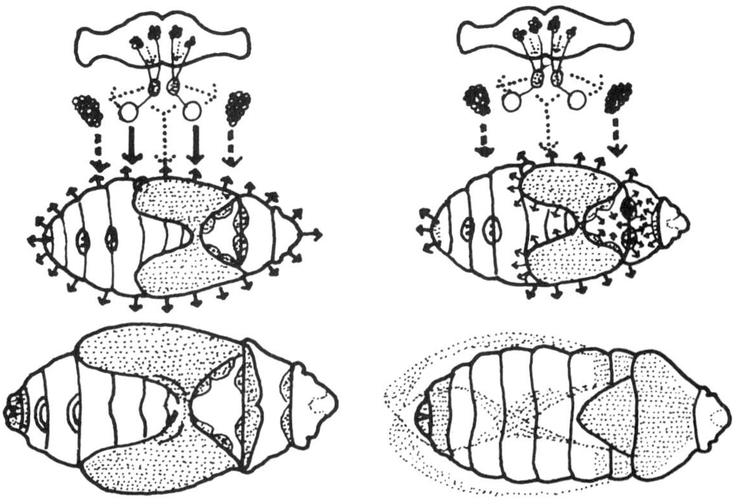

FIGURE 1.2. Diagram of the mode of action of the metamorphosis hormones according to the gradient factor theory. Dotted area = structures containing the GF; short arrows = growing parts of the body; left-hand side, with JH; right-hand side, without JH. (After Novák, 1956.)

penultimate instar nymph with a high JH concentration to a last-instar nymph, immediately after molting, resulted (unlike after active CA transplantation) in a specimen without any increase of the imaginal parts (wing pads) (Wigglesworth, 1936). It could thus be better described as a second fifth-instar nymph than a sixth instar. The examination of the shape of pronotum and distribution of the black pigmentation of such a series of transitory forms between the last-instar nymph and imago in *Oncopeltus* allows the following conclusions:

(1) The nymphal parts of the body can only grow in the presence of an active JH concentration.

(2) The growth of the imaginal parts takes place irrespective of the presence of JH. In the absence of JH, they increase in comparison with nymphal (= larval) parts.

(3) In the presence of JH, the imaginal parts grow at the same rate as the body as a whole, thus resulting in isometric growth.

(4) The growth of the imaginal parts is inversely proportional to the growth of the nymphal parts—the more the nymphal parts grow, the less is the increase of the imaginal parts (cf. the shape and pigmentation of the pronotum in Fig. 1.3).

(5) The nymphal parts may partly or fully degenerate during a longer

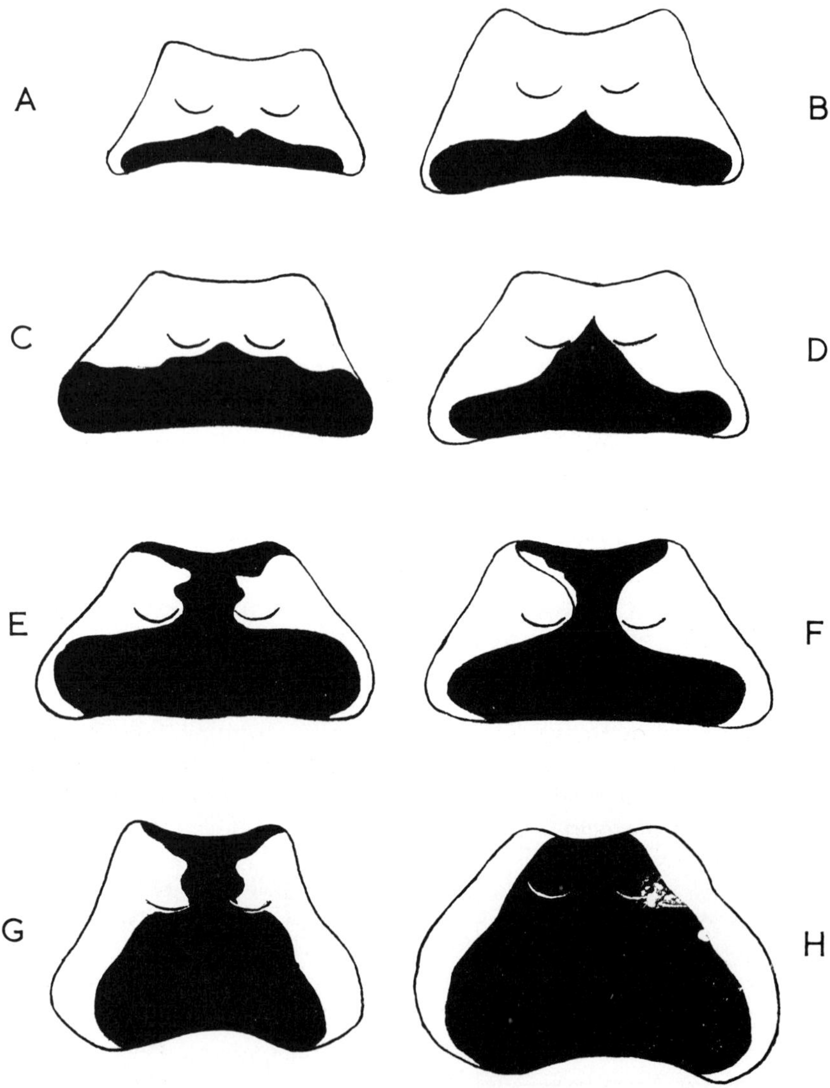

FIGURE 1.3. The development of the shape of pronotum and its black pattern in the lygaeid bug *Oncopeltus fasciatus:* (A) A normal fifth(last)-instar nymph; (B) supernumerary giant, sixth-instar nymph (result of corpus allatum implantation before the onset of adult differentiation); (C–G) transitory forms between nymph and adult (result of corpus allatum implantation during adult differentiation); (H) normal adult. (From Novák, 1951b.)

absence of JH, as they do in the course of complete metamorphosis in Holometabola. Their further growth is reversible only during a certain period, up to a particular moment. What has been observed in the pronotum of *Oncopeltus* is also true of any other surface structure of *Oncopeltus* (e.g., other sclerites, genital appendages). It is equally so for any other surface structure of any other insect where it was seriously investigated (Sehnal and Novák, 1969). And the same was also found in various internal organs, e.g., in *Galleria* (Novák, 1967; Sehnal, 1965, 1968). The foregoing conclusions may thus be viewed as well demonstrated in any structure of insects, both by observation in normal development and experimentally. (See figs. 1.2, 1.3, 1.4)

1.2. The Gradient Factor (GF) Hypothesis

To understand why the larval structures fail to grow in the absence of JH, two observations should be taken into account: First, the larval parts do not lack the ability to grow from the onset of ontogenesis. This is apparent because during the embryonic period they differentiate and grow to their definitive shape (in newly hatched larva) in relation to the other parts of the body, although no JH is supposed to be acting at that time. Secondly, from the late postembryonic period onward, they continue to grow without further differentiation but only as long as JH is present. As soon as the JH concentration in hemolymph fails to reach the active level, their growth is stopped and they gradually disintegrate, being replaced by the growing imaginal parts. In normal development, this occurs during metamorphosis, but this state may be induced experimentally from the beginning of the larval period (first larval instar). In such a case, a precocious metamorphosis is caused and a miniature adult insect results.

The fact that the ability of the larval structures to grow is completely restored by supplying them a single substance, the JH, suggests that their loss of this ability toward the end of embryogenesis has been caused by the loss or inactivation of just one chemical substance also, a tissue-bound factor, whereas it continues to be contained in the imaginal parts of the body in all other structures, and all other tissues are able to grow (and differentiate) in the absence of JH. No definite experimental data are available on the chemical character of this factor so far. Its assumption, however, has been very helpful—as we shall see in this chapter—for a consistent analysis and understanding of insect metamorphosis. As noted earlier, because this hypothetical factor is contained in all growth gradients, it has been called GF (Novák, 1951a,b, 1956, 1975).

Even if no data on the chemical character of GF are available so far, a number of conclusions can be drawn as to its properties and mode of action and three levels of morphogenesis can be recognized: (1) GF is a factor conditioning any growth (in the absence of JH). From this it follows that it is present in the body from the very beginning of ontogenesis, in the fertilized egg (zygote), and is only gradually lost in a species-specific pattern in the course of further morphogenesis), starting with the differentiation of the yolk and the animal pole from which the future embryo develops. This is the macroscopic form of differentiation or morphogenesis. (2) Such macroscopic differentiation is based on differentiation at the cellular level and closely interconnected with it. It is realized by cells that either retain or lose active GF and, in consequence of this, either continue to grow and reproduce (divide) or cease to grow and eventually disintegrate. It may concern either cells as wholes or just their parts. This, then, is the microscopic form of differentiation. Besides the two types of cells, there is a third type, where each cell behaves as the body does; that is, some of its parts contain GF, whereas others do not. Typical of this case are the epidermal cells of hemimetabolous insects (Exopterygota). They survive but change their form in metamorphosis (Novák, 1975a). (3) Another case, again closely interconnected and in causal relation with the preceding two, is molecular differentiation. It primarily concerns the DNA molecules, which may equally either (a) continue to divide or (b) die with the cell and disintegrate, and with them also the corresponding protein and other molecules.

1.2.1. *Macroscopic Morphogenesis*

Experiments with denucleated egg cells reimplanted with the nuclei of cells from later ontogenetic stages in frogs have shown that all ontogenetic differentiation is due to action of the environment, both internal and external, without any hereditary change [NPC = nonhereditary phylogenetic changes (Novák, 1981, 1985a,b)]; this means that most, if not all, cells of the animal body remain genetically identical, as in plants during their whole ontogenesis (Dubinin, 1970). It has been known that insect eggs are centrolecithal, i.e., some of the first cells after fertilization, the so-called yolk nuclei, move to the yolk surface, where by rapid cell divisions they form a continuous surface layer, the monolayered blastoderm, whereas further growth and division of those that remain inside the yolk (the yolk nuclei) are practically stopped. This may be viewed as the beginning of the GF inactivation process.

In later stages of embryogenesis, the process of gradual GF inactivation continues; we observe the same process during the whole of morphogenesis. Some parts of the body continue to grow, whereas growth of other parts ceases and they are gradually more or less decomposed (Fig. 1.4). An example is formation of the dorsal germ band and its gradual transformation into the embryo, followed by its segmentation (metamerization), which has been shown to be the process of strobilation

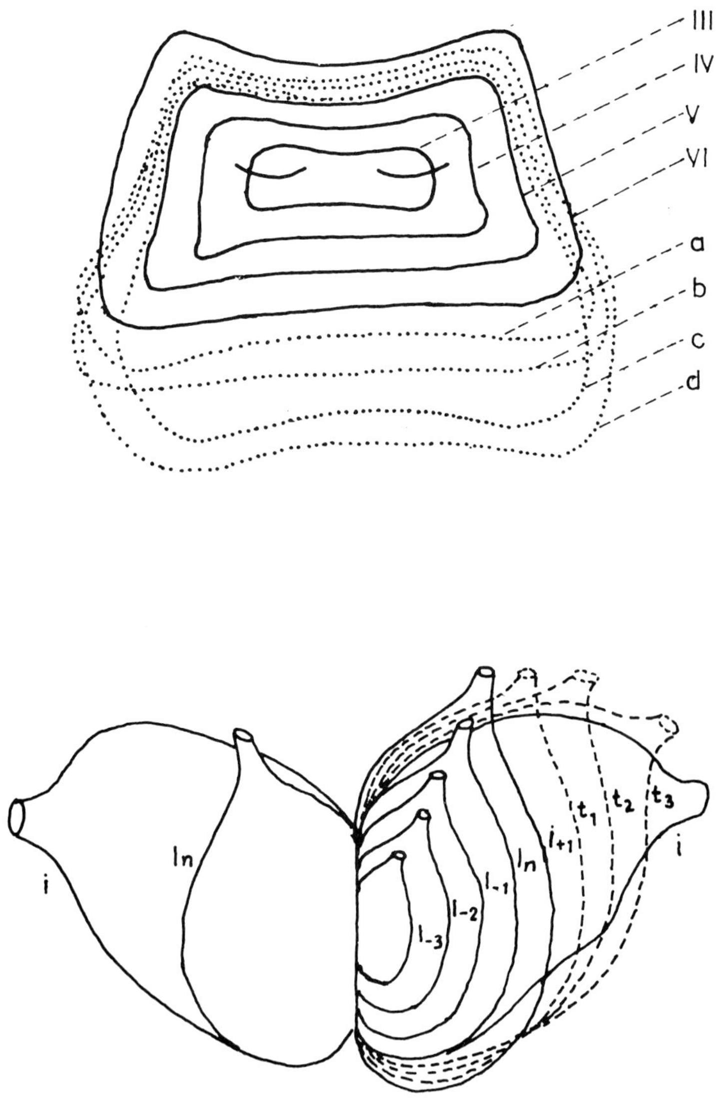

FIGURE 1.4. The effect of JH on the shape of insect body in postembryonic development: (*above*) effect on the shape of the pronotum in *Oncopeltus fasciatus* from third- to sixth (supernumerary)-instar nymph to adult (d) and three transitory forms between the last instar nymph and adult (a,b,c); (*below*) effect on the shape of the cerebral ganglion of *Galleria mellonella* in three penultimate larval instars (l_{-1}, l_{-2}, l_{-3}), the last larval instar (l_n) and in three supernumerary larvae (l_{+1}, l_{+2}, l_{+3}) and in pupa (l). (From Novák, 1975a.)

(Novák, 1985), and then stepwise organogenesis. Thus, we have four stages of Berlese–Jezhikov that follow the apodous state, i.e., the protopod, polypod, oligopod, and postoligopod stages (Novák, 1975a). In further development, the corresponding newly formed growth gradients either fully disappear in the course of further morphogenesis or may produce structures specialized for another function in adult insects. Examples are the paired abdominal extremities of the polypod stage forming the external genital appendages or some specific organs of locomotion in later larval stages, such as the pseudopods in caterpillars or the thickened bristled rings on each body segment and similar adaptations in larval Diptera: Cyclorrhapha (Novák, 1975). Contrary to Hinton's (1955) view, all of them are homologous, originating from the same growth gradients. In all cases, the adaptive character of most of the corresponding structures should be noted; they all have the character of NPC (see above and Novák, 1981, 1985a,b, 1989).

1.2.1.1. JH IN EMBRYOGENESIS

Strong evidence in support of the principal identity of embryonic and postembryonic processes was furnished by the discovery that JH is also active during the embryonic period. This was first shown by Pflugfelder as early as 1939, and was later confirmed by Novák (1951b) in his experiments with *Oncopeltus fasciatus* bugs. A rapid advancement of knowledge on this subject has been brought about by the discovery of juvenoids or JH analogues (JHA). Their application on egg surfaces has caused a great variety of effects on the structure of embryos corresponding to those known in the postembryonic period. A detailed investigation of some 1500 *Schistocerca gregaria* embryos, affected in varying ways by the JHA application in various doses and at different times after egg laying (Novák, 1969), led to the following conclusions: Embryonic development responds to different JHA in the same way but more readily and less species specifically than does postembryonic development. It was found that the juvenilizing effect increased with the amount of applied substance and decreased with time after the beginning of morphogenesis. Many diversely juvenilized and teratological specimens were obtained, some of them asymmetrical and at first glance quite fortuitous. A detailed analysis of all these forms, however, yielded an explanation fully consistent with that obtained from the findings on JH action in postembryonic development. Quite surprising was the occurrence of miniature, fully differentiated embryos, which, at first glance, appear just the opposite of what we observe by JH action in larvae.

The result can, however, be interpreted as follows: If the substance starts to act after differentiation of the germ band has begun but before embryonic differentiation is finished, then differentiation is not completed, the structure of the undifferentiated cells is preserved, and isometric growth of both the germ band and the surrounding undifferentiated blastoderm cells continues. If the applied dose of the substance is sufficiently large, this state may continue unless the reserve substances in the yolk are consumed. In this way, the various structures of embryogenesis may be preserved with the corresponding amount of growth. When, however, only a small dose has been applied, the substance ceases to act after a time (being consumed or inactivated by the tissue) and gradient growth and thus morphogenesis are resumed. Now, of course, only cells whose differentiation has been realized before further development has been stopped by the JHA can develop further, whereas those cells whose differentiation did not start have evidently passed their critical period so that they have to remain in their preceding

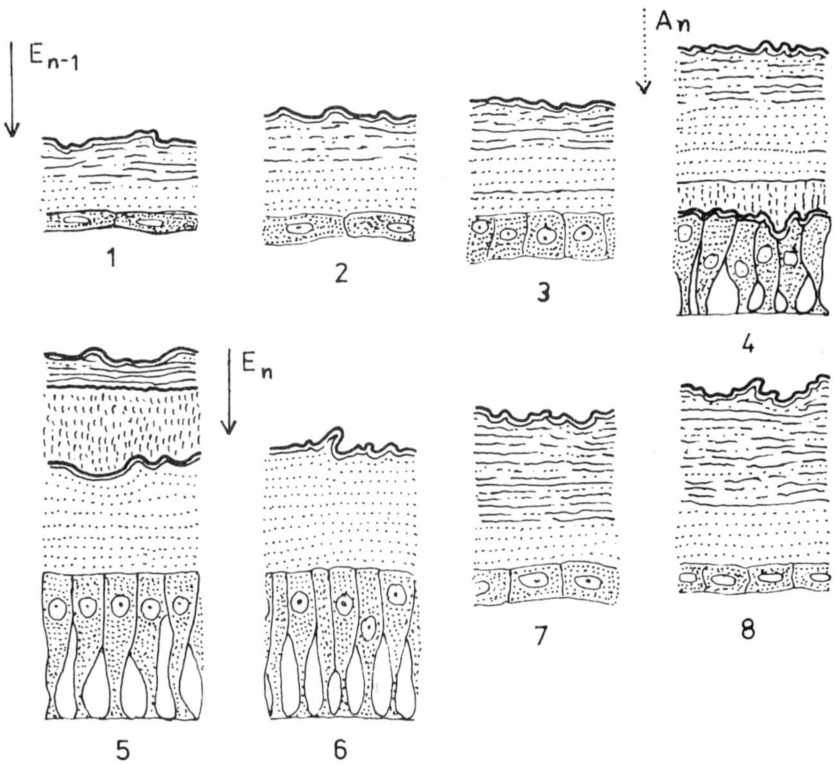

FIGURE 1.5. Molting cycle of *Rhodnius, prolixus*. *Key: E* = ecdysis; *A* = apolysis (1–8) eight stages of the molting process from the beginning of the instar (*I*), through apolysis and ecdysis, up to shortly before the next apolysis (8); $n - 1$ = penultimate; n = last nymphal instar. (From Novák, 1975a; after Jenkin, 1966, modified.)

state. The embryo is thus able to develop only from the limited number of cells forming the germ band in that time. However, as the locust eggs are regulatory, they can still achieve complete differentiation, as in Seidel's (1924) ligation experiments in ephemerids, which, however, may not concern all the organs. Their limited size decreases their requirements for food supply, so that part of the yolk remains and with it also its connection through the dorsal opening in the head (Figs. 1.5, 1.6, and 1.7). No such results were obtained in experiments with phylogenetically higher insects like bugs (Sláma and Williams, 1966; Polivanova, 1968) and *Drosophila*, probably owing to their lack of regulatory capacity.

1.2.1.2. GF TYPES OF STRUCTURES

The existence of structures that do not belong strictly either to larval or to imaginal parts of the body was observed by Geigy as early as 1941. These were his so-called larvo-imaginal structures, which differentiate with other larval structures during embryogenesis and do not change during the larval period, but do not disintegrate with them; rather, they survive and continue to function during the adult period. This is the case of CA, the source of JH activation of such larval structures, as well as of the ovarian follicle cells of adult females and the accessory glands of adult males (Wigglesworth, 1936). We can thus distinguish the following main parts of the body, differing in the time of their GF inactivation and thus of their active part in morphogenesis:

(1) *Embryonic structures* whose GF is inactivated before the embryonic JH starts to act so that they do not survive to postembryonic period
(2) *Larval structures,* which lose their GF in late embryogenesis and survive and continue to function and grow to the beginning of metamorphosis owing to JH action

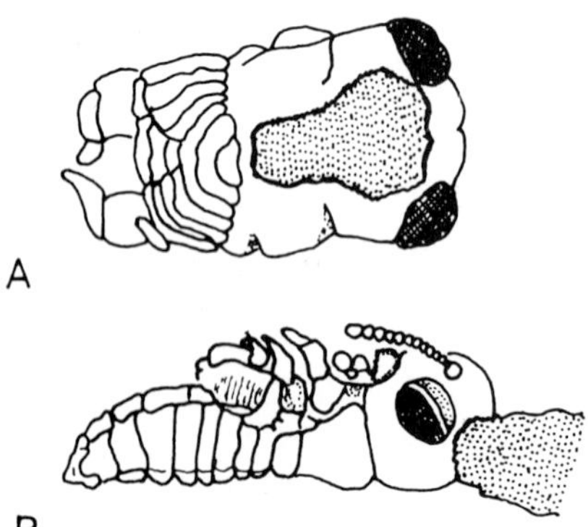

FIGURE 1.6. Two JH-treated *Schistocerca gregaria* embryos at time of hatching of controls. (A) Embryo with six abdominal segments contracted; dorsal view with the open connection with yolk sack. (B) Less influenced embryo with all abdominal segments; yolk sack opening reduced to the head, with normal structure but much diminished. (From Novák, 1969.)

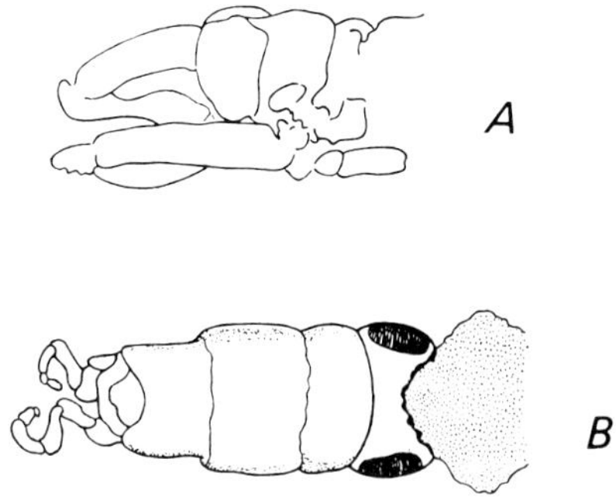

FIGURE 1.7. Two embryos of *Schistocerca gregaria* with abdomen reduced to one (A) or two (B) sclerites following JH_a treatment in late phase of embryogenesis. (From Novák, 1969.)

(3) *Late larval structures,* which differentiate in the postembryonic period, in the gradient growth periods at the beginning of each first larval instar, but lose their GF and do not survive later than other larval structures in metamorphosis (e.g., various integumental structures distinguishing individual larval instars of Saturniidae)
(4) *Larvo-imaginal structures* in the sense of Geigy (1941) as discussed earlier
(5) Most of the *epidermal cells* in larvae of Hemimetabola and some in the pupae of Holometabola, some of whose parts continue to grow, whereas other parts disintegrate in the course of metamorphosis; in addition to these, there are epidermal cells of larval character that cease to grow and disintegrate during metamorphosis, and there are imaginal epidermal cells, which start to grow, divide, and proliferate during metamorphosis when they replace the larval ones
(6) *JH-dependent imaginal structures,* developing only during metamorphosis, whose function in the adult period, however, is dependent on the presence of JH (e.g., the ovarian follicle cells and the male accessory glands in some species)
(7) *Imaginal structures,* which mostly remain in an undifferentiated state in the larval period and start to differentiate and grow intensively during metamorphosis, e.g., the nymphal wing pads in Hemimetabola and the imaginal disks of holometabolous larvae
(8) *Sex cells,* which differentiate and develop at the same time as do the imaginal structures, or later, in the adult insect, but which continue to grow and differentiate only in individuals of the next generation

The imaginal wing disks of insects with regressive metamorphosis, as in the wingless females of some Geometridae and Psychidae or in some species of fleas, might be considered as special cases between the categories of larvo-imaginal structures (item 4, above) and the epidermal cells (item 5): they develop slowly like the imaginal structures in the larval period; their growth is accelerated with the onset of metamorphosis, but they disintegrate like the larval structures partly or completely before its end (see Section 1.3, below).

Another exception to be placed between categories of JH-dependent imaginal structures and normal imaginal structures (items 6 and 7, above) are the indirect wing muscles in female ants. They develop as normal imaginal structures until the adult stage and are fully active during the mating flight, but they disintegrate completely shortly afterward, thus contributing as a food reserve for development of the ovaries and some female glands (Janet, 1899). A similar case is that of the wing muscles of crickets and certain other female insects (Chudakova et al., 1974). All these categories have a strictly adaptive character in the given environment.

1.2.2. *Microscopic (Cellular) Differentiation*

The macroscopic morphogenesis we have spoken about has its necessary basis and counterpart in microscopic morphogenesis, i.e., cell and tissue differentiation. Although experimental evidence on the effect of JH and hence also GF at the cellular level is as yet very limited, thus making a detailed analysis of morphogenesis at this level rather difficult, here too the assumption of GF may be helpful and stimulating. There are detailed histological studies of the imaginal wingpads in fourth- and fifth-instar nymphs of *Pyrrhocoris apterus* in the presence and absence of JH (Sláma, 1965), a study on the action of JHAs in the embryogenesis of *Schistocerca gregaria* (Novák, 1969), and papers on JH action on the prothoracic glands in the metamorphosis of *Galleria mellonella* at microscopic and ultrastructural levels (Malá et al., 1974; Blazsek et al., 1975).

From these observations, it seems evident that the action of JH at the cellular level is principally the same as in macroscopic morphogenesis. In this case also the JH produces its structure-preserving effect, and thus we can assume the interaction of the GF here also. Even the earliest embryonic cells preserve their original structure, increasing in size without any visible differentiation after the treatment with juvenoids (Novák, 1969). Here, of course, many questions remain open, and more experimental evidence will be necessary to clear various old and new questions in this connection (Novák, 1969, 1975a). One example is the question as to the gradual differentiation of one and the same cell; another, as to the critical period during which a cell destined to die in the process of morphogenesis can be preserved (Lockshin and Williams, 1965). These questions again are experimentally best accessible in the period of metamorphosis.

Three types of cells can be distinguished from this point of view, in the case of macroscopic differentiation or morphogenesis: (1) larval cells, which neither grow nor divide and usually disintegrate in the absence of JH; (2) imaginal cells, which grow more rapidly in the absence than in the presence of JH and which divide, differentiate, and proliferate only in the absence of JH; (3) differentiating larval cells, each of which behaves like an organism as a whole, i.e., it partly survives and changes in form, and partly disintegrates, as is most often the case in nymphal (hemimetabolous) and pupal (holometabolous) epidermal cells. During their differentiation, however, these cells preserve their function, which means that as long as they live, they perform the molting process and other functions of average epidermal cells. The type of cuticle depends on the structure of the given epidermis cell. Moltings occur in some insects, however, where even minimum morphogenesis does not take place between individual moltings. These are the so-called stationary moltings in workers of some termite species (Lüscher, 1972). Moltings with minimum growth or morphogenesis may also be induced experimentally by high doses of ecdysone, as shown by Bückmann (1969).

In terms of the GF hypothesis, it may be assumed that there are three types of cells: (1) those lacking GF, i.e., larval cells of the postembryonic period, which disintegrate in the absence of JH; (2) those possessing it in full degree (even if it is gradually inactivated inside them in the course of their differentiation), i.e., imaginal cells of the postembryonic period; and (3) those possessing it only in specific parts. The species-specific pattern of tissue formed by these cells determines the aforementioned macroscopic pattern of morphogenesis. The GF is thus a determinant of changes in form, corresponding to both hereditary and environmental internal (chemical, physical) and external morphogenetic factors, which interfere and integrate their morphogenetic action in the species-specific way.

1.2.3. *Molecular Differentiation or Morphogenesis*

From the previous discussion, it follows that the site of action of GF is at the molecular level and that it is concerned with some fundamental growth processes. The site in question has two essential features: its indispensability for growth (on all three—macroscopic, cellular, and molecular—levels), and its ability to be inactivated by various internal and external influences. On this basis, the concept of *morphogenetic factors* (including organizers, inductors, etc.) may now be introduced and their modes of action further elucidated. A morphogenetic factor (MF) is defined as any factor that inactivates or replaces the GF in the course of regular development. Morphogenetic factors may be internal or external, chemical or physical, hereditary or nonhereditary, and tissue (cell)-bound or blood-carried (hormonal). In any case, their activity is concerned with suppressing (stopping) growth at a given spot. In rare

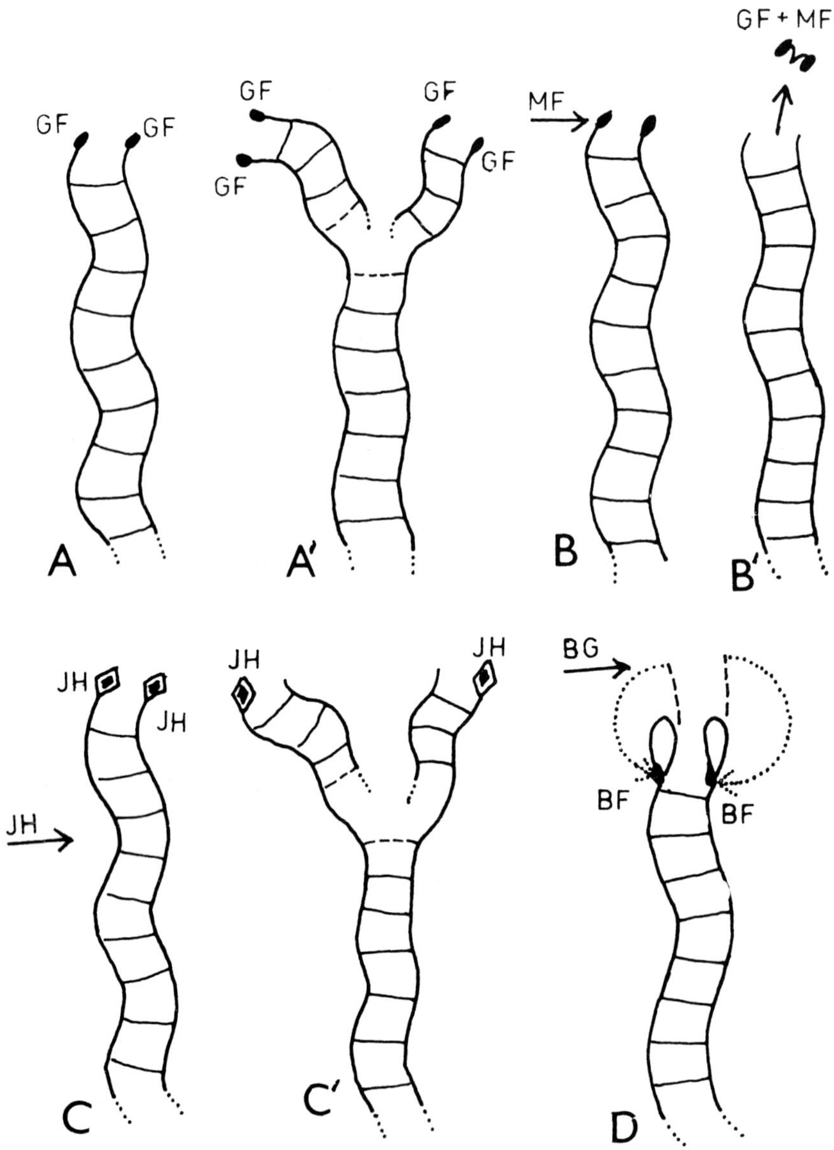

FIGURE 1.8. Model of the GF theory at the molecular level: (A) normal DNA molecule capable of replication, with its gradient factor; (A′) commencement of replication (the GF is also reproduced); (B) action of inactivating morphogenetic factor (MF); (C) replacement of lost gradient factor (GF) by JH; (C′) DNA molecule replication not accompanied by reproduction of JH molecules; (D) origination of DNA molecule with blastomofactor (BF) through binding of the GF to a component of the same DNA chain. (From Novák, 1967.)

cases, they may work in the contrary manner by replacing the inactivated GF, as in the case of JH (Fig. 1.8).

1.2.4. *Chemical Character of GF*

Little definitive can be said as yet regarding the chemical character of the GF. Its specific tissue-bound character does not allow one even to imagine some convincing direct experimental way to obtain the necessary data. There is, however, a rather rich supply of indirect evidence in this respect. First, the experiments concerning the mode of action of JH should be mentioned. It works, as most of its analogues do, throughout ontogeny, stimulating growth, and thus preventing morphogenesis, beginning with egg segmentation up to the beginning of metamorphosis. Further, it is not species- and even subclass-specific within the whole class of Insecta. There is a loss of further growth ability of those parts of the body (like the larval parts) whose GF is presumed to have been inactivated; however, such loss of ability to grow is still reversible by inducing a morphogenetically active JH concentration up to a certain point in time. In most insects, JH production starts again during the adult period when it conditions the activity of ovarian follicles and male accessory glands.

From all these observations it seems most probable that the tissue-bound factor conditioning the growth activity of the imaginal parts of the body in the postembryonic period (and their equivalents in earlier embryonic and in subsequent periods of ontogeny), i.e., the hypothesized GF, has the character of a nucleic acid, most probably DNA. This conclusion is corroborated by findings that the observed gradient growth is common not only to all animals but in a specific form also to all plants and single-celled microorganisms (Novák, 1975a). It may thus be assumed that the first step of the aforementioned inactivation is caused by some general means of blocking the activity of the DNA molecule in the growth process. During this first inactivation phase the inhibited GF may be completely replaced by JH, whereas the second step, consisting of the GF's destruction, is irreversible. Both steps are species-, time-, and place-specific.

From this, as a secondary hypothesis, it may be concluded that the phenomenon of GF is based in a nonhereditary feature of the DNA molecule (of a particular atom group therein) conditioning its ability to react with various internal and external morphogenetic factors. It gradually loses its ability to work as growth-conditioning factor in a regular species-specific way. In this way, the form that growth takes is continuously modified. This ability of the DNA molecule to fulfill its role in replication, transcription, and translation processes probably originated at the subcellular level in the protobionts (as in present-day viruses), before the first cellular organism arose.

From what has been presented above, it follows that GF is necessarily an intracellular growth factor, working at the molecular level. Further, it has been shown that GF is also the main morphogenetic regulator of the cell structure (Novák, 1975a), e.g., in the epidermal cells of Hemimetabola. This is undoubtedly its phylogenetically earlier function; later, as its distribution was conditioned by various morphogenetic factors, it started to regulate morphogenesis at the macroscopic level (see Section 1.2.2, above). It can thus be viewed as a mechanism through which the morphogenetic activity of the external environment is effected by means of its interaction with factors of the internal medium—the individual morphogenetic factors. All this happens, however, on the basis of nonhereditary phylogenetic changes (NPC), as discussed earlier.

1.2.5. *"Competence of Tissue to Undergo Metamorphosis"*

A number of further conclusions can be drawn from various extraordinary observations on morphogenetic phenomena concerning insect metamorphosis and its hormonal regulation. This is first of all the "competence of tissue to undergo metamorphosis" (Novák, 1975a). It was found by Pflugfelder (1939) in *Carausius morosus* and by Scharrer (1946) in *Leucophaea maderae* that allatectomy in the earlier nymphal instars (in penultimate and earlier nymphs) does not result in immediate metamorphosis, as in *Rhodnius* (Wigglesworth, 1936), in *Bombyx mori* (Bounhiol, 1937), and other Holometabola. Instead, two further molts are necessary after allatectomy for the adult characters to develop. The first molt leads to what has been called a "preadultoid" with incomplete imaginal differentiation, and only the second one results in a practically complete miniature adult with degenerated prothoracic glands, so that it does not molt any more. Differing from the explanation of the authors cited above, there is another more probable explanation that agrees in all respects with the GF hypothesis and contributes to a deeper understanding of the GF mode of action from both the ontogenetic and the phylogenetic perspectives.

This is based on the fact that both the aforementioned species belong to the phylogenetically lowest groups of Hemimetabola (phasmids and cockroaches). It is well known that Apterygota continue to molt in the adult stage, because their prothoracic glands persist in adults and so they continue to produce the molting hormone in the absence of JH, unlike the same glands in Pterygota. It is, therefore, quite natural that the dependence on this factor in the lower Hemimetabola is less complete, so that an extra molt subdividing the precocious metamorphosis into two instars becomes possible. This agrees with what we know of the pupal instar in Holometabola or the subimago in Ephemeroptera.

This is still more understandable in terms of the Berlese–Jezhikov theory (see Novák, 1955). The Apterygota, such as Thysanura, reach adulthood in the oligopod stage of morphogenesis. In contrast, the hypothesized inactivation of the GF of the prothoracic glands in pterygote insects occurs, followed by their disintegration, somewhere between the oligopod stage and the advanced postoligopod stage (judging from its occurrence in the last larval instar in Hemimetabola and in the pupa of Holometabola). This stage is obviously not reached in either Phasmida or Blattoidea, so that the molting hormone production with the consequent molting process is induced before the adult morphogenesis is finished. Hence there is an extra molt, and the preadultoid instar results in these groups.

This observation shows clearly the phylogenetic character and conditions of the GF mode of action and inactivation. First, the GF inactivation of a certain structure (here the prothoracic glands)—and so its dependence on JH—originates in a specific morphogenetic stage to which it remains bound in all the subsequent stages of phylogenesis. This supports the hypothesis that GF inactivation by a specific substance appears at a specific moment of ontogenetic chemical differentiation in close association with other components of this process in the body. And, secondly, it shows how the adaptive character of GF inactivation is linked with the selective value of the process, e.g., its coordination with the origin of wings in Pterygota: after their expansion, the next molting would be destructive for the given specimen, so that a successful molting of the membranous wings with no epidermis preserved is not possible. This observation may serve as a key to a deeper understanding of the mechanism of both the phylogenetic and the ontogenetic morphogenesis, including organogenesis.

1.2.6. GF and the Molting Process

An important help for a better understanding of the mechanism of GF action in particular and morphogenesis in general are observations on the histological aspects of the molting process as summarized by Jenkin (1965, 1966) and Jenkin and Hinton (1966), as well as findings on the interaction of GF and the growth hormones (Novák, 1975a). The repeated shedding of the outermost sclerotized layer of the cuticle in insects is a prerequisite for any morphogenesis and, within certain limits, also for growth itself. At the same time, the less sclerotized internal layer is gradually dissolved in the less sclerotized and mostly thicker layer, the endocuticle. The molting process is a series of interdependent morphophysiological processes beginning with molt (apolysis), culminating in the sloughing of exuvia (ecdysis), and continuing by gradual deposition of the new cuticle by the epidermis until this is stopped by subsequent apolysis.

The epidermal cells are thus typical secretory cells with uninterrupted secretory activity. The composition of their secretion changes continuously, however, beginning as the molting fluid, which dissolves the internal layers of the endocuticle, followed by deposition on the surface of the epidermal cells, first the proteinaceous epicuticle, then the exocuticle, later tanned and sclerotized by the phenolic substances, and ending with the deposition of the endocuticle. The chitin-dissolving and -producing activities are most probably ruled by one and the same enzyme—chitinase. The individual phases of the continuous molting process can be subdivided into the following 12 steps (Novák, 1975a):

(1) External stimulation of the neurosecretory cells to produce AH, transmitted by the nervous system (resumption of feeding, etc.)
(2) Activation of MH production by AH in apolytic glands, together with activation of other temporarily inhibited processes in the body
(3) Activation of the epidermis and other epithelia of ectodermal origin; their swelling and initiation of the secretion of the molting fluid
(4) Molt (apolysis) brought about by dissolution of the innermost layer of the endocuticle, thus loosening it from the epidermal cells
(5) Anisometric growth of the imaginal growth gradients, including mitotic activity and differentiation of their epidermal cells, resulting in the production of the molting fluid
(6) Attainment of a morphogenetically active JH concentration (which is not reached in the last larval and pupal instar, inducing the growth of larval parts of the body lacking gradient factor)
(7) Change of the disproportionate imaginal growth into a proportionate larval growth in this way
(8) Gradual dissolution of the internal layers of the endocuticle by an increasing amount of the molting fluid
(9) Gradual deposition of a new cuticle, starting with the epicuticle, followed by the exocuticle, and finally by a thick endocuticle with its highest content of chitin
(10) The new cuticle, progressively suppressing further growth, gradually reduces further metabolism and other physiological processes, and with them also hormone production
(11) Ecdysis, i.e., the discarding of exuvia with a series of interconnected processes, including swallowing of air and peristaltic movement of the gut and body walls
(12) Continued deposition of the endocuticle in the beginning of the next instar, interrupted by the next apolysis

Postembryonic insect development can thus be shown to consist of a continuous stream of cyclical physiological events organized around individual growth cycles, the instars, each of which follows the preceding one without interruption. It starts with the deposition and molting of the embryonic cuticle, whose ecdysis mostly coincides with hatching and ends (in pterygote insects) with imaginal molting (whereas it continues at a decreasing rate in Apterygota). The sequence can be illustrated by a helix with increasing dimensions of each turn, in which each turn represents one instar and in which each of the partial processes (e.g., molt, morphogenesis, fledging, etc.) lies in the same vertical line parallel to the axis (Fig. 1.9).

There are other experimental results that can be interpreted to agree with the GF hypothesis. These include various experiments concerning embryonic and postembryonic development and gonogenesis, as well as implants and tissue culture, involving other morphogenetic phenomena (Laufer, 1968; Naton, 1960, 1962; Schaller, 1959, 1971; Pener, 1965; Biellmann, 1964; Bhaskaran, 1972; Urvoy, 1963; Matz, 1966; and many others).

1.2.7. *Transdetermination*

A specific relation of what has been called transdetermination, shown in a long series of very interesting experiments by Hadorn (1965, 1966), to the concept of GF, as well as what has been observed by Wigglesworth (1965) and by Steinberg (1938, 1956) earlier, is also important from this point of view. These authors have shown by transplantations of imaginal disks and various other structures into various hosts that the results of such treatment may vary in a complicated but regular way. Were we to extrapolate from these results and interpret them as suggested by Hadorn (1965), we might arrive at several diverse hypotheses, but the one that accords with the GF concept seems to be rather simple and rational and therefore probable. It may be summarized as follows:

Hadorn has shown in a large number of experiments that the imaginal disks of *Drosophila*, if transplanted to another larva at the beginning of metamorphosis, shows autotypic, i.e., normal, differentiation; in other words, they differentiate into those organs for which they were originally "programmed" corresponding to their position in the body. When, however, they are implanted into adult fly abdomen, they may show allotypic differentiation, i.e., development into organs quite different from those for which they were programmed (e.g., the disks of antennae into the leg or the anal plate), or their development may stop at a less differentiated stage (either autotypic or allotypic), or there may be no differentiation at all. This may be understood quite simply as follows. The implantation into the last instar, i.e., the metamorphosis period larva, where no JH hormone is present, does not change anything regarding determination. In the adult fly, on the other hand, the

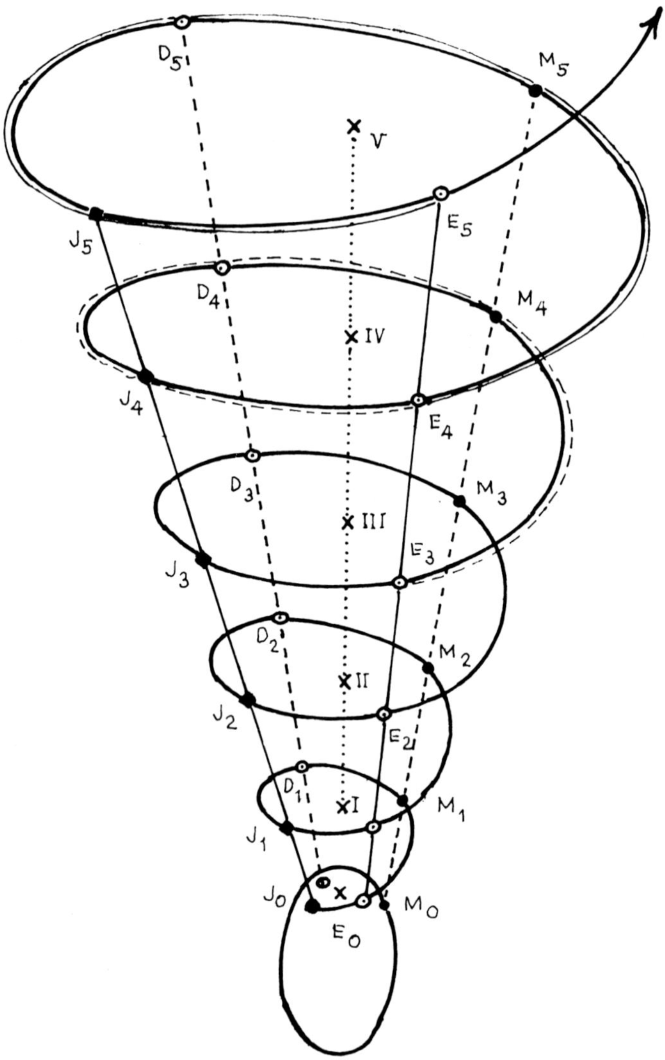

FIGURE 1.9. Continuous sequence of molting processes, from prelarval (embryonic) molt (M_0) and ecdysis (E_0) to adult ecdysis (E_5). *Key:* I–V, first to fifth instar; M_0–M_5, series of molts (apolysis) between embryonic and imaginal molt; D_0–D_5, corresponding moments of initiation of cuticle deposition; J_0–J_5, instants of inhibition of morphogenesis by freshly deposited cuticle; E_0–E_5, ecdyses. (From Novák, 1956.)

active concentration of JH is assumed to be present in a certain period. And, as we know, either the active JH concentration may prevent morphogenesis, replacing it with proportionate growth, or the morphogenesis may start before the active JH concentration has been reached, which may then interrupt it at any time during its course. In such a case, the arrested morphogenesis may resume if the JH concentration should drop again beneath the active level concomitant with the physiological state of the host (e.g., egg laying). As the experiments by Wigglesworth (1965) suggest, however, the transdetermination itself is also a process limited to a certain length of time (with a certain critical period during which each step can be influenced). Thus, if a specific period is missed owing to JH interaction, then only a more general structure (e.g., a leg instead of an antenna) can develop.

As shown earlier, the mode of action of the JH and all its analogues is the same in the embryonic as in the postembryonic period. And the same is true for the period of gametogenesis (sex maturation), as shown by JH dependence of ovarian follicles and male accessory glands in species where they are JH dependent. Less clear as yet is the role of JH and thus also GF in *metagenesis,* i.e., the last period of ontogenesis next to gametogenesis, including the maturation and involution of various other tissues, such as the indirect wing musculature in many female insects. Metagenesis also includes gradual biochemical changes and morphological involutions in the body associated with aging and gradual deterioration, and decline of the body's functions most probably connected with GF inactivation or exhaustion. Little, if anything, is known on influencing metagenesis by inducing JH.

Many papers have been concerned with the maturation of indirect flight musculature as in the cricket *Gryllus domesticus* and in the beetle *Leptinotarsa decemlineata,* or their degeneration in the female ant *Lasius niger* (Janet, 1899). In some species, JH seems to be indispensable for full development of flight muscles, whereas in others the same hormone is essential for their degeneration. The degeneration does not depend directly on the presence of ovaries, since it also takes place in castrated females, whereas both ovaries and flight muscles ripen independently of allatectomy (Chudakova and Bocharova-Messner, 1968; Chudakova et al., 1974). This is one of the unsolved paradoxical effects of the JH and thus also of GF that remain to be studied experimentally. Anyway, they do not contradict the GF hypothesis.

In summarizing, it may be observed that the results of all the available experimental and other papers dealing with insect morphogenesis not only do not contradict the assumption of GF, but many of them bring new evidence in its favor. Furthermore, their interpretation in terms of the GF hypothesis opens many new questions that remain to be studied experimentally.

1.3. GF Theory and the Origin of Insect Metamorphosis

The hypothesis concerning the role of GF in the gradient growth characteristic for insect metamorphosis, based on experimental findings on the JH, provided me with an insight into the principle of insect metamorphosis unresolved previously. It may be outlined briefly as follows: Morphogenesis during the insect embryonic period is the result of gradient growth, resulting in a continuous change in form. Since the JH starts to be produced in effective concentration toward the end of embryogenesis, the structure of the body is more or less preserved during this period until the beginning of metamorphosis. The stimulation of growth of the larval parts of the body converts the gradient growth into a proportionate growth characteristic of the larval period of Pterygota. In this way, the last portion of embryonic morphogenesis is inhibited and takes place in the form of metamorphosis only at the end of postembryonic development, when JH does not reach the effective concentration. Thus, what is phylogenetically new in insects is not the gradient growth period of metamorphosis but the larval proportionate growth period. This is then subdivided into a series of separate molting processes, the larval instars.

1.3.1. Principle and Evolution of Insect Metamorphosis

Individual larval instars of a given species are not morphologically identical: each later shows a small increase of the adult structures, so that the last instar is distinctly more advanced toward metamorphosis than the first one. This has been shown to depend on a short period of disproportionate growth that takes place at the beginning of each instar before the active concentration of JH has been reached (Wigglesworth, 1940a; Novák, 1951b) (Fig. 1.10). With the growth of the larval body, this always happens later after the preceding molting in each subsequent instar. Metamorphosis can thus be viewed as the instar in which the active JH concentration has not been reached at all, so that only disproportionate growth, and hence morphogenesis, can take place.

In Hemimetabola, metamorphosis is performed in one instar (to say nothing of the sum of the very small increases at the beginning of the individual larval instars). In Holometabola, with their internal imaginal disks, however, two instars are necessary to reach the adult structure of these undifferentiated organs. Differing from Poyarkoff's (1914) and Hinton's (1948) theories, I have shown that there are two main reasons why two moltings are needed to perform metamorphosis in phylogenetically higher insects (Endopterygota): (1) the increase in the amount of morphogenesis that is necessary to reach the adult state in

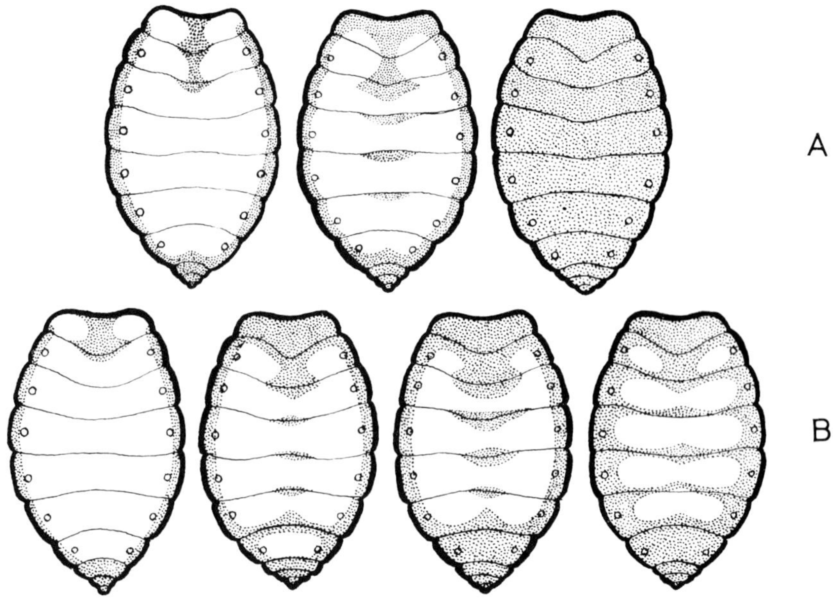

FIGURE 1.10. The occurrence of mitoses in the epidermis of the dorsal surface of the abdomen in *Rhodnius prolixus* in various periods of the penultimate (A) and last (B) nymphal instars: dotted area, with mitoses; white area, no mitoses. (From Novák, 1959; after Wigglesworth, 1940a, modified.)

connection with hatching in gradually earlier and earlier stages of morphogenesis (and thus the increase of the morphological difference between the last larval instar and the adult); and (2) development of the internal imaginal disks, first of all that of the wings, to develop into their functional adult structure. Two moltings are required for the undifferentiated internal structures to evaginate on the surface of the body, which is only possible in a rather undeveloped state of differentiation, before the wing has reached its final length and membranous form. By then, however, there is not enough time to achieve full differentiation within the given instar. Thus, the coincidence with the new extra preimaginal molting was the precondition, and later the main selection value, for survival of a species with internal imaginal disks. In this way undoubtedly the pupal instar originated, giving rise in time to various other evolutionary improvements of the pupa. All other features of the pupal instar are phylogenetically secondary achievements, often again bound together (e.g., the immobility of the pupa and its thick and hard cuticle with its conspicuous coloration; cocoon spinning in many species; the specific position of the head and appendages characteristic for the moment of the fledging (ecdysis), as observed only in higher Hemimetabola).

A more detailed comparison of various morphogenetical stages of Hemimetabola and Holometabola shows no principal morphological differences between both groups except that the same stage may perform a different ecological role in either of them and be secondarily adapted for it (Novák, 1975a). Conclusions following from these observations provide a better understanding of how the evolution of various characters proceeds on the basis of natural selection. In this connection, the often discussed question of whether Holometabola developed directly from Ametabola or whether they developed through the grade of Hemimetabola can be solved. Irrespective of what can be said in favor of the second concept, from the foregoing discussion it seems obvious that the phylogeny of all the recent Holometabola must necessarily have passed through a stage with one metamorphosis instar with external wing pads, which we should classify as Hemimetabola (even if not necessarily through any of their recent orders).

1.3.2. *Phylogenesis of the Metamorphosis Hormones*

All the above observations and phylogenetic considerations also help to solve the questions of the phylogenetic origin of the three metamorphosis hormones.

The oldest of them is undoubtedly the *activation hormone.* Groups of neurosecretory cells similar to those found in the pars intercerebralis of the insect brain, as well as in the frontal organs of crustaceans and corresponding structures of other arthropods, have also been described in the nervous system of various annelids (Clark, 1956; Gersch, 1964) and even in Turbellaria (Gersch, 1970) and Coelenterata. This also corresponds to its earliest appearance in embryogenesis before the other two hormones (Jones, 1956), confirming it as the primary hormone on which the production of the other two is dependent.

Also phylogenetically very old is the *molting hormone* (MH), or ecdysone, and its derivatives (MHd) and related substances of steroid character. Their existence has been shown in practically all the major groups of arthropods, and they seem to be present in annelids as well, evidently starting there as a condition for the synchronization of the molting process over the entire body surface. In arthropods, the source of MH may differ. It may be one of the various apolytic glands in insects, or (as suggested by Sláma, 1983) it may be some other secretory organ—like the Y-organ in crustaceans or Schneider's glands in spiders—but MH (both ecdysone and 20-hydroxyecdysone) has been shown to work without specificity within the whole phylum. Also, the dependence of its production on AH seems to be quite general. The other effects of MH seem to be secondary from the phylogenetic point of view.

Phylogenetically the youngest of all three is the *juvenile hormone* (JH), or neotenin (Wigglesworth, 1935), only known so far from insects, with its manifold and often seemingly contradictory effects. It is a growth hormone, inducing growth first of all in those parts of the body that have lost it due to the hypothesized GF inactivation. Even though it is the most active morphogenetic hormone, its activity in this direction has been shown to be indirect. In phylogeny, it has been first observed in higher Apterygota, where it is only concerned with the activation of ovarian follicle stimulation (Rohdendorf and Sehnal, 1972). Later, with an increase in its production, it spreads gradually back toward earlier and earlier larval stages, as in Ephemeroptera (the larvula), ultimately reaching the embryonic period (Novák, 1964). In this way, undoubtedly, its main function in inducing metamorphosis originated. In Holometabola, this process is continuing, so that embryonic morphogenesis is broken off at earlier and earlier morphogenetic stages, thereby increasing the degree of morphogenesis in metamorphosis (the Berlese–Jezhikov theory; cf. Novák, 1975a).

1.4. GF and Morphogenesis in Other Arthropods and Articulates

In Annelida and various groups of Arthropoda, the different morphogenetic factors affecting morphogenesis are less known than those of Insecta. This is undoubtedly owing to the lower economic importance of annelids and the fact that they are rather less accessible for rearing and experimentation than are insects. The relationship between ontogenetic development of metabolous insects and other groups of arthropods and annelids is now compared. To understand the morphological relationship between individual morphogenetic stages in embryonic and postembryonic periods in insects and those in other groups, the primary features must be distinguished from those which are phylogenetically secondary. First of all are those changes called *adaptiomorphoses* (in the sense of Shmalgauzen, 1946 = *idioadaptations* of Severtsov, 1931), i.e., various adaptations to the living conditions and habits of the particular ontogenetic stage, which do not, however, change the general organization level and vitality, unlike arogeneses (Severtsov, 1931; Paramonov, 1967). When primary features are considered separately, it becomes evident that the individual stages of morphogenesis are principally identical in different groups, whether they are reached during embryogenesis or in some postembryonic period, or in adults. This fully agrees with the law of recapitulation of Charles Darwin and Ernst Haeckel (see Fig. 1.11).

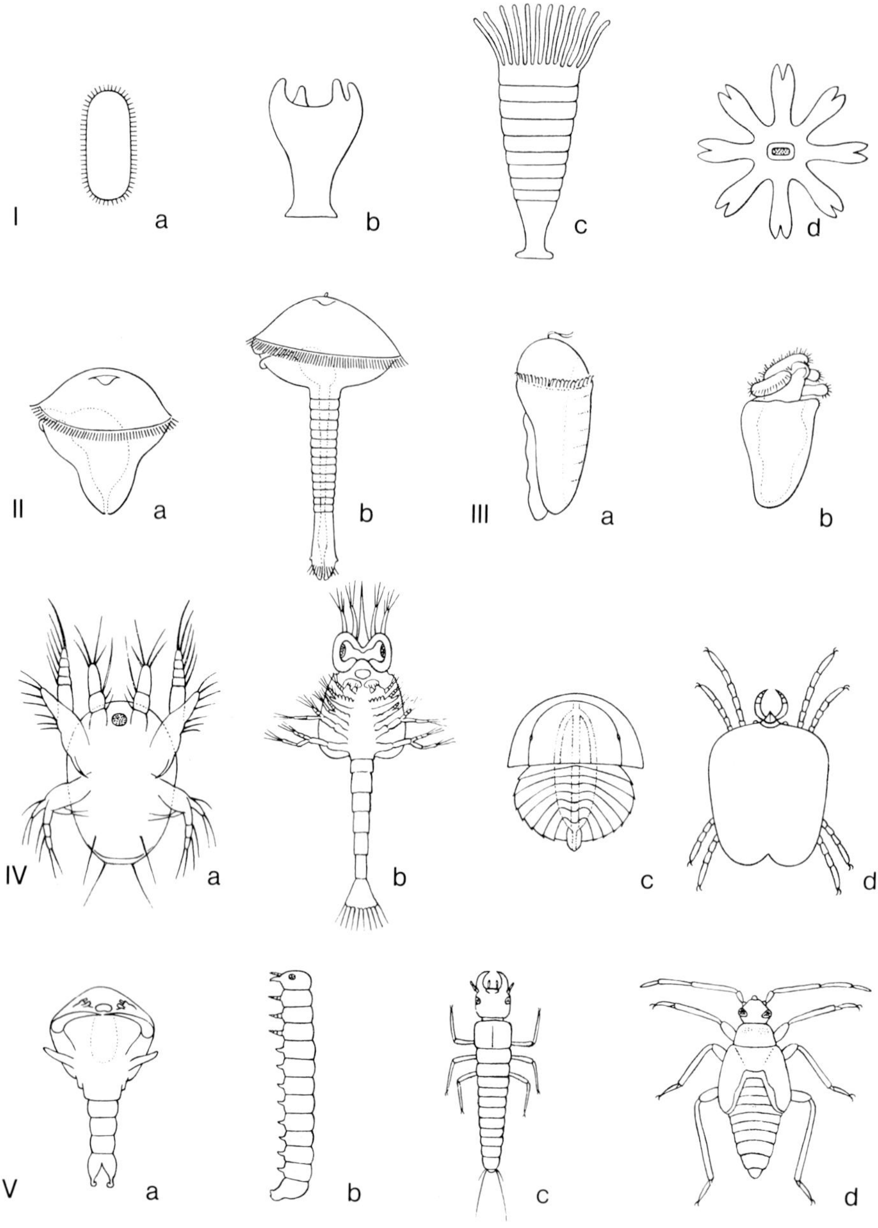

FIGURE 1.11. Main morphogenetic stages in ontogeny of invertebrates. (I) Vegetative reproduction by strobilation in Scyphomedusae: (a) planula; (b) early polypus; (c) scyphostoma; (d) ephyra (early medusa). (II) Strobilation in Archiannelida: (a) trochophora; (b) advanced trochophora. (III) Larvae in Mollusca: (a) larva of *Chiton* (Placophora); (b) veliger of *Cavolonia* (Gastropoda, Euthyneura). (IV) Arthropoda. (a) nauplius of *Cyclops* (Copepoda); (b) zoea of *Spirontocaris* (Malacostraca, Carididae); (c) larva

1.4.1. *Morphogenetic Stages of Arthropods*

The arthropod morphogenetic stages, as shown by Berlese (1913) and Jezhikov (1940), are as follows: In primitive annelids like Archiannelida, the earliest well-defined morphological stages, which are in some species free living, are trochophora (or trochosphera) and metatrochophora. In arthropods, the corresponding stages occur in embryogenesis, where the trochus has disappeared, but the stage is distinct by the beginning of metamerizatin. Then gradually the appendages appear on the first three or, less often, four metameres. This is the *protopod* stage, which may have very different ecological and functional adaptation in different groups of arthropods. It may be a free-living larva, often in the original form of the nauplius in many crustaceans, but more often the nauplial form may only come by the process of embryonization further into the embryonic period and it may also become secondarily free living (eventually within the hemolymph of the host), like the first larvae of some entomophagous parasites. And it may exist even in later stages, as is true principally in larvae and adults of Acarina; this, however, may be a secondary achievement phylogenetically on the basis of neoteny.

Next is the *polypod* stage in which each somite carries a pair of extremities; these may be very tiny and simple, like the embryonic appendages of many insects, or very big and strong, like the legs of large crustaceans (Malacostraca). Primarily, this is the adult stage, as in many annelids (the parapodia each usually with a dorsal outgrowth, the notopodium, and a ventral outgrowth, the neuropodium, each sometimes with a special threadlike cirrus). The polypod stage later reappears in adult Onychophora, Trilobita, and Myriapoda. In crustaceans, as in other higher arthropods, these appendages are usually adapted for various functions like antennae, legs, or genital appendages, or they may more or less disappear. In practically all arthropods, they appear during embryonic period, but they may also exist in some of the larval forms, such as the cypris larva of some parasitic groups (e.g., cyprid larva of some Rhizocephala).

The *oligopod* stage is the case of secondary disappearance or transformation of appendages in the abdominal metameres and their full development in specialized form as antennae, mouthparts, and legs in the head and thorax of both larvae and adults. The abdominal appendages,

---

(trilobitestage) of *Limulus polyphemus* (Xipho sura); (d) *Trombidium* (Acarina). (V) Morphogenetic stages of insect ontogeny: (a) protopod stage of *Trichacis remulus* (Chalcididae); (b) polypod larva of *Panorpa* (Mecoptera); (c) oligopod stage triungulid of *Epicauta* (Coleoptera, Meloidae); (d) postoligopod stage of *Plasiocoris* (Hemiptera, Capsidae). (Adapted from various authors.)

or some of them at least, may later reappear in specialized form, such as the genital appendages in pterygite insects. The oligopod stage is the nymphal stage of many hemimetabolous insects, and the adult stage of most Crustacea, Arachnoidea, and apterygote insects. In pterygote insects, the oligopod stage occurs in either larval or embryonic periods (Hinton, 1955).

In some cases, like *hypermetamorphosis* of some Coleoptera, Strepsiptera, Hymemoptera, and Diptera, the protopod stage of the newly hatched larva like the campodeiform larva (triungulid) of Meloidae, Bombylidae, or some entomophagous Hymenoptera may be followed by a polypod (or secondarily apodous) larva that only later transforms into an oligopod stage as a pupa. Similarly, as in case of larvula of ephemerids, the explanation in terms of the GF theory is very simple: the larvae appear to hatch in these cases before the effective level of JH has been reached; the larvae, therefore, continue their morphogenesis, with eventually two or three further moltings, before the JH starts to act, and only then the normal larval growth period begins.

The *postoligopod* stage (described by Jezhikov, 1936) differs from the oligopod stage in that there are now more developed imaginal parts of the body, such as the wing pads in larval Hemimetabola and pupal Holometabola. In Hemimetabola, it is subdivided by larval moltings into several instars with gradually increasing differentiation of the imaginal structures. Other examples of postoligopod could be the morphogenetic stage reached by the so-called zoeal larva of some Crustacea and adult stages of most Crustacea and Arachnida.

The *winged* staged, as the morphogenetic structure of adult metabolous insects, which developed through epigenesis (in the sense of Severtsov, 1931) from the oligopod stage of Apterygota. Metabolous insects constitute the only group of arthropods where it occurs.

When comparing the morphogenesis of arthropods from the point of view of their *phylogenesis* with that of other invertebrates on the basis of the GF theory, the course of their evolution may be well understood. We can then distinguish the common basis of their morphogenesis as the main mechanisms through which the various deviations developed concomitantly with their adaptation to a specific environment, way of life, and hereditary basis.

1.4.2. *GF in Parasitic Arthropods*

Among the most complicated cases of parasitism are those in arthropods. First of all are various adaptations of the parasitic way of life that take place independently in several groups of crustaceans. In some of them, the phylogenetic degeneration goes so far that the characteristics of the class and even of the phyllum Arthropoda are virtually lost.

This is, e.g., the case of *Lernaea* among copepods, which hatches as a more or less normal nauplius larva. It attaches itself on the gills of a flat fish (Heterosomata) and feeds on its blood by means of its suctorial mouthparts. Here, the parasite passes through several stages, including the immobile "pupa," after which it leaves the host and copulates in water. Whereas the male does not develop further, the female attaches itself to the gills of a new host (a fish of the family Codidae). There, the genital tract grows into the body of the host, whereas the rest of the body of the parasite completely degenerates. A complete series of transitions in degree of degeneration between this extreme case and the normal body structure is known. A similar series of transitions is known in parasitic Isopoda (*Cryptoniscus* and *Bopyrus,* parasites and hyperparasites of other crustaceans).

The highest degree of parasitic degeneration, however, is reached by some species of the order Rhizocephala, relatives of Cirripedia, e.g., the genera *Sacculina* and *Thompsonia.* Here, after the more or less normal nauplius stage, the cyprid larva attaches itself by its antenulla to a seta of a crab, and while the body of the parasite degenerates, a small group of cells continues to grow and penetrates through the antenulla and the seta into the body of the host and is transported by the blood circulation to the vicinity of the intestine. Here, it forms a rudiment of the parasite body, a nucleus, from which many rootlets grow out and permeate the body of the crab. At the end, rudimentary genital organs are formed on the underside of the intestine that press on the hosts epidermis on the ventral side of the abdomen and, after its next molting, come to project freely from the ventral side of the abdomen, where the impregnation of eggs by sperm takes place.

All these striking morphogenetic processes can be understood on a common basis now, i.e., a gradual GF inactivation in various parts of the body up to the body as a whole, except a small group of cells, as in the adult female or *Lernaea* or the cyprid larva of *Sacculina,* where it then continues to grow under the optimum conditions inside the host body. In principle, on the most gradual level, when the adult body perishes, only the sex cells (fertilized eggs) continue to grow and develop. With the solving of such old problems, of course, many new questions arise that deserve experimental research.

1.4.3. GF in Other Articulates

In all articulates, morphogenesis begins with egg segmentation in the morula-like structure, which in free-living state appears in planula-like larvae and the stage of gastrula in adult coelenterates. The next step consists of a special type of vegetative reproduction of the gastrula-like individuals (polyps) by transverse constrictions in the longitudinal body axis, as first occurs in the polyps of Scyphomedusae. This then becomes a general type of vegetative reproduction in tapeworms (Cestodes) and, at the same time, the basis for the origin of the metameric body on the basis of sociogenesis (Novák, 1982a, 1985b), as first appears in Annelida. The beginning of this process is best observed in the segmentation of the trochophora (trochosphera) larvae in Archiannelida, the phylogenetically oldest annelids, which grow to metatrochophora in this way. On principal, the same stage as metatrochophora, the larva of the aquatic crustaceans, the nauplius, forms in a similar way (corresponding to the strobilation of Scyphomedusae) and becomes a *metanauplius* with an increasing number of metameres. The development of more than three functional appendages is the median stage, which corresponds to the higher phylogenetic position of the group. Like the change in body size, the variation in number of appendages itself does not change anything, on the principle of its corresponding to the same morphogenetic stage. The same concerns the veliger larva of Mollusca. This is also true of the protopod stage of insect metamorphosis with various specific adaptations, regardless of whether it is reached in larvae or embryos. In the adult stage, this is reached in Rotifera (in form near to trochophora) as well as in Turbellaria. (See fig. 1.11)

In a similar way, the other morphogenetic stages correspond to each other in different groups of invertebrates, irrespective of their secondary hereditary, ecological, and ethological features. The polypod stage, as it occurs in all insect embryos or larvae, appears in adult Onychophora, Trilobita, and Myriapoda and in many larval and some adult crustaceans (Branchippoda, e.g., *Branchipus estheria;* Malacostraca; Isopoda), and also in larvae of many Mollusca (Hadzhi, 1963), which do not develop further in this direction. And similarly in the oligopod stage limited to arthropods: these stages, being polyphyletic and appearing parallely in various invertebrates, correspond to what has been defined as grades by Ayala and Valentine (1979). The highest (most general) ones in the hierarchy of *grades* are those of *sociogenesis* (the grades of individuals like prokaryotic, monocellular (eukaryotic), lower multicellular plants and animals, higher (cormus) plants and metameric animals, and tufts of higher plants and societies of higher (metameric) animals (Novák, 1982a, 1985b). The same is true of individual phases of sociogenesis within the above grades (Novák, 1982a, 1985b). In all these cases, the approach from the point of view of the GF theory helps to clear the seeming mysteries and helps resolve many new questions.

1.5. General GF Theory of Morphogenesis

The above observations and experiments on metamorphosis in insects, including the metamorphosis hormones and their modes of action, lead to a series of conclusions regarding *morphogenesis of animals in general.*

This experience has been further developed and verified now in arthropods and other groups of animals on a broader phylogenetic basis. It can be concluded that the very principle of animal morphogenesis is anisometric (disproportionate) growth and first of all one of its types—the gradient growth.

Anisometric (disproportionate, differential) growth of animals bodies (and the same is true for all plants; see Novák, 1971, 1985b) may be viewed as a general principle through which morphological structures arise in any organism. Of its two forms, *gradient growth*, during which various parts of the body gradually lose their ability for further growth so that growth occurs always only in specific gradients, is by far the most important. With the loss of growth potential, there is usually a decrease of vitality and resistance that may result in partial or complete removal of the affected tissue by histolysis or phagocytosis. By this, the changes in form become more rapid and conspicuous. The morphogenesis of any animal is composed of an uninterrupted series of growth processes in each of which a part of the body ceases to grow (and is eventually decomposed) whereas other parts continue to grow. The extreme case of such a process is, in principle, also individual death and the beginning of a new individual of the next generation (Fig. 1.12).

A special case of gradient growth is metameric growth, based on the original vegetative reproduction of the type of strobilation (Fig. 1.1D,E) (Hatchek, 1878). The growth gradient is placed along the hind margin of each segment (metagenesis) i.e., the original individual, and it constricts after reaching a corresponding size, without separating, however, in the case of metamerization. In this way the body size of metameric animals, including all arthropods, increases (Novák, 1982a).

The mode of action of *morphogenetic factors* in both ontogeny and phylogeny can now be better understood on the basis of the GF theory. A series of factors with marked morphogenetic activity has been described in experimental embryology and in the physiology of postembryonic development. Many of them have been characterized chemically (Needham, 1942); others have just been given various names and described by their effects, such as organizers and inductors (Speman, 1952), templates and antitemplates (Weiss and Kovanan, 1957), and the morphogenetically active hormones (Wigglesworth, 1934, 1936, 1940a,b; Kühn, 1944; etc.), or only discussed in terms of general principles (Shmalgauzen, 1946, 1982). The GF theory presents a simple and widely applicable concept of the mechanism of morphogenetic activity. This may be now understood as a regular-, species-, space-, and time-specific process of inactivation of the hypothetical GF in various parts of the body, which determines their gradient growth and thus morphogenesis. The mode of action of various environmental factors, such as air (oxygen),

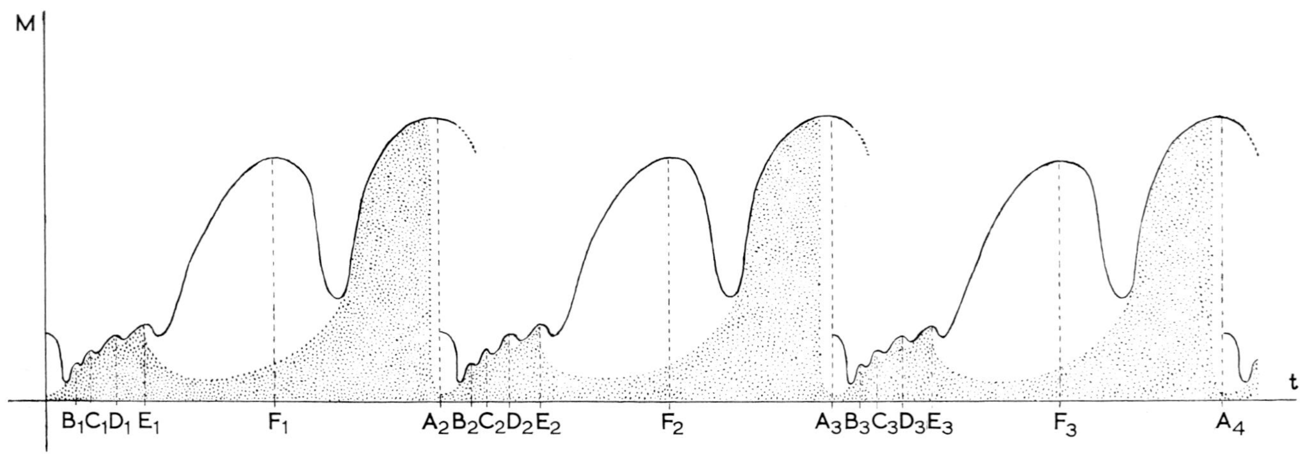

FIGURE 1.12. The relative amount of growth in metabolous insects during three subsequent generations. *Key:* abscissa, time; ordinate, amount of growth; A_1A_2 = first generation; A_2A_3 = second generation; A_3A_4 = third generation; A_1B_1 = early embryonic period (egg segmentation, etc.); B_1E_1 = embryonic morphogenesis (composed of a series of morphogenetic process: B_1C_1, C_1D_1, D_1E_1); E_1F_1 = larval period; F_1G_1 = period of metamorphosis; G_1A_2 = period of gonad development in adults; A_2B_2 = early embryonic period in second generation, etc. Each of the morphogenetic processes depends on that part of the body losing the ability for further growth and disappearing (white area), whereas the remaining parts continue to grow (dotted area). (From Novák, 1955.)

gravitation, surface tension, or humidity, can now be explained in a similar way by their reaction with the morphogenetically active substances inhibiting the GF.

1.6. Summary

When considering the justification, scope, and meaning of the GF theory, it is important first to distinguish facts, based on direct observations or experiments, from conclusions, based on these facts. Underlying the concept are the experiments on postembryonic development in insects and the JH effect on it, which I started at the suggestion of Prof. V. B. Wigglesworth, Cambridge, England, in 1949. It was the period when the first group of pioneer workers in insect endocrinology had published their papers (no more than about a dozen of them (Novák, 1951a,b, 1966, 1975): S. Kopeć, V. B. Wigglesworth, D. Bodenstein, G. Fraenkel, J. J. Bounhiol, M. Raabe, H. Piepho, E. and M. Thomsen, B. and E. Scharrer, B. Hanström, and perhaps two or three others. It was the time when the *"Entwicklungsmechanik"* approach still prevailed in experimental biology, claiming that one should draw conclusions only on the basis of one's own experiments and avoid hypotheses and conclusions from experiments of others. This rule was more or less adhered to by most of the aforementioned pioneer workers, and so a wrong idea originated and spread that perhaps each order or even lower units of the class Insecta had its own distinct hormones.

It was only my findings of the order and subclass nonspecificity of the JH of *Periplaneta americana,* which was found to be effective on the nymphs of the bug *Oncopeltus fasciatus* and on the last-instar larvae of *Bombyx mori,* that demonstrated the necessity to alter this view (Novák, 1951a,b). It also provided the stimulus to attempt a synthesis of all available data on the metamorphosis hormones of insects by various authors, which proved very fruitful. In this way, supporting Wigglesworth's findings (1934, 1936, 1940a,b), the function of the CA was fully explained; and on this basis, the first draft of the GF theory of insect metamorphosis was prepared (Novák, 1951a,b). The rapidly accumulating experimental data on these questions have further confirmed its validity.

The conclusions on the mode of action of the JH, which have been reaffirmed not only by a large number of further transplantation and extirpation experiments but also with JHA applications, may thus be viewed as well supported. Similarly documented are many observations on gradient growth and its reversibility by JH action, as well as the conclusions on morphogenesis on the basis of gradient growth in both plants and animals (Novák, 1982a). On the other hand, the conclusion as to the chemical character of GF, i.e., its identification with a specific part of the DNA molecule, which is likely to be inactivated by various morphogenetic factors, both external and internal, in specific parts of the body in a species-specific pattern, remains just a hypothesis, even though a probable one. Anyway, the GF theory based on this assumption seems to be a useful tool for analyzing various morphogenetic phenomena.

References

Ayala, F. J. and J. V. Valentine. 1979. *Evolving*. Benjamin-Cummings, Menlo Park, California.

Berlese, A. 1913. Intorno alle metamorphosis degli insetti. Redia 9: 121–136.

Bhaskaran, G. 1972. Inhibition of imaginal differentiation in *Sarcophaga bullata* by juvenile hormone. J. Exp. Zool. 182: 127–142.

Biellmann, G. 1964. Effects des rayons X sur des larves de *Locusta migratoria* L. Bull. Soc. Zool. Fr. 87: 569–588.

Blazsek, I., A. Balász, V. Novák, and J. Malá. 1975. An ultrastructure study of the prothoracic gland of *Galleria mellonella* in the penultimate, last larval and pupal stage. Cell Tissue Res. 158: 269–280.

Bounhiol, J. J. 1937. Métamorphose prématurée par ablation des corpora allata chez le jeune ver à soie. C. R. Acad. Sci. Paris 205: 175–177.

Bückmann, D. 1969. Die hormonale Entwicklungssteuerung der Arthropoden. Zool. Anz. (Suppl.) 33: 215–239.

Butenandt, A. and P. Karlson. 1954. Über die Isolierung eines Metamorphosehormones der Insekten in kristallisierter Form. Z. Naturforsch. 9B: 389–391.

Chudakova, I. V. and O. M. Bocharova-Messner. 1968. Endocrine regulation of the condition of the wing musculature in the imago of the house cricket, *Achaeta domestica*. Proc. Acad. Sci. USSR 179: 157–159 (in Russian).

Chudakova, I. V., O. M. Bocharova-Messner, and V. Novák. 1974. Analogues of the juvenile hormone and their action on the morphogenesis of the wing musculature in the cricket, *Achaeta domestica*. Ontogenesis 5: 130–138 (in Russian).

Clark, R. B. 1956. On the origin of neurosecretory cells. Ann. Sci. Nat. Zool. 18: 199–207.

Dubinin, N. P. 1970. *General Genetics*. Nauka, Moscow (in Russian).

Geigy, R. 1941. Die Metamorphose als Folge gewebsspezifischer Determination. Rev. Suisse Zool. 48: 483–494.

Gersch, M. 1964. *Vergleichende Endocrinologie der wirbellosen* Tiere. Akademishe Verlagsgesellschaft Geest & Portig KG, Leipzig.

Gersch, M. 1970. Generelle Probleme der Neuroendokrinologie der wirbellosen Tiere. Biol. Rundsch. 8: 77–89.

Hadorn, E. 1965. Problems of determination and transdetermination: genetic control of differentiation. Brookhaven Symp. Biol. 18: 148–161.

Hadorn, E. 1966. Konstanz, Wechsel und Typus der Determination und Differenzierung *in vivo*. Dev. Biol. 13: 424–509.

Hadzhi, I. 1963. *The Evolution of the Metazoa*. Pergamon Press, Oxford and Elmsford, New York.

Hatchek, B. 1878. Studien über Entwicklungsgeschichte der Anneliden. Arb. Zool. Inst. Wien No. 1.

Hinton, H. E. 1948. On the origin and function of the pupal stage. Trans. Roy. Entomol. Soc. Lond. 99: 395–409.

Hinton, H. E. 1955. On the structure, function and distribution of the prolegs of the Panorpoidea with a criticism of the Berlese–Imms theory. Trans. Roy. Entomol. Soc. Lond. 106: 455–556.

Janet, Ch. 1899. Sur les nerfs céphaliques, les corpora allata et le tentorium de la fourmi *Myrmica rubra*. Mém. Soc. Zool. Fr. 12: 295–335.
Jenkin, P. M. 1965. A comparative view of the hormonal control of molting and ecdysis in arthropods. Gen. Comp. Endocrinol. 5: 688–689.
Jenkin, P. M. 1966. Apolysis and hormones in the molting cycles of arthropods. Ann. Endocrinol. (Paris) 27: 331–341.
Jenkin, P. M. and H. E. Hinton. 1966. Apolysis in arthropod molting cycles. Nature (Lond.) 211: 871.
Jezhikov, J. J. 1936. Metamorphose, Cryptometabolie und direkte Entwicklung. Zool. Anz. 114: 451–452.
Jezhikov, J. J. 1940. Über frühe Embryonalstadien und ihren Zusammenhang mit den Typen der postembryonalen Entwicklung bei den Insekten. C. R. Acad. Sci. USSR 28: 574–576 (in Russian).
Jones, B. M. 1956. Endocrine activity during insect embryogenesis. J. Exp. Biol. 33: 174–185.
Kopeć, S. 1917. Experiments on metamorphosis of insects. Bull. Int. Acad. Sci. Cracow B pp. 57–60.
Kopeć, S. 1922. Studies on the necessity of the brain for the inception of insect metamorphosis. Biol. Bull. (Woods Hole) 42: 323–342.
Kühn, A. 1944. *Grundriss der allgemeinen Zoologie*. Thieme Verlag, Leipzig.
Laufer, H. 1968. Developmental interactions in the dipteran salivary gland. Am. Zool. 8: 257–271.
Lockshin, R. A. and C. M. Williams. 1965. Programmed cell death. I. Cytology and degeneration in the intersegmental muscles of the pernyi silkmoth. J. Insect Physiol. 11: 123–133.
Lüscher, M. 1972. Environmental control of juvenile hormone secretion and caste differentiation in termites. Gen. Comp. Endocrinol. (Suppl) 3: 509–514.
Malá, J., V. Novák, I. Blaszek, and A. Balázs. 1974. The effect of juvenile hormone on the prothoracic glands in *Galleria mellonella*. I. Morphology of the glands in the course of postembryonic development. Acta Biol. Hung. 25: 85–95.
Matz, G. 1966. Étude du cancer expérimental chez *Locusta migratoria* L. Doctoral Thesis, University of Strasbourg.
Naton, E. 1960. Über die Entwicklung des schwarzbraunen Mehlkäfers *Tribolium destructor*. I. Z. Entomol. 46: 233–244.
Naton, E. 1962. Über die Entwicklung des schwarzbraunen Mehlkäfers *Tribolium destructor*. IV. Zool. Beitr. 8: 95–123.
Needham, J. 1942. *Biochemistry and Morphogenesis*. Cambridge University Press, London and New York.
Novák, V. J. A. 1951a. New aspect of the metamorphosis in insects. Nature (Lond.) 167: 132.
Novák, V. J. A. 1951b. The metamorphosis hormones and morphogenesis in *Oncopeltus fasciatus* Dal. Acta Soc. Zool. Česk. 15: 1–48.
Novák, V. J. A. 1955. The question of the origin and evolution of the metamorphosis in insects from the point of view of the findings on the metamorphosis hormones. Acta Soc. Entomol. Česk. 52: 31–43.
Novák, V. J. A. 1956. Versuch einer zusammenfassenden Darstellung der postembryonalen Entwicklung der Insekten. Beitr. Entomol. 6: 205–493.
Novák, V. J. A. 1959. The characters of artificially produced diapause in silkworm *Bombyx mori*. Acta Symp. Evol. Inst. Prague pp. 237–242.
Novák, V. J. A. 1964. Phylogenetic considerations on the juvenile hormone and other morphogenetically active substances in insects. Proc. 12th Int. Congr. Entomol., London pp. 215–216.
Novák, V. J. A. 1966. *Insect Hormones*, 3rd ed. Methuen, London.

Novák, V. J. A. 1967. The juvenile hormone and the problem of animal morphogenesis. Pp. 119–132 *in* J. Beament and J. Treherne (eds.), *Insect and Physiology*. Oliver & Boyd, Edinburgh.

Novák, V. J. A. 1969. Morphological analysis of the effects of juvenile hormone analogues and other morphogenetically active substances on embryos of *Schistocerca gregaria* Forsk. J. Embryol. Exp. Morphol. 21: 1–21.

Novák, V. J. A. 1971. Juvenile hormone action as a special type of DNA-derepression. Pp. 139–148 *in* V. Novák (ed.), *Insect Endocrines*. Academia, Prague.

Novák, V. J. A. 1975a. *Insect Hormones*, 4th ed. Chapman & Hall, London.

Novák, V. J. A. 1975b. Das Prinzip der Soziabilität als eines der Hauptprinzipien der Evolution der Organismen. Nova Acta Leopold. (Evolution) 42: 475–488.

Novák, V. J. A. 1981. The non-hereditary phylogenetic features and the principle of morphogenesis. Pp. 247–261 *in* V. Novák and J. Mlíkovský (eds.), *Evolution and Environment*. ČSAV, Prague.

Novák, V. J. A. 1982a. *The Principle of Sociogenesis*. Academia, Prague.

Novák, V. J. A. 1982b. Neoteny as one of the general laws of evolution. Pp. 329–345 *in* V. Novák and K. Zemek (eds.), *General Questions of Evolution* (*Proc. Int. Colloq.*). Academia, Prague.

Novák, V. J. A. 1985a. The dialectics of the relation between hereditary and non-hereditary changes. Pp. 89–95 *in* V. Novák and J. Mlíkovský (eds.), *Proc. Int. Symp. Evolution and Morphogenesis, Plzeň, 1984*. Academia, Prague.

Novák, V. J. A. 1985b. The principle of sociogenesis and its morphogenetic aspects. Pp. 141–185 *in* V. Novák and J. Mlíkovský (eds.), *Proc. Int. Symp. Evolution and Morphogenesis, Plzeň, 1984*. Academia, Prague.

Paramonov, A. A. 1967. The ways of laws of evolutionary process. Pp. 342–436 *in Recent Problems of Evolutionary Theory*. Nauka, Leningrad (in Russian).

Pener, M. P. 1965. On the influence of corpora allata on maturation and sexual behavior of *Schistocerca gregaria*. J. Zool. (Lond.) 147: 119–136.

Pflugfelder, O. 1939. Weitere experimentelle Untersuchungen über die Funktion der Corpora allata von *Dixippus morosus*. Z. Wiss. Zool. 152: 384–408.

Piepho, H. 1938. Über die Auslösung der Raupenhäutung, Verpuppung und Imaginalentwicklung an Hauptimplantaten von Schmetterlingen. Biol. Zentralbl. 58: 481–495.

Polivanova, E. N. 1968. Histophysiological investigations of the differentiation of the morphological elements in the haemolymph of *Eurygaster integriceps* embryos. Vopr. Funkc. Morfol. Embryol. Nasekomych pp. 133–140 (in Russian).

Poyarkoff, E. 1914. Essai d'une théorie de la nymphe des insectes homométaboles. Arch. Zool. Exp. Gén. 54: 221–265.

Rohdendorf, E. B. and F. Sehnal. 1972. The induction of ovarian disfunctions in *Thermobia domestica* by the cecropia juvenile hormones. Experientia (Basel) 28: 1100–1101.

Seidel, F. 1924. Organisation of the insect egg. Z. Morphol. Oekol. Tiere 1: 6–64.

Schaller, F. 1959. Controle humoral du développement post-embryonaire d'*Aeschna cyanea*. C. R. Acad. Sci. Paris 248: 2525–2527.

Schaller, F. 1971. Rôle de l'ecdysone dans le régénération de l'epithélium mésentérique des insectes odonates. Arch. Zool. Exp. Gén. 112: 695–704.

Scharrer, B. 1946. The relationship between corpora allata and reproductive organs in adult *Leucophaea maderae*. Endocrinology 38: 46–55.

Sehnal, F. 1965. Einfluss des Juvenilhormons auf die Metamorphose des Oberschlundganglions bei *Galleria mellonella* L. Zool. Jahrb. Physiol. 71: 659–664.

Sehnal, F. 1968. Influence of the corpus allatum on the development of internal organs in *Galleria mellonella* L. J. Insect Physiol. 14: 73–85.

Sehnal, F. and V. Novák. 1969. Morphogenesis of the pupal integument in the wax moth *Galleria mellonella* L. and its analysis by means of juvenile hormone. Acta Entomol. Bohemoslov. 66: 137–145.

Severtsov, A. N. 1931. *Morphologische Gesetzmässigkeiten der Evolution*. Fischer, Jena.
Shmalgauzen, I. I. 1946. *The Factors of Evolution*. Dokl. Akad. Nauk SSSR, Moscow (in Russian).
Shmalgauzen, I. I. 1982. *The Organism as a Whole in Individual and Historical Development*. Nauka, Moscow (in Russian).
Sláma, K. 1965. The effects of hormones' mimetic substances on the ovarian development and oxygen consumption in allatectomized adult females of *Pyrrhocoris apterus* L. J. Insect Physiol. 11: 1121–1129.
Sláma, K. 1983. Illusive functions of a prothoracic gland in *Galleria mellonella*. Acta Entomol. Bohemoslov. 80: 161–176.
Sláma, K. and C. M. Williams. 1966. The juvenile hormone. V. The sensitivity of the bug *Pyrrhocoris apterus* to a hormonally active factor in American paper-pulp. Biol. Bull. (Woods Hole) 130: 235–246.
Speman, H. 1952. Über das Verhalten embryonalen Gewebes im erwachsenen Organismus. Wilhelm Roux' Arch. Entwicklungsmech. Org. 141: 693–769.
Steinberg, D. M. 1938. The embryonic territories of wing and leg in the hypodermis of the caterpillars of *Galleria mellonella*. Biol. Zentralbl. 7: 993–1012 (in Russian).
Steinberg, D. M. 1956. Analyse morphogénétique du développement des organes imaginaux chez les Insectes holométaboles. Proc. 10th Int. Congr. Entomol. Vol. 2, pp. 261–266.
Urvoy, J. 1963. Transplantation de patte sur les tergites de la Blatta *Blabera craniifer* Burm. Bull. Soc. Sci. Bretagne 37: 113–118.
Weiss, P. and J. L. Kovanan. 1957. A model of growth and growth control in mathematical terms. J. Gen. Physiol. 41: 1–47.
Wigglesworth, V. B. 1934. *Insect Physiology*. Methuen, London.
Wigglesworth, V. B. 1935. Function of the corpus allatum in insects. Nature (Lond.) 136: 338.
Wigglesworth, V. B. 1936. The function of the corpus allatum in the growth and reproduction of *Rhodnius prolixus*. Q. J. Microsc. Sci. 79: 91–119.
Wigglesworth, V. B. 1939. *The Principles of Insect Physiology*, 1st ed. Methuen, London.
Wigglesworth, V. B. 1940a. The determination of characters at metamorphosis in *Rhodnius prolixus*. J. Exp. Biol. 17: 201–222.
Wigglesworth, V. B. 1940b. Local and general factors in the development of "pattern" in *Rhodnius prolixus*. J. Exp. Biol. 17: 180–200.
Wigglesworth, V. B. 1952. Hormone balance and the control of metamorphosis in *Rhodnius*. J. Exp. Biol. 29: 620–631.
Wigglesworth, V. B. 1965. *The Principles of Insect Physiology*, 6th ed. Methuen, London.
Williams, C. M. 1952. Morphogenesis and the metamorphosis in insects. Harvey Lect. 47: 126–155.

Roles of Morphogenetic Hormones in Embryonic Cuticle Deposition in Arthropods

2

GIOVANNI SBRENNA

2.1. Introduction

The processes of arthropod development and growth are generally punctuated by periodic molts, even though some insects do not necessarily molt in order to grow (Williams, 1980). Ecdysteroids (Es) and in some cases also juvenile hormone (JH) are involved in these molts and processes of metamorphosis (Matsuda, 1979; Hoffmann and Charlet, 1985). In recent years, research has focused on the presence and possible role of Es and JH during arthropod oogenesis and embryonic development, as well as on their biosynthetic metabolic pathways. Numerous reviews have been published in this field (Schooley et al., 1984; Spindler et al., 1984; Hoffmann and Lagueux, 1985).

Cuticle secretion at the end of embryonic development is a vital factor for the arthropod larva, but the cuticle with which the young begin life is not the first cuticular layer laid *in ovo*. In fact, in various groups other cuticular envelopes may be laid, either form the embryonic epidermis or during the first stages of development. "Egg envelopes" will not be considered in this chapter; for information on them, the reader is referred to reviews by Anderson (1973), Furneaux and Mackay (1976), and Hinton (1981). Here, we shall report on the "embryonic cuticles" in Arthropoda and show their relationship with the hormones present *in ovo* in order to shed some light on their role in embryonic cuticulogenesis.

2.2. Embryonic Cuticles

2.2.1. *Insecta*

The presence of a provisional embryonic cuticle has been reported in several insects, and Sikes and Wigglesworth (1931) and Miller (1940) provided us with a detailed classification which, for the most part, is still valid. It was only after 1970 that a series of ultrastructural observations provided further information on the embryonic cuticle structure and on the chronology of the events of synthesis.

Apterygota

Unfortunately, not much is known about the embryonic cuticles in Apterygota. The information available seems to indicate that there are two cuticles, which are synthesized *in ovo*. In Diplura, the first cuticle is thin (Tiegs, 1944) and is shed by the embryo before abdominal segmentation has been completed. In Collembola, Tamarelle (1972) emphasized the fact that the embryonic cuticle at hatching is thick, pigmented, and has some bristles.

Pterygota Hemimetabola

Three embryonic cuticles have been observed in Hemimetabola: the first cuticle is very thin and membranous; the second is thicker; and the third, with which the insect hatches, is similar to the larval cuticles in appearance.

The first cuticle (C1). The C1 (often 5–8 nm thick) is precociously synthesized by the germ band hypoderm, and its appearance in *Leucophaea maderae* occurs simultaneously with the serosal apolysis (Rinterknecht and Matz, 1983). It appears as an electron-dense material deposited at apices of the microvilli (F. Bullière, 1973), which then becomes a continuous layer (Fig. 2.1A,B) and is clearly trilaminar (Louvet, 1974; Dorn, 1976).

In *Acheta domesticus* (Edwards and Chen, 1979), C1 detaches itself from the epidermis during the first movements of katatrepsis. However, this would not generally be considered as a true apolysis, because it takes place at different times in the various regions and hence would not be under endocrine control (Louvet, 1974). The fate of C1 is then unclear. In *Schistocerca gregaria* (Sbrenna, 1974) it remains visible until hatching, whereas in other insects it can no longer be observed at the dorsal closure; this could be owing to its extreme fragility or to inadequate manipulation (Louvet, 1974).

The second cuticle (C2). The outer epicuticle of C2 is deposited in the form of trilaminar patches (Fig. 2.1A); even though this deposition does not occur simultaneously all over the embryo surface (Louvet, 1974; Dorn and Hoffmann, 1981), all authors agree that it takes place once the dorsal closure is completed. In general, its trilaminar structure is clearly identifiable and its thickness, of approximately 10–15 nm, is greater than that of C1.

The inner epicuticle is constituted of a dense fibrous material, deposited at the apices of the microvilli, and its thickness varies from 0.1 to 0.3 μm. In the majority of insects, a lamellar procuticle is secreted under

FIGURE 2.1. Embryonic cuticles of *Schistocerca gregaria.* (A) Underneath C1 the C2 outer epicuticle patches (oe) overlying the dense microvilli tips. ×106,000. (B) The high resolution shows the trilaminar aspect of C1. ×155,000. (C) C2 before apolysis: outer (oe) and inner (ie) epicuticle; subcuticle (sc); lamellar procuticle (lp). ×32,000. (D) The embryo surface at SEM reveals the thinness of C1 and the C2 microsculpture pattern. ×320. (E) The C3 trilaminary outer epicuticle (oe) over the epidermal surface: arrows indicate inner epicuticle in deposition. ×96,000. (A & C from Sbrenna, 1974.)

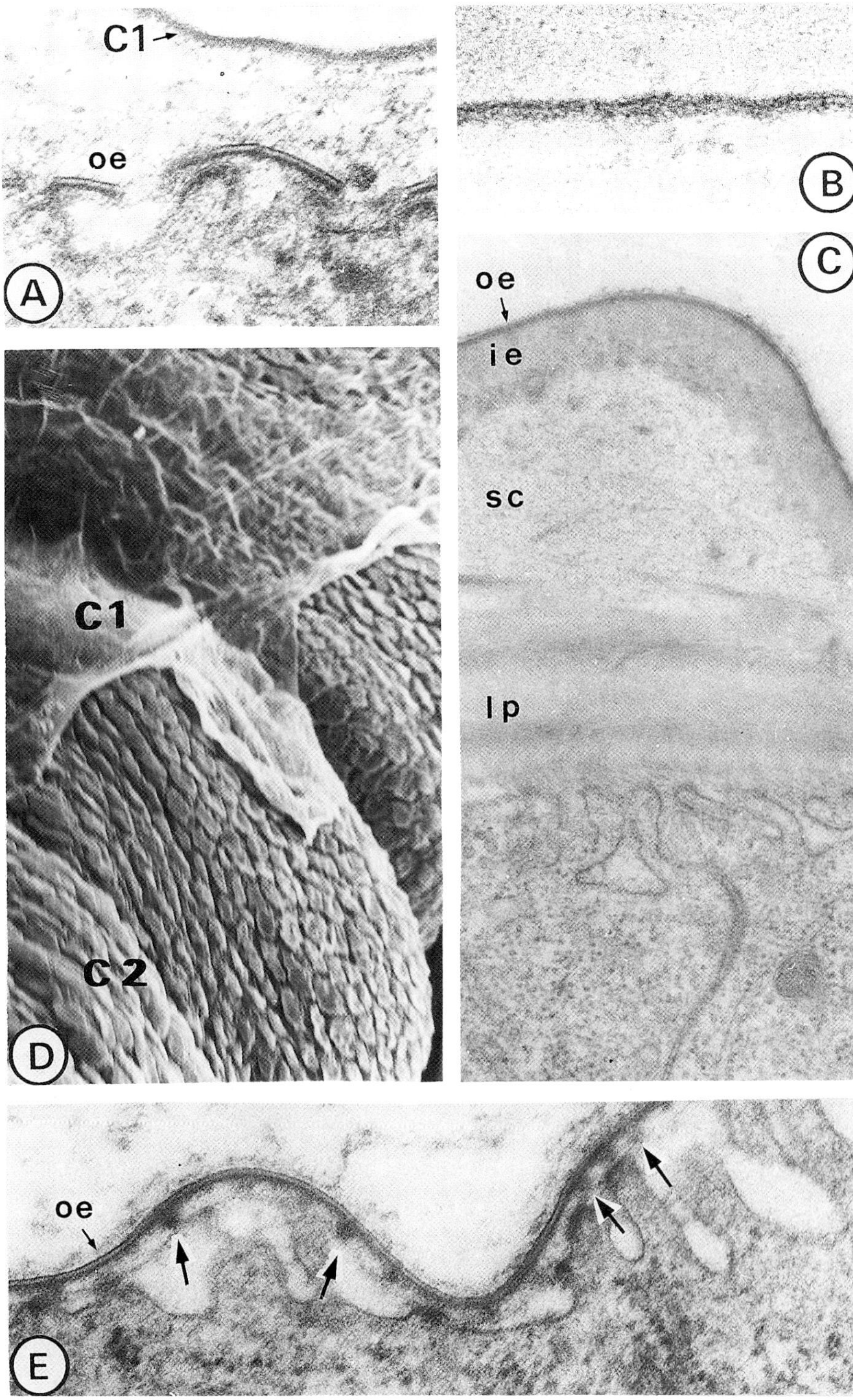
C1
oe
A
B
C
oe
ie
sc
lp
C1
C2
D
oe
E

the inner epicuticle and its thickness varies in the embryos of different species, just as it varies in the different body regions of the same embryo. In *Carausius morosus* (Louvet, 1974), procuticular lamellae are not detectable under the inner epicuticle but there is an area rich in very fine fibers.

Between the inner epicuticle and the first procuticular lamella, both in *S. gregaria* (Sbrenna, 1974) and in *Locusta migratoria* (Lagueux et al., 1979), there is a clear "subepicuticular zone" with a few fibers (Fig. 2.1C). This zone gives rise to a sinuous pattern of surface microsculptures (Fig. 2.1D), which—also in *A. domesticus* (Edwards and Chen, 1979)—would facilitate the embryo's escape from the egg envelopes at hatching.

The C2 has no hairs or pegs, but in the antennae of *L. migratoria* and in the maxillary palps of the cockroach *Periplaneta americana* (Altner and Ameismeier, 1986) tubular bodies can be observed projecting from anlagen of nymphal terminal pore sensilla toward the C2. These authors hypothesized that the nodules could be rudiments of bristles or pegs and that C2, since it is equipped with rudimentary sensilla, can be considered as an ancestral nymphal cuticle.

The C2 undergoes a true apolysis with the appearance of an exuvial space occupied by the molting gel (Dorn and Hoffmann, 1981) and is abandoned at hatching during the coinciding ecdysis.

The C2 is secreted by the fore- and hindgut cells and by the tracheae (Dorn, 1976). These different secretory patterns among various embryos and embryonic regions suggest that at the time of dorsal closure a hormonal pattern is already well defined and completely operative both from the standpoint of the specific species and of the specific tissue.

The third cuticle (C3). The outer epicuticle of C3 is deposited in the form of triple-layered patches at the apices of the microvilli (Fig. 2.1E). The inner epicuticle, which has wax canals, is secreted immediately underneath; it is followed by the procuticle, made up of several procuticular lamellae traversed by pore canals (Fig. 2.2A). The C3 is characterized by the presence of spines and bristles and can be considered to be a true nymphal cuticle. C2–C3 ecdysis occurs upon hatching, which signals the transition from embryonic to nymphal life.

Pterigota Holometabola

Ultrastructural data on embryonic cuticles of the Holometabola are scarce; hence the cuticologenetic pattern in the various orders is unclear.

FIGURE 2.2. (A) *Schistocerca gregaria* C3 before hatching. The lamellar procuticle (lp) shows pore canals (pc) with numerous filaments. ×9000. (B) An 18- to 20-h-old *Ceratitis capitata* embryo showing the C1 detaching from epidermal cells. ×34,000. (C) A 48-h-old *C. capitata* embryo. The three C2 layers are present: the outer (oe) and inner epicuticle (ie) and

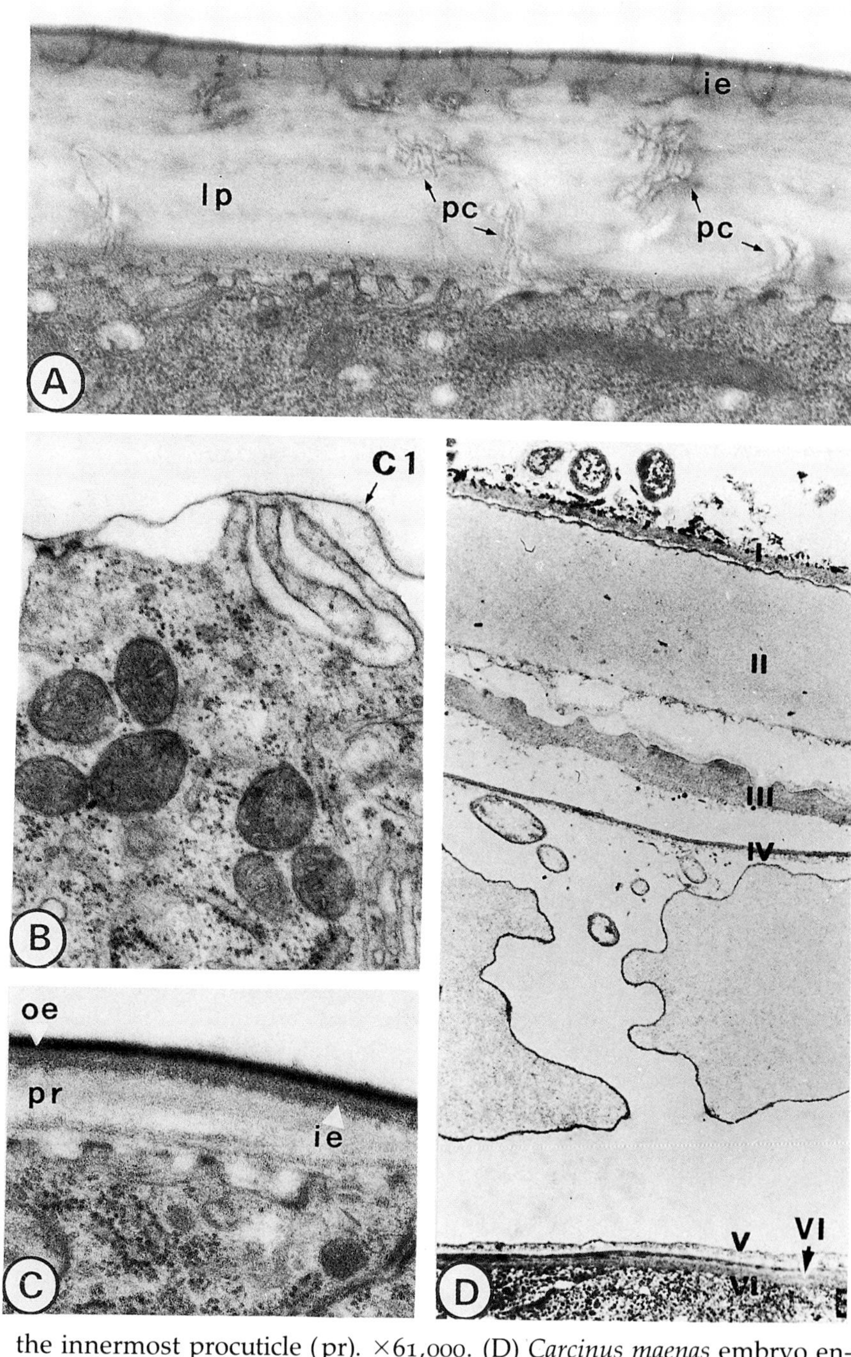

the innermost procuticle (pr). ×61,000. (D) *Carcinus maenas* embryo envelopes just before hatching. From outside to inside: fertilization envelopes (I & II); envelopes (III); embryonic cuticles (IV, V, & VI). ×11,500. (B & C, courtesy of Prof. R. Dallai; D, from Goudeau and Lachaise, 1983.)

The Lepidoptera, for instance, in which we have mainly OM (optical microscope) observations, would merit a more detailed investigation. All three cuticles have been detected in *Manduca sexta* (Dorn et al., 1987) and in *Bombyx mori* (Otsuki et al., 1976) following deposition of the serosal cuticle, but the interpretation of the first two cuticles is different. However, there does not seem to be any discordance as to the C3, designated as the "larval cuticle," with setae, crochets, and sensilla.

Not many data are also available on Diptera, but their cuticulogenetic pattern seems clearer. In fact, two cuticles have recently been detected by TEM (transmission electron microscopy) in *Calliphora erythrocephala* (Bordes-Alléaume and Sami, 1987) and in *Ceratitis capitata* (Callaini and Dallai, 1987). The very thin C1 (Fig. 2.2B) is secreted when the germ band is fully extended, it does not appear to be trilaminar and disappears before the C2 deposition. Upon dorsal closure the C2, with chitinous setae, is secreted. It consists of a trilaminar outer epicuticle, an inner epicuticle, and a procuticle (Fig. 2.2C), and becomes the first larval cuticle.

The thin structure of C1 and its constant, rapid disappearance could explain both (i) Oelhafen's report (1961), according to which there are no "embryonic molts" in *Culex pipiens,* and (ii) why only one embryonic cuticle, the "first larval cuticle," has been detected in *Drosophila melanogaster* (Hillman and Lesnik, 1970). Therefore, in the Holometabola, a third cuticulogenesis is not detected *in ovo* before hatching, as in the Hemimetabola. However, the C2 with which the insect hatches shows well-developed integumental organules. In conclusion, it would seem that not only does Williams's hypothesis (1980) regarding the evolutionary pressures for economizing on the number of molts apply during larval development but also during the embryonic period.

2.2.2. *Myriapoda*

There are no ultrastructural images or detailed descriptions at OM of the embryonic cuticle in Myriapoda. Both in the chilopod *Scutigera coleoptrata* (Knoll, 1974) and in the diplopod *Glomeris marginata* (Dohle, 1964), the embryonic epidermis deposits a thin "embryonalcuticula" (C1?), which then is shed. At the same time a "larvalcuticula" (C2?) is secreted, which thickens as it develops.

2.2.3. *Crustacea*

There is great variation in the developmental stage in which a crustacean emerges from the egg (Green, 1971; Anderson, 1982). This remarkable variability (anamorphic and epimorphic development) would imply that during development *in ovo* there are also different cuticulogenetic patterns. Unfortunately, there are not many reports on embryonic cuticulogeneses, and the very different cuticle patterns are currently difficult to schematize, given the scarcity of information on the basic embryology of many crustacean groups (Anderson, 1982).

Klepal and Barnes (1978) carried out a detailed study by TEM on the exoskeleton of different *Balanus* species (Cirripedia) during embryonic development. The epidermis of "nauplius I" does not show any evidence of exoskeletal secretory activity at the pre-limb-bud stage, but a trilaminate "blebbed" epicuticle (15–22 nm thick) and a procuticle (approximately 0.15 μm thick) composed of four or five laminae are then secreted over the embryo. Before hatching, this crustacean C1 separates itself from the epidermis and begins a second cuticular cycle. Nauplius I hatches surrounded by C1, and with the C2 ("nauplius II exoskeleton" 0.25 μm wide) in position.

This ultrastructural study clarifies previous information on the anostracan *Artemia salina* (Morris and Afzelius, 1967) and on the ostracod *Notodromas monacha* (Tetart, 1970) according to which two cuticles are secreted *in ovo* when a nauplius larva emerges at hatching.

Investigations carried out at OM on the epimorphic development of the anaspidacean *Anaspides tasmaniae* (Hickman, 1937) and of the cladoceran *Daphnia carinata* (Murugan and Venkataraman, 1977) also demonstrate that C1 appears after the three naupliar limbs and all the metameres are formed; the embryo then secretes another cuticle. Hence, in all these Crustacea, C1 is secreted *in ovo* at the end of a definitive developmental stage, the nauplius stage, in which the first three appendages appear, and its morphology would seem different from C1 of the Insecta.

Embryonic cuticulogenesis is known in Phyllocarida (Manton, 1934) and in Peracarida crustaceans, which deposit the egg in a brood pouch, or marsupium (Nair, 1956; Graf and Michaut, 1975; Wittmann, 1981). Unfortunately, the only TEM investigation on the peracarid *Hemioniscus balani* embryonic cuticles (Goudeau, 1976), in which five embryonic envelopes are described, does not clearly define the cuticular cycles.

Several cuticulogeneses occur before hatching in Decapoda, in which successive embryonized larval stages appear. In TEM studies, three embryonic cuticles ("envelopes IV, V, and VI") have been clearly demonstrated (Fig. 2.2D) in the condensed embryogenesis of *Carcinus maenas* (Goudeau and Lachaise, 1983). According to these authors, their secretory processes recall the elaboration of the trilaminate epicuticles reported for insects. The first two cuticles, composed of an epicuticle and a subjacent procuticle with a fine granular texture, are respectively synthesized during the metanauplius and protozoea stages. "Envelope VI," which is secreted before hatching during the prezoea stage, shows a laminated procuticle and after hatching constitutes the future zoeal cuticle.

Therefore, it appears evident that in Decapoda, in which an embryonalization occurs, the capacity to synthesize the nauplial cuticle is lost. Consequently, the epidermis effects three cuticular syntheses during *in ovo* development, as in the Hemimetabola. This number of cuticles seems to be the maximum permitted during the entire embryonic life span of Arthropoda.

2.2.4. *Chelicerata*

Because of the partial or scarce information on chelicerate embryology, data on *in ovo* cuticulogeneses are also lacking in this group; however, there are some brief mentions of it in OM studies dealing with the entire embryonic development (Anderson, 1973; Yoshikura, 1975; Aeschlimann and Hess, 1984). On the other hand, embryonic cuticulogeneses of the Acarina are known. In *Ornithodorus moubata* (Vogel, 1975; Dotson et al., 1991), the C1 consists of a dense layer (22 nm) and a fibrous layer (200–700 nm), and the C2 of a "cuticulin" layer and a lamellar procuticle whose deposition continues even after hatching. These two cuticulogeneses, also observed in embryos of *Boophilus microplus* and *Amblyomma hebraeum* (E. M. Dotson, unpublished data), seem to be a general feature of embryonic tick development. These data complete previous information on *in ovo* cuticulogeneses in Acarina. In fact, the "tritovum" (Claparède, 1869) and "embryonal apodermata" (Henking, 1882), with which the six-legged larva hatches, would correspond to C2, the last cuticle secreted *in ovo*, which goes into apolysis shortly after hatching (Dotson et al., 1991).

2.3. Ecdysteroid Hormones During Embryogenesis

2.3.1. *Ecdysteroids in Insecta*

Studies of ecdysteroids (Es) present during insect embryonic development have progressed remarkably during the last decade after it was proved that the adult ovaries synthesize large quantities of Es (Hagedorn et al., 1975; Hetru et al., 1978) and after their *in vitro* synthesis by the follicle cells was demonstrated (Goltzené et al., 1978). Interesting reviews have been published on the presence of Es in the eggs of various insect species (Hoffmann et al., 1980; Hagedorn, 1983; Hoffmann and Lagueux, 1985).

To date, apart from ecdysone (E) and 20-OH-E, at least 15 other Es have been isolated from insect embryos (see Chapter 1 by Gupta and Chapter 10 by Thompson et al. in Part 1). This is undoubtedly owing to the fact that the egg is a closed system (Agrell and Lundquist, 1973) and therefore permits the identification of both the precursors and the eventual metabolites present within it. In order to outline the data obtained in the various species the following points are reported:

(a) The eggs of *B. mori, Galleria mellonella, L. migratoria, Macrotermes bellicosus, Macrotermes subhyalinus, S. gregaria, Thermobia domestica* (see Hoffmann and Lagueux, 1985), and *M. sexta* (Warren et al., 1986) contain conjugated Es, mainly as 22-phosphate esters at very high concentrations and various free deoxy-E, E, and 20-OH-E at micromolar concentrations (Fig. 2.3). The conjugated fraction represents 80–85% of the total amount of the Es in *G. mellonella* (Hsiao and Hsiao, 1979), 95% in *S. gregaria* (Scalia and Morgan, 1982; Rees and Isaac, 1984), and 95–98% in *L. migratoria* embryos (Lagueux et al., 1984). Consequently, considerable amounts of the conjugated Es are still present at the end of embryonic development: at hatching, *S. gregaria* eggs contain conjugated E (140 ng/egg), 2-deoxy-E (100 ng/egg), and 20-OH-E (40 ng/egg) (Scalia et al., 1987).

(b) Maternal conjugated Es have not been evidenced in the eggs of *Blaberus craniifer, Calliphora vicina, Clitumnus extradentatus, C. morosus, D. melanogaster, L. maderae, Nauphoeta cinerea, Oncopeltus fasciatus* (see Hoffmann and Lagueux, 1985), *Sarcophaga bullata* (Wentworth and Roberts, 1984), and *C. erythrocephala* (Bordes-Alléaume and Sami, 1987). In these species, on the other hand, free Es and their metabolites are present *in ovo*.

(c) The free forms of the Es differ in the various species and are present in low concentrations, exhibiting fluctuations during embryonic development (Fig. 2.4). While 20-OH-E is generally the predominant ecdysteroid during the larval stages, E is the main free hormone in *L. migratoria* (Fig. 2.5) and *S. gregaria* eggs, its level varying from 4 to 8 μM (Lagueux et al., 1984) and from 30 to 79 ng/egg (Scalia et al., 1987), respectively. On the contrary, in *N. cinerea* embryos, 20-OH-E is the major ecdysteroid (Lanzrein et al., 1984). Furthermore, the predominant ecdysteroid in the developing eggs of *O. fasciatus* (Dorn, 1983) is makisterone A (mA), the ratio E:20-OH-E:mA being 1.0:0.8:0.07 in newly laid eggs, becoming 1.0:0.8:2.3 after blastokinesis.

(d) It has been hypothesized that the free hormone peaks could be due to the hydrolysis of the maternal conjugates. The latter might therefore serve as inactive storage forms that would release the free Es at precise times during embryogenesis (Mizuno and Ohnishi, 1975; Hoffmann et al., 1980). In *S. gregaria* embryos, the phosphatase enzyme involved in conjugate hydrolysis has the highest relative specific activity in the mitochondrial–lysosomal fraction (Rees and Isaac, 1984). Nevertheless, clear-cut conversion relationships between the conjugated and free forms have not been observed during embryogenesis in *L. migratoria* (Sall et al., 1983), in *B. mori* (Ohnishi, 1986), and in *S. gregaria* (Scalia et al., 1987) (Fig. 2.3).

(e) Besides the conjugate hydrolysis in *L. migratoria,* the capacity to perform the hydroxylation steps of free E precursors present in micromolar concentrations in the newly laid eggs has been observed (Meister et al., 1985; Dimarcq et al., 1987). Unfortunately, similar research has not yet

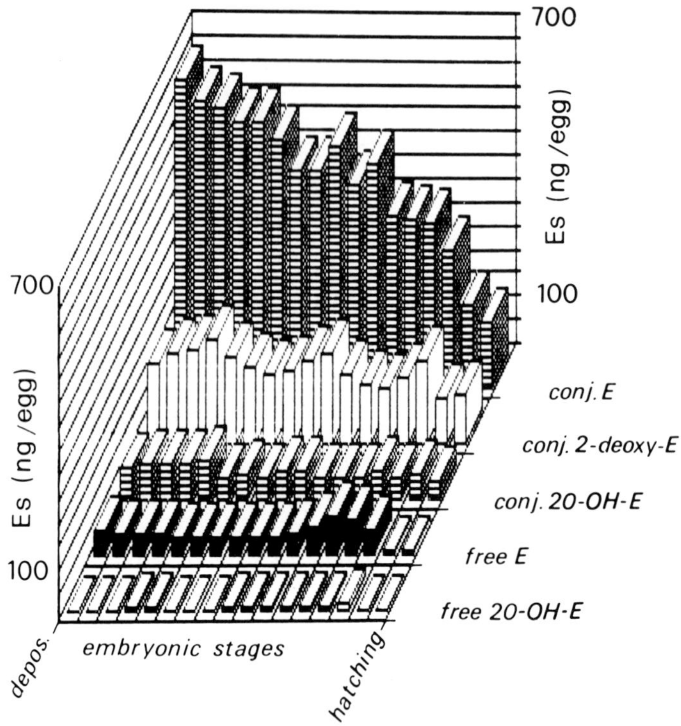

FIGURE 2.3. Fluctuations of free and conjugated Es in *S. gregaria* eggs during embryonic development. (Es titers from Scalia et al., 1987.)

been carried out on species lacking in maternal conjugated forms, in which the increments of free forms presuppose a *de novo* embryonic synthesis.

(f) Apart from conjugates of maternal origin, highly polar conjugates composed of a variety of Es metabolites that accumulate in eggs during embryogenesis have been isolated and identified (Isaac and Rees, 1984, 1985; Lagueux et al., 1984).

(g) The eggs contain not only free and polar hydrolyzable conjugates but also a variety of polar Es metabolites (i.e., ecdysonic acid and 20-OH–ecdysonic acid), whose quantities increase toward the end of embryogenesis (Isaac and Rees, 1984, 1985).

In conclusion, it can be stated that there is an active metabolism in the developing eggs and the conjugates are an integral part of this dynamic system (Hagedorn, 1983), representing either a storage form of the free hormones or their inactivation products.

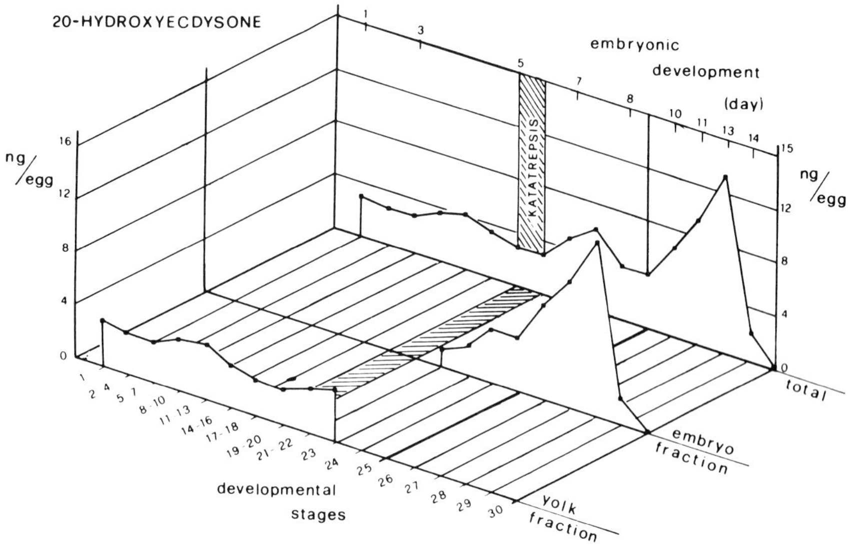

FIGURE 2.4. Titers of free 20-OH-E in the embryo and yolk fractions of *S. gregaria* developing eggs. (From Scalia et al., 1987.)

2.3.2. *Ecdysteroids in Myriapoda*

It has been shown by high-pressure liquid chromatography (HPLC) that the ovaries of the chilopod *Lithobius forficatus* produce *in vitro* both E and 20-OH-E (Leubert et al., 1982). Moreover, the presence of Es has been demonstrated by these authors in mature eggs of the same species by means of thin-layer chromatography (TLC) and radioimmunoassay (RIA). There are no reports of Es hormones in diplopod eggs.

2.3.3. *Ecdysteroids in Crustacea*

Ecdysteroids were first shown to be present in *C. maenas* egg extracts in 1977 by Lachaise and Hoffmann using RIA. Successive studies carried out by combining RIA with HPLC analysis and preceded, in some cases, by enzymatic hydrolysis have demonstrated the presence of free 20-OH-E and E during the embryogenesis of *Orchestia gammarella* (Blanchet et al., 1979), *Callinectes sapidus* (McCarthy and Skinner, 1979), *Achantonyx lunulatus* (Chaix and De Reggi, 1982), *C. maeanas* (Lachaise and Hoffmann, 1982), and *Palaemon serratus* (Spindler et al., 1987) and also in the eggs of *Astacus leptodactylus* and *A. salina* (Spindler et al., 1984). Moreover, ponasterone A is present in *C. sapidus, C. maenas, A. salina,* and *P. serratus,* almost always representing the predominant form of the total RIA activity.

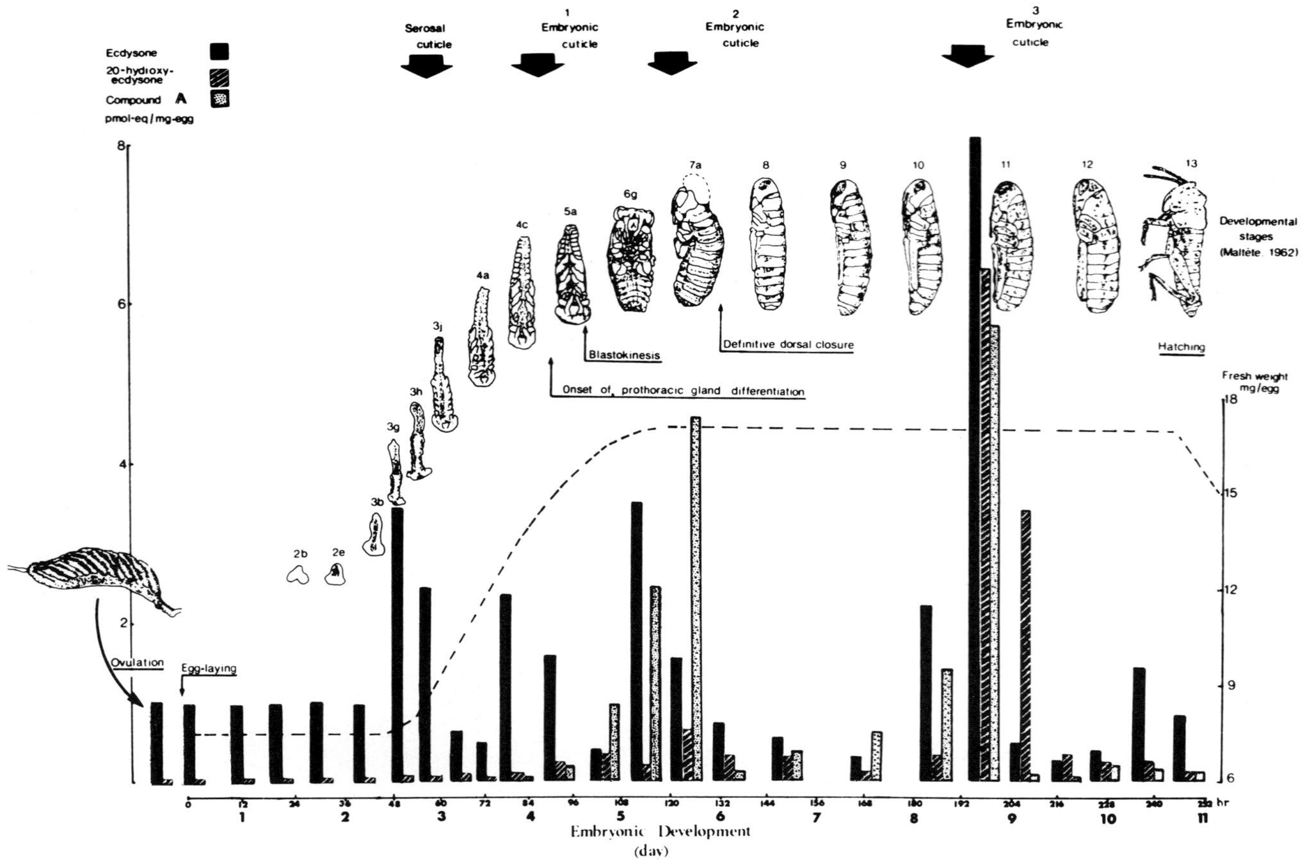

FIGURE 2.5. Titers of E, 20-OH-E, and compound A in relation to the embryonic development of *Locusta migratoria*. Arrows indicate onset of deposition of the different cuticles listed at the top of the figure. (From Lagueux et al., 1979.)

Although marked qualitative and quantitative differences exist during embryogenesis among the various Es that could be traced to the different specificity of several antisera used and to the diverse extraction methods employed, several aspects emerge which we believe deserve to be studied more closely:

(a) The production of free Es in the ovaries of reproducing females before deposition (Lachaise and Hoffmann, 1977; Lachaise et al., 1981; Blanchet et al., 1979) and the presence of free Es in newly laid eggs suggest a maternal origin for these hormones; however, experimental proof of this is lacking.

(b) The concentration of the diverse free Es varies during *in ovo* development, either by increasing from deposition to hatching (*O. gammarella, C. sapidus, C. maenas,* and *P. serratus*) or by undergoing fluctuations (*A. lunulatus*) (Fig. 2.10). This would indicate that there is a *de novo* biosynthesis of these substances. With regard to this, it is relevant that in *P. serratus* the Y-organ appears and differentiates only shortly before the prezoea stage (Le Roux, 1983).

(c) In addition to the free forms, Es conjugates are found in "dormant" *A. salina* cysts, lending weight to the hypothesis of their role as storage compounds (Spindler et al., 1984). On the other hand, the considerable increase of free ponasterone A during the early stages of the embryonic development of *C. maenas* cannot be explained by the hydrolysis of conjugated ponasterone A. Indeed, this compound becomes noticeable only after the free form concentration peak (Lachaise and Hoffmann, 1982).

(d) In *C. maenas* (Lachaise and Hoffmann, 1982) the appearance during embryogenesis of highly polar conjugates acting as inactivation products confirms the existence *in ovo* of an active Es metabolism in Crustacea as well.

(e) Lastly, the low titer of the crustacean Es could be explained by the presence (in *Orconectes limosus* eggs and in *A. salina* cysts) of binding proteins (receptors) both for Es and ponasterone A, with a higher affinity for Es than that observed for insects (Spindler et al., 1984).

2.3.4. *Ecdysteroids in Chelicerata*

In Chelicerata, the presence of Es has been detected almost exclusively during postembryonic development (reviewed by Bonaric, 1987). However, in the course of the last decade, knowledge of Es in Acarina has increased considerably (reviewed by Solomon et al., 1982; Diehl *et al.*, 1986) and a certain amount of attention has also been devoted to embryonic Es.

Using RIA before and after enzymatic hydrolysis and HPLC analysis, Connat et al. (1985) have identified free E and mainly free 20-OH-E in newly laid *A. hebraeum* eggs. Unspecified quantities of these free

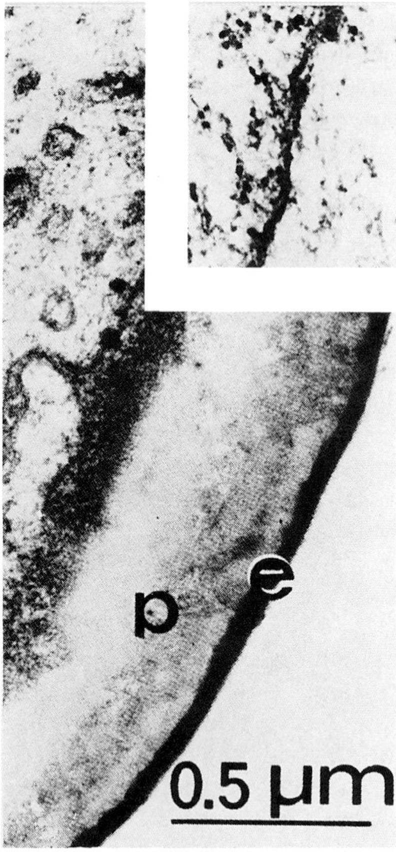

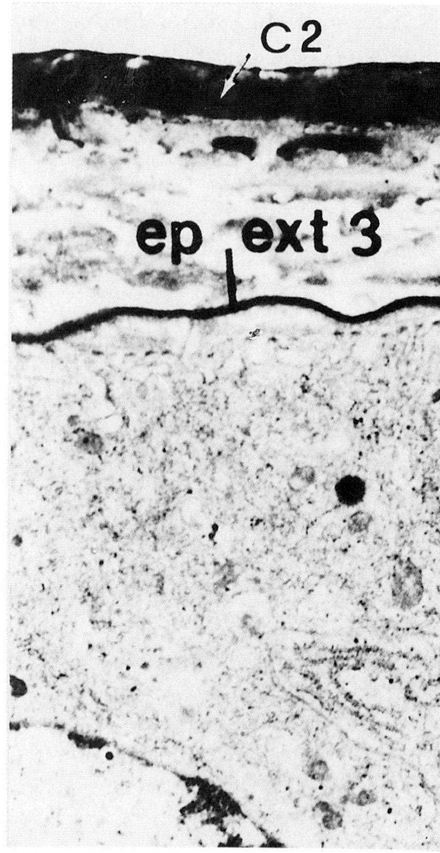

FIGURE 2.6. *Blaberus craniifer* embryonic tibiae, cultured in a medium containing 50 μg/ml of inokosterone, after 8 days of explanation. C2 has been secreted: e = epicuticle; p = procuticle. The inset shows that the control epidermis, explanted in a medium without an exogenous hormone, has not deposited cuticle. (From F. Bullière and D. Bullière, 1977.)

FIGURE 2.7. Cephalic cuticle of *Clitumnus extradentatus* embryo injected with 40 nmol/g of 20-OH-E at stage VI6b. Both C2 and C3 were secreted after 34 days: ep ext 3 = C3 outer epicuticle. ×15,500. (From Kadiri, 1983, courtesy of Prof. J. P. Louvet.)

protein-bound hormones are also present in *Ripicephalus appendiculatus* eggs (Whitehead et al., 1986). In these species, no polar Es conjugates corresponding to the maternal substances present in the insect eggs were found.

In the eggs of *O. moubata* (Connat et al., 1984), *B. micropilus* (Wigglesworth et al., 1985), *R. appendiculatus, Hyalomma dromedari,* and *Ixodes ricinus* (J. L. Connat and associates, unpublished data) ovarian apolar products are incorporated into the eggs. This maternal hormone represents a new class of apolar esterase-labile conjugates of E and 20-OH-E with long-chain fatty acid esters, bound at C-22). The role of these apolar products, which are also found in other arthropod groups (Connat and Diehl, 1986), has not yet been elucidated.

2.4. Role of Embryonic Ecdysteroids

While the role of Es as molting hormones during postembryonic arthropod development is generally accepted, no precise function for these steroids during embryogenesis has been established. This is probably owing to both the qualitative complexity of Es present in eggs and to the fact that these Es may have various physiological roles. Moreover, for some of them (2-deoxy-E, makisterone A, and ponasterone A), their function as hormonal messengers has not yet been confirmed.

2.4.1. *Role of Ecdysteroids in Insecta*

In 1963, N. S. Mueller and D. H. Bucklin (cited in Boohar and Bucklin, 1963) hypothesized that the "embryonic moults" of *Melanoplus differentialis* were controlled by E through a change of the "threshold of reaction" in the epidermal cells. However, Dorn and Romer (1976) were the first to correlate an Es (free E) peak with the second cuticologenesis in *O. fasciatus.* After this observation, other studies have clearly emphasized the correlation between the free Es fluctuations and the onset of embryonic cuticle synthesis in *L. migratoira* (Lagueux et al., 1979) (Fig. 2.5), *B. craniifer* (D. Bullière et al., 1979), *L. maderae* (Matz, 1980), *C. extradentatus* (Cavallin and Fournier, 1981), and *C. morosus* (Fournier, 1985) (see Fig. 2.9, below, in Section 2.4.1.1).

With regard to *S. gregaria,* in which the concentration variations of E and 20-OH-E have been studied in carefully staged embryos, a relationship between the titer of 20-OH-E and the cuticular depositions has been shown, especially for the C3 (G. Sbrenna, unpublished data).

On the other hand, note that in *B. mori* the Es fluctuations do not correspond with cuticle formation (Mizuno et al., 1981; Ohnishi, 1986) and in *C. erytrocephala* only one Es peak, preceding the second cuticulogenesis, is detected (Bordes-Alléaume and Sami, 1987).

The relationship between the rise of Es and cuticle synthesis has also been investigated *in vitro.* F. Bullière (1973) and F. Bullière and D. Bullière (1977), while culturing embryonic tissue in the presence of inokosterone, caused an acceleration of the last cuticologenesis with deposition of a trilaminar cuticle and a procuticle, which appears with a less defined stratification than in the controls (Figs. 2.6 and 2.8). The addition of inokosterone also provokes an accelerated uptake of proline and glucosamine, which coincides with the accelerated inner epicuticular synthesis and the beginning of procuticular synthesis (Bullière, 1977).

After diverse quantities of 20-OH-E are injected into *C. extradentatus* eggs at various stages of development, the embryonic epidermal cells respond by synthesizing a cuticle no matter what developmental stage they are in (Kadiri, 1983). This study, besides establishing the role of the hormone as the trigger of cuticular syntheses, shows that the embryos treated before the synthesis of C2 accelerate this process, but the interval between this and the following cuticulogenesis remains unchanged (Fig. 2.7). The embryos treated shortly before or immediately after the deposition of C2 accelerate the synthesis of C3.

It would appear from these first experimental attempts that the Es, in particular 20-OH-E, play a role in activating a sequence of biochemical events. Its presence, at developmental stages where it should be absent, determines nothing more than the accelerated synthesis of the successive cuticulogeneses and does not seem to influence the cuticular deposition program in any way (Fig. 2.8). Hence, as in postembryonic development (Riddiford, 1985), 20-OH-E evokes also in the embryos the secretory pattern characteristic of the epidermic cells at that precise moment of development.

However, this does not exclude the fact that in several species different Es can be used as signal molecules. When the last *M. sexta* cuticulogenesis (Warren et al., 1986) occurs, E and 20-OH-E do not exhibit fluctuations and do not appear to coordinate embryonic cuticular events as they do during postembryonic development (Riddiford, 1985); on the contrary, a high titer of 20,26-E is detected in correspondence with this larval-like C3.

2.4.1.1. CONTROL MECHANISM OF THE CUTICULAR SYNTHESES

The relationship between the presence of Es and the triggering of embryonic cuticulogeneses poses the problem of the relation between neuroendocrine centers (NCs) and the cuticular syntheses in insect embryos.

On the basis of histological observations and legation experiments, Jones (1956a,b) affirmed that the NCs of *L. migratoria* and *Locustana pardalina* control embryonic molting and pigmentation. However, investigations on embryos of various species such as *Leptinotarsa decemlineata* (Haget, 1953, 1957), *M. differentialis* (Mueller, 1963), *B. mori* (Takami, 1963), *Hyalophora cecropia* (Mueller and Bucklin, 1965), *Scapsipedus marginatus* (Grellet, 1965), and *S. gregaria* (Sbrenna-Micciarelli and Sbrenna, 1972) demonstrate that, unlike what occurs in larvae, cuticulogenesis and integumental organule differentiation of the embryo are not con-

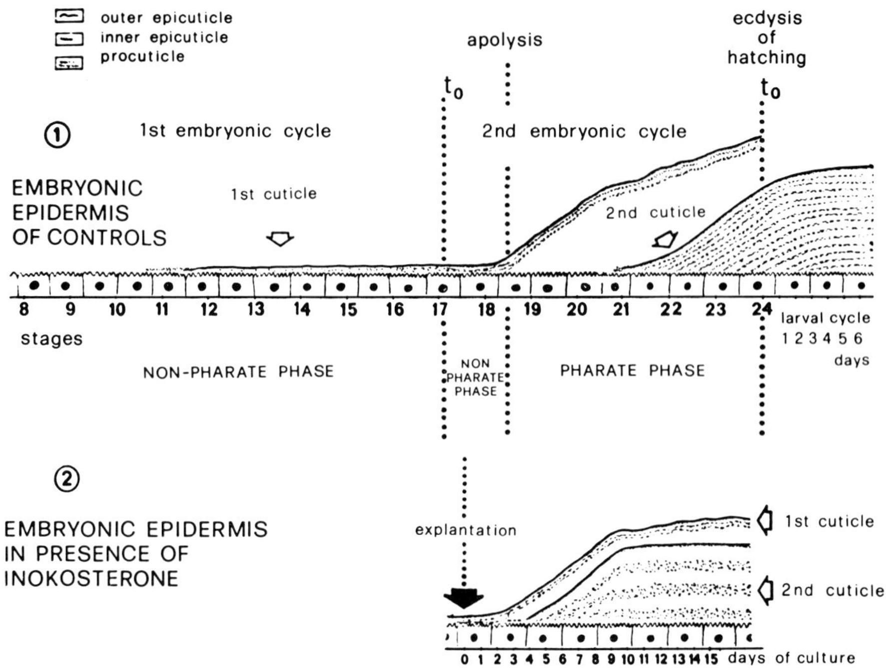

FIGURE 2.8. (1) *B. craniifer* cuticular depositions *in vivo* of the embryonic epidermis. The C2 deposition starts at stage 21. (2) Cuticular depositions in culture in the presence of inokosterone At the end of the third day of culture the accelerated synthesis of C2 begins. (From F. Bullière, 1973.)

trolled by the prothoracic glands (or ring gland) and the NCs. In fact, if the embryonic anterior and posterior halves are isolated before the cuticular synthesis, the normal cuticulogenetic succession patterns occur.

An attempt has been made to explain the marked contrast between these results and those obtained by Jones (1956a,b) under different experimental conditions (Hoffmann and Lagueux, 1985). In fact, the eggs used in such experiments contain a high quantity of conjugated Es of maternal origin and the fragments are provided *in vitro* with an adequate yolk supply. However, experiments excluding the NCs carried out recently on *C. morosus* (Fournier, 1985) and on *C. erythrocephala* (Bordes-Alléaume and Sami, 1987), which are characterized by eggs with minimal or negligible quantities of conjugated Es in the initial developmental period, provide evidence that the capacity to effect Es biosynthesis (Fig. 2.9), and thus to deposit cuticles, is maintained in the absence of stored maternal conjugates.

Furthermore, Meister et al. (1985), by incubating various *L. migratoria*

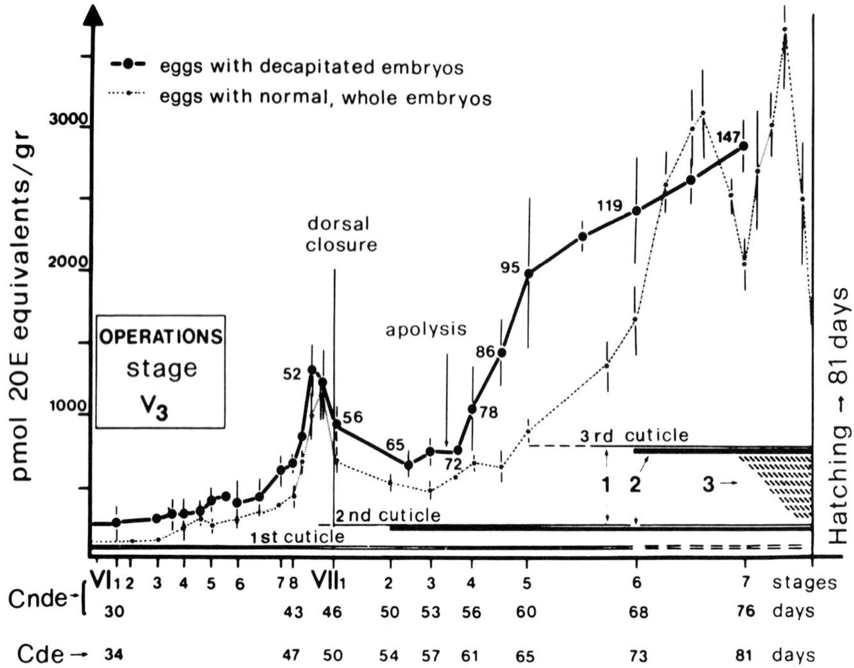

FIGURE 2.9. Titer evolution of total Es in *Carausius morosus* eggs containing both decapitated and normal whole embryos. *Key:* Cde = controls, intact or perforated dechorionated eggs (= sham operated); Cnde = controls, non-dechorionated eggs. The chronology of the cuticular depositions is shown. 1, 2, & 3 = outer epicuticle, inner epicuticle, and procuticle, respectively. (From Fournier, 1985.)

embryonic tissues in the presence of a labeled E precursor (C2,22,25-trideoxy E) normally present in newly laid eggs, have shown that the capacity to hydroxylate deoxy-E to E is present not only in the PG but also in the embryonic epidermal cells and that the enzymatic systems exist before PG differentiation.

Recent results of microsurgical reduction *in ovo* of *C. morosus* embryos (Radallah and Fournier, 1988) support the above evidence. The biosynthesis of Es is in fact more similar to the normal pattern when two regions or "compartments" are retained *in ovo* (either head-thorax, thorax-abdomen, or head-abdomen) rather than a single "compartment" (either head, thorax, or abdomen). The cuticular depositions, which are strictly dependent on the Es, also vary according to the scale of the *in ovo* reductions carried out. In the presence of two compartments the cuticologeneses prove to be normal, whereas with a single compartment only one cuticle is synthesized.

These data indicate that the insect embryo not only has a surprising capacity for Es biosynthesis but also is precociously endowed with an efficient system controlling the activity of the enzymatic molecules involved in Es biosynthesis.

2.4.2. *Role of Ecdysteroids in Noninsect Arthropoda*

The only hypothesis that can be made regarding the physiological significance of *in ovo* Es during the embryonic development of noninsect arthropods is that their major function concerns embryonic cuticle deposition.

In Crustacea, this role has been suggested in *O. gammarella* by Blanchet et al. (1979) and in *C. maenas* by Lachaise and Hoffmann (1982). According to the latter authors, ponasterone A would also be involved in the cuticulogeneses. On the other hand, in *P. serratus*, the functional relevance of ponasterone A is of minor importance, given the lack (in crustaceans) of experimental proof for any of its activities (Spindler et al., 1987).

A reasonable temporal correlation between the Es levels and cuticulogenesis has been shown in *A. lunulatus* (Chaix and De Reggi, 1982), where four hormonal peaks are detected. While it is difficult to link the first peak to any developmental event, the second seems to coincide with the "molt" that the embryo (metanauplius III) undergoes at the end of metamerization (Fig. 2.10). Similarly the other two peaks in hormone concentration seem to be related to the "molts" occurring at the protozoea and prezoea embryonic stages and to the relative cuticular syntheses.

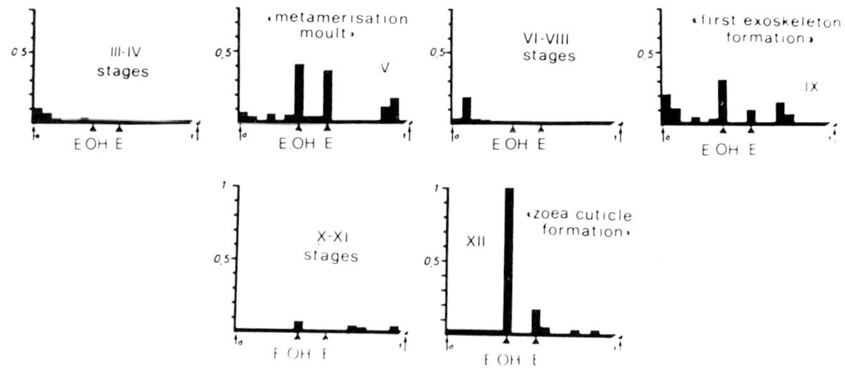

FIGURE 2.10. Changes in Es levels in *Acanthonyx lunulatus* eggs during embryonic development in correspondence with cuticle deposition: III–XII, developmental stages (IX & X, protozoea and prezoea stages, respectively). (Modified from Chaix and De Reggi, 1982.)

In Acarina (Chelicerata), it has been suggested that the Es present during embryogenesis control cuticle deposition. In *O. moubata* (Dotson et al., 1991) a peak was detected (10 pg 20-OH-E equivalents/embryo) in correlation with the epicuticle deposition of the last embryonic cuticulogenesis (C2). This hypothesis is supported by the fact that two or three days after hatching another hormonal peak occurs (32 pg 20-OH-E equivalent/larva) when the first epicuticle of postembryonic development is synthesized. In contrast, no hormonal titer fluctuations were evidenced during the deposition of C1 and the blastodermic cuticle. According to Dotson et al. (1991), this difference could be traced to the brief duration of the peaks or to the lack of sensitivity of the method.

2.5. Juvenile Hormone

While considerable research has been carried out on embryonic Es, very little work has been done on juvenile hormones (JHs) present in eggs. It is not yet clear if JHs are involved in the normal development of certain groups of noninsect arthropods (Laufer et al., 1987).

In Crustacea, the results of some research, prompted by Schneiderman and Gilbert's (1958) experiments, have supported the hypothesis that JH or a JH-like compound, methyl farnesoate (Borst et al., 1987; see also Chapter 2 by Borst and Laufer in Part 1), also plays an important role in the embryonic physiology of these arthropods. Examples of this are the effects of JH I and JH analogues on egg development of the isopod *Armadillium vulgare* (G. Greer, cited in Pihan, 1975) and on the reproduction of the cladoceran *Daphnia magna* (Templeton and Laufer, 1983). In particular, the study on *D. magna* has shown that especially in the first developmental stages, the treatment causes the appearance of either abnormal, small, or imperfectly developed embryos with some well-differentiated integumental parts with bristles.

No studies of a possible relationship between JH and the embryonic cuticles of Myriapoda and Chelicerata have been carried out (Diehl et al., 1982; Solomon et al., 1982; Booth et al., 1986).

2.5.1. *Juvenile Hormone in Insect Embryos*

By means of mainly bioassay methods, JH-active material has been observed during embryonic development in *B. mori, H. cecropia, L. migratoria, M. subhyalinus,* and *O. fasciatus* (see Hoffmann and Lagueux, 1985). However, the insufficient sensitivity of the methods used and the difficulty of identification have led to observations that have subsequently proved to be imprecise. Advances in isolation methods and structure elucidation using coupled gas chromatography/mass spectroscopy (GC/MS) have allowed Bergot et al. (1981) to determine the levels of JH titers from selected embryos of various insect species and to demonstrate the presence of JH I, JH II, and the previously unknown JH 0 and iso-JH 0 in *M. sexta* embryos. Successive qualitative and quantitative determinations of JHs by the GC/MS method have shown that only JH III is present in the majority of insect orders (Schooley et al., 1984; see also chapter 12 by Baker in Part 1).

Current information on JH during embryogenesis derives principally from the studies carried out on *N. cinerea* (reviewed in Lanzrein et al., 1984) in which JH III accumulates in the hemolymph after dorsal closure. As development takes its course, its concentration increases markedly, reaching 800 ng/g, approximately 17 pmol/embryo (Bürgin and Lanzrein, 1988), and then decreases while hatching approaches (Fig. 2.11). During the early embryonic development of *N. cinerea*, i.e., before corpora allata (CA) differentiation, JH III is absent, hence excluding the existence, as for *O. fasciatus* (Dorn, 1983), of maternal storage forms (Brüning et al., 1985).

Large quantities (about 800 ng/g) of methyl farnesoate (MF) have been observed in *N. cinerea* embryos after dorsal closure (Lanzrein et al., 1984), and this precursor of JH (Fig. 2.11) has been observed in the embryonic hemolymph shortly before JH III appears (Brüning et al., 1985). MF was found to be present specifically in the embryos; indeed, it is lacking in *N. cinerea* nymphs and adults (Lanzrein et al., 1984) owing to the low *in ovo* titer of MF epoxidase and to its scarce activity (Brüning

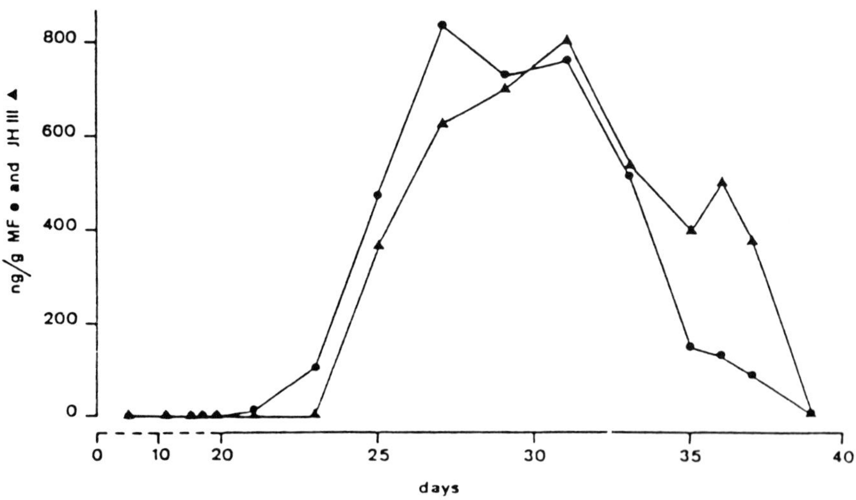

FIGURE 2.11. JH III (triangles) and MF (dots) titers throughout *Nauphoeta cinerea* embryonic development. *Abscissa:* age of embryos in days. (From Brüning et al., 1985.).

et al., 1985). However, this precursor has been found to be absent in *L. migratoria* embryos (Pener et al., 1986).

Recently, JH III variations using the GC/MS method have also been studied in developing *L. migratoria* eggs (Temin et al., 1986). The analyses of the entire embryonic period (Fig. 2.12) complete the partial data provided by Pener et al. (1986). A minimal quantity of the hormone is present in the deposition period (15 pg/egg, corresponding to at least 50 hormonal molecules). Thus, in this species there is no active transfer of JH from the mother to the egg, as is true for the maternal Es. Shortly before the third cuticulogenesis and the last free Es peak (Lagueux et al., 1979), JH III reaches a maximum (25–30 μM/egg), and it is thought that the hormone is exclusively of embryonic origin also in this insect.

The elevated concentrations of JH in *N. cinerea* embryos, although in accord with recent data on *Blattella germanica* (Brüning and Lanzrein, 1987), differ from the hormonal values reported for *L. migratoria,* as well as from those of embryos of other species such as *M. sexta* (6.4 ng/egg; Bergot et al., 1981) and *Teleogryllus commodus* (0.05 μg/g; Loher et al., 1983). It is still difficult to explain such differences in JH titers both among diverse species and different stages of the developmental time span; they could be due to the different sensitivity of the target tissues or to different functions carried out by the hormone. Moreover, to date we do not even know whether the JH levels evidence in the developing

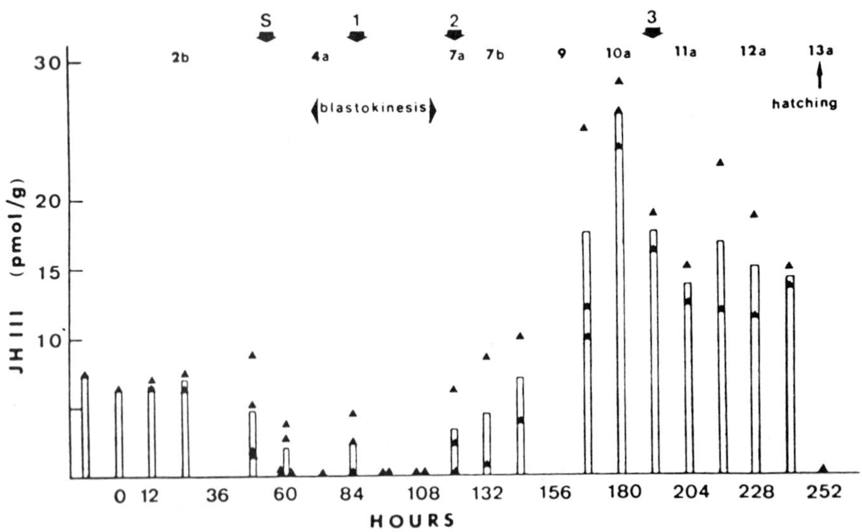

FIGURE 2.12. JH III titer fluctuations in *L. migratoria* eggs during embryogenesis. Large arrows on top indicate the periods of deposition of serosal cuticle (S), C1 (1), C2 (2), and C3 (3). (From Temin et al., 1987.)

eggs are strictly linked to a threshold concentration or whether they exceed what is necessary for a normal morphogenesis. In *M. sexta*, for example, normal embryogenesis and hatching in newly laid eggs occurs even if the *in vivo* embryonic endogenous JH levels are reduced by sixfold by means of topical treatment with fluoromevalonate (Bergot et al., 1981). Phenomena commonly associated with JH deficiency are expressed in *M. sexta* only in subsequent postembryonic development.

The fact that embryonic CA are capable of synthesizing JH seems to be substantiated by (i) the aforementioned fluoromevalonate experiments (Bergot et al., 1981); (ii) the investigations carried out on *N. cinerea* (Brüning et al., 1985; Bürgin and Lanzrein, 1988) in which corpora cardica–corpora allata (CC–CA) complexes synthesize and release *in vitro* both JH III (0.5 pmol/6 h) and MF (some pmol/6 h); and (iii) *in vitro* studies by embryonic CA on *Diploptera punctata* (Kikukawa and Tobe, 1987) in which the glands synthesize exclusively detectable quantities of JH III (0.2–1.6 pmol/h per pair).

Precocene III (P III) treatments of the *N. cinerea* "egg cases" (Brüning et al., 1985) and of *L. migratoria* eggs (Pener et al., 1986) confirm the active participation of the CA in the biosynthesis of JH III, since they determine a drastic reduction in its titer.

2.5.2. *Role of Juvenile Hormone*

Given that our attention is essentially focused on the role of JH in the embryonic cuticulogenesis, this subsection will deal chiefly with hemimetabolous insects, as there is an almost total lack of data not only on noninsect arthropods but also on holometabolous insects.

The potential practical use of the JH analogues as "insecticides of the third generation" (Williams, 1956) provided the impetus that spurred many researchers, from the mid-1960s to the mid-1970s, to investigate the effects of the administration of exogenous JH and JH analogues to insect eggs (see Smith and Arking, 1975; Enslee and Riddiford, 1977). However, in these studies the lack of optimal treatment procedures, the use of high dosages, and the impossibility of establishing the quantity of the substance actually absorbed have rendered such reports, though interesting from many points of view, somewhat inconclusive regarding the JH role during embryogenesis and especially regarding embryonic integumental differentiation and its cuticulogeneses. It is our opinion, nonetheless, that the studies undertaken to show both variations of JH and its biosynthesis *in ovo* are much more pertinent to the subject of this chapter. In both investigations Hemimetabola embryos were used.

Should the fluctuations of JH during embryonic development of *N. cinerea* (Brüning et al., 1985) and *L. migratoria* (Temin et al., 1986) be correlated with the different morphologies of the three cuticles secreted *in ovo* (see Section 2.2.1, above), then JH might well have a role in the third hemimetabolous cuticulogenesis (Figs. 2.11 and 2.12).

Moreover, the titer analyses clearly indicate the absence of JH III at stages of intense organogenesis and tissue differentiation during (i) blastokinesis, (ii) the developmental period preceding dorsal closure, and (iii) the secretion of C2. In *A. domesticus,* in fact, Roe et al. (1987) have shown a maximum JH esterase activity in newly laid eggs. Not only does this confirm Enslee and Riddiford's hypothesis (1977) regarding the relationship between the presence of JH in the first embryogenetic phases (following exogenous administration) and the subsequent absence of the embryo–larva transition, but it also indicates that its presence *in ovo* is indispensable for inducing this transition after dorsal closure.

The foregoing points are still largely speculative, as many aspects require further study. We still do not know, for example, the nature of the embryonal interactions between JH and the Es that during larval life play a regulatory role in biosynthesis (Laufer and Borst, 1983; Granger et al., 1987); neither do we know the physiological significance of the fall in JH III titer observed before hatching.

2.5.2.1. JUVENILE HORMONE AND EMBRYO–LARVA REPROGRAMMING

JH I and JH III have been administered in the periembryonic fluid of *C. extradentatus* embryos (Kadiri, 1983). In the presence of JH I (880 pmol/g) before dorsal closure and the synthesis of C2, instead of forming C2 and C3 as in normal development (Louvet, 1974), the embryo synthesizes only one cuticle, whose epicuticular and procuticular layers appear very similar (at TEM) in thickness and structure to those of C3 (Fig. 2.13A). The presence of hairs and pigmented bristles gives this cuticle a distinctly larval character. However, when the hormone is administered

FIGURE 2.13. (A) *C. extradentatus* tibiae of a JH I–treated embryo (880 pmol/g) at the VI6b stage. Only the C1 and a second cuticle (cut 2), with procuticle showing a larval structure, are present. ×10,500. (B) Cephalic cuticle of a JH I–treated *C. extradentatus* embryo at the VI8b stage (shortly before the C2 deposition). The inner epicuticle (ep int) secreted after treatment shows a stratified structure. ×57,000. (C) The abdominal sternite of *S. gregaria* embryos after topical farnesyl methyl ether treatment following blastokinesis. The only cuticle formed shows a stratified inner epicuticle (ie). ×54,000. (D) Irregular outer epicuticle synthesis in *S. gregaria* sternite embryos in presence of farnesyl methyl ether during the C2 synthesis. ×104,000. (E) Section at OM of farnesyl methyl ether–treated *S. gregaria* embryo. The C2, like the C3, shows hairs and bristles. (A & B, from Kadiri, 1983, courtesy of Prof. J. P. Louvet; C, from Sbrenna-Micciarelli, 1977.)

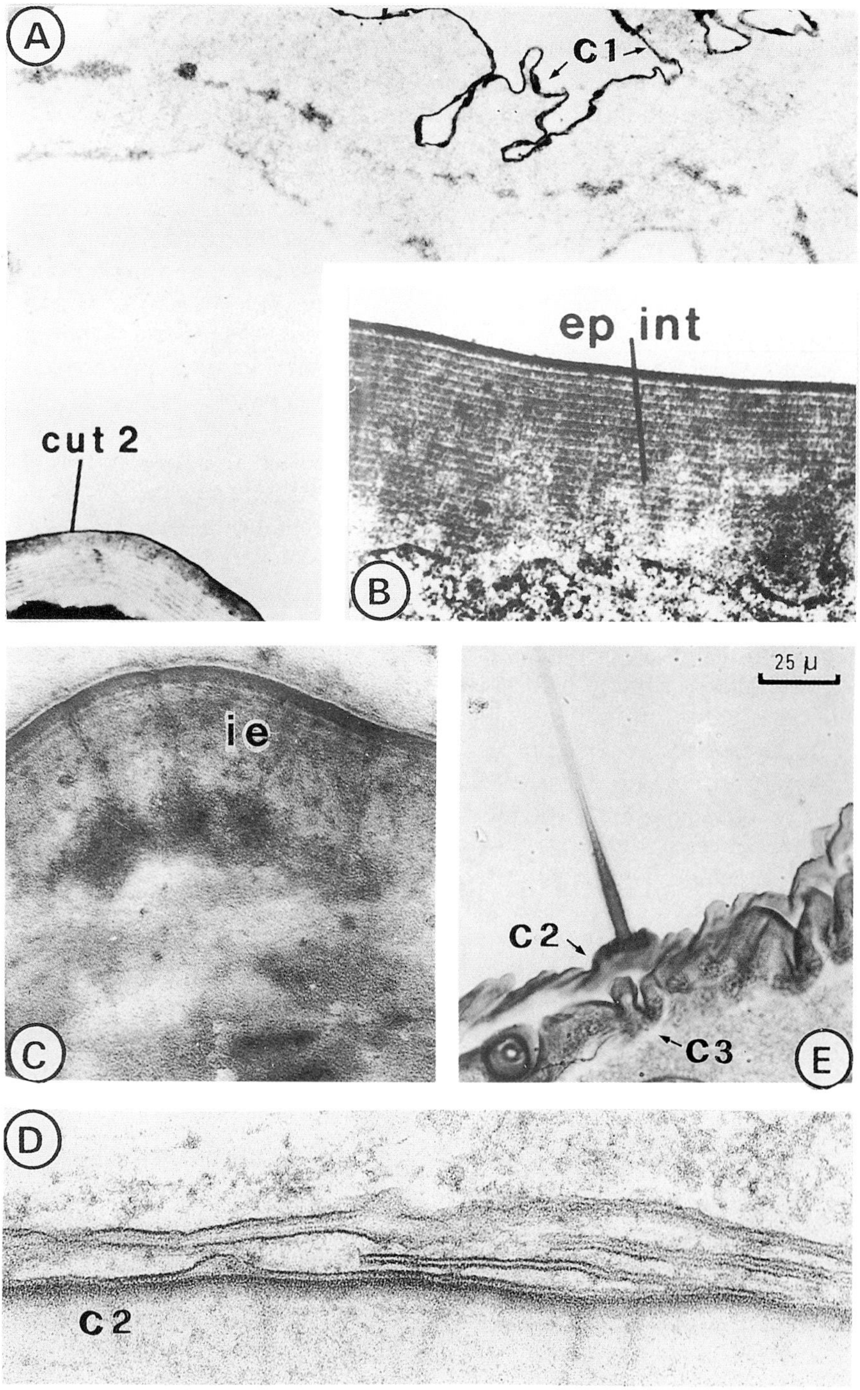
A
c1
cut 2
ep int
B
ie
C
25 µ
c2
c3
E
D
c2

during the synthesis of C2, even though only one cuticulogenesis occurs the ultrastructural patterns show irregular secretions or a thick periodic inner epicuticle (Fig. 2.13B). Similar responses (Fig. 2.13C,D) are also induced in *S. gregaria* embryos by means of topical administration of farnesyl methyl ether (G. Sbrenna, unpublished data).

In *C. extradentatus,* these extensive alterations in number and type of the embryonic cuticles were found to be dependent on the doses administered. A gradual diminution in the dosage of JH results in a return to the deposition of two cuticles as in the controls; however, the deposited C2 is always altered and shows a certain degree of larval-type epicuticular and procuticular layers even if a minimal dose (70 pmol/g) is administered. A larval-like C2 endowed with hairs and bristles (Fig. 2.13E) is also observed in *S. gregaria* embryos whose eggs were topically treated with farnesyl methyl ether immediately before blastokinesis (Sbrenna-Micciarelli, 1977). Injeyan et al. (1979) also revealed an interference of JH III with the normal development of the "provisional cuticle" (C2) when applied to the hydropyle end of *S. gregaria* eggs. During embryo–nymph ecdysis, this cuticle, perhaps because of its thicker epicuticle, is shed with difficulty by the "vermiform larva."

Apart from altering the normal pattern of embryonic cuticulogenesis, the administration of JHs or of farnesyl methyl ether induces, in accordance with studies on JH and larval pigmentation control (Goodman et al., 1987), the anticipation of mandible and tegumental pigmentation in *S. gregaria* (Sbrenna-Micciarelli, 1977; Injeyan et al., 1979), in *O. fasciatus* (Dorn, 1982), and in *C. extradentatus* (Kadiri, 1983).

The role of JH in embryonic morphogenetic events can be demonstrated particularly in the last cuticulogenesis by causing its absence experimentally using allatocidal agents on developing eggs. Following treatment of the eggs with P III in *N. cinerea,* the absence of JH III immediately after dorsal closure (i.e., when the CA become active and the JH III and MF titers increase) delays normal development (Brüning and Lanzrein, 1987). The cuticle in particular lacks larval characteristics: hairs and bristles may either be absent or, like the mandibles, may not be pigmented. Exogenous JH III applications confirm that the hormone performs an important role in the last embryonic morphogenesis, since one application is sufficient to mitigate the aforementioned alterations.

Lastly, when P III is applied to *L. migratoria* eggs (Aboulafia-Baginsky et al., 1984), after CA organogenesis and dorsal closure (thus at more advanced stages than those treated in *N. cinerea*), the embryos are found to be completely insensitive to CA degeneration and to the consequent JH III titer reduction; only during postembryonic development can prothetelic morphogenetic disturbances be observed (Pener et al., 1986). These experimental results clearly indicate that during the prehatching phase the target tissues have already been determined and are no longer modifiable.

In conclusion, it can be speculated that JH plays a role in the reprogramming of the hemimetabolous insect epidermis from embryonic to larval tissue at the third cuticulogenesis. In fact, its presence *in ovo* seems necessary when the Es evoke this cuticular synthesis to select the various genes (Willis and Cox, 1984) that trigger the larval characteristics.

The hormone appears to have an effect during embryogenesis contrary to that observed in postembryonic development (Laufer and Borst, 1983). In the larva, its function is to maintain the *status quo* (Williams, 1953); in the embryo, it is the *aditus* of larval characteristics. Indeed, in the presence of Es, depriving the larva of JH results in an anticipated metamorphosis (Granger and Bollenbacher, 1981), with the appearance of precocious adultiforms of reduced dimensions; by administering JH to young embryos, an anticipated larvalization is obtained, with the appearance of precocious larval forms of reduced dimensions (see Smith and Arking, 1975 and Enslee and Riddiford, 1977). Actually, at both the embryonic and larval developmental stages, JH always affects the target tissue (epidermis) in the same manner, i.e., by selecting the genes inducing larval characteristics during cuticulogenesis.

2.6. Conclusions

Although there is considerable detailed knowledge of insect embryonic cuticles, very little is known about those of Crustacea, Chelicerata, and Myriapoda. In agreement with Anderson's statement [in his review (1979) on embryos, fate maps, and the phylogeny of arthropods] that "new embryological research is required," we think that new studies on embryonic cuticulogenesis are necessary.

In relation to hormones involved in the control of such cuticular syntheses, it has been clearly established that arthropod eggs (Insecta, Crustacea, Chelicerata Acarina) contain both free and conjugated Es and that variations, particularly of the free forms, occur in correspondence with the embryonic molts. Investigations have demonstrated (unfortunately only with regard to insects), by means of both *in vivo* and *in vitro* Es administrations, the role of Es is triggering embryonic cuticle synthesis. Thus, it would seem that the major function of embryonic free Es is similar to that of Es during postembryonic development. It would be opportune to gather information also on the embryonic Es of the other Chelicerata and Myriapoda to determine if the biosynthesis, metabolism, and role of the Es have remained the same during the course of arthropod evolution.

The fact that successive cuticular syntheses, each one differing from the other, occur *in ovo* before endocrine gland differentiation (Y-organ and PG), and that these depositions also take place in pieces of insect embryos experimentally deprived of the NCs, leads to the suggestion that (a) there is a stage-dependent modulation of target cell responsiveness and (b) a control mechanism, either the maternal conjugate hydrolysis or of integumental biosynthesis of the Es, already exists *in ovo* during the first developmental period. Nothing certain is known in relation to this.

Although the results of research on JH reported here are very interesting, they do not completely clarify the role of endogenous JH either in embryonic cuticulogenesis or in the embryogenesis taken as a whole. In fact, the available data are on hemimetabolous insects, whereas very little is known of the holometabolous insects and practically nothing of the other arthropods. Nevertheless, it is speculated that in the Hemimetabola JH controls the ectodermal development *in ovo* and induces the epidermis to change its determinative state. Indeed, studies substantiate its obligatory absence during early embryogenesis, whereas its presence seems to be necessary during late embryogenesis in concomitance with an Es peak and with the prehatching determinative events. Thus, for JH, it is hoped that further studies will strengthen its role as *aditus* at the larval stage and, moreover, that they will elucidate the incidence of this hormonal form in the various *in ovo* developmental phases of the noninsect arthropods.

2.7. Summary

During arthropod embryonic development, both blastodermic and embryonic cuticles are secreted. Three embryonic cuticles are present in hemimetabolous insects whose morphology and synthetic processes are well known. Recent studies on Diptera embryos confirm that two embryonic cuticles are secreted in this order. No ultrastructural data are available on myriapod and chelicerate embryonic cuticles apart from some species (Acarina), whereas for the Crustacea detailed information is available on the *in ovo* cuticulogeneses of the Cirripedia and Decapoda.

A wealth of data exists on the biosynthesis and metabolism of the Es in insect eggs, whereas there is no detailed knowledge about it in noninsect arthropods apart from some Crustacea and Chelicerata (Acarina).

The temporal correlations evidenced in the majority of species studied (in insects and also in some crustaceans and ticks) between cuticulogenesis and hormonal fluctuations suggest that the Es hormones regulate the cuticular syntheses *in ovo* as they do in the larvae. Experimental investigations by means of exogenous administrations confirm the capacity of Es to evoke a cuticulogenetic response in the embryonic epidermal cells.

The extensive knowledge of embryonic Es contrasts with the lack of data on JHs. In practice, particularly in the Hemimetabola, JH is present in the embryos only after dorsal closure, closely linked to the last pulse of 20-OH-E and the last cuticulogenesis. On the basis of results obtained with exogenous administrations and allatotoxins in Hemimetabola embryos, it is hypothesized that JH induces the larval characteristics of the last embryonic cuticulogenesis.

Acknowledgments

I am grateful to Drs. J. L. Connat, B. Fournier, B. Lanzrein, and M. Walker for providing me with unpublished data and to numerous colleagues for sending me their papers. I am most indebted to Professor J. P. Louvet and Dr. Z. Kadiri for allowing me to reproduce their unpublished electron micrographs. I would like to thank my wife, A. Sbrenna-Micciarelli, as well as Dr. S. Scalia and Mrs. K. Geiger for their help and assistance in the preparation of the manuscript.

References

Aboulafia-Baginsky, N., M. P. Pener, and G. B. Staal. 1984. Chemical allatectomy of late *Locusta* embryos by a synthetic precocene and its effect on hopper morphogenesis. J. Insect Physiol. 30: 839–852.

Aeschlimann, A. and E. Hess. 1984. What is our current knowledge of acarine embryology? Pp. 90–99 *in* D. A. Griffiths and C. E. Bowman (eds.), *Acarology VI*, Pt. 1. Ellis Horwood, Chichester, England.

Agrell, I. P. S. and A. M. Lundquist. 1973. Physiological and biochemical changes during insect development. Pp. 159–247 *in* M. Rockstein (ed.), *The Physiology of Insecta*, Vol. 1. Academic Press, Orlando, Florida.

Altner, H. and F. Ameismeier. 1986. Tubular bodies in dendritic outer segments projecting to second embryonic cuticle from anlangen of contact chemoreceptors in *Locusta migratoria* L. (Orthoptera: Acrididae) and *Periplaneta americana* (L.) (Dictyoptera: Blattidae). Int. J. Insect Morphol. Embryol. 15: 253–262.

Anderson, D. T. 1973. *Embryology and Phylogeny in Annelids and Arthropods*. Pergamon Press, Oxford and Elmsford, New York.

Anderson, D. T. 1979. Embryos, fate maps, and the phylogeny of arthropods. Pp. 59–105 *in* A. P. Gupta (ed.), *Arthropod Phylogeny*. Van Nostrand Reinhold, New York.

Anderson, D. T. 1982. Embryology. Pp. 1–41 *in* D. E. Bliss (ed.), *The Biology of Crustacea*, Vol. 2. Academic Press, Orlando, Florida.

Bergot, B. J., F. C. Baker, D. C. Cerf, G. Jamieson, and D. A. Schooley. 1981. Qualitative and quantitative aspects of juvenile hormone titers in developing embryos of several insect species: discovery of a new JH-like substance extracted from eggs of *Manduca sexta*. Pp. 33–45 *in* G. E. Pratt and G. T. Brooks (eds.), *Developments in Endocrinology*, Vol. 15: *Juvenile Hormone Biochemistry*. Elsevier/North-Holland Publ., Amsterdam and New York.

Blanchet, M. F., P. Porcheron, and F. Dray. 1979. Variations du taux des ecdystéroïds au cours des cycles de mue et de vitellogenese chez le Crustacé Amphiphode, *Orchestia gammarellus.* Int. J. Invertebr. Reprod. 1: 133–139.

Bonaric, J. C. 1987. Moulting hormones. Pp. 111–118 *in* E. Nentwig (ed.), *Ecophysiology of Spiders.* Springer-Verlag, Berlin and New York.

Boohar, R. and D. H. Bucklin. 1963. Possible ecdysone activity in the embryo of the grasshopper, *Melanoplus differentialis.* Am. Zool. 3: 496.

Booth, T. F., D. J. Beadle, and R. J. Hart. 1986. The effects of precocene treatment on egg wax production in Gene's organ and egg viability in the cattle tick *Boophilus microplus* (Acarina Ixodidae): an ultrastructural study. Exp. Appl. Acarol. 2: 187–198.

Bordes-Alléaume, N. and L. Sami. 1987. Ecdysteroid titres and cuticle depositions in embryos of the Dipteran *Calliphora erythrocephala.* Int. J. Invertebr. Reprod. 11: 109–122.

Borst, D. W., H. Laufer, M. Landau, E. S. Chang, W. A. Hertz, F. C. Baker, and D. A. Schooley. 1987. Methyl farnesoate and its role in crustacean reproduction and development. Insect Biochem. 17: 1123–1127.

Brüning, E. and B. Lanzrein. 1987. Function of juvenile hormone III in embryonic development of the cockroach, *Nauphoeta cinerea.* Int. J. Invertebr. Reprod. 11: 29–44.

Brüning, E., A. Saxer, and B. Lanzrein. 1985. Methyl farnesoate and juvenile hormone III in normal and precocene-treated embryos of the ovoviviparous cockroach *Nauphoeta cinerea.* Int. J. Invertebr. Reprod. 8: 269–278.

Bullière, D., F. Bullière, and M. de Reggi. 1979. Ecdysteroid titres during ovarian and embryonic development in *Blaberus craniifer.* Wilhelm Roux's Arch. Dev. Biol. 186: 103–114.

Bullière, F. 1973. Cycles embryonnaires et sécrétion de la cuticle chez l'embryon de blatte, *Blabera craniifer.* J. Insect Physiol. 19: 1465–1479.

Bullière, F. 1977. Effects of moulting hormone on RNA and cuticle synthesis in the epidermis of cockroach embryo cultured *in vitro.* J. Insect Physiol. 23: 393–401.

Bullière, F. and D. Bullière. 1977. Règénération, différenciation et hormones de mues chez l'embryon de blatte en culture *in vitro.* Wilhelm Roux's Arch. Dev. Biol. 182: 255–275.

Bürgin, C. and B. Lanzrein. 1988. Stage-dependent biosynthesis of methyl farnesoate and juvenile hormone III and metabolism of juvenile hormone III in embryos of the cockroach, *Nauphoeta cinerea.* Insect Biochem. 18: 3–9.

Callaini, G. and R. Dallai. 1987. Cuticle formation during the embryonic development of the dipteran *Ceratitis capitata* Wied. Boll. Zool. 54: 221–227.

Cavallin, M. and B. Fournier. 1981. Characteristics of development and variations in ecdysteroid levels in *Clitumnus* embryos deprived of their cephalic endocrine glands. J. Insect Physiol. 27: 527–534.

Chaix, J. C. and M. De Reggi. 1982. Ecdysteroid levels during ovarian development and embryogenesis in the spider crab *Acanthonyx lunulatus.* Gen. Comp. Endocrinol. 47: 7–14.

Claparède, E. 1869. Studien an Acariden. Z. Wiss. Zool. 18: 445–546.

Connat, J. L. and P. A. Diehl. 1986. Probable occurrence of ecdysteroid fatty acid esters in different classes of arthropods. Insect Biochem. 16: 91–97.

Connat, J. L., P. A. Diehl, H. Gfeller, and M. Morici. 1985. Ecdysteroids in females and eggs of the ixodid tick *Amblyomma hebraeum.* Int. J. Invertebr. Reprod. 8: 103–116.

Connat, J. L., P. A. Diehl, and M. Morici. 1984. Metabolism of ecdysteroids during the vitellogenesis of the tick *Ornithodoros moubata* (Ixodoidea, Argasidae): accumulation of apolar metabolites in the eggs. Gen. Comp. Endocrinol. 56: 100–110.

Diehl, P. A., A. Aeschlimann, and F. D. Obenchain. 1982. Tick reproduction: oogenesis and oviposition. Pp. 277–350 *in* F. D. Obenchain and R. Galum (eds.), *Physiology of Ticks.* Pergamon Press, Oxford and Elmsford, New York.

Diehl, P. A., J. L. Connat, and E. Dotson. 1986. Chemistry, function and metabolism of tick ecdysteroids. Pp. 165–193 *in* J. R. Sauer and J. A. Hair (eds.), *Morphology, Physiology and Behavioral Biology of Ticks*. Ellis Horwood, Chichester, England.

Dimarcq, J. L., E. Walcher, and M. F. Meister. 1987. Conversion studies of tritiated 2-deoxyecdysone during the embryonic development on *Locusta migratoria*. Insect Biochem. 17: 871–875.

Dohle, W. 1964. Die Embryonalentwicklung von *Glomeris marginata* (Villers) im Vergleich zur Entwicklung anderer Diplopoden. Zool. Jahrb. Anat. Ontog. 81: 241–310.

Dorn, A. 1976. Ultrastructure of embryonic envelopes and integument of *Oncopeltus fasciatus* Dallas (Insecta, Heteroptera). I. Chorion, amnion, serosa, integument. Zoomorphologie 85: 111–131.

Dorn, A. 1982. Precocene-induced effects and possible role of juvenile hormone during embryogenesis of the milkweed bug *Oncopeltus fasciatus*. Gen. Comp. Endocrinol. 46: 42–52.

Dorn, A. 1983. Hormones during embryogenesis of the milkweed bug, *Oncopeltus fasciatus* (Heteroptera: Lygaeidae). Entomol. Gen. 8: 193–214.

Dorn, A., S. T. Bishoff, and L. I. Gilbert. 1987. An incremental analysis of the embryonic development of the tobacco hornworm, *Manduca sexta*. Int. J. Invertebr. Reprod. 11: 137–158.

Dorn, A. and P. Hoffmann. 1981. The "embryonic moults" of the milkweed bug as seen by the S.E.M. Tissue & Cell 13: 461–473.

Dorn, A. and F. Romer. 1976. Structure and function of prothoracic glands and oenocytes in embryos and last larval instar in *Oncopeltus fasciatus* Dallas. Cell Tissue Res. 171: 331–350.

Dotson, E. M., J. L. Connat, and P. A. Diehl. 1991. Cuticle deposition and ecdysteroid titres during embryonic and larval development of the argasid tick *Ornithodoros moubata* (Murray, 1877, *sensu* Walton, 1962) (Ixodoidea: Argasidae). Gen. Comp. Endocrinol. (in press).

Edwards, J. S. and S. W. Chen. 1979. Embryonic development of an insect sensory system, the abdominal cerci of *Acheta domesticus*. Wilhelm Roux's Arch. Dev. Biol. 186: 151–178.

Enslee, E. C. and L. M. Riddiford. 1977. Morphological effects of juvenile hormone mimics on embryonic development in the bug *Pyrrhocoris apterus*. Wilhelm Roux's Arch. Dev. Biol. 181: 163–181.

Fournier, B. 1985. Sécrétions cuticulaires et ecdystéroïdes chez les embryons decapités du phasme *Carausius morosus* Br. Int. J. Invertebr. Reprod. 8: 349–362.

Furneaux, P. J. S. and A. L. Mackay. 1976. The composition, structure and formation of the chorion and the vitelline membrane of the insect egg-shell. Pp. 157–176 *in* H. R. Hepburn (ed.), *The Insect Integument*. Elsevier, Amsterdam and New York.

Goltzené, F., M. Lagueux, M. Charlet, and J. A. Hoffmann. 1978. The follicle cell epithelium of maturing ovaries of *Locusta migratoria:* a new biosynthetic tissue for ecdysone. Hoppe-Seyler's Z. Physiol. Chem. 359: 1427–1434.

Goodman, G. W., G. Tatham, D. J. Nesbit, H. Bultmann, and R. D. Sutton. 1987. The role of juvenile hormone in endocrine control of pigmentation in *Manduca sexta*. Insect Biochem. 17: 1065–1069.

Goudeau, M. 1976. Secretion of embryonic envelopes and embryonic molting cycles in *Hemioniscus balani* Buchholz, Isopoda Epicaridea. J. Morphol. 148: 427–452.

Goudeau, M. and F. Lachaise. 1983. Structure of the egg funiculus and deposition of embryonic envelopes in a crab. Tissue & Cell 15: 47–62.

Graf, F. and P. Michaut. 1975. Chronologie du développement et évolution du stockage de calcium et des cellules a urates chez *Niphargus schellenbergi* Karaman. Int. J. Speleol. 7: 247–272.

Granger, N. A. and W. E. Bollenbacher. 1981. Hormonal control of insect metamorphosis. Pp. 105–137 *in* L. I. Gilbert and E. Frieden (eds.), *Metamorphosis, a Problem in Developmental Biology*. Plenum Press, New York.

Granger, N. A., L. R. Whisenton, W. P. Janzen, and W. E. Bollenbacher. 1987. Interendocrine control by 20-hydroxyecdysone of the corpora allata of *Manduca sexta*. Insect Biochem. 17: 949–953.

Green, J. 1971. Crustaceans. Pp. 312–362 *in* G. Reverberi (ed.), *Experimental Embryology of Marine and Fresh-water Invertebrates*. Elsevier/North-Holland Publ., Amsterdam and New York.

Grellet, P. 1965. Culture *in vitro* d'embryons de *Scapsipedus marginatus* Afz. et Br. (Orthoptères, Grillides). C. R. Acad. Sci. Paris 260: 5100–5103.

Hagedorn, H. H. 1983. The role of ecdysteroids in the adult insect. Pp. 271–304 *in* H. Laufer and R. G. H. Downer (eds.), *Endocrinology of Insects*. Liss, New York.

Hagedorn, H. H., J. D. O'Connor, M. S. Fuchs, B. Sage, D. A. Schlaeger, and M. K. Bohm. 1975. The ovary as a source of α-ecdysone in an adult mosquito. Proc. Natl. Acad. Sci. USA 72: 3255–3259.

Haget, A. 1953. Analyse experimentale des facteurs de la morphogénèse embryonnaire chez le Coléoptère *Leptinotarsa*. Bull. Biol. Fr. Belg. 87: 123–217.

Haget, A. 1957. Recherches expérimentales sur l'origine embryonnaire du crâne d'un coléoptère: le Doryphore (*Leptinotarsa decemlineata*, Say). Bull. Soc. Zool. Fr. 82: 269–295.

Henking, H. 1882. Beiträge zur Anatomie, Entwicklungsgeschichte und Biologie von *Trombidium fuliginosum* Herm. Z. Wiss. Zool. 37: 553–663.

Hetru, C., M. Lauex, B., Luu, and J. A. Hoffmann. 1978. Adult ovaries of *Locusta migratoria* contain the sequence of biosynthetic intermediates for ecdysone. Life Sci. 22: 2141–2154.

Hickman, V. V. 1937. The embryology of the syncarid crustacean *Anaspides tasmaniae*. Pap. Proc. R. Soc. Tasmania 1936: 1–36.

Hillman, R. and L. H. Lesnik. 1970. Cuticle formation in the embryo of *Drosophila melanogaster*. J. Morphol. 131: 383–395.

Hinton, H. E. 1981. *Biology of Insect Eggs*. Pergamon Press, Oxford and Elmsford, New York.

Hoffmann, J. A. and M. Charlet. 1985. Les ecdystéroïdes chez les Invertebres. Ann. Sci. Nat. Zool. Biol. Animal 7: 215–228.

Hoffmann, J. [A.] and M. Lagueux. 1985. Endocrine aspects of embryonic development in insects. Pp. 435–460 *in* G. A. Kerkut and L. I. Gilbert (eds.), *Comprehensive Insect Physiology, Biochemistry and Pharmacology*, Vol. 8. Pergamon Press, Oxford and Elmsford, New York.

Hoffmann, J. A., M. Lagueux, C. Hetru, M. Charlet, and F. Goltzené. 1980. Ecdysone in reproductively competent female adults and in embryos of insects. Pp. 431–465 *in* J. A. Hoffmann (ed.), *Progress in Ecdysone Research*. Elsevier/North-Holland Publ., Amsterdam and New York.

Hsiao, T. H. and C. Hsiao. 1979. Ecdysteroids in the ovary and the egg of the greater wax moth. J. Insect Physiol. 25: 45–52.

Injeyan, H. S., S. S. Tobe, and E. Rapport. 1979. The effects of exogenous juvenile hormone treatment on embryogenesis in *Schistocerca gregaria*. Can. J. Zool. 57: 838–845.

Isaac, R. E. and H. H. Rees. 1984. Isolation and identification of ecdysteroid phosphates and acetylecdysteroid phosphates from developing eggs of the locust *Schistocerca gregaria*. Biochem. J. 221: 459–464.

Isaac, R. E. and H. H. Rees. 1985. Metabolism of maternal ecdysteroid-22-phosphates in developing embryos of the desert locust *Schistocera gregaria*. Insect Biochem. 15: 65–72.

Jones, B. M. 1956a. Endocrine activity during insect embryogenesis: function of the ventral head glands in locust embryos (*Locustana pardalina* and *Locusta migratoria*, Orthoptera). J. Exp. Biol. 33: 174–185.

Jones, B. M. 1956b. Endocrine activity during insect embryogenesis: control of events in development following the embryonic moult (*Locusta migratoria* and *Locustana pardalina*, Orthoptera). J. Exp. Biol. 33: 685–696.

Kadiri, Z. 1983. Action de l'ecdystérone et des hormones juvéniles sur la morphogenèse, la cuticulogenèse et la cytodifferenciation (hypoderme et intestin moyen) chez l'embryon du phasme *Clitumnus extradentatus* Br. Doctoral thesis, University of Bordeaux I, 89 pp.

Kikukawa, S. and S. S. Tobe. 1987. *In vitro* biosynthesis of juvenile hormone III by the embryonic corpora allata of the cockroach, *Diploptera punctata*. Proc. 1st Cong. Asia & Oceania Soc. Comp. Endocrinol. (Nagoya) pp. 12–13.

Klepal, W. and H. Barnes. 1978. An ultrastructural study of the formation of the exoskeleton of stage I and II cirripede nauplii during embryonic development. J. Exp. Mar. Biol. Ecol. 32: 241–257.

Knoll, H. J. 1974. Untersuchungen zur Entwicklungsgeschichte von *Scutigera coleoptrata* L. (Chilopoda). Zool. Jahrb. Anat. Ontog. 92: 47–132.

Lachaise, F., M. Goudeau, C. Hetru, C. Kappler, and J. A. Hoffmann. 1981. Ecdysteroids and ovarian development in the shore crab, *Carcinus maenas*. Hoppe-Seyler's Z. Physiol. Chem. 362: 521–529.

Lachaise, F. and J. A. Hoffmann. 1977. Ecdysone et développement ovarien chez un Décapode, *Carcinus maenas*. C. R. Acad. Sci. Paris 285: 701–704.

Lachaise, F. and J. A. Hoffmann. 1982. Ecdysteroids and embryonic development in the shore crab, *Carcinus maenas*. Hoppe-Seyler's Z. Physiol. Chem. 363: 1059–1067.

Lagueux, M., H. Hetru, F. Goltzené, C. Kappler, and J. A. Hoffmann. 1979. Ecdysone titre and metabolism in relation to cuticulogenesis in embryos of *Locusta migratoria*. J. Insect Physiol. 25: 709–723.

Lagueux, M., J. A. Hoffmann, F. Goltzené, C. Kappler, G. Tsoupras, C. Hetru, and B. Luu. 1984. Ecdysteroids in ovaries and embryos of *Locusta migratoria*. Pp. 168–180 *in* J. A. Hoffmann and M. Porchet (eds.), *Biosynthesis, Metabolism and Mode of Action of Invertebrate Hormones*. Springer-Verlag, Berlin and New York.

Lanzrein, B., H. Imboden, C. Burgin, E. Brüning, and H. Gfeller. 1984. On titers, origin, and functions of juvenile hormone III, methylfarnesoate, and ecdysteroids in embryonic development of the ovoviviparous cockroach *Nauphoeta cinerea*. Pp. 454–465 *in* J. A. Hoffmann and M. Prochet (eds.), *Biosynthesis, Metabolism and Mode of Action of Invertebrate Hormones*. Springer-Verlag, Berlin and New York.

Laufer, H. and D. W. Borst. 1983. Juvenile hormone and its mechanism of action. Pp. 203–216 *in* R. G. H. Downer and H. Laufer (eds.), *Endocrinology of Insects*. Liss, New York.

Laufer, H., M. Landau, E. Homola, and D. W. Borst. 1987. Methyl farnesoate: its site of synthesis and regulation of secretion in a juvenile crustacean. Insect Biochem. 17: 1129–1131.

Le Roux, A. 1983. Histogenèse de l'organe Y (glande du mue) chez l'embryon de la crevette *Palaemon serratus* (Pennant) (Crustace Decapode Natantia). IFREMER (Inst. Fr. Rech. Exploit. Mer) Actes Colloq. 1: 255–262.

Leubert, F., H. Eibisch, H. Kroschwitz, and H. Scheffel. 1982. Ecdysteroid biosynthesis by the ovary of *Lithobius forficatus* (L.) (Chilopoda). Zool. Jahrb. Physiol. 86: 465–476.

Loher, W., L. Ruzo, F. C. Baker, C. A. Miller, and D. A. Schooley. 1983. Identification of the juvenile hormone from the cricket, *Teleogryllus commodus*, and juvenile hormone titre changes. J. Insect Physiol. 29: 585–589.

Louvet, J. P. 1974. Observation en microscopie électronique des cuticles édifiées par l'embryon, et discussion du concept de "mue embryonnaire" dans le cas du phasme *Carausius morosus* Br. (Insecta, Phasmida). Z. Morphol. Tiere 78: 159–179.

Manton, S. M. 1934. On the embryology of the crustacean *Nebalia bipes*. Philos. Trans. R. Soc. Lond. B 223: 163–238.

Matsuda, R. 1979. Abnormal metamorphosis and arthropod evolution. Pp. 137–256 *in* A. P. Gupta (ed.), *Arthorpod Phylogeny*. Van Nostrand Reinhold, New York.

Matz, G. 1980. Variations du taux des ecdystéroïdes au cours de l'embryogenèse de *Leucophaea maderae* Fabr. (Insecte Dictyoptère). C. R. Acad. Sci. Paris 291: 501–504.

McCarthy, J. F. and D. M. Skinner. 1979. Changes in ecdysteroids during embryogenesis of the blue crab, *Callinectes sapidus* Rathbun. Dev. Biol. 69: 627–633.

Meister, M. F., J. L. Dimarcq, C. Kappler, C. Hetru, M. Lagueux, R. Lanot, B. Luu, and J. A. Hoffmann. 1985. Conversion of a radiolabelled ecdysone precursor, 2,22,25-trideoxyecdysone, by embryonic and larval tissues of *Locusta migratoria*. Mol. Cell. Endocrinol. 41: 27–44.

Miller, A. 1940. Embryonic membranes, yolk cells, and morpogenesis of the stonefly *Pteronarcys proteus* Newman (Plecoptera: Pteronarcidae). Ann. Entomol. Soc. Am. 33: 437–477.

Mizuno, T. and E. Ohnishi. 1975. Conjugated ecdysone in the eggs of the silkworm, *Bombyx mori*. Dev. Growth & Differ. 17: 219–226.

Mizuno, T., K. Watanabe, and E. Ohnishi. 1981. Developmental changes of ecdysteroids in the eggs of the silkworm, *Bombyx mori*. Dev. Growth & Differ. 23: 543–552.

Morris, J. E. and B. A. Afzelius. 1967. The structure of the shell and outer membranes in encysted *Artemia salina* embryos during cryptobiosis and development. J. Ultrastruct. Res. 20: 244–259.

Mueller, N. S. 1963. An experimental analysis of molting in embryos of *Melanoplus differentialis*. Dev. Biol. 8: 222–240.

Mueller, N. S. and D. H. Bucklin. 1965. Developments and molting of *Hyalophora cecropia* embryos *in vitro*. Ann. Soc. Zool. 5: 206–212.

Murugan, N. and K. Venkataraman. 1977. Study of the *in vitro* development of the parthenogenetic egg of *Daphnia carinata* King (Cladocera: Daphnidae). Hydrobiologia 52: 129–134.

Nair, S. G. 1956. On the embryology of the Isopod *Irona*. J. Embryol. Exp. Morphol. 4: 1–33.

Oelhafen, F. 1961. Zur Embryogenese von *Culex pipiens:* markierungen und extripationen mit UV-strahlenstich. Wilhelm Roux' Arch. Entwicklungsmech. Org. 153: 120–157.

Ohnishi, E. 1986. Ovarian ecdysteroids of *Bombyx mori:* retrospect and prospect. Zool. Sci. 3: 401–407.

Otsuki, Y., S. Mori, T. Takeda, and T. Kitazawa. 1976. Morphological observation on the embryonic moult in the silkworm *Bombyx mori*. J. Sericult. Sci. Tokyo 45: 225–231.

Pener, M. P., D. Dessberg, P. Lazarovici, C. C. Reuter, L. W. Tsai, and F. C. Baker. 1986. The effect of a synthetic precocene on juvenile hormone III titre in late *Locusta* eggs. J. Insect Physiol. 32: 853–857.

Pihan, J. C. 1975. Utilisation des effets tératogènes des hormones juvéniles et de leurs mimétiques dans la lutte contre les insectes. Ann. Biol. 14: 29–44.

Radallah, D. and B. Fournier. 1988. Ecdysteroids and cuticulogeneses in partial embryonic systems following microsurgery on eggs of *Carausius morosus*. J. Insect Physiol. 34: 37–46.

Rees, H. H. and R. E. Isaac. 1984. Biosynthesis of ovarian ecdysteroid phosphates and their metabolic fate during embryogenesis in *Schistocerca gregaria*. Pp. 180–195 *in* J. A. Hoffmann and M. Porchet (eds.), *Biosynthesis, Metabolism and Mode of Action of Invertebrate Hormones*. Springer-Verlag, Berlin and New York.

Riddiford, L. M. 1985. Hormone action at the cellular level. Pp. 37–84 *in* G. A. Kerkut and

L. I. Gilbert (eds.), *Comprehensive Insect Physiology, Biochemistry and Pharmacology*, Vol. 8. Pergamon Press, Oxford and Elmsford, New York.

Rinterknecht, E. and G. Matz. 1983. Oenocyte differentiation correlated with the formation of ectodermal coating in the embryo of a cockroach. Tissue & Cell 15: 375–390.

Roe, R. M., C. L. Crawford, C. W. Clifford, J. P. Woodring, C. T. Sparks, and B. D. Hammock. 1987. Role of juvenile hormone metabolism during embryogenesis of the house cricket, *Acheta domesticus*. Insect Biochem. 17: 1023–1026.

Sall, C., G. Tsoupras, C. Kappler, M. Lagueux, D. Zachary, B. Luu, and J. A. Hoffmann. 1983. Fate of maternal conjugated ecdysteroids during embryonic development in *Locusta migratoria*. J. Insect Physiol. 29: 491–507.

Sbrenna, G. 1974. The fine structure and formation of the cuticles during the embryonic development of *Schistocerca gregaria* Forskal (Orthoptera, Acrididae). J. Submicrosc. Cytol. 6: 287–295.

Sbrenna-Micciarelli, A. 1977. Effects of farnesyl methyl ether on embryos of *Schistocerca gregaria* (Orthoptera). Acta Embryol. Morphol. Exp. 3: 295–303.

Sbrenna-Micciarelli, A. and G. Sbrenna. 1972. The embryonic apolyses of *Schistocerca gregaria* (Orthoptera). J. Insect Physiol. 18: 1027–1037.

Scalia, S. and E. D. Morgan. 1982. A re-investigation on the ecdysteroids during embryogenesis in the desert locust, *Schistocerca gregaria*. J. Insect Physiol. 28: 647–654.

Scalia, S., A. Sbrenna-Micciarelli, G. Sbrenna, and E. D. Morgan. 1987. Ecdysteroid titres and location in developing eggs of *Schistocerca gregaria*. Insect Biochem. 17: 227–236.

Schneiderman, H. A. and L. I. Gilbert. 1958. Substances with juvenile hormone activity in Crustacea and other invertebrates. Biol. Bull. (Woods Hole) 115: 530–535.

Schooley, D. A., F. C. Baker, L. W. Tsai, C. A. Miller, and G. C. Jamieson. 1984. Juvenile hormones 0, I and II exist only in Lepidoptera. Pp. 373–383 *in* J. A. Hoffmann and M. Porchet (eds.), *Biosynthesis, Metabolism and Mode of Action of Invertebrate Hormones*. Springer-Verlag, Berlin and New York.

Sikes, E. K. and V. B. Wigglesworth. 1931. The hatching of insects from the egg and the appearance of air in the tracheal system. Q. J. Microsc. Sci. 74: 165–192.

Smith, R. F. and R. Arking. 1975. The effects of juvenile hormone analogues on the embryogenesis of *Drosophila melanogaster*. J. Insect Physiol. 21: 723–732.

Solomon, K. R., C. K. A. Mango, and F. D. Obenchain. 1982. Endocrine mechanisms in ticks: effects of insect hormones and their mimics on development and reproduction. Pp. 399–438 *in* F. D. Obenchain and R. Galun (eds.), *Physiology of Ticks*. Pergamon Press, Oxford and Elmsford, New York.

Spindler, K. D., L. Dinan, and M. Londershausen. 1984. On the mode of action of ecdysteroids in crustaceans. Pp. 255–264 *in* J. A. Hoffmann and M. Porchet (eds.), *Biosynthesis, Metabolism and Mode of Action of Invertebrate Hormones*. Springer-Verlag, Berlin and New York.

Spindler, K. D., V. Wormhoudt, D. Sellos, and M. Spindler-Barth. 1987. Ecdysteroid levels during embryogenesis in the shrimp, *Palaemon serratus* (Crustacea Decapoda): quantitative and qualitative changes. Gen. Comp. Endocrinol. 66: 116–122.

Takami, T. 1963. *In vitro* culture of embryos in the silkworm *Bombyx mori*. J. Exp. Biol. 40: 735–739.

Tamarelle, M. 1972. Contribution a l'embryologie des Collemboles Arthropleones. Thesis, University of Bordeaux I, 87 pp.

Temin, G., M. Zander, and J. P. Roussel. 1986. Physico-chemical (GC–MS) measurements of juvenile hormone III titres during embryogenesis of *Locusta migratoria*. Int. J. Invertebr. Reprod. 9: 105–112.

Templeton, N. S. and H. Laufer. 1983. The effects of a juvenile hormone analog (Altosid ZR-515) on the reproduction and development of *Daphnia magna* (Crustacea: Cladocera). Int. J. Invertebr. Reprod. 6: 99–110.

Tetart, J. 1970. L'éclosion des oeufs des Ostracodes d'eau douce: étude de l'évolution des pontes, de l'ultrastructure des membranes de l'oeuf et du processus d'éclosion. Trav. Lab. Hydrobiol. 61: 189–209.

Tiegs, O. W. 1944. The dorsal-organ of the embryo of *Campodea*. Q. J. Microsc. Sci. 84: 35–47.

Vogel, B. E. 1975. Morphologische Untersuchungen über Integument und Häutung sowie Entwicklung des vierten Beinpaares bei *Ornithodoros moubata* (Murray) 1877 (Acarina, Ixodoidea, Argasidae). Ph.D. Thesis, University of Basel, 112 pp.

Warren, J. T., B. Steiner, A. Dorn, M. Pak, and L. I. Gilbert. 1986. Metabolism of ecdysteroids during the embryogenesis of *Manduca sexta*. J. Liq. Chromatogr. 9: 1759–1782.

Wentworth, S. L. and B. Roberts. 1984. Ecdysteroid levels during adult reproductive and embryonic developmental stages of *Sarcophaga bullata* (Sarcophagidae: Diptera). J. Insect Physiol. 30: 157–163.

Whitehead, D. L., E. W. Osir, F. D. Obenchain, and L. S. Thomas. 1986. Evidence for the presence of ecdysteroids and preliminary characterization of their carrier proteins in the eggs of the brown car tick *Rhipicephalus appendiculatus* (Neumann). Insect Biochem. 16: 121–134.

Wigglesworth, K. P., D. Lewis, and H. H. Rees. 1985. Ecdysteroid titre and metabolism to novel apolar derivatives in adult female *Boophilus microplus* (Ixodidae). Arch. Insect Biochem. Physiol. 2: 39–54.

Williams, C. M. 1953. Morphogenesis and the metamorphosis of insects. Harvey Lect. 47: 126–155.

Williams. C. M. 1956. The juvenile hormone of insects. Nature (Lond.) 178: 212–213.

Williams, C. M. 1980. Growth in insects. Pp. 369–383 *in* M. Locke and S. D. Smith (eds.), *Insect Biology in the Future*. Academic Press, Orlando, Florida.

Willis, J. H. and D. L. Cox. 1984. Defining the anti-metamorphic action of juvenile hormone. Pp. 466–474 *in* J. A. Hoffmann and M. Porchet (eds.), *Biosynthesis, Metabolism and Mode of Action of Invertebrate Hormones*. Springer-Verlag, Berlin and New York.

Wittmann, K. J. 1981. Comparative biology and morphology of marsupial development in *Leptomysis* and other mediterranean Mysidacea (Crustacea). J. Exp. Mar. Biol. Ecol. 52: 243–270.

Yoshikura, M. 1975. Comparative embryology and phylogeny of Arachnida. Kumamoto J. Sci. Biol. 12: 71–142.

Zacharuk, R. Y. 1976. Structural changes of the cuticle associated with moulting. Pp. 299–321 *in* H. R. Hepburn (ed.), *The Insect Integument*. Elsevier, Amsterdam, and New York.

Roles of Morphogenetic Hormones in Embryonic Diapause

3

OKITSUGU YAMASHITA
AND KOICHI SUZUKI

3.1. Introduction

As a means of adaptation to changing environmental conditions, a definitive dormant stage occurs in the life cycle of many organisms. Although the occurrence of this state of arrested growth is particularly frequent in the Arthropoda, it is by no means confined to this group. A dormant stage in the life cycle of some species is often critical to their existence, because it ensures survival of individuals during periods unfavorable for growth, development, reproduction, and propagation of the species by synchronizing the growth rates of the population.

The phenomenon of dormancy represents a spectrum of development ranging from "arrested" (diapause) to "retarded" (quiescence). It is important to make the distinction between diapause and quiescence, because the natures of the two dormancy states are very different. *Diapause* is a complex adaptative response expressed as arrested development. Characteristically, the stimulus that triggers the expression of diapause acts prior to the onset of environmental adversity, thereby enabling an individual to have sufficient time to adapt. The diapause response may involve biochemical, physiological, and endocrinological adjustment. *Quiescence,* on the other hand, is a state of retarded development that is induced immediately by adverse environmental conditions. Individuals become quiescent at any developmental stage with no prior acclimation, and they resume development without a lag phase after conditions become favorable.

There is tremendous variety in the type of dormancy expressed by plants and animals, so that it is often difficult to make a clear delineation of diapause. For most species, diapause is a developmental option, and individuals have the capacity to decide whether or not they enter diapause (facultative diapause). In a few species that complete one generation annually (univoltine life cycle), diapause occurs in each generation regardless of environmental conditions (obligatory diapause). The term *oligopause* was proposed by Mansingh (1971) to account for dormancies that appear to be intermediate between diapause and quiescence. Although within given individuals examples exist of variable intensity of diapause, that state is most consistently expressed as the threshold character. Usually, individuals can clearly be categorized as diapause or nondiapause.

Diapause may occur in any of the major stages of the life history, especially in arthropods, in which the life cycle is completed through successive metamorphoses. In insects, examples abound of diapause in each stage of development: egg, larva, pupa, and adult. The stage of diapause correlates poorly with taxonomic groupings at the family level and above, but the habitat and other components of the species niche seem to be more important in determining the diapause state.

Diapause represents an amalgam of behavioral, physiological, and biochemical characteristics, all of which enhance survival during long periods of dormancy. The key to survival is a low metabolic rate that enables diapausing individuals to stretch their nutrient reserves and gives rise to specific metabolic pathways unique to diapausing individuals.

Despite the diversity of diapause stages, the programming of diapause is not nearly so capricious. The factors that induce diapause are not the adverse environmental conditions themselves but rather various environmental signals that reliably foretell the advent of an unfavorable season. Day length is certainly the signal most widely utilized, but subtle temperature changes and seasonal variations in food quality are also useful indicators for some species. Diapause implies the capacity to receive and transduce such environmental signals into a developmental program that can be stored and retrieved at a later date.

The neuroendocrine system serves as the crucial transducer of these ambient signals into the many facets of the diapause program. The idea that diapause is mediated by the neuroendocrine system has been firmly established by pioneering experiments on insects.

The aim of this review is to place the work on embryonic diapause of arthropods in the context of physiological responses, biochemical characteristics, and hormonal regulation. Owing to the variety of species and developmental stages investigated and considerable differences in the methodology employed, it is difficult to present a clear synthesis of the experimental results. The most information on embryonic diapause to date is from the brine shrimp, *Artemia salina,* and the silkworm, *Bombyx mori.* We shall review the work on embryonic diapause of these animals as model systems with which we will, step by step, compare the data obtained in other species.

3.2. Embryogenesis and Stages of Diapause

3.2.1. *Overview of Embryogenesis of Arthropods*

Most multicellular eukaryotes go through a standard sequence of reproduction during their life cycle. It begins with gametogenesis, followed by fertilization, which results in the formation of the zygote. The zygote nucleus replicates repeatedly to form a mass of cells; in most animal groups, this occurs by cleavage of one whole cell into two, two into four, and so on. However, in many arthropods a similar result is achieved by multiplication of the nuclei within an initially undivided cytoplasm. The result of either process is usually a hollow blastula, composed of a wall of cells, the blastoderm, surrounding a central cavity.

During gastrulation, the blastoderm gives rise to two or more layers of cells known as the germinal layers—the ectoderm, mesoderm, and endoderm. Dermal differentiation is the first step leading to cytodifferentiation of embryonic cells. Following gastrulation, the basic body plan, organogenesis, becomes laid out by the process of neurulation and by the modeling of germinal layer. Internal organs are mainly derived from the endoderm together with the mesoderm, while the ectoderm usually combines with the mesoderm to form peripheral structure.

Organogenesis is followed by a period of growth and histogenesis during which the cells of organs acquire the unique structure and properties appropriate to their roles in the larva or young adult. When this has been achieved, the animal is ready to embark upon an independent existence and procure its food from the outside world.

Insect embryogenesis is characterized by adaptation to the terrestrial mode of life. Terrestrial propagation requires protection of the egg stages against desiccation, and it requires well-provisioned storage reserves. The typical insect egg is large both in absolute dimensions and by comparison to maternal body size (some 0.2–20 mm long, with average dimensions of 1.0 × 0.5 mm; Legay, 1977), and it is protected by egg covers, the chorion and vitelline envelopes, which rank among the most impenetrable and resistant in the animal kingdom. The fertilized egg cell contains stored macromolecules providing both the raw materials and energy for building the larval body, but it also contains the zygote genome and localized extrakaryotic signals (ooplasmic determinant) (Boswell and Mahowald, 1985). Development is dependent on appropriate physical conditions in the environment, and external influences releasing or directing developmental steps are sometimes important in diapausing eggs (Sander et al., 1985).

A brief outline of *Bombyx* embryogenesis is now described to provide an outline of insect embryogenesis (Fig. 3.1) (Sakaguchi, 1978; Miya, 1984). In newly laid eggs, the chromosomes, which are arrested in metaphase, resume the first meiotic division, followed by the production of the first polar nucleus. The second meiotic division takes place about 60 min after oviposition. The female pronucleus becomes active and begins to move toward the sperm nucleus, which has penetrated the egg and stayed in its anterior region. The sperm meanwhile separates from its tail and becomes the male pronucleus. Once the egg pronucleus is completed, both pronuclei approach each other and finally unite, which occurs usually some 1–2 h after deposition at 25°C. Fusion of these pronuclei does not take place at this time, however, and gonomery occurs at the interphase of the subsequent cleavage (Kawamura, 1978).

The fused nuclei immediately undergo mitosis, and nuclear cleavage proceeds through a succession of synchronous mitoses. One cycle of mitoses takes 60 min, and synchronous mitosis repeats 10 times (Ohtsuki and Murakami, 1968; Miya, 1984). The cleavage nuclei migrate

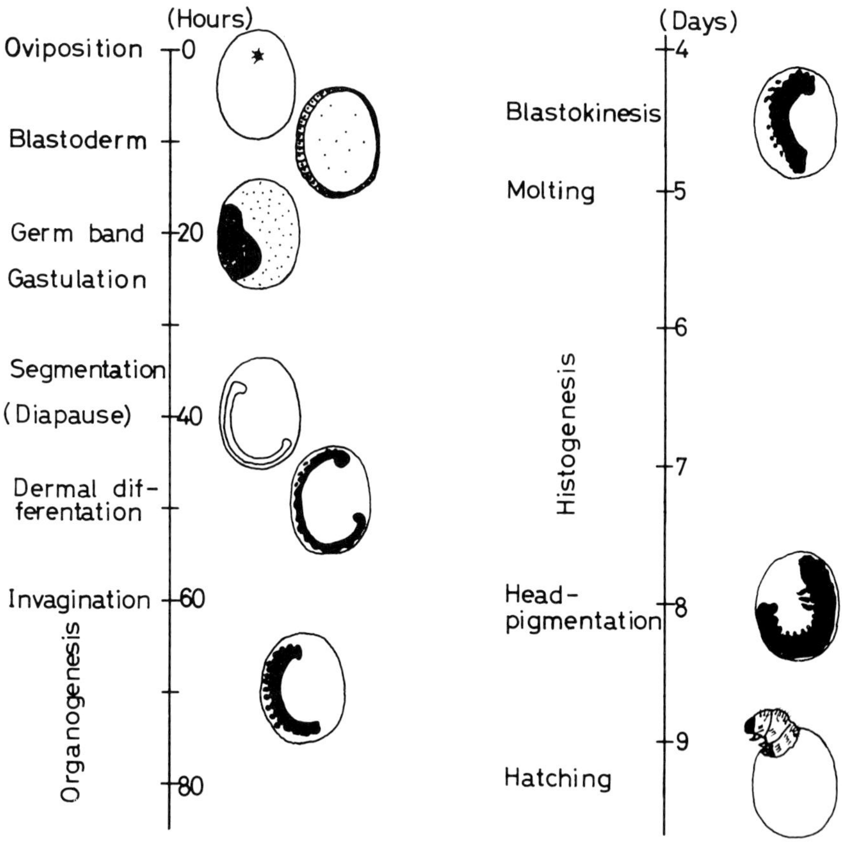

FIGURE 3.1. Diagrammatic representation of embryonic development in the silkworm, *Bombyx mori.* Embryogenesis of nondiapause eggs at 25°C is shown. When the egg programmed to enter diapause is incubated at 25°C, embryogenesis ceases completely before dermal differentiation and diapause occurs. (Modified from Takami, 1969.)

to the periplasm surrounding the yolk mass and initiate the formation of the syncytial blastoderm, the remaining few nuclei within the yolk mass becoming vitellophages.

The nuclei that migrated to the periplasm are separated from each other by plasmalemma folds, and the radial infolding of the plasmalemma expands tangentially to separate an external layer of mononuclear cells (Keino and Takesue, 1982). A uniform cellular blastoderm appears about 12 h after oviposition.

The blastoderm soon separates into two different areas, the germ anlage (embryonic rudiment) and the extraembryonic blastoderm. The margin of the inner embryonic envelope develops the amnion, which sequentially covers the germ anlage. The extraembryonic blastoderm transforms into the outer embryonic envelope, or serosa.

At about 15 h after oviposition, the cells of the germ anlage increase in thickness and become concentrated, still increasing in number through cell division. This peripheral cell layer is called the germ band. The germ anlage becomes two layers by invagination along the middle or by multipolor migration of cells. These changes occur at 25 h after oviposition and are called gastrulation. The cell layer left in contact with the amniotic cavity is the ectoderm, and the inner layer bordering the yolk system is essentially the mesoderm.

The two-layered stage leads to the germ band stage, which is externally characterized by segmental organization. The head lobes forming the anterior region of the germ band transform into the pro- and deutocerebrum, eyes, labia, and antennae, as well as a large portion of the head capsule. The typical germ band segment is characterized by a pair of ganglion rudiments that develop from the ventral ectoderm and lateral mesoderm. Finally, the germ band divides into 17 segments.

Organogenesis is a multifaceted process that starts during segmentation of the germ band. From the rudiments of mesodermal organs, the muscles and fat body are derived, while the ectoderm gives rise to the sense organs, tracheae, integument, nervous system, foregut, and Malpighian tubules. The midgut is formed from the edge of the elongating mesoderm, not from the endoderm; the vitellophages and yolk cells integrate initially with the ectoderm. The conversion of the organ anlagen into tissue-specific cell types and cell arrangements follows in the wake of organogenesis.

Organogenesis is accompanied by a conspicuous translocation of the embryonic rudiment within the egg. This embryonic movement, called blastokinesis, occurs on days 4–5 after oviposition and leads to an inverted dorsoventral orientation. During blastokinesis, the embryo quickly ingests much yolk material and expands into the entire egg space. After completion of blastokinesis, histogenesis proceeds actively and continues until larval hatching. The first embryonic molt occurs just after blastokinesis, followed by the second molt, immediately after which the true larval cuticle becomes secreted by the epidermal cells (Ohtsuki et al., 1976). Coinciding with the progress of cuticle deposition, external processes, bristles, and taenidia in the spiral band are sequentially differentiated and become deeply colored. Muscular contractions set in and gas is secreted into the tracheae. Indeed, the embryo becomes separated from the egg shell at this stage and is able to move actively in an *in vitro* condition. The dorsal closure finally completes at this stage, and the serosa is ingested by the embryos. One day later, the larva hatches by mechanical disruption of the chorion after peristaltic movement of the abdomen.

While cleavage in the Crustacea provides no information on the interrelationship of the several crustacean classes, it is possible to discern a basic pattern and a number of adaptative trends in relation to the yolk mass. The basic pattern is a modified spiral pattern of total cleavage, retained in the small eggs of those thoracican cirripeds that hatch into planktotrophic nauplii. Traces of the same pattern can be discerned in the egg of certain branchiopods, copepods, and malacostracans. In most groups of crustaceans, however, the outcome of cleavage is a blastoderm around a unitary yolk mass. Some of these groups retain traces of the ancestral spiral cleavage division (Cladocera and Eucarida), but others proceed directly through intralecithal cleavage to blastoderm formation (Leptostraca, Hoplocarida, and Paracarida).

The subsequent development of the crustacean embryo to a nauplius is already specified at the blastula stage in the pattern of the presumptive ectoderm. The first event of gastrulation is an inward and forward migration of the germ cells surrounding the yolks. Gastrulation movements remain much more distinct throughout the crustaceans in respect of the differentiation of ectoderm and mesoderm. In any crustacean that hatches as a nauplius larva, the external development of the embryo before hatching is, in general, a relatively simple process. The first notable change is the development of three pairs of nauplial limb buds. The early development of the buds proceeds, leaving short preantennulary and postnauplial regions at the end of the embryo. As development of the embryo continues, from the naupliar buds all postmandibular segments are developed anamorphically (D. T. Anderson, 1973). Figure 3.2 outlines embryonic development of *Panulirus japonius* as an example of the crustaceans (Shiino, 1950).

Embryogenesis in the egg of the spider *Floronia bucculenta* proceeds in a similar way as that of other arthropods (Schaefer, 1976). Fertilization is followed by several mitoses, after which cleaved nuclei migrate to the periphery to form the blastoderm. The germ band appears through active cell division of the outer layer of the germ disk and elongates along with the ventral surface of eggs. When the germ band occupies almost the whole periphery of the egg, segmentation occurs and outgrowth of the future appendages takes place. Involution of the embryo commences as a result of an attenuation of the body wall along the midventral line to form the ventral sulcus, whereupon the appendages lengthen and bend. After involution, the embryo continues to elongate and finally hatches to an incomplete postembryonic stage by shedding the chorion and vitelline membrane.

3.2.2. *Stages of Diapause*

In embryos, where morphogenesis is continuous, the system that receives the environmental signal is still in a premature state, but diapause is confined to a definitive stage of embryonic development. Embryonic

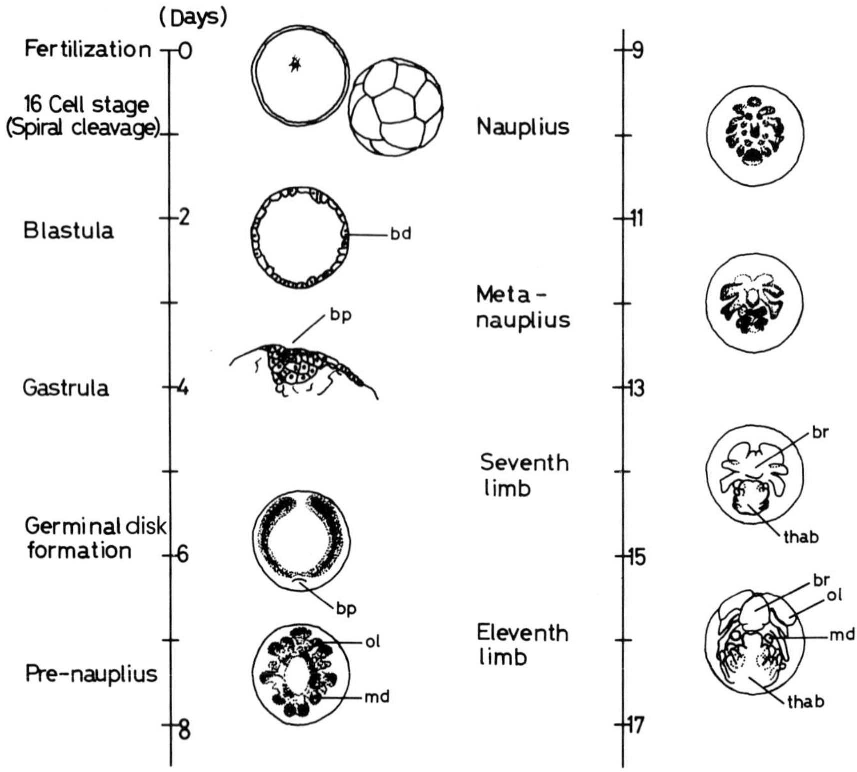

FIGURE 3.2. Diagrammatic representation of embryonic development in *Panulirus japonicus*. Embryogenesis proceeds at 15–20°C. *Key:* bd = blastoderm; bp = blastopore; br = brain; md = mandible; ol = optic lobe; thab = thoracico-abdominal process. (Modified from Shiino, 1950.)

development of insects may be divided into three stages from the viewpoint of the occurrence of diapause. The first phase sees the successive division of the cleavage nuclei and ends with the formation of the blastoderm. No species are known in which diapause supervenes during nuclear cleavage after fertilization.

The arrest occurs in some species once the cleaved nuclei are cellularized into the germ band (Umeya, 1946; see Fig. 3.3). The eggs of *Metatetranychus ulmi, Austroicetes cruciata, Notolopus thyellina, Dendrolinus undans excellans, Archips xylosteanus,* and *Homoeogryllus japonicus* are known to enter diapause at this early stage. In some species, growth may cease at a slightly later stage when the embryonic rudiment is elongated but as yet unsegmented, e.g., *Bombyx mori, B. mandarina,* and *Teleogryllus emma*. In many species, morphogenesis is suspended at the close of anatrepsis, during which many larval tissues and organs are differentiated. The eggs of several Orthoptera, including *Gryllus commodus, Melanoplus differentialis,* and *Locusta migratoria,* and those of the lepidopteran *Orgyia antiqua* are some examples. Finally, development may also halt very late in embryogenesis when the larva is fully formed and seemingly ready to hatch. This is the stage of arrest in *Antheraea yamamai, Lymantria disper, Malacosoma testacea,* and *Malacosoma disstria,* and also in certain Orthoptera (*Gampsoclei buergeri* and *Melanoplus bivittatus*) and Coleoptera (*Timarcha tenebricosa* and *Timarcha violacea-nigra*). These results imply that diapause cannot be induced during the phase of

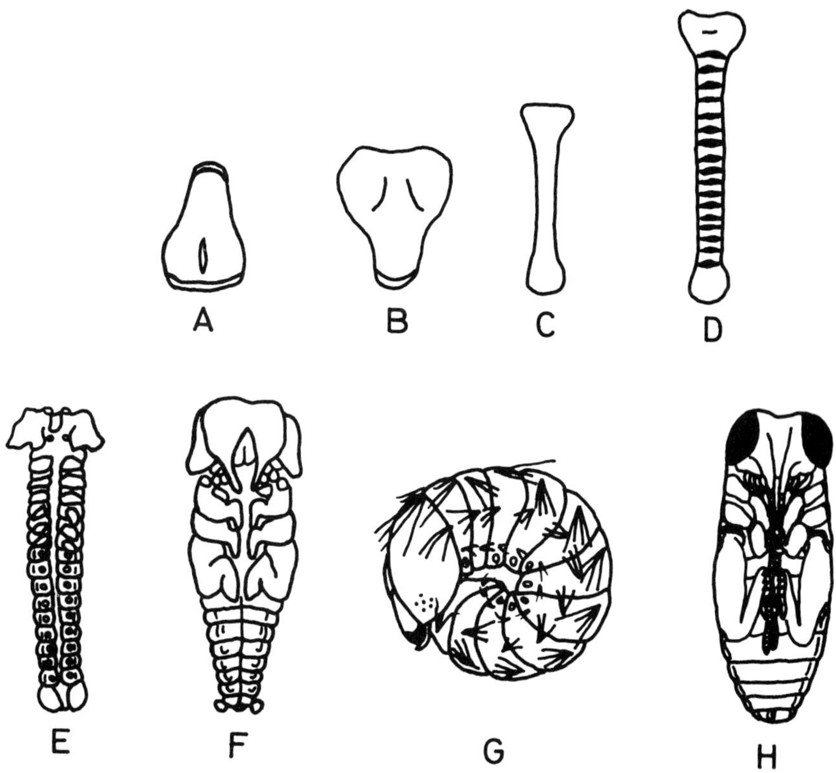

FIGURE 3.3. Embryonic stages entering diapause of various insects: (A) blastoderm embryo (some Lepidoptera); (B) reversed blastoderm embryo (some Orthoptera); (C) late germ band embryo (*Bombyx mori*); (D) segmented embryo (*Rhopobata maevana*); (E) embryo with appendages (some Lepidoptera); (F) ditto (*Locusta migratoria*); (G) pharate larva (*Lymantoria dispar, Antheraea yamamai,* and other Lepidoptera); (H) ditto (*Gampsoclei buergeri*). (From Umeya, 1946.)

intense mitotic activity that accompanies blastoderm formation or during the phase of active differentiation.

Many species of *Tettigoniidae* have the capacity to enter diapause in two or more developmental stages. The bush cricket, *Ephippiger cruciger,* diapauses as a young embryo (unsegmented primordium) and again as a nearly fully developed embryo (Hartley and Warne, 1972; Dean and Hartley, 1977a,b).

Despite the diversity of potential stages of diapause for each species, the developmental stage in which diapause occurs is quite specific. Among heteropteran insects, there are at least 16 different stages of embryonic diapause (Cobben, 1968). This great diversity of diapausing stages among closely related species clearly implies that diapause is a convergent trait that has evolved independently many times.

Among the many species of crustaceans that enter diapause in the embryonic stage, the brine shrimp, *Artemia salina,* has attracted considerable attention as an interesting object for study of various aspects of development, physiology, and biochemistry (Clegg and Conte, 1980). Female brine shrimp produces a brood of eggs about every 4 days. Each brood will develop either a thin or thick protective shell. Thin-shelled eggs hatch within a few days and are released ovoviparously, whereas thick-shelled eggs stop embryogenesis immediately and are released as cysts. The dormant cyst contains an embryo that has completed gastrulation and has just barely initiated organogenesis (Benesch, 1969). The dormant embryo is at gastrulal stage and survives 10 or more years under desiccated conditions. Following rehydration, the dormant cyst eventually resumes development into a free-swimming nauplius that usually occurs after some 12–24 h of incubation. During this period, organogenesis and histogenesis are completed without any increase in the number of nuclei or without cell division (Nakanishi et al., 1962; Olson and Clegg, 1978).

The embryonic diapause of the spider *Floronia bucculenta* seems to be unique when compared to that of other arthropods. At 5° and 0°C, normal embryogenesis without dormancy is possible from fertilization to hatching of spiderlings. From 7.5°C upward, embryogenesis is more and more delayed, especially after segmentation of the germ band. Complete arrest of embryogenesis occurs at the germ band stage when the laid egg is acclimated at 12.5°C or higher temperatures (Schaefer, 1976). In the egg of another spider, *Baetis vernus,* diapause intervenes in later stages when the germ band has formed (Bohle, 1969).

3.2.3. Morphological Characteristics of Diapause Eggs

During the dormant stage, no appreciable morphogenetic development occurs in embryos. The prolonged survival of dormant eggs associated with a specialized morphological architecture that does not develop in nondiapause eggs. Eggs of the brine shrimp, *A. salina*, are typical, showing a morphological differentiation correlated with dormancy (E. Anderson et al., 1970). The dormant eggs are covered with a thick shell consisting of a tertiary envelope. The egg shell is formed from a secretion fabricated by the shell gland that consists of lipoproteins rich in tyrosine. The secretory glanules are released into the uterus and become deposited around each egg as a dense homogeneous layer within which there develops an architecturally complex pattern at the time when the embryo is at the gastrula stage. Just how this organization is achieved or how the material hardens is unknown, although it is supposed that during egg shell formation the lipoproteins are sclerotized, presumably by a process involving spontaneous oxidation of tryosine residues to quinones as demonstrated in the process of egg-shell hardening of some insects (Brunet, 1967). It is certainly tempting to suppose that cryptobiosis of *Artemia* embryos is dependent on some degree of impermeability of the egg shell provided by the thickness, density, microstructure, and chemical composition of particular shell layers (Morris and Afzelius, 1967). Perhaps the situation will be found to be comparable to that in the grasshopper, in which it has been shown that diapause of the embryo depends on waterproofing of the special protein of the egg shell termed *hydropyle* (Slifer, 1958).

In the dormant cyst of *A. salina*, the mitochondrion seems morphologically immature and has very few cristae. A significant proportion of the mitochondria is stored inside the yolk granules. Immediately following rehydration that triggers resumption of development, there is considerable morphological maturation of the mitochondria. The maturation process is closely associated with the breakdown of yolk granules (Vallejo et al., 1979).

Electron microscopic observation of embryonic cells of diapause eggs of silkworms has shown that the fine structure of some organelles is adaptively modified (Okada, 1970): the nucleolus becomes small and compact with no projections on is surface in diapausing embryos; chromatin is observed as clumps disposed along the inner surfaces of the nuclear membrane, reflecting an inactive stage of the chromosomes in their transcription activity. The granular endoplasmic reticulum shows complicated changes during diapause. At the commencement of diapause, granular endoplasmic reticulum with five to eight layers of lamellae appears and the number of lamellae becomes reduced when diapause is initiated. At this stage, another type of granular endoplasmic reticulum develops. In the diapausing embryo, both granular endoplasmic reticulae become rather regressive as regards the size and number of their lamellae. Soon after the initiation of diapause, mitochondria gradually show a loss in electron density of their matrix, and they start swelling markedly on day 30 after oviposition. The swelling persists until embryos resume their development. These structural changes in the cellular organelles of embryos are not correlated directly to the progress of diapause, since the structure of the extraembryonic cells, such as the yolk cells, amniotic cells, and serosal cells, seem also to be modified (Miya et al., 1972). Nevertheless, the initiation and termination mechanisms of diapause might be related in some way to some changes in the fine structure of the embryo without any accompanying morphogenesis.

3.3. Metabolic Aspects of Embryonic Diapause

Although knowledge of the metabolic aspects of development and differentiation in arthropod eggs is as yet rudimentary and fragmentary, some investigations of the biochemical events occurring during embryogenesis have been reported for various arthropods. These provide promising data as the first steps in the biochemical approach to dormancy.

3.3.1. *Energetics and Related Metabolisms*

In several crustaceans (Green, 1965), O_2 uptake remains generally at a relatively low level until blastoderm formation. A gradual increase occurs during embryogenesis, and the maximum values are obtained at larval hatching (Green, 1965). The respiratory quotient (RQ) is close to 1 at cell cleavage and falls to 0.72 by blastoderm formation. Dormant embryos of *A. salina* reduce O_2 uptake to one-tenth that of developing embryos owing to the immature organization of the mitochondria (Dutrieu, 1960).

A similar phenomenon is observed in diapausing eggs of silkworms (Takami, 1969; Miura and Shimizu, 1987), the field cricket *Teleogryllus emma* (Izumiyama and Suzuki, 1986), and the spider *Floronia bacculenta* (Schaefer, 1976).

Trehalose and glycerol play significant roles in dormant embryos of *A. salina* (Clegg, 1964, 1965). The metabolic fate of carbohydrates is also influenced by the osmotic pressure of the external medium. Polyhydroxy alcohols such as glycerol, sorbitol, and mannitol are also demonstrated to accumulate in relation to diapause of insects (Asahina, 1969; Sømme, 1982; Storey and Storey, 1983; Yamashita and Hasegawa, 1985). These polyols are assumed to act as cryoprotectants, supercooling agents, or antifreezing agents through interaction with cellular free water.

In the metabolic pathway responsible for polyol metabolism in silkworm eggs, glycogen phosphorylase and NAD(nicotinamide adenine dinucleotide)–dependent sorbitol dehydrogenase are respectively the rate-limiting enzymes in the pathway from glycogen to sorbitol and in glycogenesis from sorbitol (Yamashita et al., 1975; Yaginuma and

Yamashita, 1979; Yamashita and Hasegawa, 1985). Still unknown is what kinds of regulation mechanisms are operating as triggers of these two key enzymes.

3.3.2. *Yolk Protein Metabolisms*

In newly laid eggs of many crustaceans, most of the proteins are localized in the yolk granules (Yamashita and Indrasith, 1988). In dormant cysts, the yolk granules are ellipsoidal and contain more than 80% of the total proteins. Electrophoretic analysis shows that lipovitellin is the main component of the yolk proteins (Vallejo et al., 1981). A more detailed characterization has been done on lipovitellin of *A. salina* (De Chaffoy De Courcelles and Kondo, 1980). The lipovitellin is a carotenoid–glycolipoprotein complex containing 3.3% carbohydrates, 8.6% lipids, and 88% proteins. Two polypeptides (190 and 68 kDa) are identified as apoprotein components of the lipovitellin complex.

Once the egg of *A. salina* initiates embryonic development, lipovitellin becomes utilizable by means of limited hydrolysis (De Chaffoy De Courcelles and Kondo, 1980). During embryonic development a large apolipovitellin component undergoes preferential proteolysis *in vivo,* giving rise to multiple fractions of intermediate molecular mass, whereas a small apoprotein component remains unchanged throughout embryonic life. The acid protease in lysosomes is thought to be involved in the *in vivo* degradation of lipovitellin (Perona and Vallejo, 1982, 1985).

Almost the same sort of protein metabolism is observable in insect eggs. The identification, biosynthesis, and hormonal control of vitellogenin (the precursor of vitelline) have been studied in more than 30 species of insects (for reviews, see Postlethwait and Giorgi, 1984; Kunkel and Nordin, 1985). Insect vitellogenins have been identified as glycophospholipoproteins having diverse molecular masses (from 50 to 600 kDa) and grouped into three types according to their molecular properties. A complicated yolk protein system is found in some lepidopteran insect eggs (Yamashita, 1986; Yamashita and Indrasith, 1988). The yolk proteins of silkworm eggs are composed of three major types of proteins: vitellin (~40%), 30-kDa proteins (~35%), and egg-specific protein (~25%) (Zhu et al., 1986). These proteins are markedly different from each other in their physicochemical properties and biological fates.

Figure 3.4 shows titer changes in yolk proteins during the embryonic development of silkworm eggs (Zhu et al., 1986). A distinct pattern of utilization is apparent for each protein: (a) egg-specific protein is utilized earlier and then disappears until larval hatching; (2) vitellin decreases significantly during the last phase of embryogenesis but remains unutilized at larval hatching; (3) 30-kDa proteins are not much used throughout embryonic life.

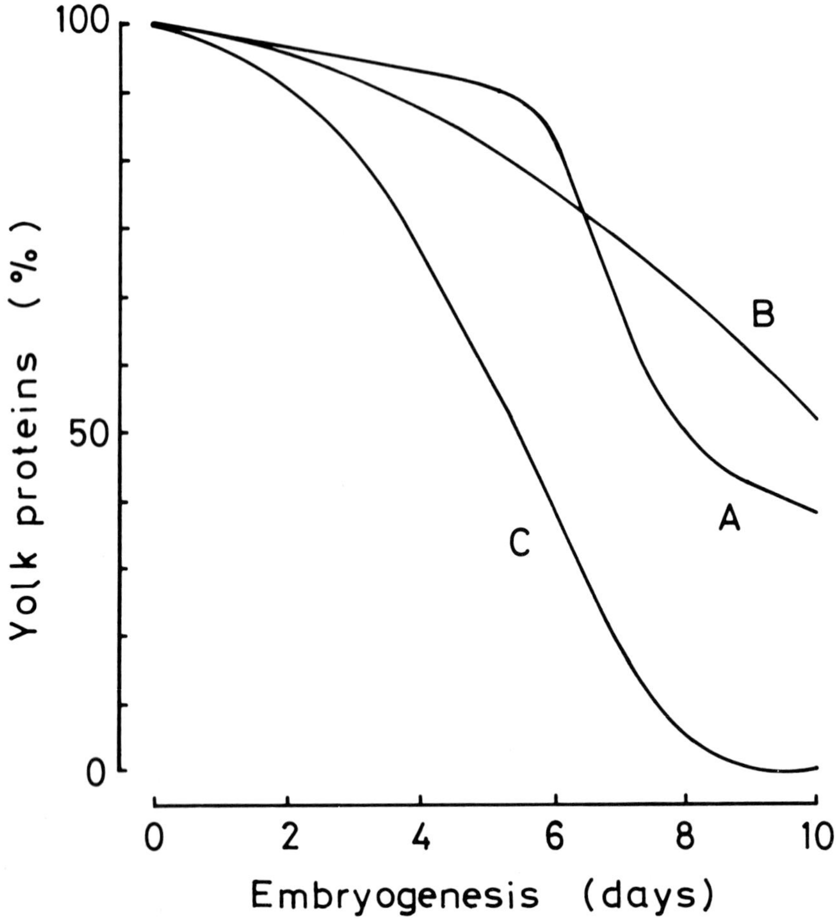

FIGURE 3.4. Developmental changes in three yolk proteins during embryonic development of silkworms. These proteins are measured by a immunochemical method and are shown as percentages of day 0 values: curve A, vitellin; curve B, 30-kDa proteins; curve C, egg-specific protein.

Using the purified egg-specific protein as the substrate, a unique protease that is responsible for limited degradation of this protein has been extracted and purified from embryonating eggs of silkworms by Indrasith et al. (1988). The protease is a trypsin-like seryl protease but prefers the egg-specific protein as a potent substrate without attacking the vitellin or the 30-kDa proteins. The native protein is specifically cleaved into three peptides at two sites; Lys^{114}–Asn^{115} and Arg^{210}–Asp^{211} of the amino acid sequence.

In dormant cysts of *A. salina*, aspartic acid, cysteine, serine, glycine, proline, tyrosine, and glutamic acid gradually accumulate (Green, 1965). In the silkworm egg, alanine begins to increase at the initiation of diapause and attains a plateau level when diapause has been established (Sonobe and Okada, 1984; Suzuki et al., 1984; Osanai and Yonezawa, 1986). The metabolic origin of this alanine remains unknown.

3.4. Hormonal Control of Embryonic Diapause

3.4.1. *Environmental Signals Controlling Diapause*

The existence of diapause eggs (resting eggs) in the life cycle of zooplankton has been postulated as the cause of the seasonal disappearance of some species from the water column (Barlow, 1955). Environmental factors such as temperature and photoperiod usually govern the induction and termination of dormancy in arthropods (Lees, 1955; Grice and Marcus, 1981). Egg dormancy may result from a physiological response of the maternal generation to a change in milieu, modifying the egg physiology. Conversely, dormancy may terminate as a response of the egg to changes in conditions.

For induction of diapause eggs in common coastal species, temperature and salinity are the predominant factors because population growth and decline are well correlated with annual changes in water temperature and salinity. Laboratory experiments on *Acartia californiensis* clearly show that diapause is only induced in eggs laid by adult females exposed to below 15°C for a few days. Nondiapause eggs spawned above 15°C may become quiescent when incubated at low temperature (Fig. 3.5). Salinity (5–25‰) does not influence the incidence of diapause eggs. These results have been interpreted as an indicator that temperature is the most important factor inducing diapause (Johnson, 1980). The response of spawning females to low temperatures by induction of diapause eggs has been shown in *Acatia fonsa* (Zillioux and Gonzalez, 1972), *Tortanus forcipatus* (Kasahara et al., 1975), and *Pontella meadi* (Grice and Gibson, 1977).

The cladoceran *Daphnia* alternates parthenogenesis with sexual reproduction in a variety of seasonal cycles. Development of the sexually derived embryo is arrested at an early stage as a diapause state. The reversal from parthenogenesis to sexual reproduction is induced by environmental changes in temperature, diet, density, and photoperiod.

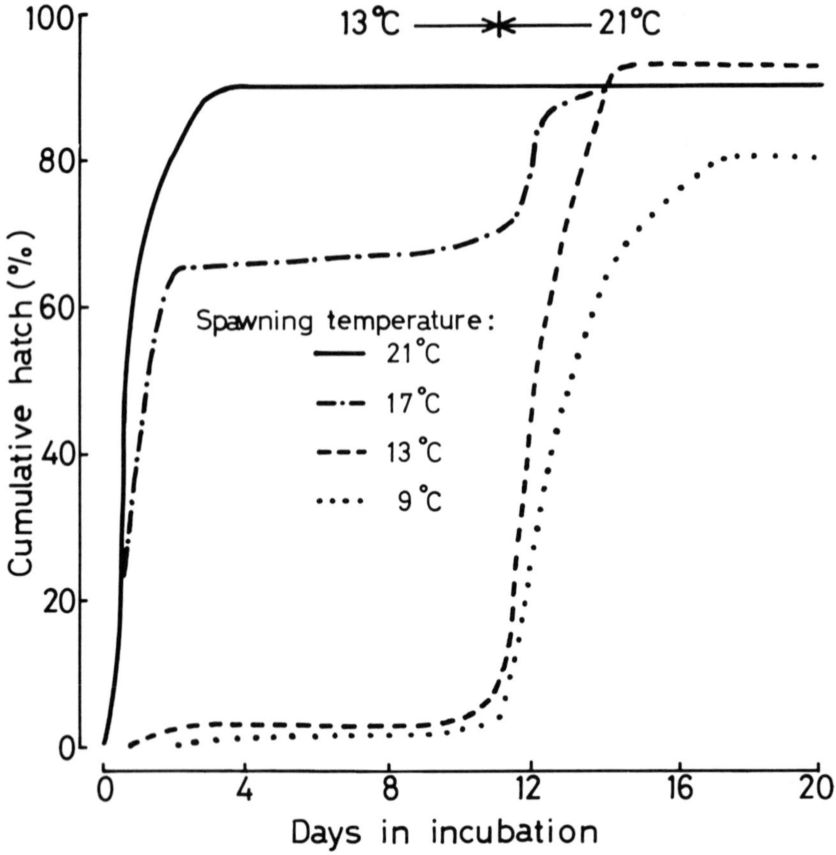

FIGURE 3.5. Effects of spawning temperatures on induction of diapause eggs of *Acarta californiensis.* Female adults are acclimated at different temperatures (9–21°C) and eggs spawned are incubated at 13°C until day 10 and then at 21°C. (Modified from Johnson, 1980.)

Stross and Hill (1965) clearly demonstrated in *D. pulex* that adults exposed to short-day photoperiod (12.75 h L : 11.25 h D) predominantly produce diapause eggs and sexual reproduction. Photoinduction of diapause eggs is modified by secondary stimuli such as population density and food availability but not influenced by temperature. In the more detailed examination of eggs produced by a single individual of *Labidocera aestiva,* the diapause nature of the eggs was determined as a result of a photoperiodic regimen to which the mature animal was exposed. Continuous exposure to adult females to 8 h L : 16 h D photoperiod induced about 90% diapause eggs (Marcus, 1980). When the animal acclimated to a short-day photoperiod was exposed to a long-day photoperiod (12 h L : 12 h D), the incidence of diapause eggs decreased gradually and finally stabilized at around 50% after 2 weeks. Thus, the induction of diapause may be determined by a cumulative input of short-day light–dark cycles (Marcus, 1982). Consequently, photoperiod is an important trigger for the induction of diapause in marine copepods as well as freshwater copepods and cladocerans (Stross, 1969a,b).

In a study of the life cycle of *Diaptomus stagnalis* in a pond in the vicinity of Chicago, Brewer (1964) noted that the egg sacs are deposited by mature adults in May. The eggs remain at the bottom of the pond from the spring through the summer, fall, and winter, until the following spring when they hatch. Thus, the annual life cycle consists of an ephemeral active phase in the spring, followed by a long inactive phase over 10 months spent in the egg stage. When the newly laid eggs were maintained for different periods of time at 25°C and then exposed to 13°C, beyond days 150–170 at 25°C, there was no acceleration in the rate of their embryonic development and the subsequent hatching at 13°C (Brewer, 1964). This stage is considered to represent the refractory phase in diapause eggs (Mansingh, 1971). Apparently, a refractory period is required for eggs to become responsive to specific environmental stimuli. Therefore, eggs that have not completed the refractory phase remain in the diapause state and never resume their embryogenesis even under favorable conditions (Grice and Gibson, 1977; Marcus, 1979; Kasahara and Uye, 1979).

To ascertain the effective stimuli for diapause termination, the percentage of eggs that hatched after incubation in different environmental conditions was monitored. When diapause eggs of *Diaptomus stagnalis* that had completed the refractory phase were exposed to various low temperatures (1°, 5°, and 13°C) for different lengths of time, nauplius hatching occurred in most eggs chilled at 1°C for 90 days. Chilling at 5°C was also effective, but the eggs took much longer to hatch. Less than 20% of the eggs were able to hatch after incubation at 13°C for 16 days (Fig. 3.6). From these experiments, it is clear that low temperature is an important stimulus for termination of diapause in this species (Brewer, 1964). Similar temperature experiments on diapause eggs of many marine copepods show that low temperature (below 10°C) is the primary cue that triggers termination of diapause (Grice and Marcus, 1981).

Although the change in day length provides an ideal cue whereby many organisms are able to predict other seasonal changes in the environment, photoperiodic stimuli do not significantly affect the termination of diapause in marine copepods (Kasahara et al., 1975; Uye and Fleminger, 1976; Uye et al., 1979; Grice and Marcus, 1981).

In some copepods, darkness completely and immediately suppresses hatching of eggs without affecting embryonic development (Landry,

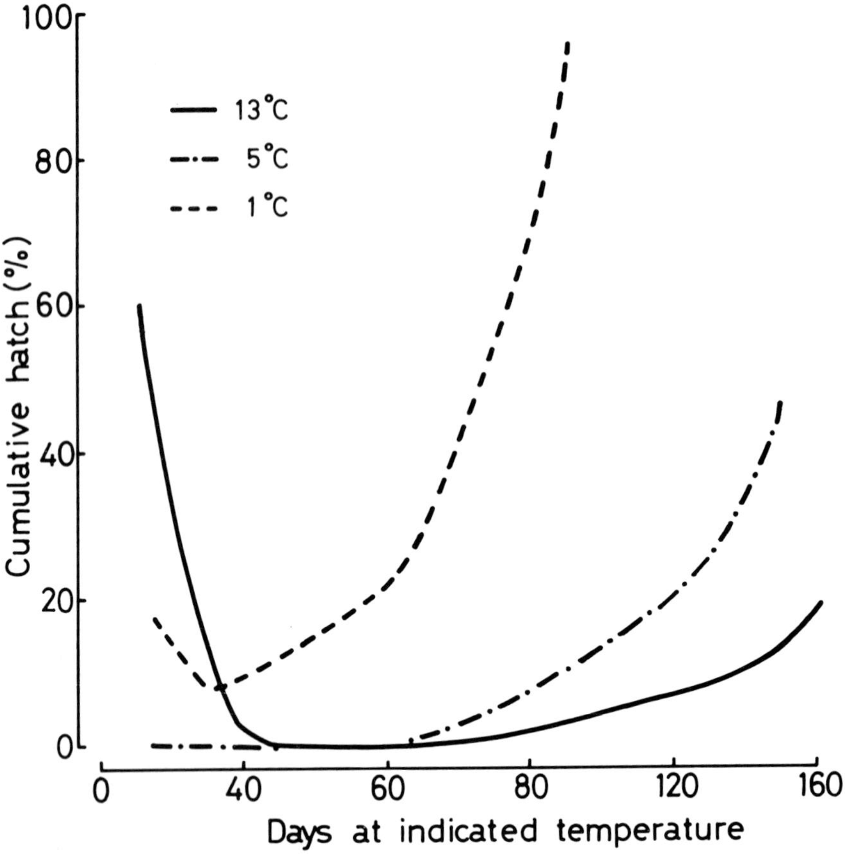

FIGURE 3.6. Effects of temperatures on termination of diapause of *Diaptomus stagnalis*. The oviposited eggs are kept at 25°C for 16 days and exposed to 1°, 5°, and 13°C for indicated days. The hatching percentage is estimated after incubation at 13°C. (Modified from Brewer, 1964.)

1975). The action spectrum for the hatching of *A. salina* cysts and the result of an illumination experiment with decapsulated cysts suggest that a heme pigment may be involved in light-induced hatching, and the screening role of hematin has also been proposed (Van der Linden et al., 1986).

In general, experiments on the effects of salinity have shown that diapause eggs of marine copepods have a wide tolerance for salinity. When diapause eggs of *Tortanus facipatus* are exposed to salinities of 0–90‰, hatching occurs at a very high rate between 18 and 54‰ (Kasahara et al., 1975). Eggs of the other calanoid copepods resume embryogenesis when exposed to the optimum salinity, but that optimum differs from species to species (Uye et al., 1979).

Another important factor affecting the development of diapause eggs is the oxygen concentration of water. None of the eggs of six species of marine calanoid copepods developed into nauplii under anoxic conditions (less than 0.4 ml O_2/litre water). However, these undeveloped eggs were able to initiate embryogenesis and to hatch when incubated later at the normal concentrations of oxygen (4.7 ml/litre) (Uye et al., 1979). Since eggs of several species of calanoid copepods exist in the diapause state in sea-bottom muds where oxygen concentration and salinity are reduced, these eggs have become adapted to use these physical factors as the crucial signals for termination of diapause.

The spider *Floronia bucculenta* completes one generation annually in nature, and its life cycle is synchronized with the annual seasons by two diapause mechanisms: prolongation of postembryonic development by long-day photoperiods, and arrest of embryonic development at high temperatures. The incidence of embryonic diapause is not influenced by different regimens of temperatures and photoperiods acting upon the maternal generation. But the complete arrest of embryogenesis is directly induced by exposure of the laid eggs to relatively high temperatures (above 12.5°C). Slow but continuous embryogenesis proceeds in an egg that is chilled at below 5°C, and the whole period of embryonic development expands to about 200 days. The prior inhibition of embryonic development by high temperatures can be removed completely by exposure to low temperatures. A temperature as low as 0°C is most effective for termination of diapause. The minimal chill period required is 10 weeks at 0°C and expands to 12 weeks at 5°C and 25 weeks at 7.5°C (Schaefer, 1976).

Despite the diversity of diapausing stages and its independent evolution in so many different insect taxa, reliance upon photoperiod and temperature as the cues for diapause induction is amazingly widespread (Lees, 1955; Danilevski, 1965; Tauber and Tauber, 1978; Denlinger, 1985). Both the quantity and quality of food can exert an effect on diapause in some insects (Clay and Venard, 1972; Scheltes, 1978). The idea prevailing in much of the diapause literature is that diapause termination, like induction, is linked directly to specific environmental cues, such as temperature and photoperiod.

The sensitive period most frequently occurs well in advance of the diapausing stage, but the nature and duration of the sensitive period vary greatly among different species. For induction of embryonic diapause in many insects, environmental stimuli are received at different stages of the maternal generation. As for silkworms, exposure of developing embryos to a high temperature (25°C) and a long-day photoperiod (16 L:8 D) plays a decisive role in programming diapause in the egg of the progeny (Yamashita and Hasegawa, 1985). When the egg is exposed to an intermediate temperature (20°C), then a high temperature and a long-day photoperiod at the larval stages, especially at the younger stages, influences diapause expression of the progeny. However, there is no conversion from nondiapause to diapause by shifting environmental conditions once the maternal egg is programmed to become nondiapause by low temperature at the embryonic stage. Thus, failure to receive the decisive cues for the determination of diapause or nondiapause will prolong the monitoring of environmental conditions until programming of diapause is set up.

In many cases of embryonic diapause, a low temperature, one lower than 15°C, acts as an environmental cue for termination of diapause. The optimum temperature varies from species to species; a still higher temperature (10–15°C) is effective for insects inhabiting a relatively high-temperature zone (Denlinger, 1985). For the eggs of silkworms (Takami, 1969) and false melon beetle, *Atrachya mentriesi* (Ando, 1978), a temperature of 7.5°C is most effective for quick and synchronous termination of diapause. For these insects, storage at a temperature as low as 0°C has been shown to render eggs inert as regards diapause termination (Ando, 1978; Furusawa et al., 1982): diapause of silkworm eggs that were preserved continuously at 0°C was only terminated after they were chilled at 5°C for more than 2 months, a procedure like that used for eggs kept at 25°C. This response of diapause eggs to 0°C has suggested a means whereby preservation of silkworm eggs can be prolonged for more than 2 years (Yamashita et al., 1988).

Although eggs of *Melanoplus differentialis* absorb water during diapause termination (Slifer, 1958), water absorption may not actually be a cue for such termination but may only be coincidental. As described previously, a diapausing egg is characterized by reduced oxygen consumption. Dechorionization of diapausing eggs of silkworms initiates embryogenesis in a culture system (Okada, 1971), suggesting that oxygen acts to terminate diapause. The permeability of oxygen through the chorion layer remains unchanged before and after diapause termination (Sonobe et al., 1980). Recently, Kim (1987) showed different results in oxygen permeability between diapause eggs and nondiapause eggs, although a very limited area of the chorion layer was examined for estimation. More detailed experiments are required to show oxygen effects.

Some chemical reagents are known to promote diapause termination. Diapause of some cricket eggs is terminated on contact with urea solution or exposure to ammonia gas (Hogan, 1962, 1964, 1965). In the egg of *M. differentialis,* xylol, toluol, carbon tetrachloride, and mineral oil are efficient and completely harmless agents for terminating diapause (Slifer, 1958). Diapause of *Atrachya menetriesi* eggs terminates when they are dipped in mercuric chloride solution (Kurihara and Ando, 1969;

Ando, 1971). The effects of asparagine and glutamine on diapause termination have been examined in *Teleogryllus emma* eggs (Tomeba et al., 1988); the precise role of these chemicals remains unknown, although permeability of the chorion is improved in the treated eggs.

Since the turn of the century, hydrochloric acid has been well known as an effective means to control the progress of diapause in silkworm eggs. This acid works in two ways: (1) it prevents the initiation of diapause; (2) it terminates diapause. Prevention can be induced by dipping the eggs in HCl solution just before the onset of diapause (specific gravity of 1.075 at 15°C) for 5 min at 46.5°C, whereas termination is effected by soaking the prechilled diapausing eggs in HCl solution (specific gravity of 1.10 at 15°C) for 5 min at 48°C. Electron probe X-ray microanalysis has recently demonstrated that Cl^- ions actively accumulate in the chorion layer, with a concomitant decrease of Ca^{2+} ions. Further, the pH of such eggs decreases from 6.6 to 5.8 during a 2-h period after HCl treatment (Yoshimi et al., 1986). Reinitiation of embryogenesis in dormant cysts of *A. salina* is accompanied by a large change in intracellular pH (Busa et al., 1982). Similar associations between intracellular pH and release from metabolic and developmental repression have been noted for germinating bacterial spores (Setlow and Setlow, 1980). Thus, such HCl treatment seems to change the intracellular pHs and hence the conditions in which various metabolic processes become active.

3.4.2. *The Diapause Hormone*

The diapause of eggs with a fully developed embryo all ready to hatch is considered to be regulated by the same endocrine system as in larval diapause (Denlinger, 1985). In eggs whose endocrine system has developed during prediapause morphogenesis, diapause is thought to be the consequence of hormonal deficiency (Neumann-Visscher, 1976); those that are neutral in respect to diapause determination are presumably regulated by some maternal effect (J. F. Anderson, 1968). Eggs entering diapause at an early stage of embryonic development are reported to have a quite different regulatory system from those with larval, pupal, or adult diapause. In this case, the program for diapause is determined by the endocrine system of the maternal generation. Such a mechanism has clearly been demonstrated in silkworms and the lymantriid *Orgyia antiqua* (Yamashita and Hasegawa, 1985).

3.4.2.1. THE ENDOCRINE SYSTEM INDUCING EMBRYONIC DIAPAUSE

Clearly, the germ band is too rudimentary to transduce environmental stimuli to internal signals. Early experiments by Umeya (1926) provided the first hint of a blood-borne factor of pupae regulating diapause of the progeny eggs. A classic series of experiments identified the subesophageal ganglion as the endocrine organ releasing diapause hormone (DH) (Yamashita and Hasegawa, 1985). The brain exerts its role for controlling the release of DH from the subesophageal ganglion. The precise interaction between the brain and the subesophageal ganglion remains to be clarified, but the brain may function as the receptor of environmental stimuli and as the programmable center of diapause by regulating the release of DH from subesophageal ganglion (Matsutani and Sonobe, 1987).

DH activity has also been detected in the subesophageal ganglia of several other species of moths (Takeda and Girardie, 1985; Yamashita and Hasegawa, 1985). When assayed on nondiapause egg producers of silkworms, ganglia from *Lymantria disper, Antheraea yamamai,* and *Dictyoploca japonica* all stimulate production of some diapause eggs. All three of these species also have an egg diapause. Results with *Antheraea pernyi, Phalaenoides glycinae, Spilosoma canescens, Leucania separata, Periplaneta americana,* and *Locusta migratoria* are more surprising: these species have no embryonic diapause, yet their subesophageal ganglion shows DH activity. DH activity is also found in the brain–corpora cardiacum–corpora allatum complexes of *L. separata* (Ogura and Saito, 1973), *P. americana* (Takeda, 1977), and *L. migratoria* (Takeda and Girardie, 1985). Thus, the active principle for diapause induction of silkworm eggs may prove to be very widespread.

3.4.2.2. CHEMISTRY OF THE DIAPAUSE HORMONE

The most direct technique for measuring the activity of DH extracts utilizes a race of polyvoltine strains as the injection recipient. In this bioassay, a DH unit is the amount of extract required to induce 50% diapause eggs (Isobe and Goto, 1980). The highly purified preparation is dissolved in a slightly alkaline solution containing bovine serum albumin as a carrier (Yamashita and Hasegawa, 1985). Each sample is routinely injected into the abdomen 4 days after pupation when the compound eyes become reddish brown. The fate of eggs can be determined with certainty 2 weeks after oviposition, but activity can be closely estimated 3 days after oviposition by observing the color of the eggs laid. Ommochrome accumulates in diapause programmed eggs and gives the egg a distinctly dark color.

Hasegawa (1957) was the first to extract the active principle from brain–subesophageal ganglion complexes of silkworm pupae and called it the diapause hormone (DH). The discovery of high activity in adult male heads eliminates the labor of dissecting organs and provides a more readily attainable extraction source. As the starting materials, 4.2 million male heads are used (Imai et al., 1982). The two crude DH fractions, DH-A and DH-B, are obtained by sequential extraction with acetone and a methanol : chloroform mixture. DH-A is then subjected to various column chromatographic separations using an organic solvent because the crude extracts easily form aggregates in an aqueous solution. The last chromatography with Merkogel OR-6000 gives a symmetrical peak in weight and activity. However, this fraction is further separated into two main fractions by the subsequent DEAE(diethylaminoethyl)–Sepharose CL-6B column chromatography, but activity is recovered in only one fraction. The active fraction is finally subjected to a high-performance liquid chromatography (HPLC). The third HPLC gives a single peak having a specific activity of 0.05 μg/DH unit.

DH-B is selectively extracted using prolonged pulverization of adult heads after extraction of DH-A. Thus, extracts of DH-B are less contaminated compared with those of DH-A. Although DH-B is purified by sequential column chromatographies like those used for purification of DH-A, the elution pattern is different, suggesting that DH-A and DH-B are similar but not identical in chemical structure (Isobe and Goto, 1980).

The apparent molecular masses of DH-A and DH-B have been estimated as 3.3 and 2 kDa, respectively, from the elution profile from gel permeation column chromatography. DH-A appears to be composed of 14 kinds of amino acids and 2 kinds of amino sugars. DH-B contains the same 14 amino acids found in DH-A, although the molar ratios of some amino acids are slightly different. DH-B lacks amino sugars, and there appear to be no free amino or carboxyl terminal groups. In both fractions, activity is readily lost after treatment with proteolytic enzymes, thus suggesting that peptide linkages are essential for hormonal activities (Isobe and Goto, 1980).

When similar methods are applied for partial purification after treatment with a detergent, diapause hormone activity is obtained from ovaries and eggs of silkworms that are programmed to diapause. Although the elution profile after chromatography with a Sephadex LH-20 column resembles that found from the subesophageal ganglia, it is now uncertain whether diapause hormone activity in eggs is of maternal origin (Kai and Kawai, 1981).

3.4.2.3. BIOCHEMICAL ACTION OF THE DIAPAUSE HORMONE

DH is actively secreted from the subesophageal ganglion throughout pupal–adult development of silkworms, during which periods ovarian development occurs to produce eggs. Induction of diapause eggs closely depends upon the time of injection of DH, and a higher incidence is obtained by injection at the vitellogenic stage of many follicles. Consequently, DH action is elicited in pharate adults even though it plays an important role in regulating the developmental fate of the laid eggs. The hormone's effect is reflected on metabolism, which provides the biochemical adjustment for diapause eggs. Eggs programmed to diapause exhibit some quantifiable biochemical characteristics: high accumulation of glycogen and 3-hydroxykynurenine. Injection of DH into pupae programmed to lay nondiapause eggs stimulates transport of 3-hydroxykynurenine from the hemolymph to the ovary, where it accumulates and is converted into the ommochrome pigment when serosal cells are differentiated (Yamashita and Hasegawa, 1985). This effect results in the dark and perhaps cryptic coloration that is unique to diapause eggs. Similarly, accumulation of glycogen in ovaries is stimulated by injection of DH extracts (Yamashita and Hasegawa, 1985); at the initiation of diapause, glycogen is converted to sorbitol, as observed in natural diapause eggs.

The elevated accumulation of glycogen in ovaries by DH is accompanied by a reduced level of hemolymph trehalose. No difference in hemolymph trehalose levels is brought about in ovariectomized female pupae after DH treatment. Hormone regulation of hemolymph trehalose level occurs even in males into which ovaries have been implanted. Thus, the ovary is seen to be a target organ of DH. Glycogen synthetic activity in follicles ranging in weight from 550 to 600 μg is selectively stimulated by DH injection, suggesting that there is a restricted sensitive stage of target cells for the hormone (Yamashita and Hasegawa, 1985). Ovarian glycogen is derived from hemolymph trehalose that originates in fat body glycogen (Yamashita and Hasegawa, 1974). Thus, glycogen accumulation in ovaries depends on the synthetic activity of the ovaries and not on the fat body.

Enzymatic analysis of the glycogen synthetic pathway in developing ovaries has demonstrated the presence of trehalase, hexokinase, phosphoglucomutase, uridine diphosphate glucose(UDPG)–pryophosphorylase, and UDPG–glycogen glucosyltransferase. The kinetic properties and subcellular localization of these enzymes are comparable with those found in other insect tissues (Yamashita, 1969). Accordingly, ovarian glycogen is synthesized from hemolymph trehalose by the following pathways:

hemolymph trehalose → glucose → glucose 6-phosphate →
glucose 1-phosphate → UDPG → glycogen

Among these enzymes, trehalase has been proposed as the rate-limiting enzyme based on the following facts: (1) the highest K_m value of trehalase is comparable to that detected with concentrations of hemolymph trehalose; (2) the lowest activity of trehalase is detected subsequent to ovarian development (Yamashita, 1969). Extirpation of the subesophageal ganglion from pupae destined to lay diapause eggs provides an ideal system by which investigators can study DH action on whole ovaries because such treatment completely switches the ovary to production of nondiapause eggs. Removal of the subesophageal ganglion reduced trehalase activity alone by one-third of the control level without affecting the other enzymes (Yamashita et al., 1972). Injection of a DH preparation into day-5 pupae whose subesophageal ganglion had been extirpated resulted in a rapid increase in trehalase activity. This hormonal effect has been clearly substantiated in an *in vitro* culture system with a DH preparation (Miyadai and Yamashita, 1980).

Absorption of trehalose from the incubation medium follows saturation kinetics, giving an apparent K_m value of 8.4 mM (Shimada and Yamashita, 1979), which resembles that of ovary trehalase acting upon trehalose (Yamashita et al., 1972). Trehalose absorption is closely correlated with the activity of ovary trehalase; the higher the trehalase activity, the greater the trehalose absorption. Accordingly, it is likely that hemolymph trehalose is taken up into the ovary by a trehalase-mediated process. Immunohistochemical and biochemical studies have disclosed that trehalose is exclusively localized on the plasma membrane of oocytes (Azuma and Yamashita, 1985). From these results we can conclude that DH does enhance trehalase activity, which is attributed to the increased uptake of trehalose from hemolymph and to the supply of glucose for glycogen synthesis (Yamashita and Hasegawa, 1985).

Recently, Takeda et al. (1988) have demonstrated that by injecting a potent and specific trehalase inhibitor, validoxylamine, into a pupa programmed to lay diapause eggs, nondiapause eggs are induced accompanied by a decrease in glycogen content in the laid eggs. This significant effect is observed only when the inhibitor is injected into day-6 pharate adults in which many follicles have become sensitive for DH. Thus, validoxylamine seems to act directly on developing follicles to inhibit trehalase activity, but not on the subesophageal ganglion to interfere with secretion of DH. It is thus quite possible that the regulation of trehalase by DH represents one of the original hormonal actions that direct the metabolic pathway responsible for diapause initiation.

3.4.3. *Ecdysteroids*

3.4.3.1. OCCURRENCE AND CHEMISTRY OF ECDYSTEROIDS

Ecdysteroids are hormones secreted by the prothoracic glands or their counterparts that control molting, metamorphosis, and reproduction in arthropods (Wigglesworth, 1985). Even in developing embryos, molts occur several times at definite stages. In some grasshoppers (*Locusta migratoria, L. pardalina,* and *Melanoplus differentialis*), four distinct cuticles are secreted: serosal cuticle, two embryonic cuticles, and larval cuticle (Mueller, 1963; see also Chapter 2 by Sbrenna herein). Embryonic molting was first considered to be controlled by neuroendocrine mechanisms in *L. migratoria* and *L. pardalina* (Johnes, 1956). When the embryonic abdomen, isolated long before the brain and prothoracic glands are differentiated, is incubated *in vitro* with sufficient amounts of yolk, it develops into the larval abdomen after molting in some insects such as *Schistocerca gregaria* (Sbrenna-Micciarelli and Sbrenna, 1972), *Bombyx mori* (Takami, 1963), and *M. differentialis* (Mueller, 1963). These data led to the conclusion that embryonic molting, in contrast to larval molting, can occur independently of the brain and prothoracic glands, but such results do not eliminate the possibility that embryonic molting might be controlled by molting hormone (ecdysteroids), provided by sources other than the prothoracic glands.

The presence of ecdysone-like activity in yolk alone has been shown in the egg of *M. differentialis* (Boohar and Bucklin, 1963). Ohnishi et al. (1971) have clearly shown relatively high ecdysteroid activity in the extracts of silkworm eggs, including diapause eggs. The greater part of such activity is due to the more polar substances, presumed to be ecdysteroid conjugates, which are obtained in their free forms upon incubation with snail juice. Particularly, ecdysteroids from diapausing eggs are almost exclusively conjugated (Ohnishi, 1981).

Research on the role of hormones in the embryonic development of insects received an unexpected impetus a few years ago when the ovaries of reproductively competent females were demonstrated to be actually capable of synthesizing ecdysteroids in several insects (Hagedorn, 1983; Hoffmann and Lagueux, 1985). These findings led to the conclusion that ecdysteroids synthesized in developing ovaries are stored in yolks as conjugated forms that are used during the subsequent embryogenesis.

Over the last decade, 16 kinds of ecdysteroids have been identified in eggs of insects (Fig. 3.7). Characteristically, egg ecdysteroids are predominantly present as conjugates in most insect species investigated. A large portion of the conjugates is esterified to phosphate at the C-22 and C-3 hydroxyl groups (Hoffmann and Lagueux, 1985; Isaac and Rees, 1985).

The apparently ubiquitous presence of ecdysteroids in insect eggs and ovaries led us to explore their occurrence in another major arthropod group, the Crustacea. Radioimmunoassay measurements of crude extracts of the shore crabs *Carcinus maenas* revealed for the first time that crustacean eggs contain ecdysteroids (Lachaise and Hoffmann, 1977). Similar findings were later reported from crude extracts of *Orchestia gammarellus* (Blanchet et al., 1979), *Acanthonyx lunulatus* (Chaix and De Reggi, 1979) and *Helleria brevicornis* (Hoarau and Hirn, 1978).

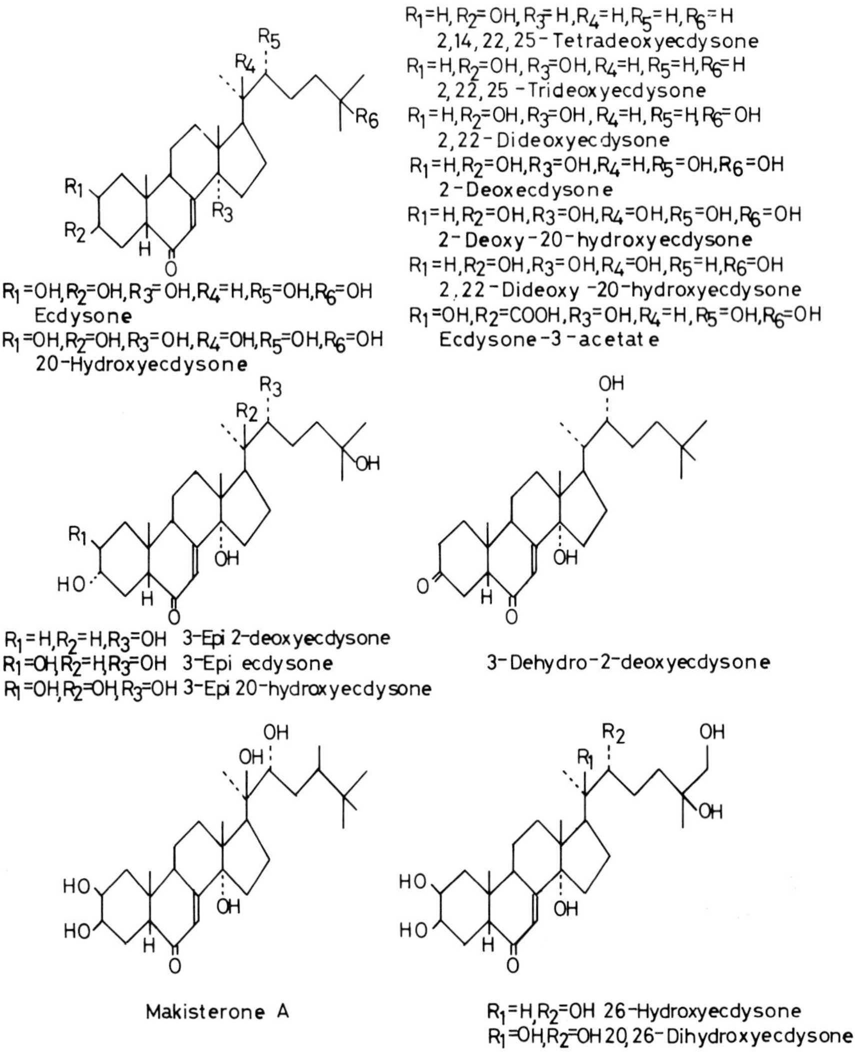

FIGURE 3.7. Structural formulas of ecdysteroids identified in insects and crustaceans.

In the eggs of the blue crab, *Callinectes sapidus*, two major ecdysteroids, 20-hydroxyecdysone and ponasterone A (2b,3b,14a,20,22-pentahydroxy-5b-cholest-7en-6one), have been identified on the basis of chromatographic, immunological, and mass spectral data (McCarthy, 1979). Ecdysone and polar conjugate of 20-hydroxyecdysone are present at a low level throughout embryonic development (McCarthy and Skinner, 1979). Eggs of *C. maenas* contain high concentrations of ponasterone

A along with lower titers of ecdysone and 20-hydroxyecdysone, and some portion of the ecdysteroids are present in their polar conjugated forms (Lachaise and Hoffmann, 1982).

Recently, materials immunoreactive to ecdysteroids have been found in eggs of the ticks *Amblyomma hebraeum* and *Rhipicephalus appediculatus* (Diehl et al., 1982). Physicochemical analysis on extracts of the ixodid tick *A. hebraeum* shows that 20-hydroxyecdysone is always the major free ecdysteroid to be found and more than 90% of the radioimmunoassay-positive materials are in the free forms (Connat et al., 1985). In nymphs of the argasid tick *Ornithodoros moubata,* the ingested ecdysteroid 20-hydroxyecdysone is rapidly transformed into a family of apolar conjugates, which are esterified at C-22 with long-chain fatty acids, $C_{16:0}$, $C_{18:0}$, $C_{18:1}$, or $C_{18:2}$. These apolar conjugates may be transported into the ovaries and then into eggs as a storage form of the ecdysteroids (Diehl et al., 1985).

3.4.3.2. ROLES OF ECDYSTEROIDS IN EMBRYOGENESIS

Changes in titers of free ecdysone and free 20-hydroxyecdysone can be followed in eggs of *L. migratoria* from oviposition to larval hatching (Fig. 3.8). Ecdysone is present at all stages, and four distinct peaks are observed during embryonic development—with each of these coincident with the deposition of a cuticle. In this insect, free 20-hydroxyecdysone remains at a low level during the first half of embryogenesis and shows two distinct peaks during later development, whereas the titers of total ecdysteroids including the conjugated forms are maintained at almost constant levels throughout embryonic life (Lagueux et al., 1981). Consequently, one of the functions of egg ecdysteroids has been proposed to be the control of embryonic molting. In several insect species, namely, *Oncopeltus faciatus* (Dorn and Romer, 1976), *Blaberus craniifer* (Bullière et al., 1979), *Leucophaea maderae* (Hoffmann and Lagueux, 1985), *Nauphoeta cinerea* (Imboden et al., 1978; Imboden and Lanzrein, 1982), and *Clitumus extradendatus* (Cavallin and Fournier, 1981), a correlation between molting events in embryos and concentration peaks of ecdysteroids is clearly evident.

Before blastokinesis in *S. gregaria,* both free and conjugated ecdysteroids are present only in the yolk and undergo a continuous and constant decrease with small fluctuations. Later, when the yolk is gradually engulfed, both forms of the hormone become detectable in the embryo. This second phase is characterized by a conspicuous rise of free ecdysone and 20-hydroxyecdysone in correspondence with mitotic activity, apolysis of the second cuticle structure, and onset of larval cuticular synthesis. Consequently, the free hormone is able to exert control of morphogenetic processes of embryonic development (Scalia et al., 1987).

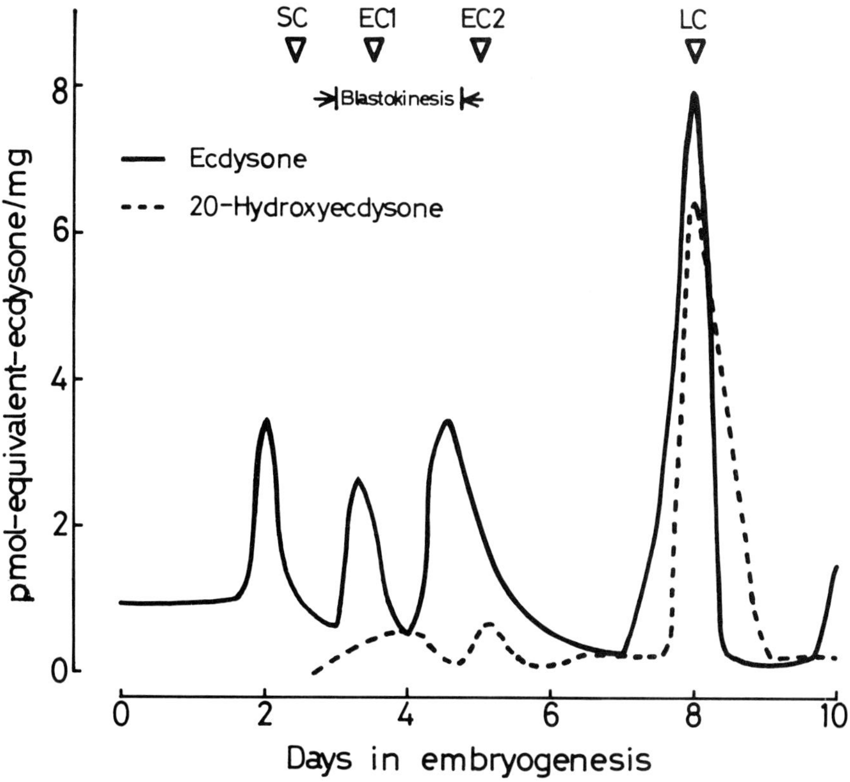

FIGURE 3.8. Developmental changes in free ecdysone and 20-hydroxyecdysone during embryogenesis of *Locusta migratoria.* Arrows indicate onset of deposition of the different cuticles: SC = serosal cuticle; EC1 and EC2 = first and second embryonic cuticles; LC = first larval cuticle. (Modified from Lagueux et al., 1981.)

In the blowfly *Calliphora erythrocephala,* the deposition of the typical (second) cuticle, the larval cuticle, is coincident with a rise in the concentration of ecdysteroids in eggs. However, the deposition of the thin, atypical (first) cuticle apparently occurs in the absence of a clearly defined surge in ecdysteroid concentration. The typical cuticle is deposited in embryos whose ring gland is experimentally removed, indicating that an ecdysteroid biosynthesis can occur in *Calliphora* embryos in the absence of ring gland (Bordes-Alléaume and Sami, 1987).

In silkworm eggs, a rather constant level of free ecdysteroids is observed during the first half of embryonic development, followed by a decrease in later periods. The content of the conjugated forms is shown to fluctuate, with several peak values, but none of the peaks corresponds in time with the differentiation of membranous structures, including serosal membrane and the first larval cuticle (Mizuno et al., 1981). Thus, it is possible that ecdysteroids functioning at embryonic stages may have different roles from those at postembryonic stages, as shown in vertebrates in which fetal steroid hormones play different roles from those working after puberty (George et al., 1981).

An interesting role of ecdysteroids of eggs is presumed in relation to embryonic diapause. Ohnishi et al. (1971) first observed in silkworm eggs that the level of ecdysteroid activity is diminished during diapause and reelevated upon resumption of embryonic development. More detailed analysis has provided evidence that a strict time correlation exists between a decrease of free ecdysteroid levels and the onset of diapause and vice versa, although the conjugate forms did not change appreciably before and after diapause. This situation apparently resembles the endocrinological mechanism responsible for pupal diapause in insects (Denlinger, 1985). Similar results have been shown by Hirn et al. (1977) and Coulon et al. (1979) by radioimmunological measurements of crude unhydrolyzed extracts of silkworm eggs. The comparisons of ecdysteroid content are made between diapause and nondiapause eggs of the Australian plague locust, *Chortoicetes terminifera,* and shows the reduced levels of ecdysteroids in diapausing eggs (Gregg et al., 1987).

The ecdysteroid role in diapause termination in silkworm eggs is supported by the fact that diapausing embryos are able to resume their embryogenesis by incubation in a medium containing relatively high concentrations of 20-hydroxyecdysone (Gharib et al., 1981). The conclusive role of ecdysteroids on regulation of embryonic diapause of this insect awaits further investigation using ecdysteroid-free eggs as controls.

Correlative studies on ecdysteroid titers and embryogenetic events have provided the information on the function of hormonal actions in crustaceans. During embryogenesis of the blue crab, *C. sapidus,* a continuous increase in ecdysteroids, especially 20-hydroxyecdysone and peak III (ponasterone A), takes place (McCarthy and Skinner, 1979). A more detailed experiment has been carried out with eggs of the shore crab *C. maenas* (Lachaise and Hoffmann, 1982). Ponasterone A is the predominant ecdysteroid and is present in both free and conjugated forms. The concentration of both forms shows two distinct peaks during the developmental periods: one approximately 40 h after egg laying, and the second at around 200 h (Fig. 3.9). The second peak precedes secretion of the first embryonic envelope by a few hours. Deposition of this envelope is followed by that of metanauplii, protozoea, and prozoea, all of which occur in the presence of high titers of ponasterone A. These observations suggest that in embryos of *C. maenas,* one of the roles of ecdysteroids is the control of deposition of embryonic envelope and cuticles.

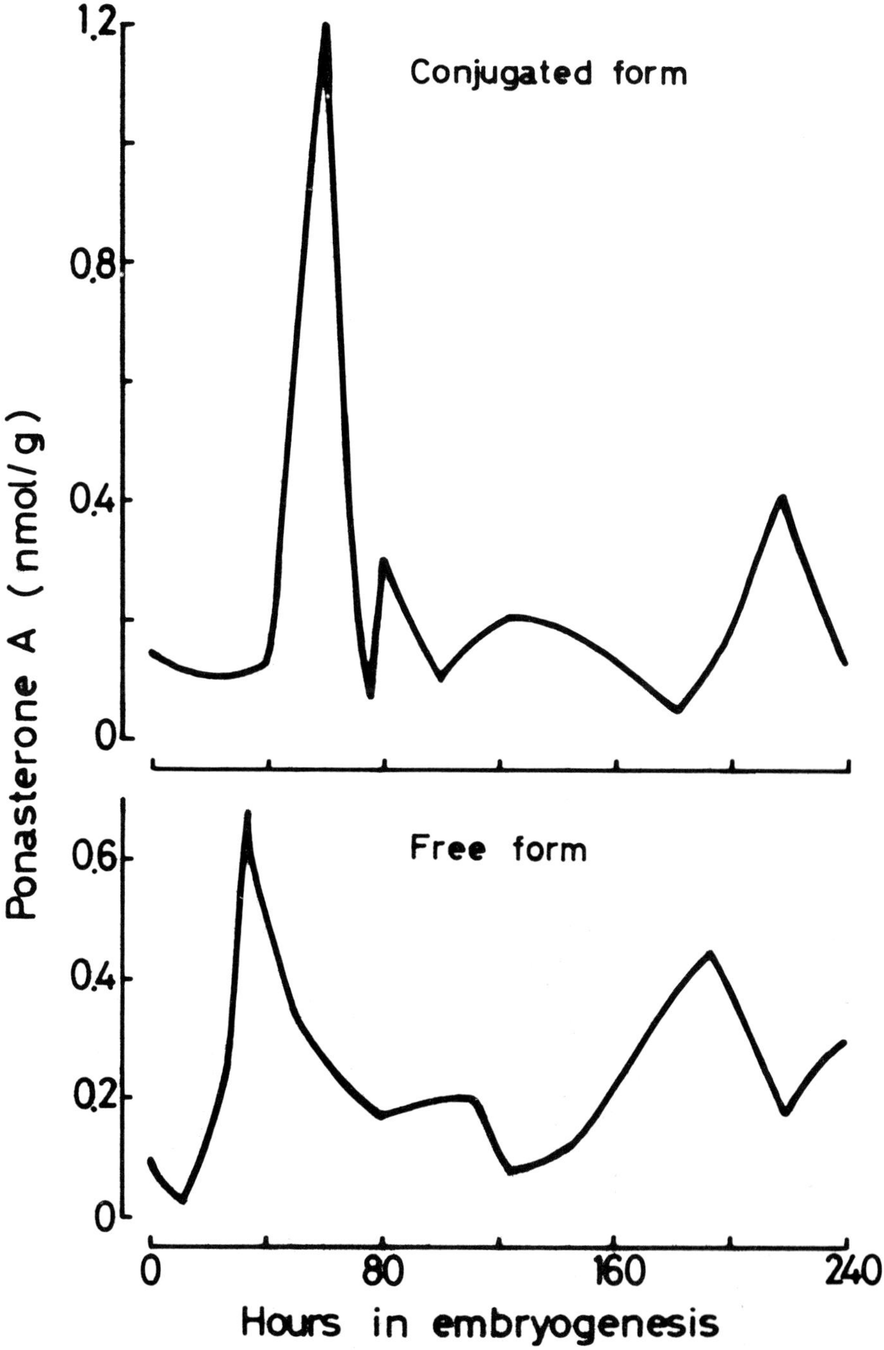

FIGURE 3.9. Changes in ponasterone A titers during embryogenesis of *Carcinus maenas*. (Modified from Lachaise and Hoffmann, 1982.)

3.4.4. *Juvenile Hormones*

3.4.4.1. OCCURRENCE AND CHEMISTRY OF JUVENILE HORMONES

In contrast to the substantial number of studies on ecdysteroids in eggs, very few papers have dealt with the presence of juvenile hormones (JHs) in eggs of insects. This is all the more surprising because as early as 1961 Gilbert and Schneiderman found that eggs of *Hyalophora cecropia* contain JH activity, as determined by the *Galleria* wax test. In their experiment, JH in newly laid eggs is shown to be of maternal origin, since no deposition of this hormone occurs in eggs laid by allatectomized females.

It was not until 1980 that these studies were continued on egg JHs, while during this period great progress has been made in determining the chemical structure, biochemical action, and physiological function of JH in postembryonic organisms. The paucity of information on embryonic JH seems to have been due to the difficulty inherent in the methodology of JH assays. However, recent developments in the field of radioimmunoassays (RIA) combined with chromatographic and mass spectrometric techniques have led to rapid advances in our knowledge of embryonic JH. In Fig. 3.10, structural formulas of JHs identified in eggs of insects and crustaceans are shown. Bergot et al. (1980, 1981) were the first to separate and identify the JHs in eggs of several insect species by using physicochemical methods. In the lepidopteran species *Manduca sexta, Heliotis virescens,* and *H. cecropia,* JH I, JH II, and JH 0 have been identified (Bergot et al., 1980, 1981). JH 0 and iso-JH 0 are exclusively present in eggs of these species. The biological activity of JH 0 was found to be higher than that of the other JHs by a substitution assay in *M. sexta* (Bergot et al., 1980). JH III has been detected in *Oncopeltus faciatus* (Bergot et al., 1980), *Telegryllus commodus* (Loher et al., 1983), *Locusta migratoria* (Temin et al., 1986; Pener et al., 1986), and *Nauphoeta cinerea* (Baker et al., 1984; Lanzrein et al., 1985).

In crustaceans, as in insects, JHs are generally thought to play regulatory roles in both metamorphosis and gametogenesis. Although Schneiderman and Gilbert (1958) extracted compounds with JH activity from crustaceans 30 years ago, until recently the chemical entity of JH in crustacea has remained unidentified. Laufer et al. (1987) extracted a compound with JH activity from hemolymph of adult spider crab *Libinia emarginata* and clearly identified it as methyl farnesoate (MF) by gas chromatography/mass spectrometry (GC/MS); see Fig. 3.10. Further, the mandibular organs have been shown to be the secretory organs of MF by comparing the secretory activities of various tissues incubated *in vitro.* The *in vitro* secretory rates of MF by these organs are closely related to

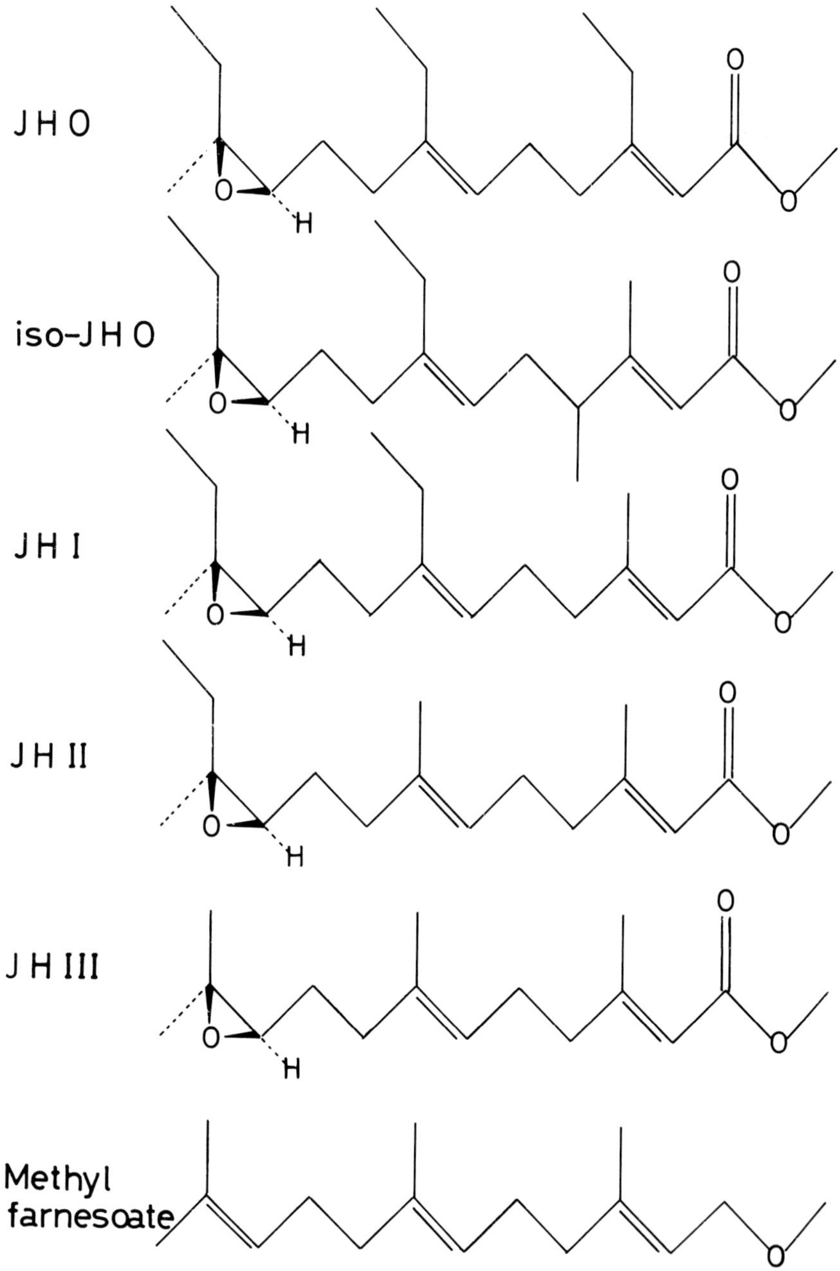

FIGURE 3.10. Structural formulas of JHs and methyl farnesoate identified in insect eggs and crustaceans.

the stages of ovarian growth. This relation between secretion of MF and vitellogenesis in *Libinia* is in all likelihood a cause-and-effect response.

The occurrence of MF as a secretory product of the mandibular organs is not unique to *L. emarginata* but is also found in other crabs and lobsters (Borst et al., 1987; see also Chapter 2 by Borst and Laufer in Part 1). It is not yet known what kinds of JH are present in eggs of crustaceans, although some circumstantial evidence is available (Templeton and Laufer, 1983).

3.4.4.2. ROLES OF JUVENILE HORMONES IN EMBRYOGENESIS

Preliminary investigations on the fluctuations of JHs in eggs of some species show that JHs remain at very low level before the onset of organ differentiation but then increase steeply through larval differentiation, with a drop prior to larval hatching (Dorn, 1982; Imboden et al., 1978; Bergot et al., 1980). In eggs of *L. migratoria* during embryonic development, four periods are apparent from the end of oogenesis until hatching as regards the fluctuation of JH titer (Fig. 3.11). From ovulation, a small amount of JH (7 n*M*) is present during germ band formation that is obviously of maternal origin. During blastokinesis, JH III is not detectable. This period corresponds to the intense stage of organogenesis when many larval tissues differentiate. The subsequent larval differentiation is characterized by a gradual rise in concentration of JH III (Temin et al., 1986). The JH III peak slightly precedes a peak of free ecdysteroids in this egg, corresponding to larval cuticle deposition (Hoffmann and Lagueux, 1985). Thus, formation of the first larval cuticle occurs in the presence of high titers of both JH III and ecdysteroid. As opposed to this, the serosal cuticle and embryonic cuticles are deposited in the presence of high ecdysteroid titers, but in the absence of significant JH III levels.

In *N. cinerea* (Bruning et al., 1985), the increase of JH III titer after dorsal closure can be prevented by the application of a precocene analogue, which has been shown to be a potent inhibitor of JH synthesis in corpora allatum in insects (Bowers, 1983). In treated eggs, the fat body disintegrates and the midgut does not differentiate normally, thus impeding yolk resorption. When a dose of JH III is applied onto the precocene-treated embryos, formation of the midgut is normalized and disintegration of the fat body is less pronounced. Consequently, in embryos, JH III plays a role in the differentiation of the midgut epithelium and probably in the development of the fat body in addition to cuticle formation (Bruning and Lanzrein, 1987).

The levels of JHs seem to be decided by the balance of synthesis and degradation. An increase in titer after dorsal closure mainly corresponds to increased synthetic activity by the newly differentiated corpora al-

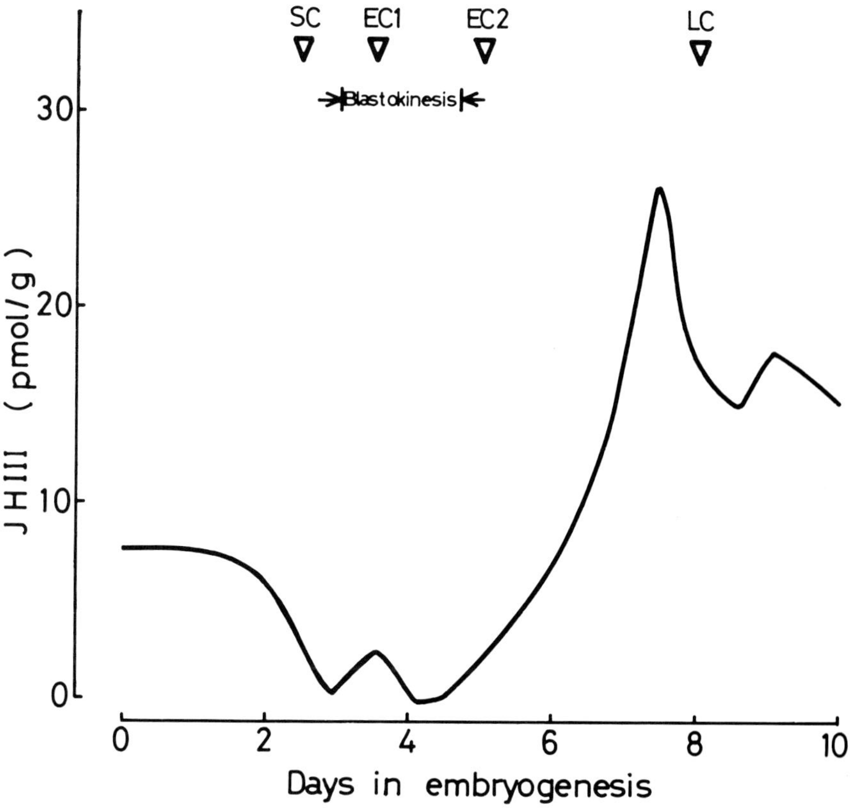

FIGURE 3.11. Changes in JH III titers in eggs of *Locusta migratoria* during embryogenesis. Arrows indicate the period of deposition of serosal (SC), first (EC1) and second (EC2) embryonic cuticles, and larval cuticle (LC). (Modified from Temin et al., 1986.)

latum, because the activity of JH esterase declines from this stage (Roe et al., 1987).

Regulatory roles of JHs during embryogenesis in insects and crustaceans have also been proposed by researchers who have topically applied JHs, their analogues, and anti-JHs onto adults and developing embryos (Sehnal, 1983; Hardie, 1987). The age- and dose-dependent effects of these compounds indicate a functional role for JH during embryogenesis. However, these results have proved difficult to interpret because the high concentrations of substances used suggest that the observed responses may be nonspecific pharmacological effects.

Concerning the effects of JHs on embryonic diapause, a preliminary experiment has been carried out on silkworm eggs (Gharib et al., 1984).

The three JHs were not found in the eggs before germ band formation, but they abruptly appeared at the level of 5 nmol/g eggs, followed by a sudden disappearance at the initiation of diapause. Although no probable conclusion has been drawn from these results on JH actions related to diapause, such information will be helpful in future elucidation of the hormonal code controlling embryonic diapause.

Recently, an interesting phenomenon was discovered regarding hormonal control of embryonic diapause of *Antheraea yamamai* (Tan et al., 1986) and *Lymantria dispar japonica* (unpublished data). Diapause in these insects intervenes after completion of larval differentiation and is terminated by exposure to low temperature for more than 60 days. Thus, this diapause seems to be under the control of endocrine mechanisms that are functional in larval diapause, maintaining a high titer of JH (Denlinger, 1985). Some imidazole derivatives such as 1-benxyl-5-[(*E*)-2,6-dimethyl-1,5-heptadienyl]imidazole (KK-42) have been synthesized and shown to have a potent anti-JH activity in silkworm larvae (Kuwano et al., 1983). Thus, KK-42 is topically applied onto diapausing eggs of *A. yamamai* under various conditions (Tan *et al.*, 1986). KK-42 terminates diapause and hatching larva in a dose-dependent fashion, a half-maximum dose being ~5 μg/egg. The effectiveness is strengthened by prechilling of diapause eggs for 30 days. A similar KK-42 effect is reproduced in terminating embryonic diapause of the gypsy moth, *Lymantria dispar japonica* (our unpublished data). Since JH titers are not measured in eggs after KK-42 treatment, it is difficult to conclude that termination of diapause is induced by reduction of JH titers through the inhibition of JH synthesis by an anti-JH compound. Further, these imidazole compounds have been demonstrated to inhibit ecdysone synthesis in prothoracic glands of silkworms (Kadono-Okuda et al., 1987). At any rate, to our knowledge this is the first time that embryonic diapause has been shown to be completely terminated by treatment with anti-JH hormone compounds.

3.4.5. *Other Hormones*

At present, little information exists on the chemistry and function of neuropeptide hormones (NHs) in eggs, although the existence of neurosecretory cells in developing embryos has been demonstrated repeatedly in several insects (Dorn, 1985). The dynamic changes in ecdysteroids and JHs correlated with embryonic development seem to be under the control of NHs, as endocrine studies have revealed for postembryonic development (e.g., Bollenbacher and Granger, 1985). In applications of the method by which prothoracicotropic hormone was purified from *Bombyx* adult heads (Nagasawa et al., 1986), neuropeptides were extracted from developing embryos of silkworms and prothoracicotropic hormone activity was assayed using the debrained pupae of *Samia cynthia ricini* (Chen et al., 1986; Fugo et al., 1987). The chemical properties of this hormone extracted from eggs resembles that from the adult heads. The activity remains at a constant level before head pigmentation and then increases markedly. There is no difference in activity before and after diapause.

Prothoracicotropic hormone activity has been extracted from head segments and brains of embryos of *Manduca sexta* and subjected to bioassay for activity on prothoracic glands from larvae and pupae (Dorn et al., 1987). Hormonal activity was noted in the head segments as early as 24 h after oviposition. Maximal activity was found in brains from embryos 117 h after oviposition, just prior to hatching. These data suggest that neurohormonal control of ecdysteroid synthesis is involved in embryonic life.

Eclosion of adults from pupal cuticles is known to be stimulated by a neurohormone, eclosion hormone, which is secreted from neurosecretory cells of brains or abdominal ganglia (Reynolds and Truman, 1984). Larval hatching from eggs is also regulated by this hormone (Truman et al., 1981). The biologically active extracts have been obtained from fully developed embryos of *M. sexta* and *B. mori* (Truman et al., 1981; Fugo et al., 1985).

During the postembryonic stages of insects, several other neurohormones function to adjust the metabolic state as homeostatic regulators (Rankin and Gade, 1988). The effects of these hormones on eggs have not yet been examined. Thus, future work must consider the regulation of diapause metabolism by neurohormones that are adaptively set up at the time of diapause initiation.

3.5. Summary

Although the occurrence of diapause (arrested growth) is particularly frequent in arthropods, it is by no means confined to this group of animals. For most species, diapause is a developmental option and individuals have the capacity to determine whether or not they enter that state by monitoring changes in environmental and internal conditions through the mediation of endocrinological mechanisms. Diapause intervenes in any of the major stages of the life cycle, and the egg stage is one of the most suitable stages for diapause.

The diapause state has not yet been identified in many species of crustaceans. Diapause occurs at the gastrulal stage in eggs of *Artemia salina*. In insects, embryonic diapause intervenes mainly at three different stages, depending on the species. Although during diapause periods no morphogenetic development occurs, the diapausing embryo undergoes specialized structural changes.

Diapause represents an amalgam of physiological and biochemical characteristics, all of which enhance survival during long-term dormancy. A unique metabolic pathway related to diapause occurs in carbohydrate metabolism. Trehalose and glycerol are largely accumulated in diapausing eggs of *A. salina*. In *B. mori* eggs, the interconversion between glycogen and sorbitol occurs precisely at the initiation and termination of diapause. The metabolic shift is introduced by activation of the key enzymes glycogen phosphorylase and NAD-dependent sorbitol dehydrogenase.

Embryonic diapause in many cases results from the physiological response of the maternal generation to a changing milieu—a response that modifies the egg physiology. For induction and termination of diapause in several species of crustaceans, low temperatures are the prodominant factors. Photoperiods provide the cue for induction of diapause in some marine copepods and many insects. In some insects, temperature is exploited as an indispensable stimulus. Many insects use a specific temperature as the signal for termination of diapause. In addition to the physical stimuli, some chemicals are known to promote diapause termination in several insects.

Endocrinological mechanisms concerning embryonic diapause have not yet been generalized. The most systematic studies have been carried out on embryonic diapause of *B. mori*, in which the DH secreted from the subesophageal ganglion induces diapause. DH is a neuropeptide hormone that is distributed in the cephaloganglionic organs of several orders of insects. The target organs of this hormone are the developing ovaries, where it stimulates trehalase activity as one of its major functions.

Ecdysteroids and JHs are well known as hormones regulating postembryonic development, including diapause in many arthropods. Ecdysteroids are stored in conjugated forms in newly laid eggs. The titer changes of free ecdysteroids coincide with embryonic molting events, so these hormones are also functional in regulating embryonic development. In a few insects, the reduced levels of ecdysteroids and increased diapause potency appear to be inversely related.

JHs have recently been found in eggs of several insects. From their titer changes during embryogenesis, some regulatory role in organogenesis of several larval tissues is proposed. Other neuropeptide hormones, i.e., prothoracicotropic hormone and eclosion hormone, identified in the postembryonic stages, have been shown to be present in eggs, especially in those ready to hatch.

Acknowledgment

We thank Professor S. Kawase, Drs. M. Kobayashi, and T. Yaginuma of Nagoya University for their stimulating discussion. We are also indebted to Professor S. Kasahara of Hiroshima University for his kind advice on preparation of the manuscript. The research from our laboratories was supported in part by a research grant from the Ministry of Education, Science, and Culture of Japan.

References

Anderson, D. T. 1973. *Embryology and Phylogeny in Annelids and Arthropods*. Pergamon Press, Oxford and Elmsford, New York.

Anderson, E., J. H. Lochhead, M. S. Lochhead, and E. Huebner. 1970. The origin and structure of the tertiary envelop in thick-shelled eggs of the brine shrimp, *Arthemia salina*. J. Ultrastruct. Res. 32: 497–525.

Anderson, J. F. 1968. Influence of photoperiod and temperature on the induction of diapause in *Aedes atropalus* (Diptera: Culicidae). Bull. Conn. Agric. Exp. Stn. 711: 1–22.

Ando, Y. 1971. Distribution of mercury in the eggs of the false melon beetle, *Atrachya menetriesi* Faldermann (Coleoptera: Chrysomelidae), treated with mercuric chloride for breaking diapause. Appl. Entomol. Zool. 6: 67–74.

Ando, Y. 1978. Studies on egg diapause in the false melon beetle, *Atrachya menetriesi* Faldermann (Coleoptera: Chrysomelidae). Bull. Fac. Agric. Hirosaki Univ. 30: 131–215.

Asahina, E. 1969. Frost resistance in insects. Adv. Insect Physiol. 6: 1–49.

Azuma, M. and O. Yamashita. 1985. Immunohistochemical and biochemical localization of trehalase in the developing ovaries of the silkworm, *Bombyx mori*. Insect Biochem. 15: 589–596.

Baker, F. C., B. Lanzrein, C. A. Miller, L. W. Tsai, G. C. Jamieson, and D. A. Schooley. 1984. Detection of only JH III in several life-stages of *Nauphoeta cinerea* and *Thermobia domestica*. Life Sci. 35: 1553–1560.

Barlow, J. P. 1955. Physical and biological processes determining the distribution of zooplankton in a tidal estuary. Biol. Bull. (Woods Hole) 109: 211–225.

Benesch, R. 1969. Zur ontogenie und morphologie von *Arthemia salina* L. Zool. Jahrb. Abt. Anat. Ontog. Tiere. 86: 307–458.

Bergot, B. J., F. C. Backer, D. C. Cerf, G. Jamieson, and D. A. Schooley. 1981. Quantitative and qualitative aspects of juvenile hormone titers in developing embryos of several insect species. Pp. 33–45 *in* G. E. Pratt and G. T. Brooks (eds.), *Juvenile Hormone Biochemistry*. Elsevier/North-Holland Publ., Amsterdam and New York.

Bergot, B. J., G. C. Jamieson, M. A. Ratcliff, and D. A. Schooley. 1980. JH zero: new naturally occurring insect juvenile hormone from developing embryos of the tobacco hornworm. Science (Wash., D.C.) 210: 336–338.

Blanchet, M. F., P. Porcheron, and F. Dray. 1979. Variations du taux des ecdysteroids au cours des cycles de mue et vitellogenese chez le crustace Amphipode *Orchestia gammarellus*. Int. J. Invertebr. Reprod. 1: 133–139.

Bohle, H. W. 1969. Untersuchungen uber die Embryonalentowicklung und die embryonale Diapause bei *Baetis vernus* Curtis und *Baetis rhodani* (Baetidae: Ephemeroptera). Zool. Jahrb. Abt. Anat. Ontog. Tiere. 86: 493–575.

Bollenbacher, W. E. and N. A. Granger. 1985. Endocrinology of the prothoracicotropic hormone. Pp. 109–151 *in* G. A. Kerkut and L. I. Gilber (eds.), *Comprehensive Insect Physiology, Biochemistry and Pharmacology*, Vol. 7. Pergamon Press, Oxford and Elmsford, New York.

Boohar, R. and D. Bucklin. 1963. Possible ecdysone activity in the embryo of the grasshopper, *Melanoplus differentialis*. Am. Zool. 3: 496.

Bordes-Alléaume, N. and L. Sami. 1987. Ecdysteroid titres and cuticle depositions in embryos of the dipteran *Calliphora erythrocephala*. Int. J. Invertebr. Reprod. Dev. 11: 109–122.

Borst, D. W., H. Laufer, M. Laudau, E. S. Chang, W. A. Hertz, F. C. Baker, and D. A. Schooley. 1987. Methyl farnesoate and its role in crustacean reproduction and development. Insect Biochem. 17: 1123–1127.

Boswell, R. and A. Mahowald. 1985. Cytoplasmic determinants in embryogenesis. Pp. 387–405 *in* G. A. Kerkut and L. I. Gilbert (eds.), *Comprehensive Insect Physiology, Biochemistry and Pharmacology*, Vol. 1. Pergamon Press, Oxford and Elmsford, New York.

Bowers, W. S. 1983. The precocens. Pp. 517–523 *in* R. G. H. Downer and H. Laufer (eds.), *Endocrinology of Insects*. Liss, New York.

Brewer, R. H. 1964. The phenology of *Diaptomus stagnalis* (Copepoda: Calanoida): the development and the hatching of the egg stage. Physiol. Zool. 37: 1–20.

Brunet, P. C. F. 1967. Sclerotins. Endeavour (Oxf.) 26: 68–74.

Bruning, E. and B. Lanzrein. 1987. Function of juvenile hormone III in embryonic development of the cockroach, *Nauphoeta cinerea*. Int. J. Invertebr. Reprod. Dev. 12: 29–44.

Bruning, E., A. Saxer, and B. Lanzrein. 1985. Methyl farnesoate and juvenile hormone III in normal and precocene treated embryo of the ovoviviparous cockroach, *Nauphoeta cinerea*. Int. J. Invertebr. Reprod. Dev. 8: 269–278.

Bullière, D., F. Bullière, and M. De Reggi. 1979. Ecdysteroid titers during ovarian and embryonic development in *Blaberus craniifer*. Wilhelm Roux's Arch. Dev. Biol. 186: 103–104.

Busa, W. B., J. H. Crowe, and G. B. Matson. 1982. Intracellular pH and the metabolic status of dormant and developing *Arthemia* embryos. Arch. Biochem. Biophys. 216: 711–718.

Cavallin, M. and B. Fournier. 1981. Characteristics of development and variations in ecdysteroid levels in *Clitumnus* embryos derived of their cephalic endocrine glands. J. Insect Physiol. 27: 527–534.

Chaix, J. C. and M. De Reggi. 1979. Ecdysteroid levels during ovarian development and embryogenesis in the spider crab *Acanthonyx lunulatus*. Gen. Comp. Endocrinol. 47: 7–14.

Chen, J.-H., H. Fugo, M. Nakajima, H. Nagasawa, and A. Suzuki. 1986. The presence of neurohormonal activities in embryos of the silkworm, *Bombyx mori*. J. Seric. Sci. Jpn. 55: 54–59.

Clay, M. and C. Venard. 1972. Larval diapause in the mosquito, *Aedes triseriatus:* effect of diet and temperature on photoperiodic induction. J. Insect Physiol. 18: 1441–1446.

Clegg, J. S. 1964. The control of emergence and metabolism by external osmotic pressure and the role of free glycerol in the developing cysts of *Arthemina salina*. J. Exp. Biol. 41: 879–892.

Clegg, J. S. 1965. The origin of trehalose and its significance during the formation of encysted dormant embryos of *Artemia salina*. Comp. Biochem. Physiol. 14: 135–143.

Clegg, J. S. and F. P. Conte. 1980. A review of the cellular and developmental biology of *Artemia salina*. Pp. 11–54 *in* G. Persoone, P. Sorgeloos, O. Roels, and E. Jaspers (eds.), *The Brine Shrimp Artemia*, Vol. 2. Universa Press, Wetteren, Belgium.

Cobben, R. H. 1968. Evolutionary trends in Heteroptera. Part I. Eggs: architecture of the shell, gross embryogenesis and eclosion. Meded. Landb. Hogesch. Wageningen. 151: 1–475.

Connat, J.-L., P. A. Diehl, H. Gfeller, and M. Morici. 1985. Ecdysteroids in females and eggs of the ixodid tick, *Amblyomma hebraeum*. Int. J. Invertebr. Reprod. Dev. 8: 103–116.

Coulon, M., B. Calvez, M. De Reggi, J. M. Legay, and M. Hirn. 1979. Variation des taux d'ecdystéroides an cours du dévelopment de *Bombyx mori:* rapport entre ces variations et les phases de crossance et de morphogénèse. Experientia (Basel) 35: 1120–1121.

Danilevsky, A. S. 1965. Photoperiodism and seasonal development of insects. Oliver & Boyd, Edinburgh.

Dean, R. L. and J. C. Hartley. 1977a. Egg diapause in *Ephippiger cruciger* (Orthoptera: Tettigoniidae). I. The incidence, variable duration and elimination of the initial diapause. J. Exp. Biol. 66: 173–183.

Dean, R. L. and J. C. Hartley. 1977b. Egg diapause in *Ephippiger cruciger* (Orthoptera: Tettigoniidae). II. The intensity and elimination of the final egg diapause. J. Exp. Biol. 66: 185–195.

De Chaffoy De Courcelles, D. and M. Kondo. 1980. Lipovitellin from the crustacean, *Artemia salina:* biochemical analysis of lipovitellin complex for the yolk granules. J. Biol. Chem. 255: 6727–6733.

Denlinger, D. L. 1985. Hormonal control of diapause. Pp. 353–412 *in* G. A. Kerkut and L. I. Gilbert (eds.), *Comprehensive Insect Physiology, Biochemistry and Pharmacology*, Vol. 8. Pergamon Press, Oxford and Elmsford, New York.

Diehl, P. A., A. Aeschlimann, and F. D. Obenchain. 1982. Tick reproduction: ooganogenesis and oviposition. Pp. 277–350 *in* F. D. Obenchain and R. Galun (eds.), *Physiology of Ticks*. Pergamon Press, Oxford and Elmsford, New York.

Diehl, P. A., J.-L. Connat, J. P. Girault, and R. Lafont. 1985. A new class of apolar ecdysteroid conjugates: esters of 20-hydroxyecdysone with long-chain fatty acids in ticks. Int. J. Invertebr. Reprod. 8: 1–13.

Dorn, A. 1982. Precocene-induced effects and possible role of juvenile hormone during embryogenesis of the milkweed bug, *Oncopeltus faciatus*. Gen. Comp. Endocrinol. 46: 42–52.

Dorn, A. 1985. Neurosecretion in the insect embryo. Proc. Arthropod Embryol. Jpn., 1984 pp. 1–16.

Dorn, A. and F. Romer. 1976. Structure and function of prothoracic glands and oenocytes in embryos and last larval instar in *Oncopeltus faciatus* Dallas. Cell Tissue Res. 171: 331–350.

Dorn, A., S. T. Bishoff, and L. I. Gilbert. 1987. An incremental analysis of the embryonic development of the tobacco hornworm, *Manduca sexta*. Int. J. Invertebr. Reprod. Dev. 11: 137–158.

Dutrieu, J. 1960. Observations biochimiques et physiologiques sur le dévelopment d'*Artemia salina* Leach. Arch. Zool. Exp. Gen. 99: 1–134.

Fugo, H., J.-H. Chen, M. Nakajima, H. Nagasawa, and A. Suzuki. 1987. Neurohormones in developing embryos of the silkworm, *Bombyx mori:* the presence and characteristics of prothoracicotropic hormones. J. Insect Physiol. 33: 243–248.

Fugo, H., H. Saito, M. Nagasawa, and A. Suzuki. 1985. Eclosion hormone activity in developing embryos of the silkworm, *Bombyx mori*. J. Insect Physiol. 31: 293–298.

Furusawa, T., M. Shikata, and O. Yamashita. 1982. Temperature dependent sorbitol utilization in diapause eggs of the silkworm, *Bombyx mori*. J. Comp. Physiol. 147B: 21–26.

George, F. W., J. E. Griffen, M. Leshin, and J. D. Wilson. 1981. Testosteron. Pp. 341–357 *in* M. J. Novey and J. A. Resko (eds.), *Fetal Endocrinology*. Academic Press, Orlando, Florida.

Gharib, B., M. De Reggi, J.-L. Connat, and J.-C. Chaix. 1984. Ecdysteroid and juvenile hormone changes in *Bombyx mori*, related to the initiation of diapause. FEBS (Fed. Eur. Biochem. Soc.) Lett. 160: 119–123.

Gharib, B., J. M. Legay, and M. De Reggi. 1981. Potentiation of developmental abilities of diapausing eggs of *Bombyx mori* by 20-hydroxyecdysone. J. Insect Physiol. 27: 711–713.

Gilbert, L. I. and H. A. Schneiderman. 1961. The content of juvenile hormone and lipid in

Lepidoptera: sexual differences and developmental stages. Gen. Comp. Endocrinol. 1: 453–472.

Green, J. 1965. Chemical embryology of the crustacea. Biol. Rev. 40: 580–600.

Gregg, P. C., B. Roberts, and S. L. Wentworth. 1987. Levels of ecdysteroids in diapause and non-diapause eggs of Australian plague locust, *Chortoicetes terminifera* (Walker.). J. Insect Physiol. 33: 237–242.

Grice, G. D. and V. R. Gibson. 1977. Resting eggs in *Pontella meadi* (Copepoda: Calanoida). J. Fish. Res. Board Can. 34: 410–412.

Grice, G. D. and N. H. Marcus. 1981. Dormant eggs of marine copepods. Oceanogr. Mar. Biol. Annu. Rev. 19: 125–140.

Hagedorn, H. H. 1983. The role of ecdysteroids in the adult insects. Pp. 271–304 *in* R. G. H. Downer and H. Laufer (eds.), *Endocrinology of Insects*. Liss, New York.

Hardie, J. 1987. Juvenile hormone stimulation of oocyte development and embryogenesis in the parthenogenetic ovaries of an aphid, *Aphis fabae*. Int. J. Invertebr. Reprod. Dev. 11: 189–202.

Hartley, J. C. and A. C. Warne. 1972. The developmental biology of the egg stage of western European *Tettigoniidae* (Orthoptera). J. Zool. (Lond.) 168: 267–298.

Hasegawa, K. 1957. The diapause hormone of the silkworm, *Bombyx mori*. Nature (Lond.) 179: 1300–1301.

Hirn, M., M. Coulon and M. De Reggi. 1977. Taux d'ecdystéroids dans l'oeuf de *Bombyx mori* et variations au cours de l'embryogénèse. C. R. Acad. Sci. Paris 284: 2147–2150.

Hoarau, F. and M. Hirn. 1978. Evolution du taux des ecdystéroïdes au cours du cycle de mue chez *Helleria brevicornis* Ebner (Isopode terrestre). C. R. Acad. Sci. Paris 286: 1443–1446.

Hoffmann, J. A. and M. Lagueux. 1985. Endocrine aspects of embryonic development in insects. Pp. 435–460 *in* G. A. Kerkut and L. I. Gilbert (eds.), *Comprehensive Insect Physiology, Biochemistry and Pharmacology*, Vol. 1. Pergamon Press, Oxford and Elmsford, New York.

Hogan, T. W. 1962. The absorption and subsequent breakdown of urea by diapausing eggs of *Acheta commodus* (Walk.) (Orthoptera: Gryllidae). Aust. J. Biol. Sci. 15: 362–370.

Hogan, T. W. 1964. Further data on the effect of ammonia on the termination of diapause in eggs of *Teleogryllus commondus* (Walk.) (Orthoptera: Gryllidae). Aust. J. Biol. Sci. 17: 752–757.

Hogan, T. W. 1965. Changes in pH associated with the application of ammonia and potassium hydroxide to diapausing eggs of *Teleogryllus comodus* (Walk.) (Orthoptera: Gryllidae). Aust. J. Biol. Sci. 18: 81–87.

Imai, K., N. Kondo, M. Isobe, T. Goto, O. Yamashita, and K. Hasegawa. 1982. The neurohormones from the subesophageal ganglion of *Bombyx mori:* separation of melanization and reddish coloration hormone from diapause hormone. J. Seric. Sci. Jpn. 51: 111–125.

Imboden, H. and B. Lanzrein. 1982. Investigation on ecdysteroids and juvenile hormones and on morphological aspects during early embryogenesis in the ovoviviparous cockroach, *Nauphoeta cinerea*. J. Insect Physiol. 28: 37–46.

Imboden, H., B. Lanzrein, J. P. Delbecque, and M. Lusher. 1978. Ecdysteroids and juvenile hormone during embryogenesis in the ovoviviparous cockroach, *Nauphoeta cinerea*. Gen. Comp. Endocrinol. 36: 628–635.

Indrasith, L. S., T. Sasaki, and O. Yamashita. 1988. A unique protease responsible for selective degradation of a yolk protein in *Bombyx mori:* purification, characterization and cleavage profile. J. Biol. Chem. 263: 1045–1051.

Isaac, R. E. and H. H. Rees. 1985. Metabolism of maternal ecdysteroid-22-phosphates in developing embryos of the desert locust, *Schistocerca gregaria*. Insect Biochem. 15: 65–72.

Isobe, M. and T. Goto. 1980. Diapause hormone. Pp. 216–243 *in* T. A. Miller (ed.), *Neurohormonal Techniques in Insects*. Springer-Verlag, Berlin and New York.

Izumiyama, S. and K. Suzuki. 1986. Nucleotide pools in the eggs of emma field cricket. *Teleogryllus emma*, and two-spotted cricket, *Gryllus bimaculatus* (Orthoptera: Gryllidae). Appl. Entomol. Zool. 21: 405–410.

Johnes, B. M. 1956. Endocrine activity during insect embryogenesis. Control of events in development following the embryonic moult (*Locusta migratoria* and *Locusta pardalina*, Orthoptera). J. Exp. Biol. 33: 685–696.

Johnson, J. K. 1980. Effects of temperature and salinity on production and hatching of dormant eggs of *Acartia californiensis* (Copepoda) in an Oregon estuary. Fish. Bull. NOAA 77: 567–584.

Kadono-Okuda, K. E. Kuwano, M. Eto, and O. Yamashita. 1987. Inhibitory action of an imidazole compound on ecdysone synthesis in prothoracic glands of the silkworm, *Bombyx mori*. Dev. Growth Differ. 29: 527–533.

Kai, H. and T. Kawai. 1981. Diapause hormone in *Bombyx* eggs and adult ovaries. J. Insect Physiol. 27: 623–627.

Kasahara, S. and S. Uye. 1979. Calanoid copepod eggs in sea-bottom muds. V. Seasonal changes in hatching of subitaneous and diapause eggs of *Tortanus forcipatus*. Mar. Biol. 55: 63–68.

Kasahara, S., T. Oube, and M. Kamigaki. 1975. Calanoid copepod eggs in sea-bottom muds. III. Effects of temperature, salinity and other factors on the hatching of resting eggs of *Tortanus forcipatus*. Mar. Biol. 31: 31–35.

Kawamura, N. 1978. The early embryonic mitosis in normal and cooled eggs of the silkworm, *Bombyx mori*. J. Morphol. 158: 57–72.

Keino, H. and S. Takesue. 1982. Scanning electron microscopic study on the early development of silkworm eggs (*Bombyx mori*). Dev. Growth & Differ. 24: 287–294.

Kim, S.-E. 1987. Changes in eggshell permeability to oxygen during early developmental stages in diapause eggs of *Bombyx mori*. J. Insect Physiol. 33: 229–235.

Kunkel, J. G. and J. H. Nordin. 1985. Yolk proteins. Pp. 84–111 *in* G. A. Kerkut and L. I. Gilbert (eds.), *Comprehensive Insect Physiology, Biochemistry and Pharmacology*, Vol. 1. Pergamon Press. Oxford and Elmsford, New York.

Kurihara, M. and Y. Ando. 1969. The effect of mercury compounds on breaking of diapause in the eggs of the false melon beetle, *Atrachya menetriesi* Falermann (Coleoptera: Chrysomelidae). Appl. Entomol. Zool. 4: 149–151.

Kuwano, E., R. Takeya, and M. Eto. 1983. Terpenoid imidazoles: new anti-juvenile hormones. Agric. Biol. Chem. 47: 921–923.

Lachaise, F. and J. A. Hoffmann. 1977. Ecdysone et dévelopment ovarian chez un decapode, *Carcinus maenas*. C. R. Acad. Sci. Paris 285D: 701–704.

Lachaise, F. and J. A. Hoffmann. 1982. Ecdysteroids and embryonic development in the shore crab, *Carcinus maenas*. Hoppe-Seyler's Z. Physiol. Chem. 363: 1059–1067.

Lagueux, M., P. Harry, and J. A. Hoffmann. 1981. Ecdysteroids during embryogenesis in *Locusta migratoria*. Am. Zool. 21: 715–726.

Landry, M. R. 1975. Dark inhibition of egg hatching of the marine copepod, *Acartia clausi*. J. Exp. Mar. Ecol. 20: 43–47.

Lanzrein, B., V. Gentinetta, H. Abegglen, F. C. Baker, C. A. Miller, and D. A. Schooley. 1985. Titers of ecdysone, 20-hydroxyecdysone and juvenile hormone III throughout the life cycle of a hemimetabolous insect, the ovoviviparous cockroach, *Nauphoeta cinerea*. Experientia (Basel) 41: 913–917.

Laufer, H., D. Borst, F. C. Baker, C. Carrasco, M. Sinkus, C. C. Reuter, L. W. Tsai, and D. A. Schooley. 1987. Identification of juvenile hormone-like compound in a crustacean. Science (Wash., D.C.) 235: 202–205.

Lees, A. D. 1955. *The Physiology of Diapause in Arthropods*. Cambridge University Press, London and New York.

Legay, J.-M. 1977. Allometry and systematics: insect egg form. J. Nat. Hist. 11: 493–499.

Loher, W., L. Ruzo, F. C. Baker, C. A. Miller, and D. A. Schooley. 1983. Identification of the juvenile hormone from the cricket, *Teleogryllus commodus*, and juvenile hormone titre changes. J. Insect Physiol. 29: 585–589.

Mansingh, A. 1971. Physiological classification of dormant in insects. Can. Entomol. 103: 983–1009.

Marcus, N. H. 1979. On the population biology and nature of diapause of *Labidocera aestiva* (Copepoda: Calanoida). Biol. Bull. (Woods Hole) 157: 297–305.

Marcus, N. H. 1980. Photoperiodic control of diapause in the marine copepod *Labidocera aestiva*. Biol. Bull. (Woods Hole) 160: 311–318.

Marcus, N. H. 1982. The reversibility of subitaneous and diapause egg production by individual females of *Laboidocera aestiva* (Copepoda: Calanoida). Biol. Bull. (Woods Hole) 162: 39–44.

Matsutani, K. and H. Sonobe. 1987. Control of diapause-factor secretion from the subesophageal ganglion in the silkworm, *Bombyx mori:* the role of the protocerebrum and tritocerebrum. J. Insect Physiol. 33: 279–285.

McCarthy, J. F. 1979. Ponasterone A: a new ecdysteroid from the embryos and serum of bracyuran crustaceans. Steroids 34: 799–806.

McCarthy, J. F. and D. M. Skinner. 1979. Changes in ecdysteroids during embryogenesis of the blue crab, *Callinectes sapidus* Rathbum. Dev. Biol. 69: 627–633.

Miura, K. and I. Shimizu. 1987. Changes of tryglyceride and glycogen content in the silkworm (*Bombyx mori*) eggs during diapause and embryogenesis. Comp. Biochem. Physiol. 86B: 719–723.

Miya, K. 1984. Early embryogenesis of *Bombyx mori*. Pp. 49–73 *in* R. C. King and H. Akai (eds.), *Insect Ultrastructure*, Vol. 2. Plenum Press, New York.

Miya, K., M., Kurihara, and I. Tanimura. 1972. Changes of fine structure of serosal cell and the yolk cell during diapause and post-diapause in the silkworm, *Bombyx mori* L. J. Fac. Agric. Iwate Univ. 11: 51–87.

Miyadai, T. and O. Yamashita. 1980. Diapause hormone action in the silkworm, *Bombyx mori* L. (Lepidoptera: Bombycidae): enhancement of trehalase activity in developing ovaries incubated *in vitro*. Appl. Entomol. Zool. 15: 439–446.

Mizuno, T., K. Watanabe, and E. Ohnishi. 1981. Developmental changes of ecdysteroids in the eggs of the silkworm, *Bombyx mori*. Dev. Growth & Differ. 23: 543–552.

Morris, J. E. and Afzelius. 1967. The structure of the shell and outer membranes in encysted *Artemia salina* embryos during cryptosis and development. J. Ultrastruct. Res. 20: 244–259.

Mueller, N. S. 1963. An experimental analysis of molting in embryos of *Melanoplus differentialis*. Dev. Biol. 8: 222–240.

Nagasawa, H., H. Kataoka, A. Isogai, S. Tamura, A. Suzuki, A. Mizoguchi, Y. Fujiwara, A. Suzuki, S. Y. Takahashi, and H. Ishizaki. 1986. Amino acid sequence of a prothoracicotropic hormone of the silkworm, *Bombyx mori*. Proc. Natl. Acad. Sci. USA 83: 5840–5843.

Nakanishi, Y. H., T. Iwasaki, T. Okigaki, and H. Kato. 1962. Cytological studies of *Artemia salina*. I. Embryonic development without cell multiplication after the blastula stage in encysted dry eggs. Annot. Zool. Jpn. 35: 223–228.

Neumann-Visscher, S. 1976. The embryonic diapause of *Aulocara elliotti* (Orthoptera: Acrididae): histological and morphometric changes during development and following experimental termination with juvenile hormone analogue. Cell Tissue Res. 174: 433–452.

Ogura, N. and T. Saito. 1973. Induction of embryonic diapause in the silkworm, *Bombyx mori* L. (Lepidoptera: Bombycidae) by implantation of subesophageal ganglia of the common armyworm larvae *Lucania separata* Walk. (Lepidoptera: Noctuidae). Appl. Entomol. Zool. 8: 46–48.

Ohnishi, E. 1981. Ecdysteroids in the silkworm, *Bombyx mori*. Sericologia 21: 14–22.

Ohnishi, E., T. Ohtaki, and S. Fukuda. 1971. Ecdysone in the eggs of *Bombyx* silkworm. Proc. Jpn. Acad. 47: 413–415.

Ohtsuki, Y. and A. Murakami. 1968. Nuclear division in the early embryonic development of the silkworm, *Bombyx mori* L. Zool. Mag. Jpn. 77: 383–387.

Ohtsuki, Y., S. Mori, T. Kanda, and T. Kitazawa. 1976. Morphological observation on the embryonic moult in the silkworm, *Bombyx mori*. J. Seric. Sci. Jpn. 45: 225–231.

Okada, M. 1970. Electron microscopic studies on diapause embryos of the silkworm, *Bombyx mori*. Sci. Rep. Tokyo Kyoiku Daigaku. 24: 95–111.

Okada, M. 1971. Role of the chorion as a barrier to oxygen in the diapause of the silkworm, *Bombyx mori* L. Experientia (Basel) 27: 658–660.

Olson, C. C. and J. S. Clegg. 1978. Cell division during the development of *Artemia salina*. Wilhelm Roux's Arch. Dev. Biol. 184: 1–13.

Osanai, M. and Y. Yonezawa. 1986. Changes in amino acid pools in the silkworm, *Bombyx mori*, during embryonic life: alanine accumulation and its conversion to proline during diapause. Insect Biochem. 16: 373–379.

Pener, M. P., D. Dessberg, P. Imzarovici, C. C. Reuter, L. M. Tsai, and F. C. Baker. 1986. The effect of a synthetic precocene on juvenile hormone III titer in late *Locusta* eggs. J. Insect Physiol. 32: 853–857.

Perona, R. and C. G. Vallejo. 1982. The lysosomal proteinase of *Artemia:* purification and characterization. Eur. J. Biochem. 124: 357–362.

Perona, R. and C. G. Vallejo. 1985. Acid hydrolases during *Artemia* development: a role in yolk degradation. Comp. Biochem. Physiol. 81B: 993–1000.

Postlethwait, J. H. and F. Giorgi. 1984. Vitellogenesis: insects. Pp. 85–125 *in* L. W. Browder (ed.), *Developmental Biology: A Comprehensive Synthesis*, Vol. 1. Plenum Press, New York.

Rankin, A. K. and G. Gade. 1988. Insect peptide nomenclature. Insect Biochem. 18: 785–787.

Reynolds, S. E. and J. W. Truman. 1984. Eclosion hormones. Pp. 196–215 *in* T. A. Miller (ed.), *Neurohormonal Techniques in Insects*. Springer-Verlag, Berlin and New York.

Roe, R. M., C. L. Crawford, C. W. Clifford, J. P. Sparks, and B. D. Hammock. 1987. Characterization of the juvenile hormone esterases during embryogenesis of the house cricket, *Acheta domestica*. Int. J. Invertebr. Reprod. Dev. 12: 57–72.

Sakaguchi, B. 1978. Gametogenesis, fertilization and embryogenesis of the silkworm. Pp. 5–29 *in* Y. Tazima (ed.), *The Silkworm: An Important Laboratory Tool*. Kodansha, Tokyo.

Sander, K., H. O. Gutzeit, and H. Jackle. 1985. Insect embryogenesis: morphology, physiology, genetical and molecular aspects. Pp. 319–385 *in* G. A. Kerkut and L. I. Gilbert (eds.), *Comprehensive Insect Physiology, Biochemistry and Pharmacology*, Vol. 1. Pergamon Press, Oxford and Elmsford, New York.

Sbrenna-Micciarelli, A. 1977. Effects of farnesyl methyl ether on embryos of *Schistocerca gregaria* (Orthoptera). Acta Embryol. Exp. 155: 295–303.

Sbrenna-Micciarelli, A. and G. Sbrenna. 1972. The embryonic apolysis of *Schistocereca gregaria* (Orthoptera). J. Insect Physiol. 18: 1027–1037.

Scalia, S., Sbrenna-Micciarelli, G. Sbrenna, and E. Morgan. 1987. Ecdysteroid titres and localization in developing eggs of *Schistocerca gregaria*. Insect Biochem. 17: 227–236.

Schaefer, M. 1976. An analysis of diapause and resistance in the egg stage of *Floronia bucculenta*. Oecologia 25: 155–174.

Scheltes, P. 1978. The condition of the host plant during aestivation-diapause of the stalk borers *Chilo parellus* and *Chilo orichalcociliella* (Lepidoptera: Pyralidae) in Kenya. Entomol. Exp. Appl. 24: 679–688.

Schneiderman, H. and L. I. Gilbert. 1958. Substances with juvenile hormone activity in crustacea and other invertebrates. Biol. Bull. (Woods Hole) 115: 530–535.

Sehnal, F. 1983. Juvenile hormone analogues. Pp. 657–672 *in* R. G. H. Downer and H. Laufer (eds.), *Endocrinology of Insects*. Liss, New York.

Setlow, B. and P. Setlow. 1980. Measurements of the pH within dormant and germinated bacterial spores. Proc. Natl. Acad. Sci. USA 77: 2474–2476.

Shiino, S. M. 1950. Studies on the embryonic development of *Panulirus japonicus*. (von Siebold). J. Fac. Fish. Prefect. Univ. Mie 1: 1–168.

Shimada, S. and O. Yamashita. 1979. Trehalose absorption related with trehalase in developing ovaries of the silkworm, *Bombyx mori*. J. Comp. Physiol. 131: 333–339.

Slifer, E. H. 1958. Diapause in the egg of *Melanoplus differentialis* (Orthoptera: Acrididae). J. Exp. Zool. 138: 259–282.

Sømme, L. 1982. Supercooling and winter survival in terrestrial arthropods. Comp. Biochem. Physiol. 73A: 519–543.

Sonobe, H. and Y. Okada. 1984. Studies on embryonic diapause of the *pnd* mutant of the silkworm, *Bombyx mori*. III. Accumulation of alanine in diapause eggs. Wilhelm Roux's Arch. Dev. Biol. 193: 414–417.

Sonobe, H., M. Ikeda, and H. Kaizuma. 1980. Oxygen premeability of the chorion in relation to diapause termination in *Bombyx* eggs. Experientia (Basel) 35: 1650–1651.

Storey, K. B. and J. M. Storey. 1983. Biochemistry of freeze tolerance in terrestrial insects. Trends Biochem. Sci. 8: 242–245.

Stross, R. G. 1969a. Photoperiodic control of diapause in *Daphnia*. II. Induction of winter diapause in the *Arctic*. Biol. Bull. (Woods Hole) 136: 264–273.

Stross, R. G. 1969b. Photoperiodic control of diapause in *Daphnia* III. Two-stimulus control of long-day, short-day induction. Biol. Bull. (Woods Hole) 137: 359–374.

Stross, R. G. and J. C. Hill. 1965. Diapause induction in *Daphnia* require two stimuli. Science (Wash., D.C.) 150: 1462–1464.

Suzuki, K., M. Hosaka, and K. Miya. 1984. The amino acid pool of *Bombyx mori* eggs during diapause. Insect Biochem. 14: 557–561.

Takami, T. 1963. *In vitro* culture of embryos in the silkworm, *Bombyx mori*. J. Exp. Biol. 40: 735–739.

Takami, T. 1969. *The Silkworm Eggs*, pp. 178–182. Japanese Association of Silkworm Eggs, Tokyo.

Takeda. S. 1977. Induction of egg diapause in *Bombyx mori* by some cephalo-thoracic organs of the cockroach, *Periplaneta americana*. J. Insect Physiol. 23: 813–816.

Takeda, S. and A. Girardie. 1985. An active principle from the corpora cardiaca, corpora allata and subesophageal ganglion of the locust, inducing egg diapause in *Bombyx mori*. J. Insect Physiol. 31: 761–766.

Takeda, S., Y. Kono, and Y. Kameda. 1988. Induction of non-diapause eggs in *Bombyx mori* by a trehalase inhibitor. Entomol. Exp. Appl. 46: 291–294.

Tan, E., K. Suzuki, E. Kuwano, S. Abe, and M. Kurihara. 1986. Effect of anti-JH (KK-42) treatment on the breaking of the diapause of the eggs of the silkmoth, *Antheraea yamamai*. J. Seric. Sci. Jpn. 55: 305–308.

Tauber, M. J. and Tauber, C. A. 1978. Evolution of phenological strategies in insects: a comparative approach with eco-physiological and genetic considerations. Pp. 53–71 *in* H. Dingle (ed.), *Evolution of Insect Migration and Diapause*. Springer-Verlag, Berlin and New York.

Temin, G., M. Zander, and J.-P. Roussel. 1986. Physico-chemical (GC-MS) measurements of juvenile hormone III titres during embryogenesis of *Locusta migratoria*. Int. J. Invertebr. Reprod. Dev. 9: 105–112.

Templeton, N. S. and H. Laufer. 1983. The effects of a juvenile hormone analog (Altosid, ZR-515) on the reproduction and development of *Daphnia magna* (Crustacea: Cladocera). Int. J. Invertebr. Reprod. 6: 99–110.

Tomeba, H., K. Oshikiri, and K. Suzuki. 1988. Changes of free amino acid pool in the eggs of the emma field cricket, *Teleogryllus emma* (Orthoptera: Gryllidae). Appl. Entomol. Zool. 23: 228–233.

Truman, J. W., P. H. Taghert, P. F. Copehaver, N. J. Tublitz, and L. M. Schwarz. 1981. Eclosion hormone may control all ecdysis in insects. Nature (Lond.) 291: 70–71.

Umeya, Y. 1926. Experiments of ovarian transplantation and blood transfusion in the silkworm with special reference to the alteration of voltinism (*Bombyx mori* L.). Bull. Seric. Exp. Stn. Chosen 1: 1–26.

Umeya, Y. 1946. Embryonic hibernation and diapause in insects from the viewpoint of the hibernating eggs of the silkworm. Bull. Seric. Exp. Stn. 12: 393–484.

Uye, S. and A. Fleminger. 1976. Effect of various environmental factors on egg development of several species of *Artemia* in southern California. Mar. Biol. 38: 253–262.

Uye, S., S. Kasahara, and T. Onbe. 1979. Calanoid copepod eggs in sea-bottom muds. IV. Effects of some environmental factors on the hatching of resting eggs. Mar. Biol. 51: 151–156.

Vallejo, C. G., R. Perona, R. Garesse, and R. Marco. 1981. The stability of the yolk granules of *Arthamia:* an improved method for their isolation and study. Cell Differ. 10: 343–356.

Vallejo, C. G., M. A. G. Sillero, and R. Marco. 1979. Mitochondrial maturation during *Artemia salina* embryogenesis. General description of the process. Cell. Mol. Biol. 25: 113–124.

Van der Linden, A., I. Vankerckhoven, R. Caubergs, and W. Decleir. 1986. Action spectroscopy of light-induced hatching of *Artemia* cysts (Branchiopoda: Crustacea). Mar. Biol. 91: 239–243.

Wigglesworth, V. B. 1985. Historical perspectives. Pp. 1–24 *in* G. A. Kerkut and L. I. Gilbert (eds.), *Comprehensive Insect Physiology, Biochemistry and Pharmacology*, Vol. 7. Pergamon Press, Oxford and Elmsford, New York.

Yaginuma, T. and O. Yamashita. 1979. NAD-dependent sorbitol dehydrogenase activity in relation to the termination of diapause in eggs of *Bombyx mori.* Insect Biochem. 9: 547–553.

Yamashita, O. 1969. Studies on the glycogen metabolism in pupal ovaries of the silkworm, *Bombyx mori.* J. Seric. Sci. Jpn. 38: 329–339.

Yamashita, O. 1986. Yolk protein system in *Bombyx* eggs: synthesis and degradation of egg-specific protein. Adv. Invertebr. Reprod. 4: 263–275.

Yamashita, O. and K. Hasegawa. 1974. Mobilization of carbohydrates in tissues of female silkworm, *Bombyx mori,* during metamorphosis. J. Insect Physiol. 20: 1749–1760.

Yamashita, O. and K. Hasegawa. 1985. Embryonic diapause. Pp. 407–434 *in* G. A. Kerkut and L. I. Gilbert (eds.), *Comprehensive Insect Physiology, Biochemistry and Pharmacology*, Vol. 1. Pergamon Press, Oxford and Elmsford, New York.

Yamashita, O. and L. S. Indrasith. 1988. Metabolic fates of yolk proteins during embryogenesis in arthropods. Dev. Growth & Differ. 30: 337–346.

Yamashita, O., K. Hasegawa, and M. Seki. 1972. Effect of the diapause hormone on trehalase activity in pupal ovaries of the silkworm, *Bombyx mori* L. Gen. Comp. Endocrinol. 18: 515–523.

Yamashita, O., M. Isobe, K. Imai, N. Kondo, and T. Goto. 1980. Serum albumin as an effective carrier for diapause hormone of the silkworm, *Bombyx mori* (Lepidoptera: Bombycidae). Appl. Entomol. Zool. 15: 90–95.

Yamashita, O., K. Suzuki, and K. Hasegawa. 1975. Glycogen phosphorylase activity in relation to diapause initiation in *Bombyx* eggs. Insect Biochem. 5: 707–718.

Yamashita, O., T. Yaginuma, M. Kobayashi, and T. Furusawa. 1988. Metabolic shift related with embryonic diapause of *Bombyx mori:* temperature-directed sorbitol metabolism. Pp. 263–275 *in* F. Sehnal and D. L. Denlinger (eds.), *Frontiers in Insect Endocrinology.* Wrocław Technical University Press, Wrocław, Poland.

Yoshimi, T., M. Shikata, T. Furusawa, and M. Tada. 1986. Chloride distribution in the shells and pH value of the extracts from *Bombyx* eggs treated with HCl. J. Seric. Sci. Jpn. 55: 197–201.

Zhu, J., L. S. Indrasith, and O. Yamashita. 1986. Characterization of vitellin, egg-specific protein and 30 kDa protein from *Bombyx* eggs, and their fates during oogenesis and embryogenesis. Biochim. Biophys. Acta 882: 427–436.

Zillioux, E. J. and J. G. Gonzalez. 1972. Egg dormancy in a neritic calanoid copepod and its implication to overwintering in boreal water. Pp. 217–230 *in* B. Battaglia (ed.), *Fifth European Marine Biology Symposium*. Piccin Editore, Padua, Italy.

Hormonal Regulation of Postembryonic Events

Roles of Morphogenetic Hormones in the Metamorphosis of Arthropods Other Than Insects

4

KLAUS-DIETER SPINDLER

4.1. Introduction

The hormonal basis of metamorphosis has been studied almost exclusively in insects or amphibians. This is documented by two recent books on metamorphosis (Gilbert and Frieden, 1981; Balls and Bownes, 1985). The former is a completely revised second edition of a book with the same title (Etkin and Gilbert, 1968), omitting the only article on metamorphosis in arthropods other than insects (Costlow, 1968). Instead, there is a survey on invertebrate metamorphosis, primarily dealing with morphological descriptions of metamorphosis in various invertebrate phyla (Highnam, 1981). This indicates the rapid progress in understanding the mechanisms of metamorphosis in insects and the rather slow progress in other invertebrate taxa.

Before going into details, we should first define *metamorphosis*, which is a difficult task, since this term has been used in either a strict or a broad sense, as outlined by Cohen (1985). The term is an old one, having been used and defined by Goethe (1790) and by Carus (1853), although in different ways. Since then, a variety of definitions have been proposed, and in modern textbooks on embryology (Fioroni, 1987) there are classifications of types of metamorphosis, such as embryonic–larval, one–multiphasic, free–intracapsular, early–late, and progressive–regressive, as well as anamorphosis, epimorphosis, and hypermetamorphosis. For insect development and metamorphosis, a new classification, including new criteria and names, has been proposed by Nüesch (1987). Throughout our present review, we will use a broader definition, covering all cases we are dealing with. Such a definition is given by Wald (1981): "What one calls metamorphosis is all a matter of how greatly and abruptly animals partition their lives"; or by Cohen (1985): "Abrupt change in an animal's form and way of life, usually with a break in vegetative activities." Of course, one should bear in mind that not only does the outer form change but also the internal organization does so, which leads to the breakdown of some larval organs; however, some others remain unchanged, leading to the building of new adult structures.

4.2. Morphogenetic Hormones

Three groups of hormones are involved in the regulation of metamorphosis, namely, ecdysteroids, juvenile hormones, and peptide hormones. The first two have been treated extensively in recent reviews (Kerkut and Gilbert, 1985a,b) and only some features will be described here.

Ecdysteroid (E) is the generic name for roughly 100 polyhydroxylated steroids whose general structure is depicted in Fig. 4.1. In contrast to the "vertebrate-type" steroid hormones, Es possess the full C_{27} skeleton,

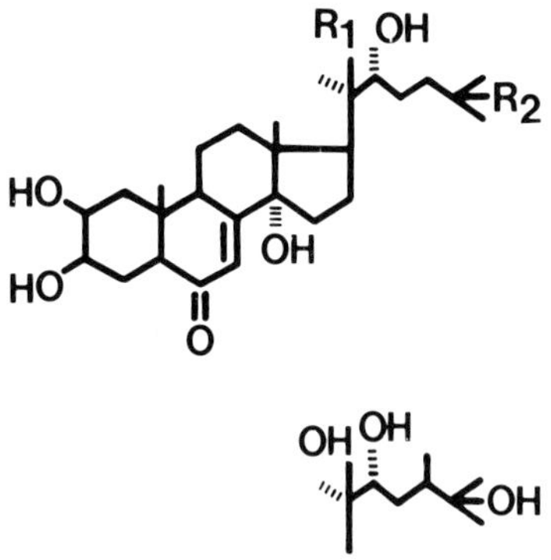

$R_1;R_2$ = OH : 20-OH-ecdysone
R_1=H;R_2=OH : ecdysone
R_1=OH;R_2=H : ponasterone A

makisterone A

FIGURE 4.1. Structure of some ecdysteroids.

and a nonplanar configuration between ring A and B. Unlike vertebrates, arthropods are not able to synthesize sterols *de novo*. Es have been found in all arthropod taxa.

Juvenile hormones (JHs) possess a dihomosesquiterpenoid skeleton, which can be synthesized (at least) by insects. The structure of the four known JHs is depicted in Fig. 4.2a. The presence of these hormones was considered a special feature of insect endocrinology, but very recently a hormone with related effects and structure (Fig. 4.2b) has been found in crustaceans (see Section 4.2.2, below).

4.2.1. *Chelicerata*

A detailed description of Es in Chelicerata and their hormonal roles has been given recently (Spindler, 1989). Much less is known about the other two classes of arthropod hormones in Chelicerata. Effects of JHs have been described in the reproduction of ticks (Connat, 1987), and there are also detailed descriptions of neurosecretory structures (e.g., Binnington, 1986), but neither JHs nor peptide hormones have been demonstrated unequivocally so far in chelicerates.

4.2.2. *Crustacea*

The presence, distribution, and function of Es in crustaceans have been summarized in several reviews (Spindler et al., 1980, 1984; Skinner, 1985; Spindler, 1989). JH activity was first found in Crustacea by Schneiderman and Gilbert in 1958, but on methodological grounds these early reports are questionable. Recently, methyl farnesoate (MF) (Fig. 4.2b), as a precursor of JH, has been identified unequivocally as a product of mandibular organs from several decapod crustaceans (Laufer et al., 1986, 1987a,b; Borst et al., 1987; see also Chapter 2 by Borst and Laufer in Part 1).

(a)

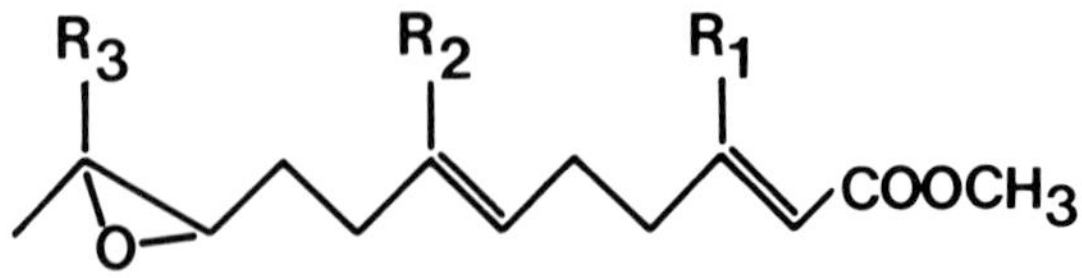

$R_1; R_2; R_3$: $-C_2H_5$ — JH 0

R_1 : $-CH_3$ $R_2; R_3$: $-C_2H_5$ — JH I

$R_1; R_2$: $-CH_3$ R_3 : $-C_2H_5$ — JH II

$R_1; R_2; R_3$: $-CH_3$ — JH III

(b)

FIGURE 4.2. Structure of juvenile hormones (a) and methyl farnesoate (b).

The most pronounced progress in discovering "new" hormones in crustaceans has resulted from work in detection of peptide hormones. Several reviews have appeared of late on this topic, including those of Kleinholz (1985), Quackenbush (1986), and Fingerman (1987), but even these recent ones have not kept up with the rapid detection of new hormones.

4.2.3. *Myriapoda*

The presence of Es and their involvement in regulation of molting, regeneration, and reproduction have recently been summarized (Scheffel, 1987; Spindler, 1989). In the former review, a description of neurosecretory centers is also included, but there are no data on the chemical nature of these neurosecretory hormones. JHs have not been detected in Myriapoda so far.

4.3. Hormonal Regulation of Metamorphosis

Our knowledge of the hormonal basis of metamorphosis is only fragmentary. This is because only a few species have been investigated, whereas there is a vast variety of developmental strategies even within one systematic group (e.g., the crustaceans), which indeed could lead to diverse means of hormonal regulation.

4.3.1. *Chelicerata*

Xiphosura

Owing to the efforts of Jegla and coworkers, we have a clear picture of the hormonal regulation of larval development of one xiphosuran, *Limulus polyphemus* (for a review see Jegla, 1982; see also Chapter 8 by Jegla in Part 1). After four embryonic molts, the so-called trilobite larva hatches and metamorphoses into a second larval instar. Metamorphosis in xiphosurans is a more gradual process. Es, predominantly 20-OH-ecdysone, were detected both in intermolt and in premolt stages, the latter showing higher concentrations. The effectiveness of exogenous Es in inducing molting greatly depends on the type of E used and on the time of application within the molting cycle. If trilobite larvae were treated with JH or JH mimics, these substances were either toxic or showed no visible effects. Only if juvenoids in high concentrations (100 μg/litre) were applied after the second embryonic molt and throughout further larval development, an abnormal development of the dorsal organ was observed which was not taken as an indication for an involvement of JH in hormonal regulation of metamorphosis.

The role of Es in the regulation of molting in the premetamorphic trilobite stage has been convincingly demonstrated, but whether other hormones are involved in metamorphosis remains to be elucidated.

Pantopoda

Among pantopods so far only one species, *Pycnogonum litorale,* and only the Es have been investigated. Despite the fact that the morphology of the adults has been unchanged for more than 370 million years (Bergström et al., 1980), there are diverse developmental strategies, even within one group of Pantopoda. For example, among the Pycnogonidae, some species molt before hatching and others do not. A detailed study of all larval stages and metamorphosis is only available for *Pycnogonum litorale,* where rearing in the laboratory was successful (Behrens, 1984). In this species, there are six larval stages with distinct morphological features (Fig. 4.3). From stage 1 to 2, typical larval characteristics are lost, and another drastic morphological and functional change occurs with the fifth molt. The last molt leads to the final form (Behrens, 1984). In this species, Es were analyzed and characterized from eggs, juvenile females and males, and from adults of both sexes (Behrens and Bückmann, 1983; Bückmann et al., 1986). Unfortunately, larval stages have not been investigated so far, and there are no studies on the effects of exogenous Es or JHs on metamorphosis in this species.

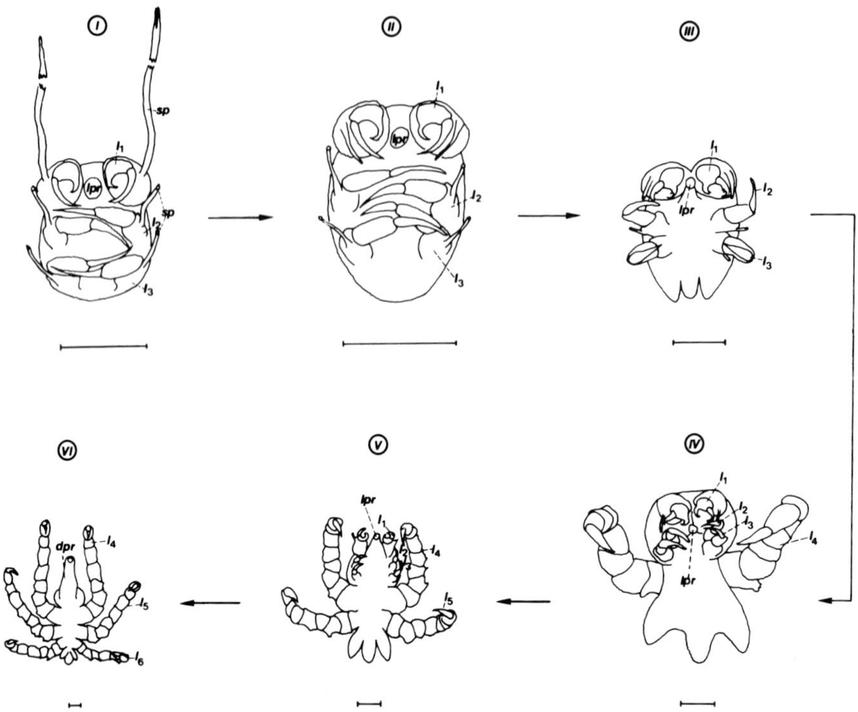

FIGURE 4.3. Larval stages (I to VI) of *Pycnogonum litorale.* The bars represent 100 μm. *Key: dpr* = definite proboscis; l_1 to l_6 = legs 1 to 6; *lpr* = larval proboscis; *sp* = *Spinndorn.* (Redrawn from Behrens, 1984.)

Arachnida

The endocrinology of the vast majority of Arachnida is mostly unknown (Bonaric, 1987; Spindler, 1989; see also Chapter 3 by Bonaric and Juberthie in Part 1). But, owing to economic considerations, e.g., the development of antiparasitic drugs (for a review see Spindler, 1988), the ectoparasitic ticks have gained much interest in the last few years (for reviews see Binnington, 1986; Diehl et al., 1986; Connat, 1987; Spindler, 1989). But again, metamorphosis and its hormonal regulation is a strongly neglected field, even in ticks.

In the spider *Pisaura mirabilis,* the molting hormone content has been studied during the intermolt cycle of the eighth nymphal stage, but the hormonal regulation of the transitions from one developmental stage to another is completely unknown.

Metamorphosis in ticks occurs when they molt from the larval to the nymphal stage. In all stages of development of ticks, Es have been detected and quantified. Preferential attention has been given to a study of nymphs and adults, especially to females. In the larval stage of *Ornithodorus moubata,* the fluctuations of E titer in relation to cuticular events are very similar to those observed in the fifth-stage nymphs of the same species (Diehl et al., 1986). Es in the larval stage were detected and quantified by radioimmunoassay. The importance of Es for ticks during metamorphosis is demonstrated by the fact that larval diapause can be terminated by application of exogenous Es, which was the first demonstrated effect exerted by Es in ticks (Wright, 1969). Several reports have been published that exogenously applied JHs either inhibit or retard molting (for a review see Solomon et al., 1982), but till now there have been no unequivocal demonstrations of JHs in tick larvae. More recently (Connat, 1987), JH activity has been demonstrated in extracts of *Boophilus microplus* females in vitellogenesis. These extracts were able to induce oviposition in the tick *Ornithodorus moubata* and to inhibit metamorphosis in the beetle *Tenebrio molitor.* The influence of this substance on metamorphosis of ticks has not been investigated so far. Whether JHs are involved in the regulation of metamorphosis is therefore not known with certainty.

4.3.2. *Crustacea*

Twenty years ago, the final statement in a review on metamorphosis in crustaceans was as follows: "The solution to the problem [metamorphosis] does not lie in isolated studies on the cellular reorganization, the physiological changes, the chemical structure of the hormones, or in the injection of insect JH into larval stages, but rather through an integrated approach that recognizes that development involves the animal in all of its parts" (Costlow, 1968). Unfortunately, we are not able to present such an integrated view, not even for a single aspect, namely, the underlying hormonal mechanism(s) of metamorphosis. Nevertheless, there is some progress in this field, as documented by a recent review on "hormonal processes in decapod crustacean larvae" (Christiansen, 1988).

The enormous variety of different larval stages and modes of development in crustaceans (Williamson, 1982) makes it quite difficult to give a unifying picture of the endocrinology of metamorphosis. We, therefore, will deal with some selected species separately.

The nauplius larva is generally considered as the primitive and characteristic larva of crustaceans. Its metamorphosis to other larval stages may be gradual or more abrupt, depending on the taxon. In copepods, the three-segmented nauplius stage with three pairs of appendages molts several times (five or six nauplius/metanauplius stages). The final nauplius stage, still containing the three pairs of appendages and rudiments of five further pairs, metamorphoses into the first copepodid stage. The hormonal basis of molting and metamorphosis

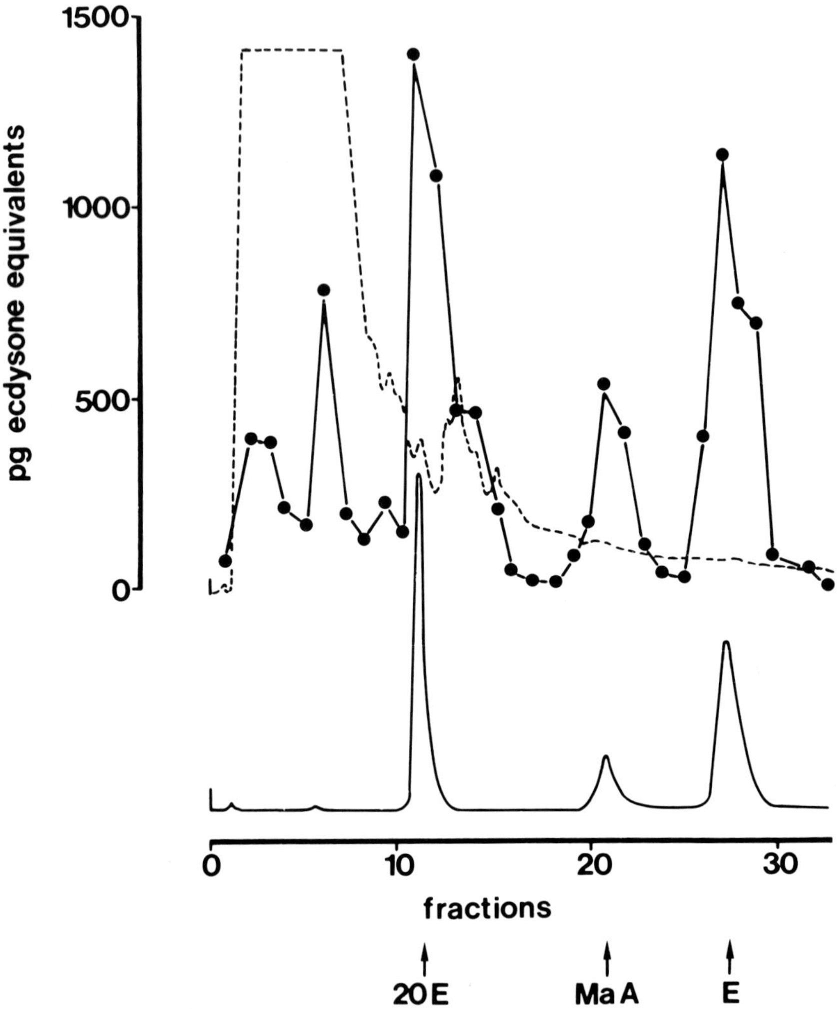

FIGURE 4.4 HPLC–RIA analyses of partially purified extracts from zooplankton (consisting of more than 60% of the freshwater copepod *Cyclops vicinus* in different developmental stages). Purification and HPLC–RIA were performed as described, except that methanol : water (33:67, v/v) was used for elution.—— = optical profile (at 254 nm) of a mixture of 20-OH-ecdysone (20E), makisterone A (MaA), and ecdysone (E); --- = absorbance of the extract at 254 nm; ●—● = concentration of ecdysteroids in the single fractions.

was unknown till now. As demonstrated in Fig. 4.4, we detected Es in a sample of zooplankton from Lake Constance containing more than 60% of the copepod *Cyclops vicinus* in different developmental stages (as well as some other freshwater copepods and some Cladocera). Owing to the small amounts of selected material, Es were only characterized according to high-pressure liquid chromatography–radioimmunoassay (HPLC–RIA) analysis, which is not sufficient for an unequivocal characterization of these hormones. However, the predominant E is 20-OH-ecdysone in addition to ecdysone and makisterone A and some polar material (fractions 1–7). Since the antibody used has about threefold higher sensitivity toward ecdysone than toward 20-OH-ecdysone, the values of 20-OH-ecdysone in Fig. 4.4 must be multiplied by 3 to show the true relation between ecdysone and 20-OH-ecdysone. For the fourth copepodid stage, which is able to enter diapause (Einsle, 1967; Spindler, 1971), the concentration of total Es per animal has been determined as about 100 pg, which is comparable to the concentrations found in larval stages of the spider crab *Hyas araneus* (see Fig. 4.5). These studies only indicate that during copepod development Es (and presumably the same Es as occur in other crustacean species) are used as molting hormones, but their exact role as well as the question of whether other hormones are involved in metamorphosis cannot be answered as yet.

The most dramatic changes in the external morphology and internal organization occur during larval development and metamorphosis of Cirripedia, leading from a free-swimming nauplial larva to a so-called cyprid larva, also named a "locomotive pupa" (Darwin, 1851), and finally to the juvenile and adult animal. This type of early development occurs both in the sessile barnacles and in the endoparasitic *Rhizocephala*. In both taxa, the final stages of metamorphosis are characterized by free-swimming instars that settle on a suitable substratum. Barnacles were the first "entomostracan" species in which morphogenetic hormones (molting hormones) were detected and characterized (Bebbington and Morgan, 1977), but only in the adults. The effects of Es on molting were first studied in adult barnacles, demonstrating a synchronization and acceleration of molting (Tighe-Ford and Vaile, 1972, 1973). Subsequently, an acceleration of molting events in cyprid larvae by 20-OH-ecdysone was demonstrated in two *Balanus* species by Cheung (1974) and Freeman and Costlow (1983a). Some cyprid larvae underwent metamorphosis but without attachment to the substrate. Freeman and Costlow (1983a) describe an interesting feature of the molting physiology of cyprids from *Balanus amphitrite*. Shortly after ecdysis to the cyprid larvae, there is an early premolt period, localized only in the anterior, posterior, and ventral regions of the carapace valves, whereas the dorsal carapace epidermis undergoes apolysis much later and only if attachment has already taken place. Only this late premolt period can be accelerated by 20-OH-ecdysone. The various parts of the epidermis must therefore respond differently to Es. One possible explanation for this phenomenon could be a differential expression of ecdysteroid receptors in different parts of the epidermis. It would be helpful to know the exact titer curve of Es during cyprid molting stage in order to see whether there are also two peaks of hormone present, as demonstrated in adult *Uca pugilator* (Hopkins, 1983). Nevertheless, it is evident that Es contribute to the regulation of metamorphosis in barnacles, but the exact mode of action is unknown. Concerning the metamorphosis of rhizocephalan cyprids, it is interesting that cyprids prefer crabs as hosts, which are late in premolt. The effect of JHs or analogues on metamorphosis was tested by Gomez et al. (1973) and Ramenofsky et al. (1974). These investigators demonstrated precocious metamorphosis without attachment of cyprid larvae from *Balanus galeatus* after addition of JH I (see Fig. 4.2a) or a JH analogue. The same substances were unsuccessful when tested on nauplial stages. In contrast to *Balanus,* nauplii from *Elminius modestus* were affected by juvenoids (Tighe-Ford, 1977). Nauplii from this species metamorphosed into abnormal cyprid larvae with some nauplial characteristics still present. Cyprids metamorphosed to unattached adults or to morphologically abnormal larvae. Induction of abnormal morphology of cyprids or adults after treatment of the foregoing developmental stage with farnesol was also described by Mortlock et al. (1984). From insect endocrinology one would suspect a retardation of metamorphosis after application of juvenoids. Unlike in insects, metamorphosis in Cirripedia is more accelerated. But we should keep in mind that all these results have to be discussed critically because of the high toxicity of these compounds in Cirripedia.

In decapod crustaceans, quite diverse types of larval developments occur—from direct development (e.g., freshwater crayfishes) to indirect development with more or less pronounced metamorphosis. The number and morphological description of larval stages before metamorphosis is given by Williamson (1982). Our knowledge of the endocrine control of metamorphosis in decapods is restricted to a few marine crabs and to the lobster *Homarus americanus.* It has evolved in three phases: (1) description of neurosecretory centers in the eyestalks and the Y-organ, combined with eyestalk ablation experiments; (2) application of hormones with known structure and function; and (3) detection of these hormones and determination of their titers. Most of these data are summarized by Christiansen (1988). From all these investigations, the conclusions can be drawn that larval molts are under the control of Es and presumably also of the molt-inhibiting hormone (MIH) as in adults.

Es have been detected and determined throughout larval development both in the lobster (Chang and Bruce, 1981) and in the spider crab

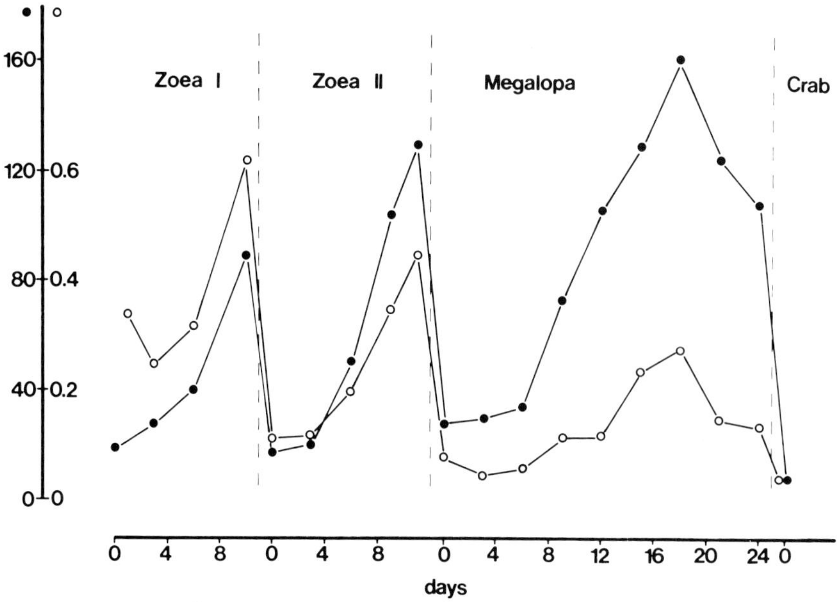

FIGURE 4.5. Titer of ecdysteroids as determined by radioimmunoassay during larval development and metamorphosis in the spider crab *Hyas araneus*. Values are expressed either as picogram ecdysone equivalents per animal (●) or as nanogram ecdysone equivalents per milligram dry weight (○); $n = 4$ to 8. (Redrawn from Spindler and Anger, 1986.)

Hyas araneus (Spindler and Anger, 1986; Anger and Spindler, 1987). The shape of the titer curves resembles those of the adults (see also Fig. 4.5). The most convincing demonstration of the involvement of MIH in controlling larval molts is given by Snyder and Chang (1986a). They demonstrated that eyestalk ablation not only accelerated molting events but also increased molting hormone titers. Replacement therapy with extracts from the sinus gland of juvenile lobsters prolonged molting cycles and decreased Es titers. From sinus glands of juvenile lobster, MIH was isolated and its amino acid composition determined (Chang et al., 1987), but unfortunately it has not yet been tested in larval lobsters. Eyestalk removal in larval lobsters leads to intermediate forms having characteristics of both the premetamorphic and the first postlarval stage (Snyder and Chang, 1986b; Charmantier and Aiken, 1987), which indicates that there is some kind of delicate hormonal balance necessary when animals enter metamorphosis. Possible candidates for this special regulatory role in metamorphosis could be either the so-called metamorphosis-inhibiting factor (MIF), as described by Freeman and Costlow (1983b) in the crab *Rhithropanopeus harrisii*, or JHs. Indeed, JH III (see Fig. 4.2a), but not the precursor farnesol, retards metamorphosis in lobsters slightly but significantly and affects also morphological characteristics (Hertz and Chang, 1986). MF, the JH-like compound from crustaceans (Laufer et al., 1987a), is also present in lobsters (in contrast to JH III) and retards metamorphosis too, but it does not induce morphogenetic effects (Borst et al., 1987). So far, the concentration of MF has only been determined from pooled late second and early third larval stages, but not yet from the first postmetamorphic stage. In both stages, there were identical concentrations, much lower than those in ovaries. The possible role of JHs in metamorphosis of decapods deserves further investigations, since the two JH analogues methoprene and hydroprene increased the length of the zoeal stages of the crab *Rhithropanopeus harrisii* but did not inhibit metamorphosis (Christiansen et al., 1977a,b).

In addition to morphogenetic hormones, a variety of other hormones interfering with reproduction, color change, and metabolism have been described in crustaceans (Kleinholz, 1985; Quackenbush, 1986; Fingerman, 1987), but their important role has been studied nearly exclusively in adults. Most of the available studies on presumably hormonal regulation during larval development are listed by Christiansen (1988). In *H. americanus,* there is a change in sodium regulation from isoionic to slightly hyperionic during metamorphosis from stage III to IV. From stage IV, sodium regulation is controlled by neuroendocrine factors. These changes between the last larval (III) and first postlarval (IV) stage are considered to be part of a true metamorphosis (Charmantier et al., 1984). In the same paper, there are listed also some morphological, ecological, behavioral, physiological, and biochemical changes associated with metamorphosis in the lobster *H. americanus.* If we add our knowledge on hormonal regulation of metamorphosis in the same species as described above, the lobster thus represents the only species where at least to a certain degree an "integrated approach" of describing metamorphosis has been pursued.

4.3.3. *Myriapoda*

The endocrine basis of development has been studied nearly exclusively in one species of centipedes, *Lithobius forficatus.* Owing to the work of R. Joly and H. Scheffel, basic features of the hormonal regulation of molting have been established and were recently summarized (Scheffel, 1987). Development of this species proceeds according to the scheme given in Fig. 4.6. The total number of molting events in adults is unknown. With the transition from larval to postlarval development typical larval organs, the anal organs, are reduced and instead coxal organs are formed and take over the function of anal organs.

Molting in *Lithobius* is under the control of neurosecretory centers and Es. Extirpation and implantation experiments demonstrated molt-

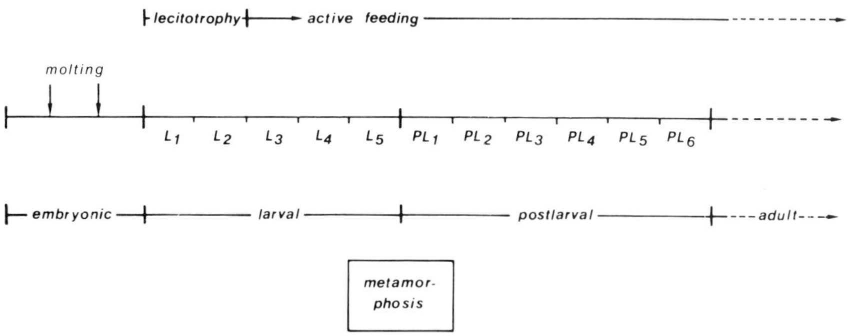

FIGURE 4.6. Scheme of development of *Lithobius forficatus:* L_1–L_5 = larval stages I to V; PL_1–PL_6 = postlarval stages. The number of adult molts is uncertain. (From Scheffel, 1987.)

inhibiting activity in a complex consisting of paired protocerebral frontal lobes, the nervus glandulae cerebralis I (II) and the cerebral gland, which shows structural similarities to the corpora allata of insects. These experiments were primarily performed with adult animals, but it could also be shown that implantation of cerebral glands into second larvae was no more effective when animals had already entered proecdysis. This obviously means that E synthesis is under neurosecretory control. Hormonal manipulations of larval stages II and III gave rise to desynchronizations between molting and morphogenetic events yielding intermediate larvae. Comparable studies have not been performed with larval stage V. It is not known whether JHs are involved in metamorphosis in this species. According to Scheffel (1979), there are no indications that JHs are present in the anamorphous stages. This author as well as Bückmann (1984) speculate that JHs play a role only in secondary metamorphosis.

4.4. Conclusions and Outlook

Until now there have not been enough facts to fulfill the demand for an integrated approach to the study of metamorphosis (Costlow, 1968). This is also true of the endocrine control of metamorphosis in arthropods other than insects. However, even if incomplete, some perspectives for future research may be pointed out:

(1) More comparative research is needed, especially in Chelicerata and Myriapoda. In no instance has metamorphosis been studied in detail in these taxa.

(2) Concerning Es, both the chemical nature and the titers of these

hormones should be determined during early development and metamorphosis. It is unknown at which developmental stage the embryo or larvae start to synthesize Es and are no longer dependent on the maternal supply. From studies on the shrimp *Palaemon serratus,* we have at least an indication of the onset of E synthesis by the embryo (Spindler et al., 1987). All these studies should enable us to determine the sensitive phases, and they might also be helpful in elucidating the nature and functions of Es, which may change during development, as shown in insects (Koolman and Spindler, 1983). The mechanisms of Es in larval development and metamorphosis in Chelicerata, Crustacea, and Myriapoda are unknown, in contrast to those of insects (Koolman and Spindler, 1983). The sensitivity toward Es during larval life suggests the presence of E receptors, but a direct proof of this hypothesis does not exist. E receptors have been demonstrated in some insects and crustaceans (See Spindler-Barth and Spindler, 1987, for a review), but with one exception (nauplii from the brine shrimp, *Artemia salina:* Spindler et al., 1984), E receptors have not been demonstrated in early larval stages. Since the techniques are available and have been successfully applied to several crustaceans (Spindler et al., 1984; Londershausen and Spindler, 1985), these studies should be possible were enough biological material present. *In vitro* studies using cell or organ cultures could also contribute to a better understanding of the regulation of metamorphosis, as has been shown in insects (Dübendorfer and Eichenberger, 1985).

(3) Even more research is necessary on JHs. The results on the application of juvenoids do not allow a clear picture. As shown in Section 4.3.2, above, there is no effect of these substances on metamorphosis, nor an acceleration or slight retardation of metamorphosis. A serious problem in evaluating these results is certainly the high toxicity of these compounds on larval crustaceans. The unequivocal demonstration of a JH-like substance in decapods (Laufer et al., 1987a,b; Borst et al., 1987) should provoke comparative research also in other arthropod taxa. A determination of the titers of JHs through larval development and metamorphosis should give a better insight in the possible functions of these hormones during early development.

(4) Finally, because antibodies against several peptide hormones from crustaceans are available, it should be possible to detect the corresponding hormones also in larval stages and to determine their titers during larval development. This may increase our knowledge of the development of neurosecretory structures. Moreover, the physiological role of these hormones must be investigated. During metamorphosis, the ecological conditions and lifestyle of these animals often change considerably. This implicates different metabolic pathways that afford hormonal regulation.

4.5. Summary

Metamorphosis in the xiphosuran *Limulus polyphemus,* the pantopode *Pycnogonum litorale,* spiders, ticks, copepods, cirripedia, crabs, the lobster *Homarus americanus,* and the myriapode *Lithobius forficatus* has been described. A short survey of the morphological changes associated with metamorphosis precedes the description of hormones and the endocrine regulation of metamorphosis. Ecdysteroids (Es) have been described for the first time in copepods. The involvement of Es in regulation of metamorphosis is shown for *Limulus,* ticks, cirripeds, and decapods and is highly probable for the other animals. In the spider crab *Hyas araneus* and in the lobster *H. americanus,* there are cyclic changes in E concentrations throughout all larval stages, resembling the situation in adults. The regulation of E synthesis is obviously under the control of a molt-inhibiting hormone from the eyestalks. The role of juvenile hormone(s) (JHs) in the regulation of metamorphosis is questionable. In *Limulus, Pycnogonum,* and *Lithobius,* an involvement of JHs in metamorphosis is excluded from the results obtained till now. The application of JHs or mimics accelerates metamorphosis in Cirripedia but slightly retards metamorphosis in the lobster. In another decapod, *Rhithropanopeus harrisii,* juvenoids are without effect on metamorphosis. Instead, a so-called metamorphosis-inhibiting factor (MIF) prevents resorption of epidermis in the dorsal spine of zoea stages, which is initiated by Es. The primary action of those hormones involved in the regulation of metamorphosis remains to be elucidated, as does the participation of hormones that regulate metabolic pathways.

Acknowledgments

I gratefully acknowledge the supply of freshwater copepods from Lake Constance by Dr. U. Einsle (Constance) as well as the unpublished work by Dr. E. S. Chang (Bodega Bay), Dr. M. E. Christiansen (Oslo), Dr. J. L. Connat (Neuchâtel), and Dr. H. Laufer (Storrs), and the critical reading of this manuscript by Dr. M. Spindler-Barth (Düsseldorf). My own work on crustacean endocrinology has been supported by the Deutsche Forschungsgemeinschaft for nearly 15 years.

References

Anger, K. and K.-D. Spindler. 1987. Energetics, moult cycle and ecdysteroid titers in spider crab (*Hyas araneus*) larvae starved after the D_0 threshold. Mar. Biol. 94: 367–375.

Balls, M. and M. Bownes (eds.), 1985. *Metamorphosis.* Oxford University Press, (Clarendon), London and New York.

Bebbington, P. M. and E. D. Morgan. 1977. Detection and identification of moulting hormone (ecdysones) in the barnacle (*Balanus balanoides*. Comp. Biochem. Physiol. 56B: 77–79.

Behrens, W. 1984. Larvenentwicklung und Metamorphose von *Pycnogonum litorale* (Chelicerata, Pantopoda). Zoomorphology 104: 266–279.

Behrens, W. and D. Bückmann. 1983. Ecdysteroids in the pycnogonid *Pycnogonum litorale* (Ström) (Arthropoda, Pantopoda). Gen. Comp. Endocrinol. 51: 8–14.

Bergström, J., W. Stürmer, and G. Winter. 1980. *Palaeoisopus, Palaeopantus* and *Palaeothea*, pycnogonid arthropods from the lower Devonian Hunsrück slate, West Germany. Palaeontol. Z. 54: 7–54.

Binnington, K. C. 1986. Ultrastructure of the tick neuroendocrine system. Pp. 152–164 *in* J. R. Sauer and J. A. Hair (eds.), *Morphology, Physiology and Behavioral Biology of Ticks.* Ellis Horwood, Chichester, England.

Bonaric, J. 1987. Moulting hormones. Pp. 111–118 *in* W. Nentwig (ed.), *Ecophysiology of the Spiders.* Springer-Verlag, Berlin and New York.

Borst, D. W., H. Laufer, M. Landau, E. S. Chang, W. A. Hertz, F. C. Baker, and D. A. Schooley. 1987. Methyl farnesoate and its role in crustacean reproduction and development. Insect Biochem. 17: 1123–1127.

Bückmann, D. 1984. The phylogeny of hormones and hormonal systems. Nova Acta Leopold. 56: 437–452.

Bückmann, D., G. Starnecker, K.-H. Tomaschko, E. Wilhelm, R. Lafont, and J.-P. Girault. 1986. Isolation and identification of major ecdysteroids from the pycnogonid *Pycnogonum litorale* (Ström) (Arthropoda, Pantopoda). J. Comp. Physiol. B156: 759–765.

Carus, J. V. 1853. *System der tierischen Morphologie.* Engelmann, Leipzig.

Chang, E. S. and M. J. Bruce. 1981. Ecdysteroids titers of larval lobsters. Comp. Biochem. Physiol. 70A: 239–241.

Chang, E. S., M. J. Bruce, and R. W. Newcomb. 1987. Purification and amino acid composition of a peptide with the molt-inhibiting activity of the lobster, *Homarus americanus.* Gen. Comp. Endocrinol. 65: 56–64.

Charmantier, G. and D. E. Aiken. 1987. Intermediate larval and postlarval stages of *Homarus americanus* H. Milne Edwards, 1837 (Crustacea: Decapoda). J. Crustacean Biol. 7: 525–535.

Charmantier, G., M. Charmantier-Daures, and D. E. Aiken. 1984. Neuroendocrine control of hydromineral regulation in the American lobster *Homarus americanus* H. Milne-Edwards, 1837 (Crustacea, Decapoda). 2. Larval and postlarval stages. Gen. Comp. Endocrinol. 54: 20–34.

Cheung, P. J. 1974. The effect of ecdysterone on cyprids of *Alanus eburneus* Gould. J. Exp. Mar. Biol. Ecol. 15: 223–229.

Christiansen, M. E. 1988. Hormonal processes in decapod crustacean larvae. In *Aspects of Decapod Crustacean Biology.* Symp. Zool. Soc. London 59: 47–68.

Christiansen, M. E., J. D. Costlow, Jr., and R. J. Monroe. 1977a. Effects of the juvenile hormone mimic ZR-512 (Altozar) on larval development of the mud-crab *Rhithropanopeus harrisii* at various cyclic temperatures. Mar. Biol. 39: 281–288.

Christiansen, M. E., J. D. Costlow, Jr., and R. J. Monroe. 1977b. Effects of the juvenile hormone mimic ZR-515 (Altosid) on larval development of the mud-crab *Rhithropanopeus harrisii* in various salinities and cyclic temperatures. Mar. Biol. 39: 269–279.

Cohen, J. 1985. Metamorphosis: introduction, usages, and evolution. Pp. 1–19 *in* M. Balls and M. Bownes (eds.), *Metamorphosis.* Oxford University Press, London and New York.

Connat, J. L. 1987. Aspects endocrinologiques de la physiologie du développement et de la reproduction chez les tiques. Doctoral thesis, University of Neuchâtel, Switzerland.

Costlow, J. D., Jr. 1968. Metamorphosis in crustaceans. Pp. 3–41 *in* W. Etkin and L. I.

Gilbert (eds.), *Metamorphosis—A Problem in Developmental Biology.* Appleton-Century-Crofts (Division of Simon & Schuster), New York.
Darwin, C. 1851. *A Monograph of the Subclass Cirripedia.* London Ray Society, London.
Diehl, P. A., J. L. Connat, and E. Dotson. 1986. Chemistry, function and metabolism of tick ecdysteroids. Pp. 165–193 *in* J. R. Sauer and J. A. Hair (eds.), *Morphology, Physiology and Behavioral Biology of Ticks.* Ellis Horwood, Chichester, England.
Dübendorfer, A. and S. Eichenberger. 1985. *In vitro* metamorphosis of insect cells and tissues: development and function of fat body cells in embryonic cell cultures of *Drosophila.* Pp. 145–161 *in* M. Balls and M. Bownes (eds.), *Metamorphosis.* Oxford University Press, London and New York.
Einsle, U. 1967. Die äusseren Bedingungen der Diapause planktisch lebender *Cyclops*-Arten. Arch. Hydrobiol. 63: 387–403.
Etkin, W. and L. I. Gilbert. 1968. *Metamorphosis—A Problem in Developmental Biology.* Appleton-Century-Crofts (Division of Simon & Schuster), New York.
Fingerman, M. 1987. The endocrine mechanisms of crustaceans. J. Crust. Biol. 7: 1–24.
Fioroni, P. 1987. *Allgemeine und vergleichende Embryologie der Tiere.* Springer-Verlag, Berlin and New York.
Freeman, J. A. and J. D. Costlow. 1983a. The cyprid molt cycle and its hormonal control in the barnacle *Balanus amphitrite.* J. Crustacean Biol. 3: 173–182.
Freeman, J. A. and J. D. Costlow. 1983b. Endocrine control of spine epidermis resorption during metamorphosis in crab larvae. Wilhelm Roux's Arch. Dev. Biol. 192: 362–365.
Gilbert, L. I. and E. Frieden (eds.) 1981. *Metamorphosis—A Problem in Developmental Biology.* Plenum Press, New York.
Goethe, J. W. von. 1790. Die Metamorphose der Pflanzen. Pp. 1–51 *in* Vol. 36 (1790) of his *Gesammelte Werke.* Bong, Berlin, 1926.
Gomez, E. D., D. J. Faulkner, W. A. Newman, and C. Ireland. 1973. Juvenile hormone mimics: effect on cirriped crustacean metamorphosis. Science (Wash., D.C.) 179: 813–814.
Hertz, W. A. and E. S. Chang. 1986. Juvenile hormone effects on metamorphosis of lobster larvae. Int. J. Invertebr. Reprod. Dev. 10: 71–77.
Highnam, K. C. 1981. A survey of invertebrate metamorphosis. Pp. 43–73 *in* L. I. Gilbert and E. Frieden (eds.), *Metamorphosis—A Problem in Developmental Biology.* Plenum Press, New York.
Hopkins, P. M. 1983. Patterns of serum ecdysteroids during induced and uninduced proecdysis in the fiddler crab, *Uca pugilator.* Gen. Comp. Endocrinol. 52: 350–356.
Jegla. T. C. 1982. A review of the molting physiology of the trilobite larva of *Limulus.* Pp. 83–101 *in* J. Bonaventura, C. Bonaventura, and S. Tesh (eds.), *Phyiology and Horseshoe Crabs: Studies on Normal and Environmentally Stressed Animals.* Liss, New York.
Kerkut, G. A. and L. I. Gilbert (eds.) 1985a. *Comprehensive Insect Physiology, Biochemistry and Pharmacology,* Vol. 7. Pergamon Press, Oxford and Elmsford, New York.
Kerkut, G. A. and L. I. Gilbert (eds.) 1985b. *Comprehensive Insect Physiology, Biochemistry and Pharmacology,* Vol. 8. Pergamon Press, Oxford, and Elmsford, New York.
Kleinholz. L. H. 1985. Biochemistry of crustacean hormones. Pp. 463–522 *in* D. E. Bliss and L. H. Mantel (eds.), *The Biology of Crustacea,* Vol. 9. Academic Press, Orlando, Florida.
Koolman, J. and K.-D. Spindler. 1983. Mechanism of action of ecdysteroids. Pp. 179–201 *in* R. G. H. Downer and H. Laufer (eds.), *Endocrinology of Insects.* Liss, New York.
Laufer, H., M. Landau, D. Borst, and E. Homola. 1986. The synthesis and regulation of methyl farnesoate, a novel juvenile hormone for crustacean reproduction. Pp. 135–143 *in* M. Porchet, J.-C. Andries, and A. Dhainaut (eds.), *Advances in Invertebrate Reproduction,* Vol. 4. Elsevier, Amsterdam and New York.

Laufer, H., D. Borst, F. C. Baker, C. Carrasco, M. Sinkus, C. C. Reuter, L. W. Tsai, and D. A. Schooley. 1987a. Identification of a juvenile hormone–like compound in a crustacean. Science (Wash., D.C.) 235: 202–205.

Laufer, H., M. Landau, E. Homola, and D. W. Borst. 1987b. Methyl farnesoate: its site of synthesis and regulation of secretion in a juvenile crustacean. Insect Biochem. 17: 1129–1131.

Londershausen, M. and K.-D. Spindler. 1985. Uptake and binding of moulting hormones in crayfish. Am. Zool. 25: 187–196.

Mortlock, A. M., J. T. R. Fitzsimons, and G. A. Kerkut. 1984. The effects of farnesol on the late stage nauplius and free swimming cypris larvae of *Elminius modestus* (Darwin). Comp. Biochem. Physiol. 78A: 345–357.

Nüesch, H. 1987. Metamorphose bei Insekten: Direkte und indirekte Entwicklung bei Apterygoten und Exopterygoten. Zool. Jahrb. Anat. 115: 453–487.

Quackenbush, L. S. 1986. Crustacean endocrinology: a review. Can. J. Fish. Aquat. Sci. 43: 2271–2282.

Ramenofsky, M., D. J. Faulkner, and C. Ireland. 1974. Effect of juvenile hormone on cirriped metamorphosis. Biochem. Biophys. Res. Commun. 60: 172–178.

Scheffel, H. 1979. Probleme der hormonalen Regulation von Häutung und postembryonaler Morphogenese bei Chilopoden. Wiss. Z. Pädagog. Hochschule "Dr. Theodor Neubauer" Erfurt–Mühlhausen. Naturwiss. Reihe 1: 111–119.

Scheffel, H. 1987. Häutungsphysiologie der Chilopoden: Ergebnisse von Untersuchungen an *Lithobius forficatus* (L.). Zool. Jahrb. Physiol. 91: 257–282.

Schneiderman, H. A. and L. I. Gilbert. 1958. Substances with juvenile hormone activity in Crustacea and other invertebrates. Biol. Bull. (Woods Hole) 115: 530–535.

Skinner, D. M. 1985. Molting and regeneration. Pp. 43–146 *in* D. E. Bliss and L. H. Mantel (eds.), *The Biology of Crustacea*, Vol. 9, Academic Press, Orlando, Florida.

Snyder, M. J. and E. S. Chang. 1986a. Effects of sinus gland extracts on larval molting and ecdysteroid titers of the American lobster, *Homarus americanus*. Biol. Bull. (Woods Hole) 170: 244–254.

Snyder, M. J. and E. S. Chang. 1986b. Effects of eyestalk ablation on larval molting rates and morphological development of the American lobster, *Homarus americanus*. Biol. Bull. (Woods Hole) 170: 232–243.

Solomon, K. R., C. K. A. Mango, and F. D. Obenchain. 1982. Endocrine mechanisms in ticks: effects of insect hormones and their mimics on development and reproduction. Pp. 399–438 *in* F. D. Obenchain and R. Galun (eds.), *Physiology of Ticks*. Pergamon Press, Oxford and Elmsford, New York.

Spindler, K.-D. 1971. Dormanzauslösung und Dormanzcharakteristika beim Süsswassercopepoden *Cyclops vicinus*. Zool. Jahrb. Physiol. 76: 139–151.

Spindler, K.-D. 1988. Parasites and hormones. Pp. 465–476 *in* H. Mehlhorn (ed.), *Parasitology in Focus: Facts and Trends*. Springer-Verlag, Berlin and New York.

Spindler, K.-D. (1989). Hormonal role of ecdysteroids in Crustacea, Chelicerata and other Arthropoda. Pp. 290–294 *in* J. Koolman (ed.), *Ecdysone*. Thieme Verlag, Stuttgart.

Spindler, K.-D. and K. Anger. 1986. Ecdysteroid levels during the larval development of the spider crab *Hyas araneus*. Gen. Comp. Endocrinol. 64: 122–128.

Spindler, K.-D., Dinan, L. and M. Londershausen. 1984. On the mode of action of ecdysteroids in crustaceans. Pp. 255–264 *in* J. Hoffmann and M. Porchet (eds.), *Biosynthesis, Metabolism and Mode of Action of Invertebrate Hormones*. Springer-Verlag, Berlin and New York.

Spindler, K.-D., R. Keller, and J. D. O'Connor. 1980. The role of ecdysteroids in the crustacean molting cycle. Pp. 247–280 *in* J. Hoffmann (ed.), *Progress in Ecdysone Research*. Elsevier/North-Holland Publ., Amsterdam and New York.

Spindler, K.-D., A. van Wormhoudt, D. Sellos, and M. Spindler-Barth. 1987. Ecdysteroid levels during embryogenesis in the shrimp, *Palaemon serratus* (Crustacea Decapoda): quantitative and qualitative changes. Gen. Comp. Endocrinol. 66: 116–122.

Spindler-Barth, M. and K.-D. Spindler. 1987. Antiecdysteroids and receptors. Pp. 497–511 *in* M. K. Agarwal (ed.), *Receptor Mediated Antisteroid Action*. De Gruyter, Berlin.

Tighe-Ford, D. J. 1977. Effects of juvenile hormone analogues on larval metamorphosis in the barnacle *Elminius modestus* Darwin (Crustacea: Cirripedia). J. Exp. Mar. Biol. Ecol. 9: 19–28.

Tighe-Ford, D. J. and D. C. Vaile. 1972. The action of crustecdysone on the cirripede *Balanus balanoides*. J. Exp. Mar. Biol. Ecol. 9: 19–28.

Tighe-Ford, D. J. and D. C. Vaile, 1973. Crustecdysone action and the effects of light during the post-breeding condition of the cirriped *Balanus balanoides* (L.). Proc. 3rd Int. Congr. Mar. Corr. & Fouling, 1972 pp. 744–756.

Wald, G. 1981. Metamorphosis: an overview. Pp. 1–39 *in* L. I. Gilbert and E. Frieden (eds.), *Metamorphosis—A Problem in Developmental Biology*. Plenum Press, New York.

Williamson, D. I. 1982. Larval morphology and diversity. Pp. 43–110 *in* D. E. Bliss and L. G. Abele (eds.), *The Biology of Crustacea*, Vol. 2. Academic Press, Orlando, Florida.

Wright, J. E. 1969. Hormonal termination of larval diapause in *Dermacentor albipictus*. Science (Wash., D.C.) 163: 390–391.

The Role of Molting Hormone in Sclerotization in Insects

5

CONSTANTINE E. SEKERIS

5.1. Introduction

Insects possess an exoskeleton that, in the form of the hardened larval, pupal, and imaginal cuticle, shelters the organism in all stages of its development. In addition, it is a permeability barrier. It must be at the same time rigid and elastic, and—in the winged insect—also light. The formation of the exoskeletal structures and their transformation into the hardened and often darkened cuticle is a complex process, involving, in most cases, interaction of specific phenolic compounds with the cuticular proteins and is hormonally controlled. This mechanism pertains specifically to insects. Crustaceans harden their skeletal system by the incorporation of minerals into their cuticle, the arachnids by organic materials, whereas the myriapods use both mineral and organic substances.

The object of this chapter is to review the current status of the control by MH of sclerotization in insects. (Note that herein MH stands for molting hormone irrespective of whether it refers to α- or β-ecdysone.) Unavoidably, I will concentrate on a few of the best studied systems and draw generalizations, fully realizing the pitfalls inherent in such an approach.

In order to analyze the effects of MH on the sclerotization process, I will have to briefly summarize our current knowledge of the components that participate in this process: the cuticle, mainly its protein components; the phenolic sclerotizing agents; the enzymes involved in their biosynthesis; and the enzyme(s) involved in the "activation" of the sclerotizing phenols. I will also briefly present the currently accepted concepts as to the sclerotization mechanisms, i.e., how the sclerotizing agents cross-link the cuticular proteins. This introductory section is not intended to be complete, but will solely touch upon the points that are important for a discussion of hormonal control. Needless to say, a personal bias, stressing some issues more than others, cannot be avoided. Many excellent, critical, and extensive reviews have appeared on various aspects of sclerotization, to which the reader is referred for more details (A. G. Richards, 1967; Hackman, 1974; Brunet, 1980; Hepburn, 1976, 1985; Andersen, 1979b, 1980; Lipke et al., 1983).

5.1.1. *Sclerotization*

The hardening, often accompanied with darkening, of the cuticle, brought about by the interaction of the sclerotizing agent(s) with the soft, flexible, cuticle, resulting in cross-linking of the protein chains and in the modification of the properties of the cuticle (see below), is referred to as *sclerotization*. Sclerotization does not necessarily involve darkening of the cuticle. The term *tanning* should be used to characterize those cases in which hardening is accompanied also by darkening of the cuticle. The color of the sclerotized cuticle varies from completely white to very dark, with all intermediate colorations. Differences in the color of the sclerotized cuticles have been ascribed both to the chemical nature of the sclerotizing agents involved and to the mechanism of sclerotization (Andersen, 1974; see Section 5.3 and 5.4, below). Tanning should be differentiated from melanization, which is due to deposition of melanin in melanin granules; this process will not be considered here. During sclerotization the solubility of the cuticular proteins decreases. In terms of biochemical changes, sclerotization involves the formation of bridges between proteins and chitin, increased molecular size of structural polypeptides, altered packing of proteins and polysaccharides, declining response to enzymatic or chemical cleavage, extrusion of water, and reorientation of chitin fibrils (Hackman, 1974; Andersen, 1985).

5.2. The Insect Cuticle

The structure of the cuticle shows large variations according to species, region, specialized areas, and developmental stage; however, the general construction pattern in all cuticles is similar.

The main part of the cuticle, the procuticle, may be regarded as a polymer network of protein and chitin, formed by cross-linking of the protein part to the chitin, both by covalent linkages and by weaker linkages such as hydrogen bonds and van der Waals forces.

A monolayer of epidermal cells underlies the procuticle. These cells are responsible for synthesizing most of the cuticle components and most of the enzymes involved in the biosynthesis of the sclerotizing agents and can be regarded as the main target cell for hormones regulating the sclerotization process (Sekeris, 1965; Karlson and Sekeris, 1966a,b).

5.2.1. *Proteins of the Cuticle*

Proteins make up a substantial amount of the insect cuticle. The protein : chitin ratio depends on the insect species; either component may predominate and account for more than half the dry weight of the cuticle. The nature of the association of protein to chitin has been a matter of discussion. There is evidence for the existence of covalent links between carboxyl groups of glutamic and aspartic acid and the ϵ-NH_2 groups of lysine, as well as of hydrogen, van der Waals, and salt bonds.

Attempts to purify and characterize individual cuticular proteins were initiated some years ago. The most serious problem in these studies is that with progressing sclerotization such proteins become exceedingly insoluble, even in very strong solvents. Extraction of cuticular proteins is usually performed with neutral buffers or 7*M* urea. The non-sclerotized cuticle shows a high content of hydrophobic amino acids with bulky side chains (tyrosine, proline, dicarboxylic acids), whereas the sclerotized cuticle contains larger amounts of smaller size amino acids (Hackman, 1976). The number of amino groups (ϵ-NH_2 of lysine) decreases in the sclerotized cuticle (Hackman and Goldberg, 1976). This important finding has led to the conclusion that the ϵ-amino group of lysine is participating in cross-linking with the sclerotizing agent. During sclerotization, the content of β-alanine in hydrolysates of cuticles increases. It has been suggested that β-alanine is incorporated in the cuticle in the form of *N*-β-alanyl dopamine (Hopkins et al., 1982; see Section 5.4, below).

Most of the work concerning identification and characterization of individual cuticular proteins refers to higher Diptera (*Drosophila*—Silvert et al., 1984; *Lucilia*—Skelly and Howells, 1987; *Sarcophaga*—Lipke et al., 1981), to *Manduca sexta* (Kiely and Riddiford, 1985a,b), and to *Schistocerca gregaria* (Højrup et al., 1986a,b; Andersen and Højrup, 1987).

Fristrom et al. (1982) and Silvert et al. (1984) demonstrated in 7*M* urea extracts of *Drosophila* larval cuticles the presence of five major nonglycosylated proteins, of low molecular weight (<20 kDa).

From pupae, two groups of proteins could be isolated—one with molecular weights of 56 kDa, and four with molecular weights between 15 and 25 kDa. The low-molecular-weight proteins are synthesized before pupation, whereas the high-molecular-weight proteins are synthesized after pupation.

Kiely and Riddiford (1985a,b) have analyzed the cuticular proteins during the fifth (last) larval instar of *M. sexta* and during production of the pupal cuticle. According to their results, the proteins can be grouped into larval-specific types, pupal-specific types, and those common to both stages. The proteins have been characterized by two-dimensional (2D) electrophoresis. The larval-specific proteins are of low molecular weight and have low isoelectric points (see Section 5.7.1, below).

From the migratory locust, *Schistocerca gregaria,* Højrup et al. (1986a,b) have identified a group of proteins with different characteristics than those of the proteins described above, from the well-defined chorion proteins (Hamodrakas et al., 1982), or from the oothecal structural proteins (Pau, 1984), although similarities in some regions of these proteins have been noted. One of these proteins with an M_r of 15 kDa (protein 38) has been totally sequenced (Højrup et al., 1986b) (Fig. 5.1).

It is worth noting that many of the scientists currently involved in the field of cuticular proteins are interested primarily in the gene regulation aspects, as this family of proteins represents coordinately regulated molecules, analogous to the well-studied chorion proteins (Kafatos, 1981). Note also that there is no information as yet concerning which of these proteins is cross-linked in the three-dimensional cuticular net, and in what position and in what way.

```
  1 GYLGGI|AAP|VGY|AAPA|V|GYA|API|AAAPVAVAHA|VA|PAAASVANTYRIS 49
     75  |AAP|AIA|AAPA|L|GYA|-RY|AAAPVAVAHA|AV|PAAASVANTYRIS 116

 50 |QTAR|V|LAAPA|------|AYAAPA|VA|AAPAIGY| 74
117 |QTAR|L|LAAPA|VAHAPV|AYAAPA|AY|AAPAIGY|GYGGLAYGAAPVAKVY 163
```

FIGURE 5.1. Primary structure of cuticle protein No. 38 of *Locusta migratoria* showing internal homology. (From Højrup et al., 1986b.)

5.3. Nature of the Cross-links Between Sclerotizing Agents and Cuticular Proteins

There are two currently accepted models of the interaction of sclerotizing agents with cuticular proteins:

(a) *Quinone tanning.* This mechanism was first proposed by Pryor (1940a,b), who regarded *o*-quinones, the oxidation product of the respective *o*-diphenols, as the immediate metabolites reacting with the proteins. It is shown in Fig. 5.2.

(b) β-*Sclerotization.* This mechanism was first suggested by Andersen and Barrett (1971) and further formulated by Andersen (see the reviews of Andersen, 1980, 1985). It is shown in Fig. 5.3A. In this model, the sclerotizing agent is activated in the β-position of the side chain to which the available amino acid residues attach (mainly the ϵ-amino group of lysine). The identification of 1,2-dehydro-*N*-acetyl dopamine from alkali-treated locust cuticles (Andersen and Roepstorff, 1982), led Andersen to modify the original β-sclerotization concept to include also the α-C of the side chain (Fig. 5.3B). Combining classical quinone tanning and β-sclerotization, Andersen and Roepstorff (1982) and Andersen (1985) propose that parallel to quinone tanning, a desaturase can introduce a double bond in the side chain of *N*-acetyldopamine, the formed dehydro-*N*-acetyldopamine being oxidized by the diphenoloxidase (or laccase) to the respective unsaturated quinone, which can then react with cuticular proteins. The unsaturated quinone can react spontaneously with two groups and thereby form cross-links without the generation of protein-bound intermediates, which would then need reoxidation, as in classical quinone tanning (Fig. 5.3C).

Recently, Sugumaran (1987), proposed an alternative mechanism of β-sclerotization (see Fig. 5.4). He postulated the initial formation from the sclerotizing agent (1) of quinone methides (42), which then react spontaneously with protein nucleophiles to generate catechol–protein adducts (47) and side-chain–hydroxylated products (11). These, upon

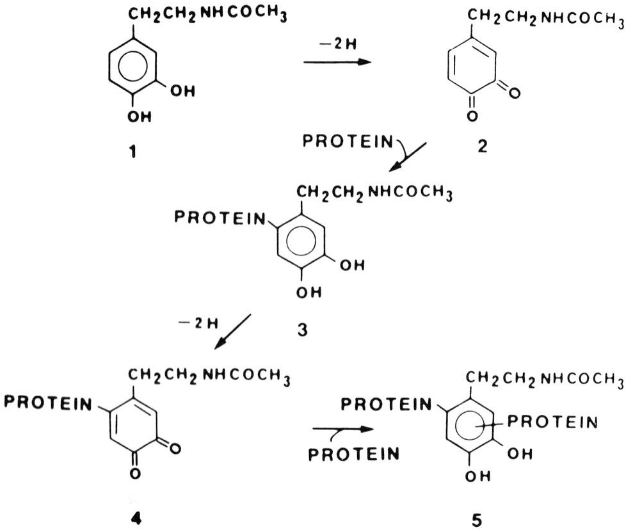

FIGURE 5.2. Quinone tanning hypothesis according to Pryor (1940a,b). *N*-Acetyldopamine (1) or another sclerotizing catechol) secreted by the epidermal cells is oxidized to the corresponding quinone (2) by the cuticular phenoloxidase. The *N*-acetyldopamine quinone undergoes a nonenzymatic Michael-1,4-addition reaction with cuticular proteins, forming *N*-acetyldopamine–protein adducts (3). Upon further oxidation (4) and coupling to another protein, these adducts yield protein–protein cross-links (5). The exact chemical structure of the cross-link and adducts are yet to be determined. Only the amino group of the proteins was considered; participation of other nucleophiles in the above process is also likely. (From Sugumaran et al., 1987.)

further oxidation (48) and coupling to other proteins (49), will result in cross-linking of the cuticular components. As regards the enzyme involved, Sugumaran and Lipke (1983) have shown that phenoloxidase can lead to the generation of quinone methides, although the presence of an additional more specific enzyme cannot be ruled out.

According to Andersen (1974, 1985), quinone tanning leads to darkly tanned cuticles whereas β-sclerotization produces white cuticles. There are indications, although not conclusive, that *N*-acetyldopamine is mainly used in sclerotization leading to white cuticles whereas *N*-β-alanyldopamine participates in quinone tanning (Hopkins et al., 1982).

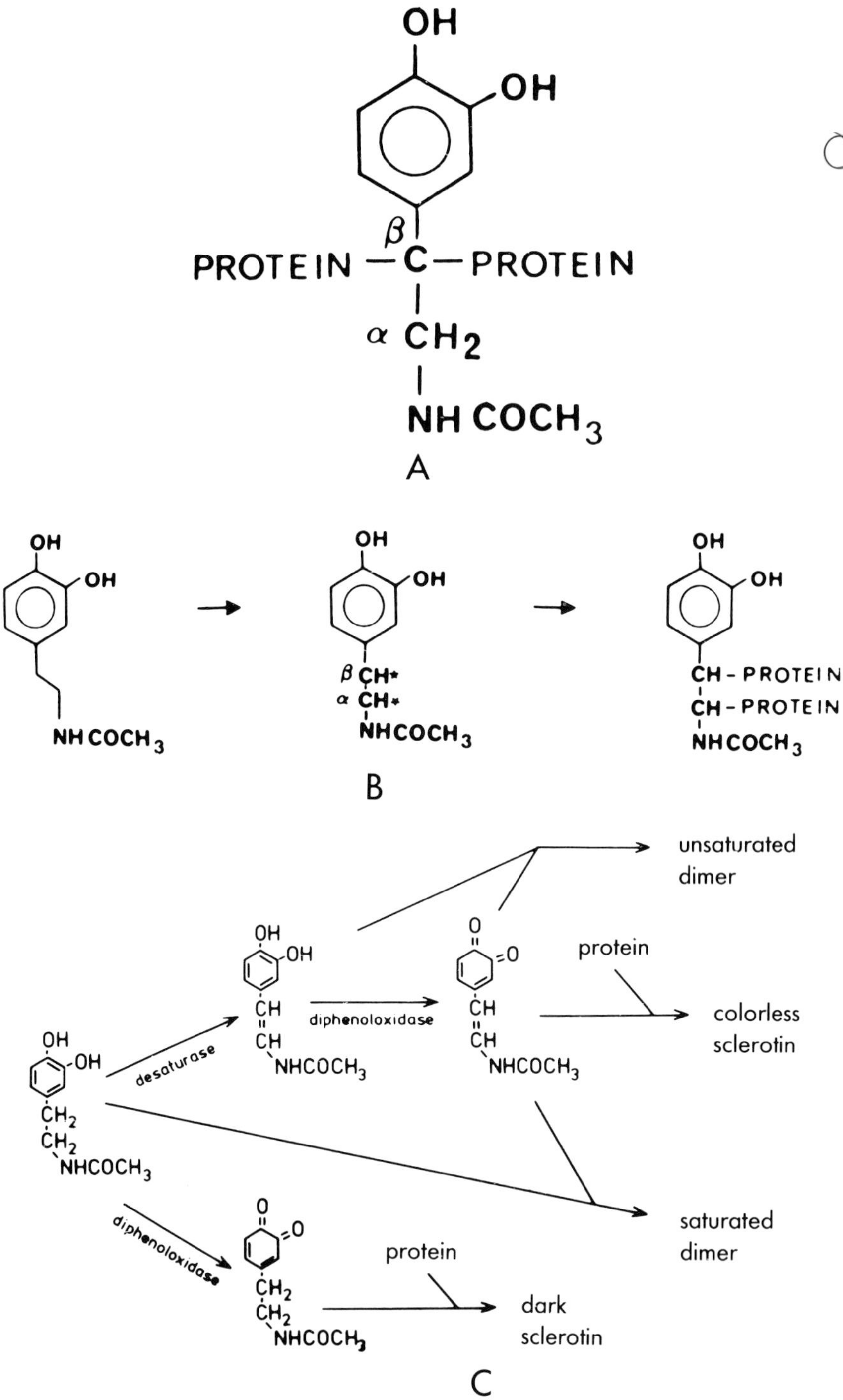

FIGURE 5.3. β-Sclerotization: (A) original scheme (Andersen and Barrett, 1971); (B) modified (Andersen and Roepstorff, 1982); (C) combined with quinone tanning. For details see the text. (From Andersen and Roepstorff, 1982; Andersen, 1985.)

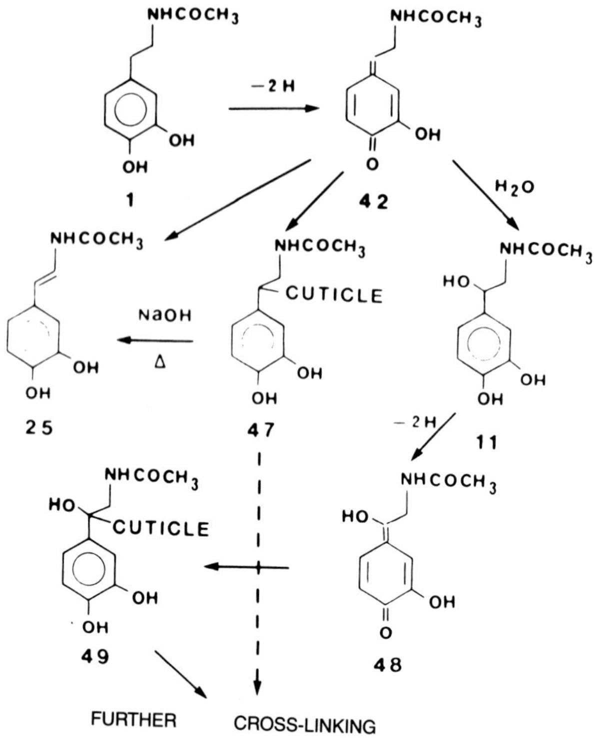

FIGURE 5.4. Quinone methide sclerotization scheme. For details see the text. (From Sugumaran, 1987.)

5.4. Biochemistry of the Sclerotizing Agents

Pryor et al., (1946, 1947) ascribed to protocatechuic acid the role of sclerotizing agent in the special case of tanning of the ootheca of *Blatta orientalis*. Brunet and Kent (1955a,b) later characterized the storage form of the acid as the 4-*O*-β-glucoside and beautifully showed the developmentally timed enzymatic release of the active agent from the collaterial glands (see Section 5.5., below). Karlson et al. (1962) isolated *N*-acetyldopamine from *Calliphora erythrocephala* (now *vicina*) larvae and proposed (Karlson and Sekeris, 1962a) that it is the tanning agent of the blow fly puparium.

Over the years, many diphenols, either in free or in conjugated form, have been isolated (see Tables 5.1 and 5.2) that have been implicated in sclerotization. One such compound, which in addition to *N*-acetyldopamine seems to play a major role in sclerotization in certain insect species, is *N*-β-alanyldopamine. Hopkins et al. (1982) isolated this

TABLE 5.1. Diphenols Identified in Insects Implicated in Sclerotization

N-Acetyldopamine	Karlson et al., 1962
N-β-Alanyldopamine	Hopkins et al., 1982
N-Acetyl noradrenaline[a]	Andersen and Roepstorff, 1978
N-β-Alanyl noradrenaline[a]	
N-(3,4-Dihydroxyphenyllactyl)dopa	Kawasaki and Sato, 1985
Dopamine	Oestlund, 1954
Noradrenaline[a]	
3,4-Dehydroxybenzoic acid	Pryor et al., 1946, 1947; Brunet and Kent, 1955a,b; Kent and Brunet, 1959
3,4-Dihydroxyphenylacetic acid	Schmalfuss and Bussmann, 1937
3,4-Dihydroxyphenylethanol	Pau and Acheson, 1968
3,4-Dihydroxybenzaldehyde	Atkinson et al., 1973
Gentisic acid (2,5-dihydroxybenzoic acid)	Brunet and Coles, 1974

[a]Noradrenaline is the British generic name for norepinephrine.

compound from hemolymph of *Manduca sexta* and observed high levels in early pharate pupae, with an 800-fold increase shortly before pupal ecdysis. During tanning of the cuticle, the levels of this compound declined. Phenoloxidase extracted from pharate pupal integument of *M. sexta* shows preference for *N*-β-alanyldopamine over *N*-acetyldopamine. *N*-β-Alanyldopamine was also found in the following species: *Periplaneta americana, Sarcophaga bullata, Diatraea grandioseila, Thyridopterya ephemeraeformis, Plodia interpunctella, Ephestia conseila,* and *Tenebrio mollitor.* It should be mentioned that Andersen and Barrett (1971) first postulated that *N*-β-alanyldopamine could be the metabolite by which β-alanine is incorporated into the cuticle to account for the observed *N*-terminal position of β-alanine in cuticles (Bodnaryk and Levenbook, 1969).

Some of the agents involved in the hardening of silk are also shown in Table 5.1.

5.5. Biosynthesis of the Sclerotizing Agents

All sclerotizing agents derive from tyrosine, with the exception of the kynurenine derivatives involved in silk hardening, which are tryptophan metabolites.

The main pathway leading from tyrosine to *N*-acetyl dopamine

TABLE 5.2. Conjugated Forms of Phenolic Metabolites Isolated from Various Insect Species.

Compound	Reference
Glucosides:	
Tyrosine-4-*O*-glucoside	Chen et al., 1978; Psarianos et al. 1985
N-Acetyldopamine-4-*O*-β-glucoside	Okubo, 1958; Karlson et al., 1962
N-Acetyldopamine-3-*O*-β-glucoside	Yago et al., 1984a
N-Malonyldopamine-3-*O*-β-glucoside	Yago et al., 1984a
N-(*N*-Acetyl-β-alanyl)dopamine-3-*O*-β-glucoside	Yago et al., 1984b
N-(*N*-Malonyl-β-alanyl)dopamine-3-*O*-β-glucoside	Yago et al., 1984b
3,4-Dihydroxybenzoic acid–5-*O*-β-glucoside	Brunet and Kent, 1955a,b
3,4-Dihydroxybenzyl alcohol–3-*O*-β-glucoside	Pau and Acheson, 1968
Phosphates:	
Tyrosine-4-*O*-phosphate	Mitchell et al., 1960; Mitchell and Lunan, 1964
N-Acetyldopamine-3-*O*-phosphate	Bodnaryk et al., 1974
Sulfates:	
Tyrosine-3-*O*-and-4-*O*-sulfate	Ahmed et al., 1983a
Dopamine-3-*O*-sulfate	Bodnaryk and Brunet, 1974; Sloley and Downer, 1987
N-Acetyldopamine-3-*O*-sulfate	Bodnaryk et al., 1974; Sloley and Downer, 1987
Dipeptides:	
β-Alanyl-tyrosine	Levenbook et al., 1969
γ-Glutamyl-phenylalanine	Bodnaryk, 1970b

involves hydroxylation of tyrosine to dopa, decarboxylation of dopa to dopamine, and *N*-acetylation of the amine (Sekeris and Karlson, 1962) (Fig. 5.5). In some insects and at certain developmental stages, decarboxylation of tyrosine to tyramine also takes place. Tyramine can then follow one of the following three pathways (Fig. 5.5): (1) hydroxylation to dopamine, which can then be acetylated to *N*-acetyldopamine (Sekeris and Karlson, 1966); (2) *N*-acetylation to *N*-acetyltyramine, a metabolite first isolated from pupae of *Bombyx mori* (Butenandt et al., 1959), followed by hydroxylation to *N*-acetyldopamine; (3) oxidative deamination to *p*-hydroxyphenylacetic acid. Pathways 1 and 2 lead to the production of the sclerotizing agent, but there is no compelling evidence that their contribution in this respect is significant (Sekeris, 1965; Brunet, 1980).

Concerning the biosynthesis of *N*-β-alanyldopamine, two possibilities can be visualized: one is the formation of this substance by coupling of dopamine with β-alanine; the other is coupling of tyramine with β-alanine and subsequent hydroxylation.

As regards biosynthesis of the phenolic acid compounds, two main pathways can be visualized—one by way of transamination of tyrosine (or dopa) to the respective pyruvic acid derivative, and another by oxidative deamination of tyramine (or dopamine). The highly intricate—indeed, Byzantine—possible interconversions are critically handled in a remarkable review by Brunet (1980).

5.5.1. *Enzymes Involved in the Formation of* N-*Acetyldopamine and in the Interaction of the Sclerotizing Agents with Cuticular Components*

Many enzymes are potentially involved in the generation of the sclerotizing agents and in their interaction with the cuticular components. The most important in this context enzymes will be considered, some of which have been definitely shown to be affected by MH and for some of which ample, although not always conclusive in the context of hormonal dependence, information has been gathered.

5.5.1.1. HYDROXYLATION OF TYROSINE TO DOPA

There is still uncertaintly regarding the insect enzyme(s) responsible for the hydroxylation of tyrosine to dopa. Many of the *o*-diphenoloxidases found in abundance in insects are also Tyrosinases, i.e., they can oxidize tyrosine as well, whereas others oxidize solely diphenols (see Section 5.5.1.4, below). Whether a specific tyrosine hydroxylase (E.C. 1.14.16.2) could be responsible for the production of dopa destined to be metabolized to *N*-acetyldopamine or whether phenoloxidases are the sole enzymes involved in this process is still an open question.

5.5.1.2. DECARBOXYLATION OF DOPA TO DOPAMINE

A key enzyme in the formation of *N*-acetyldopamine, probably also of *N*-β-alanyldopamine, is dopa decarboxylase (DDC) (E.C. 4.1.1.26). It has been detected in all insects examined (Sekeris and Karlson, 1962; Sekeris, 1963; Sekeris and Herrlich, 1966; T. Chen and Hodgetts, 1976;

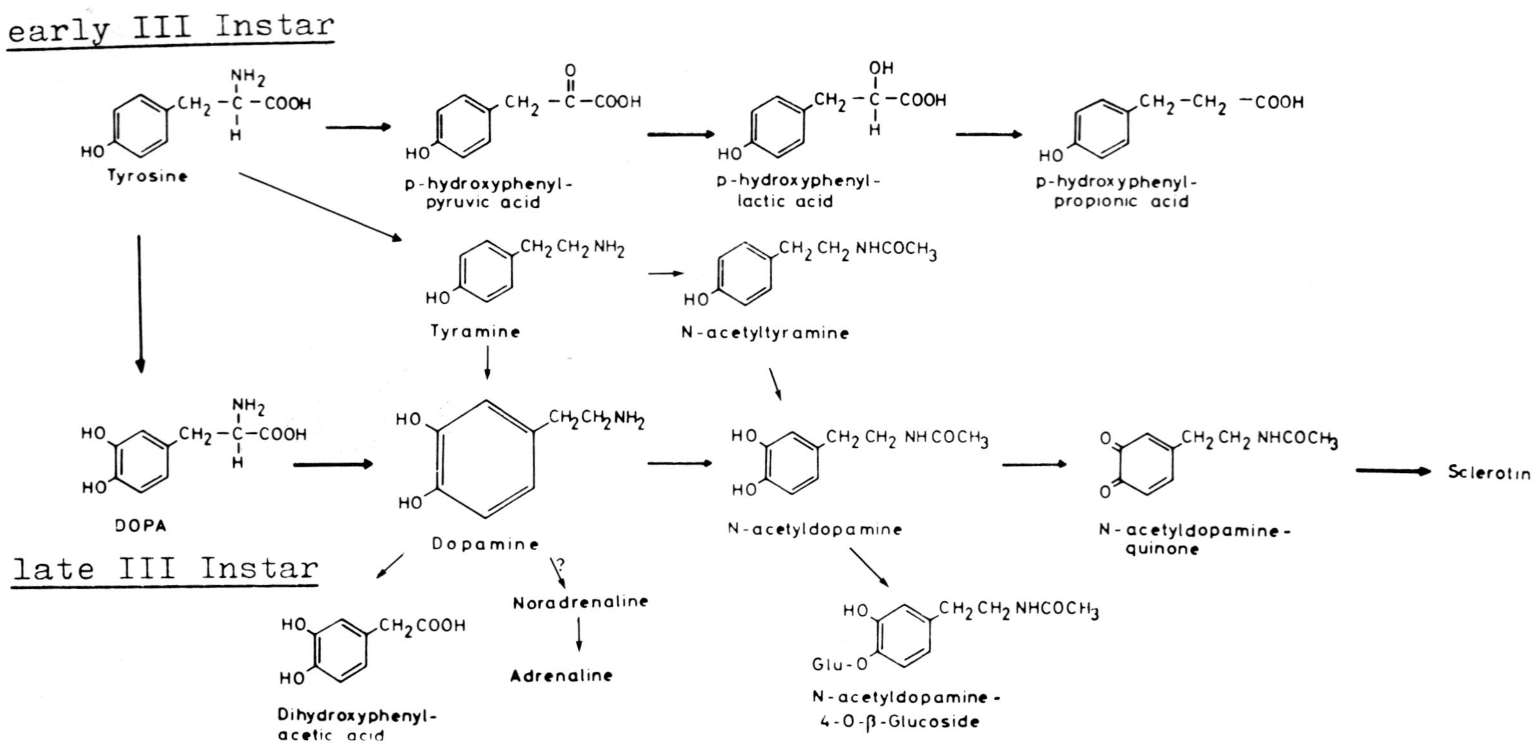

FIGURE 5.5. Metabolic transformations of tyrosine in *Calliphora vicina*. The pathway tyrosine → dopa → dopamine → *N*-acetyldopamine is activated in late-third-instar larvae. (Modified from Sekeris and Karlson, 1966.)

Clark et al., 1978; Kang et al., 1980; G. Thalassinos and E. G. Fragoulis, cited in Sekeris and Fragoulis, 1985).

The DDC involved in *N*-acetyldopamine formation is localized in the integument and is synthesized by the epidermis cells (Fragoulis and Sekeris, 1975a,b,c). DDC is also present in the ovaries (Schlaeger and Fuchs, 1974) and in the central nervous system (CNS) (Marsh and Wright, 1980), being responsible for the formation of catecholamines, needed as neurotransmitters (Oestlund, 1954). DDC has been purified from the integument of white prepupae of *Calliphora vicina* (Fragoulis and Sekeris, 1975a), from mature larvae of *Drosophila melanogaster* (Clark et al., 1978), from the ovaries of gravid female *Aedes aegypti* (Kang et al., 1980), and from white prepupae of *Ceratitis capitata* (G. Thalassinos and E. G. Fragoulis, cited in Sekeris and Fragoulis, 1985). Some characteristics of the enzymes are shown in Table 5.3).

5.5.1.3. N-ACETYLATION OF DOPAMINE

The enzyme responsible for the acetylation of dopamine to *N*-acetyldopamine, acetyl-CoA-*N*-acetyl transferase, or transacetyl transferase (E.C. 2.3.1.5) was first detected in *C. vicina* (Karlson and Ammon, 1963). It transfers the acetyl moiety of acetyl-CoA to dopamine and tyramine, but also to histamine and serotonin. The enzyme has been found as well in *Schistocerca gregaria* (Karlson and Herrlich, 1965), *D. melanogaster* and *T. mollitor* (Sekeris and Herrlich, 1966; Maranda and Hodgetts, 1977; Marsh and Wright, 1980), and pupae of the mosquito *Aedes togoi* (Shampengtong, 1987).

An enzyme transferring acetyl groups solely to tyramine has been detected in the nervous system of *Drosophila* (Dewhurst et al., 1972).

5.5.1.4. THE PHENOLOXIDASE SYSTEM AND RELATED ENZYMES

Phenoloxidases (E.C. 1.14.18.1) are the enzymes involved in tyrosine metabolism that have been most extensively studied in insects. Nevertheless, many ambiguities still exist as regards the number of enzymes, substrate specificity, activation, localization, site of synthesis, interrelationships, biological role, and hormonal regulation (for general reviews of phenoloxidases see Mason, 1955; Boyer, 1975; for more specialized reviews see Cottrell, 1964; Brunet, 1980; Andersen, 1985).

Phenoloxidases are enzymes oxidizing dopa and other *o*-diphenols to dopachinones (and further to indole derivatives). Many of the phenoloxidases also hydroxylate tyrosine to dopa (i.e. they are also tyrosinases), whereas others cannot. This, in part, could reflect no real differences of the enzymes per se, but rather in the way the enzymes were prepared, namely, the starting material, the presence of subcellular components, the aggregation state, the presence of trace metabolites, etc. (see below).

TABLE 5.3. Some Characteristics of DDC[a] from Insect Sources

Source	*Mr* (kDa)	Subunits	Effect of ions	Substrate specificity	Reference
Integument of white *Calliphora*	96	2	Fe^{2+} weakly inhibitory Cu^{2+}, Hg^{2+} strongly, inhibitory Mn^{2+}, Al^{2+}, weakly stimulatory	dopa	Fragoulis and Sekeris (1975a)
Mature larvae of *Drosophila melanogaster*	112	2	Cu^{2+}, Zn^{2+}, Hg^{2+} strongly inhibitory	dopa	Clark et al. (1978)
Female gravid *Aedes aegypti*	112		Fe^{2+}, Zn^{2+}, Cu^{2+}, Ca^{2+}, Mg^{2+} weakly inhibitory	dopa	Kang et al. (1980)

SOURCE: Modified from Sekeris and Fragoulis (1985).
[a]DDC = dopa decarboxylase.

If the enzyme can oxidize *p*-diphenols, it is characterized as a laccase; laccases are unable to oxidize tyrosine. Phenoloxidases have been detected both in the hemolymph and in the cuticle. The enzyme in the hemolymph is usually present as an inactive proenzyme that can be activated (Bodine and Allen, 1941; Ohnishi, 1953; Schweiger and Karlson, 1962) by a protein activator present in the integument (Sekeris and Mergenhagen, 1964; Karlson et al., 1964; Dohke 1973a,b; Hughes and Price, 1975a,b). Activator is also present in the hemolymph (Ashida and Ohnishi, 1967). Activation involves limited proteolysis of the proenzyme (Schweiger and Karlson, 1962; Ashida and Dohke, 1980).

The substrate specificity of the phenoloxidase depends on the source of the enzyme: the phenoloxidase from *C. vicina* larvae preferentially oxidizes dopa, dopamine, and *N*-acetyldopamine (Karlson and Liebau, 1961) but not phenolic carboxylic acids, whereas the enzyme from the collaterial glands of *Periplaneta americana* preferentially attacks phenolic carboxylic acids such as protocatechuic acid (Brunet and Kent, 1955a,b). The specificity apparently reflects the nature of the sclerotizing agent in the respective system, i.e., *N*-acetyldopamine in the blow fly and protocatechuic acid in the cockroach. Most hemolymph phenoloxidases also oxidize monophenols, e.g., tyrosine (Hackman and Goldberg, 1967), with a lag phase that is shortened in the presence of electron donors such as dopa or tetrahydropteridine.

The hemolymph phenoloxidases have been extensively studied in a variety of insect species (Ohnishi, 1954a,b,c, 1958; Karlson et al., 1964; Hackman and Goldberg, 1967; Mitchell et al., 1967; Preston and Taylor, 1970; Sin and Thompson, 1971, Hughes and Price, 1975a,b; Pau and Eagles, 1975; Pau and Kelly, 1975; Aso et al., 1985; Ashida, 1971; Tsukamoto et al., 1986). There are indications that the hemolymph of *Calliphora* contains two enzymes, an *o*-diphenoloxidase (which cannot oxidize tyrosine) and a tyrosinase (Sin and Thompson, 1971; Pau and Kelly, 1975), whereas Mitchell and Weber (1965) reported the presence of three enzymes in *Drosophila* hemolymph, two dopa oxidases and a tyrosinase.

Study of the cuticular enzymes has been hampered by difficulties in their extraction, in particular at periods in which knowledge of their characteristics would be most illuminating for sclerotization, i.e., during the sclerotization process per se. As known, the cuticular proteins at that period are refractive to extraction; the same holds true for the cuticular enzymes. It is possible that the enzyme proteins have also been tanned by the quinones they have generated ("suicide enzymes"). In spite of these difficulties, both *o*-diphenol oxidases as well as laccases have been isolated from cuticles of various species. Most cuticular diphenoloxidases do not oxidize tyrosine (Bhagvat-Richter, 1938; Dennell, 1947; Ohnishi, 1954a,b,c; Shaaya and Sekeris, 1965; Hackman and Goldberg,

1967; Mills et al., 1968; Yamazaki, 1969, 1972; Hughes and Price, 1975a,b; Andersen, 1978; Barrett and Andersen, 1981; Aso et al., 1984; Barrett, 1987).

From cuticles of some species (e.g., *S. gregaria*—Andersen, 1979a) only laccase activity could be detected, leading Andersen to regard this enzyme as the only one involved in sclerotization. Others, however (Aso et al., 1984), could detect *o*-diphenoloxidases from pupal cuticle of *M. sexta* that becomes sclerotized. Similar results have been obtained in other species. Yamazaki (1969, 1972) demonstrated laccase activity in *Drosophila virillis* and *B. mori*. The *Bombyx* enzyme reaches maximal activity some hours after ecdysis and 24 h thereafter completely disappears, whereas the *Drosophila* enzyme appears just before puparium formation with a maximum 3 h thereafter but with high activity during the pupal and pharate periods.

From the cuticles of *Calliphora* and of the sheep fly, *Lucilia cuprina*, three enzymes have been identified (Barrett and Andersen, 1981; Barrett, 1987): enzyme A is a typical phenoloxidase present throughout the wandering stage and final instar, declining at pupation; enzyme B is a laccase whose activity dramatically increases at pupation and decreases as sclerotization takes place; enzyme C is also a laccase. The developmental curve of enzyme B shows a correlation to the ecdysone titer (see Section 5.7.3, below), strongly suggesting its involvement in the sclerotization process and leaving the question of *p*-phenol participation in sclerotization still open (Dennell, 1958).

In addition to the well-known reactions catalyzed by phenoloxidases and laccases, two other reactions of potential importance for sclerotization have been ascribed to these enzymes. Andersen and Roepstorff (1982) reported the formation of dehydro-*N*-acetyldopamine upon incubation of *N*-acetyldopamine with cuticular extracts. As *o*- and *p*-diphenols will competitively inhibit the release of ^{3}H from side-chain ^{3}H-labeled *N*-acetyldopamine, Andersen (1972, 1979b, 1980) and Andersen and Roepstorff (1982) regard the laccase responsible for the side-chain oxidation. Lipke et al. (1983) and Sugumaran and Lipke (1983) reported that the cuticular phenoloxidase of *Sarcophaga bullata* can catalyze the formation of quinone methides from 4-alkylcatechols. Furthermore, the mushroom tyrosinase can oxidatively decarboxylate 3, 4-dihydroxymandelic acid to the respective benzaldehyde, very probably by way of formation of quinone methides (Sugumaran, 1986). According to Sugumaran (1986) and Sugumaran et. al. (1987a,b,c), β-sclerotization is due to the formation of quinone methides as intermediates of the sclerotizing agent.

What are the relationships between the hemolymph and the cuticle phenoloxidase? Hackman and Goldberg (1967) reported phenoloxidase activity in the cuticle of six species of flies, in each case distinct from that of the hemolymph, but there are indications that the hemolymph phenoloxidase enters the cuticle during certain developmental stages, either as a proenzyme or perhaps activated in the hemolymph prior to translocation (Hughes and Price, 1975a,b). We still do not know much about the site of synthesis of the components of the phenoloxidase system. The epidermal cells could be the site of synthesis of both the phenoloxidase and the activator, although the blood cells, where phenoloxidase activity has been reported (Leonard et al., 1985), have also been implicated in this context.

In the foregoing brief review, I have only considered the phenoloxidases in the context of sclerotization and have refrained from discussing their role in melanization, wound healing, and cellular defenses.

5.6. Hormonal Control of Molting

A characteristic of insect development is the periodic molting initiated by the molting hormone (MH). In the presence of the juvenile hormone (JH), MH induces larval molts; in the absence of JH, metamorphosis is initiated (Wigglesworth, 1940; Piepho, 1942; Williams, 1961).

The molting cycle encompasses a series of events beginning with a preparatory phase—including synthesis by the epidermal cells of RNA, protein, and in some cases also of DNA, their apolysis, followed by the secretion of the ecdysial membrane, the molting gel, and the new cuticle, starting with the epicuticle and continuing with the various layers of the procuticle (Truman, 1985). The molting gel is then activated, resulting in partial digestion of the old cuticle, the molting fluid is absorbed, ecdysis takes place, possibly also expansion, and lastly sclerotization of the new cuticle. Depending on the insect species, the epi- and exocuticle may already be sclerotized at ecdysis.

During the preparatory phase of a larval–larval molt, mRNA and protein synthesis for endocuticular proteins stops (Riddiford et al., 1981, 1982, 1984; Kiely and Riddiford, 1985), so the larvae can redirect their biosynthetic potential toward the production of molting fluid and the new epicuticle and exocuticle. Later, mRNA for the proteins of the endocuticle reappears. At metamorphosis, however, a reprogramming to produce a new type of cuticle must take place (see Section 5.7.1., below).

5.7. Hormonal Control of Sclerotization

As described in the preceding sections sclerotization is a well-orchestrated sequence of events, starting with the deposition of the cuticular proteins, followed by the biosynthesis of the sclerotizing agents, involving many enzyme systems, culminating in the availability of enzyme(s) "activating" the tanning agent and catalyzing its interaction with the cuticular proteins. We will, therefore, concentrate on hormonal control of synthesis of cuticular proteins, of synthesis of the sclerotizing phenols, and of synthesis of enzymes "activating" the sclerotizing phenols, leading to interaction of the sclerotizing agent with the cuticular components. In these processes, the molting hormone plays a primary role. In addition to MH, *bursicon,* the tanning hormone, assisted by protein factors such as puparium-tanning factor (PTF) (see Žďárek, 1985, for a review), also participates in the control of certain aspects of sclerotization. I will describe the processes that are, or seem to be, regulated by the MH but will not refer to the biochemical effects of bursicon. PTF and other related factors will be discussed in Chapter 18 by Žďárek and Sivasubramanian herein.

5.7.1. *Hormonal Control of Synthesis of Cuticular Proteins*

The type of cuticular proteins synthesized depends on the hormonal status of the insect. In the presence of both MH and JH, the proteins synthesized are characteristic of the larval cuticle. In the absence of JH, new batteries of proteins appear that characterize the pupa or the imago. How this switch in protein biosynthesis is brought about is currently receiving due attention.

With the exception of higher Diptera and Hymenoptera, the epidermal cells of all insects are polymorphic, i.e., the same cells synthesize the larval and the pupal proteins. In this case both MH and JH interact with the same cell, variably modulating gene expression, depending on the ratio of the two hormones. This is an excellent system indeed for study of hormone interactions at the molecular level. In higher Diptera and Hymenoptera, however, different cells are mobilized to produce new batteries of proteins. In these species, the pupal and imaginal cuticles are produced by the imaginal disks, as regards the head and thorax cuticle, and by the larval epithelium and the abdominal histoblasts, as regards the abdominal cuticle. Accordingly, the pupal cuticle is the product of a mosaic epithelium. At metamorphosis, and in insects with polymorphic epidermal cells, therefore, the production of larval proteins must be shut off in one type of cells and the genes responsible for producing the pupal and imaginal cuticle must be activated in the other type.

Fristrom et al. (1982) exploited the *in vitro* systems introduced by various workers (Mandaron, 1970; Marks, 1972; Fristrom et al., 1973; Oberlander, 1976; Mitsui and Riddiford, 1976; Nardi and Willis, 1979) to study cuticle formation and apolysis. When imaginal disks are incubated for 4–6 h with 0.5–10 μg MH/ml, further for a period of at least 5 h without MH, and then with 0.05 μg MH/ml, optimal procuticle deposition occurs. Continuous incubation with high concentrations of MH does not result in cuticle deposition. Pupal cuticle proteins (2, 3, and 4) were synthesized under the *in vitro* conditions maximal for cuticle deposition commencing 6 h after hormone withdrawal. In another paper the same group (Doctor et al., 1985), using both cultured imaginal disks and pupal integument from different regions of *Drosophila* and a combination of biochemical and immunocytochemical techniques, showed that a set of low-molecular-weight pupal cuticle proteins (15–25 kDa) are synthesized in the epithelium before pupation. After pupation, synthesis of these proteins ceases and new high-molecular-weight proteins (40–82 kDa) are synthesized and incorporated into the cuticle. Similar biphasic accumulation of low- and high-molecular-weight pupal cuticle proteins is also seen in cultured imaginal disks.

For the first wave of low-molecular-weight protein synthesis, a pulse of MH is needed, the proteins appearing 6 h after hormone withdrawal. This approximates the changes in MH titer in *Drosophila*, which reaches a concentration of 0.1–1.0 μM some 6 h before pupariation and drops to intermolt levels during the mid-prepupal period (Richards, 1981). *In vivo*, the deposition of the pupal procuticle beings during the mid-prepupal period, some 8–9 h after pupariation.

The second wave of high-molecular-weight protein synthesis, involving mainly the 82- and 40-kDa proteins, is induced by readdition of MH to the culture medium 8–9 h after hormone withdrawal. At the same time, incorporation of precursors in low-molecular-weight pupal cuticular protein is inhibited. A transient increase in MH titer during pupation has been shown by Handler (1982) and Klose et al. (1980).

Note that the cuticular lamellae deposited before pupation (outer lamellae) and after pupation (inner lamellae) are ultrastructurally distinct. Strong evidence is provided that the low-molecular-weight pupal cuticular proteins are structural components of the outer lamellae, whereas the high-molecular-weight proteins are components of the inner lamellae.

The parallelism of the *in vitro* experiments using imaginal disks and the *in vitro* experiments with salivary gland chromosomes (Richards, 1976a,b) should be stressed, in which the removal of MH from the medium is required for induction of prepupal puffs whereas the addition of hormone is necessary to observe the late prepupal puffs.

In the tobacco hornworm, *Manduca sexta*, under the influence of MH and in the presence of JH, the epidermis at the larval molt produces larval cuticle throughout the four feeding larval stages. During the final (fifth) instar, JH is no longer present and a small MH peak appears, initiating metamorphosis. In the absence of JH, MH acts on the epidermis, stopping the production of larval proteins and committing the epidermal cells to pupal differentiation. Riddiford and coworkers, in a series of papers (Riddiford, 1982, 1985, 1986; Wolfgang and Riddiford, 1986; Riddiford et al., 1986), have studied the change in commitment of the epidermal cells and the reprogramming of cuticular protein synthesis in *Manduca*. The epidermis of growing *M. sexta* larvae continuously deposits endocuticle. The major endocuticular proteins (class I) are produced during each larval instar but not during the molting period. They are also synthesized during the first day of the final (fifth) instar, contributing to the formation of the 1-μm thick lamellae. On the final day of feeding (day 3), synthesis of class I proteins stops and a new class of endocuticular proteins (class II) is formed, contributing to the structure of the thin (0.1–0.2 μm) cuticular lamellae. In the pupally committed cell, no such proteins are synthesized. Instead, several new high-molecular-weight cuticular proteins are formed—a 34-kDa "pupal commitment" protein and several unidentified species. The fall of JH observed during the final larval instar is followed by two small peaks of MH and a larger one (Fig. 5.6). The small MH peaks come at a period in which JH is totally absent for the first time in the life history of the animals. Riddiford and coworkers demonstrated that commitment to pupal synthesis is due to the small MH peaks whereas the large MH surge acts as a trigger for the expression of the pupal cuticle genes,

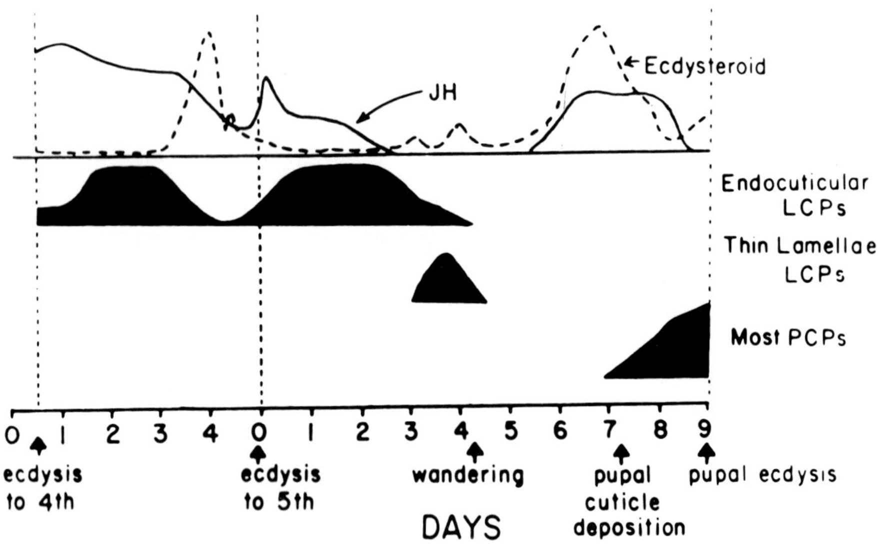

FIGURE 5.6. Summary of the relationship between the changing cuticular gene expression and the hormonal titers during the fourth and fifth larval instars and pupal development of *Manduca*. *Key:* LCP = larval cuticular proteins; PCP = pupal cuticular proteins. (From Riddiford, 1986.)

independent of whether JH is present or not. The rise of MH titer then induces synthesis of pupal cuticular proteins in the following order: epicuticular, exocuticular, and endocuticular.

Studies involving mRNA translation, incorporation of radioactively labeled amino acids into protein, and cloning of some cuticular protein gene (Rebers et al., 1987) have contributed toward a deeper insight into the mechanism of sequential expression of the cuticular protein genes. Three class I cuticle genes (Rebers et al., 1987; Riddiford, 1987) are expressed during the feeding phase of both the fourth and fifth larval instars, but not during the molt. The cessation of mRNA synthesis seems to be dependent on the rising titer of MH, since high levels of MH *in vitro* depressed these levels in fourth-instar epidermis, a situation resembling the response of pupal cuticle genes of *Drosophila* to MH (Doctor et al., 1985). Before the onset of metamorphosis, expression also rapidly declines. Two of the clones encode a 14.6-kDa polypeptide, expressed in pharate pupal and adult epidermis, producing the flexible cuticle of the intersegmental membrane regions. The class II cuticle genes were first expressed before the onset of metamorphosis, causing the formation of a stiffer cuticle. The expression of this class of genes is initiated by low MH concentrations and suppressed by high MH titer. Sequencing of these cloned genes and isolation and sequencing of still other cuticle protein genes will define putative regulatory regions and hormone-binding segments possibly necessary for the coordinate and sequential hormonal regulation of these genes.

5.7.2. *Regulation by MH of* N-*Acetyldopamine Formation*

In most insect species studied, *N*-acetyldopamine and β-*N*-alanyldopamine serve as sclerotizing agents (see Section 5.4, above). Most of the accumulated data dealing with the effect of MH on biosynthesis of sclerotizing agents refer to *N*-acetyldopamine, to which I will now turn.

Three enzymes participate in *N*-acetyldopamine biosynthesis: tyrosine hydroxylase and dopa decarboxylase (discussed next), and dopamine transacetylase (discussed in Section 5.7.2.3, below).

5.7.2.1. TYROSINE HYDROXYLASE.

As mentioned in Section 5.5.1.1, above, the demonstration of a specific tyrosine hydroxylase in insects has not yet been achieved. Many of the phenoloxidase preparations also show monophenoloxidase activity and could be involved in the hydroxylation of tyrosine to dopa. Therefore, the hydroxylase activity and its modulation by MH will be discussed in the section on regulation of phenoloxidase activity by MH (see Section 5.7.3, below).

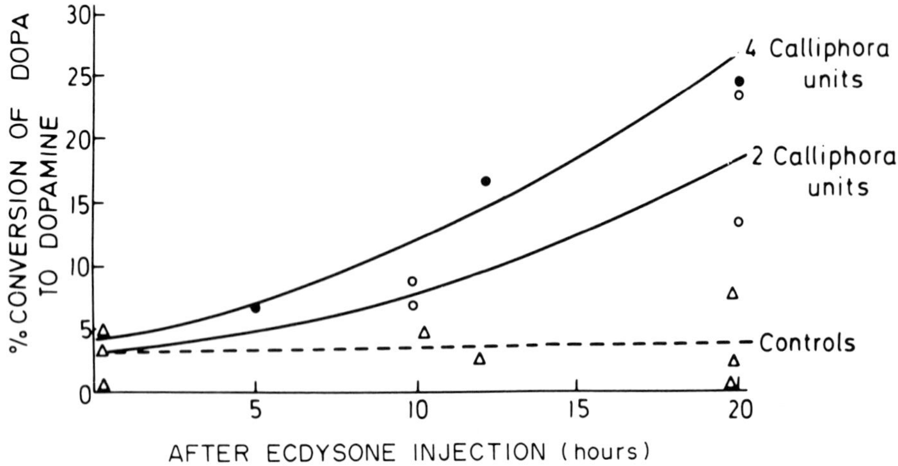

FIGURE 5.7. Induction of DDC in ligated *C. vicina* larvae by ecdysone. DDC activity is expressed as percentage transformation of dopa to dopamine. Ligated larvae were microinjected in the abdomen with 2 or 4 *Calliphora erythrocephala* units (CU) of ecdysone (a highly purified preparation, containing traces of 20-hydroxyecdysone). (From Karlson and Sekeris, 1962b.)

5.7.2.2. DOPA DECARBOXYLASE.

The dependence of dopa decarboxylase (DDC) activity on MH was first demonstrated in head-ligated larvae of *Calliphora vicina* (Karlson and Sekeris, 1962b). In these larvae MH does not reach the hind part, which therefore does not pupate; DDC activity remains low. Injection of MH preparations to the nonpupated hind part of these larvae leads within 12–20 h to increase in the activity of DDC and to pupation (Fig. 5.7). The developmental appearance of the enzyme in comparison to the titer of MH (Fig. 5.8) strongly suggested the hormonal dependence of enzyme activity (Shaaya and Sekeris, 1965), as regards the enzyme peak observed at pupariation. There is, however, no such correlation concerning an enzyme peak observed at eclosion of the imago (see Section 5.7.2.3, below).

The increased activity of DDC after MH administration is due to increased titers of DDC–mRNA, and its translation to enzyme protein (Fragoulis and Sekeris, 1975c) conforms with the hormone–gene activation hypothesis proposed by Karlson (see Clever and Karlson, 1960; Karlson, 1961, 1963; Karlson and Sekeris, 1966a,b). This had been suggested by early studies using RNA and protein synthesis inhibitors (Karlson and Sekeris, 1964; Shaaya and Sekeris, 1971) (Fig. 5.9), which blocked appearance of DDC activity and which also defined a critical

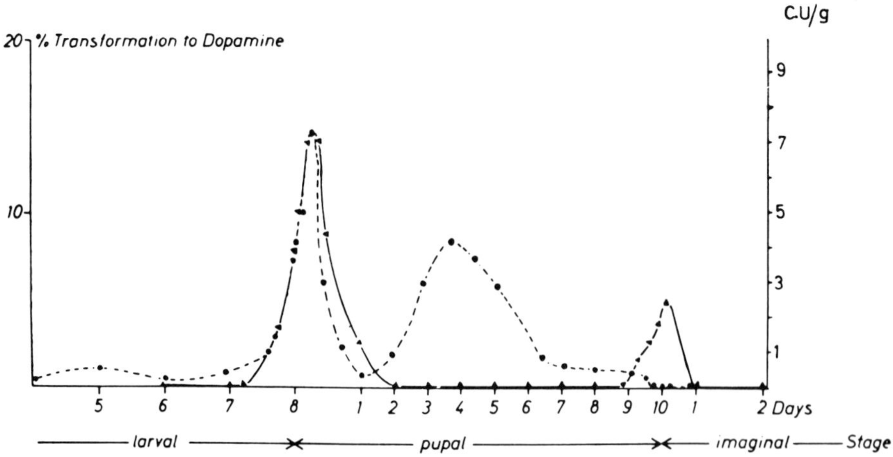

FIGURE 5.8. Development profile of DDC activity and of ecdysteroid titer of *C. vicina*. Ecdysteroid titer expressed as *Calliphora erythrocephala* units (CU) per gram of tissue (Sekeris and Karlson, 1962). (Data from Shaaya and Sekeris, 1965.)

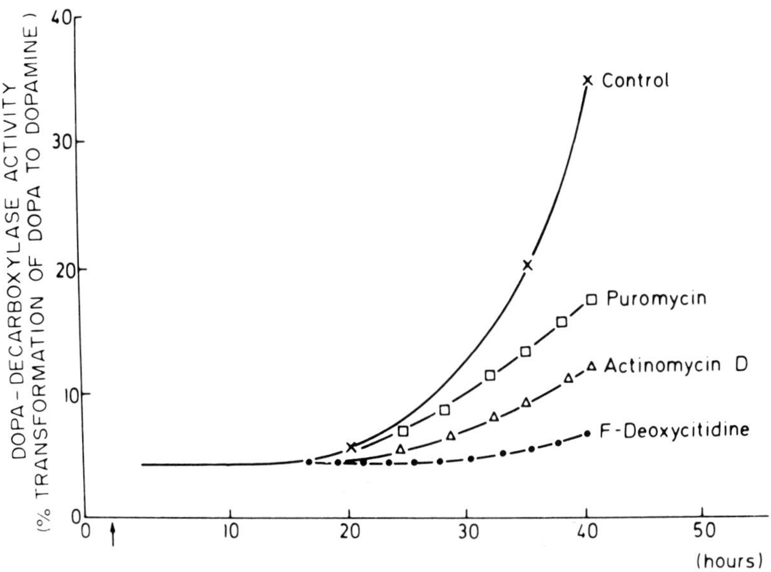

FIGURE 5.9. Influence of different inhibitors on DDC activity of *C. vicina* larvae. The respective inhibitors were microinjected into the abdomen of *Calliphora* larvae at the time period indicated by the arrow. Groups of larvae at the time period indicated on the abscissa were homogenized, and DDC activity was determined in the low-speed supernatant. Enzyme activity was expressed as percentage of transformation of dopa to dopamine. (From Sekeris and Karlson, 1964.)

period of sensitivity of pupariation to these transcriptional and translational inhibitors (Fig. 5.10). Translation of poly(A) + RNA stemming from larvae of two different developmental periods (Fragoulis and Sekeris, 1975c)—one characterized by low and one by high DDC activity, and one by low and one by high MH titer, respectively (Fig. 5.11; Table 5.4)—as well as measurement of amino acid incorporation into enzyme protein in larvae of different developmental stages (Table 5.5), in ligated animals and in ligated, MH-injected ones (Table 5.6), (Fragoulis and Sekeris, 1975b), demonstrated the following: increased DDC–mRNA and enzyme synthesis in the epidermal cells of *C. vicina* white prepupae, as compared to early wandering stage larvae; and increased DDC synthesis in ligated, MH-injected animals, as compared to the ligated controls.

A very thorough molecular and genetic approach for the study of DDC induction in *Drosophila* has been initiated by Kraminsky et al. (1980), Hirsh and Davidson (1981), and Wright et al. (1981a,b). In *D.*

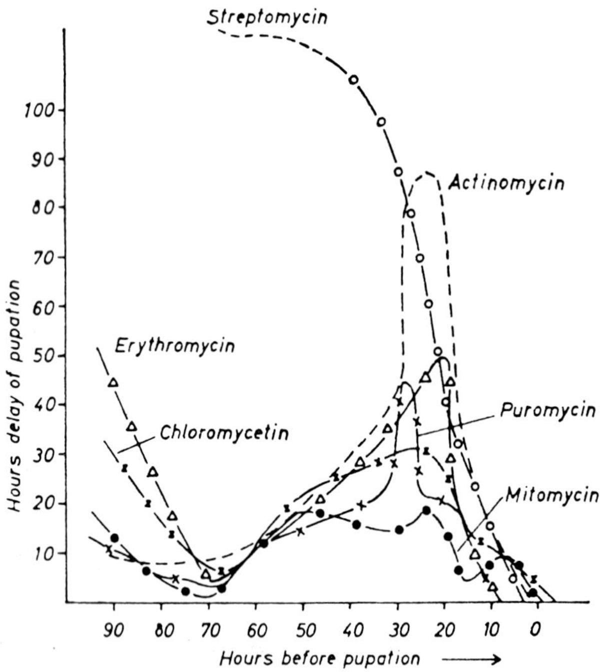

FIGURE 5.10. Influence of injection of inhibitors on the pupation of *Calliphora* larvae. The delay of pupation in hours is plotted against the time of injection of the inhibitors before pupation. (Data from Sekeris and Karlson, 1964.)

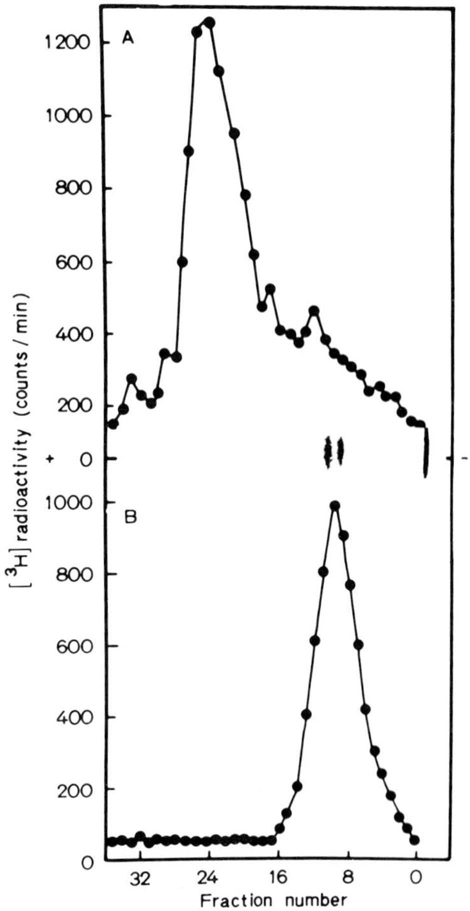

FIGURE 5.11. *In vitro* synthesis of DDC under the direction of mRNA from white prepupae (ecdysone-induced larvae of *C. vicina*. mRNA from the integument of white prepupae was added to the heterologous protein-synthesizing system (see Fragoulis and Sekeris, 1975c, for experimental details). After incubation at 36°C for 25 min, the incubate was treated with immungoglobulin G (IgG) against DDC: (A) sodium dodecyl sulfate (SDS)–polyacrylamide gel electrophoresis of the supernatant obtained after immunoprecipitation; (B) SDS-polyacrylamide gel electrophoresis of the pellet. DDC was also submitted to electrophoresis under the same conditions. The bromophenol tracking dye migrated with fractions 32–34. (From Fragoulis and Sekeris, 1975c.)

melanogaster, five peaks of DDC activity have been observed during postembryonic development (Fig. 5.12) (Kraminsky et al., 1980). One is seen during hatching, two during the larval molts, one at pupariation, and one at adult eclosion. Of the five DDC peaks, only the one at pupariation showed direct positive correlation to the MH titer; the other four peaks appearing after the MH titer had decreased to basal levels (Fig. 5.11). This will be discussed in Section 5.7.2.3. Here, I will summarize the extensive work concerning the DDC peak appearing at pupariation of *Drosophila*. The dependence of this peak on MH was elegantly shown by Kraminski et al. (1980) using the *ecd1* mutant isolated by Garen et al. (1977). The larvae of this mutant do not produce MH at the restrictive temperature of 29°C; they thus do not pupate and have a very low DDC activity (Fig. 5.12). If the animals are retuned to the permissive temperature, DDC activity rapidly begins to increase and the animals

TABLE 5.4. *In Vitro* Synthesis of DDC under the direction of mRNA from the Integument of White Prepupae and of 6- to 7-Day-Old Blow Fly Larvae[a]

mRNA source	Incorporation of [^{3}H]leucine (cpm) into: Total cell-free product (A)	Incorporation of [^{3}H]leucine (cpm) into: DDC (B)	100B/A
White prepupae	35,000 ± 2988	250 ± 23	0.70
6- 7-day-old larvae	48,000 ± 2111	88 ± 11	0.18

[a] Protein synthesis was carried out in the heterologous system of Schreier and Staehelin (1973) under the direction of 2.6-μg mRNA from the integument of white prepupae or 6- to 7-day-old larvae with [^{3}H]leucine as the radioactive amino acid. Immunoprecipitation was preformed with immunoglobulin G (IgG) against DDC. Blank values (10 cpm) and the values of the immunoprecipitate obtained with preimmune IgG (30–45 cpm) were subtracted (Fragoulis and Sekeris, 1975c).

TABLE 5.5. Incorporation of [^{14}C]Leucine into Total Protein and into DDC of Epidermis During Development of Blow Fly Larvae[a]

Developmental age	Incorporation into total epidermal protein (cpm/mg protein)	Incorporation into enzyme protein (cpm in immunoprecipitate)	Percentage radioactivity in enzyme/percentage radioactivity in total protein
6-day-old larvae	7,562	240	0.15
White prepupae	11,981	1,200	0.40
Brown pupae (2 h after pupation)	4,734	1,240	1.25
Black pupae (24 h after pupation)	25,793	1,900	0.34

[a] Groups of 40 animals were injected with [^{14}C]leucine; 1 h thereafter the animals were homogenized, and subsequently the homogenate was centrifuged at 30,000g/15 min. In the supernatant, incorporation of [^{14}C]leucine into total trichloracetic acid (TCA)–precipitable material and in material immunoprecipitable with antibody to DDC was determined (Fragoulis and Sekeris, 1975b).

TABLE 5.6. Effect of 20-Hydroxyecdysone on [14C]Leucine Incorporation into Total Protein and into DDC of Epidermis of Ligated Blow Fly Larvae[a]

Hours after treatment	Incorporation into total protein (cpm/mg protein)	Radioactivity in immunoprecipitate (cpm)
Control	5004	380
1	4516	250
3	5023	350
5	6290	1860

[a] [14C]Leucine was injected for 1 h into control ligated larvae and into three groups of ligated larvae treated for 1, 3, and 5 h with 20-hydroxyecdysone. The epidermis was then extracted with buffer, and the incorporation of [14C]leucine into total extracted protein and in material immunoprecipitated with antibody to DDC was determined (Fragoulis and Sekeris, 1975b).

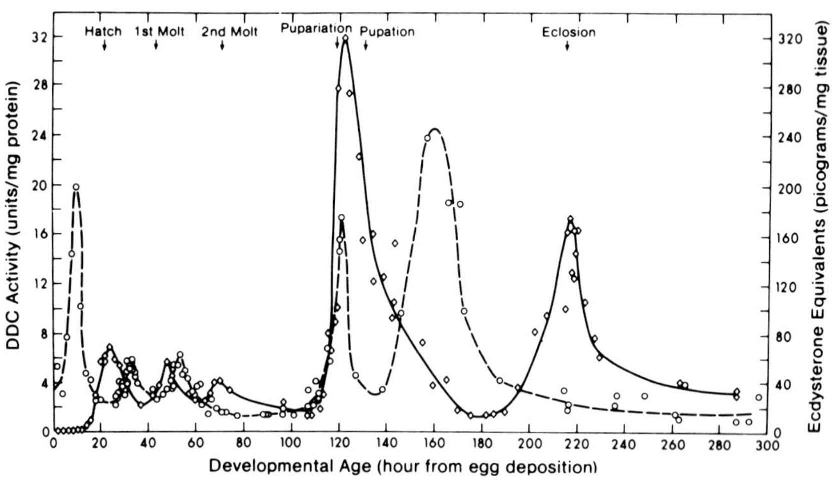

FIGURE 5.12. Developmental profile of DDC activity and of ecdysteroid titer in *Drosophila melanogaster. Key:* O---O = ecdysone titer expressed as picograms of 20-hydroxyecdysone equivalent per milligram of tissue (Hodgetts et al., 1977); O——O = DDC activity, expressed in units per milligram protein (Lunan and Mitchell, 1969, modified according to Hodgetts et al., 1977). (From Kraminsky et al., 1980.)

pupate normally. The same is seen if the animals, instead of being returned to the permissive temperature, are fed MH: 4–6 h later DDC activity begins to rise, increasing after 16 h to levels 50% of the maximal activity observed at pupariation during normal development (Fig. 5.13). At this time period, the first puparia also appear. Feeding the larvae with [^{35}S]methionine at the restrictive temperature, in the presence or absence of MH, and using immunoprecipitation with antibody to wild-type DDC (Clark et al., 1978), sodium dodecyl sulfate(SDS)–polycrylamide gel electrophoresis and fluorography, Kraminsky et al. (1980) showed that radioactivity was incorporated into a polypeptide with M_r of 53–54 kDa, corresponding to DDC (Fig. 5.14). This incorporation, hardly detectable in larvae not fed hormone, increased significantly over an 8-h period in the MH-fed larvae, in good correlation with the increasing enzyme activity. Titration of DDC–mRNA by translation of poly(A) + RNA in a rabbit reticulocyte lysate system (Pelham and Jackson, 1976), precipitation of the products synthesized with antibody to

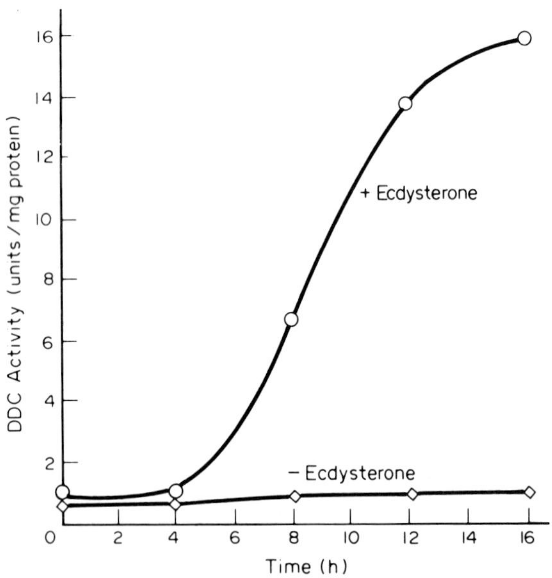

FIGURE 5.13. Induction of DDC activity in larvae of the *ecd1* mutant of *D. melanogaster* fed 20-hydroxyecdysone. Eggs were collected over 4-h intervals, and cultures were maintained at the permissive temperature of 20°C for 150 h. These were then transferred to the restrictive temperature, and after 48 h at 29°C larvae were transferred to petri dishes containing 20-hydroxyecdysone. Crude extracts of larval samples collected at the times indicated were prepared, and DDC activity was measured. (From Kraminsky et al., 1980.)

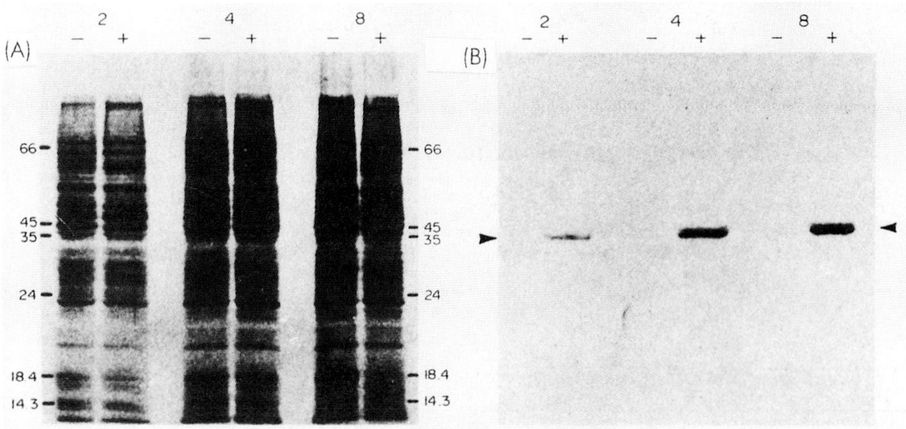

FIGURE 5.14. *In vivo* synthesis of DDC in the epidermis of *ecd1* larvae of *D. melanogaster* fed 20-hydroxyecdysone. Eggs were collected over 4-h intervals. Cultures were maintained at the permissive temperature 20°C for 150 h and then transferred to the restrictive temperature of 29°C; 48 h thereafter they were fed [^{35}S]methionine + 20-hydroxyecdysone for 2-, 4-, or 8-h periods prior to the preparation of epidermis and extraction of proteins from it. Pairwise comparisons of labeled polypeptides displayed on dried SDS-polyacrylamine gels are shown (−, control; +, 20-hydroxyecdysone-fed larvae). The positions of molecular weight makers ($\times$ 10^{-3}) are indicated. The position of the internal marker rabbit IgG heavy chain (M_r: 53 kDA) is indicated by the arrow. The subunit molecular weight of DDC is 54 kDA (Clark et al., 1978): (A) autoradiography of total *in vivo*–labeled products; (B) fluorography of the polypeptide immunoprecipitates from the labeled products. (From Kraminsky et al., 1980.)

DDC, SDS–polyacrylamide gel electrophoresis, and fluorography revealed the presence of DDC–mRNA in 8-h hormone-fed animals but not in the noninduced controls (Fig. 5.15).

Cloning of the DDC gene (Hirsh and Davidson, 1981) has initiated studies aimed at a better understanding of its expression during development. Using genomic DDC subclones and Northern blot hybridization, Hirsh and Davidson (1981) detected three DDC–RNA species. The concentration of a 2.1-kb RNA, which on the basis of its size corresponded to DDC–mRNA, varied through development, high levels being observed at pupariation and eclosion. This DDC–mRNA is concentrated in the integument, corroborating the findings of Fragoulis and Sekeris (1975b,c) that the epidermal cells are the site of synthesis of DDC.

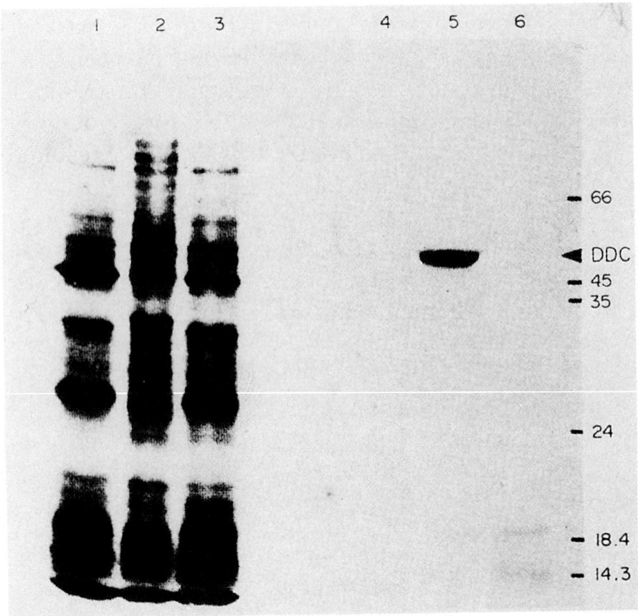

FIGURE 5.15. Induction of DDC–mRNA in *ecd1* larvae of *D. melanogaster* after feeding with 20-hydroxyecdysone. Large batches of *ecd1* larvae raised at 20°C were transferred to 29°C at mid-third instar and after 2 days were treated for 2 and 8 h with or without 20-hydroxyecdysone. Poly(A) + RNA was prepared from each group and used for *in vitro* translation in the reticulocyte lysate system (Pelham and Jackson, 1976), with [^{35}S]methionine as the labeled precursor. The fluorogram of the SDS–polyacrylamide gel shows the total products of translation reactions in the presence of poly(A) + RNA from larvae fed on hormones for 2 h (slot 1), 8 h (slot 2), and the control (slot 3), as well as the immunoprecipitates from each reaction, respectively (slots 4, 5, and 6). The epidermal DDC activity in each group of larvae was determined and found to be constant in the control, 1.2 times greater in the 2-h-induced sample, and 7.8 times greater in the 8-h-induced sample. (From Kraminsky et al., 1980.)

Beal and Hirsh (1984) later showed the presence of five size classes of DDC–RNA, which appear later in embryogenesis, coincident with the induction of DDC enzyme activity. Although all DDC–RNA species appeared at the same developmental period, there were marked differences in te subsequent developmental profiles of the larger DDC–RNA species. The investigations of Hirsh and collaborators (see Morgan et al., 1986) and of Hodgetts and associates (see Gietz and Hodgetts, 1985; Eveleth et al., 1986; Z.-A. Chen and Hodgetts, 1987) have resulted in the identification of the nature of the various DDC–RNA species and in the elucidation of their developmental appearance.

Of the five size classes of DDC–RNA with molecular weights of 4.0, 3.0, 2.7, 2.3, and 2.0 kb, the 2.3-kb species is found solely in late embryogenesis, in evaginating imaginal disks, and in adult flies. The other four RNAs appear at multiple stages during which DDC is expressed. Sequencing of the DDC gene (Eveleth et al., 1986; Morgan et al., 1986) had led to a detailed structure of the DDC–RNAs and to the recognition of alternative splicing reactions. Four exons were identified (Fig. 5.16).

The 4.0-kb RNA is the fully unprocessed RNA, the 3.0-kb RNA has the third intron spliced, the 2.3-kb molecule is the fully spliced RNA containing all four exons, whereas the 2.1-kb RNA is also a fully spliced species but does not contain the second exon (Fig. 5.16).

As mentioned in Section 5.5.1.2, two DDCs are present in insects—one in the integument, participating in the production of the sclerotizing agent, and one in the CNS, involved in the biogenesis of catecholamine neurotransmitters. The results of Eveleth et al. (1986) and Morgan et al. (1986) can now account for the appearance of the 2.3-kb mRNA in late embryogenesis and in adult flies, as it codes for brain DDC and is the

FIGURE 5.16. The structures of the DDC–RNAs. Both the 2.1- and 2.3-kb RNAs are formed from the same 4.0-kb primary transcript. The 2.3-kb RNA retains four exons (filled boxes A, B, C, and D), whereas the 2.1-kb RNA retains only exons A, C, and D. The 3.0-kb RNA is a partially processed precursor that has spliced exons C and D together. (From Morgan et al., 1986.)

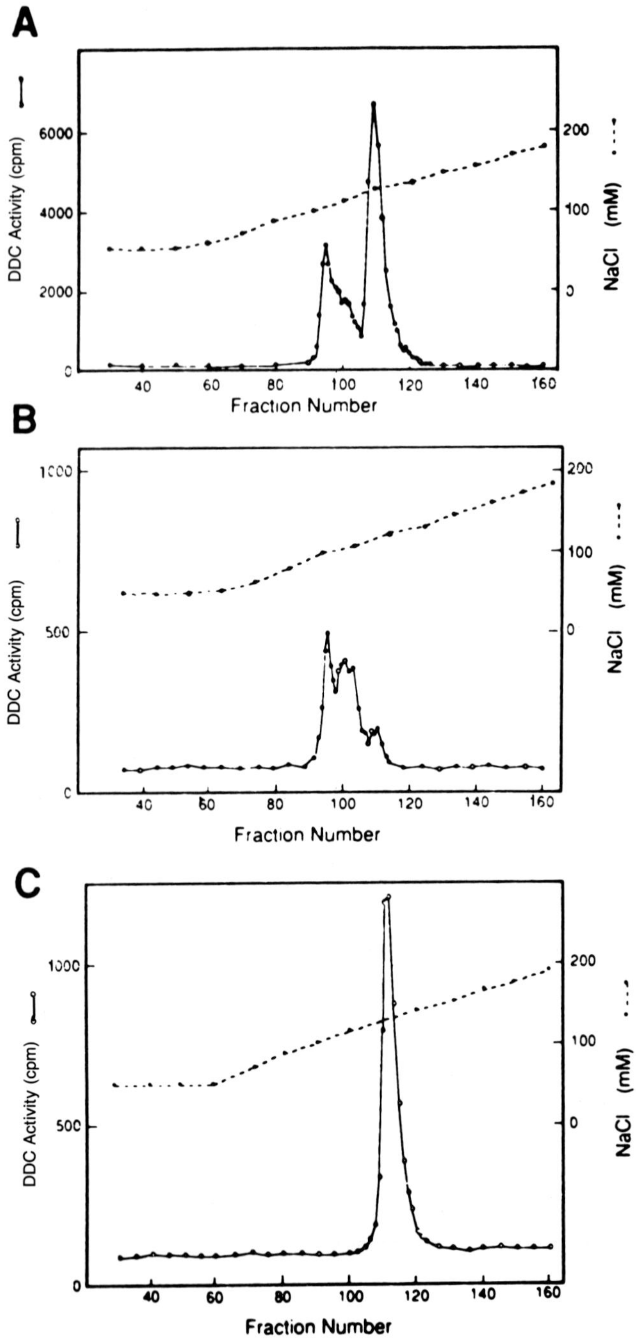

FIGURE 5.17. Elution profiles of DDC activity from a diethylaminoethyl (DEAE) high-pressure liquid chromatography (HPLC) column. Crude homogenates of heads from wild-type flies (A), hand-dissected brains

only DDC–mRNA found in the CNS. The 2.1-kb mRNA, which codes for the epidermal DDC, appears during pupariation and eclosion of the imago, and is enriched in the integument, the site of synthesis of the DDC involved in sclerotization. Experiments using $DDC^{\Delta(-208,-59)}$ flies, which are homologous for a deletion that eliminates DDC expression in the CNS, showed the absence of the 2.3-kb mRNA in the CNS (Morgan et al., 1986). These findings have been corroborated at the enzyme protein level by HPLC chromatography, which separates two major DDC peaks—one somewhat heterogenous, stemming from the CNS, and one from the integument. The aforementioned strain that carries the gene lacking the sequences required for expression in the CNS ($DDC^{\Delta(-208,-59)}$) shows on HPLC chromatography only the second peak, i.e., that of the integument enzyme (Fig. 5.17), whereas mutant flies, transformed with a gene that confers normally regulated DDC expression, exhibit a normal DDC elution profile.

5.7.2.3. REGULATION OF DOPA DECARBOXYLASE ACTIVITY AT ECLOSION OF THE IMAGO

Shortly before eclosion of *C. vicina* (Sekeris, 1964; Shaaya and Sekeris, 1965) and of *D. melanogaster* (Kraminsky et al., 1980), a new DDC peak appears, at a period in which the MH titer has decreased to low or undetectable values. A first explanation of this apparent discrepancy, when compared to DDC dependence on MH titer observed during pupariation, was that the DDC gene is more sensitive to MH during the last days of the pharate imago than at pupariation (Sekeris, 1964). The work of Hiruma and Riddiford (1984, 1985) and Hiruma et al. (1985) on *Manduca sexta* showed that the appearance of the DDC peak during the last half of the last larval molt, characterized by falling MH titer, is prevented by infusion of MH into the larvae, implying that the fall of MH is necessary for DDC induction (see also Fristrom et al., 1982; Section 5.7.1, above). However, it is the MH peak, which precedes DDC induction and which induces the molt, that determines the subsequent induction of DDC. A very interesting finding was that presence of JH 4–8 h after head capsule slippage, during the period of maximum MH titer, completely prevented DDC activity. *In vitro* experiments with epidermis from allatectomized larvae explanted 7 h before the onset of melanization corroborated the *in vivo* inhibitory effects of MH on DDC: in the

---

from wild-type flies (B), or heads from $DDC^{\Delta(-208,-59)}$ flies (C) were bound to the column and eluted with a linear salt gradient. DDC activity in cpm per assay (●—●) and ionic strength (●---●) were measured for each fraction. Values are plotted as a function of the fraction number. (From Morgan et al., 1986.)

presence of MH, the observed increase of DDC in the explanted epidermis was suppressed. Similarly, α-amanitin and cycloximide also suppress DDC appearance, strongly suggesting that the increase of enzyme activity is owing to *de novo* synthesis of protein. Thus, MH determines a particular pattern of gene expression for a later development stage, which can be elicited when the hormone titer declines. JH influences only the level of the subsequent synthesis, not the presence or absence of DDC. As Hiruma and Riddiford (1985) point out, this type of hormonal effect and the already studied direct effect of MH on DDC transcription, can be tested in *Drosophila* and *Callipora* (and in other Diptera, e.g., *Ceratitis capitata*) (G. Thalassinos and E. G. Fragoulis, cited in Sekeris and Fragoulis, 1985), where both types of control of DDC could be effective. On these lines, Hodgetts and coworkers (see Clark et al., 1986; Hodgetts et al., 1986a,b) found that isolated imaginal disks cultured in the continuous presence of MH did not accumulate DDC–mRNA nor showed increased enzyme activity. However, exposure of the imaginal disks to MH followed by culture in MH-free medium resulted in appearance of DDC transcript and activity 6 h after hormone withdrawal.

On the basis of the above findings, we can conclude that expression of the 2.1-kb mRNA coding for the epidermis DDC during pupariation is dependent on high MH titers, whereas in other developmental stages (e.g., larval molts; eclosion of the imago) its appearance is inhibited by MH but needs a preceding burst of MH in order to be inducible later on by still unknown factors.

5.7.2.4. MOLECULAR MECHANISM OF DOPA DECARBOXYLASE GENE REGULATION

Knowledge of the primary structure of the gene, in particular its flanking sequences, and of factors associated with putative regulatory sequences will no doubt advance our understanding of the mechanism of DDC regulation.

Eveleth et al. (1986) defined a sequence beginning 540 bases upstream of the DDC gene (ATGAAAATAATGCCTTT) having strong homology to a sequence (ATGGAAA-3-TACCTTT) similarly located on the hormonally regulated SGs-4 gene encoding the glue polypeptide sgs-4 (Muskavitch and Hogness, 1982) and to the sequence TTTGCAT-13-ATGGAAC of the puff 74EF–DNA, which is similarly MH regulated. The homology to the canonical enhancer sequence of mammalian viruses [Weiher et al., 1983: (G)TGGAA(G)], in particular that of the altered DDC+4 overproducing variant of Estelle and Hodgetts (1984a,b) and Hodgetts et al. (1986a,b) cannot be overlooked (see below). In addition, the DDC gene at position 846 contains the TGTTCT sequence, which is characteristic of the glucocorticoid-responsive elements found in glucocorticoid inducible genes (Scheidereit et al., 1983).

Estelle and Hodgetts (1984a,b) studied the DDC+4 *Drosophila* mutant, which has 20% more DDC activity at adult eclosion than the wild-type animal and underproduces DDC at pupariation. They showed the overproduction segregated with the second chromosome and mapped to a position within 0.15 map units from the DDC structural gene. Interestingly, the genetic element responsible for underproduction mapped at an identical position to that causing overproduction. The authors suggested that alteration in a single genetic element, lying in proximity to the structural gene, is responsible for the variation in DDC seen at different stages. Using the P-element–mediated genetic transformation (Rubin and Spradling, 1982; Scholnick et al., 1983), Chen and Hodgetts (1987) demonstrated patterns of DDC expression characteristic of the strain from which the transforming DNA had originally been derived and concluded that the essential information of the variant DDC+4 phenotype was included in a fragment that extended 2.9 kb upstream of the cap site of the DDC–mRNA and 0.9 kb downstream of the poly(A) addition site. As regards tissue-specific expression of DDC, Scholnick et al. (1986) have demonstrated multiple regulatory elements between −106 and −33 necessary for normal hypodermal expression and one between −83 and −59 required for normal expression in the CNS. No sequences upstream of −208 are necessary for normal levels of expression in the hypodermis and CNS or for the normal developmentally regulated DDC expression (Hirsh et al., 1986). Beall and Hirsh (1987) have recently defined a *cis* 16-bp (base pairs) regulatory element (I) localized between −83 and −59 necessary for normal neuronal expression and demonstrated the presence of a factor in embryonic nuclear extracts specifically protecting element I in DNase I footprinting assays (Bray et al., 1988). At least one additional regulatory element is required for normal neuronal but not for glial expression of DDC (Beall and Hirsh, 1987), located 800–2200 bp upstream, whereas element I is required for expression in both cell types. These studies, which are currently being intensively pursued in the aforementioned laboratories, will certainly shed light on the still unknown mechanisms involved in the differential expression and splicing of the DDC gene in the integument and in the brain, and on the hormonal and developmental stimuli differentially affecting expression of DDC in the epidermis tissue during the life cycle of the insect.

5.7.2.5. DOPAMINE N-ACETYLTRANSFERASE

Maranda and Hodgetts (1977) have followed *N*-acetyltransferase activity during development of *D. melanogaster*. They demonstrated considerable activity in second- and third-instar larvae, a maximum at pupariation, very little or no activity in the pupal and pharate adult stage, and an increase during the first day following eclosion to levels comparable to these seen at pupariation. Marsh and Wright (1980) also found marked changes in *N*-acetyltransferase activity during *Drosophila* development, which, however, was not correlated to MH titers (see also Schloerer et al., 1970).

Owing to the distribution of enzyme activity in more than one tissue, it would be worthwhile to follow during development the enzyme activity in the various tissues. It is possible that, in analogy to DDC, more than one *N*-acetyltransferase is present, each with its own tissue distribution and hormonal regulation.

5.7.3. *MH and the Phenoloxidase System*

The characteristics of the phenoloxidase system have already been described in Section 5.5.1.4. The possibility discussed was that the blood phenoloxidase may be a different enzyme than that of the cuticle. Furthermore, the cuticle appears to contain two or three different enzymes, although the lack of a thorough biochemical characterization of the enzymes should lead to caution in accepting multiple enzyme entities in the cuticle.

Phenoloxidase was the first enzyme system involved in tyrosine metabolism to be studied from the aspect of hormonal regulation. Due, in part, to its complexity, however, it is not yet possible to definitely state which, if any, of its components is under the direct control of the MH. In most of the studies (see Section 5.5.1.4, above), an increase in the activity of phenoloxidase in hemolymph or in the cuticle has been noted during ecdysis and sclerotization, either during the larval to pupal or the pupal to imaginal transition, suggesting hormonal control of the enzyme. Next, some of the relevant papers will be reviewed in which, directly or indirectly, correlations between MH and components of the phenoloxidase system have been presented.

One of the early studies dealt with the phenoloxidase system of the blow fly. In Fig. 5.18, the developmental appearance of phenoloxidase during the late third larval instar up to the early posteclosion period of *Calliphora* is shown, in relation to MH titer (Shaaya and Sekeris, 1965). During the third instar, total extractable phenoloxidase activity is high, showing a small peak at pupariation and a sharp decline thereafter (as regards the *Drosophila* enzyme, see also Ohnishi, 1953; Mitchell, 1966). During the pharate imago stage enzyme activity is almost undetectable. One day before eclosion of the imago, phenoloxidase activity reappears, attaining levels approximately 25% of those during pupariation, falling again to background levels a day after eclosion.

In blow fly larvae, Karlson and Schweiger (1961) followed the activity of total extractable prophenoloxidase and of total extractable activator during development. They found a low concentration of both proenzyme and activator in early third-instar larvae, both activities increas-

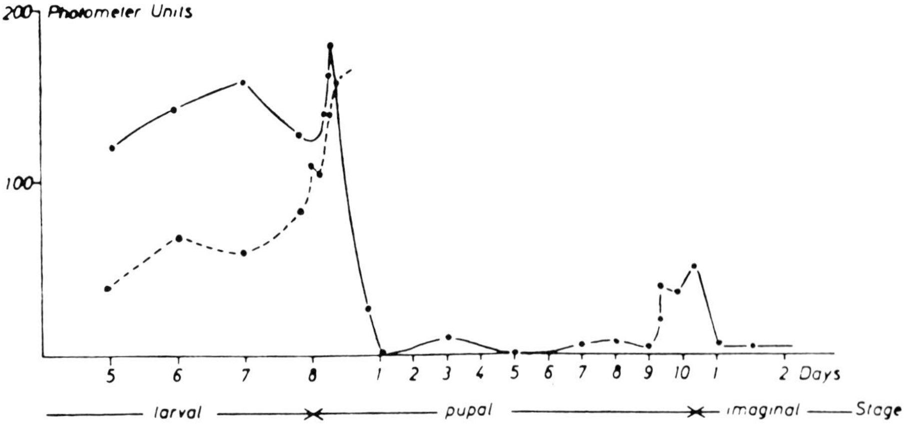

FIGURE 5.18. Phenoloxidase activity in the whole body during development of *Calliphora erythrocephala*. Phenoloxidase activity, expressed in photometer units, is plotted against days of insect life: ●---● = monophenoloxidase activity; ●—● = diphenoloxidase activity. (From Shaaya and Sekeris, 1965.)

ing thereafter, reaching a maximum before pupariation and then abruptly declining (Fig. 5.19A,B). Somewhat different curves are obtained if the activity of the hemolymph prophenoloxidase of the cuticular phenoloxidase and of the activator of the cuticle are separately titrated (Shaaya and Sekeris, 1965) (Fig. 5.20A,B). The concentration of proenzyme in the hemolymph reaches maximal levels at days 6–7, then falls to background levels; the cuticle phenoloxidase shows an increase in late third-instar larvae with a small peak before pupariation; the activator, although present in 6- to 7-day-old larval cuticles, shows a discernible peak shortly before pupariation, correlating with the increased MH titer observed at that time period (see below). Similarly, Hackman and Goldberg (1967) have shown that the cuticle phenoloxidase of *Lucilia cuprina* increases rapidly as the larvae approaches the time for puparium formation, but enzyme activity falls to zero just when puparium formation begins (Table 5.7). The increase in total phenoloxidase activity of the larvae and the decrease of enzyme activity in the hemolymph has been interpreted as a translocation of phenoloxidase from the hemolymph to the cuticle. Measurement, however, of total phenoloxidase activity in the cuticle showed an increase during the prepupal period, but not of such intensity as to account for the amount of prophenoloxidase disappearing from the hemolymph (see Fig. 5.19A,B). The difficulty of extracting the phenoloxidase from cuticles during the period of sclerotization should be kept in mind when one is interpretating such experiments.

In another fly, the flesh fly *Sarcophaga bullata*, Hughes and Price

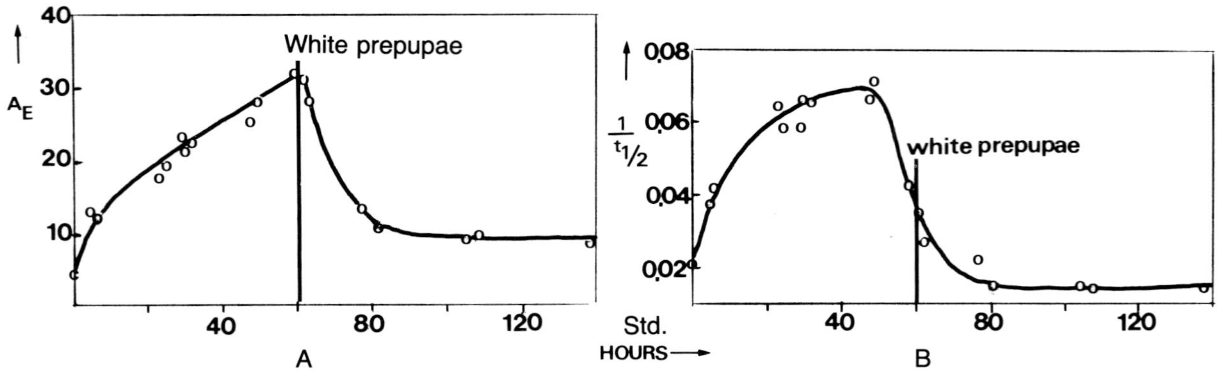

FIGURE 5.19. Change of activity of (A) prophenoloxidase and (B) activator of developing third-instar *Calliphora vicina* larvae. The abscissa denotes time period in hours after the larvae have left the meat; the ordinate denotes activity (A) in dopa units (B) in reverse half activation period in min^{-1}). (From Karlson and Schweiger, 1961.)

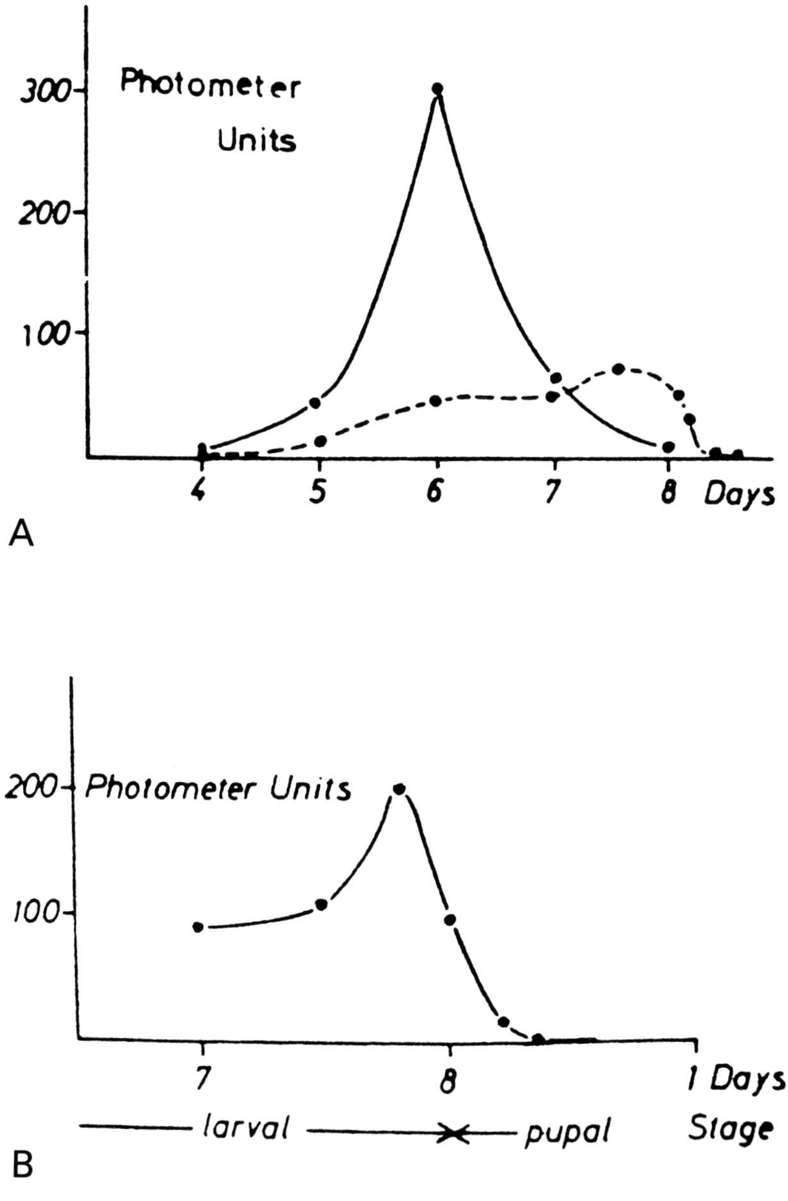

FIGURE 5.20. Activity of (A) prophenoloxidase in hemolymph and cuticle and (B) activator in cuticle of developing blow fly larvae. (A) Prophenoloxidase activity expressed in photometer units, is plotted against days of larval life: ●---● = activity in the cuticle; ●—● = activity in the hemolymph. (B) Activity of activator, expressed in photometer units, is plotted against days of larval life.

TABLE 5.7. *o*-Diphenoloxidase in the Cuticle of *Lucilia cuprina* Larvae During Development[a]

Time after migration from food (h)	Absorption of oxygen, initial velocity (μl/10 min per 3 cuticles)
0–5	17.9
5–11	25.9
11–17	42.2
17–23	54.0
23–29	60.8
29–35	66.9
35–41[b]	65.3
41–47	62.0
Just-formed puparia	0.8

SOURCE: From Hackman and Goldberg (1967).
[a]*o*-Diphenoloxidase activity was assayed by oxygen absorption with conventional Warburg techniques and dopamine as substrate.
[b]Puparia present.

(1975a,b) followed both prophenoloxidase and active phenoloxidase activity of hemolymph in 3- to 8-day-old larvae and during pupariation. They found (Fig. 5.21) minimal concentrations of active phenoloxidase in 3- to 8-day-old larvae that increased dramatically during pupariation, whereas the amount of proenzyme increased progressively in 3- to 7-day-old larvae, reaching a plateau at days 7–8. At pupariation, all of the hemolymph prophenoloxidase has been activated. This could be owing to increased activator concentrations, a decrease in the amount of the inhibitor of activation in the hemolymph (see Section 5.5.1.4, above), or an increase in the activator concentration of the cuticle, which could then diffuse to the hemolymph. Hughes and Price (1975a,b) found that the titer of total phenoloxidase activity remained fairly constant during the third larval instar but fell at the rounded off white puparium stage [similar findings to those of Hackman and Goldberg (1967) for *Lucilia*]. The question therefore arises as to the fate of the hemolymph phenoloxidase in *Sarcophaga*. As in *Calliphora*, the possibility that the enzyme enters the cuticle to take part in the sclerotization process during pupariation should be considered. Hughes and Price (1975a,b) report a constant level of cuticular enzyme in 3- to 8-day-old larvae, but they do not give data as to what happens during pupariation.

The possible dependence of either the phenoloxidase or the activator

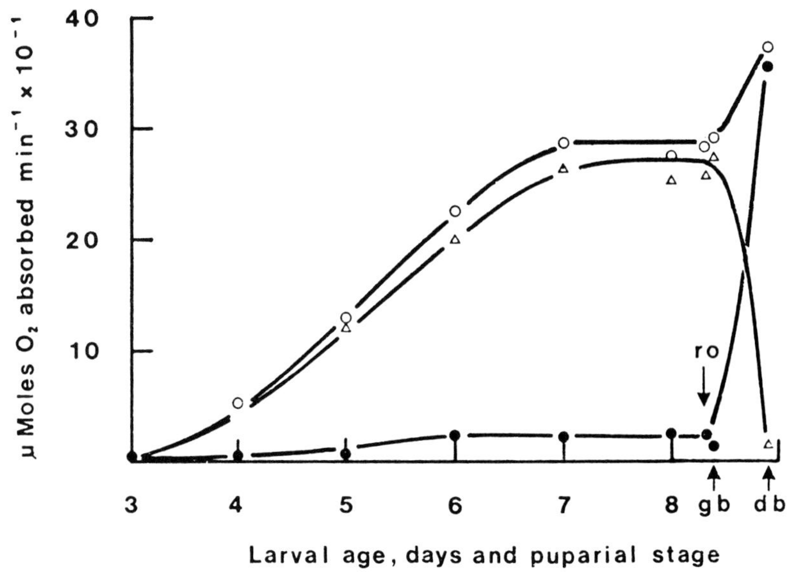

FIGURE 5.21. Level of proenzyme (△—△) and of active enzyme (●—●) in plasma fractions during the third larval instar and early puparial stages of *Sarcophaga bullata*. The other curve (○—○) is the sum of the activities present in all the fractions after activation. *Key:* db = dark-brown puparial stage; gb = golden-brown puparial stage; ro = rounded-off white puparial stage. (From Hughes and Price, 1975a.)

on MH was evaluated by Karlson and Schweiger (1961) in *Dauernlarven* of *C. vicina,* i.e., in third-instar larvae whose ring gland, the site of production of MH, was destroyed by thermocauterization and which retain their larval characteristics many days after the operation. In these animals, prophenoloxidase activity did not appreciably change even weeks after the operation, whereas the activator titer dropped to zero values 2 weeks after destruction of the ring gland, suggesting a dependence only of the latter component on MH. Injection of MH to these larvae at the period of zero activator activity restores within 24 h the level of the activator to initial preoperation values (Fig. 5.22). After 48 h the activator titer is even higher.

Although these experiments strongly suggest induction of activator by MH, they have not been further pursued on a molecular level, something that should indeed be done. Important in this respect is the biochemical characterization of the activator and of the other factors involved in the activation process, as well as—among others—

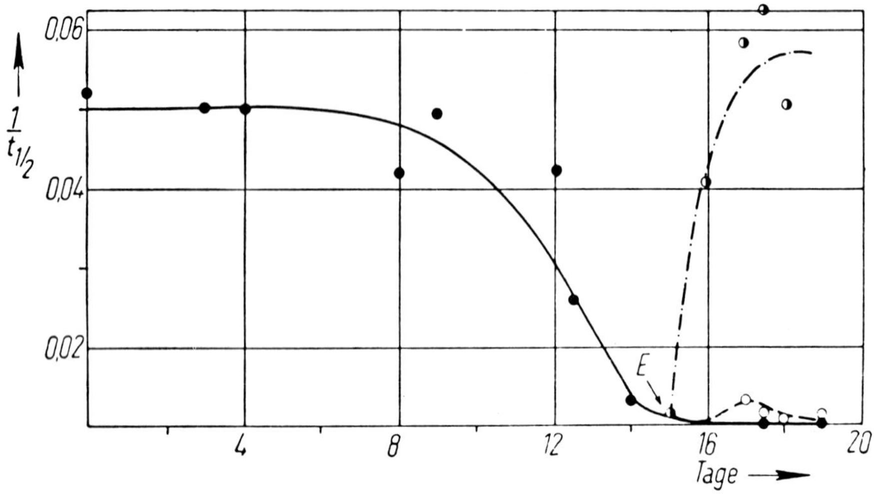

FIGURE 5.22. Activator and prophenoloxidase of *C. vicina* larvae after cauterization of the ring gland and subsequent MH injection. *Ordinate:* reverse half activation period in min^{-1}. *Abscissa:* time (in days) after ring gland cauterization. *Key:* E = time of MH injection; ● = without MH; ◑ = after MH; ○ = water-injected control. (After Karlson and Schweiger, 1961.)

understanding of the relation of the cuticular to the hemolymph activator and recognition of the site of synthesis of the proteins.

5.7.3.1. MH AND THE PHENOLOXIDASE SYSTEM AT ECLOSION OF THE IMAGO

As seen from Fig. 5.18, a day before eclosion of the imago, phenoloxidase activity appears again. Shaaya and Sekeris (1965) have suggested that the disappearance of phenoloxidase activity during the pupal and pharate imago stage is due to lack of both prephenoloxidase and activator. The increase, therefore, of phenoloxidase activity at eclosion denotes, very probably, a *de novo* synthesis of both components. Owing to the complexity of the system, many other possibilities might be considered, such as removal of the inhibitors or activation of the components. As the increase of phenoloxidase activity occurs before eclosion, bursicon (see Section 5.7) cannot be responsible for the effect. As regards the titer of MH at the period of appearance of phenoloxidase activity, it is very low or undetectable. As already noted in connection with cuticle deposition (Fristrom et al., 1982) and with the appearance of DDC activity at eclosion (Hiruma and Riddiford, 1985), a preceding MH peak is responsible for reprogramming of the pattern of protein synthesis; furthermore, we have seen that the ensuing decrease of MH permits the switching on of new genes. The same should hold true for the possible induction of the phenoloxidase system at eclosion. Hiruma et al. (1985) have shown that the activation of prophenoloxidase in the premelanine granules of the larval cuticle of *Manduca sexta* is inhibited by MH.

5.7.3.2. MH AND LACCASE

Another enzyme that appears to be involved in sclerotization is laccase, isolated from cuticles of various insect species. In Section 5.5.1.4 some papers showing developmental profiles of laccase have been reviewed, suggesting MH dependence of this enzyme. I should like to stress the paper by Barrett (1987), who followed the activities of phenoloxidase (enzyme A) and the laccase (enzyme B) from cuticles of *L. cuprina* during the wandering stage, up to the final instar and pupation. As seen from Fig. 5.23, laccase activity shows a developmental curve parallel to the curve of MH titer, whereas phenoloxidase shows high activity throughout. Both enzyme activities A and B abruptly decrease at pupation, perhaps due to cross-linking by the sclerotizing agents. As both enzymes A and B have been purified, this system should, in the near future, be the object of more detailed studies as regards regulation by MH.

5.7.4. *Other Enzymes Involved in Tyrosine Metabolism and Their Possible Control by MH*

A series of hydrolases releasing phenol substances from their conjugates seem to be affected by MH. In the following subsections, we will consider respectively the glycosidases, the phosphatases and sulfatases, and the dipeptides.

5.7.4.1. GLYCOSIDASES

A series of glycosides isolated from insects serve as storage forms for many of the phenolic substances involved in sclerotization, such as tyrosine, dopa, dopamine, *N*-acetyldopamine, *N*-β-alanyldopamine, and phenolic carboxylic acids (P. S. Chen et al., 1978; Kramer et al., 1980; Ahmed et al., 1983a,b; Psarianos et al., 1985; Sekeris and Karlson, 1962; Sekeris, 1964; Brunet and Kent, 1955a,b; Kent and Brunet, 1959; see also Brunet, 1980).

The timing of the release of the phenolic component is coordinated in such a way as to enable it to participate in the sclerotization process, either as a precursor or as the immediate tanning agent. It was, therefore, natural to follow the effects of MH on the enzymes involved in glycoside hydrolysis.

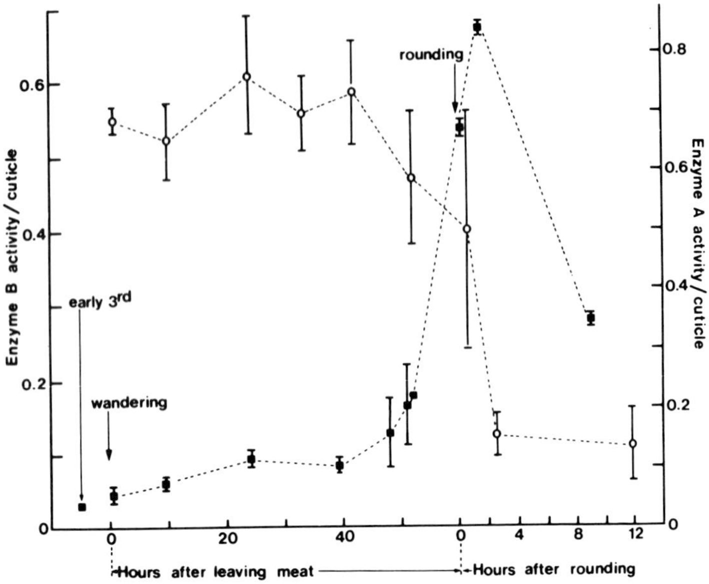

FIGURE 5.23. Changes in activity of enzymes A and B in intact cuticle of *Lucilia cuprina* during the third instar and following the onset of puparium formation. Values for enzyme A (open circles) are means + SD of four determinations (n = 2 for 12 h after rounding) using three cuticles for assay following 3-h preincubation at pH 8.9. Values for enzyme B (closed squares) are means + SD of five determinations using five cuticles per assay following 2- to 3-h preincubation at pH 8.9. All values corrected for low levels of activity present in the absence of added substrate. (From Barrrett, 1987.)

One well-studied case is that of the tyrosine-*O*-glucoside found in *M. sexta* (Ahmed et al., 1983a,b). The concentration of the glucoside has been followed in various developmental stages (Fig. 5.24). The glucoside appears in 2-day-old fifth-instar larvae and reaches maximum concentrations in pharate pupae. Shortly before larval–pupal ecdysis, the glucoside titer falls, reaching low levels during tanning. The glucoside titer remains low during pharate adult development. Some days before eclosion glucoside concentration gradually increases and falls to very low levels after eclosion and during the sclerotization of the adult cuticle.

A similar developmental curve to that of the tyrosine glucoside has been shown for the *O*-glucoside of *N*-acetyldopamine in the blow fly (Sekeris, 1964, and unpublished data). Sekeris found that the activity of

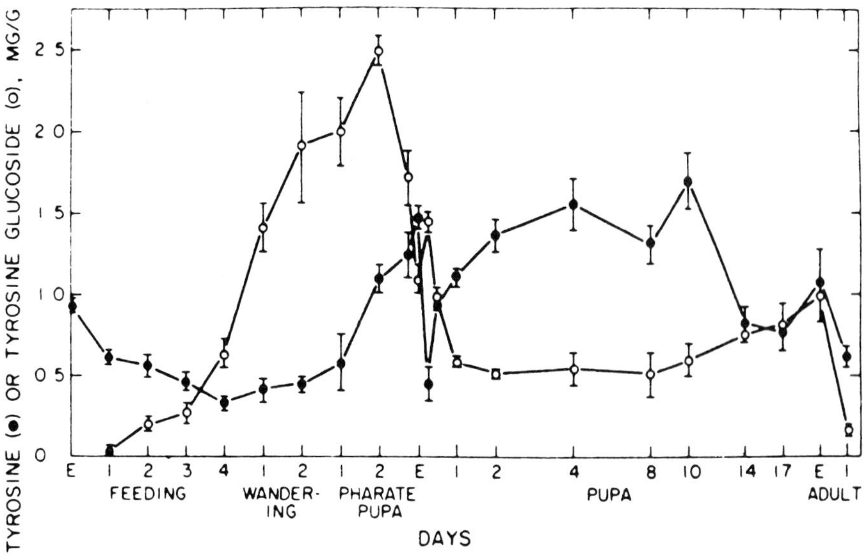

FIGURE 5.24. Titers of tyrosine and tyrosine glucoside in the whole body during the development of *Manduca sexta*. *Key:* bars = SEM; n = 3–6; E = ecdysis. (From Ahmed et al., 1983a.)

the transglucosidase responsible for formation of the glucoside remains constant throughout development, the glucoside formed depending on the amount of available phenolic agent. Thus, the changing titers of the glucoside apparently reflect the activity of the glucosidase, which seems to be under hormonal control.

Ahmed et al. (1983b) found that the activity of the hydrolase of the tyrosine glucoside, localized in the fat body of *M. sexta,* starts to increase a few hours before pupal ecdysis, at which period MH titers are very high (Bollenbacher et al., 1981) (Fig. 5.25) Furthermore, injection of ecdysone to ligated abdomens of wandering larvae of *M. sexta* led to decrease of tyrosine glucoside concentration in the hemolymph, with concomitant increase of tyrosine (Fig. 5.26A,B) (Ahmed et al., 1983b). As the site of synthesis and hydrolysis of the glucoside is the fat body (Ahmed et al., 1983a), transport of the glucoside or of the respective phenol between the fat body and the hemolymph must take place also in a coordinated way, and in synchrony with all processes involved in sclerotization.

5.7.4.2. PHOSPHATASES AND SULFATASES

Mitchell and coworkers (1960; also Mitchell and Lunan, 1964) reported the existence of tyrosine-4-*O*-phosphate in *Drosophila* and Seligman et al.

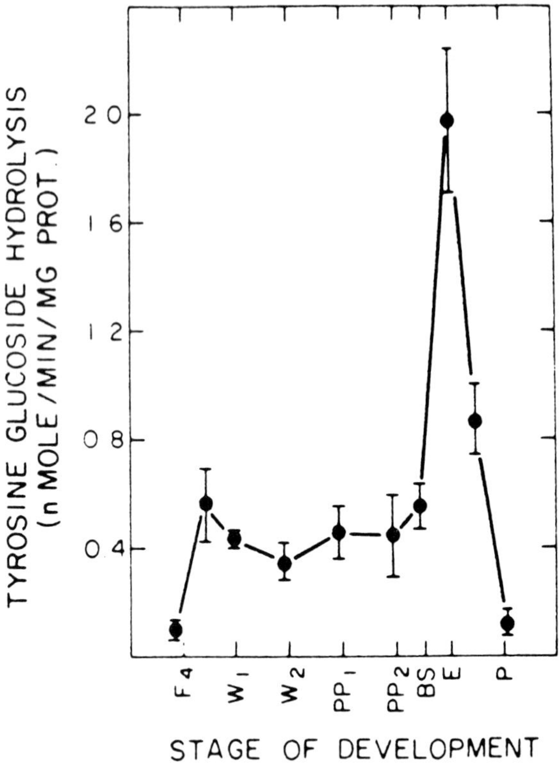

FIGURE 5.25. Tyrosine glucoside hydrolase activity in fat body tissue during larval–pupal development of *M. sexta*. *Key:* F = feeding larva; W = wandering larva; PP = pharate pupa; BS = brown thoracic bar stage; E = pupal ecdysis; P = pupa; bars = SEM; *n* = 2–3. (From Ahmed et al., 1983b.)

(1969a,b) demonstrated its presence in *Sarcophaga*. The concentration of this ester increases in growing larvae with a maximum at purariation and a rapid fall thereafter. As with the glucoside of tyrosine, the phosphate ester is regarded as a storage form serving a dual role: making tyrosine unavailable for the phenoloxidase and rendering the amino acid water soluble, thus permitting the high concentrations of tyrosine seen before pupation (Mitchell and Lunan, 1964; Bodnaryk et al., 1974; Sloley and Downer, 1987).

N-Acetyldopamine 3-*O*-phosphate, dopamine 3-*O*-sulfate, and *N*-acetyldopamine 3-*O*-sulfate (Bodnaryk et al., 1974) were identified in the cockroach *Periplaneta americana* and their role in the sclerotization

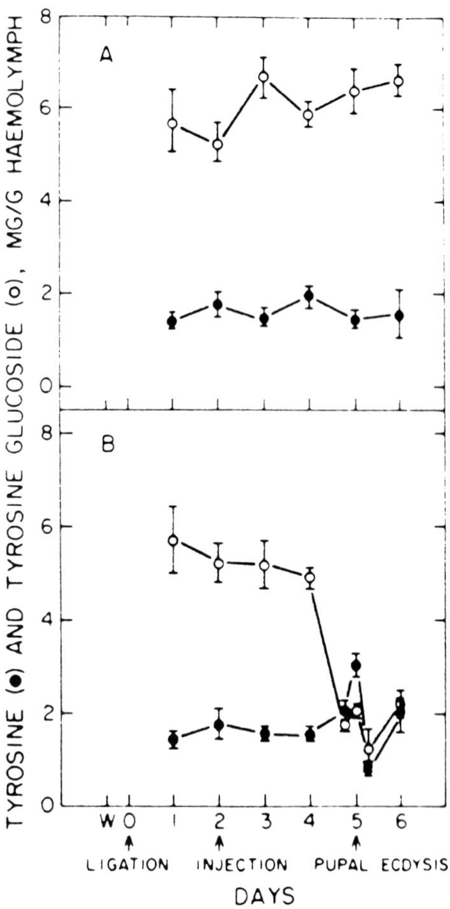

FIGURE 5.26. The effect of 20-hydroxyecdysone on tyrosine and tyrosine glucoside titers in hemolymph of isolated larval abdomens of *M. sexta:* (A) solvent-injected control; (B) abdomens injected with 20-hydroxyecdysone. *Key:* W = wandering stage; bars = SEM; $n = 3$–6. (From Ahmed et al., 1983b.)

process determined. The crucial finding was the injection of the aforementioned compounds, 1-^{14}C, ^{32}P, or ^{35}S labeled, in newly ecdysed larvae led after 7–12 days to incorporation into the sclerotized cuticle only of the ^{14}C radioactivity and not that of the phosphate or sulfate, indicating that prior to incorporation the phosphate and sulphate esters are hydrolyzed (Bodnaryk et al., 1974; Bodnaryk and Brunet, 1974). The role of the phosphate or of the sulfate group has not yet been elucidated. A transport function was proposed by Bodnaryk et al. (1974), taking into account the findings of Koeppe and Gilbert (1973). They found binding of dopamine or its metabolites to blood proteins of *P. americana,* which then entered the cuticle during sclerotization (see also Koeppe and Gilbert, 1974), carrying the phenolic agents. As in other cases in which a hydroxyl group of a diphenol is blocked, a protective role against oxidation by the omnipresent phenoloxidase(s) should also be considered.

As the concentration of these esters decreases abruptly during sclerotization, at the period of high MH titer, a possible role of ecdysone in controlling the respective esterases is plausible but has yet to be proven.

5.7.4.3. DIPEPTIDASES

Two peptides, β-alanyl-L-tyrosine and γ-glutamyl-L-phenylalanine, have been isolated from *Sarcophaga bullata* and *Musca domestica,* respectively (Levenbook et al., 1969; Bodnaryk, 1970a). All members of the genus *Sarcophaga* possess the tyrosine-containing dipeptide, and all of the genus *Musca,* the phenylalanine dipeptide (Bodnaryk, 1972). Although, peptides had been detected as constituents of insects, the two aforementioned dipeptides have characteristics differentiating them from the previously described ones. Their concentration increases continuously during larval growth, reaching maximal values before the onset of pupariation, and then decreases abruptly at the white pupa stage to reach almost undetectable levels approximately 12 h after initiation of sclerotization (see Fig. 5.27A,B). As both peptides contain an amino acid that can serve as a precursor to tanning agents, Levenbook et al. (1969) and Bodnaryk (1970a) have considered the role of these dipeptides to be as storage forms of amino acids to be released and used during pupariation. The question as to why a storage form of these two amino acids would be needed has been discussed (Bodnaryk, 1970a). As far as the tyrosine dipeptide is concerned, it is far more soluble in water than is tyrosine (α-alanyl-tyrosine is not), and its amount in hemolymph exceeds the amount of free tyrosine, which could be dissolved by two orders of magnitude (Bodnaryk and Levenbook, 1969). The solubility factor does not seem to play a role as regards γ-glutamyl-phenylalanine, for its concentration in hemolymph does not exceed the solubility of free phenylalanine.

The strongest argument for the participation of MH in the control of dipeptide metabolism and for the role of dipeptides in the sclerotization process are the results of ligation experiments (Bodnaryk, 1970a). Ligation of third-instar house fly larvae prior to the critical period, i.e., that of release of ecdysone from the ring gland, and analysis of the posterior nonpupated part of the animal revealed that the concentration of the dipeptide remained at the same level as that of larvae prior to ligation. This implies that MH is involved in the mobilization of phenylalanine, very probably by inducing a specific dipeptidase. The levels of glutamic acid and phenylalanine during house fly pupation do not increase, whereas that of tyrosine decreases. It is still not possible from such data to conclude in what form the dipeptide is further processed. However, data concerning β-alanyltyrosine in *S. bullata* (Bodnaryk and Levenbook, 1969) suggest that the dipeptide is hydrolyzed to its individual components, which are then incorporated into the puparium. Thus, it is logical

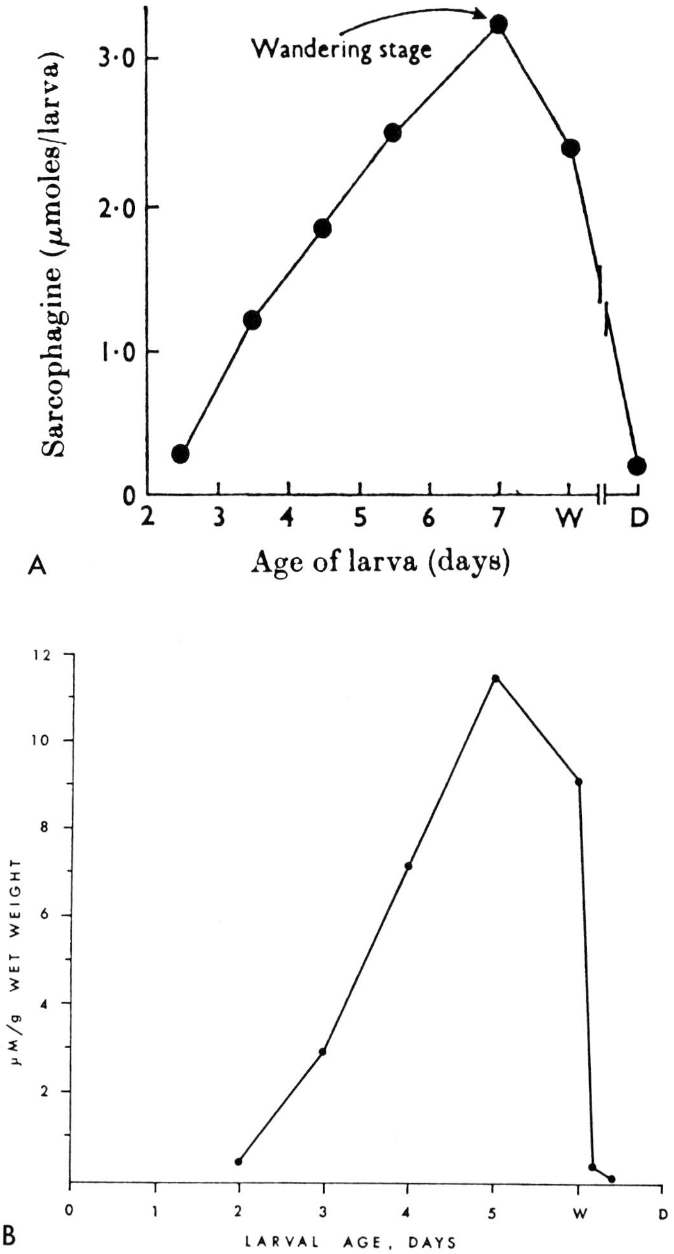

FIGURE 5.27. Changes in (A) β-alanyl-L-tyrosine and (B) γ-L-glutamyl-L phenylalamine content during larval growth and pupation of *Sarcophaga bullata* and *Musca domestica*, respectively: (A) D = pupa, 12 h; W = white pupa. (From Levenbook et al., 1969.) (B) D = 24 h after initiation of tanning; W = white pupa. (From Bodnaryk, 1970c.)

to assume that the released phenylalanine is rapidly hydroxylated to tyrosine by the phenylalanine hydroxylase, which at that time period shows maximal activity (Schloerer et al., 1970: experiments with blow fly larvae) and that tyrosine then also rapidly enters the pathway leading to formation of the sclerotizing agent.

5.8. Epilogue

A series of metabolic transformations and anabolic processes are required for sclerotization, some of which are controlled by MH. It was not possible to critically and in depth discuss all of these MH-regulated steps, nor was it feasible to elaborate on the many facets of sclerotization; much greater attention was focused on puparium formation, a special case of cuticle tanning restricted to cyclorrhaphous Diptera, compared with larval and imaginal postecdysial sclerotization processes. Of all MH-controlled metabolic steps, the best studied is induction of DDC during pupariation. The rapid progress in cloning of the DDC gene, combined with extensive genetic studies, have opened this field to molecular analysis. Provided that the MH receptor and gene are purified and characterized, the DDC system will contribute significantly to elucidating the molecular mechanism of action of steroid hormones. Intriguing is the reverse relationship of MH titer to DDC induction in all developmental stages with the exception of pupariation. It should be interesting to identify the factor(s) that are the immediate inducers of DDC at these stages. Cloning of the DDC gene has also led to understanding the existence of two DDC species—one in the epidermal cells, the other in the CNS. The two enzymes result from the differential splicing of the DDC gene, and knowledge is beginning to accumulate on the cause of the differential expression and MH control of the epidermis and CNS–DDC. The availability of insect cell lines demonstrating DDC induction by MH should facilitate future relevant studies (Swiderski and O'Connor, 1986).

Another enzyme system that has also received considerable attention is phenoloxidase. Earlier work demonstrated an effect of MH on the activator of phenoloxidase. These studies, hampered by inadequate data on the characteristics of the several components involved in tyrosine oxidation, have for a long time been discontinued, but they should certainly be started again, taking into account the recent progress in the purification of the protein activator(s) of the phenoloxidase.

The identification of a laccase in the cuticle whose activity shows a remarkable parallelism to MH titer should also focus interest on this system. As already mentioned, more basic biochemical work is needed to investigate the specific effects of MH on the cuticular enzymes. Effects of MH on enzymes hydrolyzing conjugated forms of the sclerotizing agent or its precursors have also been demonstrated. These are potentially interesting systems, which also need more fundamental biochemical work to make them appealing to molecular analysis.

A field of explosive progress and current interest is MH control of cuticular protein gene expression. The change in the pattern of cuticular proteins during development has attracted to this highly promising field the attention of molecular biologists interested in the control of expression of multigene families. As a result, cuticular proteins are now beginning to be characterized in terms of primary structure, which, in addition to significantly delineating MH regulation of gene expression, is a first step toward understanding the intricate details of the interaction of sclerotizing agents with cuticular proteins.

5.9. Summary

Sclerotization of insect cuticles results from interaction of certain diphenolic compounds mainly with the cuticular proteins. A prerequisite for the onset of this process is the availability of the sclerotizing agents and of the suitable protein substrate.

To this end a series of biosynthetic processes are activated at the appropriate time period, in part due to the action of MH. This hormone induces the synthesis of novel cuticular proteins encoded by multigene families, either by reprogramming polymorphic epidermal cells or by inducing the respective genes in cells newly drafted for cuticle protein synthesis, such as those of imaginal disks.

The biosynthesis of the principal sclerotizing agent, *N*-acetyldopamine, starts from tyrosine by way of hydroxylation to dopa, decarboxylation to dopamine, and *N*-acetylation. A similar pathway is very probably involved in the biosynthesis of another major sclerotizing agent, *N*-β-alanyldopamine.

"Activation" of the sclerotizing diphenols involves phenoloxidase(s) and laccase(s) of the cuticle. Although the literature is rich in publications concerning these enzymes, there is no definite picture as to which concrete enzyme(s) is (are) involved in this process or as to the exact mechanism (quinone tanning, β-sclerotization, or quinone methide sclerotization) of activation.

During pupariation, MH induces the *de novo* synthesis of DDC of the epidermal cells by way of gene activation. This enzyme is not induced by MH in other developmental periods in which DDC activity is increased; on the contrary, in these developmental stages MH acts as an anti-inducer. Cloning of the DDC gene has resulted in the recognition of regulatory regions necessary for MH induction and has also revealed alternative splicing reactions, explaining the presence of another DDC form in the CNS whose activity is not regulated by MH.

Phenoloxidases are present as inactive proenzymes, activated by a protein activator(s), localized mainly in the integument. There are indications from experiments with *Dauernlarven* that MH affects the "activity" of the protein activator, although the lack of biochemical information concerning the components of the highly complex phenoloxidase system has hampered progress in understanding the role of MH in this respect. The presence of a laccase in the cuticle, whose developmental appearance during metamorphosis shows parallelism to MH titer, makes this enzyme a possible target of regulation by MH.

Sclerotizing phenols or their precursors are stored as conjugated metabolites—glucosides, phosphates, sulfates, dipeptides—released at the appropriate time periods by the respective hydrolases, to be used for the sclerotization process. Mostly indirect evidence, e.g., comparison of metabolite to MH titers, suggests a dependence of the hydrolases on MH.

Acknowledgments

I thank Dr. E. Skantzo for invaluable help in the preparation of the manuscript, Ms. V. Delaki for help in preparing the references and indexes, and Ms. A. Hatzistili for competent secretarial assistance.

References

Ahmed, R. F., T. L. Hopkins, and K. J. Kramer. 1983a. Tyrosine and tyrosine glucoside titres in whole animals and tissues during development of the tobacco hornworm *Manduca Sexta* (L.). Insect Biochem. 13: 369–374.

Ahmed, R. F., T. L. Hopkins, and K. J. Kramer. 1983b. Tyrosine glucoside hydrolase activity in tissues of *Manduca Sexta* (L.): effect of 20-hydroxyecdysone. Insect Biochem. 13: 641–645.

Andersen, S. O. 1972. An enzyme from locust cuticle involved in the formation of cross-links from *N*-acetyldopamine. J. Insect Physiol. 18: 527–540.

Andersen, S. O. 1974. Evidence for two mechanisms of sclerotization in insect cuticle. Nature (Lond.) 251: 507–508.

Andersen, S. O. 1978. Characterization of a trypsin-solubilized phenoloxidase from locust cuticle. Insect Biochem. 8: 143–148.

Andersen, S. O. 1979a. Characterization of the sclerotization enzymes(s) in locust cuticle. Insect Biochem. 9: 233–239.

Andersen, S. O. 1979b. Biochemistry of insect cuticle. Annu. Rev. Entomol. 24: 29–61.

Andersen, S. O. 1980. Cuticular sclerotization. Pp. 185–215 *in* T. A. Miller (ed.), *Cuticle Techniques in Arthropods.* Springer-Verlag, Berlin and New York.

Andersen, S. O. 1985. Sclerotization and tanning of the cuticle. Pp. 59–74 *in* G. A. Kerkut and L. I. Gilbert (eds.), *Comprehensive Insect Phisiology, Biochemistry and Pharmacology,* Vol. 3. Pergamon Press, Oxford and Elmsford, New York.

Andersen, S. O. and F. M. Barrett. 1971. The isolation of ketocatecols from insect cuticle and their possible role in sclerotization. J. Insect. Physiol. 17: 69–83.

Andersen, S. O. and P. Højrup. 1987. Extractable proteins from abdominal cuticle of sexually mature locusts, *Locusta migratoria.* Insect Biochem. 17: 45–51.

Andersen. S. O. and P. Roepstorff. 1978. Phenolic compounds released by mild acid hydrolysis from sclerotized cuticle: purification, structure and possible origin from cross-links. Insect Biochem. 8: 99–104.

Andersen, S. O. and P. Roepstorff. 1982. Sclerotization of insect cuticle. III. An unsaturated derivative of *N*-acetyldopamine and its role in sclerotization. Insect Biochem. 12: 269–276.

Ashida, M. 1971. Purification and characterization of prophenoloxidase from hemolypmph of the silkworm *Bombyx mori*. Insect Biochem. 10: 37–47.

Ashida, M. and K. Dohke. 1980. Activation of prophenoloxidase by the activating enzyme of the silkworm, *Bombyx mori*. Insect Biochem. 10: 37–47.

Ashida, M. and E. Ohnishi. 1967. Activation of prophenol oxidase in hemolymph of the silkworm *Bombyx mori*. Arch. Biochem. Biophys. 122: 411–416.

Aso, Y., K. J. Kramer, T. L. Hopkins, and G. Lookhart. 1985. Characterization of haemolymph protyrosinase and a cuticular activator from *Manduca sexta*. Insect Biochem. 15: 9–17.

Aso, Y., K. J. Kramer, T. L. Hopkins, and S. Z. Whetzel. 1984. Properties of tyrosinase and DOPA quinone imine conversion factor from pharate pupal cuticle of *Manduca sexta* L. Insect Biochem. 14: 463–472.

Atkinson, P. W., W. V. Brown, and A. R. Gilby. 1973. Phenolic compounds from insect cuticle: identification of some lipid antioxidants. Insect Biochem. 3: 309–315.

Barrett, F. M. 1987. Phenoloxidases from larval cuticle of the sheep blow fly *Lucilia cuprina:* characterization, developmental changes and inhibition by antiphenoloxidase antibodies. Arch. Insect Biochem. Physiol. 5: 99–118.

Barrett, F. M. and S. O. Andersen. 1981. Phenoloxidase in larval cuticle of the blow fly, *Calliphora vicina*. Insect Biochem. 11: 17–24.

Beall, C. J. and J. Hirsh. 1984. High levels of intron-containing RNAs are associated with expression of the *Drosophila* DOPA decarboxylase gene. Mol. Cell. Biol. 4: 1669–1674.

Beall, C. J. and J. Hirsh. 1987. Regulation of the *Drosophila* DOPA decarboxylase gene in neuronal and glial cells. Genes 1: 510–520.

Bhagvat, K. and D. Richter. 1938. Animal phenolases and adrenaline. Biochem. J. 32: 1397–1406.

Bodine, J. H. and T. Allen. 1941. Enzymes in ontogenesis (Orthoptera). XV. Some properties of protyrosinases. J. Cell. Comp. Physiol. 18: 151–160.

Bodnaryk, R. P. 1970a. Effect of dopa-decarboxylase inhibition on the metabolism of β-alanyl-L-tyrosine during puparium formation in the fleshfly *Sarcophaga bullata* Parker. Comp. Biochem. Physiol. 35: 221–227.

Bodnaryk, R. P. 1970b. Biosynthesis of γ-L-glutamyl-L-phenylalanine by the larva of the housefly *Musca domestica*. J. Insect Physiol. 16: 919–929.

Bodnaryk, R. P. 1970c. Levels of free glutamic acid, phenylalanine and γ-glutamyl-phenylalanine during pupal sclerotization in the housefly *Musca domestica*. Comp. Biochem. Physiol. 35: 499–502.

Bodnaryk, R. P. 1972. A survey of the occurence of β-alanyl-tyrosine: γ-glutamyl-phenylalanine and tyrosine-*O*-phosphate in the larval stage of flies (Diptera). Comp. Biochem. Physiol. 43B: 587–592.

Bodnaryk, R. P. and P. C. J. Brunet. 1974. 3-*O*-Hydrosulphato-4-hydroxyphenethylamine (dopamine 3-*O*-sulphate), a metabolite involved in the sclerotization of insect cuticle. Biochem. J. 138: 463–469.

Bodnaryk, R. P., P. C. J. Brunet, and J. K. Koeppe. 1974. On the metabolism of *N*-acetyldopamine in *Periplaneta americana*. J. Insect Physiol. 20: 911–923.

Bodnaryk, R. P. and L. Levenbook. 1969. The role of β-alanyl-L-tyrosine (sarcophagine) in puparium formation in the fleshfly *Sarcophaga bullata*. Comp. Biochem. Physiol. 30: 909–921.

Bollenbacher, W. E., S. L. Smith, W. Goodman, and L. I. Gilbert. 1981. Ecdysteroid titer during larval–pupal–adult development of the tobacco hornworm *Manduca sexta*. Gen. Comp. Endocrinol. 44: 302–306.

Boyer, P. D. (ed.) 1975. *The Enzymes*, Vol. 12: *Oxidation–Reduction*. Academic Press, Orlando, Florida.

Bray, S. J., W. A. Johnson, J. Hirsh, U. Heberlein, and R. Tjian. 1988. A *cis*-acting element and associated binding factor required for CNS expression of the *D. melanogaster* dopa decarboxylase gene. EMBO (Eur. Mol. Biol. Org.) 7: 177–188.

Brunet, P. C. J. 1980. The metabolism of the aromatic amino acids concerned in the cross-linking of insect cuticle. Insect Biochem. 10: 467–500.

Brunet, P. C. J. and B. C. Coles. 1974. Tanned silks. Proc. Roy. Soc. Lond. B187: 133–170.

Brunet, P. C. J. and P. W. Kent. 1955a. Mechanism of sclerotin formation: the participation of a beta-glucoside. Nature (Lond.) 175: 819.

Brunet, P. C. J. and P. W. Kent. 1955b. Observations on the mechanism of a tanning reaction in *Periplaneta* and *Blatta*. Proc. R. Soc. Lond. B144: 259–274.

Butenandt, A., U. Gröschel, P. Karlson, and W. Zillig. 1959. Über *N*-acetyltyramin, seine Isolierung aus *Bombyx*-puppen und seine chemischen und biologischen Eigenschaften. Arch. Biochem. Biophys. 83: 77–83.

Chen, T. and R. B. Hodgetts. 1976. Some properties of dopa decarboxylase from *Sarcophaga bullata*. Comp. Biochem. Physiol. 53B: 415–418.

Chen, P. S., H. K. Mitchell, and M. Neueg. 1978. Tyrosine glucoside in *Drosophila buskii*. Insect Biochem. 8: 279–286.

Chen, Z.-A. and R. B. Hodgetts. 1987. Functional analysis of a naturally occurring variant dopa decarboxylase gene in *Drosophila melanogaster* using P element mediated germ line transformation. Mol & Gen. Genet. 207: 441–445.

Clark, W., P. Pass, B. Vankataraman, and R. Hodgetts. 1978. Dopa decarboxylase from *Drosophila melanogaster*. Mol. & Gen. Genet. 162: 287–297.

Clark, W. C., J. Doctor, J. W. Fristrom, and R. B. Hodgetts. 1986. Differential responses of the dopa decarboxylase gene to 20-OH-ecdysone in *Drosophila melanogaster*. Dev. Biol. 114: 141–150.

Clever, U. and P. Karlson. 1960. Induktion von Puff-Veränderungen in den Speicheldrüsen-chromosomen von *Chironomus tentans* durch Ecdyson. Exp. Cell Res. 20: 623–626.

Cottrell, C. B. 1964. Insect ecdysis with particular emphasis on cuticular hardening and darkening. Adv. Insect Physiol. 2: 175–218.

Dennell, R. 1947. A study of an insect cuticle: the formation of the puparium of *Sarcophaga falculata* Pand. (Diptera). Proc. R. Soc. Lond. B134: 79–110.

Dennell, R. 1958. The hardening of insect cuticles. Biol. Rev. 33: 178–196.

Dewhurst, S. A., S. G. Croker, K. Ikeda, and R. E. McCaman. 1972. Metabolism of biogenic amines in *Drosophila* nervous tissue. Comp. Biochem. Physiol. 43B: 975–981.

Doctor, J., D. Fristrom, and J. W. Fristrom. 1985. The pupal cuticle of *Drosophila*: biphasic synthesis of pupal cuticle proteins *in vivo* and *in vitro* in response to 20-hydroxyecdysone. J. Cell Biol. 101: 189–200.

Dohke, K. 1973a. Studies on pro-phenoloxidase-activating enzyme from cuticle of the silkworm *Bombyx mori*. I. Activation reaction by the enzyme. Arch. Biochem. Biophys. 1576: 203–209.

Dohke, K. 1973b. Studies on pro-phenoloxidase-activating enzyme from cuticle of the silkworm *Bombyx mori*. II. Purification and characterization of the enzyme. Arch. Biochem. Biophys. 157: 210–224.

Estelle, M. A. and R. B. Hodgetts. 1984a. Genetic elements near the structural gene modulate the level of dopa decarboxylase during *Drosophila* development. Mol. & Gen. Genet. 195: 434–441.

Estelle, M. A. and R. B. Hodgetts. 1984b. Insertion polymorphisms may cause specific variation in mRNA levels for dopa decarboxylase in *Drosophila*. Mol. & Gen. Genet. 195: 442–451.

Eveleth, D. D., Jr., R. D. Gietz, C. A. Spencer, F. F. Nargang, R. B. Hodgetts, and J. I. Marsh. 1986. Sequence and structure of the dopa decarboxylase gene of *Drosophila:* evidence for novel RNA splicing variants. EMBO (Eur. Mol. Biol. Org.) J. 5: 2663–2672.

Fragoulis, E. G. and C. E. Sekeris. 1975a. Purification and characteristics of DOPA decarboxylase from the integument of *Calliphora vicina* larvae. Arch. Biochem. Biophys. 168: 15–25.

Fragoulis, E. G. and C. E. Sekeris. 1975b. Induction of dopa (3,4-dihydroxyphenylalanine) decarboxylase in blow fly integument by ecdysone. Biochem. J. 146: 121–126.

Fragoulis, E. G. and C. E. Sekeris, 1975c. Translation of mRNA for 3,4-dihydroxyphenylalanine decarboxylase isolated from epidermis tissue of *Calliphora vicina* R.-D. in an heterologous system. Eur. J. Biochem. 51: 305–316.

Fristrom, J. W., W. R. Logan, and C. Murphy. 1973. The synthetic and minimal culture requirements for evagination of imaginal discs of *Drosophila melanogaster in vitro*. Dev. Biol. 33: 441–456.

Fristrom, J. W., J. Doctor, D. K. Fristrom, W. R. Logan, and D. J. Silvert. 1982. The formation of the pupal cuticle by *Drosophila* imaginal discs *in vitro*. Dev. Biol. 91: 337–350.

Garen, A., L. Kauvar, and J. A. Lepeasant. 1977. The role of ecdysone in *Drosophila* development. Proc. Natl. Acad. Sci. USA 74: 5009–5103.

Gietz, R. D. and R. B. Hodgetts. 1985. An analysis of dopa decarboxylase expression during embryogenesis in *Drosophila melanogaster*. Dev. Biol. 107: 142–155.

Hackman, R. H. 1974. Chemistry of insect cuticle. Pp. 216–270 *in* M. Rockstein (ed.), *The Physiology of Insecta*, Vol. 6. Academic Press, Orlando, Florida.

Hackman, R. H. 1976. The interactions of cuticle proteins and some comments on their adaption to function. Pp. 107–120 *in* H. R. Hepburn (ed.), *The Insect Integument*. Elsevier, Amsterdam and New York.

Hackman, R. H. and M. Goldberg. 1967. The *o*-diphenoloxidases of fly larvae. J. Insect Physiol. 13: 531–544.

Hackman, R. H. and M. Goldberg. 1976. Comparative chemistry of arthropod cuticle proteins. Comp. Biochem. Physiol. 55B: 201–206.

Hackman, R. H., M. G. M. Pryor, and A. R. Todd. 1948. The occurrence of phenolic substances in arthropods. Biochem. J. 43: 474–477.

Hamodrakas, S. J., C. W. Jones, and F. C. Kafatos. 1982. Biochem. Biophys. Acta 700: 42–51.

Handler, A. M. 1982. Ecdysteroid titers during pupal and adult development in *Drosophila melanogaster*. Dev. Biol. 93: 73–82.

Hepburn, H. R. (ed.) 1976. *The Insect Integument*. Elsevier, Amsterdam and New York.

Hepburn, H. R. 1985. Structure of the integument. Pp. 1–58 *in* G. A. Kerkut and L. I. Gilbert (eds.), *Comprehensive Insect Physiology, Biochemistry and Pharmacology*, Vol. 3. Pergamon Press, Oxford and Elmsford, New York.

Hirsh, J. and N. Davidson. 1981. Isolation and characterization of the dopa decarboxylase gene of *Drosophila melanogaster*. Mol. Cell. Biol. pp. 475–485.

Hirsh, J., B. Morgan, and S. Scholnick. 1986. Delimiting regulatory sequences of the *Drosophila melanogaster* Ddc gene. Mol. Cell Biol. 6: 4548–4557.

Hiruma, K. and L. M. Riddiford. 1984. Regulation of melanization of tobacco hornworm larval cuticle *in vitro*. J. Exp. Zool. 230: 393–403.

Hiruma, K. and L. M. Riddiford. 1985. Hormonal regulation of Dopa-decarboxylase during a larval molt. Dev. Biol. 110: 509–513.

Hiruma, K., L. M. Riddiford, T. L. Hopkins, and T. D. Morgan. 1985. Roles of dopa decarboxylase and phenoloxidase in the melanization of the tobacco hornworm and their control by 20-hydroxyecdysone. J. Comp. Physiol. B155: 659–669.

Hodgetts, R. B., W. C. Clark, D. D. Eveleth, Jr., R. O. Gietz, C. A. Spencer, and L. J. Marsh. 1986a. Pp. 221–234 *in* H. C. Slavkin (ed.), *Progress in Developmental Biology*, Pt. A. Liss, New York.

Hodgetts, R. B., W. C. Clark, R. D. Gietz, B. A. Sage, and J. D. O'Connor. 1986b. Stage-specific mechanisms regulate the expression of the dopa decarboxylase gene during *Drosophila* development. Arch. Insect Biochem. Physiol. 1: 97–104.

Hodgetts, R. B., B. Sage, and J. D. O'Connor. 1977. Ecdysone titers during postembryonic development of *Drosophila melanogaster*. Dev. Biol. 60: 310–317.

Højrup, P., S. O. Andersen, and P. Roepstorff. 1986a. Isolation, characterization, and *N*-terminal sequence studies of cuticular proteins from the migratory locust, *Locusta migratoria*. Eur. J. Biochem. 154: 153–159.

Højrup, P., S. O. Andersen, and P. Roepstorff. 1986b. Primary structure of a structural protein from the cuticle of the migratory locust, *Locusta migratoria*. Biochem. J. 236: 713–720.

Hopkins, T. L., T. D. Morgan, Y. Aso, and K. J. Kramer. 1982. *N*-β-alanyldopamine: major role in insect cuticle tanning. Science (Wash., D.C.) 217: 364–366.

Hughes, L. and G. M. Price. 1975a. Tyrosinase activity in the larva of the fleshfly *Sarcophaga barbaria*. J. Insect Physiol. 21: 1373–1384.

Hughes, L. and G. M. Price. 1975b. Tyrosinase activity in larvae of the fleshfly *Sarcophaga barbaria* [sic]. Biochem. Soc. Trans. 3: 72–75.

Kafatos, F. C. 1981. Structure, evolution and developmental expression of the silkmoth chorion multigene families. Am. Zool. 21: 707–714.

Kang, S.-H., M. S. Fuchs, and P. M. Webb. 1980. Purification and characterization of dopa decarboxylase from adult gravid *Aedes aegypti*. Insect Biochem. 10: 501–508.

Karlson, P. 1961. Biochemische Wirkungweise der Hormone. Dtsch. Med. Wochenschr. 86: 668–674.

Karlson, P. 1963. New concepts on the mode of action of hormones. Perspect. Biol. Med. 6: 203–214.

Karlson, P. and H. Ammon. 1963. Zum Tyrosinstoffwechsel der Insekten. XI. Biogenese und Schicksal der Acetyl-gruppe des *N*-Acetyl-dopamins. Hoppe-Seyler's Z. Physiol. Chem. 330: 161–168.

Karlson, P. and P. Herrlich. 1965. Zum Tyrosinstoffwechsel der Insekten. XVI. Der Tyrosinstoffwechsel der Heuschrecke *Schistocerca gregaria* Forsk. J. Insect Physiol. 11: 79–89.

Karlson, P. and H. Liebau. 1961. Zum Tyrosinstoffwechsel der Insekten. V. Reindarstellung, Kristallisation und Substratspezifität der *o*-Diphenoloxydase aus *Calliphora erythrocephala*. Hoppe-Seyler's Z. Physiol. Chem. 326: 135–143.

Karlson, P., D. Mergerhagen, and C. E. Sekeris. 1964. Zum Tyrosinstoffwechsel der Insekten. XV. Weitere Untersuchungen über das *o*-Diphenoloxydase-system von *Calliphora erythrocephala*. Hoppe-Seyler's Z. Physiol. Chem. 338: 42–50.

Karlson, P. and A. Schweiger. 1961. Zum Tyrosinstoffwechsel der Insekten. IV. Das Phenoloxydase-system von *Calliphora* und seine Beeinfluessung durch das Hormon Ecdyson. Z. Physiol. Chem. 323: 199–210.

Karlson, P. and C. E. Sekeris. 1962a. *N*-Acetyl-dopamine as sclerotizing agent of the insect cuticle. Nature (Lond.) 195: 183–184.

Karlson, P. and C. E. Sekeris. 1962b. Zum Tyrosinstoffwechsel der Insekten. IX. Kontrolle des Tyrosinstoffwechsels durch Ecdyson. Biochem. Biophys. Acta 63: 489–495.

Karlson, P. and C. E. Sekeris. 1964. Biochemistry of insect metamorphosis. Pp. 221–243 *in* M. Florkin and H. S. Mason (eds.), *Comparative Biochemistry*, Vol. 6. Academic Press, Orlando, Florida.

Karlson, P. and C. E. Sekeris. 1966a. Ecdysone, an insect steroid hormone, and its mode of action. Recent Prog. Horm. Res. 22: 473–501.

Karlson, P. and C. E. Sekeris. 1966b. Biochemical mechanisms of hormone action. Acta Endocrinol. 53: 505–518.

Karlson, P., C. E. Sekeris, and K. E. Sekeris. 1962. Zum Tyrosinstoffwechsel der Insekten. VI. Identifizierung von *N*-Acetyl-3,4-diydroxy-β-phenäthylamin (*N*-Acetyl-dopamine) als Tyrosinmetabolit. Hoppe-Seyler's Z. Physiol. Chem. 327: 86–94.

Kawasaki, H. and H. Sato. 1985. The tanning agent in the silk of the Japanese giant silkmoth, *Dictyoploca japonica* butler. Insect Biochem 15: 681–684.

Kent, P. W. and C. J. Brunet. 1959. The occurence of protocatechuic acid and its 4-*O*-β-D-glucoside in *Blatta* and *Periplaneta*. Tetrahedron 7: 252–256.

Kiely, M. L. and L. M. Riddiford. 1985a. Temporal programming of epidermal cell protein synthesis during the larval–pupal transformation of *Manduca sexta*. Wilhelm Roux's Arch. Dev. Biol. 194: 325–335.

Kiely, M. L. and L. M. Riddiford. 1985b. Temporal patterns of protein synthesis in *Manduca* epidermis during the change to pupal commitment *in vitro:* their modulation by 20-hydroxyecdysone and juvenile hormone. Wilhelm Roux's Arch. Dev. Biol. 194: 336–343.

Klose, W. E., G. H. Emmerich, and H. Beikirch. 1980. Developmental studies on two ecdysone deficient mutants of *Drosophila melanogaster*. Wilhelm Roux's Arch. Dev. Biol. 189: 51–57.

Koeppe, J. K. and L. I. Gilbert. 1973. Immunochemical evidence for the transport of haemolymph protein into the cuticle of *Manduca sexta*. J. Insect Physiol. 19: 615–624.

Koeppe, J. K. and L. I. Gilbert. 1974. Metabolism and protein transport of a possible pupal cuticle tanning agent in *Manduca sexta*. J. Insect Phyisol. 20: 981–992.

Kramer, K. J., T. L. Hopkins, R. F. Ahmed, D. Mueller, and G. Lookhart. 1980. Tyrosine metabolism for cuticle tanning in the tobacco hornworm, *Manduca sexta* (L.) and other Lepidoptera: identification of β-D-glucopyranosyl-*O*-L-tyrosine and other metabolites. Arch. Biochem. Biophys. 205: 146–155.

Kraminsky, G. P., W. C. Clark, M. A. Estelle, D. R. Gietz, B. A. Sage, J. D. O'Connor, and R. B. Hodgetts. 1980. Induction of translatable mRNA for dopa decarboxylase in *Drosophila*: an early response to ecdysterone. Proc. Natl. Acad. Sci. USA 77: 4175–4179.

Leonard, C., K. Soderhall, and N. A. Ratcliffe. 1985. Studies on prophenoloxidase and protease activity of *Blaberus craniifer* haemocytes. Insect Biochem. 15: 803–810.

Levenbook, L., R. P. Bodnaryk, and T. F. Spande. 1969. β-Alanyl-L-tyrosine: chemical synthesis, properties and occurrence in larvae of the fleshfly *Sarcophaga bullata* Parker. Biochem. J. 113: 837–841.

Lipke, H., K. Stout, W. Henzel, and M. Sugumaran. 1981. Structural proteins of sarcophid larval exoskeleton: composition and distribution of radioactivity derived from [7-^{14}C]dopamine. J. Biol. Chem. 256: 4241–4246.

Lipke, H., M. Sugumaran, and W. Henzel. 1983. Mechanisms of sclerotization in dipteras. Adv. Insect Physiol. 17: 1.

Lunan, K. D. and H. K. Mitchell. 1969. The metabolism of tyrosine-*O*-phosphate in *Drosophila*. Arch. Biochem. Biophys. 132: 450–456.

Mandaron, P. 1970. Development *in vitro* des disques imaginaux de la Drosophile: aspects morphologiques et histologiques. Dev. Biol. 22: 298–320.

Maranda, B. and R. Hodgetts. 1977. A characterization of dopamine acetyltransferase in *Drosophila melanogaster*. Insect Biochem. 7: 33–43.

Marks, E. P. 1972. Effects of ecdysterone on the deposition of cockroach cuticle *in vitro*. Biol. Bull (Woods Hole) 142: 293–301.

Marsh, J. L. and T. R. F. Wright. 1980. Developmental relationship between dopa decar-

boxylase, dopamine acetyl-transferase and ecdysone in *Drosophila*. Dev. Biol. 80: 379–387.
Mason, H. S. 1955. Comparative biochemistry of the phenolase complex. Adv. Enzymol. 16: 105–184.
Mills, R. R., S. Androuny, and F. R. Fox. 1968. Correlation of phenoloxidase activity with ecdysis and tanning hormone release in the American cockroach. J. Insect Physiol. 14: 603–612.
Mitchell, H. K. 1966. Phenoloxidase and *Drosophila* development. J. Insect Physiol. 12: 755–765.
Mitchell, H. K. and K. D. Lunan. 1964. Tyrosine-*O*-phosphate in *Drosophila*. Arch. Biochem. Biophys. 106: 219–222.
Mitchell, H. K., P. S. Chen, and E. Hadorn. 1960. Tyrosine phosphate on paper chromatograms of *Drosophila melanogaster*. Experientia (Basel) 16: 410–411.
Mitchell, H. K. and U. M. Weber. 1965. *Drosophila* phenoloxidases. Science (Wash., D.C.) 148: 964–965.
Mitchell, H. K., U. M. Weber, and G. Schaar. 1967. Phenoloxidase characteristics in mutants of *Drosophila melanogaster*. Genetics 57: 357–368.
Mitsui, T. and L. M. Riddiford. 1976. Pupal cuticle formation by *Manduca sexta* epidermis *in vitro:* patterns of ecdysone sensitivity. Dev. Biol. 54: 172–186.
Morgan, B. A., W. A. Johnson, and J. Hirsh. 1986. Regulated splicing produces different forms to dopa decarboxylase in the central nervous system and hypoderm of *Drosophila melanogaster*. EMBO (Eur. Mol. Biol. Org.) J. 5: 3335–3342.
Muskavitch, M. A. T. and D. S. Hogness. 1982. An expandable gene that encodes a *Drosophila* glue protein is not expressed in variants lacking remote upstream sequences. Cell 29: 1041–1051.
Nardi, J. B. and J. H. Willis. 1979. Control of cuticle formation by wing discs *in vitro*. Dev. Biol. 68: 381–395.
Oberlander, H. 1976. Hormonal control of growth and differentiation of insect tissues cultured *in vitro*. In Vitro (Rockville) 12: 225–235.
Oestlund, E. 1954. The distribution of catechol amines in lower animals and their effect on the heart. Acta Physiol. Scand. 31(Suppl. 112): 1–67.
Ohnishi, E. 1953. Tyrosinase activity during puparium formation in *Drosophila melanogaster*. Jpn. J. Zool. 11: 69–74.
Ohnishi, E. 1954a. Tyrosinase in *Drosophila virilis*. Annot. Zool. Jpn. 27: 33–39.
Ohnishi, E. 1954b. Melanin formation in the mutant "ebony" of *Drosophila melanogaster*. Annot. Zool. Jpn. 27: 76–81.
Ohnishi, E. 1954c. Activation of tyrosinase in *Drosophila virilis*. Annot. Zool. Jpn. 27: 188–193.
Ohnishi, E. 1958. Tyrosinase activation in the pupae of the housefly, *Musca vicina* M. Jpn. J. Zool. 12: 179–188.
Okubo, S. 1958. Occurence of *N*-acetylhydroxytyramine glucoside in *Drosophila melanogaster*. Med. J. Osaka Univ. 9: 327–333.
Pau, R. N. 1984. Cloning in cDNA for a juvenile hormone–regulated oothecin mRNA. Biochem. Biophys. Acta 782: 422–428.
Pau, R. N. and R. M. Acheson. 1968. The identification of 3-dydroxy-4-*O*-β-D-glucosidobenzyl alcohol in the left collaterial gland of *Blaberus discoidalis*. Biochem. Biophys. Acta 158: 206–211.
Pau, R. N. and P. A. M. Eagles. 1975. The isolation of an *o*-diphenol oxidase from third-instar larvae of the blow fly *Calliphora erythrocephala*. Biochem. J. 149: 707–712.
Pau, R. N. and C. Kelly. 1975. The hydroxylation of tyrosine by an enzyme from third-instar larvae of the blow fly *Calliphora erythrocephala*. Biochem. J. 147: 565–573.
Pelham, H. R. B. and R. J. Jackson. 1976. An efficient mRNA-dependent translation system from reticulocyte lysates. Eur. J. Biochem. 67: 247–256.

Piepho, H. 1942. Untersuchungen zur Entwicklungsphysiologie der Insekten-Metamorphose: über die Puppenhäutung der Wachsmotte *Galleria mellonella* L. Wilhelm Roux Arch. Entwicklungsmech. Org. 141: 500–583.

Preston, J. W. and R. L. Taylor. 1970. Observations on the phenoloxidase system in the haemolymph of the cockroach. *Leucophaea maderae*. J. Insect Physiol. 16: 1729–1744.

Pryor, M. G. M. 1940a. On the hardening of the ootheca of *Blatta orientalis*. Proc. R. Soc. Lond. B128: 378–393.

Pryor, M. G. M. 1940b. On the hardening of the cuticle of insects. Proc. R. Soc. Lond. B128: 393–407.

Pryor, M. G. M., P. B. Russell, and A. R. Todd. 1946. Protocatechuic acid, the substance responsible for the hardening of the cockroach ootheca. Biochem. J. 40: 627–628.

Pryor, M. G. M., P. B. Russell, and A. R. Todd. 1947. Phenolic substances concerned in hardening the insect cuticle. Nature (Lond.) 159: 399–400.

Psarianos, C. G., V. J. Marmaras, and J. N. Vournakis. 1985. Tyrosine-4-*O*-β-glucoside in the mediterranean fruit-fly *Ceratitis capitata*. Insect Biochem. 15: 129–135.

Rebers, J., F. Horodyski, R. Hice, and L. M. Riddiford. 1987. Genomic cloning of developmentally regulated larval cuticle genes from the tobacco hornworm, *Manduca sexta*. Mol. Entomol. 201–210.

Richards, A. G. 1967. Sclerotization and localization of brown and black colors in insects. Zool. Jahrb. Anat. 84: 25–62.

Richards, G. 1976a. Sequential gene activation by ecdysone in polytene chromosomes of *Drosophila melanogaster*. IV. The mid-prepupal period. Dev. Biol. 54: 256–263.

Richards, G. 1976b. Sequential gene activation by ecdysone in polytene chromosomes of *Drosophila melanogaster*. V. The late prepupal puffs. Dev. Biol. 54: 264–275.

Richards, R. 1981. The radioimmune assay of ecdysteroid titres in *Drosophila melanogaster*. Mol. Cell. Endocrinol. 211: 181–197.

Riddiford, L. M. 1981. Hormonal control of epidermal cell development. Am. Zool. 21: 751–762.

Riddiford, L. M. 1982. Changes in translatable mRNAs during the larval–pupal transformation of the epidermis of the tobacco hornworm. Dev. Biol. 92: 330–342.

Riddiford, L. M. 1984. Hormonal control of sequential gene expression in insect epidermis. Pp. 265–272 *in* J. A., Hoffmann and M. Porchet (eds.), *Biosynthesis, Metabolism and Mode of Action of Invertebrate Hormones*. Springer-Verlag, Berlin and New York.

Riddiford, L. M. 1985. Hormone acation at the cellular level. Pp. 37–84 *in* G. A. Kerkut and L. I. Gilbert (eds.), *Comprehensive Insect Physiology, Biochemistry and Pharmacology*, Vol. 8: *Endocrinology*, Pt. II. Pergamon Press, Oxford and Elmsford, New York.

Riddiford, L. M. 1986. Hormonal regulation of sequential larval cuticular gene expression. Arch. Insect Biochem. Physiol. Suppl. 1: 75–86.

Riddiford, L. M. 1987. Hormonal control of sequential gene expression in insect epidermis. Pp. 211–222 *in* J. H. Law (ed.) *UCLA Symposia on Molecular and Cellular Biology*, N. S. 49, *Molecular Entomology*. Alan R. Liss, Inc, New York.

Riddiford, L. M., A. C. Chen, B. J. Graves, and A. T. Curtis. 1981. RNA and protein synthesis during the change to pupal commitment of *Manduca sexta* epidermis. Insect Biochem. 11: 121–127.

Riddiford, L. M., A. Bechman, R. H. Hice, and J. Rebers. 1986. Developmental expression of three genes for larval cuticular proteins of the tobacco hornworm, *Manduca sexta*. Dev. Biol. 118: 82–94.

Rubin, G. M. and A. C. Sprading. 1982. Genetic transformation of *Drosophila* with transposable element vectors. Science (Wash., D.C.) 218: 348–353.

Saul, S. J. and M. Sugumaran. 1986. Phenoloxidase activation, proteases and proteases inhibitors in insect hemolymph. Fed. Proc. 45: 1860.

Scheidereit, C., S. Geisse, H. M. Westphal, and M. Beato. 1983. The glucocorticoid receptor binds to defined nucleotide sequences near the promoter of mouse mammary tumor virus. Nature (Lond.) 304: 749–752.

Schlaeger, D. A. and M. S. Fuchs. 1974. Effects of dopa decarboxylase inhibition of *Aedes aegypti* eggs: evidence of sclerotization. J. Insect Physiol. 20: 349–357.

Schloerer, J., C. E. Sekeris, and P. Karlson. 1970. Zum Tyrosinstoffwechsel der Insekten. XVIII. Ueber die Aktivitaet der Phenylalanin-4-hydroxylase und des *N*-acetyl-dopamin-glucosid-bildenden Systems *in vivo* im Verlauf der Entwicklung von *Calliphora erythrocephala* Meigen. Z. Physiol. Chem. 351: 1035–1040.

Schmalfuss, H. and G. Bussmann. 1937. Das dunkeln des lebenden Mehlkäfers *Tenebrio molitor* L. I. Z. Vergl. Physiol. 24: 493–508.

Scholnick, S. B., S. J. Bray, B. A. Morgan, C. A. McCormick, and J. Hirsh. 1986. Distinct central nervous system and hypoderm regulatory elements of the *Drosophila melanogaster* dopa decarboxylase gene. Science (Wash., D.C.) 234: 998–1022.

Scholnick, S. B., B. A. Morgan, and J. Hirsh. 1983. The cloned dopa decarboxylase gene is developmentally regulated when reintegrated into the *Drosophila* genome. Cell 34: 37–45.

Schreier, M. and T. Staehelin. 1973. Initiation of protein synthesis: the importance of ribosome and initiation factor quality for the efficiency of *in vitro* systems. J. Mol. Biol. 73: 329–347.

Schweiger, A. and P. Karlson. 1962. Zum Tyrosinstoffwechsel der Insekten. X. Die Aktivierung der Präphenoloxydase und das Aktivator-Enzym. Hoppe-Seyler's Z. Physiol. Chem. 329: 210–221.

Sekeris, C. E. 1963. Zum Tyrosinstoffwechsel der Insekten. XII. Reinigung, Eigenschaften und Substratspezifität der Dopa decarboxylase. Hoppe-Seyler's Z. Physiol Chem. 332: 70–78.

Sekeris, C. E. 1964. Sclerotization in the blow fly imago. Science (Wash., D.C.) 144: 419–420.

Sekeris, C. E. 1965. Action of ecdysone on RNA and protein metabolism in the blow fly. *Calliphora erythrocephala*. Pp. 149–167 *in* P. Karlson (ed.), *Mechanisms of Hormone Action*. Thieme Verlag, Stuttgart, and Academic Press, Orlando, Florida.

Sekeris, C. E. and E. G. Fragoulis. 1985. Control of dopa decarboxylase. Pp. 147–164 *in* G. A. Kerkut and L. I. Gilbert (eds.), *Comprehensive Insect Physiology, Biochemistry and Pharmacology*, Vol. 8. Pergamon Press, Oxford and Elmsford, New York.

Sekeris, C. E. and P. Herrlich. 1966. Zum Tyrosinstoffwechsel der Insekten. XVII. Der Tyrosinstoffwechsel von *Tenebrio molitor* und *Drosophila melanogaster*. Hoppe-Seyler's Z. Physiol Chem. 344: 267–275.

Sekeris, C. E. and P. Karlson. 1962. Zum Tyrosinstoffwechsel der Insekten. VII. Der katabolische Abbau des Tyrosins und die Biogeneses der Sklerotisierungssubstanz, *N*-acetyldopamin. Biochem. Biophys. Acta 62: 103–113.

Sekeris, C. E. and P. Karlson. 1964. On the mechanism of hormone action. II. Ecdysone and protein biosynthesis. Arch. Biochem. Biophys. 105: 483–487.

Sekeris, C. E. and P. Karlson. 1966. Biosynthesis of catecholamines in insects. Pharmacol. Rev. 18: 89–94.

Sekeris, C. E. and D. Mergerhagen. 1964. Phenoloxidase system of the blow fly, *Calliphora erythrocephala*. Science (Wash., D.C.) 145: 68–69.

Seligman, M., S. Friedman, and G. Fraenkel. 1969a. Bursicon mediation of tyrosine hydroxylation during tanning of the adult cuticle of the fly *Sarcophaga bullata*. J. Insect Physiol. 15: 553–562.

Seligman, M., S. Friedman, and G. Fraenkel. 1969b. Hormonal control of turnover of tyrosine and tyrosine phosphate during tanning of the adult cuticle in the fly *Sarcophaga bullata*. J. Insect Physiol. 15: 1085–1101.

Shaaya, E. and C. E. Sekeris. 1965. Ecdysone during insect development. III. Activities of some enzymes of tyrosine metabolism in comparison with ecdysone titer during the development of the blow fly *Calliphora erythrocephala*. Gen. Comp. Endocrinol. 5: 35–39.

Shaaya, E. and C. E. Sekeris. 1971. Inhibitory effects of α-amanitin on RNA synthesis and induction of dopa decarboxylase by β-ecdysone. FEBS (Fed. Eur. Biochem. Soc.) Lett. 16: 333–336.

Shampengtong, L. 1987. *N*-Acetylation of dopamine and tyramine by mosquito pupae (*Aedes togoi*). Insect Biochem. 17: 111–116.

Silvert, D. J., J. Doctor, L. Quesada, and J. W. Fristrom. 1984. Pupal and larval cuticle proteins of *Drosophila melanogaster*. Biochemistry 23: 5767–5774.

Sin, Y. T. and J. A. Thompson. 1971. Developmental changes in the prophenoloxidases of larval haemolymph in the fly *Calliphora*. Insect Biochem. 1: 56–62.

Skelly, P. J. and A. J. Howells. 1987. Larval cuticle proteins of *Lucilia cuprina:* electrophoretic separation, quantification and developmental changes. Insect Biochem. 17: 625–633.

Sloley, B. D. and R. G. H. Downer. 1987. Dopamine, *N*-acetyldopamine and dopamine-3-sulphate in tissues of newly ecdysed and fully tanned adult cockroaches (*Periplaneta americana*). Insect Biochem. 17: 591–596.

Sugumaran, M. 1986. Tyrosinase catalyzes an unusual oxidative decarboxylation of 3,4-dihydroxymandelate. Biochemistry 25: 4489–4492.

Sugumaran, M. 1987. Quinone methide sclerotization: a revised mechanism for β-sclerotization of insect cuticle. Bioorg. Chem. 15: 194–211.

Sugumaran, M. and H. Lipke, 1983. Quinone methide formation from 4-alkylcatechols: a novel reaction catalyzed by cuticular polyphenoloxidase. FEBS (Fed. Eur. Biochem. Soc.) Lett. 155: 65–68.

Sugumaran, M., B. Hennigan, and J. O'Brian. 1987a. Tyrosinase-catalyzed protein polymerization as an *in vitro* model for quinone tanning of insect cuticle. Arch. Insect Biochem. Physiol. 6: 9–25.

Sugumaran, M., B. Hennigan, and J. O'Brien. 1987b. Tyrosinase-catalyzed protein polymerization as an *in vitro* model for quinone tanning of insect cuticle. Arch. Insect Biochem. Physiol. 6: 9–25.

Sugumaran, M., D. Hemalata, V. Semensi, and B. Hennigan. 1987c. Tyrosinase-catalyzed unusual oxidative dimerization of 1,2-Dehydro-*N*-acetyldopamine. J. Biol. Chem. 262: 10546–10549.

Swiderski, R. E. and J. D. O'Connor. 1986. Modulation of novel-length dopa decarboxylase transcripts by 20-OH-ecdysone, in a *Drosophila melanogaster* Kc cell subline. Mol. Cell Biol. 6: 4433–4439.

Truman, J. W. 1985. Hormonal control of ecdysis. Pp. 413–440 *in* G. A. Kerkut and L. I. Gilbert (eds.), *Comprehensive Insect Physiology, Biochemistry and Pharmacology*, Vol. 8. Pergamon Press, Oxford and Elmsford, New York.

Tsukamoto, T., M. Ishiguro, and M. Funatsu. 1986. Isolation of latent phenoloxidase from prepupae of the housefly *Musca domestica*. Insect Biochem. 16: 573–581.

Weiher, H., M. Konig, and P. Gruss. 1983. Multiple-point mutations affecting the Simian virus 40 enhancer. Science (Wash., D.C.) 219: 626–631.

Wigglesworth, V. B. 1940. The determination of characters at metamorphosis in *Rhodnius prolixus* (Hemiptera). J. Exp. Biol. 17: 201–222.

Williams, C. M. 1961. The juvenile hormone II: its role in the endocrine control of molting, pupation and adult development in the cecropia silkworm. Biol. Bull. (Woods Hole) 121: 572–585.

Wolfgang, W. J. and L. M. Riddiford. 1986. Larval cuticular morphogenesis in the tobacco hornworm *Manduca sexta* and its hormonal regulation. Dev. Biol. 113: 305–316.

Wright, T. R. F., R. B. Hodgetts, and A. F. Sherald. 1976. The genetics of dopa decarboxylase in *Drosophila melanogaster.* I. Isolation and characterization of deficiencies that delete the dopa decarboxylase dosage sensitive region and the alpha-methyl-dopa–hypersensitive locus. Genetics 84: 267–285.

Wright, T. R. F., R. Steward, K. W. Bently, and P. N. Adler. 1981a. The genetics of dopa decarboxylase in *Drosophila melanogaster.* III. Effects of a temperature sensitive dopa decarboxylase deficient mutation on female fertility. Dev. Gen. 2: 223–235.

Wright, T. R. F., W. Beermann, J. L. Marsh, C. P. Bishop, R. Steward, B. C. Black, and E. Y. Wright. 1981b. The genetics of dopa decarboxylase in *Drosophila melanogaster.* IV. The genetics and cytology of the 37B10-7D1 region. Chromosoma 83: 45–58.

Yago, M., H. Sato, and H. Kawasaki. 1984a. The identification of five *N*-acetyldopamine glucosides in the left collateral gland of the praying mantid, *Hierodula patellifera* Serville. Insect Biochem. 14: 487–489.

Yago, M., H. Sato, and H. Kawasaki. 1984b. The identification of *N*-acetyldopamine glucosides in the left collateral gland of the praying mantid, *Mantis religiosa, Statilia maculata* Thunberg, and *Tenodera augustipensis* Saussure. Insect Biochem. 14: 7–9.

Yamazaki, H. I. 1969. The cuticular phenoloxidase in *Drosophila virilis.* J. Insect Physiol. 15: 2203–2211.

Yamazaki, H. I. 1972. Cuticular phenoloxidase from the silkworm *Bombyx mori:* properties, solubilization and purification. Insect Biochem. 2: 431–444.

Žďárek, J. 1985. Regulation of pupariation. Pp. 301–334 *in* G. A. Kerkut and L. I. Gilbert (eds.), *Comprehensive Insect Biochemistry, Physiology and Pharmacology,* Vol. 8. Pergamon Press, Oxford and Elmsford, New York.

Role of Morphogenetic Hormones in Morphological Color Changes in Arthropods

6

A. BOUTHIER
AND P. Y. NOËL

6.1. Introduction

Pigment cells may have more than one physiological function. The chromatophores (almost exclusively represented in Crustacea) and eye pigment cells form the only kind that have primarily a chromatic function. The others have a different primary function, such as skeletal (cuticle-producing pigment cells), digestive (hepatocytes), or reproductive (oocytes). Among these diverse pigment cell categories, only chromatophores and cuticle-producing pigment cells play a significant role in color changes and therefore will be considered in this chapter.

Morphological color changes were first reported to occur during color adaptation processes in vertebrates (Bàbàk, 1913; Odiorne, 1933a; Sumner, 1940) and in invertebrates (Keeble and Gamble, 1905; Brown, 1934, 1935; Chassard-Bouchaud, 1965). Some other color changes (i.e., those related to the nychthemeral rhythm) do not clearly appear as adaptive. Moreover, the adaptive function of the aforementioned color changes is not well defined. General information on arthropod color changes is presented in Parker (1948).

6.1.1. *Morphogenetic Hormones in Arthropods*

Generally, the term *morphogenetic hormone* is applied to an internal secretion that influences growth and nutrition of organs or organisms. In arthropods, it is classically applied to ecdysteroids and juvenile hormones (true JHs and JH-like substances). Other hormones, namely, the pigment hormones (or chromatophorotropins), affect the external appearance of animals and control their apparent morphology; they may therefore be called *pseudomorphogenetic* hormones, as compared with the *eumorphogenetic* hormones. The chemical nature of the pigment hormones was reviewed by Mordue and Stone (1976).

According to Carlisle and Knowles (1959), morphogenetic hormones "are hormones which act upon sequences of growth, differentiation and maturation of the whole body or specific organ systems." They differ from metabolic and kinetic hormones, which act immediately on metabolism and organs, respectively; but a hormone may belong to more than one category, and it appears that chromatophorotropins are involved in kinetic and metabolic as well as morphogenetic changes (Nagabhushanam and Vasantha, 1971).

Hormonal control of color changes differs according to the groups of arthropods, but there is evidence of some relationships between hormones of different groups. For instance, extracts from different parts of the central nervous system (CNS) of insects have effects on pigment retraction within red chromatophores of decapods (Brown and Meglitsch, 1940; Thomsen, 1943; Mordue and Stone, 1977).

6.1.2. *What Are Morphological Color Changes?*

Two types of processes are involved in color adaptation: those involving pigment migration within pigment cells are faster and require 24 h at the utmost for their completion; those involving pigment formation, deposition, and destruction are slow and require weeks to adapt the animal to the environment. The former are called "physiological color changes," and the latter, "morphological color changes" (Mégusar, 1912; Odiorne, 1933a; Bowman, 1942; Brown, 1934, 1935, 1961; Chassard, 1962; Chassard-Bouchaud, 1965; Vuillaume, 1969).

Characteristics of morphological color changes are as follows:

(1) *slowness*
(2) *prolongation of effect*
(3) *variations* (increase or decrease) *in pigment cell number*
(4) *variations in pigment quantity* (production or destruction, storage or elimination)
(5) *variations in pigment quality*
(6) *control by light* (intensity of incident light; light reflectance of background color), *temperature, humidity, carbon dioxide, and ethology* (group and crowding effects; nongenetic parental influence)

In arthropods, the coloration depends both on pigments deposited in the cuticle and on those present in the epidermis and/or in deeper tissues. Eumorphogenetic hormones are well known only in Insecta and were recently demonstrated in Crustacea (Borst et al., 1985; Laufer et al., 1986). In Crustacea, they are still poorly known and nothing is known about their role in pigmentation. Arthropods present a great variety of colors and pigmentation. Some of these colorations have a role in the modification of the apparent shape of animals (mimicry).

6.1.3. *Chromatic Effectors*

Most of the pigment cells playing a prominent role in arthropods' color display are located in the epidermis. We shall discuss only the cells and structures that play a major role in the external appearance of animals, and so only cells with a chromatic function, main or accessory, will be considered (for eye pigments see Bouthier, 1981). General information is given in Brown (1973) and Bagnara and Hadley (1973).

Morphological color changes originate from modification of effectors, directly (pigment cells) or indirectly (cuticle, through epidermal cells). We shall adopt the definitions given earlier (Noël, 1983a), and introduce the term *chromoepidermal cells* for epidermis pigmented cells playing a role in colors and color change in arthropods, including those called "chromatophores" by other authors in Odonata (see Section 6.2.4, below). We believe that the term *chromatophore* should be restricted to the subepidermal (or internal) multibranched pigment cells found in various phyla (vertebrates, echinoderms, arthropods) (Table 6.1).

Recent reviews are very cautious as to the role of hormones on arthropod morphological color changes (Fingerman, 1963; Bellon-Humbert, 1970a; Noël, 1981; Bückmann, 1985; Fuzeau-Braesch, 1985). The distribution and control of pigment cells are presented in Table 6.2.

6.1.4. *Pigments*

Arthropod epidermis pigmentation depends on eight main chemical families:

- *biliary pigments*—green, blue, yellow, red, or violet, of tetrapyrrolic form derived from δ-aminolevulinic acid and common in many arthropods; respiratory pigments related to this group are found in some arthropod classes
- *carotenoids and carotenoproteins*—yellow, orange, red, blue, or green, widespread in all arthropod classes, are of nutritional origin
- *flavonoids*—yellow, red, or brown benzochromanes, also originating in plant foods, essentially present in insects
- *melanins*—black, brown, or reddish, high polymers of modified aromatic amino acids (tyrosine), present in most arthropods, infiltrated in the cuticular proteic matrix
- *ommochromes*—yellow, red, brown, purple, or violet condensed nitrogenous or sulfurous heterocycles derived from the tryptophan nucleus, prevalent in arthropods
- *pterins, flavins, and lumazines* (deaminated pterins)—white, yellow, orange, or red nitrogenous heterocycles derived from puric nucleic bases, widely distributed
- *purinic pigments* (guanine and the like)—are simple whitish derivatives of uric acid, sometimes crystallized, very common
- *quinones*—yellow or red aromatic diacetones, complex, derived from tyrosine

More information is available in recent reviews on insects (Kayser, 1985) or crustaceans (Ghidalia, 1985).

6.2. Morphological Color Changes in Insects and Their Hormonal Control

Many aspects of endocrinology have been widely investigated in insects. Almost all insect orders have been studied because of their role as economic pests or as convenient biological material for laboratory experiments. However, the larger species are the best known because experimentation is easier on them.

TABLE 6.1. Main Categories of Pigment Cells in Arthropods

Main function	Cell category	Example	Pigments
digestion/excretion	digestive epithelium	hepatopancreas	carotenoids
reproduction	oocytes	ovary	carotenoproteins
vision:	ocelli		
photoreception	retinular cells	axon ending with proximal pigment nauplius eye	rhodopsin (primary visual pigment)
photoprotection	accessory pigment cells in eyes	basal red pigment cell dark distal pigment cell light distal pigment cell	carotenoids ommochromes
photofocalization	reflecting pigment cells	compound eye	pterins, flavins, guanin

photoemission	photoluminous organs	glowworm, firefly	riboflavin
chromatic only	euchromatophores	Decapoda, Mysidacea	carotenoids, ommochromes, pterins, flavins
	parachromatophores	Isopoda	carotenoids, ommochromes, pterins, flavins
	pseudochromatophores	Diptera larvae	ommochromes ?
Chromatic and protective or enzymatic	chromoepidermal cells	phasmids, Odonata	ommochromes, pterins
		Crustacea	carotenoproteins
	pigment cells	Lepidoptera, Crustacea, Orthoptera	flavins, pterins

TABLE 6.2. Distribution and Control of Pigment Cells and Color Changes in Arthropods[a]

Groups	Chromatophores	Chromo-epidermal cells	Control			Color changes occurrence	
			Nervous	*Hormonal*	*Direct*	*Morphological*	*Physiological*
Crustacea							
Decapoda	eu[b]	+	+ (?)	++	+	++	++
Amphipoda	para	+	–	+ (?)	+	+?	+
Isopoda	para	+	–	+	–	++	+
Mysidacea	eu[b]	–	–	+	+	+	++
Stomatopoda	eu	+	–	+	–	+	+
Branchiura	eu	–	–	+?	?	?	+
Copepoda	?	+	?	?	?	?	?
Insecta							
Diptera larvae	pseudo	–	–	+?	–	–	+
Odonata	–	+	?	?	?	?	?
Most insects	–	+	–	+	?	+	+
Myriapoda							
most species	?	+	?	?		?	?
Chelicerata							
Merostomata	–	–	?	?	?	?	–
Pycnogonida	+?	+	?	?	?	?	?
Arachnida	+	+	?	?	?	+	+?

[a] *Key:* eu = euchromatophores; para = parachromatophores; pseudo = pseudochromatophores; – = absent; + = present; ++ = frequent; ? = unknown or uncertain.

Numerous reviews have been published on pigments and color changes in insects (Rowell, 1971; Fuzeau-Braesch, 1972, 1985; Bückmann, 1974a, 1985; Raabe, 1982, 1983b; Hoffmann, 1985; Kayser, 1985). Hormonal control of such color changes has been considered—more or less—in these reviews. From year to year, new results have been fitted in with those already obtained. Yet insects for which the complete hormonal control of color change has been elucidated remain very scarce; indeed, the lepidopterans *Cerura vinula*, *Pieris brassicae*, and *Manduca sexta* are the only ones we have found in the literature.

6.2.1. *Definitions and General Features*

Several kinds of morphological color changes affecting various developmental stages and under different physiological controls exist in insects. Bückmann differentiated physiological color change from morphological ones several years ago (1974a) and again more recently (1985): the former color change is "caused by pigment movement within the living epidermal cells," and the latter is "caused by change in pigment content within the living cells and/or melanin secreted into the cuticle." Bückmann adds: "While the former can be changed at all times the latter can be changed only through a moult." He suggested the following classification:

- ontogenetic color change (successive developmental stages pigmented differently)
- morphological color adaptation (only one developmental stage variable in pigmentation; irreversible change)
- morphological color change *sensu stricto* (pigmentation modified by external factors; reversible)
- biochemical color change (chemical change in the pigments without change in quantity).

In the present chapter, we prefer to adopt a quite different classification based on the mode of hormonal control of color change. When morphological color changes belong to normal individual development and the unfolding of obligatory steps of its program, they are dependent almost exclusively on morphogenetic hormones. In contrast, when the color changes are subordinate to the random and unforeseeable alteration of agents exterior to the animal, they are not registered in the genetic program of specific development and cannot therefore be regulated by morphogenetic hormones alone, whose levels are not adaptable. These environment-dependent color changes are then induced by neurohormones, which are produced with more or less adaptiveness according to the strength of the stimulus.

In each class of color change, however, there is a frequent conjunction of the peculiar action of both types of endocrine mechanism. We can therefore differentiate the two major kinds of color changes according to their predominent hormonal control, and we must consider moreover the color changes caused mainly by neurohormones acting as morphogenetic hormones (pseudomorphogenetic) as well as those actually caused by authentic morphogenetic (eumorphogenetic) hormones.

6.2.1.1. INSECT HORMONE-PRODUCING CENTERS

The eumorphogenetic hormones belong to two chemical kinds:

- ecdysteroids such as ecdysone (E) and ecdysterone (20-E) produced by the prothoracic gland (PTG) (or by the ventral gland)
- juvenile hormones (JHs, namely, JH I, II, III, and 4-methyl–I), sesquiterpenoids produced by the corpora allata (CA)

Neurohormones, including pseudomorphogenetic hormones, are produced by various areas of the nervous system, as follows (for additional details, see the review articles in Borhovek and Gelman, 1986):

- brain [median and lateral neurosecretory cells (NSC)]
- subesophageal ganglia
- corpora cardiaca (CC)
- sometimes ganglia of the nervous ventral chain

6.2.1.2. NEUROHORMONES INVOLVED IN MORPHOLOGICAL COLOR CHANGES

These peptidic hormones are the following:

- prothoracicotropic hormones (4- and 22-kDa PTTH: Suzuki, 1986), produced by the brain
- melanization and reddish coloration hormone (MRCH), produced by the complex brain/subesophageal ganglia showing sequence similarities with the eclosion hormone implicated in color change determinism (Suzuki, 1986)
- pupal melanization reducing factor (PMRF: Bückmann and Maisch, 1987), an hydrophobic peptide
- another hormone apparently controls the carotenoid pigmentation in *Papilio xuthus* (Awiti and Hidaka, 1982)

In addition, we might report the production of the CC of several peptide hormones (adipokinetic hormones, or AKHs) present in various insect species (Goldsworthy and Wheeler, 1986). Their amino acid sequences differ little from one another, or from that of the red pigment concentrating hormone (RPCH) isolated from Crustacea. Insect AKHs manifest the same activity as RPCH on crustacean chromatophores and also on certain insects.

Recently, the structure of several peptidic chromactivating neurohormones was determined; these peptides belong to the so-called RPCH/AKH family (for data and references see Schaffer, 1986; Goldsworthy and Wheeler, 1986).

Other pigment hormones present in Crustacea (such those mentioned in Section 6.3.1.2, below) are not yet reported in insects.

6.2.1.3. PIGMENTED STRUCTURES: CUTICLE AND PIGMENT CELLS

The integument of insects is constituted of two parts: the outer one, or cuticle, an exoskeleton forming a protective layer against dehydration and atmospheric oxygen; and the inner one, the epidermis, consisting of ordinary integumental cells and their accompanying cells (Fig. 6.1.).

The cuticle consists of a matrix containing chitin (a polyacetyl glucosamine) and proteins (often sclerotized by phenolic tanning), with associated proteins conferring suppleness to it. The cuticle may be mineralized in a few species. It has three layers:

- epicuticle, consisting of waxes
- exocuticle, the most rigid, deposited before ecdysis
- endocuticle, the most supple and thick, deposited after ecdysis

The scleroproteins are frequently colored by phenolic tanning; eumelanins are also present in both endocuticle and exocuticle and nowhere else. The epidermis is formed of pavement (or cylindric) cells producing the cuticular layers and containing pigment granules (ommochromes, pterins), pigment crystals (uric acid or alcaline urates), pigment vesicles (carotenoproteins), or soluble pigments (biliary pigments, flavonoids, etc.).

Apparently, none of the different kinds of chromatophores described below in Sections 6.1.3. and 6.3.1.3. (Fig. 6.2.) is present in insect epidermis. An equivalent of chromatophores sometimes occurs in some species where chromoepidermal cells' pigment can migrate through the cell from the apex to the base and vice versa along the cytoskeleton under proper stimulation.

Phasmids (Berthold, 1977, 1980) and perhaps some Odonata (Veron et al., 1974) are known to have such pigment migration in their epidermis, but they are the only ones reported in Insecta.

The pseudochromatophores on the dorsal side of the tracheal bladders of the phantom midge larva (*Corethra plumicornis*) represent the only chromatophoric structure; their extension or retraction state affects

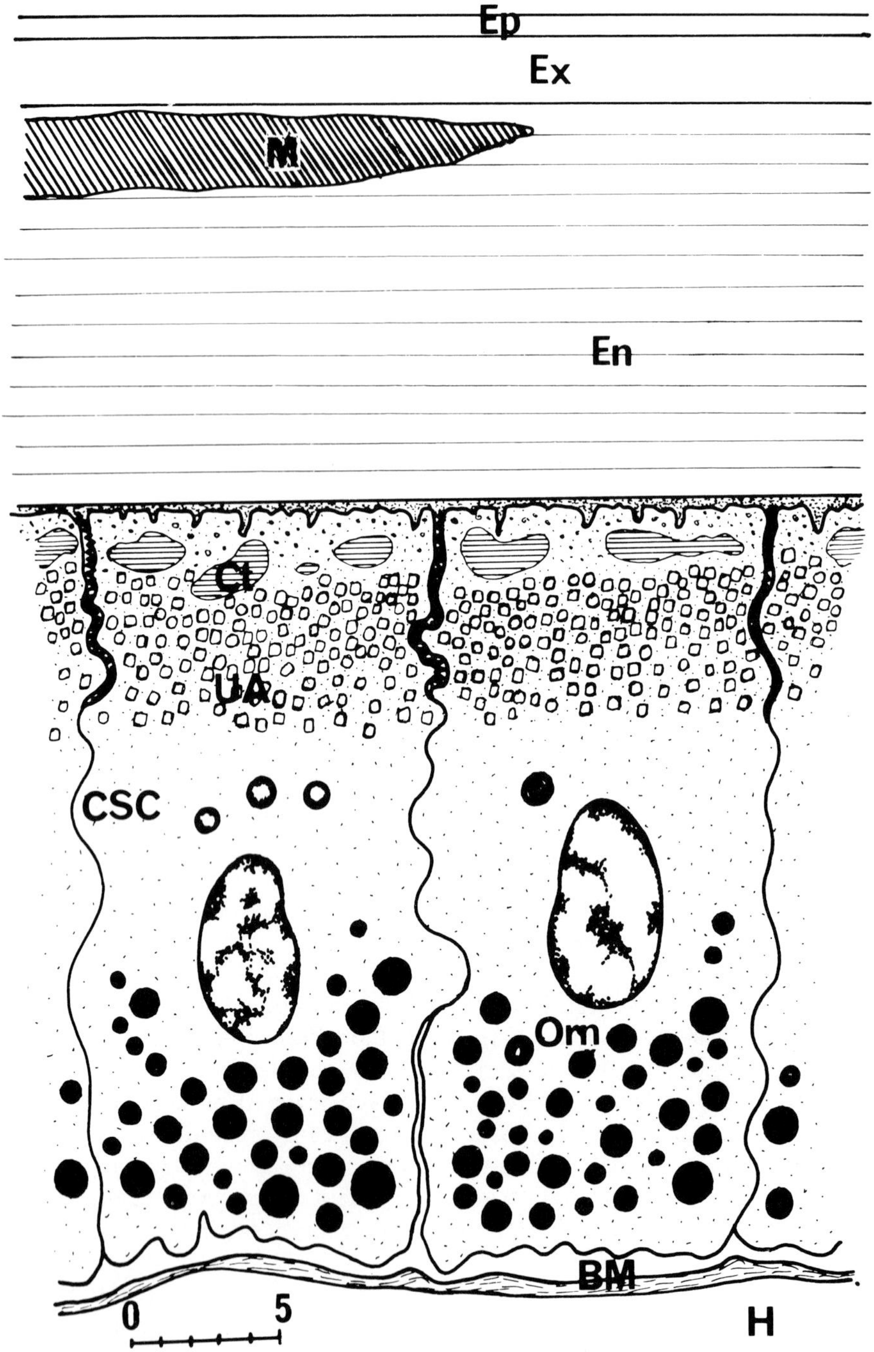

FIGURE 6.1. Section of *Locusta migratoria* fifth-instar integument, showing the situation of pigmented structures involved in morphological color changes. Melanin (M) is present in external part of cuticle [epicuticle

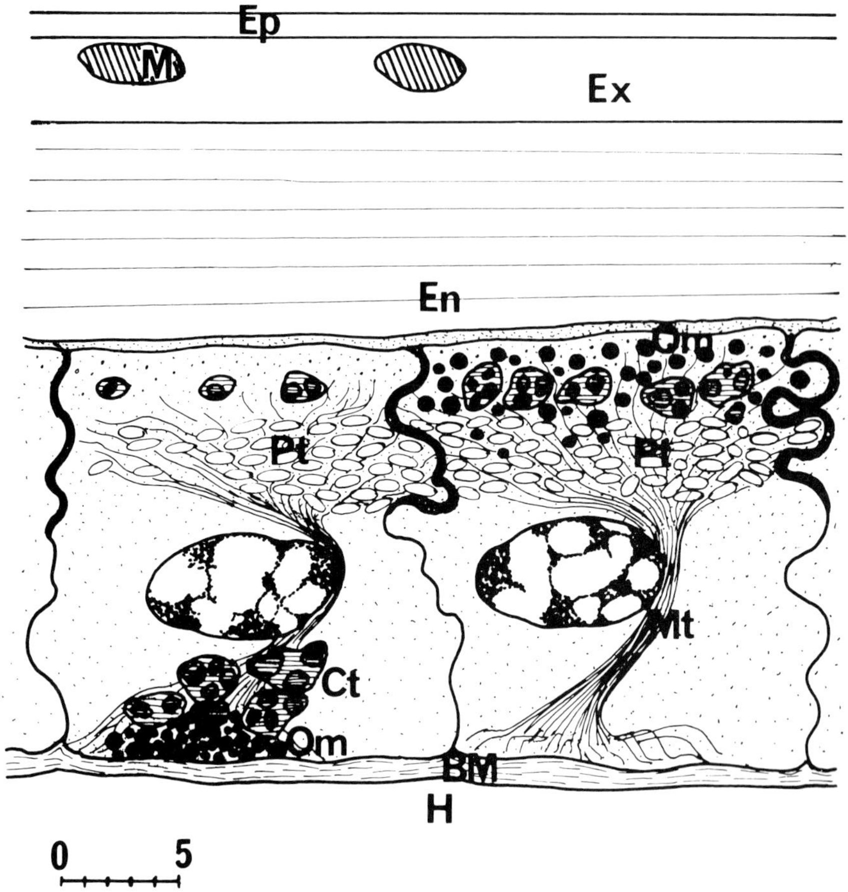

FIGURE 6.2. Section of *Carausius morosus* larval integument. Left part of the drawing schematizes a light-adapted cuticle-secreting cell, and the right part a dark-adapted one. Microtubules bundles (Mt) help the migration of pigment granules: pterinosomes (Pt), ommochromes granules (Om), and carotenoids (Ct); pigments are close to the basement membrane (BM) on dark adaptation; epicuticle (Ep), exocuticle (Ex), and endocuticle (En) are not much pigmented. Bar = 5 μm. (After Berthold, 1977, 1980; Berthold and Seifert, 1977.)

(Ep), exocuticle (Ex), endocuticle (En)]; cuticle-secreting cell (CSC) contains carotenoids (Ct) and ommochromes (Om) chromogranules, together with uric acid microcrystals (UA). Basement membrane (BM) limits epidermis from hemocoel (H). Bar = 5 μm.

the general aspect of the larva because of its complete body transparency, but they are not epidermis cells.

Wings are integumental parts peculiar to insects and are often pigmented (in Odonata, Dictyoptera, Phasmoptera, Orthoptera, Heteroptera, Homoptera, Neuroptera, Diptera, Coleoptera, and obviously Lepidoptera), constituting the major illustration of morphological color change since they characterize the adults and appear at the imaginal ecdysis. In all these orders, the wings are membranous and have no pigmentary differentiation. Lepidoptera wings have specialized surface structures such as scales, hairs, or bristles formed by dead cells with enclosed dry pigmented organelles (e.g., pterinosomes) or amorphous pigments (flavonoids, ommochromes, melanins, etc.). These surface structures confer physical colors.

6.2.2. *Morphological Color Changes Controlled Mainly by Eumorphogenetic Hormones*

Numerous kinds of morphological color change are present in insects affecting various developmental stages and submitted to different physiological determinisms. Larval or adult stages are the only ones concerned by this type of color change.

6.2.2.1. LARVAL INDIVIDUAL COLOR CHANGE

The more common kind of color change takes place around the end of larval life, corresponding in holometabolous insects to a basic change in environment, mode of life, and feeding practices (Table 6.3). In heterometabolous insects, this is less clear because modifications in ethological habits surrounding the imaginal ecdysis are not as common. This important behavioral change occurs at the pupal stage (or the last larval instar in heterometabolous insects) and is already prepared for during the last larval instar (or the penultimate larval instar in heterometabolous insects).

All cases of larval individual color change may be comprised under the category of so-called green/brown polymorphism, or green/black polymorphism, which is widespread in many insect orders. Everywhere the JHs play an important but variable role in this change. It is known that JH secretion is interrupted well before imaginal (or pupal) ecdysis, therefore orienting the following molt in a nonlarval direction.

Orthoptera

In Orthoptera, especially in all Acrididae, JH (i.e., JH III) produced by the CA is obviously the most efficient factor in development of the green coloration (Pfeiffer, 1945; P. Joly 1951, 1952a,b; L. Joly, 1958, 1960; Rowell, 1967; for specific details, see Table 6.4). The appearance of biliverdin-IX-α, at first in the hemolymph (even during intermolts) and

TABLE 6.3. Larval Individual Color Change[a]

Groups and species	20-E ommochromes	JH *Ommochromes*	JH *Turning green*	JH *Clearing*	References
Orthoptera					
Acrididae					
3 species[b]	ND	ND	+	ND	Rowell, 1967
Locusta migratoria cinerascens	ND	ND	+	ND	P. Joly, 1952b
Acrida turrita	ND	ND	+	ND	P. Joly, 1952a
Melanoplus differentialis	ND	ND	+	ND	Pfeiffer, 1945
Gryllidae					
Gryllodes sigillatus	ND	ND	ND	+	Yaragamblimath and Mathad, 1978
Gryllus bimaculatus	ND	ND	ND	+	Roussel, 1967; Fuzeau-Braesch, 1968
Lepidoptera					
Pieridae					
Pieris brassicae	D	D	ND	ND	Beydon et al., 1980
Nymphalidae					
Hestina japonica	=	=	ND	ND	Osanai and Arai, 1962
Sasakia charonda	=?	=?	ND	ND	Osanai, 1966
Bombycidae					
Bombyx mori	ND	I	ND	ND	Kiguchi, 1973; Ohashi et al., 1983

(continued)

TABLE 6.3. *(Continued)*

Groups and species	20-E ommochromes	JH *Ommochromes*	JH *Turning green*	JH *Clearing*	References
Noctuidae					
Mamestra brassicae	ND	I	ND	ND	Hiruma et al., 1984
Notodontidae					
Cerura vinula	I	D	ND	ND	Bückmann, 1974b; Hintze, 1969
Pyralidae					
Diatraea grandiosella	ND	ND	ND	ND	Yin and Chippendale, 1974
Sphingidae					
Cephonodes hylas	I	=	+	ND	Ikemoto, 1975, 1983
Deilephila nerii	ND	ND	=	ND	Chang and Jang, 1980
Manduca sexta	ND	D	ND	ND	Riddiford and Hori, 1981; Hori and Riddiford, 1982
Smerinthus ocellata	ND	ND	ND	ND	Hintze-Podufal, 1970, 1975
Sphinx ligustri	ND	I	ND	ND	Hintze-Podufal, 1970, 1975

[a] *Key:* D = decrease; I = increase; JH = juvenile hormones and analogues; ND = no data; 20-E = 20-hydroxyecdysone; - = inhibition of depigmentation; + = positive action; ? = uncertain.
[b] *Acanthacris ruficornis, Humbe tenuicornis,* and *Gastrimargus africanus.*

TABLE 6.4. Species Affected by a Density-Dependent Larval Darkening[a]

Groups and species	Voltinism	Diapause	Developmental stage affected by darkening	MRCH action	JH action	References
Orthoptera						
Pyrgomorphidae						
2 species[b]	1	ND	L	ND	ND	Uvarov, 1966, 1977
Acrididae						
Calliptaminae						
Calliptamus italicus	ND	ND	LA	ND	ND	Della Beffa, 1948
Cyrtacanthacris tatarica	ND	ND	L	ND	ND	Uvarov, 1966
Nomadacris septemfasciata	1	A	LA	ND	ND	Faure, 1932, 1935
Schistocerca gregaria	1–S	A	LA	ND	ND	Uvarov, 1966, 1977
S. obscura	ND	ND	L	ND	ND	Uvarov, 1966
S. paranensis	ND	ND	LA	ND	ND	Uvarov, 1966
Catantopinae						
Melanoplus bivittatus	1	E	LA	ND	ND	Uvarov, 1966, 1977
M. sanguinipes	1–2	E	LA	ND	ND	Uvarov, 1966, 1977

(continued)

TABLE 6.4. (*Continued*)

Groups and species	Voltinism	Diapause	Developmental stage affected by darkening	MRCH action	JH action	References
Oedipodinae						
Gastrimargus affinis	ND	ND	LA	ND	ND	Ritchie, 1978
G. musicus	ND	ND	LA	ND	ND	Uvarov, 1966
Locusta migratoria	1–S	EA	LA	ND	ND	Uvarov, 1966, 1977
Locustana pardalina	1–3	E	LA	ND	ND	Faure, 1932
Oedaleus senegalensis	1	E	LA	ND	ND	Ritchie, 1978
Pyrgodera armata	ND	ND	LA	ND	ND	Ritchie, 1978
Gomphocerinae						
Chorthippus albomarginatus	1	ND	LA	ND	ND	Uvarov, 1966, 1977
Dociostaurus maroccanus	1	E	LA	ND	ND	Uvarov, 1966, 1977
Tettigoniidae						
3 species[c]	1	E	LA	ND	ND	Della Beffa, 1948
Lepidoptera						
Pieridae						
Pieris brassicae	S	P	LP	ND	ND	Long, 1953

Noctuidae						
Aedia leucomelas	S	Ø	L	ND	ND	Iwao, 1968
Anomis sabulifera	S	+ND	L?	ND	ND	Dutt, 1969
Diataraxia oleracea	S	P	L	ND	ND	Long, 1953
Leucania loreyi	ND	ND	L	D+	ND	Iwao, 1962 Matsumoto et al., 1981
Leucania separata	S	Ø	L	D+	Ø	Ogura, 1975a,b; Matsumoto et al., 1981
Lithacodia stygia	ND	+ND	L	ND	ND	Iwao, 1962
Malacosoma pluviale	ND	ND	L	ND	ND	Iwao, 1968
Mamestra brassicae	S	P	L	D+	Om+	Hiruma et al., 1984 Matsumoto et al., 1981
Naranga aenescens	ND	PF	L	ND	ND	Iwao, 1962
Persetania ervingi	ND	PF	L	ND	ND	Doull, 1953
Plusia gamma	S	Ø	L	ND	ND	Long, 1953
Spodoptera (Laphygma) exempta	S	Ø	L	D+	ND	Faure, 1943b; Matsumoto et al., 1981
2 species[d]	S	Ø	L	ND	ND	Faure, 1943a
S. (Prodenia) litura	S	Ø	L	D+	Ø	Matsumoto et al., 1981; Tojo et al., 1985
Trachea atriplicis	ND	+ND	ND	ND	ND	Iwao, 1962

(continued)

TABLE 6.4. *(Continued)*

Groups and species	Voltinism	Diapause	Developmental stage affected by darkening	MRCH action	JH action	References
Geometridae						
Ennomos subsignarius	1	E	LP	ND	ND	Iwao, 1968
Notodontidae						
Exaereta ulmi	1	P	L	ND	ND	Iwao, 1968
Saturnidae						
Saturnia (Eudia) pavonia	1	P	L	ND	ND	Long, 1953; Hintze-Podufal, 1977
Sphingidae						
Erinnyis ello	ND	ND	L	ND	ND	Schneider, 1973
Cephonodes hylas	S	ND	L	ND	Ø–S	Ikemoto, 1983; Matsumoto et al., 1981

Key: [a] *Voltinism:* 1, 2, . . . = one, two, . . . generation(s) per year; S = several generations per year; ND = no data. *Diapause:* Ø = no diapause; E = embryo; P = pupa; A = adult; F = facultative; + = diapausing stage not indicated; ND = no data. *Darkening:* L = larva; P = pupa; A = adult; D+ = darkening increase; Om+ = ommochrome synthesis increase; S = solitarization; Ø = no action.

[b] *Zonocerus variegatus* and *Phymateus* sp.

[c] *Barbitistes fischeri (B. berenguieri), Decticus verrucivorus,* and *Orphania denticauda* (var. *azami*).

[d] *Spodoptera (Laphygma) exempta* and *S. copicola.*

then in the epidermis (only at molt), seems to be the only cause of this color change. The act of turning green must be considered here to be a phenomenon entirely independent of the crowding effect even if the two processes are superimposed on one another (see Section 6.2.3.1, below). Therefore, as noted by Staal (1961), the green coloration should not be considered as the main character of the solitary phase (the morphometric numbers are better but little affected by CA implantation).

Implantation of extra CA in *Locusta* hoppers (P. Joly, 1952b; P. Joly and L. Joly, 1953; L. Joly, 1960; Staal, 1961; Girardie, 1967; Roussel, 1975; Bouthier, 1979), like JH injection (Roussel, 1975, 1976) provokes a substantial disturbance in their further development.

An extra larval molt instead of the imaginal molt gives rise to either supernumerary larvae (also called "adultoids," in reality sixth-instar larvae), or pseudoadults (imperfect), or adults according to their morphology. A notable proportion of implanted animals die without molting or during the molt. The pseudoadults are seen in various color types (personal observations and data from experiments by Bouthier, 1979): a green epidermal color type (27–45%), a beige or brown epidermal color type with green hemolymph (9–22%), and a normal "gregaricolored" type (45–50%). Some of these animals achieve a sixth molt; many die at this time. Pseudoadults have an acridiommatin level much higher than that of the supernumerary larvae, but the latter, when they do not molt, increase their acridiommatin content between twice and thrice that measured at the beginning of this sixth "permanent larval" instar. This extra larval instar therefore behaves like normal instars, and JH does not lessen ommochrome metabolism whereas it indubitably enhances that of biliverdin. Natural anti-JH products (precocenes I and II) induce precocious imaginal molts in *Locusta* and *Schistocera* hoppers, which become normally pigmented adults (Pener et al., 1978; Nêmec et al., 1978; Kruse-Pedersen, 1978; Unnithan et al., 1980). JH II or III acts on *in vitro* cultivated larval epidermis of *Schistocerca*, delaying the normal disappearance of the ommochrome granules, whereas JH I or ecdysteroids (20-E) hurries this process along (Caruelle, 1976; Caruelle and Cassier, 1979).

Thus, the impact of JH on the ontogenetic program seems to belong to an all-or-none mechanism switching on development to a larval or an adult differentiation, if we consider only pigmentation. More generally, JH action on pigmentation is inseparable from its general influence on larval–adult development.

The rearing of Acrididae in wet conditions induces the epidermis to become green (Faure, 1932; Okay, 1956), but there is no experimental proof of an effective role of JH here.

A similar action (epidermal clearing without green pigment occurrence) was reported after CA implantation in *Gryllus bimaculatus* by Roussel (1967), as well as after synthetic JH application on the same species

(Fuzeau-Braesch, 1968) or after action of JH analogues on *Gryllodes sigillatus* (Yaragamblimath and Mathad, 1978). The multiplicity of geographic races and the extreme individual variability in *Locusta migratoria* subspecies make it hard to draw definite conclusions as to the hormonal determinism of color change from experimental results. Many experiments have been made on the laboratory strain *Locusta migratoria migratorioides* R. & F. issued from an African stock of the Anti-Locust Research Center, but some have been performed on the circummediterranean strain *L. migratoria cinerascens* Fabr., which is less amenable to crowding.

Low temperatures provoke an increase of epidermal ommochrome synthesis in many species that persists in adults (Okay, 1954; Bouthier, 1979), but its hormonal control remains unknown. Daily CO_2 action induces an epidermal darkening in *Locusta* gregarious adults, restoring in gregarious hoppers the homochromic response to background color present in solitary hoppers (Nicolas and Fuzeau-Braesch, 1968; Nicolas, 1972).

Lepidoptera

In Lepidoptera, caterpillars that are uniformly or largely green exhibit a spectacular color change during their last larval instar, turning brown or orange-yellow or grey: they do not become green again until the pupal stage. Caterpillars having another basic coloration more rarely show a noticeable color change. In Lepidoptera, the action of JH on larval color change seems to be determining in most species studied. The color change is generally characterized by a transitory increase of the ommochrome epidermis content (often xanthommatin): in *Cerura vinula*, this ommochrome increase is explained by the liberation of tryptophan from hydrolyzed proteins the amino acids of which are reutilized for the synthesis of the fibrous proteins needed for cocoon spinning (Linzen and Bückmann, 1961). This attractive explanation is unfortunately not applicable to caterpillars other than that of *Cerura*, except perhaps the silkworm *Bombyx mori* (Ohashi et al., 1983): according to Linzen (1974), the amounts of kynurenine, 3-hydroxykynurenine, 3-hydroxykynurenine glucoside, and xanthommatin are together exactly equivalent to that of the tryptophan released.

The modes of action of the two kinds of eumorphogenetic hormones are often antagonist even in closely related species. Injected 20-E causes an epidermal ommochrome content increase in *Cerura* (Bückmann, 1974b), but a decrease in *Pieris brassicae* (Beydon et al., 1980) and *Cephonodes hylas* (Ikemoto, 1975).

In another species, *Deilephila nerii* (Chang and Jang, 1980), the color change consists in the melanization of the integument, which passes within 5 h from green to dark black dorsally and orange ventrally some

36–48 h before the larval–pupal ecdysis. Here 20-E and JH-like substances injected or applied at high levels during the precritical period block melanization but have no effect on pupation. The active hormonal factor is ecdysone (E), which initiates the melanization process in synergy with a brain neurohormone if secreted during the sensitive period of the epidermal cells. Ligation at a specific moment (at least 24 h before the color change but before the pharate pupal stage) blocks this phenomenon; ligations made later are inactive.

We must distinguish between the hormonal action on synthesis and degradation of epidermal pigments: in *Pieris brassicae* (Beydon et al., 1980), 20-E is ineffective on epidermal xanthommatin synthesis (whereas JH I inhibits them) but causes the rapid (within 24 h) vanishing of xanthommatin granules in the epidermis (JH I shows an antagonistic effect); Beydon et al. (1980) evoke another possibility of control by a neuroendocrine mechanism. The same scheme is also true for *Cephonodes hylas* (Ikemoto, 1983) except for the action of 20-E on ommochrome synthesis (Ikemoto, 1975). Generally only pigment synthesis is considered.

In *Mamestra brassicae* (Hiruma et al., 1984), allatectomy inhibits the xanthommatin synthesis, which may be restored by application of JH-like substances. JH seems to be necessary to provoke ommochrome synthesis, even if the subesophageal neurohormonal secretion (MRCH) acts synergically, whereas MRCH is indispensable to melanization. A similar mode of action was demonstrated in *B. mori:* the JH-like substances injected just before the fourth molt induce xanthommatin synthesis with a dose-dependent effect (Kiguchi, 1973; Ohashi et al., 1983) and have no effect on melanization induced by 20-E (Ohashi et al., 1983) or by allatectomy (Kiguchi, 1973). JH enhances the yellow color of the *lemon* mutant of the silkworm due to xanthopterin (Kiguchi and Kimura, 1981); allatectomy has the opposite effect.

In *Manduca sexta,* extensively investigated by Riddiford and her coworkers, a quite different mechanism exists (see Riddiford and Hiruma, 1984); the synthesis or degradation of melanin, ommochrome, or insecticyanin (a chromoprotein with a biliverdin-IX-γ chromophore) is dependent only on the balance JH/20-E (MRCH has no effect on these pigments):

Prior to a certain critical period, if JH is absent, injection of 20-E acts dose-dependently on melanization, increasing the tyrosine level in the hemolymph (Ahmed et al., 1983); after that critical period, even high doses of 20-E are ineffective, i.e., premelanin granules are deposited in the cuticle only in the absence of JH during head capsule slippage preceding the last larval molt (Curtis et al., 1984). The decrease of the 20-E level induces activation of the prophenoloxidase in these granules (Hiruma et al., 1985). At the same time JH diminishes almost twofold the activity of dopa decarboxylase, which is completely inhibited if the 20-E level reaches or exceeds 2.6 μg/ml in the hemolymph (Hiruma and Riddiford, 1985, 1986). Thus, 20-E acts dose-dependently directly on epidermal cells (Hiruma and Riddiford, 1985).

In all larval instars, and there only, insecticyanin is synthetized during the entire instar and the mRNAs responsible for this synthesis are absent only around the molting time; during larval instars, JH induces a dose-dependent increase of epidermal insecticyanin whereas ecdysteroids show an antagonistic effect (Goodman et al., 1987). Allatectomy (5–6 h before head cap apolysis) inhibits the synthesis of new insecticyanin mRNAs, inducing in consequence the appearance of a pink color due to ommochrome (xanthommatin) synthesis; insecticyanin mRNAs reappear to a reduced extent thereafter; then JH acts as a partial repressor on the insecticyanin mRNA (larval) gene, and a lack of JH causes an enhancement of kynurenine hydroxylase activity (about three times higher) and therefore an increase in the ommochrome content. The two isoelectric forms of insecticyanin (a and b) appear in a sequential mode during the larval instar (namely, the fifth): insecticyanin a appears and disappears first, and insecticyanin b appears after the a form but persists longer; the nature of the hormonal determinism of this sequence is still unknown (Riddiford et al., 1986).

The essential role of JH in the synthesis of the biliary blue/green pigments would be clarified if the mechanism demonstrated in *Manduca* was revealed to be applicable to other insects (Orthoptera, for instance).

In *Cerura vinula,* the color change corresponds to a succession of biochemical processes controlled only by the eumorphogenetic hormones (E and JH). At first, the brown saddleback spot (xanthommatin) becomes red (dehydroxanthommatin) at the beginning of the starvation period (10 days before pupation); then, from cocoon-spinning onward the whole epidermis turns red, with a rapid increase of the dehydroxanthommatin content, which disappears quickly by transformation (esterification) into rhodommatin and ommatin D (5 days later) (Linzen and Bückmann, 1961; see the color plates in Bückmann, 1959, and Hintze, 1969). Ligation of the larval abdomen before the change in color of the dorsal spot inhibits this color change and prevents reddening of the abdomen. After this, the color change occurs in spite of ligation and persists without evolution (Bückmann, 1952, 1953). This process is E-dependent: low doses induce xanthommatin synthesis; middle-sized doses directly provoke the appearance of esterified pigment forms; and high doses cause ecdysis without color change (Bückmann, 1959). E seems to act on ommochrome synthesis itself via release of the protein tryptophan and synthesis of 3-hydroxykynurenine (Bückmann, 1974b). JH has an antagonistic effect, blocking the ommochrome synthesis only on the treated half of the larva (Hintze-Podufal and Fricke, 1971).

In *Hestina japonica* (and perhaps *Sasakia charonda*), the control mechanism is different again: E and JH are ineffective; the active factor seems to be neurohormonal and secreted by the posterior part of the ventral nerve cord (Osanai and Arai, 1962, with color plate).

Hymenoptera

The caterpillar-like larvae of sawflies (Symphyta) undergo a drastic color change at the last larval molt, losing their green larval coloration. The mechanism of control, certainly hormonal, is unknown.

Odonata

Odonata nymphs (*Aeschna cyanea, Agrion puella,* and various other species: Krieger, 1954) present a morphological color change occurring only at the time of ecdysis (with a critical period ending 3 days before molt) according to the background color: on a white background (as on a yellow, red, or green one) they get lighter; on a black background (as on a blue one) they darken. Lightening or darkening seems to be caused only by a variation in the cuticular melanin content. The stimulus, which is perceived by the eyes, is transmitted by an unknown mechanism and requires 2 days to become effective.

6.2.2.2. ADULT INDIVIDUAL COLOR CHANGE

This kind of color change is very scarce in holometabolous insects but exists in heterometabolous insects and seems to be related to the acquisition of sexual maturity.

Orthoptera

In *Locusta migratoria* (Orthoptera, Acrididae), adult males in crowded conditions, beige or light brownish colored just after the imaginal ecdysis, become deep yellow when they acquire sexual maturity. This yellowing is strictly dependent on the JH III released by the CA after action of a cerebral neurohormone (Pener et al., 1972). As both sexes age, browning of the abdominal epidermis occurs (acridiommatin accumulation: Bouthier, 1979). In *Schistocerca gregaria,* another gregarious acridian, the color change taking place in crowded animals at the beginning of adult life is more complex: animals of both sexes pass through pink coloration, then violet, before yellowing some 10–12 days after the imaginal ecdysis. The epidermis of both males and females shows the same reciprocal competence to yellowing (Fogal, 1968). Extirpation of the CA from young adults maintains the pink coloration (Cassier and Joulié-Delorme, 1976), whereas the same experiment performed on old adults causes rapid vanishing of the yellow color (within a week) (Pener, 1965, 1967), which is restored by topical application or injection of JH I or III (Amerasinghe, 1978), a dose-dependent effect (Pener and Lazarovici, 1979). The yellowing is so induced by topical application of JH analogue in precocious adults that were CA-deprived after anti-JH treatment (Unnithan et al., 1980). The ephemeral pinkish and purplish coloration is due to a transitory accumulation of acridiommatin granules; when they disappear, mineral inclusions replace them in epidermal cells before the advent of yellow carotenoids (Cassier and Delorme-Joulié, 1976). In *Nomadacris septemfasciata,* only the hind wings develop a red coloration prior to sexual maturity, and after that the body color becomes yellow (Faure, 1935, with color plates). Numerous acridians undergo a marked reddening of their abdominal epidermis associated with sexual maturity, a characteristic that may be fully acquired either before (*Omocestus ventralis, Chorthippus vagans*) or after (*C. brunneus, Stenobothrus lineatus*) maturation (Richards and Waloff, 1954). *Atractomorpha similis* undergoes wing and abdominal epidermis reddening that is clearly associated with sexual maturation (Bernays and Chapman, 1973).

Odonata

Other examples, comparable to that of *S. gregaria,* are described in Odonata (Zygoptera) of the genus *Ischnura:* in *I. elegans* (Europe and North Africa), the young adults are purplish-pink, turning to blue in males, and to blue, violet, orange, then green or brown in females, of more intense hue on the eighth abdominal segment, although with considerable variability, after which sexual maturity is reached. In Australian species exhibiting physiological color changes, the newly emerged males show no color change and remain yellowish brown to purplish orange until the appearance in the epidermis of "chromatophores" (Veron et al., 1974), which we shall call chromoepidermal cells (as defined earlier, in Section 6.1.3). According to Veron et al., "attainment of mature coloration probably coincides with onset of sexual maturity."

From abdominal ligature experiments, Veron (1973) postulated the existence of a darkening factor migrating from the posterior to the anterior part of the abdomen along the ventral nerve cord and controlling pigment migration in the chromoepidermal cells. Could this neurohormonal control be the cause of the color change in young adults?

6.2.3. *Morphological Color Changes Controlled Mainly by Neurohormones*

6.2.3.1. DENSITY-DEPENDENT LARVAL DARKENING

The crowding effect, related or not to migratory behavior, is widespread among insects (Odonata, Orthoptera, Phasmoptera, Homoptera, Psocoptera, Coleoptera, Lepidoptera), but this group effect acts on body color only in Orthoptera and Lepidoptera (see the reviews of Iwao, 1968;

Nolte, 1974; Nijhout and Wheeler, 1982; Pener, 1983; Hardie and Lees, 1985).

Hormonal Control

Numerous species are known with density-dependent larval darkening (see Table 6.4), but few have been seriously studied regarding hormonal control of their color change as it relates to crowding. The migratory species of Acrididae on which swarming confers an important economic impact are evidently the best investigated, as are also many pest moths, but we must take into account the preliminary discussion and conclusion of Pener (1983) in a chapter dealing with endocrines and phase polymorphism in locusts.

Generally, the crowding effect amounts to a more or less complete darkening of the integument, namely, a melanin deposition in the endocuticular layers. Concurrently with this melanization, a synthesis of ommochromes occurs in epidermis cells with the same dispersion pattern as that of the superimposed cuticular melanic pigment. This darkening appears, and sometimes intensifies, with the successive larval instars; and because of its cuticular localization, it takes place immediately after the time of ecdysis. This temporal coincidence has inspired the idea of hormonal control by eumorphogenetic hormones. In fact, there is some experimental proof of ecdysteroid action on tyrosine metabolism just after ecdysis (Karlson and Schweiger, 1961; Karlson and Sekeris, 1962; Sekeris and Fragoulis, 1985; see also Section 6.2.2.1, above), but this mechanism does not seem to be widespread in most insects.

In both orders the aforementioned mechanism essentially affects the nymphs, frequently also the adults in Orthoptera (Ritchie, 1978, with color plates), rarely the pupae and never the adults in Lepidoptera. In Orthoptera, darkening can also result from the influence of a burnt background on the adult integumental color (Burtt, 1951). In Lepidoptera, certain noctuid and geometrid moths are affected by "industrial melanism," which is a pigmentary response to the environment.

Orthoptera

Some Acridian species regularly manifest gregarious behavior, with repercussions on the morphology of both the hoppers and adults (*Locusta migratoria, Schistocerca gregaria, Locustana pardalina, Nomadacris septemfasciata, Dociostaurus maroccanus*). Many others less regularly or only occasionally exhibit gregarious behavior, such as *Calliptamus italicus*, American *Melanoplus* or *Dissosteira* sp. (Uvarov, 1966), *Oedalus senegalensis* (Ritchie, 1978), and some species of Tettigoniidae (Della Beffa, 1948). A few species undergo the gregarious transformation only in hoppers (*Phymateus* or *Zonocerus variegatus:* Uvarov, 1966) or only under crowded rearing conditions (*Cyrtacanthacris tatarica* or *Schistocerca obscura:* Uvarov, 1966). Rowell (1971) distinguished three main kinds of color changes in Acridians: (1) a homochromic reaction to a colored background; (2) "green/brown polymorphism"; and (3) a color change related to phase. *Locusta* and *Locustana* (Faure, 1932) manifest these three kinds together or one at a time or associated in pairs. *Schistocerca* and *Nomadacris* are mainly subject to phase-related color change (for additional details see Pener, 1983; also color plates in Faure, 1932, 1935). The first kind will not be considered here, as its hormonal determinism is not at all clear. The second was considered in Section 6.2.2.1. We can, however, add that in *Locusta* (L. Joly et al., 1977), as in *Schistocerca* (Injeyan and Tobe, 1981), the JH level is higher in solitary hoppers than in gregarious ones, especially at molting time (L. Joly et al., 1977). The third kind will now be discussed.

In locusts, melanization related to gregariousness has been successively attributed to diverse hormonal, neurohormonal, or even pheromonal factors:

- an unknown hemolymph factor (Nickerson, 1954)
- absence of PTG (Ellis and Carlisle, 1961)
- a neurohormone of the CC (Staal, 1961)
- a pheromone present in feces (Nolte, 1963; Nolte et al., 1970) known as locustol (Nolte et al., 1973),
- a neurohormone produced by the median NSC of the pars intercerebralis (C cells) and later elaborated by the CC (Girardie and Cazal, 1965; Girardie, 1967, 1974),
- a neurohormone released by the nerves posterior to metathoracic ganglia (in hatchlings of *S. gregaria* just after the first ecdysis: Padgham, 1976)

It seems that the most effective hormonal factor in the control of melanization is the neurohormone demonstrated by Staal (1961) and later by Girardie and Cazal (1965). Our conclusions are not consistent with those drawn by Pener (1983), but we consider here only the "melanization" component within the gregarization multicomponent complex. Girardie's laboratory experiments were performed on *L. migratoria cinerascens* (Girardie and Cazal, 1965; Girardie, 1967), then on *L. migratoria migratorioides* (Girardie, 1974). They consist of the selective electrocoagulation of the different NSC groups of the pars intercerebralis. The central AB zone controls (inhibits) the activity of the CA, whereas the peripheral C zone counteracts this effect. Moreover the C cells have an intrinsic melanotropic action. The lateral NSC have an action similar to that of the C cells (Girardie, 1974). Fifth-instar *Locusta* hoppers issued from NSC

C-deprived fourth instar have a lighter appearance (melanin deficiency) than controls, while their ommochrome content seems to remain unaffected (Bouthier, 1976, 1979). This neurohormone is effective only on melanin synthesis and not on that of ommochromes.

Lepidoptera

The crowding effect accompanied by a color change, namely, darkening, can affect some species more and other closely related ones less, within the same genus. For example, the noctuid *Orthosia incerta* manifests a general blackening when crowded, whereas *O. cruda* shows only a dorsal browning, *O. gothica* a grayish color, and *O. stabilis* almost no change (Long, 1953).

Superimposition of the crowding effect on the frequent normal individual color variation, as in the noctuid *Diataraxia oleracea* (Long, 1953) or in the migratory locust (*L. migratoria:* P. Joly and L. Joly, 1953; Nicolas, 1972), should be taken into consideration. For instance, Iwao (1962) distinguishes five larval color morphs in the common armyworm (*Leucania separata*): type 1 is yellow green to dull orange or reddish brown, where the reddish component (xanthommatin) (H. Ikemoto, 1968, cited in Ogura and Saito, 1972) is not affected by density and the brown component is liable to vary independently of the density; type 5 is intensively melanized (cuticular black melanin) (Ikemoto, 1971); and types 2, 3, and 4 are intermediates with a less or more melanized cuticle. In isolation, the level of color variation never goes beyond type 3. The darkening increases in proportion to the density. The color of a given instar is dependent on that of the foregoing instars: the darkening that occurs at the end of larval life is determined by the rearing density during the last three instars, but it is also affected by that of the first two even though no color difference (and no darkening) can be observed during the initial instars under crowded conditions.

The reversibility of the crowding effect decreases with the duration of conditioning. In *Leucania loreyi*, a density-dependent color change also occurs in the latter half of larval life, but it is not so conspicuous as in *L. separata* (Iwao, 1962). In *L. separata*, ligation applied 24 h before the fourth ecdysis inhibits darkening in the posterior part of the resulting fifth-instar gregarious larvae (and reddening in solitary ones). Removal of the subesophageal ganglia in gregarious larvae is more effective than that of the brain, whereas excision of the PTG has no effect (as in controls). Implantation of subesophageal ganglia or brain increases the melanic pigmentation (or the reddish brown pigmentation in solitary larvae). Implantation of brain–CA–CC complexes or subesophageal ganglia into ligated abdomens induces a black coloration in gregarious larvae and a reddish brown one in solitaries (Ogura et al., 1971; Ogura and Saito, 1972).

Implantation of silkworm subesophageal ganglia into isolated abdomens of *Leucania* induces cuticular melanization (Ogura, 1975b). Therefore, the same hormonal agent causes darkening in gregarious larvae and ommochrome synthesis in solitary ones. Transplantation experiments between gregarious and solitary larvae demonstrate that this factor exists at a similar level in the two populations (Ogura, 1975a); thus, they differ from one another only in the competence of epidermal cells to synthesize either pigment. The causal agent, MRCH, was extracted from the brain–subesophageal ganglia complex and subsequently from ganglia cultivated *in vitro* (Ogura and Mitsuhashi, 1978). It is effective on *L. loreyi, Spodoptera litura, S. exempta,* and *Mamestra brassicae.* MRCH is secreted 21 h before the molt, and melanization begins 9 h before ecdysis (Ogura and Mitsuhashi, 1979). JH has no action in tested animals, nor in controls.

In *S. litura,* JH is present at a higher level in solitary larvae than in gregarious ones (Yagi and Kuramochi, 1976), but topical application of JH or JH-like substances has no effect on either color or behavior (feigned death). It seems that MRCH is more effective on the specific darkening observed in the gregarious phase (Matsumoto et al., 1981) than the CA–JH system even if implicated in phase variation (Pener, 1983).

In *Mamestra brassicae,* MRCH induces a cuticular melanization characteristic of crowded larvae (Ogura and Mitshuhashi, 1979). When cultured *in vitro* in an anhormonal medium, larval integument develops a cuticular melanization that is inhibited in the presence of 20-E. Melanization normally provoked by subesophageal ganglia is also prevented by a previous conditioning with 20-E (Mochizuki and Agui, 1976).

The rearing conditions during the two preceding instars are determining for the darkening reaction in fourth and fifth instars of *Eudia pavonia,* and neurohormones are possibly involved in this regulation (Hintze-Podufal, 1977).

The action of JH on color change due to the crowding effect is clearer in larvae of *Cephanodes hylas:* topical JH application on crowded larvae causes a return to the solitary state. Cervical ligation applied to fourth-instar solitary larvae induces darkening at the next molt, a process inhibited by topical application of JH-like substance just after ligation with a dose-dependent effect. In this species, the darkening of crowded larvae is caused by the increase of both ommochrome and melanin content and not only by melanin as in *Leucania* (Ikemoto, 1983).

6.2.3.2. PUPAL COLOR CHANGE RELATED TO DIAPAUSE

Frequently, pupal diapause is correlated with a marked color change (Table 6.5), and environmental control (day length or light wavelength) has been demonstrated. The best studied example is *Pieris brassicae,* where melanization occurs sequentially during the first 6 h after pupa-

TABLE 6.5. Species Exhibiting a Pupal Color Change Related to Diapause[a]

Family and species	20-E	JH	PMRF	References
Pieridae				
Pieris brassicae	ND	Cl	+	Kayser-Wegmann, 1975; Ressin, 1980
P. napi	ND	ND	?	Smith, 1980
P. rapae crucivora	ND	Cl±	?	Ohtaki and Ohnishi, 1967
Nymphalidae				
Aglais urticae	ND	ND	+	Bückmann, 1960
6 species[b]	ND	ND	+	Koch and Bückmann, 1982/83
Inachis io	Ø	Ø	+	Bückmann and Maisch, 1987
Papilionidae				
Battus philenor	ND	ND	ND	Hazel and West, 1979
Papilio macilentus	ND	ND	+?	Shimada, 1985
P. machaon	ND	ND	ND	Hidaka, 1961
P. polyxenes	ND	ND	ND	Hazel, 1977
2 species[c]	ND	ND	+?	Hidaka, 1961

[a] *Key:* Cl = clearing; JH = juvenile hormone; PMRF = pupal melanization–reducing factor; ND = no data; 20-E = 20-hydroxyecdysone; Ø = no action; + = induction of pupal color change; ? = uncertain or not proved; Cl± = less clean clearing.
[b] *Araschnia levana, Limenitis camilla, Nymphalis antiopa, N. polychloros, Polygonia c-album, Vanessa atalanta,* and *V. cardui.*
[c] *Papilio protenor demetrius* and *P. xuthus.*

tion. Sectioning of the nervous chain between the brain and subesophageal ganglia provokes strong melanization, which is less obvious if the section is made between the subesophageal ganglia and PTG. Ligation between the first and second thoracic ganglia demonstrates the existence of two critical phases during the prepupal stage at 3 and 16 h of age, respectively (Kayser-Wegmann, 1975). JH I injected during these critical periods inhibits melanization (Ressin, 1980).

A similar control seems to apply to the related species *P. rapae crucivora* (Hidaka and Ohtaki, 1963) and *P. napi* (Smith, 1980).

Recently, a new hormonal factor, called PMRF, was isolated from heads and prothorax of *Inachis io* prepupae (Bückmann and Maisch, 1987). Its production by the cephalic nervous system induces the decrease or absence of cuticular melanin synthesis and of epidermal lutein incorporation (Maisch and Bückmann, 1987; Bückmann and Maisch, 1987). The existence of PMRF is presumed in *P. brassicae* and in numerous nymphalid species (Koch and Bückmann, 1982/83; see Table 6.5) and perhaps in papilionids.

In papilionids, some authors (e.g., Hazel and West, 1979) attribute the pupal color change to environmental conditions at the time immediately preceding pupation, whereas others advance a genetic determinism (Hazel, 1977) in other species; however, Hidaka (1961, with a color plate) supposes a humoral transmission mechanism implicating brain and PTTH.

In all these species, the background color to which prepupae are exposed (3–4 h after the spinning onset in *I. io:* Maisch and Bückmann, 1987) seems to have a determining influence on the resulting pupal color. In other species, such as *Malacosoma neustria testacea* (species with "brown pupal monochromatism"), melanization is strictly related to pupation and accompanies it obligatorily (Hidaka, 1959).

6.2.3.3. SEASONAL POLYMORPHISM RELATED TO PUPAL DIAPAUSE

Many Lepidoptera (Papilionidae, Pieridae, Nymphalidae, Geometridae, Notodontidae, Noctuidae, Sphingidae) develop with two or more (up to four or five) generations in a year. This situation implies a great plasticity in the adaptive mechanisms toward day length, warmth, and other conditions variable over the year. Often, they present a pupal diapause affecting the overwintering generation, preceded by a color change. Some species show a simple wing color polymorphism (as *Phytometra chrysitis* or *Phasiane clathrata*) or a regular succession of seasonal morphs identical year after year and affecting only wing coloration (e.g., species of the genus *Selenia;* see Table 6.6) or both color and form of the wings (*Polygonia c-aureum*).

Existence of seasonal morphs in Lepidoptera is always related to overwintering diapausing pupae (see Table 6.6). However, it seems that diapause is not necessary for the occurrence of spring morphs (cf. *Papilio xuthus:* Endo et al., 1985). The factor determining this seasonal polymorphism, when sufficiently elucidated, seems to be at first a neurohormone produced by the median nerve cells of the brain of nondiapausing pupae. It induces the appearance of summer forms in *Polygonia c-aureum* (Endo, 1984), *Lycaena phlaeas daimio* (Endo and Kamata, 1985), *P. xuthus* (Endo and Funatsu, 1985; Endo et al., 1985; see Table 6.6). An action of 20-E by mediation of diapause has been reported in *Araschnia burejana,* with opposite results according to the time of the intervention: induction of summer forms at the beginning of diapausing stage or of spring forms later in the pupal stage (Keino and Endo, 1973).

This fact has been recently confirmed in *Araschnia levana* (Koch and Bückman, 1987): summer forms result from the presence of ecdysteroids 3 days after larval–pupal ecdysis, whereas 7 days later or after a diapausing period ecdysteroids induce the realization of the spring form. The brain–subesophageal ganglia–CC complex seems to have no influence in this control.

No general feature can be given to characterize these seasonal morphs: in some species the summer forms are browner; in others they are lighter or lack the orange hue.

Some examples of seasonal polymorphism have been reported in tropical African Lepidoptera, and these are related to the annual alternation of rainy and dry seasons. The two species mentioned in Table 6.6 belong to the family Nymphalidae, but probably examples just as good could be found in other families and other tropical regions. Mell (1931) tied the dry-season form of *Hestina assimilis nigrivena* (the lighter form) to a diapausing form. Experiments on pupae of *H. assimilis* (Mell, 1931) exposed to wet or dry conditions showed that the appearance aspect of the resulting adults is modified in accordance with the causal agent, but no indication is given as to the nature of the physiological mechanism involved.

Rare cases are known in other insect orders. In *Chrysopa carnea* (Neuroptera), seasonal color polymorphism affects both body color and wing color. Honêk (1973) clearly established a determinism by both temperature and photoperiod and a facultative relation with winter diapause (Honêk, 1976), but he postulated, in addition, the action of another intrinsic factor of unknown nature. The biochemical cause of the color change has been demonstrated by Rüdiger and Klose (1970): the xanthommatin content increases, while the biliverdin-IX-α amount decreases simultaneously during the year. The autumnal color change of *Nezara viridula*, and perhaps of other Heteroptera such as *Palomena prasina* or *P. viridissima* or Miridae, represents another case and is also related to winter diapause caused by temperature and photoperiodic variations, but the hormonal mechanism is just as unknown, even if the cause (accumulation of yellow or red flavonoids) has been elucidated (Gogala and Michieli, 1962, 1966).

6.2.4. *Insect "Chromatophores" and Their Control*

Insect "chromatophores," or rather chromoepidermal cells with pigment migrations (see Section 6.1.3. for our justification of this view), are present in the epidermis of only two groups of insects: Phasmoptera (Fig. 6.2) and certain odonate Zygoptera (Veron, 1973).

In Phasmoptera, while many species show a color change adapted to

TABLE 6.6. Adult Seasonal Polymorphism Related to Pupal Diapause[a]

Groups Family and species	Seasonal forms	Diapause	Decap. decereb.	Impl. cer.	20-E	References
Lepidoptera						
Nymphalidae						
Araschnia burejana	yellow spots/ reddish yellow spots	P	ND	ND	1–2	Keino and Endo, 1973
A. levana	red colored	P	Ø	Ø	1–2	Koch and Bückmann, 1987
gen. est. *A. prorsa*	black & white					
Polygonia c-aureum	reddish yellow/ brownish black	ND	2	2	ND	Fukuda and Endo, 1966; Endo, 1984
	brownish yellow/ dark yellow					
Precis octavia	rainy season, red; dry season, dark blue	ND	ND	ND	ND	Stanêk, 1985
Hestina assimilis assimilis	rainy season, melanized + little red					
Hestina (Diagora) nigrivena	dry season, light	ND	ND	ND	ND	Mell, 1931

Lycaenidae						
Lycaena phlaeas daimio	red/reddish brown	P	2	1	1	Endo and Kamata, 1985
Papilionidae						
Papilio xuthus	smaller, orange col.; larger, without orange	P	Ø–2	1	Ø	Endo and Funatsu, 1985 Endo et al., 1985
Geometridae						
Selenia lunaria gen. aest. *delunaria*[b]	larger, brownish; smaller, light	P	ND	ND	ND	Herbulot, 1949
Neuroptera						
Chrysopa carnea	ultragreen to yellow-green salmon	A	ND	ND	ND	Honêk, 1973, 1976
Heteroptera						
3 species[c]	green-brown	A	ND	ND	ND	Gogala and Michieli, 1962, 1966

[a] *Key:* A = presence of an adult winter diapause; Decap. decereb. = decapitation, decerebration; gen. aest. = specific name of the summer generation without diapause [for each species spring form is described first and summer form second (in *Polygonia c-aureum* the colors of upper and lower wing sides of each seasonal form are given)]; Impl. cer. = implantation of cerebrum; P = presence of a pupal winter diapause; Ø = no action; 1 = enhancement of sumer forms; 2 = enhancement of spring forms; 20-E = action of 20-hydroxyecdysone.

[b] Two other species: *S. bilunaria* gen. aest. *illunaria* and *S. tetralunaria* gen. aest. *aestiva*.

[c] *Nezara viridula*, *Palomena prasina*, and *P. viridissima*.

the dominant background color, the most studied species are *Carausius (Dixippus) morosus* and *Clitumnus extradentatus*. These animals are able to become green, beige, brown, or black according to the background. The pigments involved in these changes are biliary pigments (biliverdin), five carotenoids (β,β-carotene-2-ol, β,β-carotene-2,2′-diol, β-carotene, lutein, and an unidentified form: Kayser, 1976), two ommochromes (xanthommatin and ommin), four pterins (leucopterin, isoxanthopterin, xanthopterin, and pterin: Berthold and Henze, 1971), and melanin. The ommochrome content remains about constant (10 μg/animal) during the first five larval instars and then triples during the sixth instar. Then it reaches 35 μg in adult males and 57.5 μg in adult females (Stratakis, 1979).

Presence of CA (and JH) has no influence on carotenoids and is not necessary to maintain viliverdin synthesis, but seems to be essential to that of ommochromes. Extra CA implantation induces an enhancement of the ommochrome content in epidermis (Berthold, 1973), whereas implantation of extra PTG leads to a darkening in beige or green animals owing to an increase in the ommochrome content only if the CA are present (Raabe, 1961). Implantation or extirpation of CA (Raabe, 1962; Dustmann, 1964) counterfeits respectively rejuvenation or aging of the larvae, reversing the chromatic reaction, because only the larvae are able to synthetize ommochromes (Berthold, 1973). Darkness and/or temperature increase induce a dark color in animals by synthesis of ommochromes and melanin, the only pigments whose level can be modified; after darkening of the ventral part of the eyes, the epidermal ommochrome content reaches 600 μg in adult animals (Bückmann, 1977, 1979), whereas the fecal excretion of kynurenic acid falls to 1.1. mg/g, instead of 2.4 mg/g in the feces of controls (Berthold and Bückmann, 1975).

In CA-deprived animals, higher temperatures no longer affect darkening, but darkness maintains the same influence (Berthold, 1973). The latter fact implies the existence of another hormonal control for this darkening. Indeed, such a mechanism was demonstrated by Raabe (1961, 1962, 1966, 1983a): implantation of subesophageal ganglia (where B cells are numerous and manifest a cyclic activity) is effective, in contrast to that of pars intercerebralis or whole brain.

Another cause of color change is superimposed on those described above: a circadian rhythm affecting the whole epidermis except in green animals. Beige individuals become brown, and gray ones black. This is owing to an active migration of black ommochrome pigment granules along epidermis cytoskeletal microtubules (Berthold, 1980), a migration whose mechanism is still unknown but seems to be under neurohormonal control.

In Odonata, color changes are quite different: weak lighting or low temperatures induce a change from the normal blue color to a dark hue (which becomes effective in 9 h); increase in either temperature and/or light intensity causes a faster return to the blue color (Veron, 1976; see color illustrations in Veron, 1973, and Veron et al., 1974). The neurohormonal factor responsible for darkening is released by the last abdominal ganglia, but the reverse color change seems to occur independently of the nervous system, as demonstrated by *in vitro* experiments (Veron, 1973).

6.3. Morphological Color Changes in Crustacea and Their Control

Morphological color changes and hormones have been much more studied in decapods than in any other group. Almost nothing is known about color changes and endocrinology in non-malacostracans, particulatly in important groups such as Amphipoda, Mysidacea, Stomatopoda, Anomura. The best studied groups are within Isopoda or Decapoda, namely, Caridea, Macrura Reptantia, and Brachyura. Thus, most of the examples are taken from these groups.

6.3.1. *Definitions and General Features*

Hormonal control of morphological color changes has curiously almost been ignored in Crustacea (see reviews of Panouse, 1946, 1947; Bellon-Humbert, 1970a; Nagabhushanam et al., 1981). Different aspects of crustacean morphological color changes and what is currently known about their hormonal control will be reviewed hereafter.

Unlike insects, where most of the morphological color changes are due either to cuticle pigments or/and cuticle-secreting cells (chromoepidermal cells), morphological color changes in malacostracan Crustacea are essentially due to specialized pigment cells, the chromatophores. However, some examples are known where, as in insects, "ordinary" cuticle secreting cells are pigmented and play a role in color changes.

Two more or less distinct groups can be distinguished within Crustacea according to dominant color changes: some species present essentially background color adaptation, whereas others have a developed nychthemeral rhythm of pigment migration. Both phenomena are hormonally controlled; this control has mostly been studied in the former group (Table 6.7).

6.3.1.1. CRUSTACEAN HORMONE-PRODUCING CENTERS

Most of the hormones affecting color changes in Crustacea originate from the central nervous system (CNS). Several studies have established

TABLE 6.7. Morphological Color Changes in Some Crustacea, as Part of Chromatic Phenomena[a]

Groups and species	Background adaptation (1)[b]	Nychthemeral rhythms (2)(3)[b]	Morphological color changes(1)[b]	References
Caridea				
Lysmata seticaudata	−	+++	+	Bhaud-Couturier, 1974
Processa edulis	−	+++	+	Noël, 1979
Periclimenes amethysteus	−	+++	++	Noël, 1983b
Palaemon serratus	+++	+	+++	(1)[b]
Palaemonetes vulgaris	+++	+	+++	(1)(3)[b]
Crangon crangon	+++	++	++	(1)[b]
Macrura Reptantia				
Cambarellus shufeldti	+	+	+	(3)[b]
Orconectes clypeatus	+	?	++	Fingerman and Aoto, 1958b

Brachyura				
Uca pugilator	++	++	+	Fingerman et al., 1966; Fingerman and Lowe, 1957b
Carcinus maenas	++	+	++	Powell, 1962b
Isopoda				
Idothea spp	+	+	+++	Lee, 1966a,b,c
Mysidacea				
Mysidacea spp.	+	+?	?	Keeble and Gamble, 1904; (1)[b]
Stomatopoda				
Chloridella empusa	+	?	?	Fingerman and Rao, 1969

[a] *Key:* − = almost absent; + = present (weak); ++ = present; +++ = well developed; ? = unknown or uncertain.
[b] (1) See references in Chassard-Bouchaud, 1965. (2) See references in Noël, 1982. (3) See references in Bellon-Humbert, 1970a.

that the different parts of the CNS contain unequal amounts of chromatophorotropins (reviewed in Noël, 1981). Therefore, two regions of the CNS are known to assume most of the control:

- the protocerebrum–X-organ–sinus gland complex, located in the eyestalk (Decapoda, Stomatopoda, Mysidacea, etc.) or head (Isopoda, Amphipoda) (Gabe, 1966; Cooke and Sullivan, 1982)
- the postcommissural organs that belong to the tritocerebrum (Knowles, 1953a,b; reviewed in Carlisle and Knowles, 1959)

6.3.1.2. CHROMATOPHOROTROPINS AND OTHER HORMONES INVOLVED IN MORPHOLOGICAL COLOR CHANGES

The biochemistry of crustacean hormones was recently reviewed by Kleinholz (1985). Other studies have shown the existence of antagonistic pigmentary hormones, one of them being a dispersing one, the other a retracting one (Fingerman and Aoto, 1958a). Two forms of the same hormone probably occur: an inactive complex and an active free form (Fingerman and Aoto, 1958b). These hormones occur in Crustacea both with (malacostracans) and without chromatophores (see references in Hanström, 1939; Noël, 1981). The structure of several peptidic chromactivating neurohormones was determined (see references in Rao, 1985); they belong either to the RPCH–AKH family (see Section 6.2.1.2, above) or to the PDH family. AKH hormones also have a RPCH activity. From the PDH family, only two structures (distal retinal pigment hormone, or DRPH) are known: the α-PDH from *Pandalus borealis* (Fernlund, 1976), and the β-PDH from *Uca pugilator* (Rao et al., 1985) and *Cancer magister* (Kleinholz et al., 1986).

1 3 5 7 9 11 13 15 17
Asn-Ser-Gly-Met-Ile-Asn-Ser-Ile-Leu-Gly-Ile-Pro-Arg-Val-Met-Thr-Glu-Ala-NH_2 α-PDH
Asn-Ser-Glu-Leu-Ile-Asn-Ser-Ile-Leu-gly-Leu-Pro-Lys-Val-Met-Asn-Asp-Ala-NH_2 β-PDH

6.3.1.3. PIGMENTED STRUCTURES: CUTICLE AND PIGMENT CELLS

Pigments are generally deposited in an amorphous state in the crustacean cuticle just before molting (exocuticle) (Figs. 6.3 and 6.4). The only pigments characterized in the crustacean cuticle belong to carotenoids. Pigment granules have never been demonstrated at that level. Changes in the coloration of the cuticle may be achieved only by molting. For this reason the anomuran *Hippa pacifica* needs several months for adaptation to a new background color (Wenner, 1972). Modifications of cuticle pigmentation occur within the isopod *Idothea* (Lee, 1966a); different pigments are present in the cuticle—the red canthaxanthin being present in

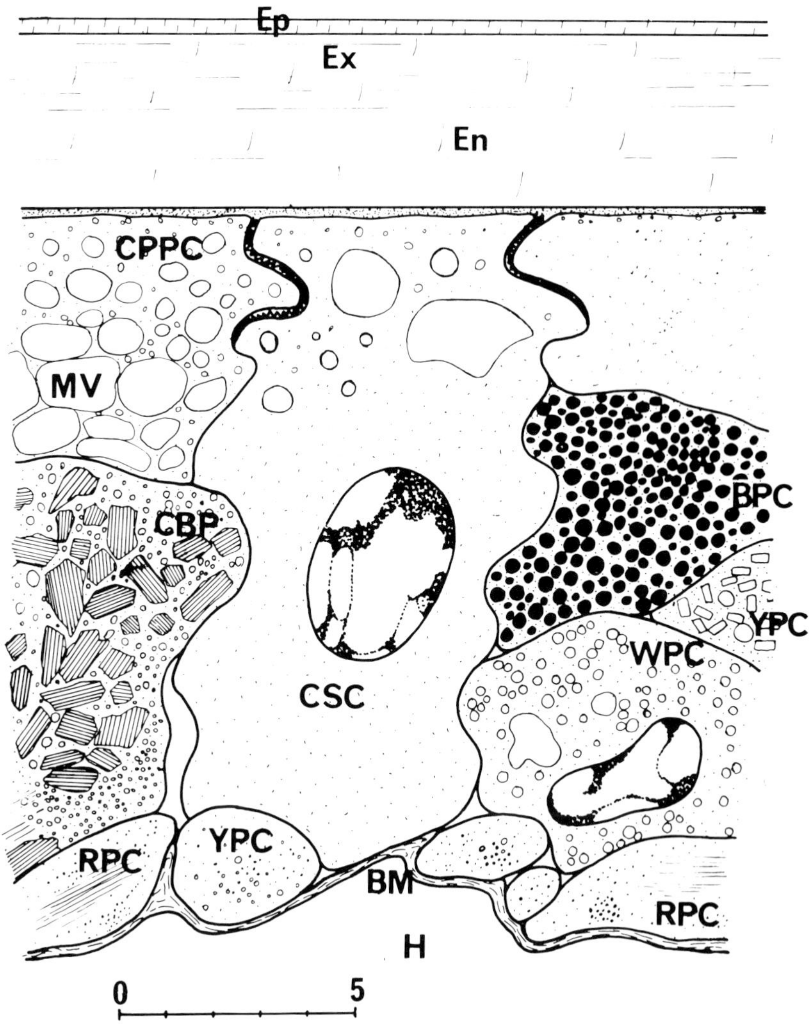

FIGURE 6.3. Shrimp integument section. Carotenoid-containing pigment cells such as yellow pigment cell (YPC)(= xanthophore), red pigment cell (RPC)(= erythrophore), cell with blue paracrystals (CBP)(= palaemonid erythrophore), and carotenoprotein pigment cell (CPPC) with pigment macrovesicles (MV) are schematized on left. On right, crangonid-dark (ommochrome) pigment cell (BPC)(= "melanophore") and light (pterins, flavins) pigment cell (WPC)(= leukophore) are represented. Pigment cells are typically associated into chromatosomes (CS). Epicuticule (Ep) and endocuticle (En) are not pigmented, whereas exocuticle (Ex) may be slightly pigmented by carotenoids. Cuticle-secreting cells (CSC) play a role in pigment transfer toward cuticle. Epidermic cells are separated from hemocoel (H) by basement membrane (BM). Bar = 5 μm.

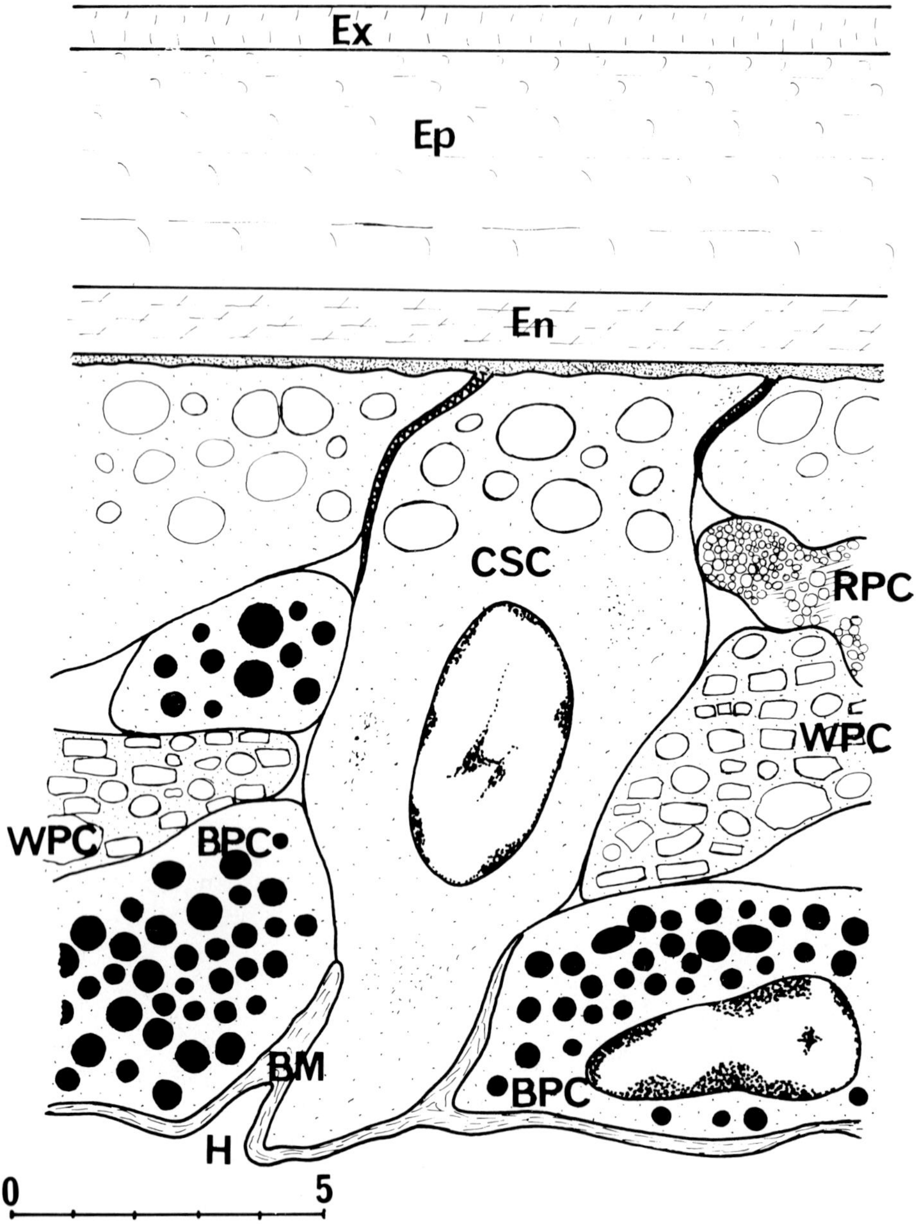

FIGURE 6.4. Section of crab integument. Epicuticule (Ep) is pigmentless; exocuticle (Ex) and endocuticle (En) contain carotenoproteins associated or not with matrix (opaque calcified cuticle). Red pigment (carotenoids) cell (RPC), dark (ommochrome) pigment cells (BPC), and white pigment cell (WPC) are present. Cuticle-secreting cells (CSC) play a role in pigment transfer toward the cuticle. Basement membrane (BM) separates epidermis from mesodermic derivatives such as hemocoel (H) and muscles. Bar = 5 μm.

the exocuticle, whereas lutein is the yellow pigment of the endocuticle (Lee, 1966c).

In epidermis, pigments can be either in cuticle-secreting cells (Bocquet, 1951; Koulish and Klepal, 1981; Barden and Koulish, 1983) or in chromatophores (references in Noël, 1979). Different types of chromatophores may assemble into chromatosomes. When these different types are present, only some chromatophores may play a role in color changes (Lee, 1972). Unlike those in vertebrates, crustacean chromatophores contain only one type of pigment.

Internal organs may contain substantial amounts of pigments that interfere with the general body coloration in more or less translucent animals. After egg laying, females may incubate eggs in their breeding pouch and the color of the egg mass or developing embryos changes the general color of the female. Crustacean pigmentation is due essentially to carotenoids and carotenoproteins (red, blue, green, brown colors), ommochromes (brown, violet, black), pterins, and flavins (white, yellow); melanins are never present in chromatophores, and may occur in cuticles (only associated with pathological processes). Color plates for illustration of morphological color changes may be found in Knowles (1955), Chassard-Bouchaud (1965), and Robison and Charlton (1973).

6.3.2. *Molting, Molting Hormones, and Color Changes*

Drach (1939) reported the first observations on color changes related to the molt cycle. Even in larvae, Costlow (1961) observed fluctuations of chromatophorotropins (black pigment–dispersing hormone, or BPDH) in the larval stages of the crab *Sesarma reticulata*. Activity is lower before and after molting. Chromatophorotropins extracted from the barnacle *Balanus eburneus* vary according to the molt cycle (Costlow, 1963). Color of the fiddler crab *Uca pugilator* carapace (Guyselman, 1953) changes during the premolt period, and these changes are associated with epidermal activity preceding ecdysis. The animals typically have brown or purple patterns, and a blue-gray color may be regarded as an indication of imminent ecdysis. No color change correlated with the molt is observed in eyestalkless crabs, indicating the possibility of an influence of hormonal factors. In Stomatopoda, a green male of *Gonodactylus breedini* became red just after molting and took 2 weeks to recover its normal green color (Morgan and Goy, 1987). Isopoda are known to present an exuviation in two steps; color change at molting was accordingly observed to occur with a 2-day interval (Lee, 1966a). In Caridea, several investigations have established that pigment dispersion varies during the molt cycle (Table 6.8).

TABLE 6.8. Pigment Dispersion Variations During Molt Cycle in Caridea (Decapoda)[a]

Species and scale used	Pigment or chromatophore	Intermolt stages (Drach, 1939)							References
		A	B	C	D_0	D_1	D_2	D_3–D_4 & E	
Lysmata seticaudata									
0 → 5	red I	4.5	4	3	3	3.5	4.5	4.5	Bhaud-Couturier, 1974
Palaemon elegans									
0 → 5	red	4		2.5	2	0.71	1.5	NG	Chassard-Bouchard and Genofre, 1974
	bl?	5.4		3	3	1.5?	?4	NG	
Leander serratus	red:								
0 → 12	stripes	NG	7.0	8	NG	8.3	8.4	8.4	Scheer and Scheer, 1954
	interspaces	NG	2.7	3.9	NG	3.8	4.1	3.8	
Leander xiphias									
0 → 4	white	2.7	2.1	2	NG	1.9	2.1	2.2	Scheer, 1960
	yellow	3.4	3	2.9	NG	2.9	2.8	2.7	
	red	2.5	2.3	1.9	NG	1.9	1.8	1.9	
Caridina weberi									
1 → 5	red	1.0	1.2	2.0	NG	2.6/2.7	2.7	2.8	Nagabhushanam and Vasantha, 1971
	white	1.0	2.0	2.5	NG	3.0	3.1	3.1	

Key: [a]NG = data not given. When more than one value is given for the same stage (for instance, subdivisions of the C or D_0 stages), only the mean is shown.

6.3.3. *Development, Juvenile Hormones, and Color Changes*

6.3.3.1. HORMONES CONTROLLING DEVELOPMENT IN CRUSTACEA

Methyl farnesoate is produced by the mandibular organs of decapods and can be converted to juvenile hormone (Borst et al., 1985; see also Chapter 2 by Borst and Laufer in Part 1); it seems to be present in the hemolymph of the spider crab *Libinia emarginata* (Laufer et al., 1986). Nothing is presently known about the roles of such hormones in Crustacea, especially in morphological color changes.

6.3.3.2. EGGS AND LARVAE

The first chromatophores differentiate usually before hatching, and the chromatophore system develops afterward as either a gradual uniform process or a stepwise one (Keeble and Gamble, 1904). Generally, larvae are colorless at the beginning of their life, particularly planktonic larvae. The chromatophore system develops progressively from hatching onward, and it may evolve or transform even in the adult. Pautsch (1967) has reviewed the first studies on larval chromatophores, and the first complete detailed study throughout development of color changes in a crustacean is that of Rao (1968) on the ghost crab, *Ocypode macrocera*. Table 6.9 summarizes current knowledge of pigmentation variations during crustacean development.

The hormonal control may appear relatively early in larval stages, as noted by Couturier-Bhaud (1975b) in the Monaco shrimp, *Lysmata seticaudata*, or somewhat later (Pautsch, 1967). In the juvenile period, the typical pattern of pigmentation is established.

Larval chromatophores could respond to chromatophorotropins (Rao, 1967). Chromatophorotropins appear early in development: BPDH was shown to be present in the first zoeal stage in the crab *Sesarma reticulata* (Costlow, 1961), but the great majority of color changes in crustacean larvae are in fact due solely to direct responses of pigment cells, particularly photoreactivity.

6.3.3.3. YOUNG AND IMMATURE ANIMALS

Evolution of the chromatophore pattern has been studied only in a few examples of higher Crustacea (Fig. 6.5); nothing is known about its control. Usually, the chromatophore number progressively increases during growth (Castrucci and Mendes, 1975), and new patches may appear even asymmetrically (Noël, 1983b). Appearance and disappearance of chromatophores may occur even in larval development (Couturier-Bhaud, 1975a). *Uca princeps* passes from a greenish-brown carapace

TABLE 6.9. Development of Color Changes in Decapoda[a]

Group and species	Stage of develop-ment	Pigment or Chroma-tophore	Reaction to illumina-tion	Diurnal rhythm	Back-ground adaptation	Response to Chromato phorotro-pins	References
Brachyura							
Uca pugilator	zoea 1–5	melanophore	NG	no	no	NG	Johnson, 1974
	megalopa	melanophore	NG	DD	yes	NG	
Sesarma reticulata	1st zoea	melanophore	dispersion	DD	no, un-certain	no VV	Nagabhushan-am, 1965
Ocypode ma-crocera	1st zoea	black	dispersion	NG	yes	yes VT; no VV	Rao, 1967
		white	dispersion	NG	no	yes	
	megalopa	black	dispersion	NG	yes	NG(5)	Rao, 1968
		red	dispersion	NG	no	NG(5)	
		white	dispersion	NG	yes	NG(5)	
	adult	black	no	yes ND	yes	NG(5)	Rao, 1968

		red	uncertain	yes ND	yes	NG(5)	
		white	dispersion	yes DD	yes	NG(5)	
Geograpsus lividus	1st zoea	black-dark	dispersion	yes DD (1,2)	no	—	Pautsch, 1967
		white	—	yes DD (1,2)	—	—	
	adults	black	—	yes	yes	—	Pautsch, 1967
		white	—	yes	no	—	
Carcinus maenas	prezoea	brown and yellow	no	—	—	—	Pautsch, 1961
	1st zoea	brown and yellow	dispersion	yes DD (3)	no	no	Pautsch, 1967
	juvenile	black	NG	yes DD (3)	NG	NG	Powell, 1962a
		red	NG	yes DD (3)	NG	NG	
		white	NG	no	NG	NG	

(continued)

TABLE 6.9. (*Continued*)

Group and species	Stage of development	Pigment or Chromatophore	Reaction to illumination	Diurnal rhythm	Background adaptation	Response to Chromato phorotropins	References
Rhithropanopeus harrisi	1st zoea	dark + yellow	dispersion	yes	yes	—	Pautsch, 1967
	later zoea & megalopa	melanophores	—	—	yes	—	Pautsch and Lawinski, 1967
	adult	black	dispersion	yes(2,3)	yes	—	Pautsch, 1967
		red	dispersion	yes(2,3)	—	—	
		white	dispersion	yes(2)	yes	—	
		yellow	dispersion	yes(2)	—	—	
Anomura							
Diogenes bicristimanus	"eggs"(4)	red	no	no	no	no	Nagabhushanam and Sarojini, 1969

	1st zoea	red	dispersion	yes DD	no	no VV	Nagabhushanam and Sarojini, 1969
	adult	no functional chromatophores					
Macrura Reptantia							
Upogebia affinis	—	red	dispersion	—	uncertain	—	Nagabhushanam, 1964
Homarus americanus	zoea 1–3	yellow/red	dispersion	yes DD	NG	NG	Hadley, 1905
	zoea 4+ juveniles 11 = adults	—	no	no	NG	NG	Hadley, 1905
Caridea							
Crangon crangon	"eggs"(4)	"embryonic"	no	NG	no	no VV	Pautsch, 1951, 1953
	1st zoea	Brown/ yellow	dispersion	NG	no	no VV	Pautsch, 1951, 1953

(continued)

TABLE 6.9. *(Continued)*

Group and species	Stage of develop-ment	Pigment or Chroma-tophore	Reaction to illumina-tion	Diurnal rhythm	Back-ground adaptation	Response to Chromato phorotro-pins	References
Hippolyte varians	zoea	red yellow-green	retraction expansion	NG NG	 uncertain	NG NG	Keeble and Gamble, 1904
Lysmata seti-caudata	zoea I–IV	yellow/red	dispersion	yes DD	NG	NG	Courturier-Bhaud, 1975b
	zoea V–IX	red type I	weakens	yes DD	NG	NG	
		red type II	weakens	yes DD	NG	NG	
Palaemonetes vulgaris	2nd zoea	—	—	—	yes	yes VT	Broch, 1960
Macrobra-chium kistnensis	1st(?) zoea	red	—	—	yes	—	Keshavan, 1974
Caridina weberi	1st(?) zoea	red	retraction VT	—	yes	—	Vasantha, 1968

[a] *Key:* DD = day dispersion; ND = night dispersion; NG = data not given by authors; VV = *in vivo;* VT = *in vitro;* ? = uncertain; (1) = abolished in constant darkness; (2) = abolished in constant light; (3) = persists in constant darkness for 3 days or more (indicating an hormonal control); (4) "eggs" = embryos with chromatophores before hatching; (5) probably yes.

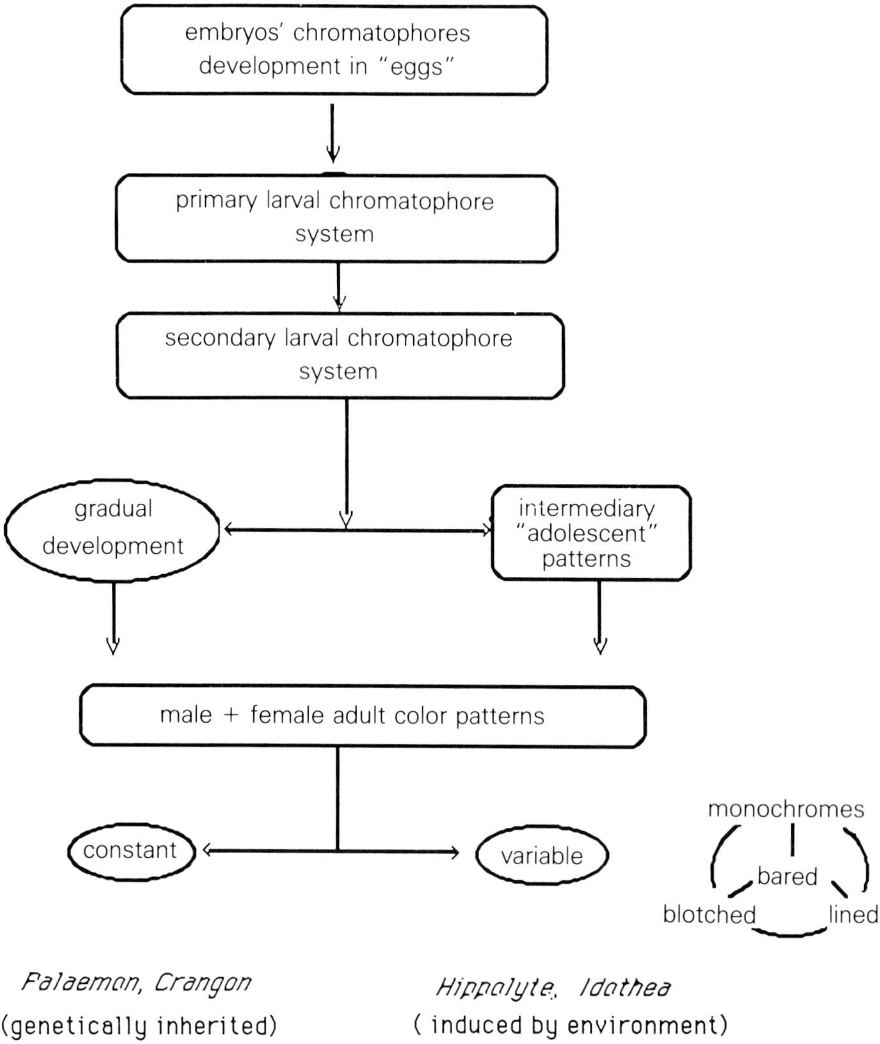

FIGURE 6.5. Evolution of chromatophore pigmentation in higher Crustacea. (Simplified from data of Keeble and Gamble, 1904, and Chassard-Bouchaud, 1965.)

through orange and then yellow phases before becoming almost wholly dazzlingly white (Crane, 1944). Chromatophorotropins present variations with aging (= size) in crabs (Rao, 1968; Rao and Fingerman, 1969), and since they are present in eggs, they appear before differentiation of target cells, i.e., chromatophores (Costlow and Sandeen, 1961; Vasantha, 1971). Since BPDH concentration is higher in small crabs than in larger ones, it has been suggested that chromatophore sensitivity to chromatophorotropins increases in proportion to size (Rao and Fingerman, 1969). A non-chromatophoral pigmentary evolution has also been described in harpacticoid copepods from larvae to adults (Bocquet, 1951), but its control is unknown.

6.3.3.4. SEX, SEXUAL MATURITY, AND COLOR CHANGES

Differences in the pigmentation between males and females may appear at sexual differentiation (Knowles and Callan, 1940). The morphology and physiology of adult pigmentation have occasionally been reported to vary according to sex and reproductive activity. Sexual dichromatism is often limited in Crustacea to minor differences. In Decapoda, it was reported to occur in portunid crabs (Hopkins, 1963). In the blue crab, *Callinectes sapidus,* the female shows bright orange fingers on the chelae, as compared to the male whose fingers are white on the outer margins and blue on the inner margins. Similar dichromatism is also observed in crabs of the genus *Ovalipes.* In fiddler crabs such as *Uca saltitanta, U. stenodactyla,* and *U. latimanus,* where the dimorphism in size is extreme, the female is always very dull and the male exceptionally brilliant in display coloration (Crane, 1944); this could well be an example of linked hormonal effects. Stomatopoda also present some examples of sex pigment dimorphism (see references in Morgan and Goy, 1987): the male of *Gonodactylus bredini,* for instance, is green, whereas the female is pinkish red or rust colored. Pigment sexual dimorphism within Isopoda is rarely reported; males of *Spheroma teissieri* have a large red spot on the pleotelson (Bocquet and Lejuez, 1967). Another type of sexual differences is found in hermaphrodites: for instance, males of *Lysmata seticaudata* correspond both to males and younger (1-year) shrimps (Bhaud-Couturier, 1974). In other decapod species, males are simply smaller and correlatively less pigmented than females, which appear darker (Keeble and Gamble, 1900a; Miyawaki and Tsuruda, 1984). Pigment sexual dimorphism does exist in amphipods. The male of the sand flea *Talitrus saltator* possesses second antennae containing astaxanthine as the major carotenoid, while female antennae are almost pigmentless. This pigmentation is under hormonal control: ablation of the male androgenic glands leads to the disappearance of pigmentation within 3–4 weeks. So the androgenic hormone may control pigmentation in male antennae (Barbier et al., 1966). This is to our knowledge the only study establishing the role of sexual hormones in pigment sexual dimorphism in Crustacea. The hormonal nature of the modification is indirectly proved in parasitized decapods where male parasitized animals present female characteristics.

In freshwater prawn *Macrobrachium rosenbergii,* three male morphotypes also exhibit three successive types of pigmentation: small pale males, orange-claw males, and dominant blue-claw males (see references in Sagi et al., 1988). Nothing is known about the control of these pigmentations.

Sex hormones seem to play a minor role in crustacean pigmentation as compared with genetic factors. Gynandromorphic specimens are known only in decapods, and such individuals have the male pigmentation on one side and the female pigmentation on the other. Bicolored specimens are known in Macrura such as *Homarus* (see references in Chace and Moore, 1959), *Palinurus, Jasus,* and *Cambarus,* Anomura such as *Eupagurus,* and Brachyura such as *Uca, Carcinus,* and *Potamon* (see references in Charniaux-Cotton, 1975). Genetic origin of pigmentation is also documented in harpacticoid copepods (Bocquet, 1951) and isopods (Bocquet et al., 1951), where a rare nongenetic determinism is also known (Bocquet, 1960).

Males do not usually exhibit modifications of pigmentation according to the reproduction season. In contrast, females often display marked color differences: white pigments in particular are frequently enhanced by vitellogenesis (Panouse, 1946). For instance, in female shrimps, variations of white chromatophore number are correlated with vitellogenesis (Chassard-Bouchaud, 1965); moreover, red pigment dispersion in the midgut erythrophores is more marked in females in reproduction than in other females (Miyawaki and Tsuruda, 1984). Incubation of eggs is also accompanied by morphological color changes and consequently pigmentation changes. Pigmentation develops at that time in oostegites of Mysidacea (Keeble and Gamble, 1904): females at the time of hatching are darker than males (Chicewicz, 1951). Some of the seasonal color changes can be attributed to acquisition of sexual maturity (Crane, 1944; McVay, 1942), but others are only attributable to chromatic adaptation (*Hippolyte varians*) depending on seasonal changes of vegetation.

Sex also modifies the variation of chromatophore number in long-term background adaptation (Table 6.10): females usually have larger variations than males. Frequently, when the chromatophore responses were studied separately in males and females, differences were evidenced. Pigment dispersion in male chromatophores is less effective than in females (Chassard-Bouchaud, 1965; Bhaud-Couturier, 1974).

6.3.4. *Pigment Cells, Pigment Hormones, and Morphological Color Changes*

6.3.4.1. MORPHOLOGY: PATTERNS, CHROMATOPHORES, AND THEIR MODIFICATIONS

The most striking morphological color change phenomena in Crustacea are probably those present in the littoral shrimp *Hippolyte varians.* The so-called sea chameleon displays many different colored forms and can

TABLE 6.10. Morphological Chromatic Adaptation: Chromatophore Number Relative Evolution After 1 Month on Black or White Background in Males and Females

Chromatophore: *Background:*	Black Black/white	White Black/white	Yellow Black/white	Red Black/white
Species and sex				
Crangon crangon				
Male	+56%/0%	−42%/+203%	+10%/0%	0%/0%
Female	+45%/+13%	−50%/+351%	+12%/+16%	+6%/+16%
Athanas nitescens				
Male	no	0%/+32%	no	+44%/0%
Female	melanophore	0%/+100%	xanthophore	+42%/0%
Palaemon serratus				
Male	no	−15%/+3%	0%/−20%	0% to +20%/−17% to −20%
Female	melanophore	−33%/+7%	+40%/−50%	0% to +40%/0% to −50%

SOURCE: Adapted from data of Chassard-Bouchaud (1965).

adapt its color quite well to a new background within a few weeks (Kröyer, 1842; Keeble and Gamble, 1900a,b, 1904, 1905; Minkiewicz, 1908; Kleinholz and Welsh, 1937). Its morphological chromatic adaptation is now well known (Chassard, 1957, 1958b; Chassard-Bouchaud, 1965). Homochromy results from modification of both the place and number of chromatophores. For instance, a heterogeneous brown pattern can be transformed into a uniform green one within a few weeks. Progressively, developing chromatophores appear in translucent parts of the body. The appearance of new pigment cells and the vanishing of others is characteristic of this complete transformation of the pigment pattern. Reversion to the initial pigment pattern is possible, but the chromatophores do not necessarily reappear at the same place. The cells able to accumulate pigment are not confined in a single place, but rather may colonize predetermined regions.

In some Crustacea, pigment patterns are primarily genetically determined. The same is true of factors controlling the possibilities of chromatic adaptation, i.e., regulating pigment hormones (Bocquet et al., 1951). In some others like *Carcinus* (Powell, 1962a), *Idothea* (Lee, 1966a,b,c), or *Hippolyte* (Chassard-Bouchaud, 1965), the great adaptability of the pattern shows that its determinism is not primarily genetic.

Experiments on regeneration and background chromatic adaptation have established that the pigment pattern may evolve and that the chromatophore number tends to change (Table 6.10): as a general rule, if the animal is kept on a white background, the leucophore number increases while other chromatophores undergo little variation in number; if kept on a black background, the leucophore number decreases, but the number of darker (black, red, or yellow) chromatophores generally increases. This was observed for Caridea (Brown, 1934; Fingerman et al., 1959; Chassard-Bouchaud, 1965; Vasantha, 1971, 1972, 1974; Keshavan, 1974), Astacura (Bowman, 1942; Fingerman and Whitsell, 1956; Fingerman and Lowe, 1957a,b; Fingerman and Aoto, 1958a,b), or Brachyura (Fingerman and Lowe, 1957a,b). Some species like *Palaemon elegans* (= *Leander squilla*), on the contrary, display a more stable pigment pattern and, after the animals are kept 1 month on a white or black background, their chromatophore number is not significantly changed (Chassard-Bouchaud, 1965).

One can wonder how the morphological color changes proceed in animals and, if the variation in cell number originates from the division or disappearance of preexistent chromatophores, whether new chromatophores appear. Interesting experiments by Chassard-Bouchaud (1965) during the regeneration of uropods (Table 6.11) show that in *Athanas nitescens* the regenerated appendage is first transluscent, then red pigment appears, and then white pigment, whereas in the sand shrimp *Crangon crangon* the order is black, white, and then yellow-red. The

TABLE 6.11. Morphological Chromatic Adaptation After 1 Month on White or Black Background, Showing Chromatophore Number Evolution on Regenerated Uropod in the Shrimp *Athanas nitescens*

Chromatophores:	Erythrophores		Leukophores	
Background:	Black	White	Black	White
Males	initial number recovered	−50%	−52%	initial number recovered
Females	initial number recovered	−49%	−50%	initial number recovered

SOURCE: Adapted from Chassard-Bouchaud (1965).

neoformation of chromatophores leads to a normal quantity of pigment within the cells.

In parallel to changes in chromatophore number, there are also changes in morphological aspects of chromatophores: typically, dark ones become larger and more branched if the animal is kept on a black background, whereas they are smaller and punctate if it is kept on a white background (Fingerman and Whitsell, 1956; Fingerman and Lowe, 1957b). Ultrastructural investigations on crustacean pigment cells (see references in Noël, 1979) have rarely given information on morphological color changes. Variations in the number of microtubules and pigment granules or their appearance might well be expected, but only indirect evidence has been obtained within Isopoda (Castrucci and Mendes, 1975). Robison and Charlton (1973) used morphological color changes in the glass shrimp *Palaemonetes vulgaris* to confirm identification of the different types of pigment granules. Almost nothing is known about the chromoepidermal cells. These cells have been investigated only in relation to cuticle production, but they often contain pigments (mostly carotenoproteins, present in vesicles) that are deposited in the cuticle just before molting. These water-soluble pigments might not be seen in electron microscopy with the usual morphological techniques.

6.3.4.2. BIOCHEMICAL VARIATIONS IN PIGMENTS AND THEIR HORMONAL CONTROL

Both qualitative and quantitative variations may occur within morphological color changes. Reports dealing with this aspect of color changes are often imprecise, being essentially founded on subjective observations of animals. Biochemical studies are too scarce to permit firm conclusions to be drawn.

True qualitative pigment variations have never been reported in Crustacea. All pigments seem to preexist in the organism before adaptation, and there are only some (more or less) important quantitative variations. Some changes in proportions of the different pigments occur during white background adaptation, where development of white pigment and reduction of dark ones (red, brown, blue, or black) are typically observed; the opposite phenomenon occurs for black background adaptation (Brown, 1934).

Quantitative data are known only for carotenoids in palaemonid prawns. In *Palaemonetes vulgaris,* total carotenoid content decreases by 65% after the animal is kept 20 days on a white background (Brown, 1934). The common European prawn *Palaemon* (= *Leander*) *serratus* kept 2 months on a white background exhibits a substantial decrease in red pigment in the chromatophores of the stripes; when it is kept on a black background, the red pigment content increases and a great amount of blue pigment appears, while the white pigment disappears (Panouse, 1947; Chassard-Bouchaud, 1965). These observations were later confirmed by biochemical investigations on epidermal pigments. Carotenoids show large quantitative variations according to the color of the substrate, while food has little or no influence (Chassard-Bouchaud et al., 1971). Hormones play a role in carotenoid metabolism: selective ablation of the sinus gland leads to decrease of xanthophylls and increase of β-carotene (Genofre, 1976). The appearance of dark carotenoproteins was demonstrated to be controlled by hormones from the sinus gland both for crab green pigment (Lenel and Veillet, 1951) and for prawn blue pigment (Genofre, 1973).

6.3.4.3. LONG-TERM VARIATIONS IN CHROMATOPHORE PHYSIOLOGICAL RESPONSE POTENTIAL

Long-term variations in chromatophore physiological response potentially concern the final state, i.e., dispersion or retraction of pigment within chromatophores, the speed with which these migrations occur, or the quantity of chromatophorotropins, for all known physiological responses of chromatophores, mainly background adaptation and nychthemeral rhythm of pigment dispersion. As a general rule, the morphological chromatic adaptation slows down more and more as the physiological color changes, and this holds true for background adaptation or nychthemeral rhythm of pigment migration (Powell, 1962b).

Fingerman and Lowe (1957b) have shown that there is a decrease in the ability of pigment retraction or dispersion (physiological adaptation) after 3 weeks of morphological adaptation to a given background for red, white, and black pigments. The speed of pigment retraction in *Palaemonetes vulgaris* (Fingerman et al., 1958) is less affected than that of pigment dispersion after 14 days of colored background adaptation, the rate of pigment migration decreases progressively, and the color change system becomes sluggish (Fingerman et al., 1959). In the crab *Ocypode macrocera,* a prolonged adaptation to a background color influences the possibilities of physiological chromatic adaptation. There is a marked decrease in the capacity of chromatophores to respond to background color changes, with corresponding variations in the CNS content in chromatophorotropins (Rao, 1969).

Chromatophores having a nychthemeral rhythm of pigment dispersion are also subjected to long-term variations. For instance, the male pea crab *Pinnotheres veterum* is unable to retract chromatophore pigment nightly after one rearing month (Atkins, 1926). The diurnal chromatophore rhythm of the fiddler crab *Uca pugnax* is also affected under different conditions of light periods and intensities (Brown et al., 1953).

6.3.4.4. HORMONAL CONTROL OF MORPHOLOGICAL COLOR CHANGES

Hormonal control of morphological color changes is indicated by numerous findings of variations in chromatophorotropins during adaptive processes and is further supported by experimental analysis. Most of the morphological color changes seem to be under hormonal control; it may happen therefore that the relative wealth of pigment (i.e., carotenoids) supplied in food may lead to long-term color modifications (Gilbert, 1977).

As a result of long-term background adaptation, one might well expect changes in the hormonal content of organs producing or storing chromatophorotropins. In fact, experiments conducted with *Palaemonetes vulgaris* have showed that the titers of red pigment-controlling hormones change in the circumesophageal connectives. A possible explanation is that the portion of the hormone not used to maintain an appropriate degree of pigment dispersion or retraction is stored in this organ (Fingerman et al., 1959). Major changes in the quantity of different chromatophorotropins in hemolymph as well as in CNS occur during background adaptation. The origin of the sluggishness observed for chromatophores might be due to modifications in the level of chromatophorotropins (Fingerman and Lowe, 1957b).

Mechanisms controlling morphological chromatic adaptation are not fully understood. Brown (1934) and Fingerman et al. (1958) supposed a control similar to that of physiological chromatic adaptation. Hanström (1939) supposed that pigment-retracting hormones act simultaneously on pigment destruction. Hormonal control for crustacean morphological color changes is supported by long-term studies on the effects of bilateral eyestalk ablation. This operation suppress the protocerebrum–neuroendocrine (neurosecretory) X-organ–sinus gland (neurohemal organ) complex responsible for elaboration, storage, and release of a number of hormones, among which are chromatophorotropins.

Generally, eyestalk ablation leads to events that may differ in the short, middle, or long term. As is well documented, soon after eyestalk removal, red pigment dispersion is maximal. It appears therefore that within a few days or a week after operation red pigment begins to retract progressively. This occurs both in species presenting a well-developed background adaptation as the main phenomenon, such as *Palaemon*, or in species with a nychthemeral rhythm, such as *Lysmata seticaudata* (Chassard-Bouchaud and Couturier-Bhaud, 1969) or *Processa edulis* (Noël, 1979). In *Palaemon paucidens*, eyestalk ablation leads to red pigment dispersion but this condition lasts less than a week; after that, the pigment becomes less expanded. Apparently, in the absence of eyestalks the PDH (pigment-dispersing hormone) cannot be put to use and is therefore stored in supraesophageal ganglia, while the PCH (pigment-concentrating hormone) in circumesophageal connectives is released more or less freely from the NSC (Aoto, 1961). Similarly, in *Lysmata seticaudata*, a progressive night retraction of red pigment in type I erythrophores reappears over a period of 90 days (Bhaud-Couturier, 1974). Eyestalk ablation in *Lysmata uncicornis* results in decoloration by loss of pigment, analogous to a long period of adaptation to a white background (Bellon-Humbert, 1970b). In *Palaemon serratus*, ablation of the sinus gland gives some modification in the level of the blue pigment found in epidermis (Genofre, 1973). Two weeks after eyestalk ablation, *Hippolyte varians* manifests a disappearance of green pigments and the appearance of the heterogeneous brown-red pattern, independently of the initial pattern. So, this pattern can be considered as the neutral (fundamental) one for the species (Chassard-Bouchaud, 1965). Note that the brown heterogeneous pattern is common in other species of small Hippolytidae such as *Hippolyte leptocerus, H. longirostris,* or *H. inermis* and therefore may represent an original pigment pattern in evolution. The establishment of pigment pattern seems to be independent of eyestalk hormones: after eyestalk removal, the chromatophores appearing in regenerated appendages do not generally show much variation from regenerated appendages in normal animals (Chassard-Bouchaud, 1965). Variations in white pigment after eyestalk removal may be at least partly due to the stimulation of vitellogenesis through the absence of VIH (vitellogenesis-inhibiting hormone) from the eyestalk.

After eyestalk ablation, variations may occur in the chromatophores according to background color: blinded males of *Crangon crangon* show a drastic increase of leukophores when kept on a white background (+260%), whereas there is a decrease when kept on a black one (−30%). In eyestalkless *Palaemon serratus* a maximum response is observed for white-red chromatosomes, whose number increases (+361%) within 2 months on a black background, and less (+50%) on a white background (Chassard-Bouchaud, 1965). The reasons for such differences in the reaction of blinded animals to white and black backgrounds remain unexplained. Direct effect of total illumination may play a role here. Ten days after eyestalk removal in the crayfish *Cambarellus shufeldtii*, there is a difference of reactivity according to the position of the chromatophores (Fingerman, 1957b). The long-term effects of eyestalk ablation have not yet been satisfactorily explained. Hormonal compensation of CNS neurosecretory structures (such as postcommissural organs) may compensate the absence of protocerebral neurosecretory structures; another possibility is the regeneration of an organ similar to the sinus gland.

The first role attributed to morphological color changes in Crustacea is a protective one and is often related to substrate color or appearance (Powell, 1962a; Kuris and Carlton, 1977). The animal, it would seem, after having undergone a morphological color adaptation to a new condition, is sometimes able (*Hippolyte varians:* Chassard-Bouchaud, 1965) and sometimes unable (*Idothea:* Lee, 1966a) to choose a new environment matching its new color. Evolutionary processes are undoubtedly fundamental to any explanation of these morphological color changes, since closely related species within the same genus may or may not show such adaptation (*Idothea:* Lee and Gilchrist, 1972). Some problems remain to be investigated. Why do the chromatophores fail to respond after a prolonged period of background adaptation when chromatophorotropins seem to be still there? The pigment cell itself seems to become refractory to adaptation (Rao, 1969).

6.4. Morphological Color Changes in Myriapoda and Chelicerata

6.4.1. *Myriapoda*

Very little is known about morphological color changes and their control in Myriapoda or Chelicerata. The chemistry and identification of pigments has been studied by various authors, especially by Needham (1960). Xanthommatin, riboflavin, benzoquinones, coproporphyrins, and bilins are present in different species. Most of these pigments are fluorescent and are to be complexed with proteins that extinguish at least partly their fluorescence. Pigments are located primarily in epidermal cells, since the cuticle is almost pigmentless. A violet pigment is present in *Lithobius* within connective tissue (Bannister and Needham, 1971). In some species, only the digestive tract is pigmented. Although it is not fully ascertained, the existence of a terminal molt in Myriapoda is not sure; molting occurs after reproduction but is apparently not associated with any special color change. Some species exhibit bright black and red aposematic colorations; the role of these is unknown. Eggs are pinkish at the time of hatching in the geophilid *Strigamia maritima* (Lewis, 1962); the pigment is located in cells of the digestive tract.

6.4.2. *Chelicerata*

In the order Araneae, many species have brownish or grayish colors. Species of some families such as Thomisidae or Salticidae show different types of bright pigmentation and color patterns (Millot, 1949; Bristowe, 1958; Jones, 1983); physiological color changes are rare (Blanke, 1972, 1975), and an abundant pigmentation is therefore present in the epidermis. A white-to-yellow color change occurs within a few days in some spiders such as *Misumena vatia* or *Thomisus onustus* (Neet, 1987:

Figs. 54 & 55); it persists even in blinded animals (Rabaud, 1923), evoking a direct photoreactivity of pigment cells. True chromatophores exist in some spiders such as the pholcid *Holocnemus pluchei*, but nothing is known about their endocrine control (Legendre and Lopez, 1973) nor about their possible role in color changes (André, 1949). Some Mygalomorphae such as *Scodra griseipes* or *Brachypelma smithi* exhibit slight color changes during development or at molt; nothing is known about the control of these morphological color changes. Acarina often show a pigmentation in relation to the background color, and their color often depends on their food; sexual activity has also been reported to affect color, but the control mechanism is unknown. Some Scorpionidea undergo slight color changes, although most species are darkly pigmented (brown or black) and not homochromatic. Pigmentation seems related to the family to which the species belongs and not to the background. The body color undergoes changes throughout development in *Androctonus australis*, and the color differences between some body parts or appendages (the end of the abdomen or chelae, for instance) persist but are more pronounced at the beginning of life (Vachon, 1952). This phenomenon is common only in Buthidae, where young are more pigmented than adults. Pigments are mostly deposited in the epidermis, and the cuticle is transparent. In some species, male and female pigmentation differs slightly. The study of American Buthidae, especially the genus *Tityus*, allows Lourenço (1983) to distinguish "coloration" (cuticle) and "pigmentation" (epidermis). The cuticle coloration varies considerably and depends largely on the age of the animal; the cuticle is sometimes colorless, especially in juveniles, as observed on pigmentless exuviae. A terminal molt exists in adults, and therefore the cuticle may be very dark, by progressive sclerotization. Epidermal cells contain pigments responsible for the typical species pattern. This pattern is evident when the cuticle is transparent enough, such as soon after molting, when neoformed cuticle is still thin. In different species of *Tityus* the pigment pattern appears to be almost constant during postembryonic development.

Pantopoda (= Pycnogonida) often exhibit a coloration similar to their nutritional animal background. The present mechanism of homochromy as well as the nature of pigmented structures are not well known (Fage, 1949). Almost nothing is known on morphological color changes in other groups of arthropods such as Phalangida, Solifuga, Palpigrada, Xiphosura, and Pseudoscorpionidea.

6.5. General Discussion and Conclusions

In arthropods, most of the knowledge on morphogenetic hormones and color changes has been obtained from insects and crustacea and very little information is available for the other groups, viz. Myriapoda and Chelicerata. It is interesting to note that the basic structure of the integument of insects and crustaceans is essentially the same. However, in most insects, the cuticle-producing cells give rise to general pigmentation and chromatophores are absent, whereas in Crustacea chromatophores are common and pigmentation of cuticle-producing cells has less importance. In both insects and crustaceans, larger animals have thick cuticles whose pigmentation hides that of internal tissues. In Crustacea the pigmented cuticle is predominantly the exocuticle, whereas in insects it is the endocuticle. In crustaceans, entomostracan pigments are not in chromatophores, and little is known on the morphology or physiology of their pigment cells. Hormonal control of morphological color changes is still obscure in Crustacea, and some data suggest that different controls could exist (Green, 1964b). Morphological (like physiological) color changes play an essential role in the mode of life of many arthropods. They correspond to an adaptation to environmental changes. Species that maintain similar behavioral habits during their larval and adult life stages in a buffered medium, as do many crustaceans, arachnids, and some insects, need not adapt to changing edaphic conditions; on the other hand, some crustaceans and most insects must adapt to changes in their environment.

6.6. Summary

The general factors such as morphogenetic hormones, morphological color changes, chromatic effectors, and pigments involved in arthropod color changes have been examined successively in insects, crustaceans, myriapods, and chelicerates.

Eumorphogenetic hormones of insects are well known and their effects on pigmentation have been particularly investigated because effectors (pigmented epidermal cells) may be steadily observed. Individual color changes affecting both larvae and adults are controlled essentially by eumorphogenetic hormones. However, the neurohormones play a more or less important role in conjunction with the eumorphogenetic hormones. The role of neurohormones is predominant in color changes related to the crowding effect, as in lepidopteran pupal color changes.

In Crustacea, morphological color changes mostly affect the chromatophores. The role of the epidermal cells, which often contain pigments, is as yet little known. Neurohormonal control of these color changes (pseudomorphogenetic hormones) has been demonstrated in only a few examples, but the role of eumorphogenetic hormones is still to be established.

Acknowledgments

We are grateful to our colleagues, particularly to Drs. J. M. Demange, R. Lafont, W. R. Lourenço, R. Stockman, and P. Testard, for their help in collecting information for the present review.

References

Ahmed, R. F., T. L. Hopkins, and K. J. Kramer. 1983. Tyrosine glucoside hydrolase activity in tissues of *Manduca sexta* (L.): effect of 20-hydroxyecdysone. Insect Biochem. 13(6): 641–645.

Amerasinghe, F. P. 1978. Effects of J.H. I and J.H. III on yellowing, sexual activity and pheromone production in allectomized male *Schistocerca gregaria*. J. Insect Physiol. 24(8/9): 603–612.

André, M. 1949. Ordre des Acariens. Pp. 794–892 *in* P. P. Grassé (ed.), *Traité de Zoologie*, Vol. VI. Masson, Paris.

Aoto, T. 1961. Chromatophorotropins in the prawn *Palaemon paucidens* and their relationship to longterm background adaptation. J. Fac. Sci. Hokkaido Univ., Ser. VI 14(4): 544–560.

Atkins, D. 1926. On nocturnal colour change in the pea-crab (*Pinnotheres veterum*). Nature (Lond.) 117(2942): 415–416.

Awiti, L. R. S. and T. Hidaka. 1982. Neuroendocrine mechanism involved in pupal colour dimorphism in swallowtail *Papilio xuthus*. Insect Sci. Appl. 3(2/3): 181–191.

Bàbàk, E. 1913. Über den Einfluss des Lichtes auf die Vermehrung der Hautchromatophoren. Pfluegers Arch. Ges. Physiol. 149:462–470.

Bagnara J. T. and M. E. Hadley. 1973. *Chromatophores and Color-change: The Comparative Physiology of Animal Pigmentation*. Prentice-Hall, Englewood Cliffs, New Jersey.

Bannister, W. H. and A. E. Needham. 1971. Connective tissue pigment of the centipede *Lithobius forficatus* (L.). Naturwissenschaften 58(1): 58–59.

Barbier, M., H. Charniaux-Cotton, and M. C. Fried-Montaufier. 1966. Présence dans les secondes antennes des mâles de *Talitrus saltator* et *Orchestia gammarella* (Crustacés Amphipodes) d'un taux relativement élevé d'astaxanthine. C. R. Acad. Sci. Paris 263D(20): 1508–1510.

Barden, H. and S. Koulish. 1983. The dark brown integumentary pigment of a barnacle (*Balanus eburneus*). Histochemistry 78: 41–52.

Bellon-Humbert, C. 1970a. Les changements de couleur des Crustacés. In *Les Hormones et le comportement*. Probl. Actuels Endocrinol. Nutrit. 14: 259–317.

Bellon-Humbert, C. 1970b. Recherches sur *Lysmata uncicornis* (Holthuis et Maurin). IV. Rôle du pédoncule oculaire dans le comportement chromatique. Bull. Soc. Sci. Nat. Phys. Maroc 50: 83–100.

Bernays, E. A. and R. F. Chapman. 1973. Changes in the coloration of *Atractomorpha similis* I. Bolivar (Orthoptera: Pyrgomorphidae). Austral. Entomol. Mag. 1(4): 41–43.

Berthold, G. 1973. Der Einfluss der *Corpora allata* auf die Pigmentierung von *Carausius morosus* Br. Wilhelm Roux' Arch. Entwicklungsmech. Org. 173:249–262.

Berthold, G. 1977. Die Wanderung der Pigmentgranula in der Epidermis beim physiologischen Farbwechsel von *Carausius morosus*. J. Insect. Physiol. 23(2): 235–240.

Berthold, G. 1980. Microtubules in the epidermal cells of *Carausius morosus* (Br.): their pattern and relation to pigment migration. J. Insect Physiol. 26(6): 421–425.

Berthold, G. and D. Bückmann. 1975. Morphologischer Farbwechsel und Kynurensäure-Exkretion bei der Stabheuschrecke *Carausius morosus* Br. J. Comp. Physiol. B100(4): 347–350.

Berthold, G. and M. Henze. 1971. Über die Beteiligung der Pterine an den Farbmodifikationen der Stabheuschrecke *Carausius morosus*. J. Insect Physiol. 17(12): 2375–2382.

Berthold, G. and G. Seifert. 1977. Elektronenmikroskopische Untersuchungen über die Wanderung der Pigment-Granula beim physiologischen Farbwechsel der Stabheuschrecke *Carausius morosus* (Phasmatodea: Phasmatidae). Entomol. Germ. 3(4): 303–315.

Beydon, P., B. Mauchamp, J. Lhonoré, A. Bouthier, and R. Lafont. 1980. The epidermal cell cycle during the last larval instar of *Pieris brassicae* (Lepidoptera). II. Endocrine control of xanthommatin content. J. Comp. Physiol. B136(1): 21–29.

Bhaud-Couturier, Y. 1974. Biologie et physiologie chromatique de *Lysmata seticaudata* Risso (Crustacé, Décapode). Doctoral thesis (Sci. Nat.), University of Paris VI.

Blanke, R. 1972. Untersuchungen zur Ökophysiologie und Ökoethologie von *Cyrtophora citricola* Forskäl (Araneae, Araneidae) in Andalusien. Forma Functio 5(2): 125–206.

Blanke, R. 1975. Die Bedeutung der Guanozyten für den physiologischen Farbwechsel bei *Cyrtophora cicatrosa* (Arachnida: Araneidae). Entomol. Germ. 2(1): 1–6.

Bliss, D. E. 1979. From sea to tree: saga of a land crab. Am. Zool. 19: 385–410.

Bocquet, C. 1951. Recherches sur *Tisbe* (= *Idyaea*) *reticulata*, n. sp.: essai d'analyse génétique du polychromatisme d' un Copépode Harpacticoïde. Arch. Zool. Exp. Gén. 87: 335–416.

Bocquet, C. 1960. Sur un phénotype "semi-pigmenté" réalisé chez de rares *Sphaeroma serratum* (Fabricius). Bull. Soc. Linn. Normandie [10] 1:210–212.

Bocquet, C. and R. Lejuez. 1967. Sur un nouveau sphérome appartenant à la faune endogée des sables de la région de Roscoff, *Sphaeroma teissieri* n. sp. C. R. Acad. Sci. Paris 265: 689–692.

Bocquet, C., C. Levi, and G. Teissier. 1951. Recherches sur le polychromatisme de *Sphaeroma serratum* (F.). Arch. Zool. Exp. Gén. 87: 245–297.

Borhovec, A. B. and D. B. Gelman (eds.). 1986. *Insect Neurochemistry and Neurophysiology*. Humana Press, Clifton, New Jersey.

Borst, D. W., H. Laufer, M. Landau, E. S. Chang, W. A. Hertz, F. C. Baker, and D. A. Schooley. 1985. Methyl farnesoate and its role in crustacean reproduction and development. Proc. 4th Int. Symp. on Juvenile Hormones: Physiology, Biochemistry & Chemistry. *Publ.* in Insect Biochem. 17(7): 1123–1128.

Bouthier, A. 1976. Action des facteurs hormonaux sur le métabolisme des ommochromes chez le criquet *Locusta migratoria* L. (Orthoptères, Acrididae). Ann. Endocrinol. 37(6): 539–540.

Bouthier, A. 1979. Biochimie et métabolisme des ommochromes chez *Locusta migratoria cinerascens* Fabr. (Orthoptères, Acrididae) et son mutant *albinos*. Publ. Lab. Zool. ENS (Ecole Normale Supérieure, Paris) No. 15.

Bouthier, A. 1981. Les ommochromes, pigments absorbants des yeux des Arthropodes. Arch. Zool. Exp. Gén. 122(3): 237–252.

Bowman, T. E. 1942. Morphological color changes in the crayfish. Am. Nat. 76: 332–336.

Bristowe, W. S. (ed.) 1958. *The World of Spiders*. Collins, London.

Broch, E. S. 1960. Endocrine control of the chromatophores of the zoeae of the prawn *Palaemonetes vulgaris*. Biol. Bull. (Woods Hole) 119(2): 305–306.

Brown, F. A., Jr. 1934. The chemical nature of the pigments and the transformations responsible for color changes in *Palaemonetes*. Biol. Bull. (Woods Hole) 67: 365–380.

Brown, F. A., Jr. 1935. Color changes in *Palaemonetes*. J. Morphol. 57: 317–334.

Brown, F. A., Jr. 1939. Background selection in crayfishes. Ecology 20(4): 507–516.

Brown, F. A., Jr. 1944. Hormones in Crustacea: their sources and activities. Q. Rev. Biol. 19(1): 32–46.

Brown, F. A., Jr. 1973. Chromatophores and color change. Pp. 915–950 *in* C. L. Prosser and F. A. Brown, Jr. (eds), *Comparative Animal Physiology*. Saunders, Philadelphia.

Brown, F. A., Jr. and A. Meglitsch. 1940. Comparison of the chromatophorotropic activity of insect *corpora cardiaca* with that of crustacean sinus-glands. Biol. Bull. (Woods Hole) 79(3): 409–418.

Brown, F. A., Jr., M. Fingerman, and M. N. Hines. 1953. Persisting modifications of the diurnal chromatophore rhythm of *Uca pugnax* by altering periods of light of different intensities. Anat. Rec. 117: 634.

Brown, F. A., Jr., H. M. Webb, and M. I. Sandeen. 1952. The action of two hormones regulating the red chromatophores of *Palaemonetes*. J. Exp. Zool. 120(3): 391–420.

Bückmann, D. 1952. Die Umfärbung von Schmetterlingsraupen von der Verpuppung (Untersuchungen an *Cerura vinula* L.). Naturwissenschaften 39(9): 213–214.

Bückmann, D. 1953. Über den Verlauf und die Auslösung von Verhaltensänderungen und Umfärbungen erwachsener Schmetterlingsraupen. Biol. Zentralbl. 72(5/6): 276–311.

Bückmann, D. 1959. Die Auslösung der Umfärbung durch das Häutungshormon bei *Cerura vinula* (Lepidoptera, Notodontidae). J. Insect Physiol. 3(2): 159–189.

Bückmann, D. 1960. Die Determination der Puppenfärbung bei *Vanessa urticae* L. Naturwissenschaften 47(24): 610–611.

Bückmann, D. 1974a. Die hormonale Steuerung der Pigmentierung und des morphologischen Farbwechsels bei den Insekten. Fortschr. Zool. 22(2/3): 1–22.

Bückmann, D. 1974b. Die Wirkung von Ecdyson und von Tryptophanüberschuss bei den Ommochrombildung von *Cerura vinula* L. vor der Verpuppung. Zool. Jahrb., Abt. Allg. Zool. Physiol. Tiere 78(3): 257–278.

Bückmann, D. 1977. Morphological colour change: stage independent, optically induced ommochrome synthesis in larvae of stick insect, *Carausius morosus* Br. J. Comp. Physiol. B115(2): 185–194.

Bückmann, D. 1979. Morphological colour change. 2. The effects of total and partial blinding on epidermal ommochrome content in the stick insect, *Carausius morosus* Br. J. Comp. Physiol. B130(4): 331–336.

Bückmann, D. 1985. Color change in Insects. Pp. 209–217 *in* J. Bagnara, S. N. Klaus, E. Paul, and M. Schartl (eds.), *Pigment Cell 1985: Biological, Molecular and Clinical Aspects of Pigmentation*. University of Tokyo Press, Tokyo.

Bückmann, D. and A. Maisch. 1987. Extraction and partial purification of the pupal melanization reducing factor (PMRF) from *Inachis io* (Lepidoptera). Insect Biochem. 17(6): 841–844.

Burtt, E. 1951. The ability of adult grasshoppers to change colour on burnt ground. Proc. R. Entomol. Soc. Lond., Ser. A 26(1): 45–48.

Carlisle, D. B. and F. Knowles. 1959. Endocrine control in Crustaceans. *Camb. Monog. Exp. Biol.* 10: 1–120.

Carlson, S. Ph. 1935. The colour changes in *Uca pugilator.* Proc. Natl. Acad. Sci. USA 21(9): 549–551.

Carstam, S. P. 1947. Endocrinologie des Arthropodes. Colour change and colour hormones in Crustaceans. Colloq. Int. Cent. Natl. Rech. Sci. 4: 139–148.

Caruelle, J.-P. 1976. Etude *in vitro* de l'évolution imaginale du tégument de *Schistocerca gregaria,* Forskål, et de son déterminisme endocrine. Thesis, 3rd cycle, University of Paris VI.

Caruelle, J.-P and P. Cassier. 1979. Actions hormonales *in vitro* et différenciation cellulaire dans le tégument du Criquet pélerin *Schistocerca gregaria.* Arch. Anat. Microsc. Morphol. Exp. 68(2): 127–138.

Cassier, P. and C. Delorme-Joulié. 1976. La différenciation imaginale du tégument chez le criquet pélerin, *Schistocerca gregaria* Forsk. III. Les différences phasaires et leur déterminisme. Insectes Soc. 23(2): 179–198.

Cassier, P. and C. Joulié-Delorme. 1976. La différenciation imaginale du tégument chez le criquet pélerin, *Schistocerca gregaria* Forsk. I. L'évolution post-imaginale et son déterminisme. Arch. Zool. Exp. Gén. 177(1): 95–116.

Castrucci, A. M. de L. and E. G. Mendes. 1975. Ultrastructure of the pigmentary system and chromatophorotropic activity in land isopods. Biol. Bull. (Woods Hole) 149: 467–479.

Chace, F. A. and G. M. Moore. 1959. A bicolored gynandromorph of the lobster *Homarus americanus.* Biol. Bull. (Woods Hole) 116(2): 226–231.

Chang, F. and E. Jang. 1980. Endocrine regulation of melanization in late larvae of the oleander hawk moth *Deilephila nerii.* Ann. Entomol. Soc. Am. 73(1): 89–92.

Charniaux-Cotton, H. 1975. Hermaphroditism and gynandromorphism in malacostracan

Crustacea. Pp. 91–105 *in* R. Reinboth (ed.), *Intersexuality in the Animal Kingdom*, Springer-Verlag, Berlin and New York.

Chassard, C. 1957. Modifications de la répartition topographique des chromatophores au cours de l'adaptation chromatique d'*Hippolyte varians* Leach (Crustacé, Décapode). C. R. Acad. Sci. Paris 245: 2406–2409.

Chassard, C. 1958. L'adaptation chromatique chez *Hippolyte varians* Leach (Crustacé, Décapode). *In* Proc. 15th Int. Congr. Zool., London, Sect. VI, Paper No. 50.

Chassard. C. 1960. L'adaptation chromatique chez *Athanas nitescens* Leach (Crustacé, Décapode). Cah. Biol. Mar. 1(4): 453–463.

Chassard, C. 1962. Changements de couleur morphologiques chez *Leander serratus* (Crustacé Décapode). C. R. Acad. Sci. Paris 255(25): 3477–3479.

Chassard, C. 1963. Effets chromatiques à long terme de l'ablation des pédoncules oculaires chez *Leander serratus* (Crustacé Décapode). C. R. Acad. Sci. Paris 256: 5630–5633.

Chassard-Bouchaud, C. 1965. L'adaptation chromatique chez les Natantia (Crustacés Décapodes). Cah. Biol. Mar. 6(5): 469–576.

Chassard-Bouchaud, C. and Y. Couturier-Bhaud. 1969. Etude des phénomènes chromatiques de *Lysmata seticaudata* Risso (Crustacé Décapode). II. Effets de l'ablation des pédoncules oculaires sur le cycle nycthéméral. Cah. Biol. Mar. 10: 173–180.

Chassard-Bouchaud, C. and G. C. Genofre. 1974. Effets de l'ablation élective de la glande du sinus sur l'adaptation chromatique physiologique de *Palaemon elegans* Rathke (Crustacé Décapode) en fonction du cycle d'intermue. C. R. Acad. Sci. Paris 278D: 893–896.

Chassard-Bouchaud, D. and P. Noël. 1972. Carotenoid metabolism in the prawn *Palaemon serratus* Pennant: qualitative and quantitative aspects. Proc. 3rd Int. Symp. on Carotenoids Other Than Vitamin A, Sept. 4–7, 1972 pp. 77–78.

Chassard-Bouchaud, C. M. Barbier, P. Noël, and Y. Bhaud-Couturier. 1971. Sur le métabolisme des pigments caroténoïdes de *Palaemon serratus* (Pennant) (Crustacé Décapode): influence de la nutrition et des conditions d'éclairement. C. R. Acad. Sci. Paris 273D: 483–486.

Chicewicz, M. 1951. A study of the colour change in *Praunus flexuosus* Müll. Bull. Int. Acad. Cracovie, Ser. B 2(7–10): 371–384.

Cooke, I. M. and R. E. Sullivan. 1982. Hormones and neurosecretion. Pp. 206–290 *in* D. E. Bliss (ed.), *The Biology of Crustacea*, Vol. 3: H. L. Atwood and D. C. Sandeman (eds.), *Neurobiology: Structure and Function*. Academic Press, Orlando, Florida.

Costlow, J. D. 1961. Fluctuations in hormone activity in Brachyura larvae. Nature (Lond.) 192(4798): 183–184.

Costlow, J. D. 1963. Molting and cyclic activity in chromatophorotropins of the central nervous system of the barnacle, *Balanus eburneus*. Biol. Bull. (Woods Hole) 124(3): 254–261.

Costlow, J. D. and M. I. Sandeen. 1961. The appearance of chromatophorotropic activity in the developing crab, *Sesarma reticulatum*. Am. Zool. 1: 191.

Couch, E. F., N. Hagino, and J. W. Lee. 1987. Changes in estradiol and progesterone immunoreactivity in tissues of the lobster, *Homarus americanus*, with developing and immature ovaries. Comp. Biochem. Physiol. A87: 765–770.

Couturier-Bhaud, Y. 1975a. Etude des phénomènes chromatiques chez *Lysmata seticaudata* Risso (Crustacé, Décapode): évolution de la livrée chromatique au cours du développement larvaire. Vie & Milieu A25(1): 59–65.

Couturier-Bhaud, Y. 1975b. Etude des phénomènes chromatiques chez *Lysmata seticaudata* Risso (Crustacé, Décapode): physiologie chromatique des stades larvaires. Vie & Milieu A25(1): 67–73.

Crane, J. (ed.) 1944. *Fiddler Crabs of the World, Ocypodidae: Genus* Uca. Princeton University Press, Princeton, New Jersey.

Curtis, A. T., M. Hori, J. M. Green, W. J. Wolfgang, K. Hiruma, and L. M. Riddiford. 1984.

Ecdysteroid regulation of the onset of cuticular melanization in allatectomized and black mutant *Manduca sexta* larvae. J. Insect Physiol. 30(8): 597–606.

Della Beffa, G. 1948. Studi su alcune infestazioni di Fasgonuridi e Locustidi in Piemonte, nel 1947, e prime esperienze di lotta. Ann. Sper. Agrar. 2(4): 567–589.

Dores, R. M. and W. S. Herman. 1980. The purification and partial characterization of the melanophore-dispersing hormone of *Uca pugilator.* Gen. Comp. Endocrinol. 42(2): 179–186.

Doull, K. M. 1953. Phase coloration in lepidopterous larvae. Nature (Lond.) 172(4383): 813–814.

Drach, P. 1939. Mue et cycle d'intermue chez les Crustacés Décapodes. Ann. Inst. Océanogr. 19(3): 103–391.

Dustmann, J. H. 1964. Die Redoxpigmente von *Carausius morosus* und ihre Bedeutung für den morphologischen Farbwechsel. Z. Vgl. Physiol. 49(1): 28–57.

Dutt, N. 1969. Phases in jute semilooper *Anomis sabulifera* Guen. Sci. Cult. 35(9): 485–487.

Ellis, P. E. and D. B. Carlisle. 1961. The prothoracic gland and colour change in locusts. Nature (Lond.) 190(4773): 368–369.

Endo, K. 1984. Neuroendocrine regulation of the development of seasonal forms of the Asian comma butterfly, *Polygonia c-aureum* L. Dev. Growth & Differ. 26(3): 217–222.

Endo, K. and S. Funatsu. 1985. Hormonal control of seasonal morph determination in the swallowtail butterfly, *Papilio xuthus* L. (Lepidoptera: Papilionidae). J. Insect Physiol. 31(9): 669–674.

Endo, K. and Y. Kamata. 1985. Hormonal control of seasonal morph determination in the small copper butterfly, *Lycaena phlaeas daimio* Seitz. J. Insect Physiol. 31(9): 701–706.

Endo, K., Yamashita I., and Y. Chiba. 1985. Effect of photoperiodic transfer and brain surgery on the photoperiodic control of pupal diapause and seasonal morphs in the swallowtail, *Papilio xuthus* L. (Lepidoptera: Papilionidae). Appl. Entomol. Zool. 20(4): 470–478.

Fage, L. 1949. Classe des Pycnogonides. Pp. 906–941 *in* P. P. Grassé (ed.), *Traité de Zoologie,* Vol. VI. Masson, Paris.

Faure, J.-C. 1932. The phases of locusts in South Africa. Bull. Entomol. Res. 23: 293–405.

Faure, J.-C. 1935. The life history of the red locust (*Nomadocris septemfasciata* (Serville)). Bull. Dep. Agric. S. Afr. No. 144: 1–32.

Faure. J.-C. 1943a. The phases of the lesser armyworm, *Laphygma exigua* (Hübn). Farming S. Afr. 18(203): 69–78.

Faure, J.-C. 1943b. Phase variation in the armyworm *Laphygma exempta* (Walk.) Dept. Agr. Farming S. Afr. Sci. Bull. No. 234: 1–17.

Fernlund, P. 1976. Structure of a light-adapting hormone from the shrimp *Pandalus borealis.* Biochem. Biophys. Acta 439(1): 17–25.

Fernlund, P. and L. Josefsson. 1972. Crustacean color-change hormone: amino acid sequence and chemical synthesis. Science (Wash., D.C.) 177: 173–175.

Fingerman, M. 1957. Physiology of the red and white chromatophores of the dwarf crayfish *Cambarellus shufeldtii.* Physiol. Zool. 30(2): 142–154.

Fingerman, M. 1963. *The Control of Chromatophores* (Pergamon Int. Ser. Monogr., No. 14), p. 184. Pergamon Press, Oxford and Elmsford, New York.

Fingerman, M. and T. Aoto. 1958a. Electrophoretic analysis of chromatophorotropins in the dwarf crayfish *Cambarellus shufeldti.* J. Exp. Zool. 138(1): 25–50.

Fingerman, M. and T. Aoto. 1958b. Chromatophorotropins in the crayfish *Orconectes clypeatus* and their relationship to longterm background adaptation. Physiol. Zool. 31(3): 193–208.

Fingerman, M. and T. Aoto. 1962. Regulation of pigmentary phenomena: hormonal regulation of pigmentary effectors in crustaceans. Gen. Comp. Endocrinol. Suppl. 1: 81–93.

Fingerman, M. and M. Lowe. 1957a. Background adaptation in the dwarf crayfish, *Cambarellus shufeldti*. Anat. Rec. 128(3): 548.

Fingerman, M. and M. Lowe. 1957b. Influence of time on background upon the chromatophore systems of two crustaceans. Physiol. Zool. 30(3): 216–231.

Fingerman, M. and K. R. Rao. 1969. Physiology of the brown-black chromatophores of the stomatopod crustacean *Squilla empusa*. Physiol. Zool. 42(2): 138–147.

Fingerman, M. and J. S. Whitsell. 1956. Background responses of the red and white chromatophores of the dwarf crayfish *Cambarellus shufeldti*. Anat. Rec. 125: 637.

Fingerman, M., R. A. Krasnow, and S. W. Fingerman. 1971. Separation, assay, and properties of the distal retinal pigment light-adapting and dark adapting hormones in the eyestalks of the prawn *Palaemonetes vulgaris*. Physiol. Zool. 44(3): 119–128.

Fingerman, M., M. I. Sandeen, and M. E. Lowe. 1958. Influence of long-term background adaptation on the liability of chromatophores and the sources of chromatophorotropins in *Palaemonetes vulgaris*. Biol. Bull. (Woods Hole) 115:351–352.

Fingerman, M., M. I. Sandeen, and M. E. Lowe. 1959. Experimental analysis of the red chromatophore system of the prawn *Palaemonetes vulgaris*. Physiol. Zool. 32(2): 128–149.

Fogal, W. H. 1968. Pigment development in male integument transplanted onto females in the desert locust, *Schistocerca gregaria*. Entomol. Exp. Appl. 11:475–476.

Fukuda, S. and K. Endo. 1966. Hormonal control of the development of seasonal forms in the butterfly, *Polygonia c-aureum* L. Proc. Jpn. Acad. 42(9): 1082–1087.

Fuzeau-Braesch, S. 1968. Action de l'hormone juvénile de synthèse dans la morphogenèse et la pigmentogenèse de *Gryllus bimaculatus* (Orthoptères). C. R. Soc. Biol. 162(5/6): 1086–1090.

Fuzeau-Braesch, S. 1972. Pigments and color changes. Annu. Rev. Entomol. 17: 403–424.

Fuzeau-Braesch, S. 1985. Colour changes. Pp. 549–589 *in* G. A. Kerkut and L. I. Gilbert (eds.), *Comprehensive Insect Physiology, Biochemistry and Pharmacology*, Vol. 9: *Behaviour*. Pergamon Press, Oxford and Elmsford, New York.

Gabe, M. 1966. *Neurosecretion*. Pergamon Press, Oxford and Elmsford, New York.

Gamble, F. W. 1910. The relation between light and pigment formation in *Crenilabrus* and *Hippolyte varians*. J. Microsc. Sci. 55(3): 541–583.

Gamble, F. W. and F. W. Keeble. 1900. *Hippolyte varians*: a study in colour change. Q. J. Microsc. Sci. 43(172): 589–698.

Genofre, G. C. 1973. Sur le métabolisme des pigments caroténoïdes de *Palaemon serratus* (Pennant)(Crustacé Décapode): influence de l'ablation de la glande du sinus, de la nutrition et des conditions d'éclairement sur les caroténoprotéines bleues. C. R. Acad. Sci. Paris 276D(15): 2269–2272.

Genofre, G. C. 1976. Efeito da glândula do seio sobre a quantidade de pigmentos carotenôides tegementares em *Palaemon serratus* (Pennant, 1777)–Crustacea–Decapoda–Macrura: estudo comparativo. An. Acad. Bras. Cienc. 48(2): 301–303.

Ghidalia, W. 1985. Structural and biological aspects of pigments. Pp. 301–394 *in* D. E. Bliss (ed.), *The Biology of Crustacea*, Vol. 9: D. E. Bliss and L. H. Mantel (eds.), *Integument, Pigments, and Hormonal Processes*. Academic Press, Orlando, Florida.

Gibert, J. 1977. Recherches sur la pigmentation de *Niphargus virei* Chevreux, 1896. Crustaceana Suppl. 4: 45–57.

Gilchrist, B. M. and W. L. Lee. 1972. Carotenoid pigments and their possible role in reproduction in the sand crab, *Emerita analoga* (Stimpson, 1857). Comp. Biochem. Physiol. B42: 263–294.

Girardie, A. 1967. Contrôle neuro-hormonal de la métamorphose et de la pigmentation chez *Locusta migratoria cinerascens* (Orthop.). Bull. Biol. Fr. Belg. 101(2): 79–114.

Girardie, A. 1974. Recherches sur le rôle physiologique des cellules neurosecrétrices latérales du protocérébron de *Locusta migratoria migratorioides* (Insecte Orthoptère). Zool. Jahrb. Abt. Allg. Zool. Physiol. Tiere 78(3): 310–326.

Girardie, A. and M. Cazal. 1965. Rôle de la *pars intercerebralis* et des *corpora cardiaca* sur la mélanisation chez *Locusta migratoria* L. C. R. Acad. Sci. Paris 261: 4525–4527.

Gogala, M. and S. Michieli. 1962. Sezonsko prebarvarje pri nekaterih vrstah stenic (Heteroptera). [Zyklische Umfärbung bei einigen Heteropteren-Arten.] Biol. Vest. 10(1): 33–44.

Gogala, M. and S. Michieli. 1966. Vpliv svetlobe in temperature na sezonsko prebarvanje pri *Nezara viridula* L. (Heteropt.). [Der Einfluss des Lichtes und der Temperatur auf die herbstliche Umfärbung bei *Nezara viridula*.] Biol. Vest. 14: 83–90.

Goldsworthy, G. and C. H. Wheeler. 1986. Structure/activity relationships in the adipokinetic hormone/red pigment concentrating hormone family. Pp. 183–186 *in* A. B. Borkovec and D. B. Gelman (eds.), *Insect Neurochemistry and Neurophysiology*. Humana Press, Clifton, New Jersey.

Goodman, W. G., G. Tatham, D. J. Nesbit, H. Bultmann, and R. D. Sutton. 1987. The role of juvenile hormone in endocrine control of pigmentation in *Manduca sexta*. Insect Biochem. 17(7): 1065–1069.

Green, J. P. 1963. An analysis of morphological color change in two species of brachyuran crustaceans. Diss. Abs. 24(3): 1294–1295.

Green, J. P. 1964a. Morphological color change in the Hawaiian ghost crab, *Ocypode ceratophthalma* (Pallas). Biol. Bull. (Woods Hole) 126(3): 407–413.

Green, J. P. 1964b. Morphological color changes in the fiddler crab, *Uca pugnax* (S. I. Smith). Biol. Bull. (Woods Hole) 127(2): 239–255.

Guyselman, J. B. 1953. An analysis of the moulting processes in the fiddler crab *Uca pugilator*. Biol. Bull. (Woods Hole) 104(2): 115–137.

Hadley, P. B. 1905. The changes in form and color in successive stages of the American lobster (*Homarus americanus*). Annu. Rep. Comm. Inland Fish., Rhode Island, 35th Rep. pp. 44–80.

Hanström, B. 1939. *Hormones in Invertebrates*. Oxford University Press (Clarendon), London and New York.

Hardie, J. and A. D. Lees. 1985. Endocrine control of polymorphism and polyphenism. Pp. 441–490 *in* G. A. Kerkut and L. I. Gilbert (eds.), *Comprehensive Insect Physiology, Biochemistry and Pharmacology*, Vol. 8: *Endocrinology*, Pt. II. Pergamon Press, Oxford and Elmsford, New York.

Hazel, W. N. 1977. The genetic basis of pupal colour dimorphism and its maintenance by natural selection in *Papilio polyxenes* (Papilionidae: Lepidoptera). Heredity 38(2): 227–236.

Hazel, W. N. and D. A. West. 1979. Environmental control of pupal colour in swallowtail butterflies (Lepidoptera; Papilioninae)—*Battus philenor* (L.) and *Papilio polyxenes* Fabr. Ecol. Entomol. 4(4): 393–400.

Heldt, J. H. 1932a. Sur quelques différences sexuelles (coloration, taille, rostre) chez deux crevettes tunisiennes: *Penaeus caramote* Risso et *Parapenaeus longirostris* Lucas. Bull. Stn. Océnogr., Salammbô, Tunisia 27: 1–20.

Herbulot, C. 1949. *Atlas des Lépidoptères de France, Belgique, Suisse. III. Hétérocères*. Boubée, Paris.

Herrick, F. H. 1895. The American lobster: a study of its habits and development. Bull. U.S. Fish. Comm. 15:1–252.

Hidaka, T. 1959. Recherches sur le déterminisme hormonal de la coloration pupale chez les Lépidoptères. 3. Cas de *Malacosoma neustria testacea*, une espèce à monochromisme pupal. Zool. Mag. 68(7): 270–273. [In Japanese, with French summary.]

Hidaka, T. 1961. Recherches sur le mécanisme de l'adaptation chromatique morphologique chez les nymphes de *Papilio xuthus* L. J. Fac. Sci. Univ. Tokyo, Sec. IV 9(2): 223–261.

Hidaka, T. and T. Ohtaki. 1963. Effet de l'hormone juvénile et du farnésol sur la coloration tégumentaire de la nymphe de *Pieris rapae crucivora* Boisd. C. R. Soc. Biol. 157(4): 928–930.

Hintze, C. 1969. Die Wirkung von Methylfarnesoatdihydrochlorid auf die Umfärbung und das Verpuppungsverhalten der Raupe von *Cerura vinula* L. (Lepidoptera). Biol. Zentralbl. 88(1): 77–83.

Hintze-Podufal, C. 1970. Farbanpassung von Sphingidenraupen. Naturwissenschaften 57(9): 460–461.

Hintze-Podufal, C. 1975. Effects of juvenile hormone and analogues on pupae and imagines of *Sphinx ligustri* after application to mature larvae (Lepidoptera, Sphingidae). Dtsch. Entomol. Z. 22(1–3): 219–228.

Hintze-Podufal, C. 1977. The larval melanin pattern in the moth *Eudia pavonia* and its initiating factor. J. Insect Physiol. 23(6): 731–738.

Hintze-Podufal, C. and F. Fricke. 1971. The effect of farnesol derivatives on the mature larva of *Cerura vinula* L. (Lepidoptera). J. Insect Physiol. 17(10): 1925–1932.

Hiruma, K. and L. M. Riddiford. 1985. Hormonal regulation of dopa decarboxylase during a larval molt. Dev. Biol. 110: 509–513.

Hiruma, K. and L. M. Riddiford. 1986. Inhibition of dopa decarboxylase synthesis by 20-hydroxyecdysone during the last larval moult of *Manduca sexta*. Insect Biochem. 16(1): 225–231.

Hiruma, K., Matsumoto S., Isogai A. and A. Suzuki. 1984. Control of ommochrome synthesis by both juvenile hormone and melanization hormone in the cabbage armyworm, *Mamestra brassicae*. J. Comp. Physiol. B154(1): 13–21.

Hiruma, K., L. M. Riddiford, T. L. Hopkins, and T. D. Morgan. 1985. Roles of dopa decarboxylase and phenoloxidase in the melanization of the tobacco hornworm and their control by 20-hydroxyecdysone. J. Comp. Physiol. B155: 659–669.

Hitchcock, H. B. 1941. The coloration and color changes of the gulf-weed crab, *Planes minutus*. Biol. Bull. (Woods Hole) 80: 26–30.

Hoffmann, K. H. 1985. Color and color changes. Pp. 206–224 *in* K. H. Hoffmann (ed.), *Environmental Physiology and Biochemistry of Insects*, Springer-Verlag, Berlin and New York.

Honêk, A. 1973. Induction of a winter coloration in *Chrysopa carnea* Steph. (Neuroptera: Chrysopidae). Vêst. Cesk. Spol. Zool. 37(4): 253–257.

Honêk, A. 1976. Maintenance and termination of winter coloration in *Chrysopa carnea* (Neur. Chrysop.): relationships to the development of diapause. Ann. Zool. Ecol. Anim. 8(3): 411–416.

Hopkins, T. S. 1963. Sexual dichromatism in three species of portunid crabs. Crustaceana 5(3): 238–239.

Hori, M. and L. M. Riddiford. 1982. Regulation of ommochrome biosynthesis in the tobacco hornworm, *Manduca sexta*, by juvenile hormone. J. Comp. Physiol. B147(1): 1–10.

Howell, J. 1897. The protective coloring of the Aesop prawns. J. Mar. Zool. Microsc. 2: 101–103.

Humbert, C. 1965. Etude expérimentale du rôle de l'organe X (pars distalis) dans les changements de couleur et la mue de la crevette *Palaemon serratus*. Trav. Inst. Sci. Chérifien, Ser. Zool. 32: 1–86.

Ikemoto, H. 1971. On the black pigment of the larval integument of the armyworm *Leucania separata*. Botyu-Kagaku [Sci. Pest Control] 36: 128–131.

Ikemoto, H. 1975. Hormonal control of the body-colour change in larvae of the larger pellucid hawk moth, *Cephonodes hylas* L. (1). Botyu-Kagaku [Sci. Pest Control] 40(2): 59–62.

Ikemoto, H. 1983. The role of juvenile hormone in the density-related color-variation in larvae of *Cephonodes hylas* L. (Lepidoptera, Sphingidae). Appl. Entomol. Zool. 18(1): 57–61.

Injeyan, H. S. and S. S. Tobe. 1981. Phase polymorphism in *Schistocerca gregaria:* assessment of juvenile hormone synthesis in relation to vitellogenesis. J. Insect Physiol. 27(3): 203–210.

Iwao, S. 1962. Studies on phase variation and related phenomena in some lepidopterous insects. Mem. Coll. Agric., Kyoto Univ. 84: 1–80.

Iwao, S. 1968. Some effects of grouping in lepidopterous insects. In *L'effet de groupe chez les animaux*. Colloq. Int. Cent. Natl. Rech. Sci. 1967: 185–212.

Johnson, D. F. 1974. The development of the chromatophore response to light in the larvae of the crab, *Uca pugilator*. Chesapeake Sci. 15(3): 165–167.

Joly, L. 1958. Comparaison des divers types d'adultoïdes chez *Locusta migratoria* L. Insectes Soc. 5(4): 373–378.

Joly, L. 1960. Fonctions des *corpora allata* chez *Locusta migratoria*. Doctoral thesis, University of Strasbourg, 107 pp.

Joly, L., J. Hoffmann, and P. Joly. 1977. Contrôle humoral de la différenciation phasaire chez *Locusta migratoria migratorioides* (R. & F.) (Orthoptères). Acrida 6(1): 33–42.

Joly, N. 1843. Recherches sur le développement et les métamorphoses d'une petite salicoque d'eau douce (*Caridina desmarestii* Nobis, *Hippolyte Desmarestii* Millet), suivies de quelques réflexions sur les métamorphoses des crustacés décapodes en général. Ann. Sci. Nat. [2] 19:34–86.

Joly, P. 1951. Déterminisme endocrine de la pigmentation chez *Locusta migratoria* L. C. R. Soc. Biol. Paris 145(17/18): 1362–1364.

Joly, P. 1952a. Déterminisme de la pigmentation chez *Acrida turrita* L. C. R. Acad. Sci. Paris 235(18): 1054–1056.

Joly, P. 1952b. Production d'adultoïdes chez *Locusta migratoria* L. (Insecte orthoptéroïde). C. R. Acad. Sci. Paris 235(23): 1555–1557.

Joly, P. and L. Joly. 1953. Résultats de greffe de *corpora allata* chez *Locusta migratoria* L. Ann. Sci. Nat., Zool. Biol. Anim. [11] 15: 331–345.

Joly, R. and M. Descamps. 1988. Endocrinology of myriapods. Pp. 429–452 *in* H. Laufer and R. G. H. Downer (eds.), *Invertebrate Endocrinology*, Vol. 2: *Endocrinology of Selected Invertebrate Types*. Liss, New York.

Jones, D. 1983. *The Country Life Guide to Spiders of Britain and Northern Europe*. Country Life Book, Feltham, Middlesex, England.

Karlson, P. and A. Schweiger. 1961. Zum Tyrosinstoffwechsel der Insekten. IV. Das Phenoloxydase-System von *Calliphora* und seine Beeinflussung durch das Hormon Ecdyson. Hoppe-Seyler's Z. Physiol. Chem. 323: 199–210.

Karlson, P. and C. E. Sekeris. 1962. Zum Tyrosinstoffwechsel der Insekten. IX. Kontrolle des Tyrosinstoffwechsels durch Ecdyson. Biochem. Biophys. Acta 63: 489–495.

Kayser, H. 1976. Identification of β,β-caroten-2-ol and β,β-caroten 2,2'-diol in the stick insect *Carausius morosus* Br.: a reinvestigation study. Z. Naturforsch. C31(11/12): 646–651.

Kayser, H. 1985. Pigments. Pp. 367–415 *in* G. A. Kerkut and L. I. Gilbert (eds.), *Comprehensive Insect Physiology, Biochemistry and Pharmacology*, Vol. 10: *Biochemistry*. Pergamon Press, Oxford and Elmsford, New York.

Kayser-Wegmann, I. 1975. Untersuchungen zur Photobiologie und Endokrinologie der Farbmodifikationen bei der Kohlweisslingspuppe *Pieris brassicae:* Zeitverlauf der sensiblen und kritischen Phasen. J. Insect Physiol. 21(5): 1065–1072.

Keeble, F. W. and F. W. Gamble. 1900a. The colour physiology of *Hippolyte varians*. Proc. R. Soc. Lond. 65: 461–468.

Keeble, F. W. and F. W. Gamble. 1900b. Physiologie de la coloration chez *Hippolyte varians*. Bull. Mus. Natl. Hist. Natl. 6(4): 185–188.

Keeble, F. and F. W. Gamble. 1904. The colour physiology of higher Crustacea. Philos. Trans. R. Soc. Lond. B196: 295–388.

Keeble, F. and F. W. Gamble. 1905. The colour physiology of higher Crustacea. Part III. Philos. Trans. R. Soc. Lond. B198: 1–16.

Keino, H. and K. Endo. 1973. Studies on the determination of seasonal forms in the butterfly, *Araschnia burejana*. Zool. Mag. (Tokyo) 82(1): 48–52.

Keller, R. and S. Webster. 1986. Isolation and functions of crustacean peptides hormones. Bull. Sox. Zool. Fr. 111(1/2): 24.

Keshavan, R. 1974. Studies on colour changes in the prawn *Macrobrachium kistnensis*. Ph.D. thesis, Marathwada Agricultural University, Maharashtra, India 257 pp.

Kiguchi, K. 1973. Hormonal control of the coloration of the larval body and the pigmentation of larval markings in *Bombyx mori*. II. Relationship between the juvenile hormone and the larval body color or the larval markings on various mutants of the silkworm. J. Sericult. Sci. Jpn. 42(4): 293–300.

Kiguchi, K. and S. Kimura. 1981. Hormonal control of larval colouration in the silkworm, *Bombyx mori:* changes in tyrosine contents during the larval moulting. J. Sericult. Sci. Jpn. 50(5): 435–443.

Kleinholz, L. H. 1985. Biochemistry of crustacean hormones. Pp. 464–522 *in* D. E. Bliss (ed.), *The Biology of Crustacea*, Vol. 9: D. E. Bliss, and L. H. Mantel (eds.), *Integument, Pigments, and Hormonal Processes*. Academic Press, Orlando, Florida.

Kleinholz, L. H. and J. H. Welsh. 1937. Colour changes in *Hippolyte varians*. Nature (Lond.) 140: 851–852.

Kleinholz, L. H., K. R. Rao, J. P. Riehm, G. E. Tarr, L. Johnson, and S. Norton. 1986. Isolation and sequence analysis of a pigment-dispersing hormone from eyestalks of the crab, *Cancer magister*. Biol. Bull. (Woods Hole) 170: 135–143.

Knowles, F. G. W. 1949. Control of pigment migration in Crustaceans. Nature (Lond.) 164(4157): 36–37.

Knowles, F. G. W. 1953a. Endocrine activity in the crustacean nervous system. Proc. R. Soc. Lond. B141: 248–267.

Knowles, F. G. W. 1953b. Neurosecretory pathways in the prawn *Leander serratus*. Nature (Lond.) 171(4342): 131–132.

Knowles, F. G. W. 1955. Crustacean colour change and neurosecretion. Endeavour 14(54): 95–104.

Knowles, F. G. W. and H. G. Callan. 1940. A change in the chromatophore pattern of Crustacea at sexual maturity. J. Exp. Biol. 17: 262–266.

Koch, P. B. and D. Bückmann. 1982/83. Vergleichende Untersuchung der Farbmuster und der Farbanpassung von Nymphalidenpuppen (Lepidoptera). Zool. Beitr. [N.F.] 28(3): 369–401.

Koch, P. B. and D. Bückmann. 1987. Hormonal control of seasonal morphs by the timing of ecdysteroid release in *Araschnia levana* L. (Nymphalidae: Lepidoptera). J. Insect Physiol. 33(11): 823–829.

Koller, G. 1927. Über Chromatophorensystem, Farbensinn und Farbwechsel bei *Crangon vulgaris*. Z. Vgl. Physiol. 5: 191–246.

Koulish, S. and W. Klepal. 1981. Ultrastructure of the epidermis and cuticle during the moult–intermoult cycle in two species of adult barnacles. J. Exp. Mar. Biol. Ecol. 49: 121–149.

Krieger, F. 1954. Untersuchungen über den Farbwechsel der Libellenlarven. Z. Vgl. Physiol. 36(4): 352–366.

Kröyer, H. 1842. Monographisk Fremstilling af slaegten *Hippolytes* nordiske Arten. K. Dan. Vidensk. Selsk. Afhandl. 9: 209–361.

Kruse-Pedersen, L. E. 1978. Effects of anti-juvenile hormone (precocene 1) on the development of *Locusta migratoria* L. Gen. Comp. Endocrinol. 36(4): 502–509.

Kuris, A. M. and J. T. Carlton. 1977. Description of a new species, *Crangon handi,* and new genus *Lissocrangon,* of crangonid shrimps (Crustacea, Caridae) from the California coast, with notes on adaptation in body shape and coloration. Biol. Bull. (Woods Hole) 153(3): 540–559.

Laufer, H., M. Landau, D. Borst, and E. Homola. 1986. The synthesis and regulation of methyl farnesoate, a new juvenile hormone for crustacean reproduction. Pp. 135–143 *in* M. Porchet, J.-C. Andries, and A. Dhainaut A. (eds.), *Advances in Invertebrate Reproduction,* Vol. 4. Elsevier, Amsterdam and New York.

Laufer, H., M. Landau, E. Homola, and D. W. Borst. 1987a. Methyl farnesoate: its site of synthesis and regulation of secretion in a juvenile crustacean. Insect Biochem. 17(7): 1129–1129.

Laufer, H., D. Borst, F. C. Baker, C. Carrasco, M. Sinkus, C. C. Reuter, L. W. Tsai, and D. A. Schooley. 1987b. Identification of a juvenile hormone-like compound in a crustacean. Science (Wash., D.C.) 235: 202–205.

Lee, W. L. 1966a. Color change and the ecology of the marine isopod *Idothea* (*Pentidotea*) *montereyensis* Maloney 1933. Ecology 47(6): 930–941.

Lee, W. L. 1966b. Pigmentation of the marine isopod *Idothea montereyensis.* Comp. Biochem. Physiol. 18: 17–36.

Lee, W. L. 1966c. Pigmentation of the marine isopod *Idothea granulosa* (Rathke). Comp. Biochem. Physiol. 19(1): 13–27.

Lee, W. L. and B. M. Gilchrist. 1972. Pigmentation, color change and the ecology of the marine isopod *Idotea resecata* (Stimpson). J. Exp. Mar. Biol. Ecol. 10: 1–27.

Legendre, R. and A. Lopez. 1973. Les chromatophores de l'araignée *Holocnemus pluchei* (Scop.)(Pholcidae).(Note préliminaire.) Bull. Soc. Zool. Fr. 98(4): 487–494.

Lenel, R., G. Nègre-Sadargues, and R. Castillo. 1978. Les pigments caroténoïdes chez les crustacés. Arch. Zool. Exp. Gén. 119(2): 297–334.

Lenel, R. and A. Veillet. 1951. Effets de l'ablation des pédoncules oculaires sur les pigments caroténoïdes du crabe *Carcinus maenas.* C. R. Acad. Sci. Paris 233: 1064–1065.

Lewis, J. G. E. 1962. The ecology, distribution and taxonomy of the centipedes found on the shore in the Plymouth area. J. Mar. Biol. Assoc. U.K. 42: 655–664.

Linzen, B. 1974. The tryptophan → ommochrome pathway in insects. Adv. Insect Physiol. 10: 117–246.

Linzen, B. and D. Bückmann. 1961. Biochemische und histologische Untersuchungen zur Umfärbung der Raupe von *Cerura vinula* L. Z. Naturforsch. B16(1): 6–18.

Long, D. B. 1953. Effects of population density on larvae of Lepidoptera. Trans. R. Entomol. Soc. Lond. 104(15): 543–585.

Lourenço, W. R. 1983. Importance de la pigmentation dans l'étude taxonomique des Buthidae néotropicaux (Arachnida, Scorpiones). Bull. Mus. Natl. Hist. Nat. [4] 5A(2): 611–618.

Maisch, A. and D. Bückmann. 1987. The control of cuticular melanin and lutein incorporation in the morphological colour adaptation of a nymphalid pupa, *Inachis io* L. J. Insect Physiol. 33(6): 393–40.

Matsumoto, S., A. Isogai, A. Suzuki, N. Ogura, and H. Sonobe. 1981. Purification and properties of the melanization and reddish colouration hormone (MRCH) in the armyworm, *Leucania separata* (Lepidoptera). Insect Biochem. 11(6): 725–734.

McNamara, J. C. 1981c. Morphological organization of crustacean pigmentary effectors. Biol. Bull. (Woods Hole) 161(2): 270–280.

McVay, J. A. 1942. Physiological experiments upon neurosecretion with special reference to *Lumbricus* and *Cambarus.* Doctoral thesis, Northwestern University, Evanston, Illinois. Cited in Brown, 1944.

Mégusar, F. 1912. Experiment über den Farbwechsel der Crustaceen. IV. *Palaemon.* Wilhelm Roux' Arch. Entwicklungsmech. Org. 33: 462–665.

Mell, R. 1931. Die Trockenzeitform als Hemmungserscheinung (*Diagora nigrivena* (*Leech*) als Trockenzeitform von *Hestina assimilis L.*) Biol. Zentralbl. 51: 187–194.

Meusy, J.-J. and Payen G. G. 1988. Female reproduction in malacostracan Crustacea. Zool. Sci. 5(2): 213–260.

Millot, J. 1949. Ordre des Aranéides. Pp. 559–743 *in* P. P. Grassé (ed.), *Traité de Zoologie*, Vol. VI. Masson, Paris.

Minkiewicz, R. 1908. Etude expérimentale du synchromatisme de *Hippolyte varians* Leach. Note préliminaire. Bull. Int. Acad. Sci. Cracovie 3: 918–929.

Miyawaki, M. and T. Tsuruda. 1984. Sexual dimorphism in the behavior of the red chromatophores on the midgut of a freshwater shrimp, *Caridina denticulata*. Proc. Jpn. Acad. B60: 353–356.

Mochizuki, S. and N. Agui. 1976. Formation of black pigment in the integument of the cabbage armyworm *Mamestra brassicae* L. *in vitro*. Appl. Entomol. Zool. 11(3): 258–260.

Mordue, W. and J. V. Stone. 1976. Comparison of the biological activities of an insect and a crustacean neurohormone that are structurally similar. Nature (Lond.) 264(5583): 287–289.

Mordue, W. and J. V. Stone. 1977. Relative potencies of locust adipokinetic hormone and prawn red pigment–concentrating hormone in insect and crustacean systems. Gen. Comp. Endocrinol. 33(1): 103–108.

Morgan, S. G. and J. W. Goy. 1987. Reproduction and larval development of the mantis shrimp *Gonodactylus bredini* (Crustacea: Stomatopoda) maintained in the laboratory. J. Crustacean Biol. 7(4): 595–618.

Nagabhushanam, R. 1964. Chromatophores physiology of the zoeae of the mud shrimp, *Upogebia affinis* (Say). Proc. Semin. on Some Aspects of Plankton Res., Mar. Biol. Stn., Porto Novo, India, March 23–25, 1964.

Nagabhushanam, R. 1965. The comparative physiology of the crustacean pigmentary effectors. XIV. Colour changes in the zoaea of the crab, *Sesarma reticulata*. J. Anim. Morphol. Physiol. 12(2): 199–204.

Nagabhushanam, R. and R. Sarojini. 1969. Chromatophore physiology of the zoaea of *Diogenes bicristimanus*. J. Anim. Morphol. Physiol. 16(2): 190–195.

Nagabhushanam, R. and N. Vasantha. 1971. Moulting and colour changes in the prawn, *Caridina weberi*. Hydrobiologia 38(1): 39–47.

Nagabhushanam, R., U. M. Farooqui, and R. Sarojini. 1981. Endocrine control of chromatophores in Decapod Crustacea. Indian Rev. Life Sci. 1: 79–104.

Nagano, T. 1949. Physiological studies on the pigmentary system of crustacea. III. The color change of an Isopod *Ligia exotica* (Roux). Sci. Rep. Tohoku Univ. [4] 18(2): 167–175.

Needham, A. E. 1960. Properties of the connective tissue pigment of *Lithobius forficatus* (L.). Comp. Biochem. Physiol. 1: 72–100.

Needham, A. E. 1974. *The Significance of Zoochromes*. Springer-Verlag, Berlin and New York.

Neet, C. 1987. *Les araignées*. Payot, Lausanne.

Nêmec, V., T. T. Chen, and G. R. Wyatt. 1978. Precocious adult locust, *Locusta migratoria migratorioides*, induced by precocene. Acta Entomol. Bohemoslov. 75(4): 285–286.

Nickerson, B. 1954. A possible endocrine mechanism controlling locust pigmentation. Nature (Lond.) 174(4425): 357–358.

Nicolas, G. 1972. Evolution vers le type solitaire chez le criquet grégaire, *Locusta migratoria cinerascens* (Fabr.), soumis à l'action périodique du gaz carbonique. Acrida 1(2): 97–110.

Nicolas, G. and S. Fuzeau-Braesch. 1968. Etude de quelques facteurs contrôlant l'homochromie chez *Locusta migratoria migratorioides* (Orthoptère). C. R. Soc. Biol. 162(5/6): 1091–1094.

Nijhout, H. F. and D. E. Wheeler. 1982. Juvenile hormone and the physiological basis of insect polymorphisms. Q. Rev. Biol. 57(2): 109–133.

Noël, P. 1972. Sur le métabolisme des pigments caroténoïdes de *Palaemon serratus* (Pennant) (Crustacé Décapode): synthèse de l'astaxanthine à partir du β-carotène. C. R. Acad. Sci. Paris 276D: 2679–2682.

Noël, P. 1979. Contribution à l'étude de la fonction chromatique de *Processa edulis* (Crustacé Natantia). Doctoral thesis (Sci. Nat.), University of Paris VI, 273 pp.

Noël, P. Y. 1981. Hormonal control of chromatophores in Crustaceans. J. Sci. Ind. Res. (India) 40(4): 267–276.

Noël, P. 1982. Circadian chromatic rhythms of some European Processidae (Crustacea, Decapoda). Proc. 13th Int. Conf. Int. Soc. Chronobiol., 1982 pp. 285–293.

Noël, P. Y. 1983a. La pigmentation animale et ses differents aspects. Bull. Soc. Zool. Fr. 108(2): 169–185.

Noël, P. Y. 1983b. Observations sur la pigmentation et la physiologie chromatique de *Periclimenes amethysteus* (Crustacea, Caridea). Can. J. Zool. 61(1): 153–162.

Nolte, D. J. 1963. A pheromone for melanization of locusts. Nature (Lond.) 200(4907): 660–661.

Nolte, D. J. 1974. The gregarization of locusts. Biol. Rev. 49(1): 1–14.

Nolte, D. J., S. H. Eggers, and I. R. May. 1973. A locust pheromone: locustol. J. Insect Physiol. 19(8): 1547–1554.

Nolte, D. J., I. R. May, and B. M. Thomas. 1970. The gregarization pheromone of locusts. Chromosoma 29(4): 462–473.

Odiorne, J. M. 1933a. Morphological color changes: fishes. Proc. Natl. Acad. Sci. USA 19: 329–332.

Odiorne, J. M. 1933b. Adrenaline and teleost leucophores. Proc. Natl. Acad. Sci. USA 19: 750–754.

Ogura, N. 1975a. Induction of cuticular melanization in larvae of armyworm, *Leucania separata* Walker (Lepidoptera: Noctuidae), by implantation of ganglia of the silkworm, *Bombyx mori* L. (Lepidoptera: Bombycidae). Appl. Entomol. Zool. 10(3): 216–219.

Ogura, N. 1975b. Hormonal control of larval coloration in the armyworm, *Leucania separata*. J. Insect Physiol. 21(3): 559–576.

Ogura, N., and J. Mitsuhashi. 1978. *In vitro* cultures of the endocrine organs secreting melanization and reddish coloration hormone in the common armyworm *Leucania separata* (Lepidoptera: Noctuidae). Appl. Entomol. Zool. 13(4): 274–277.

Ogura, N. and J. Mitsuhashi. 1979. Melanization of integuments cultured *in vivo* and *in vitro*. Appl. Entomol. Zool. 14(1): 118–120.

Ogura, N. and T. Saito. 1972. Hormonal function controlling pigmentation of the integument in the common armyworm larva, *Leucania separata* Walker. Appl. Entomol. Zool. 7(4): 239–241.

Ogura, N., S. Yagi, and M. Fukaya. 1971. Hormonal control of larval coloration in the common armyworm *Leucania separata* Walker. Appl. Entomol. Zool. 6(2): 93–95.

Ohashi, M., M. Tsusue, and K. Kiguchi. 1983. Juvenile hormone control of larval colouration in the silkworm, *Bombyx mori:* characterization of epidermal brown colour induced by the hormone. Insect Biochem. 13(2): 123–128.

Ohtaki, T. and E. Ohnishi. 1967. Pigments of the pupal integuments of two colour types of cabbage white butterfly *Pieris rapae crucivora*. J. Insect Physiol. 13(10): 1569–1574.

Okay, S. 1954. Further investigations on colour change in Orthoptera. Commun. Fac. Sci. Univ. Ankara 4:31–43.

Okay, S. 1956. The effect of temperature and humidity on the formation of green pigment in "*Acrida bicolor*" (Thunb.). Arch. Int. Physiol. Biochim. 64(1): 80–91.

Osanai, M. 1966. Pigmente der Raupen von *Hestina japonica* und *Sasakia charonda*. Hoppe-Seyler's Z. Physiol. Chem. 347(4–6): 145–155.

Osanai, M. and Y. Arai. 1962. Effect of ligation at different levels on the change of body color in the larvae of the nymphalid butterfly, *Hestina japonica*, at the beginning of wintering. Zool. Mag. (Tokyo) 71(6): 202–205.

Padgham, D. E. 1976. Control of melanization in first instar larvae of *Schistocerca gregaria*. J. Insect Physiol. 22(10): 1409–1419.

Panouse, J. B. 1946. Recherches sur les phénomènes humoraux chez les Crustacés: l'adaptation chromatique et la croissance ovarienne chez la crevette *Leander serratus*. Ann. Inst. Océanogr. 23: 65–147.

Panouse, M. L. 1947. Variations de la teneur en pigment de *Leander serratus* au cours de l'adaptation au fond. Arch. Zool. Exp. Gén. 85: 44–48.

Parker, G. H. 1948. *Animal Colour Changes and Their Neurohumours: A Survey of Investigations, 1910 to 1943*. Cambridge University Press, London and New York.

Pasteur-Humbert, C. 1962b. Chromatophoric responses of eyestalkless shrimps. Nature (Lond.) 196(4856): 790.

Pautsch, F. 1951. Colour adaptation of the zoea of the shrimp *Crangon crangon* L. Bull. Acad. Pol. Sci. Let., Cl. Sci. Math. Nat. B2: 511–523.

Pautsch, F. 1953. The colour change of the zoea of the shrimp, *Crangon crangon* L. Experientia (Basel) 9(7): 274–276.

Pautsch, F. 1961. The larval chromatophoral system of the crab, *Carcinus maenas* (L.). Acta Biol. Med. (CG-dansk) 5(6): 105–119.

Pautsch, F. 1967. Pigmentation and colour change in decapod larvae. J. Mar. Biol. Assoc. India 3: 1108–1123.

Pautsch, F. and L. Lawinski. 1960. On some responses of the chromatophores in the larvae and megalops of the crab *Rhithropanopeus harrisi* (Gould) subsp. *tridentatus* (Maitland). Abstr. 4th Conf. Eur. Comp. Endocrinol. p. 53.

Pener, M. P. 1965. On the influence of *corpora allata* on maturation and sexual behaviour of *Schistocerca gregaria*. J. Zool. 147(2): 119–136.

Pener, M. P. 1967. Effects of allatectomy and sectioning of the nerves of the *corpora allata* on oöcyte growth, male sexual behaviour and colour change in adults of *Schistocerca gregaria*. J. Insect Physiol. 13(5): 665–684.

Pener, M. P. 1983. Endocrine aspects of phase polymorphism in locusts. Pp. 379–394 *in* R. G. H. Downer and H. Laufer (eds.), *Invertebrate Endocrinology*, Vol. 1: *Endocrinology of Insects*. Liss, New York.

Pener, M. P. and P. Lazarovici. 1979. Effect of exogenous juvenile hormones on mating behavior and yellow colour in allatectomized adult male desert locusts. Physiol. Entomol. 4(3): 251–261.

Pener, M. P., A. Girardie, and P. Joly. 1972. Neurosecretory and *corpus allatum* controlled effects on mating behavior and color change in adult *Locusta migratoria migratorioides* males. Gen. Comp. Endocrinol. 19(3): 494–508.

Pener, M. P., L. Orshan, and J. De Wilde. 1978. Precocene II causes atrophy of *corpora allata* in *Locusta migratoria*. Nature (Lond.) 272(5651): 350–352.

Perez, C. 1929a. Caractères sexuels chez un crabe Oxyrhynque (*Macropodia rostrata* L.). C. R. Acad. Sci. Paris 188: 91–93.

Perez, C. 1929b. Différences sexuelles dans l'ornementation et dans le système pigmentaire chez un crabe Oxyrhynque (*Macropodia rostrata* L.). C. R. Acad. Sci. Paris 188: 271–273.

Pfeiffer, I. W. 1945. The influence of the *corpora allata* over the development of nymphal characters in the grasshopper *Melanoplus differentialis*. Trans. Conn. Acad. Arts Sci. 36: 489–515.

Powell, B. L. 1962a. Types, distribution and rhythmical behaviour of the chromatophores of juvenile *Carcinus maenas* (L.). J. Anim. Ecol. 31: 251–261.

Powell, B. L. 1962b. The responses of the chromatophores of *Carcinus maenas* (L., 1758) to light and temperature. Crustaceana 4(2): 93–102.

Raabe, M. 1961. Recherches sur le déterminisme des genèses de pigments chez un Phasme, *Carausius morosus.* C. R. Acad. Sci. Paris 252(23): 3663–3665.

Raabe, M. 1962. Rôle des glandes endocrines dans la genèse des pigments chez le Phasme *Carausius morosus:* comparaison avec les faits connus chez les Acridiens. Symp. Genet. Biol. Ital. 10: 235–249.

Raabe, M. 1966. Recherches sur la neurosecrétion dans la chaîne nerveuse ventrale du Phasme, *Carausius morosus:* liaisons entre l'activité des cellules B1 et la pigmentation. C. R. Acad. Sci. Paris 263D(4): 408–411.

Raabe, M. 1982. Morphological and physiological color change. Pp. 141–162 *in* M. Raabe (ed.), *Insects Neurohormones.* Plenum Press, New York.

Raabe, M. 1983a. Chromatophorotropic factors. Pp. 485–491 *in* R. G. H. Downer and H. Laufer (eds.), *Invertebrate Endocrinology,* Vol. 1: *Endocrinology of Insects.* Liss, New York.

Raabe, M. 1983b. Pigment metabolism. Pp. 493–502 *in* R. G. H. Downer and H. Laufer (eds.), *Invertebrate Endocrinology,* Vol. 1: *Endocrinology of Insects.* Liss, New York.

Rabaud, E., 1923. Recherches sur la variation chromatique et l'homochromie des arthropodes terrestres. Bull. Biol. Fr. Belg. 62(1): 1–69.

Rao, K. R. 1967. Responses of crustacean larval chromatophores to light and endocrines. Experientia (Basel) 23(3): 231–232.

Rao, K. R. 1968. Variations in the chromatophorotropins and adaptive color changes during the life history of the crab, *Ocypode macrocera.* Zool. Jahrb., Abt. Allg. Zool. Physiol. Tiere 74(3): 274–291.

Rao, K. R. 1969. Influence of prolonged adaptation to a background upon the chromatophorotropins and physiological responses of the chromatophores in the crab, *Ocypode macrocera.* Gen. Comp. Endocrinol. 12: 574–585.

Rao, K. R. 1985. Pigmentary effectors. Pp. 395–462 *in* D. E. Bliss (ed.), *The Biology of Crustacea,* Vol. 9: D. E. Bliss and L. H. Mantel (eds.), *Integument, Pigments, and Hormonal Processes,* Academic Press, Orlando, Florida.

Rao, K. R. and M. Fingerman. 1969. The influence of size on the response of melanophores in the fiddler crab, *Uca pugilator,* to eyestalk extracts. Z. Vgl. Physiol. 62: 86–92.

Remane, A. 1931. Farbwechsel, Farbrassen, Farbanpassung bei der Meerassel *Idothea tricuspidata.* Verh. Dtsch. Zool. Ges. 34: 109–114.

Ressin, W. J. 1980. The effect of juvenile hormone on pupal pigmentation of *Pieris brassicae* L. J. Insect Physiol. 26(5): 295–302.

Richards, O. W. and N. Waloff. 1954. Studies on the biology and population dynamics of British grasshoppers. Anti-Locust Bull. 17: 182 pp.

Riddiford, L. M. and K. Hiruma. 1984. Hormonal control of pigmentation in *Manduca sexta.* Pp. 723–731 *in* H. G. Schlossberger, W. Kochen, B. Linzen, and H. Steinhart (eds.), *Progress in Tryptophan and Serotonin Research.* De Gruyter, Berlin and New York.

Riddiford, L. M. and M. Hori. 1981. Control of larval cuticle formation and pigmentation by juvenile hormone. Pp. 241–250 *in* G. E. Pratt and G. T. Brooks (eds.), *Juvenile Hormone Biochemistry.* Elsevier, Amsterdam and New York.

Riddiford, L. M., M. L. Kiely, and W. J. Wolfgang. 1986. Hormonal regulation of larval-specific protein synthesis in *Manduca* epidermis. Pp. C203/1–203/11 *in* E. Kurstak and H. Oberlander (eds.), *Techniques in* In Vitro *Invertebrate Hormones and Genes* Elsevier, Amsterdam and New York.

Riehm. J. P. and K. R. Rao. 1982. Structure–activity relationships of a pigment-dispersing crustacean neurohormone. Peptides (N.Y.) 3(4): 643–647.

Ritchie, J. M. 1978. Melanism in *Oedaleus senegalensis* and other Oedipodines (Orthoptera, Acrididae). J. Nat. Hist. 12(2): 153–162.

Robison, W. G., Jr. and J. S. Charlton. 1973. Microtubules, microfilaments and pigment movement in the chromatophores of *Palaemonetes vulgaris* (Crustacea). J. Exp. Zool. 186(3): 279–304.

Roussel, J. P. 1967. Fonction des *corpora allata* et contrôle de la pigmentation chez *Gryllus bimaculatus* de Geer. J. Insect Physiol. 13(1): 113–130.

Roussel, J. P. 1975. Actions juvénilisante, chromatotrope, gonadotrope et cardiotrope de JH-III sur *Locusta migratoria*. J. Insect Physiol. 21(5): 1007–1015.

Roussel, J. P. 1976. Hormones juvéniles: actions sur diverses fonctions physiologiques chez le Criquet Migrateur Africain (*Locusta migratoria*). Bull. Soc. Zool. Fr. 101: 1093–1094.

Rowell, C. H. F. 1967. *Corpus allatum* implantation and green/brown polymorphism in three African grasshoppers. J. Insect Physiol. 13(9): 1401–1412.

Rowell, C. H. F. 1971. The variable coloration of the acridoid grasshoppers. Adv. Insect Physiol. 8: 145–198.

Rüdiger, W. and W. Klose. 1970. Über die Pigmente der Florfliege *Chrysopa carnea*, Experientia (Basel) 26(5): 498.

Sagi, A., Y. Milner, and D. Cohen. 1988. Spermatogenesis and sperm storage in the testes of the behaviorally distinctive male morphotypes of *Macrobrachium rosenbergii* (Decapoda, Palaemonidae). Biol. Bull. (Woods Hole) 174: 330–336.

Schaaning, H. T. L. 1929. En eiendommelig varietet av. hummer (*Homarus vulgaris*). Stavanger Mus. Aarsh. 5: 1–3.

Schaffer, M. H. 1986. Functional and evolutionary relationships among the RPCH–AKH family of peptides. Am. Zool. 26: 997–1005.

Scheer, B. T. 1960. Aspects of the intermoult cycle in natantians. Comp. Biochem. Physiol. 1: 3–18.

Scheer, B. T. and M. A. R. Scheer. 1954. The hormonal control of metabolism in crustaceans. VII. Moulting and colour change in the prawn *Leander serratus*. Pubbl. Stn. Zool. Napoli 25(3): 397–418.

Scheer, B. T. and M. A. R. Scheer. 1955. Relations of colour change to moulting in prawns. Nature (Lond.) 175(4454): 473–474.

Schneider, G. 1973. Über den Einfluss verschiedener Umweltfaktoren auf den Färbungspolyphänismus der Raupen des tropisch-amerikanischen Schwärmers *Erinnyis ello* L. (Lepidopt.,Sphingid.). Oecologia 11(4): 354–370.

Schneiderman, H. A. and L. I. Gilbert. 1958. Substances with juvenile hormone activity in Crustacea and in other invertebrates. Biol. Bull. (Woods Hole) 115: 530–535.

Sekeris, C. E. and Fragoulis E. G. 1985. Control of DOPA decarboxylase. Pp. 147–164 *in* G. A. Kerkut and L. I. Gilbert (eds.), *Comprehensive Insect Physiology, Biochemistry and Pharmacology*, Vol. 8: *Endocrinology*, Pt. II. Pergamon Press, Oxford and Elmsford, New York.

Shibley, G. A. 1968. Eyestalk function in chromatophore control in a crab, *Cancer magister*. Physiol. Zool. 41(3): 268–279.

Shimada, K. 1985. Reduction in the critical number of short days for pupal diapause in *Papilio machaon* with precocious metamorphosis. J. Insect. Physiol. 31(9): 683–688.

Smith, A. G. 1980. Environmental factors influencing pupal colour determination in Lepidoptera. II. Experiments with *Pieris rapae, Pieris napi* and *Pieris brassicae*. Proc. R. Soc. Lond. B207(1167): 163–186.

Staal, G. B. 1961. Studies on the physiology of phase induction in *Locusta migratoria migratorioides* R. & F. Publ. Fonds. Landbouw Exp. Bur. 40: 1–125.

Stahnke, H. L. 1971. Some observations of the genus *Centruriodes* Marx (Buthidae, Scorpionida) and *C. sculpturatus* Ewing. Entomol. News 82: 281–307.

Stanêk, V. J. (ed.) 1985. *Encyclopédie des Papillons*. Gründ, Paris.

Stephens, G. J., F. Halberg, and G. C. Stephens. 1964. The blinded fiddler crab: an invertebrate model of circadian desynchronisation. Ann. N.Y. Acad. Sci. 117: 386–406.

Stewart, D. M. 1988. Endocrinology of Arachnids. Pp. 415–428 *in* H. Laufer and R. G. H. Downer (eds.), *Invertebrate Endocrinology*, Vol. 2: *Endocrinology of Selected Invertebrate Types*. Liss, New York.

Stratakis, E. 1979. Ommochrome synthesis and kynurenic acid excretion in relation to metamorphosis and allatectomy in the stick insect, *Carausius morosus*. J. Insect Physiol. 25(12): 925–929.

Sumner, F. B. 1940. Quantitative changes in pigmentation, resulting from visual stimuli in fishes and amphibia. Biol. Rev. 15: 351–378.

Suzuki, A. 1986. Prothoracicotropic hormones and neurohormones in *Bombyx mori*. Pp. 29–51 *in* A. B. Borhovec and D. B. Gelman (eds.), *Insect Neurohormones and Neurophysiology*: Humana Press, Clifton, New Jersey.

Thomsen, M. 1943. Effect of *corpus cardiacum* and other insect organs on the colour-change of the shrimp *Leander adspersus*. Kl. Dan. Vidensk. Selsk. Biol. Medd. 19(4): 1–38.

Tojo, S., M. Morita, and K. Hiruma. 1985. Effects of juvenile hormone on some phase characteristics in the common cutworm, *Spodoptera litura*. J. Insect Physiol. 31(3): 243–249.

Unnithan, G. C., K. K. Nair, and A. Syed. 1980. Precocene-induced metamorphosis in the desert locust *Schistocerca gregaria*. Experientia (Basel) 36(1): 135–136.

Uvarov, B. (ed.) 1966 & 1977. *Grasshoppers and Locusts: A Handbook of General Acridology*, Vol. 1 (1966) & Vol. 2 (1977). Cambridge University Press, London and New York.

Vachon, M. (ed.) 1952. *Etudes sur les Scorpions*. Institut Pasteur d'Algérie, Algiers.

Vasantha, N. 1968. Some aspects of endocrinology of the freshwater prawn, *Caridina weberi*. Ph.D. thesis, Marathwada Agricultural University, Maharashtra, India.

Vasantha, N. 1971. Chromatophores and chromatophorotropins during development in the freshwater prawn *Caridina weberi*. Marathwada Univ. J. Sci. Sect. B, Biol. Sci. 10(3): 213–217.

Vasantha, N. 1972. Morphological colour changes in the freshwater prawn, *Caridina weberi*. Marathwada Univ. J. Sci. 11: 177–181.

Vasantha, N. 1974. Effect of long trem background adaptation on the red chromatophores and chromatophorotropins in the freshwater prawn, *Caridina weberi* (de Man). Marathwada Univ. J. Sci. 13(6): 51–58.

Veron, J. E. N. 1973. Physiological control of the chromatophores of *Austrolestes annulosus* (Odonata). J. Insect Physiol. 19(8): 1689–1703.

Veron, J. E. N. 1976. Responses of Odonata chromatophores to environmental stimuli. J. Insect Physiol. 22(1): 19–30.

Veron, J. E. N., A. F. O'Farrell, and B. Dixon. 1974. Fine structure of Odonata chromatophores. Tissue Cell 6(4): 613–626.

Vuillaume, M. 1969. Les pigments des invertébrés; biochimie et biologie des colorations. In *Les grands problèmes de la biologie*. Masson, Paris.

Walker, R. J. 1988. Endocrinology of merostomates. Pp. 395–414 *in* H. Laufer and R. G. H. Downer (eds.), *Invertebrate Endocrinology*, Vol. 2: *Endocrinology of Selected Invertebrate Types*. Liss, New York.

Wenner, A. M. 1972. Incremental color change in an anomuran decapod, *Hippa pacifica* Dana. Pac. Sci. 26(3): 346–353.

Yagi, S. and K. Kuramochi. 1976. The role of juvenile hormone in larval duration and spermiogenesis in relation to phase variation in the tobacco cutworm, *Spodoptera litura* (Lepidoptera: Noctuidae). Appl. Entomol. Zool. 11: 133–138.

Yaragamblimath, J. S. and S. B. Mathad. 1978. Pigmentation in relation to juvenile hormone analogue application in the cricket *Gryllodes sigillatus*. Comp. Physiol. Ecol. 3(3): 115–116.

Yin, C. M. and G. M. Chippendale. 1974. Juvenile hormone and the induction of larval polymorphism and diapause of the Southwestern cornborer, *Diatraea grandiosella*. J. Insect Physiol. 20(9): 1833–1847.

Termite Polymorphism and Morphogenetic Hormones

7

CHARLES NOIROT AND
CHRISTIAN BORDEREAU

7.1. Introduction

Polymorphism in termites is a developmental phenomenon. All the individuals of the society share the same genetic constitution, except the genes determining the sex. Thus, social polymorphism is in fact a matter of genetic expression.

From the earliest works on insect endocrinology, a role of hormones in polymorphism was suspected. In 1938, Pflugfelder observed the pronounced hypertrophy of the corpora allata (CA) in the physogastric queen of *Odontotermes redemanni*. Later (1947), he provided the first description of the molt gland (ventral gland) of termites; he noted in *Microcerotermes* the presence of this gland in all the larval instars (a termite nymph without wing buds), as well as in workers and soldiers, and its degeneration at the time of imaginal molt. Lüscher's experiments on differentiation of soldiers after implantation of CA were first published in 1958(a,b), followed by numerous other studies on the role of hormones on caste formation; however, owing to the difficulties linked to the nature of the material, we still cannot provide a complete interpretation of the endocrinological conditions triggering caste differentiation.

Termites are frequently categorized into "lower" groups (Mastotermitidae, Kalotermitidae, Termopsidae, Hodotermitidae, and Rhinotermitidae) and "higher" groups (Termitidae).

7.2. Morphogenesis and Polymorphism

An essential morphological marker is the more or less precocious appearance of wing buds on the meso-and metanotum. It is thus a common practice to distinguish between *larvae* without wing buds and *nymphs* with wing buds.

7.2.1. *Imaginal Development*

7.2.1.1. DEVELOPMENT OF SEXUAL ALATES

Imaginal development must be considered as *normal development;* the others proceed through some deviations.

The newly hatched larvae never possess wing buds. These appear at some later stage: at the third instar in *Reticulitermes* (Rhinotermitidae: Hare, 1934; Buchli, 1958); at the fifth instar in *Kalotermes flavicollis* (Kalotermitidae), but at the second in *K. aemulus* (Sewell and Watson, 1981); at the last instar in *Prorhinotermes* (Miller, 1942; Roisin, 1988). In the higher termites the wing buds appear at the second instar (Bathellier, 1927; Noirot, 1955; Kaiser, 1956).

The number of nymphal instars varies among the lower termites depending on the species. On the other hand, in the higher termites five nymphal instars are always observed. Imaginal development is thus of a classical type for heterometabolous insects, except for the stationary and regressive molts (see Section 7.2.2, below).

7.2.1.2. THE PHYSOGASTRY OF THE QUEENS

During development of the colony, the female undergoes an abdominal enlargement called physogastry. It is absent or very small in the Kalotermitidae and Termopsidae, present in the Hodotermitidae and Rhinotermitidae, and maximal in the Termitidae. For example, in *Cubitermes fungifaber*, the weight is increased 20 times, but as much as 2000 times in *Cephalotermes rectangularis* (Bordereau, 1982).

The ovaries enlarge progressively with a posteroanterior maturation of the ovarioles. The abdominal growth involves also a multiplication of the epidermal cells and an increase in tracheae both in length and in diameter (Bordereau, 1971, 1975). The fat body is deeply transformed ("royal" fat body) (Han and Bordereau, 1982a,b). The midgut is also greatly enlarged. Most of the abdominal organs, such as the Malpighian tubules, heart, and muscles are profoundly modified. Hence, the physogastry of the queens involves intensive morphological modifications, exceptional in an adult insect.

In the king, the modifications are much more limited. The dilation of the abdomen remains very small, without any obvious growth of organs except the testes. The fat body remains of the classical type (Han and Bordereau, 1982a). However, the cuticle is notably thickened (Bordereau, 1982).

7.2.2. *Development of Neuters*

7.2.2.1. DIFFERENTIATION OF A NEUTER LINE

In many lower termites (Kalotermitidae, Termopsidae, and *Prorhinotermes* in the Rhinotermitidae), no distinct neuter line exists (an opposite interpretation was put forward by Sewell and Watson, 1981, but refuted by Noirot and Pasteels, 1987). The soldiers differentiate from larvae and/or nymphs of various instars. There are no true workers, the social tasks being performed by the aged larvae, the nymphs, and the pseudergates. [Some larvae and nymphs may undergo *stationary* molts (without growth of the wing buds) or *regressive* molts (with a shortening or disappearance of the wing buds), giving rise to the *pseudergates* (Grassé and Noirot, 1947), which may later differentiate into soldiers, neotenics, or imagoes. These special molts give considerable flexibility to postembryonic development.] In the higher termites, in contrast, the first larval instar is undifferentiated and the first molt allows the separation of two lines: the imaginal or sexual line, terminating in the alates, and the neuter line where the wing buds never appear and the individuals differentiate into workers and soldiers. This separation is observed in some lower termites: in *Reticulitermes*, the separation occurs at the second molt but the nymphs may occasionally undergo regressive molts giving rise to pseudergates. In the Hodotermitidae, the separation also occurs at the second molt but seems irreversible.

Two exceptional cases must be noted in which the neuter line is unisexual: the genus *Anacanthotermes*, where all the neuters are males; and *Schedorhinotermes*, where they are females (reviewed in Noirot, 1985a). These facts have been well demonstrated but remain totally unexplained.

In most termites the production of nymphs (sexual line) follows a seasonal cycle, and in the Termitidae the first instar nymphs appear each year for only a few weeks.

7.2.2.2. DIFFERENTIATION OF WORKERS

The morphogenesis of workers is mainly characterized by the absence of development of several organs, resulting in persistence of numerous *larval* traits, such as absence of wings and the associated structures of the pterothorax, lack of compound eyes and optic lobes in the protocerebrum (except in the Hodotermitidae), and only rudimentary development of genital organs. All these features are the consequence of an arrest of development of these organs after the molt that separates the neuter line. The gut, however, is well developed and the head voluminous, with powerful mandibular muscles. On the whole, the workers appear as simplified nonspecialized insects (Noirot, 1982; Noirot and Pasteels, 1987).

The development of larvae into workers is progressive in the lower termites (when the workers are present): during the successive larval instars, sclerotization increases, the digestive tract develops, and the intestinal fauna is progressively completed. In the Termitidae, in contrast, the transformation is abrupt: the larvae are devoid of any pigmentation or sclerotization and are completely dependent on the workers. The molt transforming them into workers does not greatly modify their morphology but profoundly affects their physiology and behavior.

The workers, however, cannot be considered as imagoes, because in most species they remain capable of molting again, either into a worker of another instar or into a soldier or (exceptionally) into a replacement reproductive. Only in a few species do completely stabilized workers exist, such as the large workers of the fungus-growing termites. Even in this case, the molt glands are still present (see Section 7.3.2, below).

In many species, the worker caste is itself polymorphic, which can result from two phenomena not mutually exclusive. First, in most cases the workers remain capable of molting into another worker instar. These worker molts most often bear very minor modifications—mainly an increase of pigmentation with a very limited or no growth. The resulting polymorphism is therefore generally inconspicuous, with some exceptions. The second phenomenon is the existence of a *sexual dimorphism* already evident after the molt that separates the neuter line. As a result, two types of workers are present—large and small—differing especially in size of the head. Depending on the species, either the males or the females may be larger (Noirot, 1955, 1969b). Remarkably, this dimorphism is much less marked, even absent, in the imaginal line.

7.2.2.3. DIFFERENTIATION OF SOLDIERS

Origin

Depending on the species, the soldiers differentiate from various instars. In *K. flavicollis,* they derive from advanced larvae, pseudergates, or all the nymphal instars, including the last (Grassé and Noirot, 1946). In the incipient colonies, they come from younger instars: second or third instar larvae (Grassé and Noirot, 1957); in older colonies, they come from individuals of the fifth instar and beyond (pseudergates and nymphs). A similar situation seems to exist in the other Kalotermitidae and the Termopsidae. In the other lower termites, the soldiers—as far as their origin is known—come from larvae or workers. The genus *Prorhinotermes,* however, where the nymphs appear very late, is similar to the Kalotermitidae (Miller, 1942, 1969; Roisin, 1988). In the Termitidae, the soldiers always derive from the neuter line, never from the imaginal line (except for very rare intercastes), and for a given species their origin may be precisely identified in a few instars or even in one: it is most often the differentiated workers; more rarely, the last instar larvae (small soldiers in Macrotermitinae); in a few cases, young larvae, identified after the first molt (small soldiers of *Trinervitermes* and *Acanthotermes*) (Noirot, 1955, 1969b).

Too often it is believed that the soldiers come from male and female individuals. Indeed, there are numerous exceptions. In the lower termites, the genera *Anacanthotermes* and *Schedorhinotermes* should be cited: all soldiers and workers are of the same sex, male in the former genus, female in the latter. The soldiers' sex in many species, however, has never been studied! In the Termitidae, soldiers of both sexes were found in some species (e.g., *Amitermes evuncifer, Leptomyxotermes doriae*), but in most species studied they belong to the *same sex:* generally male in the Nasutitermitinae, female in the Termitinae and Macrotermitinae (Noirot, 1955, 1969b).

In many species, the soldiers are polymorphic. This is a consequence of their origin from diverse developmental stages, not (with a few exceptions) from sexual dimorphism, unlike the workers.

Morphogenesis

Termite soldiers are without an equivalent in other insects. Indeed, they differentiate through two successive molts: the first gives rise to the presoldier; the second, to the final soldier. The intermediate presoldier is always unpigmented, unsclerotized, and has a larval aspect and behavior. This instar is always short—from 8 to 15 days—and is comparable to the pupa of holometabolous insects (Deligne, 1970). All the modifications of this "metamorphosis" may be considered as adaptations to defense, which is their essential function. Whereas the workers and imagoes have undergone only minor modifications during their long phylogenetic evolution, the soldiers have achieved much more varied adaptations either in shape and function of the mandibles or by differentiation of a chemical weapon (frontal or salivary gland) or both (reviewed in Deligne et al., 1981). The soldier caste in the Isoptera really appears to be a unique evolutionary development.

The soldier is a terminal instar (but see Section 7.5.1, below) and is definitely sterile. The "fertile soldiers" described in the literature (Myles, 1986) are, in fact, intercastes. However, the soldier cannot be considered as an imago: its morphology is very different, ant it retains its molt glands (see Section 7.3.2, below).

7.2.3. *Differentiation of Neotenic Reproductives*

Replacement or supplementary reproductives were observed in many termites, most often as a consequence of the elimination of the royal pair. Depending on the species, they may originate from various individuals. Sometimes they are simply imagoes maturing into a new royal pair in their own nest, but most often they are *neotenics* differentiating through one or several special molts and thus with a new morphogenesis. Their origin may be aged larvae, pseudergates, nymphs, or workers. In the latter case, they are called *ergatoids* and must be considered as neotenics, for the workers are not imagoes (see Section 7.2.2.2, above). In addition to the development of the sex organs, various morphological changes occur (see reviews in Miller, 1969; Lüscher, 1974; Noirot, 1956, 1985b). An increase in pigmentation is general. The compound eyes are often conspicuous (but in ergatoids are without the development of optic lobes of the brain). The last abdominal sternites become imaginal in type. The wing buds deserve special attention, because they are generally reduced when the starting insect is a nymph but may appear during the differentiation of some ergatoids.

7.2.4. *Conclusion*

The phylogenetic evolution of termite polymorphism (Noirot and Pasteels, 1987) can be summarized as follows: in the prototermites, the development was probably not different from that of the solitary orthopteroids, except for the soldier caste. The workers were not yet differentiated, their functions being assumed by old larvae and nymphs from which arose the soldiers. Then through stationary molts, the larval or nymphal condition (pseudergates) was prolonged, which increased the "worker force" of the society—a phenomenon becoming more noticeable with the regressive molts. This is the status in Kalotermitidae and Termopsidae, where the postembryonic development shows maximum flexibility, since every individual (old larva or nymph) can, according to the status of the society, develop toward imaginal morphogenesis or soldier, pseudergate, or neotenic morphogenesis.

An important evolutionary leap occurred when a precocious molt separated one neuter line from the imaginal line. Workers and soldiers arose from the neuter line, alate imagoes from the imaginal line, and the separation between these two lines is never entirely reversible. In *Reticulitermes* the development remains flexible (Buchli, 1958), but it becomes more rigid in the Termitidae. Moreover, the larva–worker transition is progressive in the lower termites, which possess true workers, whereas it is rough in the Termitidae. These developmental steps are concluded at about the same time that sex-related differences appear: sexual dimorphism of the workers; specialization of one sex into soldier production. These differences are most marked when all neuters are of the same sex—male in *Anacanthotermes,* female in *Schedorhinotermes.*

Finally, it must be underscored that an individual can successively belong to different castes in the course of development: in *Kalotermes,* an individual can successively be a larva, a nymph, a pseudergate, then a soldier or a neotenic. Many workers of Termitidae can transform into soldiers or even in ergatoid sexuals. We suggest the term *temporal polymorphism* for this phenomenon.

7.3. Morphological and Histophysiological Data

7.3.1. *Anatomy of the Endocrine Glands*

7.3.1.1. NEUROSECRETORY CELLS AND NEUROHEMAL ORGANS

The neurosecretory cells (NSC) of termites are generally of small or medium size (6–18 μm: Gillott and Yin, 1972). They are mainly observed in the protocerebrum, in the median part of the pars intercerebralis (median cells), and in the lateral parts, near the optic ganglia (Noirot,

1957, 1969a; Gillott and Yin, 1972). NSC have also been noted in other parts of the nervous system (reviewed in Noirot, 1969a; Lebrun, 1983).

The corpora cardiaca (CC) are triangular in shape and medially anastomosed in their posterior regions. They are situated just behind the brain and enclosed within the wall of the aorta, on each side of the hypocerebral ganglion.

7.3.1.2. CORPORA ALLATA

The corpora allata (CA), spherical or ovoid in shape, lie just behind the CC, with which they are connected by the allatocardiac nerves. The CA are composed of cells generally arranged radially, with nuclei usually positioned at the periphery. Their histological structure and nuclear-cytoplasmic ratio, however, vary during development.

7.3.1.3. MOLT GLANDS

The molt glands, first noted by Jucci (1924) as "tentorial glands," then by Pflugfelder (1947) as "ventral glands," are composed of cellular cords attached to the posterodilatator muscles of the pharynx (Herlant-Meewis and Pasteels, 1961). In the lower termites, they have a cephalic part, which extends into the collum and the prothorax, where they form an X-shaped structure between the digestive tube and the ventral nervous chain (Fig. 7.1); thus, these glands may be called "prothoracic glands." This arrangement is observed in *Mastotermes, Kalotermes, Zootermopsis,* and *Anacanthotermes* (Bernardini-Mosconi, 1958; Mosconi-Bernardini and Vecchi, 1964; Lüscher, 1960; Gillott and Yin, 1972). In the Termitidae, in contrast, these glands are entirely cephalic, situated in the posterior part of the head capsule, on both sides of the digestive tube (Noirot, 1969a). The molt glands of the Rhinotermitidae (*Reticulitermes* or *Schedorhinotermes*) appear to be of the same type (Jucci, 1924; Mosconi-Bernardini and Vecchi, 1964).

7.3.2. *Development of the Endocrine Glands During Caste Differentiation*

7.3.2.1. SEPARATION OF THE NEUTER AND SEXUAL LINES IN THE TERMITIDAE

The main observation was made by Kaiser (1956), especially in *Anoplotermes pacificus.* This author has shown that at the end of the first larval instar (during which the larvae remain morphologically undifferentiated), the future sexuals can be recognized by their molt glands, which are more voluminous (Fig. 7.2). This enlargement, which doubles the volume of the gland, is presumed to be the result of cellular

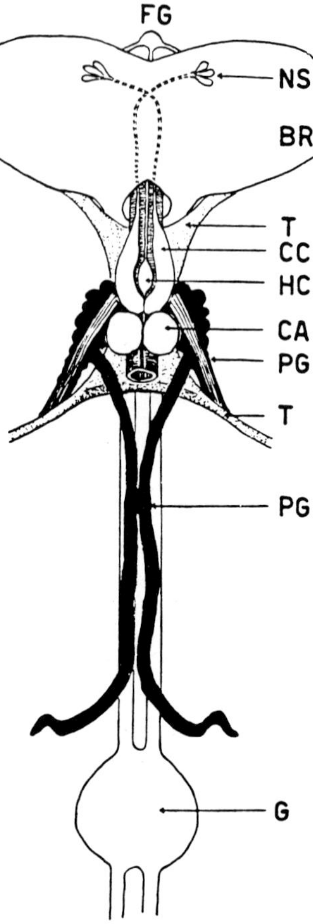

FIGURE 7.1. Endocrine glands of *Kalotermes flavicollis*. *Key:* BR = brain; CA = corpora allata; CC = corpora cardiaca; FG = frontal ganglion; G = first thoracic ganglion; HC = hypocerebral ganglion; NS = neurosecretory cells; PG = prothoracic glands; T = tentorium. (From Lüscher, 1960.)

growth, since there is no increase in cell number. During the successive nymphal instars, this enlargement of the molt glands increases markedly. In contrast, such growth is slight during development of the workers (Fig. 7.2); this suggests that the sexual–neuter separation is closely related to a difference in the ecdysteroid level. Some observations carried out on *Termes hospes* (Noirot, 1969a) confirmed Kaiser's data concerning the enlargement of the molt glands during nymphal devel-

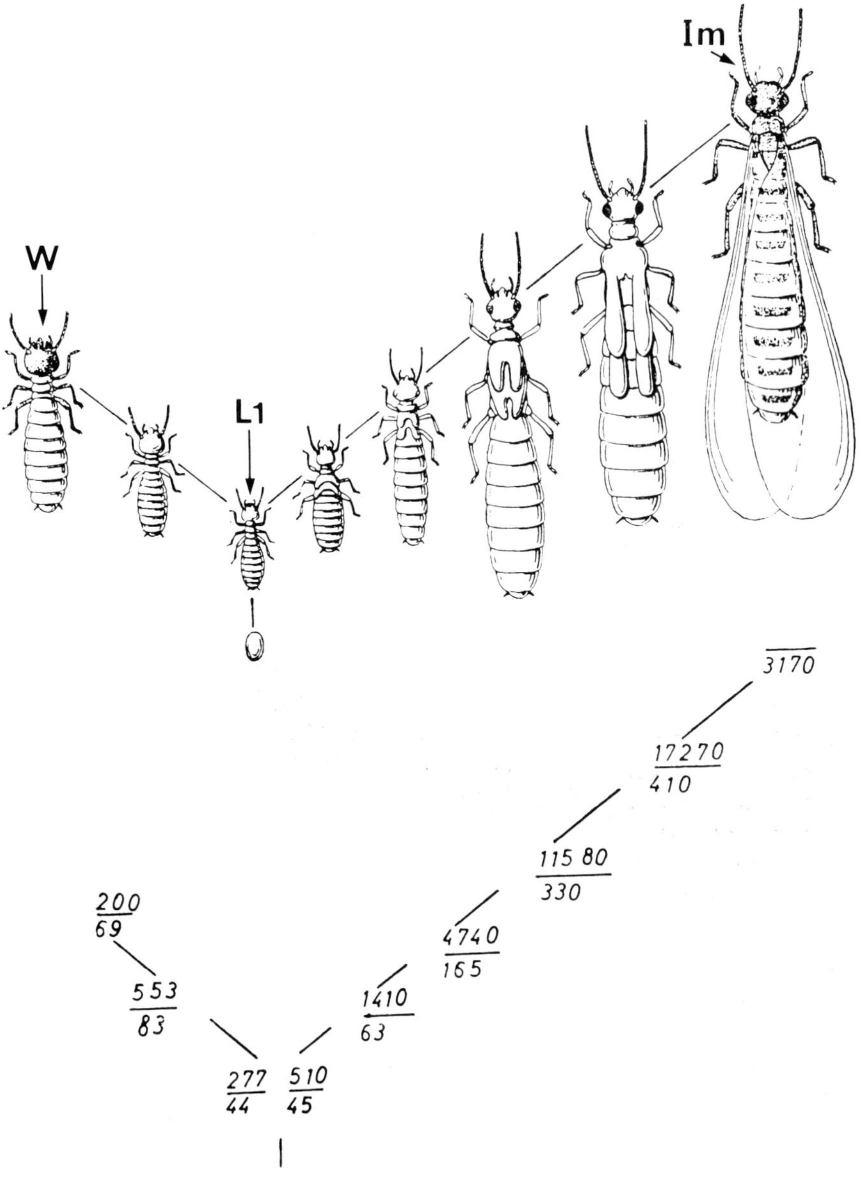

FIGURE 7.2. Postembryonic development in *Anoplotermes pacificus*. The numbers indicate for each instar the volume (expressed in units of 100 μm^3 of the molt glands (above) and the corpora allata (below). *Key:* Im = imago; L_1 = first larval instar; W = worker. At the end of the first larval instar, the individuals are morphologically undifferentiated. However, according to Kaiser, some of them—the future sexuals—have much more voluminous molt glands. In contrast, there are no differences in the corpora allata volume in the first-instar larvae. (From Kaiser, 1956.)

opment but could not establish (because of a lack of material) that the difference with the larvae of the workers started at the undifferentiated first instar. And this is essential, because it is not certain that Kaiser could distinguish between the old larvae of the first instar and the youngest nymphs in *A. pacificus;* indeed, he observed only four nymphal instars, while five nymphal instars seem to be present in all Termitidae, and in his scheme (Fig. 7.2) the first nymphal instar appears to be missing. Obviously these larvae of the first stage require new investigation.

In contrast, at the undifferentiated first instar, the CA do not appear different in both lines (Kaiser, 1956), and their growth seems similar. Our observations on *Termes hospes* (Noirot, unpublished data) support this. Then, if there is a relationship between the volume of the CA and their secretory activity, it is unlikely that these glands play a major role in the differentiation between the sexual and the neuter lines. Once the separation of both lines is achieved, the growth of the CA may be greater in the sexual than in the neuter line, but the differences are far less than for the molt glands (Fig. 7.2).

7.3.2.2. DIFFERENTIATION OF WORKERS

In the Termitidae, the workers are differentiated through two or three larval instars (the undifferentiated first instar included). During this development, the endocrine glands do not change conspicuously; molt glands and CA remain moderate in size. In many Termitinae and Nasutitermitinae, the workers molt two or more times; during these molts the growth is very limited and sometimes even insignificant. During the period of premolt, no marked changes of the endocrine glands are seen, except a temporary enlargement of the molt glands (after our observations on *Nasutitermes arborum* and *Termes hospes*). More visible are the modifications occurring during the transformation of workers into soldiers (see Section 7.3.2.3, below).

The molt glands are always present in the workers, even in the species where these workers never molt (Noirot, 1969a). This finding is in accord with the larval character of the workers.

We have no information about the endocrine glands in workers of other families where workers are present (Mastotermitidae, Hodotermitidae, or Rhinotermitidae).

7.3.2.3. DIFFERENTIATION OF SOLDIERS

The most significant and constant modification of the endocrine glands is the enlargement of the CA during the molt inducing the presoldier. It has been observed in the lower termites (Lüscher, 1958a,b; Lebrun, 1967; Gillott and Yin, 1972; Yin and Gillott, 1975a), as well as in the Termitidae (Kaiser, 1956; Noirot, 1969a; Lüscher, 1976; Okot-Kotber, 1980a, 1982).

During the presoldier–soldier transformation, the volume of the CA generally decreases. In *Termes hospes,* however, the CA volume of the soldiers is greater than that of the presoldier (Noirot, 1969a).

According to Yin and Gillott (1975a), the CC of *Zootermopsis angusticollis* increase in volume during the presoldier differentiation, suggesting an activation of neurosecretion.

Modifications of the molt glands have been much less studied. Nevertheless, it seems that they also enlarge noticeably before the presoldier molt (Kaiser, 1956; Okot-Kotber, 1982). Here, an important fact must be stressed: *the molt glands are always present in soldiers* (Noirot 1969a), although their histological structure suggests an inactive state.

7.3.2.4. DIFFERENTIATION OF NEOTENIC REPRODUCTIVES

In all cases, this differentiation is correlated with a hypertrophy of the CA that is often pronounced. This growth is precocious, noticeable before the molt of differentiation as early as the first signs of the molt (gut emptying). Such growth was observed in neotenics arising from pseudergates or nymphs in *Kalotermes flavicollis* (Lüscher, 1957, 1960; Lebrun, 1967) and in ergatoid neotenics in the Termitidae *Termes hospes* (Noirot, 1956) and *Nasutitermes corniger* (Thorne and Noirot, 1982). The CA may enlarge again after the molt: in *Zootermopsis angusticollis,* for example, the CA volume of the neotenics 2 days after the molt is 7 times that of the larvae from which they arise, and 11 times that of the larvae 5 days later (Yin and Gillott, 1975a).

Little is known about the development of the NSC; Noirot (1957) reported that new NSC appear in *Kalotermes* during the differentiation of the neotenics, but Yin and Gillott (1975a) did not observe the same phenomenon even though the enlargement of the CC can be interpreted as an increase in neurosecretory activity.

In the same way, our data concerning the molt glands remain imprecise; it is difficult to know whether the hypertrophy of these glands occurring before the differentiating molt(s) is linked to preparation of the molt, as in all pterygote insects, or to precocious sexualization. On the contrary, degeneration of the molt glands after differentiation clearly indicates that these sexual individuals are in ultimate instars and unable to molt again. A few exceptions have been observed (*Neotermes connexus:* Myles and Chang, 1984; *Nasutitermes corniger:* Thorne and Noirot, 1982), but they concern nonfunctional neotenics.

7.3.2.5. DEVELOPMENT OF PHYSOGASTRY

In the Termitidae, the physogastry of the queen is correlated with a high growth of the CA, as already observed by Pflugfelder (1938), and which in *Macrotermes subhyalinus* (Fig. 7.3) corresponds to an increase in volume of about 100 times with regard to the CA of the winged imago

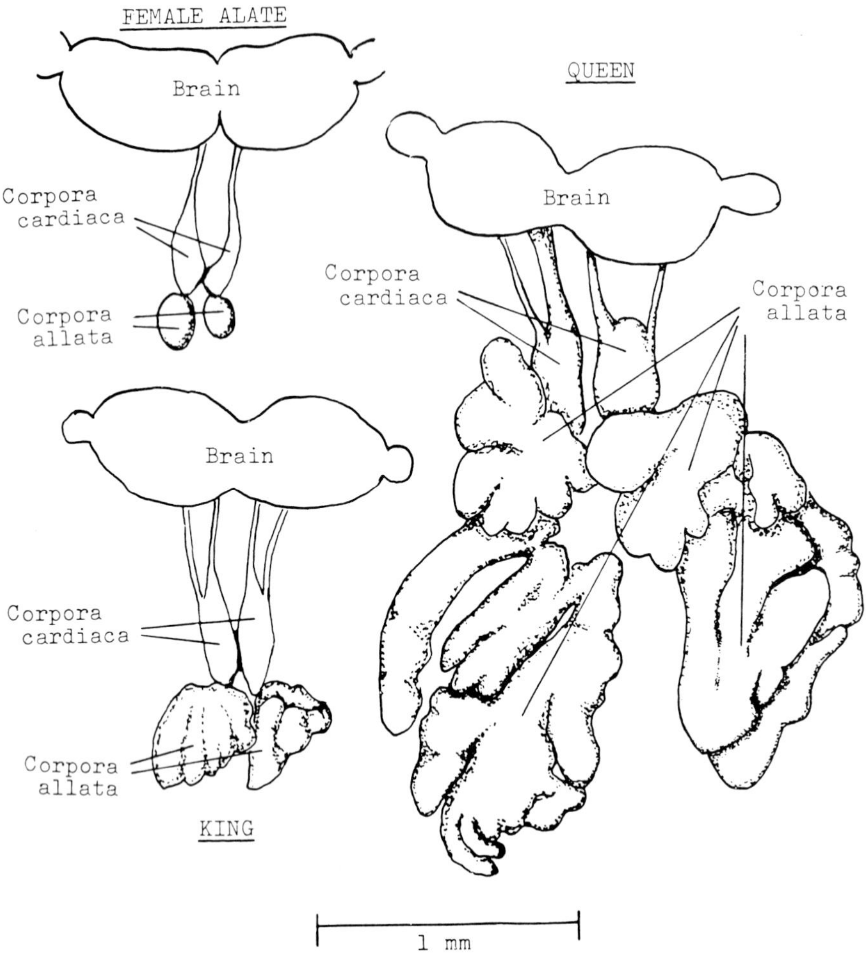

FIGURE 7.3. Endocrine glands of adult sexuals of *Macrotermes subhyalinus*. Note the hypertrophy of the corpora allata in the physograstric queen, which are 100 times more voluminous than in the alate. This enlargement is correlated with a very high secretory activity. ×35. (From Lüscher, 1976.)

(Lüscher, 1976). The gland becomes very irregular in shape and is so large that it extends itself in the collum. In the king, the enlargement of the CA is more limited, although their volume is multiplied by 5.

The protocerebral neurosecretion, histologically visible, reaches its highest peak in the physogastric queens, and the CC are markedly dilated (Pasteels and Deligne, 1965; Noirot, 1969a).

7.3.3. *Conclusion*

In spite of the as yet fragmentary documentation, some important morphological and histophysiological data suggest the action of the two hormones involved in insect development.

The CA become, often temporarily, hypertrophied at the time of the soldier differentiation. They also show an enlargement at the time of sexualization either of the neotenics or of the imaginal sexuals. In this case, however, it is not known whether this hypertrophy is related to maturation of the genital apparatus or to the peculiar morphogenesis or both.

The molt glands degenerate in imagoes and neotenics but persist in workers and soldiers. According to Kaiser's observations (1956), they could intervene in the separation between the neuter and the sexual lines, since they are much more enlarged in the second line than in the first.

The morphological data, however, must be carefully interpreted, since the volume of a gland is not necessarily in direct relationship to its secretory activity.

7.4. Hormones: Identification and Titer Determinations of Ecdysteroids and JH

7.4.1. *Hormones in Higher Termite Queens and Kings*

7.4.1.1. ECDYSTEROIDS

The first measurements were made by radioimmunoassay (RIA) by Bordereau et al. (1976) in *Macrotermes bellicosus*. Physogastric queens contain high levels of ecdysteroids: about 500 ng/g of fresh weight in *M. bellicosus*, 1500 ng/g in *Cephalotermes rectangularis*, and 1600 ng/g in *Cubitermes fungifaber* (Bordereau et al., 1977). Gas chromatography/mass spectrometry (GC/MS) demonstrates ecdysone as well as 20-OH ecdysone, the ecdysone being quantitatively the most important (Delbecque et al., 1978).

In *Macrotermes*, ecdysteroids are preferentially located in ovaries (up to 96%), but noticeable quantities are also found in hemolymph (Lanzrein et al., 1977; Delbecque et al., 1978). They probably originate from follicular cells, as occurs in other adult insects (Hoffmann et al., 1980; Hagedorn, 1985). The ovarian ecdysteroids are accumulated in the eggs to high levels that can fluctuate greatly according to the nature of the colonies (Delbecque et al., 1978), without correlation with the reproductive cycle (Lanzrein et al., 1985).

7.4.1.2. JUVENILE HORMONES

JH III is the only GC/MS-identified JH in termites, in the hemolymph of the queen of *Macrotermes michaelseni* (Meyer et al., 1976). The queen CA also synthesize JH III *in vitro* (Lanzrein et al., 1978). All the other titer determinations were made with the *Galleria* test after thin-layer chromatography (TLC) or high-pressure liquid chromatography (HPLC).

In correlation with CA hypertrophy (Fig. 7.3), the physogastric queen of *Macrotermes* contains high levels of JH, up to 13 μg/ml of hemolymph (Lanzrein et al., 1977, 1985). A part of this JH is found in the newly laid eggs in variable concentrations up to 250 ng/g of fresh weight (Lanzrein et al., 1985). Lüscher (1976) suggested that the seasonal cycle of nymphal production should be due to a seasonal decrease of JH in queen hemolymph and therefore in eggs. The JH levels, however, fluctuate in queen hemolymph as well as in eggs without correlation with a seasonal cycle (Lanzrein et al., 1985). Nevertheless, as indicated by the latter authors themselves, the titer determinations were made on pools of 1000 eggs and—even at the season of maximal nymphal production—only a small part of these eggs will develop into the nymphal line. Hence, Lüscher's hypothesis cannot be definitely rejected. The JH rise in queens is very precocious: their JH titer rises rapidly after colony foundation when oogenesis starts, a long time before the first signs of physogastry (N. Abo-Khatwa, cited by Lüscher, 1976). The *M. michaelseni* king also shows a high level of JH in the hemolymph, which is unusual in male insects (Lanzrein et al., 1985). Titers are, however, lower than in the queen and are only vaguely correlated to the titers of the queen of the same nest.

7.4.2. *Hormones During Postembryonic Development*

7.4.2.1. ECDYSTEROIDS

The only available data were obtained by RIA and expressed in ecdysone or 20-hydroxyecdysone equivalent. We should emphasize that as yet nothing is known about nymphal or imaginal development in termites, and therefore it is not known whether the difference in size of the molt glands between nymphs and larvae (see Section 7.3.2.1, above) is correlated to a difference in ecdysteroids titers.

In *M. michaelseni*, the third-instar female larvae can develop either into a worker or a presoldier. Both morphogeneses show two peaks of ecdysteroids: a precocious first peak (about 125 ng/g) is common to both types, but the second peak is later and higher in the future presoldier (550 ng/g) than in the future worker (310 ng/g) (Okot-Kotber, 1983). The difference in the chronology of the second peak is probably more important than the quantitative difference. In *Kalotermes flavicollis*, no significant differences were noted in ecdysteroid levels during the molt of neotenics from pseudergates, in comparison with normal molts of pseudergates (Bordereau, unpublished data).

7.4.2.2. JUVENILE HORMONES

The only available data are those of Greenberg and Tobe (1985) on *Z. angusticollis*. *In vitro*, the CA of the orphaned larvae secrete much more JH than do larvae reared with reproductives. Thus, the presence of functional reproductives seems to slow down the activity of the larval CA.

7.5. Experimental Data

7.5.1. *Corpora Allata and Juvenile Hormones*

7.5.1.1. STIMULATION OF SOLDIER PRODUCTION

Lüscher (1958a,b) pioneered research in this field by implantation of CA into pseudergates of *K. flavicollis;* these experiments were extended by Lüscher and Springhetti (1960), and then further investigations were conducted by Lebrun (1967, 1978). The results depend on the physiological condition of the donors, the recipients, and the colony.

CA from functional sexuals (imagoes and neotenics) and nymphs just before the imaginal molt can induce the soldier differentiation. In contrast, CA from pseudergates or from nymphs away from a molt are ineffective. This difference seems to be related to the quantity of JH produced and not (as first hypothesized by Lüscher and Springhetti, 1960) to the secretion of two separate hormones. Besides, Lebrun (1967) showed that CA from adult cockroaches in vitellogenesis (*Periplaneta americana*) could induce soldier differentiation.

The competence of the recipient varies during an instar: for soldier differentiation, it is maximal in the second half of the instar in *Kalotermes flavicollis* (Springhetti, 1972). Just after a molt, the competence for neotenic differentiation is very high, but only in orphaned colonies. This might well explain the aforementioned results of Lebrun, who obtained soldier–neotenic intercastes by implanting CA into newly molted pseudergates reared in orphaned colonies.

Such experiments often induce imperfect individuals intermediate between soldiers and another caste. A whole series of intermediates between soldiers and sexuals (Fig. 7.4A) was thus obtained by Lebrun (1970) after implantation of cockroach CA into last-instar nymphs near the imaginal molt.

These previous data obtained by CA studies were confirmed on *K. flavicollis* by the use of exogenous JH by Lüscher (1969) and then extended to other species. JH or JH analogue (JHA) is given to termites, by

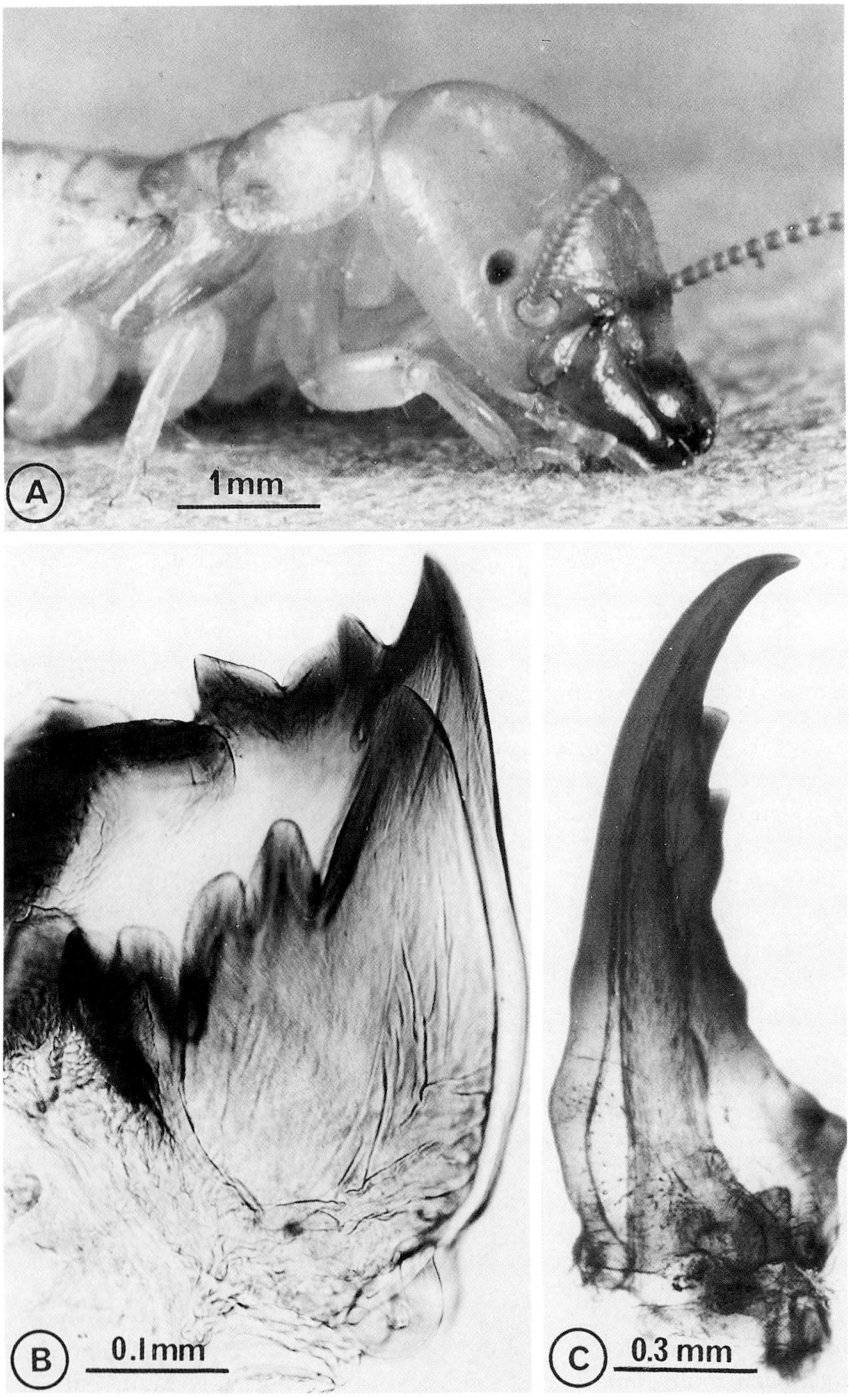
A
1mm
B
0.1mm
C
0.3 mm

injection, topical application, feeding of impregnated substrate, or by vaporization (JHA is applied as an acetone solution to the surface of a petri dish). These various types of application induce qualitatively similar results but make comparison of doses and interpretation of the dose–response relation difficult.

Soldier induction by JHAs was mostly obtained in the lower termites (Hrdy et al., 1979; Howard and Haverty, 1979; Dai, 1980; Afzal and Ahmad, 1982a,b; Howard, 1983, 1984; Doki et al., 1984; Jones, 1984; Myles and Chang, 1984; Varma, 1985; Tsunoda et al., 1986). In higher termites, only the genera *Nasutitermes* (French, 1974; Lenz, 1976; Lenz and Westcott, 1985) and *Macrotermes* (Okot-Kotber, 1980b,c) were tested.

Results show great variability that comes not only from the techniques used (types of JHAs, mode and duration of application, doses, and the like) but also from the competence of the treated insects (stage of developement and age within the instar) and from the species. Normal presoldiers and soldiers can be induced; but intercastes are more often observed.

For soldier induction by JHAs, a high threshold dose is required in the range of 1 μg per insect (Lüscher, 1969; Okot-Kotber, 1980b; Varma, 1985). Above this threshold dose a dose–effect relationship is sometimes observed, but 100% soldiers are very rarely obtained even with the highest doses (which often induce high mortality).

Regarding the competence of the treated insects, it seems extensive in the lower termites in the same way as in the natural colonies in which soldiers arise from various stages such as old larvae, pseudergates, workers, and nymphs. JHAs, however, can induce soldier differentiation of individuals that do not naturally transform into soldiers. For example, in *Z. angusticollis*, farnesyl methyl ether (FME) or ZR-512 (Zoecon) induced soldier formation from last nymphal instars (Wanyonyi

---

FIGURE 7.4. (A) Presoldier–imago intercaste of *Kalotermes flavicollis* obtained after implantation of cockroach corpora allata in a last nymphal instar. This result clearly shows the role of JH in soldier morphogenesis. (From Lebrun, 1970.) (B) Mandible of a neotenic of *K. flavicollis* after implantation of cockroach molt glands. The neotenic is an ultimate instar, i.e., one which does not normally molt. It can, however, be artificially induced to molt, and probably under the influence of high JH levels the new morphogenesis is orientated toward the soldier type, as indicated by the mandible observed inside the exuvium. (From Lebrun, 1967.) (C) Mandible of a soldier of *K. flavicollis* after implantation of cockroach molt glands. The soldier does not normally molt, but the artificially induced molt emphasizes the soldier morphogenesis. (From Lebrun, 1967.)

and Lüscher, 1973), contrary to what happens in field colonies (Miller, 1969). In *Prorhinotermes,* Hrdy et al. (1979) obtained soldiers from second-instar larvae, whereas in natural colonies soldier differentiation occurs only from third-instar larvae and beyond (Miller, 1942). In *Reticulitermes speratus,* JHAs induced presoldier differentiation from all developmental stages (Tsunoda et al., 1986). In the higher termites, JHAs can also induce soldiers from naturally incompetent instars, but to a less extensive degree. In *Nasutitermes nigriceps* (Lenz, 1976; Lenz and Westcott, 1985), ZR-512 induces soldier formation from large workers, whereas they arise naturally from small workers. With ZR-515, Okot-Kotber (1980b) obtained soldier formation from female and male third-instar larvae, whereas only female larvae produce soldiers under natural conditions. Thus, JHA could trigger production of soldiers or intercastes of the heterologous sex. It is probably the same in *N. nigriceps.*

Within the JHA-responsive instars, there are periods of competence. As in *Kalotermes,* the pseudergates of *Zootermopsis* are the most competent during the second half of the instar (Wanyonyi and Lüscher, 1973; Wanyonyi, 1974). In contrast, competence is maximal at the beginning of the intermolt in *Reticulitermes flaviceps* (Dai, 1980) and in *Macrotermes michaelseni* (Okot-Kotber, 1980c).

JHAs indisputably have a prothoracicotropic action: not only do they act on the morphogenesis of the molting individuals but also on the induction of the molts. In *Zootermopsis nevadensis,* molts are three times more numerous in ZR-515–treated populations than in controls (Wanyonyi, 1974). In *Nasutitermes exitiosus,* from French's data (1974), the increase of soldier production after JHA treatment also appears to result from a stimulation of the molts. In *N. lujae,* the phenomenon is still more clear: in populations of small workers, which naturally develop into soldiers, 72% of the individuals treated with ZR-515 molt and differentiate into presoldiers in a 30-day period, whereas only 2.4% produce presoldiers in the same period in controls (Bordereau, unpublished data). On the other hand, Wanyonyi (1974) observed an increase in the volume of the molt glands in pseudergates of *Z. nevadensis* treated with ZR-512. In *M. michaelseni,* the first peak of enlargement of the CA in the larvae developing into presoldiers is followed by an increase in volume of the molt glands (Okot-Kotber, 1982). Finally, JHA application was shown to influence the precocious peak of ecdysteroids in larvae of *Zootermopsis* (Konig and Lanzrein, cited by Lanzrein et al., 1985).

7.5.1.2. INHIBITION OF NEOTENIC DIFFERENTIATION

This inhibition was first shown by Lebrun (1967) by implantation of CA in newly molted pseudergates; 95% of such individuals reared in orphaned groups would normally differentiate into neotenics, but the CA implantation induced a transformation into soldiers or soldier–neotenic intercaste. The same conclusions result from the JHA applications: in *K. flavicollis* (Springhetti, 1974), *Bifiditermes beesoni* (Afzal and Ahmad, 1982b), *Neotermes connexus* (Myles and Chang, 1984), *Z. nevadensis* and *Z. angusticollis* (Wanyonyi and Lüscher, 1973; Wanyonyi, 1974; Yin and Gillott, 1975b), and *Reticulitermes flaviceps* (Dai, 1980), low doses of synthetic JH or JHAs inhibit neotenic differentiation.

7.5.2. *Molt Glands and Ecdysteroids*

As early as 1958, however, Lüscher and Karlson injected ecdysone into newly molted pseudergates of *K. flavicollis* and obtained normal molts, instead of neotenics as expected. Hence, it is not possible to induce an anticipated molt to obtain neotenics. In other respects the injection of ecdysone into last-instar nymphs of *K. flavicollis* induces an anticipated imaginal molt producing pseudimagos, intermediate between nymphs and adults (Lüscher, 1960).

By implantation of cockroach or *Locusta* molt glands into *K. flavicollis*, Lebrun (1967, 1978) succeeded in obtaining molt of instars considered to be ultimate instars: imagoes, soldiers, and neotenics. The sclerotization of the integument prevents exuviation; the new morphogenesis, however, can be recognized inside the exuvium. The imagoes molt without transformation; in contrast, the neotenics show mandibles clearly modified toward the soldier type (Fig. 7.4B). As for soldiers, the molt induces new soldiers but their mandibles are thinner and have very reduced lateral teeth, which corresponds to an emphasizing of the soldier type (Fig. 7.4C). Thus, these results mean that the epidermis of the imagoes, soldiers, and neotenics remains sensitive to ecdysteroids. Probably under the influence of a high level of endogenous JH, neotenics and soldiers differentiate soldier characters or reinforce them. The imagoes appear unable to realize a new morphogenesis and only secrete a new cuticle.

Pseudergates are induced to molt by implantation of molt glands, as expected. When the experiment was done on pseudergates destined to become neotenics (newly molted insects in orphaned colonies), however, the neotenic transformation was inhibited in 54%. Only 14% of neotenics and 32% of pseudergate–neotenic intercastes were obtained (Lebrun, 1967). These results complement those of Lüscher and Karlson (1958), since here the operated individuals were already induced toward neoteny and, despite implantation, some individuals could realize transformation completely or not at all.

7.6. Interpretation: Polymorphism, Hormones, and Pheromones

Termite polymorphism is clearly not due to genetic differences but to different expressions of a common genetic pool. Differences in polymorphism between males and females of the same species, however, have a genetic basis, but nothing more can be said at present. In particular, the system of multiple sex chromosomes (reviewed in Luykx, 1985) does not seem involved, because it attains its maximum in some lower termites where no difference is observed between the polymorphism of the males and that of the females.

The phenotypic realization of polymorphism is certainly regulated by endocrine factors, which induce or inhibit one kind of morphogenesis. The endocrine equilibrium, however, is itself subjected to extraindividual influences, among which *social* factors are of prime importance, especially in advanced species. The primitive societies (Kalotermitidae, Termopsidae), which nest in wood, are extremely dependent on climatic fluctuations and particularly on temperature. In the most evolved species, on the other hand, the complexity of the nest and behavior leads to precise regulation of the environment of the termitarium, which in effect becomes air-conditioned. In parallel, the interindividual relationships become more and more complex, either for food exchange by trophallaxis or for communication, notably by means of pheromones.

7.6.1. *Soldiers*

The soldier caste is most probably monophyletic and very ancient in the evolution of Isoptera, as attested by the characteristic development through two molts.

All the endocrinological studies point out the major role of JH, and it is certain that soldier differentiation requires a high level of this hormone. We must not forget, however, that the morphogenetic role of JH is expressed at the time of the molts, which are mostly regulated by the ecdysteroids. Thus, it would be essential to improve our knowledge of the interactions between JH and ecdysteroids. For termites, experimental conditions that stimulate soldier production generally increase the frequency of molts. That is the case for JH or JHA applications, which suggests a double action of these exogenous hormones: prothoracicotropic and morphogenetic. The presoldier differentiating molt thus is a molt induced in a competent individual, involving at the same time activation of the molt glands and of the CA. The timing of this double action is probably as important as the hormonal levels, if not more so, and on this basis one might venture to explain the numerous intercastes induced by JH or JHA applications.

These endocrinological phenomena are subjected to multiple regulations that lead to a proportion of soldiers that hold approximately constant for a given species. The action of climatic factors (temperature) is probable but remains unexplored. Various social factors appear to be essential and help to explain the constancy of the soldier proportion (see the discussion of social regulation in Grassé, 1982, 1986). Among these factors, two must be considered: nutrition, which remains almost unexplored; and the exchange of information, especially through pheromones, which is better known.

Some observations have shown that soldier production is significantly reduced by soldiers present in the colony (see the review by Bordereau, 1985). The pheromonal nature of this inhibition was demonstrated in *Nasutitermes lujae* by Lefeuve and Bordereau (1984), where the inhibitory pheromone is found in the secretion of the frontal gland of soldiers. It is probably the same situation in *Schedorhinotermes lamanianus* (Renoux, 1975). Soldier formation is inversely stimulated by the reproductives (*Kalotermes:* Springhetti, 1970, 1985; *Prorhinotermes:* Miller, 1942; *Cubitermes* and *Nasutitermes:* Bordereau and Han, 1986), but the pheromonal nature of this influence has not yet been demonstrated.

The mode of action of these primer pheromones is not known. Regarding the endocrinological effects, the target organ can be the molt gland or the CA or both. In *N. lujae* for example, the slowing down of soldier production by soldiers present in the nest is principally due to the slowing down of molts of the small workers that give rise to soldiers. The presence of the royal pair inversely stimulates such molts and therefore favors soldier production.

7.6.2. *Workers*

In contrast to the soldier caste, the worker caste is *polyphyletic* and appeared relatively recently in termites, in an independent way in several different evolutionary lines (Noirot and Pasteels, 1987).

In the Kalotermitidae and Termopsidae, there are no true workers; their functions are assumed by old larvae and nymphs, as well as by pseudergates, individuals developing through stationary or regressive molts but remaining capable of differentiating into any caste, including the imago. We do not know anything about the endocrinological factors regulating the stationary and regressive molts. They probably depend on the quantitative and especially temporal balance between the secretion of JH and that of ecdysteroids. The society also has an influence on these molts: the regressive and stationary molts are much more numerous in small populations, which explains their high frequency in laboratory colonies.

A decisive evolutionary leap occurred in the course of development with the separation of two lines: one neuter, one sexual. A particular molt leads to the individualization of these two lines: the sexual line in which the wing buds develop progressively, and the neuter line with no wing buds. The neuter line is principally characterized by an absence of development of imaginal characters, and the workers retain an essentially larval and very uniform morphology. Let us remember that workers are not imagoes.

The endocrinological mechanisms can only be inferred from morphological data, especially those by Kaiser (1956) (see Section 7.3.2.1, above). As a working hypothesis, the differentiation of the neuters would result from a deficiency in ecdysteroid hormones compared with normal sexual development. Not only the ecdysteroid level, however, but also the JH level and the respective chronology of the secretion of both hormones must be taken into account.

In the Termitidae, the separation of the two lines is visible after the first molt; the determination may occur either during the first larval instar or during embryonic development or oogenesis. This phenomenon follows a very seasonal cycle, nymphal production being restricted to a short period, often at the beginning of the rainy season. A determination in the egg was hypothesized by Lüscher (1976) with a role of the maternal JH, but we have seen (Section 7.4.1.2) that this has not yet been confirmed. A determination during the first larval instar was suggested by the experiments of Bordereau (1975): the removal of the royal pair in nests of *Macrotermes bellicosus* during the season of neuter production induced unseasonal nymphal production. This suggests that the royal pair may exert an inhibitory influence on the sexual production. Results, however, varied from one nest to another and were not confirmed in *M. michaelseni* (Sieber and Darlington, 1982; Sieber 1985). In conclusion, we suggest a mechanism in two steps as already shown in some ants: (1) an initial determination occurs in the egg, but (2) that determination may be confirmed or reversed at the beginning of the first larval instar. This is a purely hypothetical scheme, but possibly the importance of the determination before hatching varies according to the species and, over the course of evolution, has become more and more pronounced. In the Rhinotermitidae and the Hodotermitidae, which possess workers and where the separation between the two developmental lines appears after the second molt, the determination probably occurs in the young larvae.

7.6.3. *Neotenic Reproductives*

One or several molts are always required for neotenic differentiation, and it is an anticipated molt, with a variable morphogenesis, often with a mixture of development and regression of the imaginal characters (see Section 7.2.3, above).

At present, elucidation of the endocrinological control of neoteny remains difficult, for as yet no neoteny molts have been experimentally induced by manipulation of the endocrines.

Enlargement of the CA occurring during neoteny may be linked more to JH gonadotropic activity than to neotenic morphogenesis. JHA applications have an inhibitory influence on neoteny (see Section 7.5.1.2, above). On the other hand, a postcephalic ligature made before the neoteny molt in *K. flavicollis* does not prevent normal neotenic morphogenesis but inhibits ovarian development (Lüscher and Springhetti, 1960).

The problem gets more complicated with the intervention of social factors. The functional reproductives clearly exert inhibition on neotenic differentiation, which for a long time was thought to be chemically mediated (Pickens, 1932). In this way Lüscher (1960, 1963, 1969) proposed the following scheme: in a normal society of *K. flavicollis,* the functional reproductives stimulate CA activity of other individuals of the colony and thus inhibit neotenic differentiation (very high stimulation induces soldier formation). This hypothesis has been taken into account by many authors with little modification (Yin and Gillott, 1975b; Nijhout and Wheeler, 1982; Myles and Chang, 1984), but it was contradicted by Greenberg and Tobe's results (1985), which indicate a CA stimulation after orphaning. The latter results, however, were obtained from individuals already orphaned for 9 days. Thus, we propose the following working hypothesis: removal of the functional reproductives would induce a decrease in CA activity of the competent individuals but also an activation of the molt glands; under these conditions, neotenic morphogenesis would be induced. The CA enlargement would be a consequence (and not the cause) of this differentiation, and these CA would later become active, assuming then their gonadotropic function. This scheme, where not only the hormonal levels but above all the chronology of their secretion intervene, could enable us to understand the particulars of neoteny molts.

The pheromones, and especially the inhibitory pheromones, certainly play a major role in neotenic differentiation. Nothing, however, is known of the origin and chemistry of these substances, and their transmission through the society is controversial: Lüscher (1961) showed transmission through the proctodaeal food, but this was disputed by Stuart (1979) and Greenberg and Stuart (1980). Lüscher (1972) suggested that the royal inhibitory pheromone could be JH itself, produced by the reproductives and transmitted by trophallaxy. Although this hypothesis was later rejected by Lüscher himself (1975), it has recently been taken up again by Myles (1982) and Myles and Chang (1984).

The surprising case of *Mastotermes* must finally be noted: according to Watson et al. (1975) and Watson and Abbey (1985), the production of new neotenics may be stimulated by the neotenics themselves.

A fundamental difference between social polymorphism in Hymenoptera and Isoptera is that in the former it is an *imaginal* polymorphism whereas in the latter the various castes (except the sexual alates) are not imagines, which is possible in hemimetabolous insects. An individual may correlatively realize successive separate differentiations during its postembryonic development, that is to say, exhibit a *temporal polymorphism*. Another peculiarity of the Isoptera is the relative independence of several characters during development (Noirot and Pasteels, 1987). A striking example is the importance of neoteny in termites, in which maturation of the sex organs is dissociated from imaginal differentiation.

All these peculiarities are most probably correlated with variations of the hormonal milieu. In spite of numerous research studies and significant results, we are far from completely understanding the endocrinological basis of the expression of several possible phenotypes (castes). A promising field of research remains open.

7.7. Summary

Social polymorphism in termites is a developmental phenomenon: different endocrinologically regulated expressions of a common genetic pool lead to the morphogenesis of various castes. In termites, as workers and soldiers are not adult insects, the polymorphism is not imaginal in contrast with social Hymenoptera.

Herein, the origin and morphogenesis of different castes (imaginal and neuter line, workers, soldiers, and neotenic reproductives) have been analyzed. As the same individual can develop into two or several phenotypes (= castes) in succession, the term *temporal polymorphism* has been proposed.

The development of endocrine glands during caste differentiation and the experimental results (endocrine implantations, hormone applications or injections) highlight the major role played by CA and JH in termite polymorphism, especially in soldier and neotenic formation. Unfortunately, data about hormonal levels during development are so scanty that the endocrinological interpretation of caste differentiation remains mostly hypothetical. To understand this phenomenon, it is important to consider not only JH but also ecdysteroid levels and above all the respective chronology of their secretion.

Finally, an interpretation of isopteran polymorphism has been proposed, with interactions of hormones and social pheromones for soldier, worker, and neotenic differentiation.

Acknowledgments

We thank Professor D. Lebrun for Fig. 7.4.

References

Afzal, M. and M. Ahmad. 1982a. Effects of juvenile hormone analogues on colony foundation and caste differentiation in *Bifiditermes beesoni* (Gardner) (Isoptera, Kalotermitidae). Mater. Org. (Berl.) 17: 35–50.

Afzal, M. and M. Ahmad. 1982b. Significance of existent castes on the future caste differentiation of *Bifiditermes beesoni* (Gardner) under the influence of a juvenile hormone analogue. Mater. Org. (Berl.) 17: 93–116.

Bathellier, J. 1927. Contribution à l'étude systématique et biologique des Termites de l'Indochine. Faune Colon. Fr. 1: 125–332.

Bernardini-Mosconi, P. 1958. Le ghiandole endocrine e le cellule neurosecretici protocerebrali di ninfa di *Zootermopsis angusticollis*. Symp. Genet. Biol. Ital. 6: 129–139.

Bordereau, C. 1971. Le système trachéen de la reine physogastre et du roi chez *Bellicositermes natalensis* Haviland (Isoptera, Termitidae). Arch. Zool. Exp. Gén. 112: 747–760.

Bordereau, C. 1975. Déterminisme des castes chez les termites supérieurs: mise en évidence d'un contrôle royal dans la formation de la caste sexuée chez *Macrotermes bellicosus* (Smeathman) (Isoptera, Termitidae). Insectes Soc. 22: 363–374.

Bordereau, C. 1982. Ultrastructure and formation of the physogastric termite queen cuticle. Tissue & Cell 14: 371–396.

Bordereau, C. 1985. The role of pheromones in termite caste differentiation. Pp. 221–226 *in* J. A. L. Watson, B. M. Okot-Kotber, and Ch. Noirot (eds.), *Caste Differentiation in Social Insects*. Pergamon Press, Oxford and Elmsford, New York.

Bordereau, C. and S. H. Han. 1986. Stimulatory influence of the queen and king on soldier differentiation in the higher termites *Nasutitermes lujae* and *Cubitermes fungifaber*. Insectes Soc. 33: 296–305.

Bordereau, C., J. P. Delbecque, M. Hirn, and M. De Reggi. 1977. Ecdysones des adultes et des oeufs de termites. Bull. Soc. Zool. Fr. 102: 314–315.

Bordereau, C., M. Hirn, J. P. Delbecque, and M. De Reggi. 1976. Présence d'ecdysones chez un insecte adulte: la reine de termites. C. R. Acad. Sci. Paris 282: 885–888.

Buchli, H. 1958. L'origine des castes et les potentialités ontogéniques des termites Européens du genre *Reticulitermes* (Holmgren). Ann. Sci. Nat. Zool. 20: 261–429.

Dai, J. D. 1980. Induction and inhibition of caste-differentiating potentials in *Reticulitermes flaviceps* (Oshima). Acta Entomol. Sin. 23: 374–380.

Delbecque, J. P., B. Lanzrein, C. Bordereau, H. Imboden, M. Hirn, J. D. O'Connor, C. Noirot, and M. Lüscher. 1978. Ecdysone and ecdysterone in physogastric termite queens and eggs of *Macrotermes bellicosus* and *Macrotermes suybhyalinus*. Gen. Comp. Endocrinol. 36: 40–47.

Deligne, J. 1970. Recherches sur la transformation des jeunes en soldats dans la société de termites (Insectes, Isoptères). Thesis, University of Brussels.

Deligne, J., A. Quennedey, and M. S. Blum. 1981. The enemies and defense mechanisms of termites. Pp. 1–76 *in* H. R. Hermann (ed.), *Social Insects*, Vol. II. Academic Press, Orlando, Florida.

Doki, H., K. Tsunoda, and K. Nishimoto. 1984. Effect of juvenile hormone analogues on caste differentiation of the termite, *Reticulitermes speratus* (Kolbe) (Isoptera: Rhinotermitidae). Mater. Org. (Berl.) 19: 175–187.

French, J. R. J. 1974. A juvenile hormone analogue inducing caste differentiation in the Australian termite, *Nasutitermes exitiosus* (Hill) (Isoptera, Termitidae). J. Aust. Entomol. Soc. 13: 353–355.

Gillott, C. and C. M. Yin. 1972. Morphology and histology of the endocrine glands of *Zootermopsis angusticollis* Hagen (Isoptera). Can. J. Zool. 50: 1537–1545.

Grassé, P. P. 1982. *Termitologia*, Pt. I: *Anatomie, Physiologie, Reproduction des Termites*. Masson, Paris.

Grassé, P. P. 1986. *Termitologia*, Pt. III: *Comportement, Socialité, Ecologie, Evolution, Systématique*. Masson, Paris.

Grassé, P. P. and Ch. Noirot. 1946. Le polymorphisme social du Termite à cou jaune (*Calotermes flavicollis*): la production des soldats. C. R. Acad. Sci. Paris 223: 929–931.

Grassé, P. P. and Ch. Noirot. 1947. Le polymorphisme social du Termite à cou jaune

(*Calotermes flavicollis* F.): les faux-ouvriers ou pseudergates et les mues régressives. C. R. Acad. Sci. Paris 224: 219–221.

Grassé, P. P. and Ch. Noirot. 1957. La société de *Calotermes flavicollis* (Insecte Isoptère) de sa fondation au premier essaimage. C. R. Acad. Sci. Paris 246: 1789–1795.

Greenberg, S. and A. M. Stuart. 1980. Control of neotenic development in a primitive termite (Isoptera, Hodotermitidae). J. N.Y. Entomol. Soc. 88: 49–50.

Greenberg, S. and S. S. Tobe. 1985. Adaptation of a radiochemical assay for juvenile hormone biosynthesis to study caste differentiation in a primitive termite. J. Insect. Physiol. 31: 347–352.

Hagedorn H. H. 1985. The role of ecdysteroids in reproduction. Pp. 205–262 *in* G. A. Kerkut and L. I. Gilbert (eds.), *Comprehensive Insect Physiology, Biochemistry and Pharmacology*, Vol. 8. Pergamon Press, Oxford and Elmsford, New York.

Han, S. H. and C. Bordereau. 1982a. Ultrastructure of the fat body of the reproductive pair in higher termites. J. Morphol. 172: 313–322.

Han, S. H. and C. Bordereau. 1982b. Origin and formation of the royal fat body of the higher termite queens. J. Morphol. 173: 17–28.

Hare, L. 1934. Caste determination and differentiation with special reference to the genus *Reticulitermes* (Isoptera). J. Morphol. 56: 267–293.

Herlant-Meewis, H. and J. M. Pasteels. 1961. Les glandes de mue de *Calotermes flavicollis* F. (Insecte Isoptère). C. R. Acad. Sci. Paris 253: 3078–3080.

Hoffmann, J. A., M. Lagueux, C. Hetru, M. Charlet, and F. Goltzène. 1980. Ecdysone in reproductively competent female adults and in embryos of insects. Pp. 431–466 *in* J. A. Hoffmann (ed.), *Progress in Ecdysone Research*. Elsevier/North-Holland Publ., Amsterdam and New York.

Howard, R. W. 1983. Effects of methoprene on binary caste groups of *Reticulitermes flavipes* (Kollar) (Isoptera, Rhinotermitidae). Environ. Entomol. 12: 1059–1063.

Howard, R. W. 1984. Effects of methoprene on laboratory colonies of *Reticulitermes flavipes* (Kollar) (Isoptera, Rhinotermitidae). J. Ga. Entomol. Soc. 19: 291–298.

Howard, R. W. and M. I. Haverty. 1979. Termites and juvenile hormone analogues: a review of methodology and observed effects. Sociobiology 4: 269–278.

Hrdy, I., J. Krecek, and Z. Zuskova. 1979. Juvenile hormone analogues: effects on the soldier caste differentiation in termites (Isoptera). Vêstn. Cesk. Spol. Zool. 43: 260–269.

Jones, S. C. 1984. Evaluation of two insect growth regulators for the bait-block method of subterranean termite (Isoptera: Rhinotermitidae) control. J. Econ. Entomol. 77: 1088–1091.

Jucci, C. 1924. Sulla differenziazione delle caste nella societa dei Termiti. I. Neotenici Reali veri e neotenici l'escrezione nei reali neotenici—la fisiologia e la biologia. Atti Accad. Naz. Lincei, Rend. Cl. Sci. Fis. Mat. Nat. 14: 269–500.

Kaiser, P. 1956. Die Hormonalorgane der Termiten in Zusammenhang mit der Entstehung ihrer Kasten. Mitt. Hamb. Zool. Mus. Inst. 54: 129–178.

Lanzrein, B., V. Gentinetta, and M. Lüscher. 1977. *In vivo* and *in vitro* studies on the endocrinology of the reproductives of the termite *Macrotermes subhyalinus*. Proc. 8th Congr. IUSSI, Wageningen, Netherlands pp. 265–268.

Lanzrein, B., V. Gentinetta, and M. Lüscher. 1978. *In vitro*–Juvenilhormonsynthese durch Corpora allata der Termite *Macrotermes subhyalinus*. Rev. Suisse Zool. 85: 10.

Lanzrein, B., V. Gentinetta, and R. Fehr. 1985. Titres of juvenile hormone and ecdysteroids in reproductives and eggs of *Macrotermes michaelseni*: relation to caste determination. Pp. 307–327 *in* J. A. L. Watson, B. M. Okot-Kotber, and Ch. Noirot (eds.), *Caste Differentiation in Social Insects*. Pergamon Press, Oxford and Elmsford, New York.

Lebrun, D. 1967. La détermination des castes du termite à cou jaune (*Calotermes flavicollis* Fabr.). Bull. Biol. Fr. Belg. 101: 139–217.

Lebrun, D. 1970. Intercastes expérimentaux de *Calotermes flavicollis* Fabr. Insectes Soc. 17: 159–176.

Lebrun, D. 1978. Implications hormonales dans la morphogénèse des castes du termite *Kalotermes flavicollis* Fabr. Bull. Soc. Zool. Fr. 103: 351–358.

Lebrun, D. 1983. Cephalic neurohaemal organs in Isoptera. Pp. 336–345 *in* A. P. Gupta (ed.), *Neurohaemal Organs of Arthropods: Their Development, Evolution, Structure, and Functions*. Thomas, Springfield, Illinois.

Lefeuve, P. and C. Bordereau. 1984. Soldier formation regulated by a primer pheromone from the soldier frontal gland in a higher termite, *Nasutitermes lujae*. *Proc. Natl. Acad. Sci. USA* 81: 7665–7668.

Lenz, M. 1976. The dependence of hormone effects in termite caste determination on external factors. Pp. 73–89 *in* M. Lüscher (ed.), *Phase and Caste Determination in Insects: Endocrine Aspects*. Pergamon Press, Oxford and Elmsford, New York.

Lenz, M. and M. Westcott. 1985. Homeostatic mechanisms affecting caste composition in groups of *Nasutitermes nigriceps* (Isoptera, Termitidae) exposed to a juvenile hormone analogue. Pp. 251–266 *in* J. A. L. Watson, B. M. Okot-Kotber, and Ch. Noirot (eds.), *Caste Differentiation in Social Insects*. Pergamon Press, Oxford and Elmsford, New York.

Lüscher, M. 1957. Ersatzgeschlechtstiere bei Termiten und die Beeinflussung ihrer Entstehung durch die Corpora allata. Verh. Dtsch. Ges. Angew. Entomol. 14: 144–150.

Lüscher, M. 1958a. Uber die Enstehung der Soldaten bei Termiten. Rev. Suisse Zool. 65: 372–377.

Lüscher, M. 1958b. Experimentelle Erzeugung von Soldaten bei der Termite *Kalotermes flavicollis* (Fabr). Naturwissenschaften 45: 69–70.

Lüscher, M. 1960. Hormonal control of caste differentiation in termites. Ann. N.Y. Acad. Sci. 89: 549–563.

Lüscher, M. 1961. Social control of polymorphism in termites. Pp. 57–67 *in* J. S. Kennedy (ed.), *Insect Polymorphism*. (Symp. R. Entomol. Soc. Lond.)

Lüscher, M. 1963. Functions of the corpora allata in the development of termites. Proc. 16th Int. Congr. Zool. Vol. 4, pp. 244–250.

Lüscher, M. 1969. Die Bedeutung des Juvenilhormons für die Differenzierung der Soldaten bei der Termite *K. flavicollis*. Proc. 6th Congr. IUSSI, Bern pp. 165–170.

Lüscher, M. 1972. Environmental control of juvenile hormone (JH) secretion and caste differentiation in termites. Gen. Comp. Endocrinol. Suppl. 3: 501–514.

Lüscher, M. 1974. Kasten und Kastendifferenzierung bei niederen Termiten. Pp. 694–739 *in* C. H. Schmidt (ed.), *Sozialpolymorphismus bei Insekten*. Wissenschaftliche Verlagsgesellschaft, Stuttgart.

Lüscher, M. 1975. Pheromones and polymorphism in bees and termites. Pp. 123–141 *in* Ch. Noirot, P. E. Howse, and G. Le Masne (eds.), *Pheromones and Defensive Secretions in Social Insects*. (Proc. Symp. IUSSI, Dijon, France.)

Lüscher, M. 1976. Evidence for an endocrine control of caste determination in higher termites. Pp. 91–103 *in* M. Lüscher (ed.), *Phase and Caste Determination in Insects*. Pergamon Press, Oxford and Elmsford, New York.

Lüscher, M. and P. Karlson. 1958. Experimentelle Auslösung von Häutungen bei der Termite *Kalotermes flavicollis* (Fabr.). J. Insect Physiol. 1: 341–345.

Lüscher, M. and A. Springhetti. 1960. Untersuchungen über die Bedeutung der Corpora allata für die Differenzierung der Kasten bei der Termite *Kalotermes flavicollis*. J. Insect Physiol. 5: 190–212.

Luykx, P. 1985. Genetic relations among castes in lower termites. Pp. 17–25 *in* J. A. L. Watson, B. M. Okot-Kotber, and Ch. Noirot (eds.), *Caste Differentiation in Social Insects*. Pergamon Press, Oxford and Elmsford, New York.

Meyer, D. R., B. Lanzrein, M. Lüscher, and K. Nakanishi. 1976. Isolation and identification of a juvenile hormone (JH) in termites. Experientia (Basel) 32: 773.

Miller, E. M. 1942. The problem of castes and caste differentiation in *Prorhinotermes simplex* Hagen. Bull Univ. Miami 15: 3–27.
Miller, E. M. 1969. Caste differentiation in the lower termites. Pp. 283–310 *in* K. Krishna and F. M. Weesner (eds.), *Biology of Termites*. Academic Press, Orlando, Florida.
Mosconi-Bernardini, P. and M. L. Vecchi. 1964. La ghiandola protoracica di *Anacanthotermes ochraceus* (Hodotermitidae) e di *Schedorhinotermes javanicus* (Rhinotermitidae). Symp. Genet. Biol. Ital. 13: 169–177.
Myles, T. G. 1982. Pheromonal JH polytiterism: the basis of termite polymorphism. Proc. 9th Congr. IUSSI, Boulder, Colorado.
Myles, T. G. 1986. Reproductive soldiers in the Termopsidae (Isoptera). Pan-Pac. Entomol. 62: 293–299.
Myles, T. G. and F. Chang. 1984. The caste system and caste mechanisms of *Neotermes connexus* (Isoptera: Kalotermitidae). Sociobiology 9: 163–321.
Nijhout, H. F. and D. E. Wheeler. 1982. Juvenile hormone and the physiological basis of insect polymorphisms. Q. Rev. Biol. 57: 109–133.
Noirot, Ch. 1955. Recherches sur le polymorphisme des Termites supérieurs (Termitidae). Ann. Sci. Nat. Zool. 17: 400–595.
Noirot, Ch. 1956. Les sexués de remplacement chez les Termites supérieurs (Termitidae). Insectes Soc. 3: 145–158.
Noirot, Ch. 1957. Neurosécrétion et sexualité chez le Termite à cou jaune *Calotermes flavicollis F*. C. R. Acad. Sci. Paris 245: 743–745.
Noirot, Ch. 1969a. Glands and secretions. Pp. 89–123. *in* K. Krishna and F. M. Weesner (eds.), *Biology of Termites*. Academic Press, Orlando, Florida.
Noirot, Ch. 1969b. Formation of castes in the higher termites. Pp. 311–350 *in* K. Krishna and F. M. Weesner (eds.), *Biology of Termites*. Academic Press, Orlando, Florida.
Noirot, Ch. 1982. La caste des ouvriers, élément majeur du succès évolutif des Termites. Riv. Biol. 75: 157–195.
Noirot, Ch. 1985a. Pathways of caste development in the lower termites. Pp. 41–57 *in* J. A. L. Watson, B. M. Okot-Kotber, and Ch. Noirot (eds.), *Caste Differentiation in Social Insects*. Pergamon Press, Oxford and Elmsford, New York.
Noirot, Ch. 1985b. Differentiation of reproductives in higher termites. Pp. 177–186 *in* J. A. L. Watson, B. M. Okot-Kotber, and Ch. Noirot (eds.), *Caste Differentiation in Social Insects*. Pergamon Press, Oxford and Elmsford, New York.
Noirot, Ch. and J. M. Pasteels. 1987. Ontogenetic development and evolution of the worker caste in termites. Experientia (Basel) 43: 851–860.
Okot-Kotber, B. M. 1980a. Histological and size changes in corpora allata and prothoracic glands during development of *Macrotermes michaelseni* (Isoptera). Insectes Soc. 27: 361–376.
Okot-Kotber, B. M. 1980b. The influence of juvenile hormone analogue on soldier differentiation in the higher termite *Macrotermes michaelseni*. Physiol. Entomol. 5: 407–416.
Okot-Kotber, B. M. 1980c. Competence of *Macrotermes michaelseni* (Isoptera: Macrotermitinae) larvae to differentiate into soldiers under the influence of juvenile hormone analogue (ZR-515, Methoprene). J. Insect Physiol. 26: 655–659.
Okot-Kotber, B. M. 1982. Correlation between larval weights, endocrine gland activities and competence period during differentiation of workers and soldiers in *Macrotermes michaelseni*. (Isoptera: Termitidae). J. Insect. Physiol. 28: 905–910.
Okot-Kotber, B. M. 1983. Ecdysteroid levels associated with epidermal events during worker and soldier differentiation in *Macrotermes michaelseni*. (Isoptera: Macrotermitidae). Gen. Comp. Endocrinol. 52: 409–417.
Pasteels, J. M. and J. Deligne. 1965. Etude du système endocrine au cours du vieillissement chez les "reines" de *Microcerotermes parvus* (Haviland) et *Cubitermes heghi* (Sjöstedt) (Isoptères Termitidae). Biol. Gabonica 1: 325–336.

Pflugfelder, O. 1938. Untersuchungen über die histologischen Veränderungen und das Kernwachstum der "Corpora allata" von Termiten. Z. Wiss. Zool. 150: 451–467.

Pflugfelder, O. 1947. Über die Ventraldrüsen und einige andere inkertorische Organe des Insektenkopfes. Biol. Zentralb. 66: 211–235.

Pickens, A. L. 1932. Observations on the genus *Reticulitermes* Holmgren. Pan.-Pac. Entomol. 8: 178–180.

Renoux, J. 1975. Le polymorphisme de *Schedorhinotermes lamanianus* (Sjöstedt) (Isoptera: Rhinotermitidae). Insectes Soc. 23: 279–494.

Roisin, Y. 1988. Morphology, development and evolutionary significance of the working stages in the caste system of *Prorhinotermes* (Insecta, Isoptera). Zoomorphology (Berl.) 107: 339–347.

Sewell, J. J. and J. A. L. Watson. 1981. Developmental pathways in Australian species of *Kalotermes* Hagen (Isoptera). Sociobiology 6: 243–323.

Sieber, R. 1985. Replacement of reproductives in Macrotermitinae (Isoptera, Termitidae). Pp. 201–208 *in* J. A. L. Watson, B. M. Okot-Kotber, and Ch. Noirot (eds.), *Caste Differentiation in Social Insects*. Pergamon Press, Oxford and Elmsford, New York.

Sieber, R. and J. P. E. C. Darlington. 1982. Replacement of the royal pair in *Macrotermes michaelseni*. Insect Sci. Appl. 3: 39–42.

Springhetti, A. 1970. Influence of the king and the queen on the differentiation of soldiers in *Kalotermes flavicollis* Fabr. (Isoptera). Monit. Zool. Ital. 4: 99–105.

Springhetti, A. 1972. The competence of *Kalotermes flavicollis F.* (Isoptera) pseudergates to differentiate into soldiers. Monit. Zool. Ital. 6: 97–111.

Springhetti, A. 1974. The influence of farnesenic acid ethyl ester on the differentiation of *Kalotermes flavicollis* Fabr. (Isoptera) soldiers. Experientia (Basel) 30: 541.

Springhetti, A. 1985. The function of the royal pair in the society of *Kalotermes flavicollis* (Fabr.) (Isoptera: Kalotermitidae). Pp. 165–176 *in* J. A. L. Watson, B. M. Okot-Kotber, and Ch. Noirot (eds.), *Caste Differentiation in Social Insects*. Pergamon Press, Oxford and Elmsford, New York.

Stuart, A. M. 1979. The determination and regulation of the neotenic reproductive caste in the lower termites (Isoptera) with special reference to the genus *Zootermopsis* (Hagen). Sociobiology 4: 223–237.

Thorne, B. L. and Ch. Noirot. 1982. Ergatoid reproductives in *Nasutitermes corniger* (Motschulsky) (Isoptera: Termitidae). Int. J. Insect Morphol. Embryol. 11: 213–226.

Tsunoda, K., H. Doki, and K. Nishimoto. 1986. Effect of developmental stages of workers and nymphs of *Reticulitermes speratus* (Kolbe) (Isoptera: Rhinotermitidae) on caste differentiation induced by JHA treatment. Mater. Org. (Berl.) 21: 47–61.

Varma, R. V. 1985. Hormonal mechanisms of soldier differentiation in *Postelectrotermes nayari*. Pp. 239–244 *in* J. A. L. Watson, B. M. Okot-Kotber, and Ch. Noirot (eds.), *Caste Differentiation in Social Insects*. Pergamon Press, Oxford and Elmsford, New York.

Wanyonyi, K. 1974. The influence of the juvenile analogue ZR-515 (Zoecon) on caste development in *Zootermopsis nevadensis* (Hagen) (Isoptera). Insectes Soc. 21: 35–44.

Wanyonyi, K. and M. Lüscher. 1973. The action of juvenile hormone analogues on caste developement in *Zootermopsis* (Isoptera). Proc. 7th Congr. IUSSI, London pp. 392–395.

Watson, J. A. L., E. C. Metcalf, and J. J. Sewell. 1975. Preliminary studies on the control of neotenic formation in *Mastotermes darwiniensis* Froggatt (Isoptera). Insectes Soc. 22: 415–426.

Watson, J. A. L. and H. M. Abbey. 1985. Development of neotenics in *Mastotermes darwiniensis* Froggatt: an alternative strategy. Pp. 107–124 *in* J. A. L. Watson, B. M. Okot-Kotber, and Ch. Noirot (eds.), *Caste Differentiation in Social Insects*. Pergamon Press, Oxford and Elmsford, New York.

Yin, C. M. and C. Gillott. 1975a. Endocrine activity during caste differentiation in *Zootermopsis angusticollis* Hagen (Isoptera): a morphometric and autoradiographic study. Can. J. Zool. 53: 1690–1700.

Yin, C. M. and C. Gillott. 1975b. Endocrine control of caste differentiation in *Zootermopsis angusticollis* Hagen (Isoptera). Can. J. Zool. 53: 1701–1708.

Roles of Morphogenetic Hormones in Caste Polymorphism in Sting Bees

8

HEINZ REMBOLD

8.1. Introduction

Insect morphogenesis is controlled by many extrinsic factors like temperature, humidity, and food quality, or by rhythmic (diurnal and seasonal) environmental changes. These often rigid stress situations can either be overcome by the physiological routes of homeostasis or, in an emergency, can switch on such alternative reactions as precocious molt and diapause. In case of a short period of starvation, the fat body, as a site of storage and synthesis, can be mobilized. However, during a more extended period of starvation, the biosynthetic activity of endocrine glands, and consequently the hormone titers of the reproductive system, may be affected. Can such general phenomena help us to better understand the mechanism that controls development of reproductive queen and sterile worker honey bees, differentiating into what we call castes? In other words, is there anything special (i.e., a new phenomenon) in the evolution of these social insects that might be explained by the activity of a unique hormone-like factor? Or, if such caste differentiation is not a unique phenomenon, does it just follow the general rules of insect morphogenesis as delineated in earlier chapters of this treatise?

What exactly is special about the honey bee as a colony-forming insect? As is found in higher organisms, a type of homeostasis exists in the honey bee colony. A sensitive thermoregulatory mechanism controls the very constant brood nest temperature of 34–35°C, which is maintained even despite considerable external temperature changes. At the same time, humidity is kept constant between 45% and 50% saturation. Food quality is controlled by interposing nurse bees for feeding the brood after assimilation of external nutrients, again in a manner rather like what occurs in the mammalian system. The nurse bees secrete a jelly from their hypopharyngeal glands and deliver it, together with honey and pollen, to the larvae at regular time intervals. The queen larvae grow in special cells outside the brood nest; maintained as such, at lower temperatures than the workers, they receive their food, the royal jelly, in abundance. Much less food, only a few milligrams, is found in the worker cells. However, the lower amount of food given to the worker larvae still ensures normal development. One fundamental difference in larval nutrition is that only the queen larvae receive a food of fairly constant quality through their entire period of morphogenesis. Worker larvae are fed a mixture of glandular secretion and honey only during their first three instars. Then food quality changes dramatically. The glandular secretions are gradually reduced and finally excluded. The food of old worker larvae then is a mixture of honey and pollen. Is this diverse nutrition of the female honey bee larvae the key to elucidation of the programming of honey bee caste formation? We shall consider this important issue in the following paragraph in more depth.

What are the effects on honey bee caste formation as a consequence of differential nutrition and brood care? More than 100 morphological caste differences have been described by Lukoschus (1956) for the two females. For an in-depth comparison of the anatomic differences, see Snodgrass (1956). As a consequence of the strictly enforced scheme of labor division, the reproductive queen bee has two huge, pear-shaped ovaries, each containing some 160–180 ovarioles, and a well-developed spermatheca. Both these organs are extremely reduced in the worker: each ovary consists of only 2–12 slender ovarioles, and the spermatheca is vestigial. On the other hand, the worker is perfectly equipped for producing wax, food, sting poison, and its specific pheromones, which are responsible for a perfectly functioning colony. The queen releases other pheromones that inform the workers of her presence.

It is an old apicultural practice to produce queens from worker larvae. They must be younger than fourth larval instar when transferred into queen cells. In other words, queens are only produced from worker larvae before the change in worker food quality. In terms of molecular biology, does this mean that the genome of the female honey bee has defined morphogenetic programs for both queen and worker characters? One can easily postulate that such developmental programs are under hormonal control. The following question then arises: how are the queen and worker hormone titers respectively switched on and modulated in a caste-specific way? The most convincing experimental approach for answering this question is to remove the sensitive worker larva from its natural environment and rear it under controlled conditions of climate and nutrition. If queen and worker bee establishment can be experimentally controlled under these conditions, any endocrine differences of queen and worker larvae that have grown within the colony will then be derived from nutritional differences.

Before approaching the role of morphogenetic hormones in sting bees, we shall discuss the type of hormones present in the honey bee and the caste-specific maturation of the endocrine system. Then we shall focus on the effect of food quality on caste formation, and on this basis compare the titers of ecdysteroids and juvenile hormone in worker and queen bees.

8.2. The Morphogenetic Hormones

It was established quite early that in insects hormonal regulation of growth and development works in a hierarchical manner. In close analogy to the hypothalamo-hypophyseal system of mammals, neuropeptides from the central nervous system control the synthesis of steroid (molting) and sesquiterpenoid (juvenile) hormones in peripheral glands. The neurosecretory cells (NSC) in the brain synthesize neurosecretory material, which through the corpora cardiaca (CC) is released into the hemolymph. However, not much is known about synthesis and control of release of these organotropic neuropeptides (primarily the prothoracicotropic hormone, PTTH, and the allatropic hormone, ATH). Again in analogy to the mammalian system, one can expect that both nervous and humoral factors are involved.

8.2.1. *The Neuroendocrine System*

NSC in the brain of the honey bee have already been described by Weyer (1935) as glandular nerve cells with granules. However, until recently not much attention has been given to the presence of any hormone-releasing factors, because most of them have simply not yet been characterized in other insect species. A detailed study of the caste-specific maturation of the preimaginal neurosecretory system has shown a clear correlation with hormonal events. This will be discussed in Section 8.3.

8.2.2. *Ecdysteroids*

Both ecdysone and 20-hydroxyecdysone (20-E) have been found in all developmental stages of the honey bee, 20-E sharing about 80% of total ecdysteroids (Rembold and Hagenguth, 1981). Like all insects, the honey bee depends for its ecdysone biosynthesis on exogenous cholesterol or dietary phytosterols, cholesterol being the usual ecdysteroid precursor (Karlson and Hoffmeister, 1963). However, honey bees are unable to dealkylate dietary C_{28} or C_{29} phytosterols as ecdysteroid precursors (Svoboda et al., 1981, 1983). Feldlaufer et al. (1985) isolated makisterone A (24-methyl-20-hydroxyecdysone) from honey bee pupae as the main ecdysteroid. This C_{28} ecdysteroid has recently also been described in a number of phytophagous insects as their principal ecdysteroid (Feldlaufer and Svoboda, 1986).

8.2.3. *Juvenile Hormones*

Four homologous sesquiterpenoids have been isolated from insects and structurally determined as C_{19}–C_{16} juvenile hormones (JH O–III). Only JH III has been found in adult bees (Trautmann et al., 1974). Using the 10-(heptafluorobutyryloxy-11-methoxy derivative as analyte and electron capture gas-liquid chromatography for quantitative determination, Hagenguth and Rembold (1979) found all developmental stages (larvae, pupae, and adults) of the honey bee to show a modulation of JH III titers between 0 and 2 ng/g body weight during imaginal development, as well as a rather constant JH III titer near 10 ng/g in the adult worker bee (Hagenguth and Rembold, 1978).

8.3. Caste-Specific Maturation of the Endocrine System

It is generally accepted that the brain plays a central role in regulating molt and development of insects (Steel and Davey, 1985). According to the classical scheme of insect metamorphosis, PTTH and ATH control as *brain hormones* the titers of ecdysone and JH, which are the real morphogenetic hormones. Metamorphosis is controlled by these latter two hormones according to their differing relations, larval–larval molts being initiated under high JH and low molting hormone titers, and larval–pupal molts under lower JH and higher molting hormone titers. It was a great surprise, therefore, when morphological studies of the endocrine system showed that the neurosecretory complex of the brain is still in an embryonic state in first-instar honey bee larvae (Dogra et al., 1977). From this study, it became evident that maturation of the larval endocrine system and caste formation must be tightly linked to each other and that this system forms the basis of caste differentiation in the honey bee, *Apis mellifera* (Rembold and Ulrich, 1982).

8.3.1. *Pars Intercerebralis*

In the first-instar honey bee larva, the NSC are embryonic and without axons (Fig. 8.1). After [^{3}H]thymidine injection, their nuclei are labeled, which indicates intensive mitotic activity. However, no DNA labeling is found in any other larval instar. In the queen larva, outgrowth of axons starts in the second instar and is completed at the end of the third. In the worker, onset (third instar) and termination (fourth instar) of axonal outgrowth is delayed for the period of one larval instar. Incorporation of [^{3}H]uridine, as an indicator of metabolic activity, into the bodies of the NSC begins in the queen larva near the end of the third instar, and in the worker larva not before the fifth instar. Stainable neurosecretory material is first detectable in the late fifth instar of the queen larva, but no stainable material can be found in the worker throughout larval development. Growth and development of the CC begins in the fourth instar only. Incorporation of [^{3}H]uridine takes place in the queen at the end of the fourth larval stage, and in the worker during the fifth larval stage. No metabolic activity was observed in the early larval instars (Rembold and Ulrich, 1982; Ulrich and Rembold, 1983).

8.3.2. *Prothoracic Glands*

The aforementioned authors followed incorporation of [^{3}H]uridine into prothoracic glands during the several stages of development of honey bee larvae. No significant difference between queen and worker was observed. By counting the silver grains located above the cell nuclei, they found that labeling of the gland is fairly constant and at a high level until the end of the third instar; it decreases during the following instars and reaches a zero level at the end of the fifth instar (Rembold and Ulrich, 1982; Ulrich and Rembold, 1983).

8.3.3. *Corpora Allata*

The retardation in maturation of the neuroendocrine system of the pars intercerebralis of worker larvae, as compared with the queen, is also

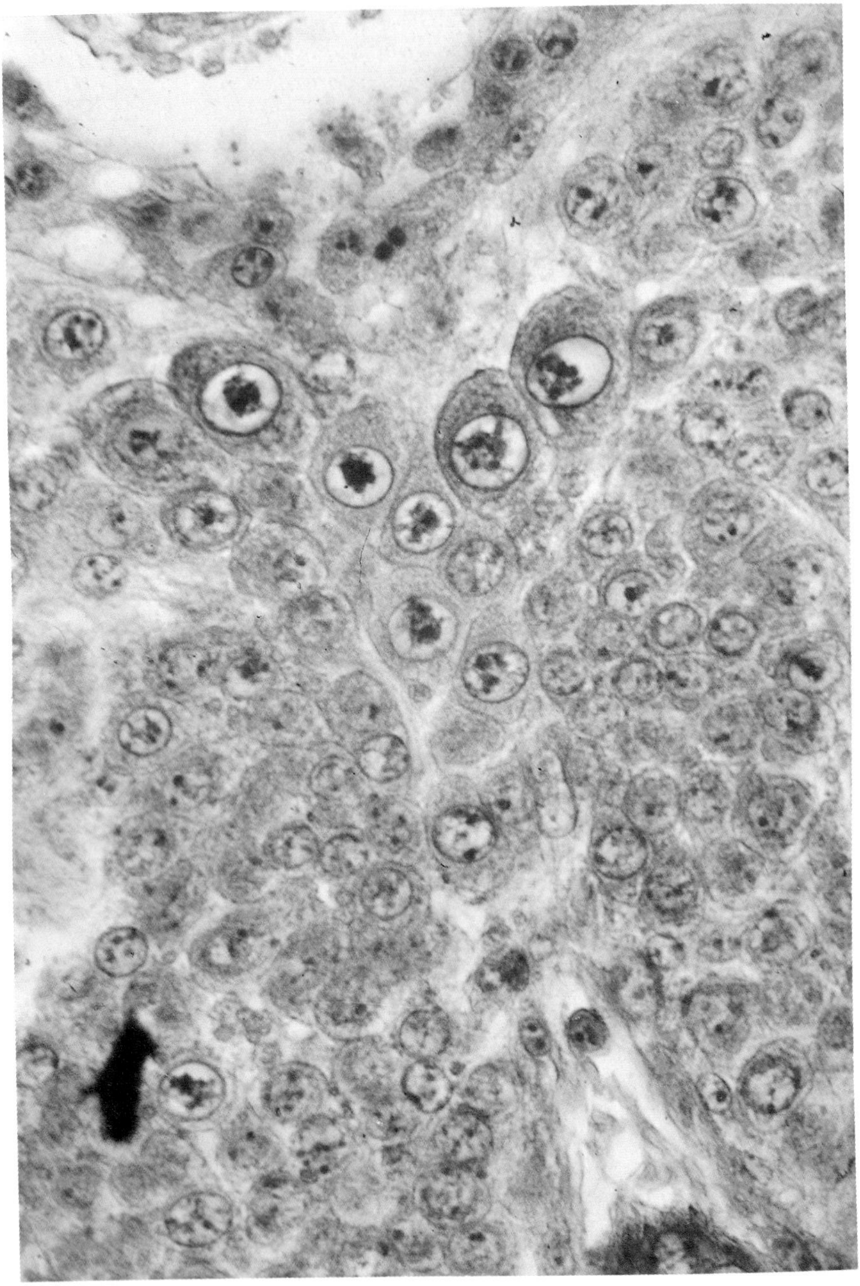

FIGURE 8.1. Neurosecretory cells of pars intercerebralis in the brain of a worker during its first larval instar (left) and a queen pupa (right). ×400.

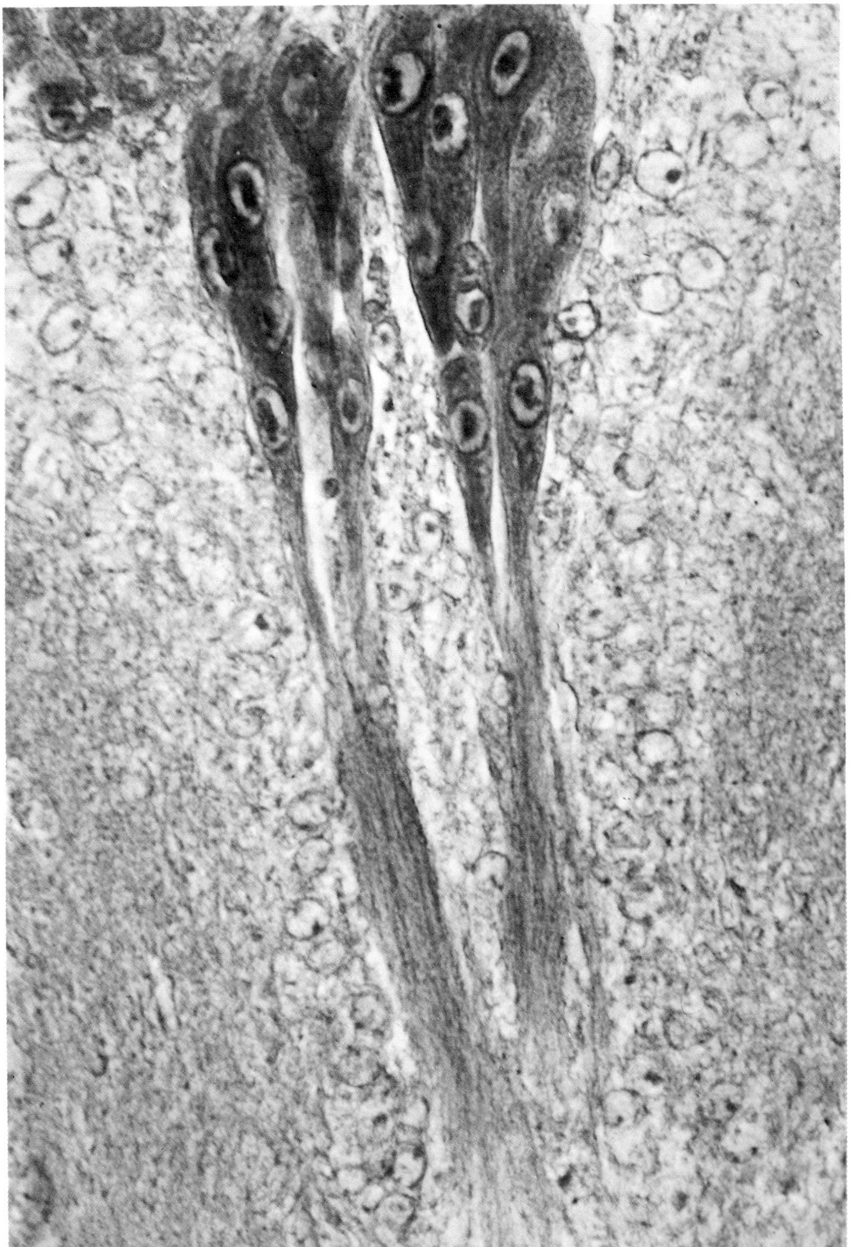

FIGURE 8.1. Continued

reflected by the development of the corpora allata (CA). From third instar onward, the queen CA are always larger than those of the worker. At its maximum in the fifth instar, the volume of a queen CA is about twice that of a worker. The CA undergo several phases of chromatin condensation and decondensation during larval development, which is correlated with an intensive [^{3}H]thymidine incorporation, reflecting endomitosis. The last phase of endomitosis is found in the queen larva at the end of the fourth instar, and in the worker during its fifth instar. During these endomitotic phases, an acceleration of the queen's larval development and a respective retardation of the worker's appears with the onset of the last (fifth) larval instar, a stage which is shorter by two-thirds in the queen. Even more accelerated is pupal development: the pupal phase of the queen takes about 5 days; that of the worker is almost twice as long (9 days) (Rembold et al., 1980; Rembold and Ulrich, 1982; Ulrich and Rembold, 1983).

8.4. Food Quality and Caste Differentiation

Are the caste-specific differences in development of the endocrine system of the honey bee larvae a consequence of their different nutrition? As already discussed, a third-instar worker larva can still be transferred into a queen cell and become a queen under colony conditions. However, at this stage, the endocrine system is already developing in different directions. The histological picture confirms the same endocrine situation for both queen and worker larvae at the onset of morphogenesis. Both queen and worker larvae originate from fertilized eggs. However, they are grown and fed under different conditions of microclimate and nutrition. Thus, is there anything in the royal jelly that must be fed to the queen larva during its entire development and might also explain its rapid growth? And is it possible to grow newly hatched bee larvae at high survival rates and under controlled conditions, thus excluding any influence of the nurse bees?

8.4.1. *Chemical Composition of Royal Jelly*

Royal jelly has been extensively analyzed, and several reviews are available on its composition and biological mode of action (Armbruster, 1960; Townsend and Shuel, 1962; Rembold, 1965, 1974, 1985, 1987a,b; Weaver, 1966; Haydak, 1970; Weiss, 1978; de Wilde, 1985). The milky, viscous material can be easily collected in gram quantities from queen cells, where it is present as a white-to-yellowish material in amounts of up to 300 mg/cell. It consists of about 40% dry matter, which has a fairly constant composition of 10% lipids, 38% proteins and glycoproteins, and 52% low-molecular, water-soluble material (sugars, amino acids, vitamins, and mineral salts). The jelly fed to the young worker larvae has a fairly similar composition. A comparison of both these foods reveals no surprising data. Neither the lipids, whose main compound is royal jelly acid (*trans*-10-hydroxy-Δ^2-decenoic acid), nor the other compounds give any indication of a special "determinator" that could be responsible for queen differentiation. Also, the only *quantitative* differences found between queen and worker jelly (pantothenic acid, biopterin, and neopterin, which are present in royal jelly in 10-fold higher concentrations than in worker jelly) did not prove to have any effect on queen bee establishment. To summarize: royal jelly is a food that is produced in extremely constant composition and from whose chemical analyses no queen-producing determinator can be derived. However, the analytical data now available make it possible to modify its composition and follow the effect of such a semisynthetic diet on caste formation *in vitro.*

8.4.2. *Effect of Food Quality on Caste Differentiation*

Since the pioneering studies of Karl von Frisch (see Lindauer, 1987), it has become clear that the honey bee colony provides a carefully controlled environment for its brood. Nevertheless, newly hatched worker larvae can be reared on royal jelly *in vitro* and, on average, 60% of them become queens. A queen-determining fraction can be removed from royal jelly by extraction with ethanol, leaving an inactive food that is still sufficient for larval survival. The queen-determining activity can be restored by adding back the ethanol extract to the inactive basic food. Rembold and colleagues concluded from this result that there must exist a defined queen bee determinator in royal jelly (see Rembold and Hanser, 1964; Rembold et al., 1974; Rembold, 1976). A more detailed study revealed, however, that a rather inactive royal jelly can be activated by addition of total yeast extract or certain fractions thereof. These results demonstrate the fundamental effect of food quality and nutritional balance on caste differentiation (Rembold and Lackner, 1981). By using a semisynthetic diet (Rembold, 1987b) composed of 50% royal jelly and 50% of a synthetic mixture of all its low-molecular components, it was proved that even minute changes in the composition of the original royal jelly are enough to reduce or even abolish queen bee formation. Extraction with ethanol therefore had simply unbalanced the food composition to such an extent that the larvae could still grow and develop into workers but not into queens (Rembold, 1987a,b).

8.5. Effect of Caste Differentiation on Hormone Titers

Social polymorphism, as exemplified by sting bees, if analyzed on the basis of our present knowledge of molecular biology, can be viewed as a sequential selection and expression of caste-specific morphogenetic programs. The genes responsible for programming the morphological expression must be switched on by the appropriate hormones. From what is known in this most speculative field, this switch involves a receptor protein. Obviously, the affinity of such a hormone receptor determines the amount of hormone necessary for releasing a signal. Concerning this hormone–receptor complex, its dissociation constant (K_D) can only be measured by the pure binding proteins, which in most cases are not yet available. The K_D value for the insect hormones is estimated to be, on average, between 10^{-8} *M* (ecdysone) and 10^{-11} *M* (juvenile hormone). This means that a JH concentration of 10^{-11} *M* may be enough to induce a 50% response of the sensitive organ. Such a concentration corresponds to about 3 pg hormone per millilitre of hemolymph! No doubt such concentrations are below the sensitivity of a biological test such as the *Galleria* assay and, because of its high cross-reactivity with components usually contained in the biological probe, also of the JH radioimmunoassay (RIA). Whereas for quantitative ecdysterone determinations a combination of high-performance chromatography with RIA is still in use, physicochemical methods, including microderivatization, are becoming more and more common for JH quantification at the same sensitivity as RIA and with unequivocal specificity. An additional and important advantage of this method is that an internal standard can be used that allows for good comparative estimations.

Even with such sensitive analytical techniques, however, short titer oscillations with a duration of some hours or less cannot be followed, owing to low synchronization and biological variation of the test animals. A scheme for the characterization of postembryonic developmental stages of the female castes of the honey bee has been published (Rembold et al., 1980). It has also been applied for titer measurements to be described below (Section 8.5.2). The developmental stages are accordingly associated with unequivocal features, such as head diameter and weight of the larvae, or colors shown by the compound eyes and the thorax of the pupae. Even with these criteria, an overlap of sometimes half a day occurs, which finally sets the physical limits of any short-term peak measurements.

8.5.1. *Ecdysteroids*

The total ecdysteroid titer in honey bee larvae is modulated in a caste-specific way (Hagenguth and Rembold, 1978; Rembold and Hagenguth, 1981). Up to the end of fourth instar, no significant differences between worker and queen larvae become evident (Fig. 8.2). Then the ecdysteroid titer decreases in concentration from 400 to less than 40 ng/g both in worker and queen larvae. In the prepupal stage, the hormone titer is then modulated in a caste-specific way, with an ecdysteroid titer

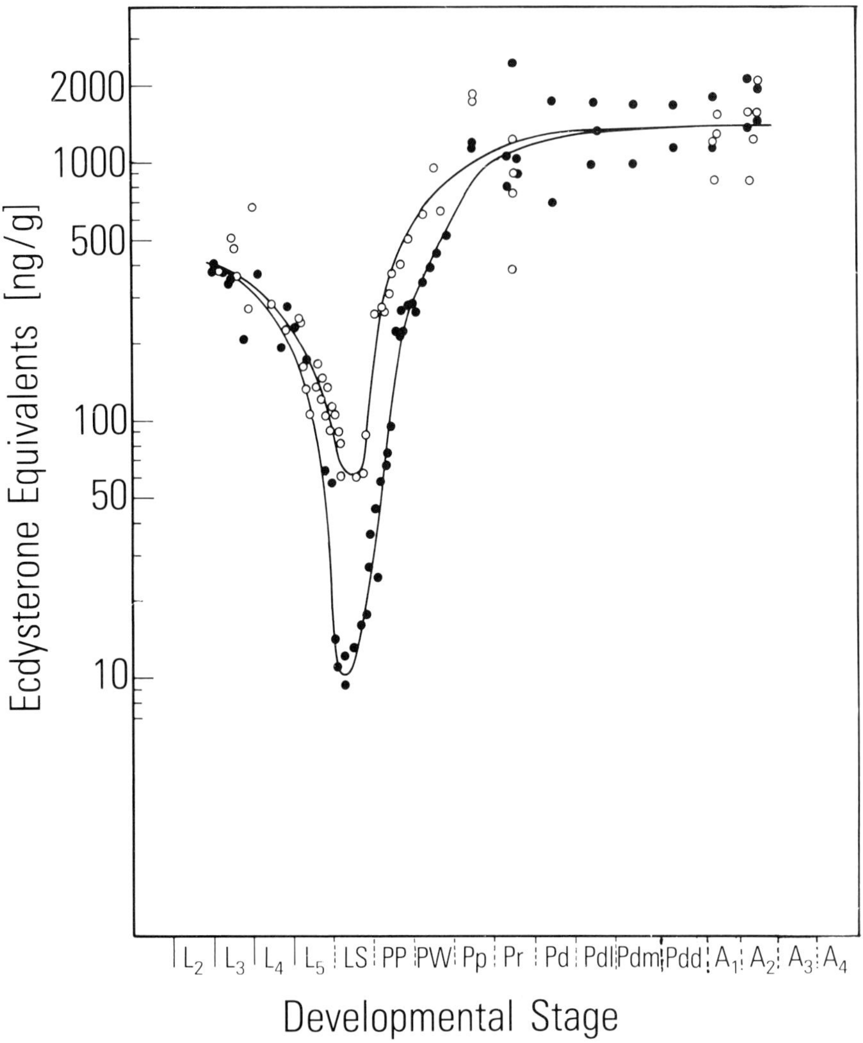

FIGURE 8.2. Ecdysterone equivalents (nanograms per gram), measured by radioimmunoassay in developmental stages of the queen (open circles) and the worker (solid circles) of the honey bee. The developmental stages are compared at the same physiological and developmental state and are therefore of different ages after L_4. For more exact characterization of the developmental stages, see Rembold et al. (1980). *Key:* L_2–L_5 = second to fifth larval instar; LS = spinning L_5; PP = pharate pupal stage of L_5; A_1–A_4 = 1- to 4-day-old adults. The pupae are characterized by eye color (w = white, p = pink, r = red, d = dark) and by degree of melanization of the cuticle (l = light, m = medium, d = dark color).

in the middle of this stage of 30 ng/g for the queen and of less than 30 ng/g for the worker. During pupal development, both the castes come to a level of around 1000 ng/g.

If the total amount of molting hormone (MH) as contained in each individual is followed, the weight gain during larval development comes into consideration. A maximum titer is reached in the fourth larval instar, which keeps increasing in the queen during the next instar (fifth). The total amount of MH comes to a minimum during the spinning period and subsequently increases again. The final value of ecdysteroids per individual is 150 ng for the worker and 300 ng for the queen pupa, owing to the higher body weight.

The MH titer follows the metabolic activity of the prothoracic glands, as shown by the incorporation of labeled uridine. This incorporation—and hence protein synthesis—decreases after the third larval instar in both the castes and reaches a zero level in the last larval instar (Rembold and Ulrich, 1982; Ulrich and Rembold, 1983).

8.5.2. *Juvenile Hormone*

Treatment of the worker brood with JH I (which is not the natural JH homologue) induces to a high extent formation of intercastes with queenlike character or even pure queens, in the colony (Wirtz and Beetsma, 1972) as well as under *in vitro* conditions (Rembold et al., 1974). Interestingly enough, this effect is much less if the natural hormone, JH III, is applied (Rembold, 1976). The queen-inducing effect of JH is especially interesting in that even after the "sensitive phase" of queen determination, i.e., after the third larval instar, such an induction is possible.

An explanation for this phenomenon came from titer estimations of queens and workers in their different developmental stages. For the extremely low JH levels during the critical developmental periods, in both the larval and especially the pupal stages, a method that is sensitive in the picogram range had to be developed. Only then could enough queen material be collected. While the gas chromatographic/mass spectrometric/multiple ion selection (GC/MS/MIS) method is sensitive enough for femtomole quantities, the problem of obtaining the hormone analyte in enough yield proved to be critical. The 10-(heptafluorobutyryloxy)-11-methoxy derivative is extremely useful for JH quantification. However, it is obtained in a maximum yield of only 20% (Rembold et al., 1980). Using this method, Rembold and Hagenguth (1981; see also Rembold, 1985) obtained a first JH III titer curve for both the castes that showed fourfold higher concentration in the queen larvae and built up further near the end of larval development. With this result, the effect of queen induction by application of JH, even during the last larval instar, now becomes intelligible as a simulation of the natural situation in the queen larva with its high JH III titer. However, only after sensitivity of the GC/MS technique had been improved by introduction of a new analyte, 10-dimethyl(nonafluorohexyl)silyloxy-11-methoxy-JH, which is obtained in total yields of more than 90% (Rembold and Lackner, 1985), was it possible to measure the individual queen and worker larvae and pupae and in that way to come to a final JH III titer curve through all developmental stages after L_2 (Rembold, 1987c).

The results from this detailed study are presented in Fig. 8.3A,B. The caste-specific titer modulation is visible during the entire developmental period. At the beginning, at the third larval instar, and at the end of the sensitive phase for queen determination, the titers differ by a factor of 6: the worker larvae have an average JH III titer of about 80 pmol/g, whereas the titer of the queen larvae at this stage is around 450 pmol/g. During the entire period of larval development, the queens have a higher JH titer than the workers. It will now be interesting to find out at which developmental state the hormone titers of queen and worker larvae first start to diverge.

Whereas the JH III titer decreases during larval development from molt to molt as in other insect species, there is another titer maximum between the late spinning phase and early stage of the pharate pupa, again in both queen and worker. Its peak coincides with the maximum of JH esterase activity fairly well, which is much higher in the queen larvae than in the respective workers (Mane and Rembold, 1977). It is remarkable from a biochemical point of view that even in the late PP stages, when all the genes and enzymes that are needed for pupation should be switched on and should be active, many animals still exhibit high JH levels. In the newly formed pupa (Pw), however, not a trace of JH is detectable. It is exactly at this stage, between PP and Pw, that the ecdysteroid titer increases steeply (Fig. 8.2). Possibly, therefore, only after removal of JH III can the signal for an increase in ecdysteroid titer be given.

Measurement of individuals and an analytical method that gives highly reliable JH III titer data allow us to take biological variation into account. The large differences within each group could reflect an uncertainty in timing the individuals within the short time periods of titer fluctuations. It seems unlikely, however, that JH, which is protected by high-affinity and receptor proteins, should be modulated in short time intervals in its titer. It seems much more probable that these titer differences reflect an individual bandwidth of hormonal regulation within a factor of 2.

An important difference between the two castes may be reflected by the relatively high JH III titer at stage LS/PP. During this developmental period, the gonads are being resorbed in the fifth-instar worker larva. It

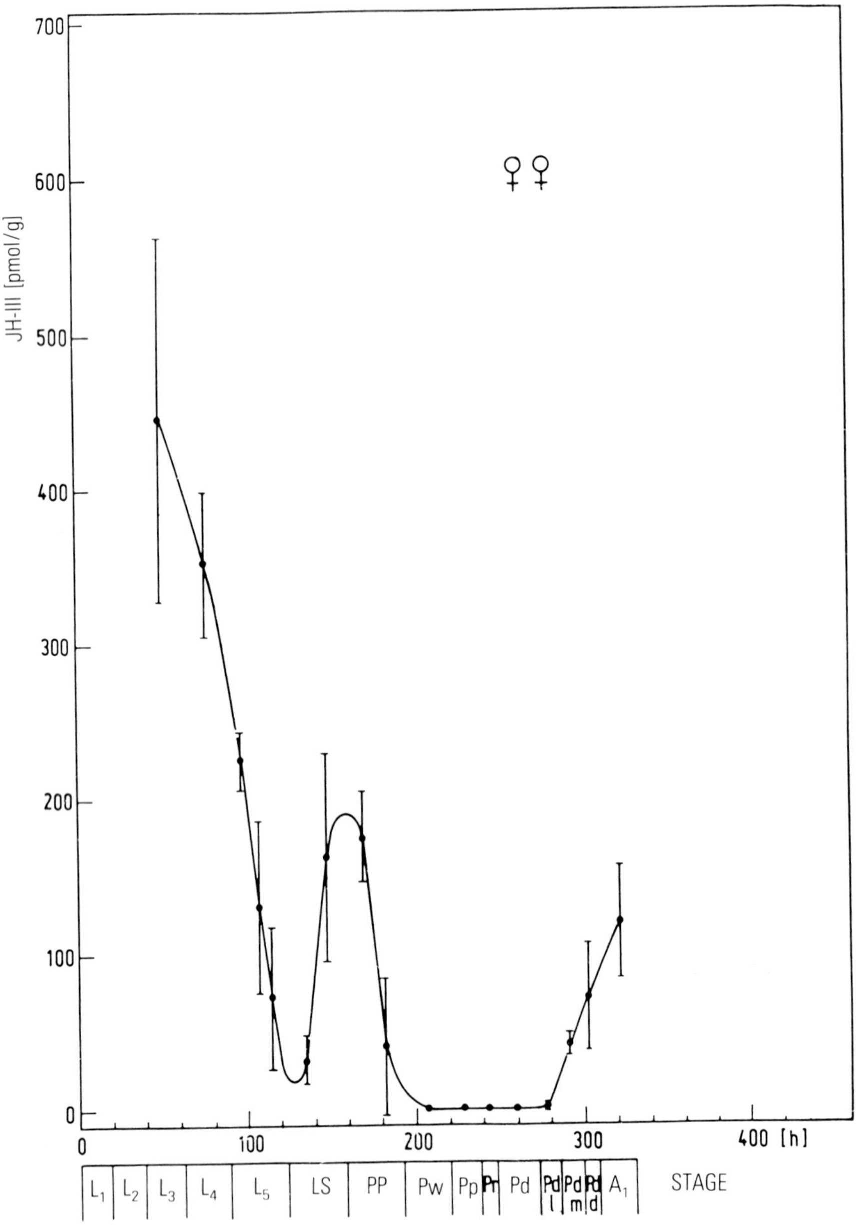

A

FIGURE 8.3. Average values of JH III titers from queen (A) and worker (B) developmental stages. (From Rembold, 1987c.) For more details see the legend of Fig. 8.2 and Rembold et al. (1980).

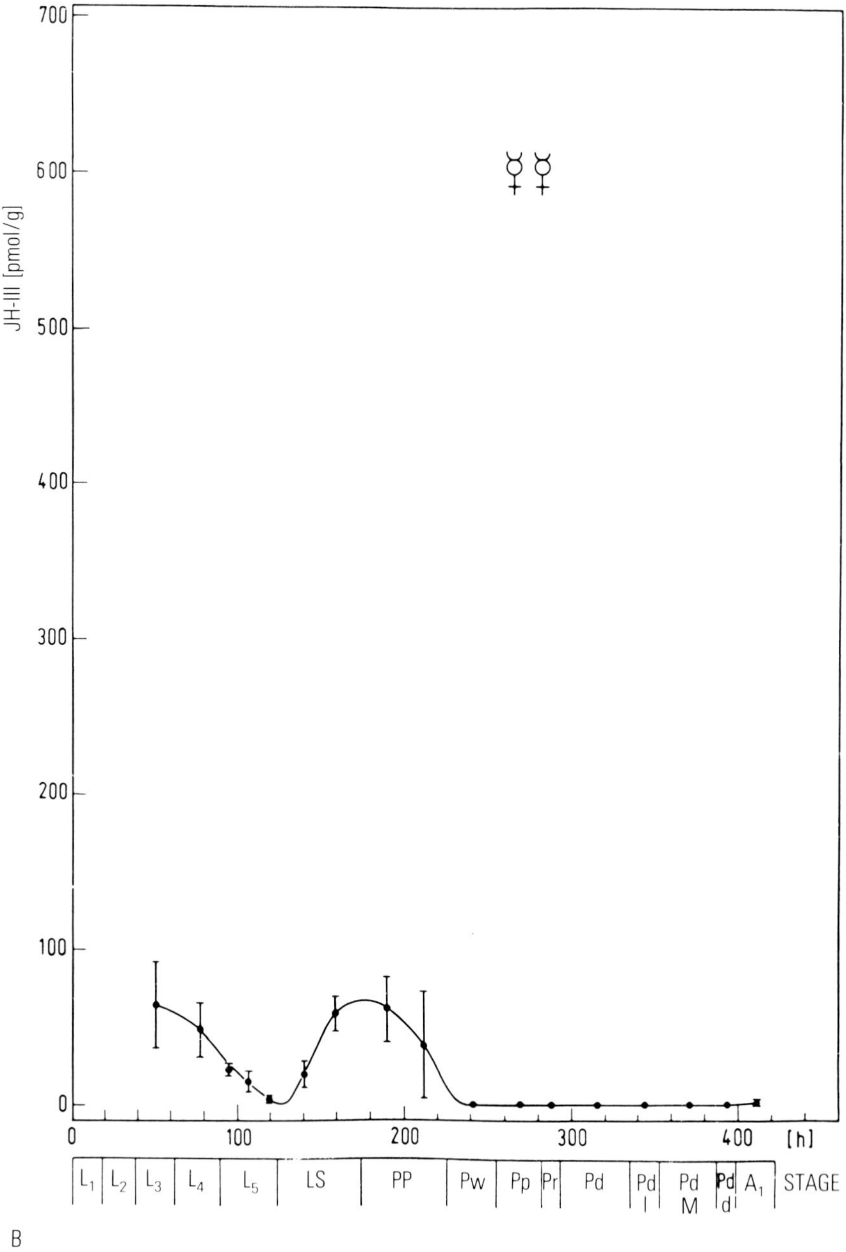

FIGURE 8.3. Continued

seems possible that the JH here has an additional function, which is to protect an organ as a maintenance hormone, as is also true of prothoracic glands (Wigglesworth, 1955; Gilbert, 1962). There is some evidence that, besides an involvement of ecdysteroids in the last-instar larvae, JH may also be needed for proliferation of cells in immature imaginal disks (Kurushima and Ohtaki, 1975). Also, a high JH esterase activity has been described for imaginal disks from *Drosophila melanogaster* (Chihara et al., 1972) and *Manduca sexta* (Hammock et al., 1975) during the ultimate larval instar. This would provide an additional argument that JH may be important in the maintenance of an organ in its immature state, as is the case for the ovaries of the prospective queen bee. The peak in the last-instar worker larva, however, may have some function in the maintenance of the imaginal disks and is not sufficient to protect the gonads from being degraded.

As a consequence of the maturation of the female gonads in the queen pupa, vitellogenin synthesis and egg production are initiated and the JH III titer rises with the onset of melanization of the cuticle from zero level slowly to about 100 pmol/g in the newly hatched adult. The worker adult, however, hatches without a trace of JH.

8.6. Caste-Specific Biochemical Effects

As already mentioned, morphogenesis is, in terms of molecular biology, the translation of a DNA sequence, which represents the blueprint of a developmental program, into a protein sequence, which may then act as an enzyme and synthesize such structural elements as lipids or polysaccharides. Gene expression is under hormonal control in higher organisms, most likely through a receptor–hormone complex of high affinity. Such a switching-on signal must be reversible, and a fairly high turnover of the hormone must be postulated. The high JH esterase activity during the JH III peak in the fifth queen instar is an example of such a rapidly inactivating catabolic enzyme.

As one example of caste-specific protein synthesis, the modulation of cytochrome *c* pools will now be considered. This respiratory protein is present in the mitochondria of the worker larvae in limited amounts only (Osanai and Rembold, 1968), a finding that may explain the well-known physiological difference in oxygen consumption between worker and queen larvae (Melampy and Willis, 1939). In order to use cytochrome *c* as an indicator for biochemical events during caste determination, cytochrome *c* must be measured in individual animals, as already discussed for JH estimations. For that purpose a viroimmunoassay was applied for the quantitative measurement of honey bee cytochrome *c* (Eder et al., 1977). The borderline of detection for this assay is at about 15 pg. With this assay, the cytochrome *c* content of worker and queen larvae was measured, as well as the body weight, W, oxygen consumption, and the respiratory quotient, R. This study showed that R times the logarithm of larval cytochrome c content, divided by W, yields a caste-specific constant, K:

$$K = R \log (\text{cyt } c)/W.$$

The queen-specific value, K_q, is about 3, and that of the worker, K_w, about 0.8 (Eder et al., 1983). In quantitative terms, queen determination results in high cytochrome c concentration and respiration throughout larval development. The difference between worker and queen in cytochrome c concentration is most pronounced in the spinning larval stage, which coincides with the JH III titer maximum, being high in the queen and much lower in the worker larva. Interestingly, among a series of experimental larvae that had been reared *in vitro*, some had the queen-specific value, some that of a worker, and some were in between, indicating that they were intercastes.

This caste-specific effect of the hormone titer on enzyme synthesis and activity is not a general phenomenon. A detailed study of carbohydrate metabolism and its glycolytic enzymes in queen and worker larvae showed a clear dependence of the enzyme activities on the developmental state. No caste-specific activity differences were found, however (Czoppelt and Rembold, 1970). This result is important in the sense that apparently the castes are sharing most of their metabolic programs and that only the special biochemical pathways, which are related to their social function, are expressed differentially.

In principle, the development of both the honey bee castes therefore follows the general scheme of a hymenopteran insect—with one exception. A change in the hormonal information transfer from the neurosecretory system to the peripheral hormone glands is used to trigger a caste-specific developmental program. This evolutionary trick allows the honey bee colony to produce an optional number of queens in case of need. The only limitation in this system is larval age. Because of the change in food quality, the worker larvae reduce their growth rate after third larval instar. According to data obtained *in vitro*, the growth rate during this period is directly linked with the capacity for queen determination (Rembold and Lackner, 1981). In other words, this means that only a well-fed larva is competent for further stimulation of its endocrine system. Taking this into consideration, the caste-specific modulation of JH titer throughout queen development has several functions. One is a caste-specific protein synthesis, as exemplified by cytochrome c biosynthesis, which in turn secures an adaptive response to special physiological situations—the queen larva growing much faster than the worker, and at a lower temperature, with a higher respiration rate. The second function of a high JH titer is in the last-instar queen larva—to protect the ovaries from degradation. And the third function is to initiate vitellogenin synthesis at the end of the queen pupal stage.

It is obvious from the preceding discussion that caste formation in the honey bee follows a sequence of several distinct steps (Rembold, 1985). In the first, the role of nutrition becomes clear. Only under optimal conditions, which are reflected by an intensive weight gain in the fourth-instar larva, is the queen-forming response of the female honey bee larva switched on during the third instar. A difference in food in the colony, or the feeding of an imbalanced food *in vitro*, can cause activation of the controlling system in the larval brain to be delayed by more than one instar in the worker. The second step, therefore, is marked by a time difference in the initiation of hormonal regulation. During this period, both worker and queen larvae start with a phase of intensive growth that is not under direct JH control. The third step is a caste-specific maturation of the peripheral hormone glands, especially the CA. Only in the last larval instar is caste induction during the third larval stage reflected by a modulated hormone synthesis, primarily of JH III. The fourth important step is, then, a queen-specific JH III peak, which coincides with the maintenance of the female gonads and with a breakdown of most of the ovarioles in the worker larva. With this important event, the caste-specific programming comes to an end. In the last step, the queen and worker developmental programs are expressed during the pupal phase.

8.7. Summary

The morphogenetic hormones play an important role in the modulation of developmental programs during imaginal metamorphosis of the honey bee, *Apis mellifera*. The biochemical basis of caste formation can be separated into a sequence of time-dependent developmental programs:

Step 1. The newly hatched female larva is fed a high-quality food in abundance through the first three larval instars. After the sensitive phase, nutrition of the workers is maintained with a food of lower quality. The queen larva is offered royal jelly throughout the entire feeding period.

Step 2. Maturation of the neurosecretory system in the pars intercerebralis is initiated in a caste-specific way. As a consequence of optimal nutrition, differentiation of the neurosecretory cells and RNA synthesis starts in the queen larva at the end of the third instar and in the worker at the fifth. The queen's CA become active at the end of the fourth instar; those of the worker, in the fifth.

Step 3. The caste-specific maturation of the endocrine organs results in titer differences of the peripheral hormones. The ecdysteroids decrease in their concentration at the end of the prepupal stage, from 400 ng/g in the fourth instar to less than 60/10 (queen/worker) ng/g in the fifth. During pupal development, both the castes maintain an ecdysteroid level of about 1000 ng/g. The titers of JH III are much more modulated in a caste-specific way. The queen builds up a higher JH titer than does the worker through its entire imaginal development. From third to fifth instar, the JH III titer decreases from 450/80 (queen/worker) to 10/0 ng/g, subsequently builds up to a peak with 200/70 ng between LS and PP, and then decreases in both the castes to zero values with the onset of pupal development. Only the queen pupa produces JH III from stage Pdl onward, and the newly hatched adult queen already has a JH III titer of more than 100 ng/g. No JH III is present in the newly hatched worker bee.

Step 4. A caste-specific modulation of hormone titers, primarily of JH III, controls the maintenance and breakdown respectively of the ovaries in the last larval instar. These high differences in JH III titer also affect biochemical and physiological regulations. As an example, the regulation of cytochrome *c* pools during the last instar was discussed herein. A caste factor *K*, connecting cytochrome *c* concentrations, respiration rate, and body weight, was also introduced.

References

Armbruster, L. 1960. Gelee royale. Arch. Bienenkd. 37: 1–39.

Chihara, C. J., W. H. Petri, J. W. Fristrom, and D. S. King. 1972. The assay of ecdysones and juvenile hormones on *Drosophila* imaginal disks *in vitro*. J. Insect Physiol. 18: 1115–1123.

Czoppelt, Ch. and H. Rembold. 1970. Vergleichende Analyse des Kohlenhydratstoffwechsels bei den Kasten der Honigbiene, *Apis mellifera*. J. Insect Physiol. 16: 1249–1264.

de Wilde, J. 1985. Extrinsic control of caste differentiation in the honey bee (*Apis mellifera* L.) and in other apidae. Pp. 361–369 *in* J. A. L. Watson, B. M. Okot-Kotber, and Ch. Noirot (eds.), *Caste Differentiation in Social Insects*. Pergamon Press, Oxford and Elmsford, New York.

Dogra, G. S., G. M. Ulrich, and H. Rembold. 1977. A comparative study of the endocrine system of the honey bee larvae under normal and experimental conditions. Z. Naturforsch. 32c: 637–642.

Eder, J., M. Osanai, S. Mane, and H. Rembold. 1977. Immunoassay for honey bee cytochrome *c* in single animals with cytochrome *c*–coated bacteriophages: a sensitive tool for the study of caste formation in the honey bee, *Apis mellifera*. Biochim. Biophys. Acta 496: 401–411.

Eder, J., J.-P. Kremer, and H. Rembold. 1983. Correlation of cytochrome *c* titer and respiration in *Apis mellifera:* adaptive response to caste determination defines workers, intercastes and queens. Comp. Biochem. Physiol. 76B: 703–716.

Feldlaufer, M. F. and J. A. Svoboda 1986. Makisterone A: a 28-carbon insect ecdysteroid. Insect Biochem. 16: 45–48.

Feldlaufer, M. F., E. W. Herbert, J. A. Svoboda, M. J. Thompson, and W. R. Lusby. 1985.

Makisterone A: the major ecdysteroid from the pupa of the honey bee, *Apis mellifera*. Insect Biochem. 15: 597–600.

Gilbert, L. J. 1962. Maintenance of the prothoracic gland by the juvenile hormone in insects. Nature (Lond.) 193: 1205–1207.

Hagenguth, H. and H. Rembold. 1978. Identification of juvenile hormone 3 as the only JH homolog in all developmental stages of the honey bee. Z. Naturforsch. 33c: 847–850.

Hagenguth, H. and H. Rembold. 1978. Kastenspezifische Modulation des Ecdysteroid-Titers bei der Honigbiene. Mitt. Dtsch. Ges. Allg. Angew. Entomol. 1: 296–298.

Hagenguth, H. and H. Rembold. 1979. Identification and quantification of insect juvenile hormones as their 10-heptafluorobutyryloxy-11-methoxy derivatives by combination of high-performance liquid chromatography and electron capture gas-liquid chromatography. J. Chromatogr. 170: 175–184.

Hammock, B. D., J. Nowock, W. Goodman, V. Stamoudis, and L. I. Gilbert. 1975. The influence of hemolymph binding protein on juvenile hormone stability and distribution in *Manduca sexta* fat body and imaginal discs *in vitro*. Mol. Cell. Endocrinol. 3: 167–184.

Haydak, M. 1970. Honey bee nutrition. Annu. Rev. Entomol. 15: 143–156.

Karlson, P. and H. Hoffmeister. 1963. Zur Biogenese des Ecdysons. I. Umwandlung von Cholesterin in Ecdyson. Hoppe-Seyler's Z. Physiol. Chem. 331: 298–300.

Kurushima, M. and T. Ohtaki. 1975. Relation between cell number and pupal development of wing discs in *Bombyx mori*. J. Insect Physiol. 21: 1705–1712.

Lindauer, M. 1987. Karl von Frisch, a pioneer in sensory physiology and experimental sociobiology. Pp. 15–19 *in* J. Eder and H. Rembold (eds.), *Chemistry and Biology of Social Insects*. Verlag Peperny, Munich.

Lukoschus, F. 1956. Untersuchungen zur Entwicklung der Kastenmerkmale bei der Honigbiene (*Apis mellifica* L.). Z. Morph. Oekol. Tiere 45: 157–197.

Mane, S. D. and H. Rembold. 1977. Developmental kinetics of juvenile hormone inactivation in queen and worker castes of the honey bee, *Apis mellifera*. Insect Biochem. 7: 463–467.

Melampy, R. M. and E. R. Willis. 1939. Respiratory metabolism during larval and pupal development of the female honey bee (*Apis mellifica* L.). Physiol. Zool. 12: 302–311.

Osanai, M. and H. Rembold. 1968. Entwicklungsabhängige mitochondriale Enzymaktivitäten bei den Kasten der Honigbiene. Biochim. Biophys. Acta 162: 22–31.

Rembold, H. 1965. Biologically active substances in royal jelly. Vitam. Horm. 23: 359–382.

Rembold, H. 1974. Die Kastenbildung bei der Honigbiene, *Apis mellifica* L., aus biochemischer Sicht. Pp. 350–403 *in* G. H. Schmidt (ed.), *Sozialpolymorphismus bei Insekten*. Wissenschaftliche Verlagsgesellschaft, Stuttgart.

Rembold, H. 1976. The role of determinator in caste formation in the honey bee. Pp. 21–34 *in* M. Lüscher (ed.), *Phase and Caste Determination in Insects*. Pergamon Press, Oxford and Elmsford, New York.

Rembold, H. 1985. Sequence of caste differentiation steps in *Apis mellifera*. Pp. 347–359 *in* J. A. L. Watson, B. M. Okot-Kotber, and Ch. Noirot (eds.), *Caste Differentiation in Social Insects*. Pergamon Press, Oxford and Elmsford, New York.

Rembold, H. 1986. Control of imaginal development by juvenile hormone. Pp. 59–68 *in* M. Porchet, J.-C. Andries, and A. Dhainaut (eds.), *Advances in Invertebrate Reproduction*, Vol. 4. Elsevier, Amsterdam and New York.

Rembold, H. 1987a. Caste differentiation of the honey bee: fourteen years of biochemical research at Martinsried. Pp. 3–13 *in* J. Eder and H. Rembold (eds.), *Chemistry and Biology of Social Insects*. Verlag Peperny, Munich.

Rembold, H. 1987b. Die Aufklärung der Kastenentstehung im Bienenstaat. Pp. 167–231 *in* H. von Ditfurth (ed.), *Mannheimer Forum 87/88*. Boehringer Mannheim Corp., Mannheim and New York.

Rembold, H. 1987c. Caste specific modulation of juvenile hormone titers in *Apis mellifera*. Insect Biochem. 17: 1003–1006.

Rembold, H. and G. Hanser. 1964. Über den Weiselzellenfuttersaft der Honigbiene. VIII. Nachweis des determinierenden Prinzips im Futtersaft der Königinnenlarven. Hoppe-Seyler's Z. Physiol. Chem. 339: 251–254.

Rembold, H. and H. Hagenguth. 1981. Modulation of hormone pools during postembryonic development of the female honey bee castes. Pp. 427–440 *in* F. Sehnal, A. Zabża, J. J. Menn, and B. Cymborowski (eds.), *Regulation of Insect Development and Behaviour.* Wrołcaw, Poland.

Rembold, H. and B. Lackner. 1981. Rearing of honeybee larvae *in vitro:* effect of yeast extract on queen differentiation. J. Apicult. Res. 20: 165–171.

Rembold, H. and G. Hanser. 1964. Über den Weiselzellenfuttersaft der Honigbiene. VIII. Nachweis des determinierenden Prinzips im Futtersaft der Königinnenlarven. Hoppe-Seyler's Z. Physiol. Chem. 339: 251–254.

Rembold, H. and B. Lackner. 1985. Convenient method for the determination of picomole amounts of juvenile hormone. J. Chromatogr. 323: 355–361.

Rembold, H. and G. Ulrich. 1982. Modulation of neurosecretion during caste determination in *Apis mellifera* larvae. Pp. 370–374 *in* M. D. Breed, Ch. D. Michener, and H. E. Evans (eds.), *The Biology of Social Insects.* Westview Press, Boulder, Colorado.

Rembold, H., Ch. Czoppelt, and P. J. Rao. 1974. Effect of juvenile hormone treatment on caste differentiation in the honeybee, *Apis mellifera.* J. Insect Physiol. 20: 1193–1202.

Rembold, H., J.-P. Kremer, and G. M. Ulrich. 1980. Characterization of postembryonic developmental stages of the female castes of the honey bee, *Apis mellifera* L. Apidologie 11: 29–38.

Rembold, H., B. Lackner, and I. Geistbeck. 1974. The chemical basis of honeybee, *Apis mellifera,* caste formation: partial purification of queen bee determinator from royal jelly. J. Insect Physiol. 20: 307–314.

Snodgrass, R. E. 1956. Anatomy of the honey bee. Comstock Publishing Associates, Ithaca, New York.

Steel, C. G. H. and K. G. Davey. 1985. Integration in the insect endocrine system. Pp. 1–35 *in* G. A. Kerkut and L. I. Gilbert (eds.), *Comprehensive Insect Physiology, Biochemistry and Pharmacology,* Vol. 8. Pergamon Press, Oxford and Elmsford, New York.

Svoboda, J. A., E. W. Herbert, M. J. Thompson, and H. Shimanuki. 1981. The fate of radiolabeled C-28 and C-29 phytosterols in the honey bee. J. Insect Physiol. 27: 183–188.

Svoboda, J. A., E. W. Herbert, and M. J. Thompson. 1983. Definitive evidence for lack of phytosterol dealkylation in honey bees. Experientia (Basel) 39: 1120–1121.

Townsend, G. F. and R. W. Shuel. 1962. Some recent advances in apicultural research. Annu. Rev. Entomol. 7: 481–500.

Trautmann, K. H., P. Masner, A. Schuler, M. Suchy, and H. K. Wipf. 1974. Evidence of the juvenile hormone methyl (2*E*,6*E*)-10,11-epoxy-3,7,11-trimethyl-2,6-dodecadienoate (JH III) in insects of four orders. Z. Naturforsch. 29c: 757–759.

Ulrich, G. M. and H. Rembold. 1983. Caste-specific maturation of the endocrine system in the female honey bee larva. Cell Tissue Res. 230: 49–55.

Weaver, N. 1966. Physiology of caste determination. Annu. Rev. Entomol. 11: 79–102.

Weiss, K. 1978. Zur Mechanik der Kastenentstehung bei der Honigbiene (*Apis mellifica* L.). Apidologie 9: 223–258.

Weyer, F. 1935. Über drüsenartige Nervenzellen im Gehirn der Honigbiene *Apis mellifica* L. Zool. Anz. 112: 137–141.

Wigglesworth, V. 1955. The breakdown of the thoracic gland in the adult insect, *Rhodnius prolixus.* J. Exp. Biol. 32: 485–491.

Wirtz, P. and J. Beetsma. 1972. Induction of caste differentiation in the honeybee (*Apis mellifera*) by juvenile hormone. Entomol. Exp. Appl. 15: 517–520.

Roles of Morphogenetic Hormones in Caste Polymorphism in Stingless Bees

9

H. H. W. VELTHUIS AND
M. J. SOMMEIJER

9.1. Introduction

9.1.1. *What Are Stingless Bees?*

Although most species of the Apoidea live solitary lives, development of social behavior can be found in five of its nine families. The highest levels of social organization are found in the subfamilies of the Apinae, the honey bees, and the Meliponinae, the stingless bees (Michener, 1974). These groups were formerly considered to be phylogenetically closely related and, together, to be distinct from the Bombinae and Euglossinae. Recent research, however, has indicated that the stingless bees separated earlier from a common bumble bee/honey bee ancestor (Winston and Michener, 1977).

Stingless bees are restricted to the tropics, where they are abundant in species and numbers. A few hundred species have been described. This is in sharp contrast to the genus *Apis*, where four (Ruttner, 1986) or six to eight (Sakagami, 1982) species are recognized.

Taxonomically, the group of stingless bees has been studied insufficiently. Moure (1951, 1961) recognizes a number of (sub)genera within the large *Trigona* group, whereas Wille and Michener (1973; see also Wille, 1979, 1983) consider these taxa mainly as species of the large genus of *Trigona*. Sakagami (1982) adopts the subgenera of Moure because they often demonstrate distinct ethological differences.

The fact that the Meliponinae and the Apinae are the only highly eusocial groups within the Apidae implies the existence of some basic biological similarities. However, there are also some principal differences, indicating their distant relatedness. These include differences in wing venation and structure of the sting apparatus (the latter being very rudimentary in Meliponinae), the stronger mandibular musculature in Meliponinae, and differences in the position of the wax glands (dorsal in Meliponinae; ventral in Apinae). While Apinae are progressive provisioners, the Meliponinae have a system whereby they mass provision their brood cells. During short periods a restricted number of bees deposit food in the cell, after which an egg is laid on top of the food. In all species, this is characterized by a pronounced rhythmicity: periods of cell-building behavior alternate with short bouts of intensive cell-provisioning behavior. Variations in this process are discussed in Sakagami and Zucchi (1974).

Another distinction is in the number of ovarioles. The ovaries of honey bee queens are composed of many more ovarioles than are those of their workers. In stingless bees worker and queen have the same number of ovarioles, namely, either four or, in a few species, eight per ovary (Sakagami, 1982). Therefore, the two groups of bees differ principally in the way they obtain the high rate of egg production by the queen (Velthuis, 1976).

Biological variability of the group of stingless bees. The important size variation within the Meliponinae ranges from *Melipona fuliginosa,* the largest stingless bee, measuring more than 13 mm to the dwarf species like *Trigonisca duckei,* measuring only about 2 mm. This size variation, combined with the great abundance of these bees, makes them the most important pollinators of the tropics (Wille, 1983).

The size of the colonies varies from a few hundred individuals in some *Melipona* species to densely populated nests of species of the genus *Trigona.* Colonies of 100,000 individuals have been reported (Lindauer and Kerr, 1958), but this might be exceptional (Wille, 1983). Most species build their nests in existing cavities like hollow trees or in the soil. Few species build their nests exposed. The narrow nest entrance of *Melipona* and other genera permits the defence of the nest by one or a few guards positioned in the mouth of the often rather elaborate tube. Within the nest a clear separation of the brood chamber from the area of food storage is always found. Storage pots are several times larger than brood cells. There are no separate types of pots, and pots containing honey are generally intermixed with those that contain pollen.

Most species normally arrange their brood cells in regular horizontal combs, but others build their brood cells in clusters, which may be very loose or compact. In all species that build horizontal combs (all *Melipona* spp. and most *Trigona* spp.), the brood chamber is surrounded by a system of waxy sheets composing the involucrum. Brood cells in a cluster arrangement are not surrounded by an involucrum. This allows for adaptation to irregularly shaped cavities. Intermediate arrangements of broodcells, e.g., irregular combs, are also found.

9.1.2. *Caste Determination Through Trophic and Genetic Factors*

The distinct morphological differences between queen and worker clearly contribute to a better functioning of the individual for a given task. Under natural circumstances there are no intermediates; this indicates the existence of two distinct developmental programs for the larva. The decision as to which of these two programs will be followed depends on information obtained from the environment and concerns the amount and/or the composition of the food. The moment at which this information must be available differs for the various social Hymenoptera. In honey bees and in the Trigonini, this information is based on the markedly different sizes of the cells from which queens or workers emerge. In honey bees, the queen cell stimulates the nurse bee to provide the larva with a different type of food (see Chapter 8 by Rembold in this volume). In the mass-provisioning system of the stingless bees, qualitative differences are probably not linked to cell size, so it is only the amount of food that determines the development into either a queen or a worker. Queen cells of the Trigonini contain some two or three (or even more) times as much food as cells that produce a worker.

In the Meliponini the two female castes emerge from the same type of cell. Kerr (1950) advanced the hypothesis that development into a queen or a worker is governed by a genetic mechanism. Only 25% of the female larvae will respond to the quantity of food available: above a certain quantitative threshold these larvae become queens; below that amount they become workers.

The question of caste differentiation is of interest because of the developmental pathways open to an individual. In addition, the ecological conditions and the processes at colony level that regulate queen production at certain times and its cessation at other times are of equal importance. In the stingless bees, this concerns the processes that regulate food production in an individual nurse bee in such a way that brood food of a supposedly constant composition becomes available, as well as the deposition of this food in a single cell by a number of provisioning bees. The physiology of individuals, their behavioral interactions, and the ecological factors influencing individual performances are fundamental to our understanding of how caste regulation contributes to the functional adaptation of a colony and of a species to its environment. We intend to discuss these various aspects in the following sections.

Three systems of rearing queens. The various stingless bee species have one of the following systems to produce queens: (1) the construction of special queen cells; (2) the confiscation of food from a neighboring cell by a larva; or (3) a genetic mechanism that prevents large proportions of the female larvae from becoming queens even if other conditions favor such a development. We shall make some brief comments on these three mechanisms:

(1) Perez (1895) and Von Ihering (1903) were the first to describe the cells from which the queens of many Trigonine species develop. Generally, such cells are constructed along the edges of the completed combs. Because of their larger size, they are very conspicuous. In 1972, de Camargo furnished evidence that the food in these queen cells is like the food in the worker and male cells. Similar but less detailed observations were made by Darchen and Delage-Darchen (1971).

(2) Another natural way of producing a queen was discovered by Terada (1973); we call it the two-cell system. It was first observed in *Frieseomelitta varia*. Cells larger than normal have never been encountered in this cluster-building species. Where brood cells are generally separated from each other by pillars of wax and resin, occasionally two brood cells are constructed contiguously. This allows the older of the two larvae in one of these cells to perforate the common cell wall and to

consume the food destined for its neighbor. If a larva consumes twice its usual ration, it becomes a queen. In *Leurotrigona muelleri,* as already noted by von Ihering (1903), larger cells are found only seldom. Terada (1973) supposed that a majority of queens are produced by the two-cell system. If these observations are correct, this species has two ways of producing queens. In *L. muelleri* and in *F. varia* the volume of the cocoon of a queen is twice that of a worker.

(3) In the genus *Melipona* no special queen cells are present. Kerr (1948, 1950) advanced the hypothesis that in these species queens are genetically predetermined. This predetermination will be expressed if the cell contains an above-average amount of food. Those female larvae that lack the genetic structure for becoming a queen will be unable to take advantage of the times when food happens to be abundantly available; they will always develop into workers. The queen-determined larvae will become workers if food does not permit queen development. This hypothesis has been refined and has become rather detailed in subsequent papers. We will deal with it more particularly in a later section.

9.2. Mass Provisioning and the Quality of Larval Food

9.2.1. *The Composition of Larval Food*

The food in the brood cell contains pollen and carbohydrates (which have been collected from flowers and have been stored separately in the honey and pollen pots), as well as proteinaceous material secreted by the hypopharyngeal glands of the nurse bees. Whether mandibular gland secretion is involved is still unclear.

The carbohydrates in the honey are mainly glucose and fructose. The ratio in which they occur depends on the nectar sources visited by the bees. The concentration of sugars in the honey is between 55% and 85% (Roubik, 1983). When the honey is transported by a bee to a brood cell, the concentration is lowered; Martinho (1975) reported a rather stable sugar concentration in the crop of *Melipona quadrifasciata.* Sugar concentration is probably also well regulated in the brood cell.

During the storage of pollen in the pollen pots, a fermentation process takes place. After an equilibrium is reached the pollen is taken by some house bees into the crop, where it becomes mixed with the honey solution. The house bees transport this mixture to a brood cell, either immediately or, more probably, after it has been regurgitated to another bee that is participating in the provisioning of a brood cell.

Some of the pollen might also be passed into the ventriculus, where degradation of the protoplast takes place. The protein, probably in a degraded form, reaches the hemolymph and may be transported to the hypopharyngeal glands where a secretion is produced that constitutes the third component of the larval food. The function of this secretion is not quite clear. Whereas in the honey bee the secretion has mainly a nutritive value, largely due to its proteinaceous character, we consider it quite possible that in other bees, including stingless bees, the enzymatic properties are more important than the direct nutritive aspects.

In any given species we can expect the composition of the larval food to be to a certain degree stabilized, so that the proportion of pollen, honey, glandular secretion, and water becomes the same in each cell. However, there seems to be a large variation in the composition among bee species. In some species, the food is rather liquid, so that the larva is almost completely submerged, whereas in other species the food has a viscous consistency. Table 9.1 gives an impression of this variation in the food composition.

The reasons for these interspecific differences are not yet understood. Are these different compositions adaptive to the respective species? And, if so, what other purpose does the specific composition serve in addition to providing the construction material for building a bee? For instance, might the absence of thermoregulation inside the nest have an effect on food composition?

In several cases, it has been possible to establish mixed colonies in

TABLE 9.1. The Composition of the Food Present in Brood Cells

Species	Dryweight (in mg/g fresh weight)	Sugars (in mg/g fresh weight)	Protein (in mg/g fresh weight)	Free amino acids (mg/ml fresh food)
Scaptotrigona depilis	590	129	19.4	10.6
S. bipunctata	541	136	17.6	12.9
Nannotrigona testaceicornis	461	51	23.8	8.6
Partamona cupira	402	96	2.8	3.8
Tetragonisca angustula	586	115	3.7	10.6
Plebeia droryana	452	82	1.3	6.9
Melipona quadrifasciata	427	64	1.1	2.1

SOURCE: After Hartfelder (1986).

which the queen was of a different species than the workers. For instance, da Silva (1973) exchanged the queens of colonies of two subspecies of *Melipona quadrifasciata*—*M. q. quadrifasciata* and *M. q. anthidioides;* the second is slightly larger than the first. In this case brood development was as in untreated colonies. If exchanges were made at the species level, also some brood was successfully reared. Often, however, exchanges at the species or higher level led to behavioral disturbances that prevented the rearing of offspring. When 100 workers of *Scaptotrigona postica* were introduced into an *M. marginata* colony, workers of the two species could be observed working together at the same cell, but the cells were opened again 6–10 days after sealing and the larvae were removed.

Interestingly, eggs laid by a *Plebeia droryana* queen in colonies of *Nannotrigona testaceicornis* produced workers and males, and eggs laid in two *Nannotrigona* queen cells developed into queens. In the case of *P. schrottkyi* queens, the normal brood cells of *Nannotrigona* led to the production of queens and males, whereas in four queen cells the larvae—apparently unable to digest the enormous amount of food—died. Such exchanges indicate that even if the food composition deviates from a species-specific pattern, development can still occur.

Hartfelder and Engels (1989) have discussed the relation between the characteristic food composition of a species and its effects in caste differentiation. In applying the conclusions of Sasaki et al. (1987) about the role of the physical properties of royal jelly in caste determination of the honey bee to stingless bees, these authors explored the possible effect of a change in food composition for the developmental program of the larva. It should be remembered that the food of a stingless bee larva consists of a liquid fraction, possibly the initial food for a larva, and a more or less solid fraction made up mainly of the pollen grains that in many species form a kind of a sediment in the cell. Few authors have considered the possibility that a large quantity of food, although of the same composition as minor quantities, may be translated into qualitatively different food in certain larval stages, leading to different developmental patterns. Hartfelder and Engels (1989) conclude that the interspecific differences in the composition of the food is correlated with the phylogenetic distance between the species.

9.2.2. *Intraspecific Variation in the Food*

Variation in the food a larva may encounter can take three forms: (1) the quantity of food in a cell; (2) the ratio in which the three components (pollen, carbohydrates, and hypopharyngeal gland secretion) are present; and (3) the concentrations of these components (in other words, the water content of the larval food). Such information, however, is very scarce and to our knowledge has never been the focus of any study. The need for such studies is indicated by the following account of data oncerning *M. quadrifasciata*. Kerr and Nielsen (1966) reported that the capacity of cells varies from 0.144 to 0.172 ml. In Kerr et al. (1966, Table 4), the amount of food deposited in a cell is given as 122 mg on average, but it varies from 87.6 to 185.8 mg. This variation results in differences in pupal weights: worker pupae may vary from about 60 to 100 mg, and queen pupae from 72 to 106 mg.

The food is deposited by some 6–14 workers, and there is an inverse relation between the number of provisioning bees and the quantity of larval food in the cell at which oviposition takes place. Apparently, the amount of food regurgitated per bee is subject to a large variation.

In 1972, de Camargo also reported the amount of food deposited per cell as 125 mg on average. With 93 mg of food, she obtained workers weighing 38.8 mg upon emergence; with 186 mg she obtained workers of 80.5 mg and queens of 71.4 mg; if 218 mg of food was available, the larvae were unable to consume it all. Those that succeeded in becoming adult weighed on average 82.2 (workers) or 79.3 mg (queens). Note that the variation in adult weight is less than in the study by Kerr et al. (1966). In 1983, da Cruz Landim reported larvae having a maximum average weight (in the fifth instar) of 168 mg, an extremely high value in comparison to values reported in the other papers cited above. Nevertheless, the larvae were collected from natural colonies.

Hartfelder (1986) gives standard deviations around the mean values for the various components of the food. For *M. quadrifasciata* the dry weight of the food, 427 mg/g fresh weight, has a standard deviation of 11 mg; with an average of 64 mg of sugars/g dry weight, the standard deviation is 9 mg. In *Scaptotrigona depilis*, these values are 590 ± 40 mg dry weight/g food and 129 ± 12.5 mg sugars/g food.

These data suggest a marked intraspecific variation in the food factor, which deserves more attention than it has been given so far.

9.3. Developmental Patterns of Worker and Queen

9.3.1. *Developmental Patterns*

The considerable variability in stingless bees is also reflected by differences in the time aspects of development. Embryogenesis takes 2 days in *Scaptotrigona* (Beig, 1971), 5.5 days in *Tetragonula* and *Trigonella* (Salmah et al., 1987), 5 days in *Melipona quadrifasciata* (Kerr, 1950), and 6 days in *M. beecheii* (M. J. Sommeijer, unpublished data). In all these species, food consumption by the larva is completed in 7–13 days; prepupal and pupal development takes between 20 and 33 days (Table 9.2). Terada (1973) mentions a developmental time of approximately 3 months for a *Frieseomelitta varia* worker. Since rather different data have been reported, even

TABLE 9.2. Some Data on the Duration of the Developmental Stages of Workers of Various Stingless Bee Species (Time in Days)

Species	Larva	Prepupa (if not included in larva)	Pupa	Total including egg stage	Reference
Melipona beecheii	12	+10	20–26	49–54	Sommeijer (unpublished data)
Melipona quadrifasciata	8		20–23	33–36	Kerr (1950)
worker	6.8	+10	18.6	40.4	Camargo (1972)
queen	7.8	+8.5	16	36.5	
male	7	+10.4	18	40.5	
Scaptotrigona depilis	10	+6	16	34	Hartfelder (1986)
Scaptotrigona postica	13		31–33	46–48	Sakagami and Akahira (in Salmah, 1987)
	14	+2	22	40	Beig (in da Cruz and Mello, 1981)
Tetragonula minankabau	9.5		27	42	Salmah et al. (1987)
Trigonella moorei	10	+1	31	46.5	Salmah et al. (1987)

for the same species or for closely related ones, we infer that developmental times are not fixed but are subject to variation due to environmental conditions such as temperature.

Generally, data on developmental stages concern workers only. In comparison, the development of a queen either takes 2–4 days more than that of a worker, as in the case in Trigonini, or takes distinctly less time than a worker, as occurs in the Meliponini. This difference is connected with the fact that the queen larvae of the Trigonini ingest more food. The *Melipona* queens, owing to their higher metabolic activity, complete their pupal development 1 day earlier than the workers.

The most detailed information concerning the time pattern has been provided for *Scaptotrigona depilis* (Hartfelder, 1986). From these data (Table 9.3) we conclude that the major difference between worker and queen development is in the longer duration of the fourth and fifth larval instar of the queen larva. In the last instar, the larger quantity of food is consumed over a longer period, but its digestion also takes more time. From defecation onward, the time aspects of development of workers and queens are almost similar again.

Darchen and Delage-Darchen (1971) report an interruption of development in the prepupal stage of queens of *Axestotrigona* and *Dactylurina*. These prepupae can stay alive without any apparent change for at least 2 months.

9.3.2. *Metabolic Aspects*

The two female castes differ in the economy by which they convert larval food into body biomass. In *Scaptotrigona* an average brood cell contains 35.4 mg of food; a worker emerges with a body weight of 19.6 mg, equivalent to 55% of the food mass. If a larva is given 105 mg of food, which is three times the quantity encountered in a worker cell, it will develop into a queen with an average body weight of 32.6 mg, which is 31% of the weight of the food (de Camargo, 1972).

In *Melipona rufiventris* 137.5 mg of food, which is equivalent to 125% of the average amount found in a brood cell, results either in the development of a worker or in a queen. The worker emerges with a mean body weight of 91.6 mg, and the queen will have an average weight of

TABLE 9.3. A Comparison of the Developmental Program of Queen and Worker in *Scaptotrigona depilis*[a]

Stage	L_1	L_2	L_3	L_4	L_5	PP	P	Emergence
Queen	0	25	59	101	147	293	423	813
Worker	0	25	57	102	132	246	376	763
Major difference					!	!		

SOURCE: After Hartfelder (1986).
[a]Time is given in hours after hatching.

73.8 mg, the conversion factors are 0.67 and 0.54, respectively. In *M. quadrifasciata*, 125 mg., the average quantity of food, results in either a worker of 76.6 mg or a queen of 49.4 mg; the conversion factors are 0.59 and 0.39, respectively. At the initiation of the pupal stage these worker pupae weigh 96.2 mg on average, and the queen pupae, 9.1 mg; this indicates that during pupal development the loss of weight in queens is more substantial than it is in workers, even though the queens' pupal stage takes less time in the *Melipona* species (de Camargo, 1972).

Hartfelder (1986) reared queens of *Scaptotrigona depilis in vitro* and compared their development with that of workers developing within the colony. He found a first difference between the two female castes in the third larval instar. Although there is no difference yet in the time pattern (compare Table 9.3), by the end of this stage a queen larva has accumulated 1.66 times the weight of a worker larva. By the time the larvae are spinning their cocoons, the difference has become a factor of 2.3 (71 vs. 31 mg), but when they emerge the difference is reduced to 1.7 again (33.3 vs. 19.1 mg in the queen and worker, respectively). These body weights compare quite well with those reported by de Camargo (1972). Hartfelder (1986) remarks, however, that queens reared *in vitro* are only two-thirds of the weight of naturally obtained queens. This could mean either that in natural conditions food quantities are even more distinct than those chosen for artificial rearing or that the exact dietary sequence and/or the resulting conversion into body weight is poorer in the experimental situation.

A large proportion of the digested material is incorporated in the larval fat body. The process has been studied by da Cruz Landim and Mello (1981) in *Scaptotrigona postica* and by da Cruz Landim (1983) in *M. quadrifasciata*. Both studies indicate that the number of cells of the fat body remains more or less the same during the various larval instars. However, the volume of each cell increases steadily until it represents a 32-fold (in *S. postica*) or even a 136-fold (in *M. quadrifasciata*) increase in the last larval stage as compared to the first.

9.3.3. *Corpora Allata and JH Production*

As in other social insects, development into a worker or a queen is regulated by the corpus allatum (CA). Kerr et al. (1975) determined the number of cells in the CA in relation to the stage of development in *M. quadrifasciata*. As soon as workers can be distinguished from queens by external characters—that is, in the early pupal stage—workers, if compared to queens, are found to possess about 75% of the number of cells in the CA. In volume units, workers show a decrease from 41 in the prepupal stage, before defecation, a stage when workers and queens cannot yet be distinguished, to 18 in the late pupa; in queens, the volume increases up to 114 in the late pupa and even up to 141 upon emergence. One can conclude that there is enhanced activity of the CA in queens, but this probably does not imply an immediate release of juvenile hormone (JH).

Bueno and Beig (1980) studied the CA during the preimaginal stages of *S. postica* and found the difference in the size of the CA between workers and queens to become important during the spinning phase of the last larval instar. At that moment CA volume of queen larvae is 3.6 times that of a worker larva. Only in the queen this volume is reduced sharply to 40% in the prepupal stage, although the CA of queens still remain 1.3 times larger than those in workers. After pupation, the workers' CA become continuously smaller: in black-eyed pupae, only 35% of the volume that was present in the spinning stage is left, whereas in queens the volume increases immediately after pupation; in white-eyed pupae, this volume is already 130% and in black-eyed pupae 150% of the volume found in the spinning larvae.

The JH synthesis in the CA has been studied by Hartfelder (1987) by an *in vitro* radiochemical method. The median parts of the brain, together with the corpora cardiaca (CC) and the CA of *S. depilis* were incubated in a [^{14}C]methylmethionine-containing medium. The JH released into the medium during 4 h of incubation was measured. In queens JH synthesis was always higher from the fourth larval instar up to the early pupal stage. At that last moment, the rate of synthesis was found to have dropped to unmeasurable levels. Major differences between queens' and workers' rates of JH synthesis were found early in the fifth larval instar; in this phase the remainder of the food is ingested. Once defecation starts and the transition to the prepupal phase takes place, the CA of queens and workers becomes inactive (Fig. 9.1).

These results are to be interpreted as a caste-specific modulation of a general pattern of JH synthesis, leading to pupation. Hartfelder (1987) poses the relevant question as to how the amount of food ingested results into the rate of hormone release. In *Scaptotrigona,* queens develop from specific cells, and therefore a difference in hormonal activity in the fourth larval instar of queens and workers becomes understandable. The situation in *Melipona* becomes the more intriguing, because here differences in the amount of food are believed to be perceived only in the final instar. The hormonal window, therefore, can be expected to be even narrower than in *Scaptotrigona*.

9.4. Mechanisms of Caste Formation

9.4.1. *The Effects of Applying JH*

The first investigators to study the effects of JH on caste determination in stingless bees were Campos et al. (1975), Velthuis and Velthuis-Kluppell (1975), and Campos (1975). Working at the same institute but

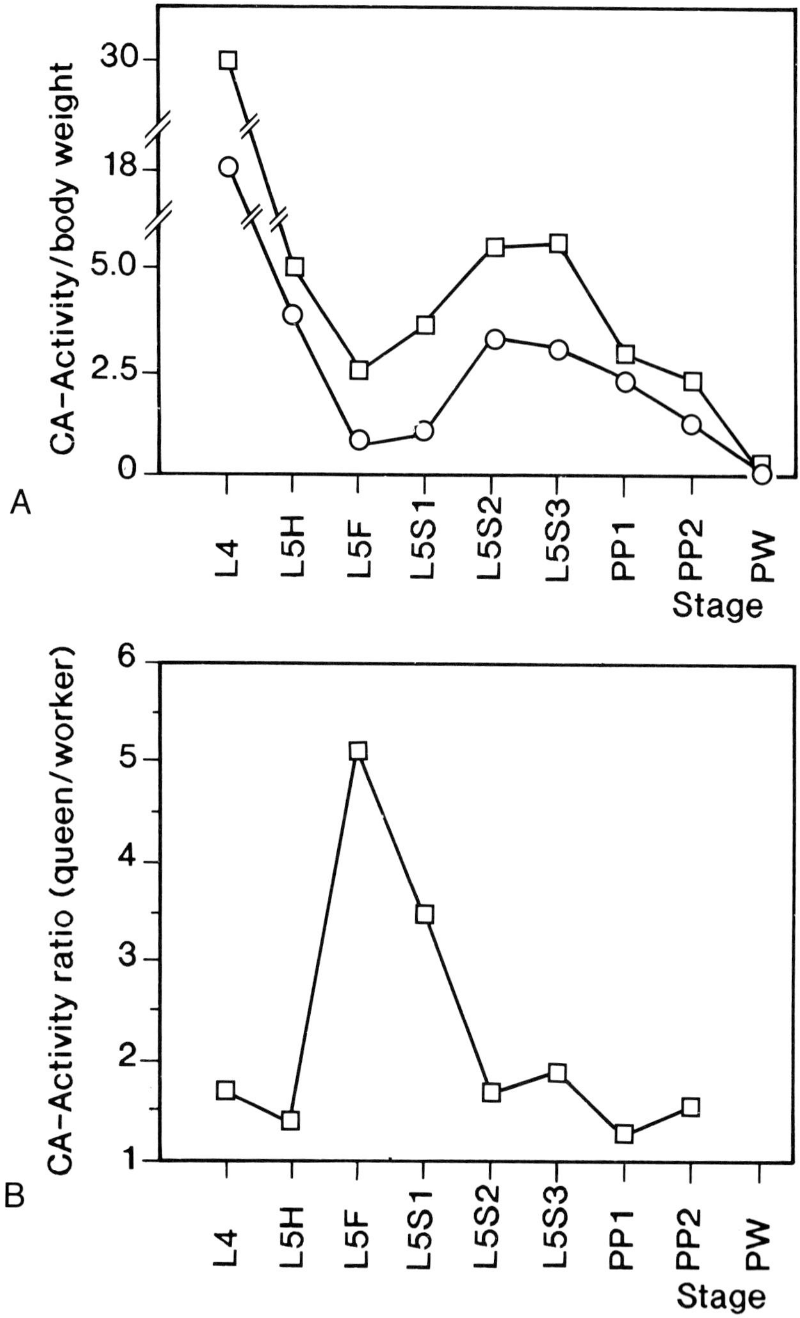

FIGURE 9.1. (A) Rates of JH synthesis of CA of *Scaptotrigona depilis* queens (□) and workers (○). (B) Ratio of CA activities of queens over workers. *Key:* L_4 = fourth larval instar; L_5 = fifth larval instar; H = horizontal feeding phase; F = vertical feeding phase; S_1–S_3 = spinning phases; PP_1, PP_2 = prepupal phases; PW = white-eyed pupa. (After Hartfelder, 1986.)

using different methods, they came to the general conclusion that if JH or an analogue was applied to a spinning larva of *M. quadrifasciata* this could lead to the emergence of a queen. The effect was strongly dose dependent.

The two methods used by these authors were the following: Campos removed larvae from their cells in the colony at the moment they had consumed their food and were starting or preparing to spin. They were transferred to petri dishes in the incubator and were subjected to one of the treatments of the experiments. Then they were allowed to continue their development. Often high mortality occurred in his experiments, even in the control treatments, possibly as a consequence of their horizontal position in the petri dish and possibly also because they were handled with forceps. This was probably one of the reasons why the effect of the treatment was recorded once the larva had developed into a pupa. At this stage workers and queens can be distinguished by the morphology of the head. A further consequence of this method is that there are no data concerning the amount of food each larva ingested, and therefore we have to accept this as an unknown source of variation.

The method of Velthuis and Velthuis-Kluppell involved the transfer of food in predetermined quantities from cells still containing an egg to artificial cells. A very young larva was placed on this food. When this larva had consumed the food and had started to prepare a cocoon it received its treatment, after which it was allowed to continue development up to the adult stage within the artificial cell. The effect of the treatment was determined on the basis of the emerged adults.

This method precludes some of the disadvantages of the other procedure, but it does have an inherent weakness in that the natural pattern of food layering in the cell (a liquid phase on top and sedimented pollen below) is disturbed. Therefore, the earlier instars probably ingested a higher proportion of pollen than would occur in the natural situation. We suppose this factor to be responsible for the fact that even with relatively high amounts of food, the control series never yielded 25% queens. The reader should examine Table 9.4, which gives further evidence that this method of *in vitro* culture in*M. quadrifasciata* does not lead to 25% queens.

Velthuis and Velthuis-Kluppell (1975) concluded that an isomer mixture of *Cecropia* JH in a quantity of approximately 10^{-7} μg was at the lower limit of effectiveness; after application of 10^{-3} μg of JH all female larvae became queens, whereas higher dosages led to increased mortality of abnormal development in the surviving animals. This was an indication that a genetic mechanism would regulate JH production. At the same time, a genetic mechanism responding to the JH titer was envisaged that would convert graded information of increasing concentrations into a discrete response, thereby preventing the production

TABLE 9.4. The Frequency of Queens in Experimental Series of *in Vitro* Cultures of Various Species of *Melipona*

Species	Segregation		Amount of food[a]	References
	Queens	*Workers*		
M. marginata	9	19	≥120	Camargo et al. (1976)
M. rufiventris	12	29	≥125	Camargo et al. (1976)
M. scutellaris	5	13	≥145	Camargo et al. (1976)
M. quadrifasicata	12	57	≥125	Camargo et al. (1976)
M. quadrifasciata	9	44	125	Engel (1979)
M. quadrifasciata	6	36	136	Velthuis and Velthuis-Kluppell (1975)
M. quadrifasciata	6	26	150	Camargo (1976)

[a]The amount of food given to each larva is given as the percentage of the average amount encountered under natural conditions.

of intercastes. Engel (1979) confirmed these results; using the same hormone preparation at 2×10^{-3} μg/larva, she obtained only queens.

Campos (1975) needed larger amounts to obtain queens in *M. quadrifasciata*. The *Cecropia* JH I applied in quantities of 0.64 μg/larva yielded 75% queens (51 queens and 19 workers); Calbiochem JH in 14.7 μg/larva yielded 10 queens and 4 workers; and the Zoecon products Altozar (9 μg/larva) and Altosid (18 μg/larva) produced 11 queens and 2 workers, and 5 queens and 6 workers, respectively. No real intercastes were encountered in these series either, but miniature individuals, especially workers, were produced.

These experiments have been extended to other *Melipona* species: *M. scutellaris* (Campos, 1975), *M. marginata* (Campos, 1975), *M. rufiventris* (Bonetti, 1982), and *M. compressipes* (Bonetti, 1982). The latter author also studied *M. quadrifasciata* again. Her data on *M. compressipes* appear especially conclusive with regard to the effects of the homologues JH I, II, and III, when applied separately or in combination, to larvae in the comb or transferred to petri dishes. She found JH I to be a good inductor of queen development; repeated treatments with 0.025 μg/day even for 4 successive days increased its effectiveness. JH II and JH III produced an average 14% and 16% queens, respectively, whereas in some untreated series 13% and 11% queens were obtained. These homologues, therefore, are to be considered less effective, presumably because the naturally occurring JH III is rapidly degraded if the titer exceeds the physiological range. The tested combinations of two or three homologues only reduced the effect of JH I alone. In accordance with this interpretation, in *S. depilis* JH III was found to be the only homologue present (K. Hartfelder et al., in preparation; cf. Hartfelder, 1987).

9.4.2. *The Genetic Basis of Caste Determination in* Melipona

Whereas in all other eusocial bees caste is regulated only by means of food supply, for the genus *Melipona* an additional mechanism has been proposed by Kerr (1946, 1950). In this subsection, we will describe the original formulation of this theory as well as its later version, without stressing those points where experimental evidence for some of its details are still lacking. Kerr's hypothesis is based on the observations that (1) queens and workers emerge from the same type of cell; (2) the proportion of queens depends on the season, being higher in periods with ample food; and (3) the proportion of queens attained an apparent maximum of 25%. This was explained by assuming two genes, X_a and X_b, each occurring in the population with two alleles. It is hypothesized that these genes regulate caste in such a way that if an animal is heterozygous for both genes (X_a^1 X_a^2/ X_b^1 X_b^2) it can develop into a queen, but if it is homozygous for one or both of these genes it will be unable to do so. In addition to this intrinsic mechanism there must be an extrinsic one, namely, the amount or quality of food available in the cell, which determines whether a "genetic queen" will become a queen or not.

In fact, the peculiar element in this proposal is that a category of individuals is believed to be intrinsically restrained from becoming a queen. Although this is an unusual situation, a specific genetic mechanism regulating morphogenesis is also found in other social Hymenoptera. In the ant *Harpagoxenus sublaevis*, Buschinger (1978) discovered a genetic factor prohibiting certain genotypes from developing into winged queens.

The inability of some of the female larvae to become queens should be distinguished from the question as to how caste differentiation is regulated by the extrinsic factors in a genotypic configuration that has retained developmental flexibility. First of all we shall concentrate on the observations that support the genetic model and we shall arrive at the latest version of it, bypassing the specifications presented from the conception of the model until its latest elaboration.

A proportion of approximately 25% queens has been reported to occur under natural conditions in the following species: *Melipona favosa, M.*

marginata, M. melanoventer, M. nigra, M. interrupta, M. quadrifasciata, and *M. rufiventris* (see Kerr and Nielsen, 1966). The same percentage has been found in experiments where larvae were reared *in vitro* on abundant food (Table 9.4).

Differences that occur in the ventral nerve chord (Kerr and Nielsen, 1966) are further evidence for the existence of two classes of genotypes. In *M. marginata* queens characteristically have four abdominal ganglia, whereas workers have either four or five. The workers segregate in such a way that workers with only four ganglia can be considered as being queens in the genetic sense. However, in *M. quadrifasciata,* where nerve chords also may have four or five ganglia, some queens were recorded as having five ganglia. This made the aforementioned authors propose another gene in the expression for the X_a/X_b system in the nerve chord character.

Another morphological trait permitting the detection of genetic queens that became workers as the consequence of unfavorable food conditions has been encountered in the extension of the tergal glands (da Cruz Landim et al., 1980). All queens of *M. quadrifasciata* have these glands on the abdominal tergites II–VII, whereas most workers have them only on tergite II. However, some of the workers have the queen type of gland distribution. The two types segregate again in a 3:1 distribution. As with the nerve chord character, however, the situation is different in other species (Bonetti, 1982).

The fact that the application of JH to sensitive larvae turns them all into queens can be interpreted as indicating that the production of JH is governed by a genetic mechanism. Campos et al. (1979) therefore attempted to specify the action of the two genes X_a and X_b in this way. The amount of food consumed by a diploid female larva is considered to be positively correlated with the number of cells developing in the CA. In these cells the genes X_a and X_b are both thought to regulate the production of a JH synthetase. The actions of these genes are supposed to be mutually independent. It is also envisaged that the action of X_a^1 X_a^2 is different from X_b^1 X_b^2, either in a quantitative or a qualitative sense. In the presence of a precursor of JH the amount of the enzymes *a* and *b* determines the amount of JH that is synthesized, which would subsequently interact with genes that program for queen caste characters. For the production of queens a specific threshold level of JH must be surpassed.

Under natural conditions development into a queen is only possible for animals having an X_a^1 X_a^2/X_b^1 X_b^2 genome; this is the case in 25% of the animals. However, since the food determines the size of the workshop where the JH is produced, manipulation of the food factor, leading to a larger than usual number of cells in the CA, could result in a higher percentage of queens. A genome in which X_a^1 X_a^2 is combined with a homozygous X_b could already be enough to make a queen, and conversely the X_b^1 X_b^2 in combination with a homozygous X_a would also suffice. The model, therefore, implies the possibility of producing queens in proportions higher than 25% under seminatural conditions.

The latter possibility was considered necessary because of an intriguing result obtained by da Silva (1973). She exchanged the queens of colonies of two subspecies of *M. quadrifasciata,* namely, *M. q. quadrifasciata (M.q.q.)* and *M. q. anthidioides (M.q.a.)*. The latter is a subspecies with a slightly larger body and presumably should provision its cells with a slightly larger amount of food. In one out of the four of these exchanges, she obtained offspring of the *M.q.q.* queen in the *M.q.a.* colony consisting of 48% queens (compare Table 9.5). Campos et al. (1979) explain this observation as follows: either exogenous JH (from pollen in the provision) or increased production of JH by the bees (owing to the larger amount of food) in conjunction with a heterozygous nature of only one of the two genes will produce a second batch of 25% queens, the first batch being the bees having the double heterozygous genome. If the categories of bees heterozygous for only one genome

TABLE 9.5. The Effect of Queen Exchange Between Colonies of *Melipona quadrifasciata quadrifasciata (M.q.q.)* and *M. q. anthidioides (m.q.a.)* on Caste of Female Offspring

	Before exchange of queens		After exchange of queens	
Workers of subspecies	Number of females	Percentage queens	Number of females	Percentage queens
M.q.q.	53	26	200	15
M.q.a.	25	28	199	48
M.q.q.	100	27	100	20
M.q.a.	100	22	100	23
M.q.q.	100	16	100	22
M.q.a.	100	18	100	19
M.q.q.	200	7.5	200	8
M.q.a.	200	10	200	16

SOURCE: After da Silva (1973).

($X_a^1\ X_a^2$ combinations and $X_b^1\ X_b^2$ combinations) were equally sensitive to this additional exogenous stimulus, a proportion of 75% queens might be expected. This was not found to be the case, however. Therefore the present model was designed in such a way that four sensitivity levels were distinguished: (1) 25% queens if only the genome $X_a^1\ X_a^2/X_b^1\ X_b^2$ develops into queens; (2) 50% queens if an additional group ($X_a^1\ X_a^2/X_b$ homozygous) develops into queens; (3) 75% queens if the other larvae being homozygous for one of the two genes can become queens too (X_a homozygous $X_b^1\ X_b^2$); and (4) 100% queens when the double-homozygous bees received enough exogenous JH to compensate for their deficiency. It should be remembered that according to this concept every increase over 25% queens is considered to be the product of unnatural conditions.

In our opinion, another explanation for the aforementioned results obtained by da Silva (1973) in the queen exchange experiment can be given. The subspecies *M.q.a.* occurs in the Ribeirão Preto area, whereas the colonies of *M.q.q.*, occurring only in distant populations, have been imported. Suppose the subspecies would have two indentical alleles for one of the *X* genes and partly different ones (X_b^1, X_b^2, X_b^3) for the other. It could be possible then that the M.q.q. queens, when mated in the experimental area would have exclusively heterozygote offspring for this gene. Queen: worker segregation is 1 : 1.

If the concept of the genetic mechanism is modified in the way Campos et al. (1979) propose, the model also implicitly explains the experiments of Darchen and Delage-Darchen (1974, 1975, 1977) on caste determination in *M. beecheii.* The clear distinction we made at the beginning of this section between females that have lost the developmental flexibility to become a queen and those that have retained this flexibility is somewhat eroded by the adoption of four different levels of sensitivity whose expression is further modified by the amount of food ingested, the availability of JH precursor, and the effects of JH on the genes regulating the expression of queen characters. The model almost approaches the hypothesis of Darchen and Delage-Darchen, according to which, in *Melipona* as in all other bees, only extrinsic factors govern the number of queens that are produced. At any rate, it is becoming difficult to discriminate between the two theories on the basis of obtained results. Conclusive experiments appear difficult to design, unless they reveal the Mendelian inheritance of any specific sensitivity level. The need for such kinds of experiment has been stressed previously (Velthuis and Velthuis-Kluppell, 1975; Campos, 1975; Velthuis, 1976).

9.4.3. *Trophic Regulation of Caste in* Melipona

Darchen hypothesized that there would be no genetic mechanism for caste in *Melipona* but that every female larva could become a queen if food conditions so permitted. In experiments with *M. beecheii,* Darchen and Delage-Darchen (1974, 1975, 1977) attempted to support this view, but their original presentations of the results were received critically (Velthuis, 1976). In their 1977 paper, more quantitative information was presented to support their statement that extra food provided to a developing larva increases the probability of its becoming a queen, even above the 25% level. In one series of overfed larvae 42 queens and 31 workers were obtained.

Darchen and Delage-Darchen (1977) correctly distinguish between quantitative and qualitative properties of the food. They assume that the liquid portion, which is probably consumed first, contains a qualitatively active factor comparable to the "determinator" in the royal jelly of the honey bee, whereas the quantitative aspect is above all in the amount of sedimented pollen. The combination of these two elements allows for the production of small queens and large workers without the intervention of a genetic regulating system.

It appears that the regulation of caste in *Melipona* remains an open question. We still do not agree with Brian (1985), who states: "in neither wasps nor bees is there yet any evidence that the egg is caste-biased and determination depends on a switch of the food supply."

9.4.4. *A New Model for the Genetic Determination of Caste in* Melipona

If we want further to explore the genetic impact on caste determination, a hypothesis that slightly differs from the original one proposed by Kerr et al. (1975) can be envisaged that will serve as an alternative to the extended model proposed by Campos et al. (1979). This alternative model includes the interaction of two independent genetic factors with food conditions (Fig. 9.2).

In this model, the first factor concerns the genetic system that regulates development in the young larva. This development could be related to the rate of food intake, which is either high or low, for instance, as the result of differences in the quality or amount of an enzyme. The enzyme system is less efficient if alleles are homozygous than if they are heterozygous. The heterozygous condition promotes queen development; the homozygous condition, worker development. However, both queen and worker disposition are not irreversibly fixed. They can still be modified later in life.

A second gene, being fully independent of the first, operates at a later larval phase. It regulates the production of JH at either a high or a low level. This gene is again, in the same way, more effective in its heterozygous form than in its homozygous constellation. Two alleles are envisaged. The physiological translation of this genetic system can have

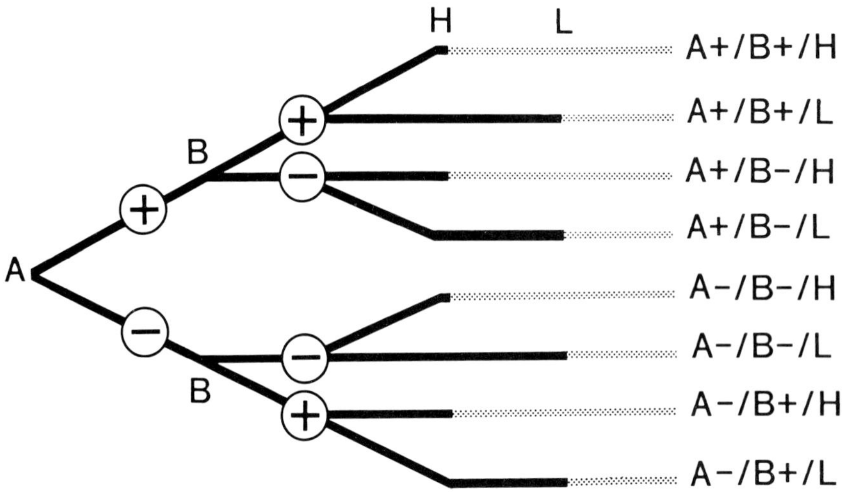

FIGURE 9.2. Diagrammatic representation of the alternative model for caste determination in the genus *Melipona:* A and B are genetic mechanisms operating early and later in preimaginal development, respectively; H and L are the food conditions as a larva experiences it, separated into high and low. See the text for further explanation.

various forms; one of them could be that the genes lead to the production of an enzyme such as a JH synthetase.

A set of two genes operating in this way allows for the existence of four classes of animals: A^+/B^+, A^+/B^-, A^-/B^-, and A^-/B^+. In connection with the further development of these four classes, a sensor mechanism is assumed that measures the amount of food ingested in the course of larval development. For the purpose of our discussion we distinguish between large and small amounts, but probably a continuous distribution exists in the amount of food ingested.

The combination of two sets of genes and two classes of food amounts leads to eight categories. An ideal developmental programme for a *Melipona* queen is $A^+/B^+/H$, for a worker bee $A^-/B^-/L$. In the latter case the lower level of JH is considered to be an adaptation to inferior food conditions. If the hormonal activity were higher, a higher metabolic activity would result, which would in turn lead to reduced body weight in the adult; this could be disadvantageous, if not fatal ($A^-/B^+/L$). However, if food amounts were higher ($A^-/B^-/H$), a larger worker or a worker that had some reserves could develop. In this part of Fig. 9.2 the category $A^-/B^+/H$ represents the larvae that are the nearest to queen development, although they still will develop into a firm worker only.

In the upper part of Fig. 9.2, A^+ animals are predestined to become queens but B^- levels preclude this in half of these larvae. By consequence all A^+/B^- animals become workers. The $A^+/B^+/L$ will become workers too, so only the $A^+/B^+/H$ type will develop into queens.

The action of B^+ is designed to be the most influential, because the artificial application of JH has stronger effects than those of increasing the amount of food. It is to be expected, however, that the various categories respond differently to such experimental treatments, and in addition to inducing queen development the effect of JH application on mortality should be considered.

The model is consistent with the observation that under natural conditions the maximum production of queens is 25%. This occurs when food conditions are excellent and, as a consequence, no L categories are present. More frequently the percentages of queens under natural conditions are lower and correspond to the ratio of H to L in the A^+/B^+ category.

9.4.5. *Consistency of the Model with Experimental Results*

The foregoing model, although highly speculative, could explain interesting aspects of some undiscussed results published by Campos et al. (1979). We shall consider these aspects in this subsection.

The model explains why in some experiments 50% of the larvae became queens. The first example is from Velthuis and Velthuis-Kluppell (1975) and detailed in Velthuis (1976). Table 9.6 is derived from the latter. We note that in all treatments the duration of larval development varied; this is supposed to reflect the action of the genes A. In the high JH treatment the effects of A^- (and of L) are apparently compensated. In the lower dosage probably only L was compensated, so the 1:1 ratio can be attributed to the effect of A^+ and A^-, respectively. In the control series, 3 queens were obtained, possibly from the 12 larvae having a short larval stage (i.e., 25%); the other larvae, supposed to be A^-, could therefore only develop into workers.

The occurrence of 50% queens in experimental situations is a central issue in the study by Campos et al. (1979), because 50% is expected in one of their models as a response to a moderate level of JH in two out of the four of their postulated genome types. As explained in our model, the application of a moderate dose of JH will likewise lead to 50% queens (all A^+ will become queens if food conditions do not prevent this), but the explanation is different. A first case is derived from their Fig. 1. That figure combines the results of JH applications by Velthuis and Velthuis-Kluppell (1975), Campos et al. (1975), Campos (unpublished data), and Velthuis (1976). The data are arranged according to the ratio of queens to workers. This arrangement suggests that with increasing JH dosage a clustering of ratios is obtained (0–25%, ±50%, ±75%, and 100%). Table

TABLE 9.6. The Relation Between the Duration of Larval Development and the Effect of JH Application on Caste Development in *Melipona quadrifasciata*

	Duration of larval development (days)				Caste of resulting adults	
Treatment	4	5	6	7	Queen	Worker
2×10^{-3} μg JH						
Grafted on:						
4/24/74		2	6	1	9	
4/25/74		7	1		8	
4/26/74	1	1	3		5	
Becoming queen	1	10	10	1	22	
2×10^{-7} μg JH						
Grafted on:						
4/24/74	1		6	1	1	7
4/25/74		8			6	2
4/26/74	2	3	1		4	2
Becoming queen	2	8	1		11	
Becoming worker	1	3	6	1		11
Acetone only						
Grafted on:						
4/24/74		1	5	2		8
4/25/74		7			2	5
4/26/74	2	2	1		1	4
Becoming queen		3			3	
Becoming worker	2	7	6	2		17

SOURCE: From Velthuis (1976).

9.7 is a version of that figure which also includes the mortality. Several interesting conclusions can be drawn in addition to those obtained by Campos et al. (1979):

(1) All control series of these experiments have, as expected, less than 25% queens, whereas all experimental series have more than 25% queens (ratio of queens to workers only). If we separate the 16 control

TABLE 9.7. Mortality as a Function of Queen Determination Among Surviving Larvae in the Control Series (Queen Determination Ranging from 0 to 20%) and in the Series Treated with a JH Preparation (Queen Determination Always Above 20%)

Effect of treatment	Degree of queen determination in surviving larvae between: 0–10%	11–20%	21–30%	31–70%	71–99%	100%
Number of larvae	117	156	0	78	50	73
Emerging—						
workers	95	120		40	5	
queens	1	22		37	32	73
Dead larvae	21	14		1	13	

SOURCE: Data derived from Campos et al. (1979, Fig. 1).

series into two groups as in Table 9.7, we note that the percentage of workers (from all larvae studied) is about the same in each group (81% and 77%, respectively) and that there are major differences between the mortality and queen numbers ($\chi^2 = 18.7$). The extremely low percentage of queens in the first column is probably due to predominantly *L* food conditions, whereas in column 2 this factor is less influential. Two explanations are possible: (1) the higher mortality in column 1 concerns specifically the bees with an A^+/B^+ genotype; (2) this higher mortality concerns specifically the A^-/B^+ genotype, the A^+/B^+ becoming workers. In the experiments of column 2 we can assume that food conditions were better (because more queens were obtained); now the A^-/B^+ survive and become workers, while the A^+/B^+ become queens. We prefer, therefore, the second explanation, because it gives further insight into the other details of Table 9.7.

(2) In the central part of Table 9.7, between 30% and 70% queen determination, mortality is extremely low; but with a further increase in determination, mortality increases again (71–99%). Along with this increase, the JH becomes more effective. Mortality appears to affect the worker genome specifically. On average, in the 30–70% region, the determination is of course 1:1. We interpret this to be A^+ vs. A^-. At the 71–99% level all A^+ bees and $A^-/B^+/H$ will become queens (5/8), all

$A^-/B^-/H$ will become workers (1/8), and both $A^-/B^-/L$ and $A^-/B^+/L$ will die (2/8). For these experiments this leads us to the following expectation: 31 queens, 6 workers, and 13 dead larvae—almost exactly the numbers found. In the last column, where determination is given as 100%, the food condition was probably much better because so many queens emerged and no workers appeared.

At this stage, the consequences of the different experimental methods applied by Campos, on the one hand, and Velthuis and Velthuis-Kluppell, on the other, need to be discussed. As already mentioned, Campos took larvae from their natural cells after they had finished their intake of food, whereas Velthuis and Velthuis-Kluppell adopted the method developed by de Camargo (1972), involving the use of artificial cells into which predetermined quantities of food and a first-instar larva had been transferred. The method of Campos precludes an estimation of the amount of food ingested. A further disadvantage could be that the few larvae that were probably left behind in the comb were of a special genotype, such as our $A^-/B^-/H$ larvae, which are expected to have a slower development because of A^-/B^- and need extra time because of H. On the other hand, transfer of food might disturb any pattern of sedimentation of pollen in it, so the young larva in its artificial cell would have to digest its pollen earlier than in the natural condition (the embryonic development of *Melipona* takes 5 days; in this time the pollen in the food settles so that the young larva probably starts consuming food that is mainly in a liquid form). This disturbance could affect the A^+ animals.

A further difference is the position in which the larvae pupate and complete their development. In Campos's experiments, the animals were placed in petri dishes and were consequently in an unnatural, horizontal position, whereas those in artificial cells were in the natural upright position during and after pupation. However, it is not known whether this has any significance, and the effects can only be of minor importance considering the series in which mortality was fully absent.

(3) In a further experiment, Campos et al. (1979) investigated whether, with slowly increasing JH amounts between series, results could be obtained around the predicted 50% determination. For this purpose, *Hyalophora cecropia* JH was administered in a range of dosages between 0.043 and 0.647 μg per bee. Their data are summarized in Figs. 9.3 and 9.4. From Fig. 9.3 we see that in the control series only 4% queens were obtained. This is again a rather low percentage; as in the previously discussed case (Table 9.7, controls), the low percentage goes with 75% workers. The explanation is the same: it is the A^-/B^+ bees that die, whereas almost all A^+/B^+ become workers.

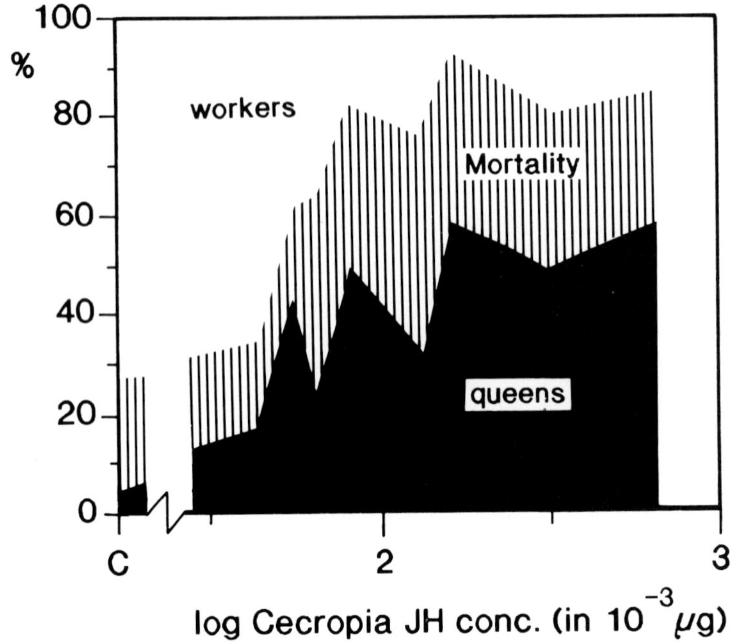

FIGURE 9.3. The relation between JH concentrations and the partition of treated larvae into surviving workers or queens and deceased larvae: C = control treatment. (Data derived from Campos et al., 1979, Table II.)

With the lowest doses of JH, 0.043 and 0.054 μg, mortality becomes somewhat lower (curve C in Fig. 9.4), but a notable increase in the percentage of queens has been found. As doses are increased, still further queen induction reaches a plateau between 70% and 90% of the surviving adults (curve A), which corresponds to 50–60% of the treated larvae (curve B). From 0.081μg onward, the total number of 251 treated larvae divide into 138 queens, 39 workers, and 74 dead individuals. This fits in well with a 4:1:2 distribution (which would yield 143, 36, 72 individuals, respectively; $\chi^2 = 0.5$). This distribution indicates that, as already explained, one of the eight types from our model probably has disappeared from the sample, namely the $A^-/B^-/H$ type. From the remaining seven types, the four having A^+ are expected to become queens, since B^- is compensated for by the exogenous JH. Workers are supposed to derive from $A^-/B^+/H$, where the increased JH compensates for A^-, whereas $A^-/B^+/L$ and $A^-/B^-/L$ groups have probably died.

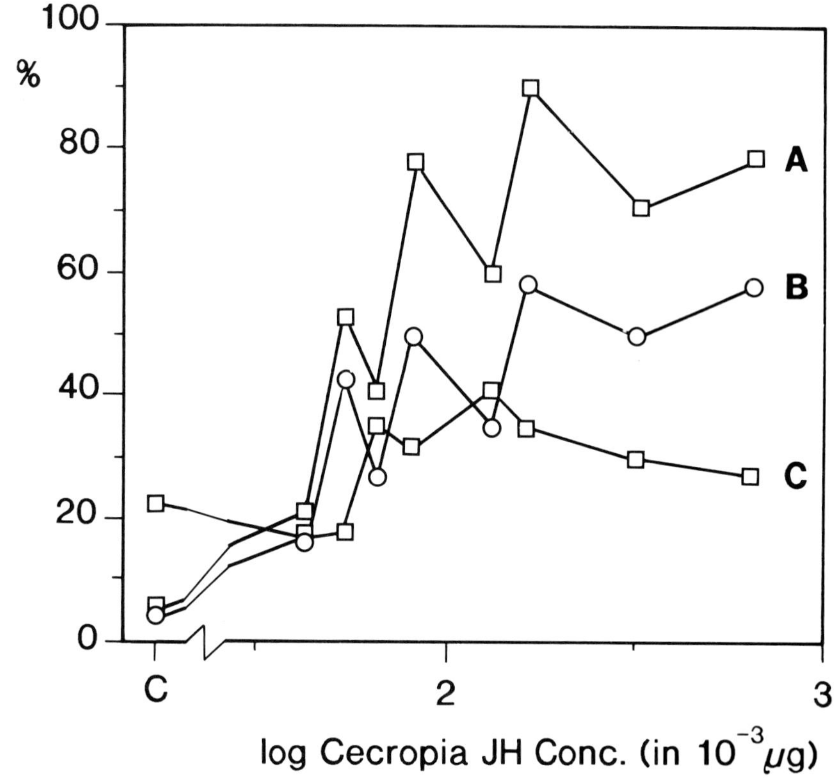

FIGURE 9.4. The effect of various JH concentrations on populations of larvae: (A) the percentage of queens among surviving larvae; (B) the percentage of queens among treated larvae; (C) mortality of larvae as related to JH concentration. Note that curve B shows a continuous increase, while A reaches a plateau. This is due to C having its maximum at 129×10^{-3} μg JH/larva. (Data derived from Campos et al., 1979, Table II/III.)

9.5. Natural Regulation of Larval Provisions

9.5.1. *Collection, Processing, and Storage of Food*

The division of labor for food collection has been studied only preliminarily in a few species, e.g., in *Melipona quadrifasciata* (Kerr and dos Santos Neto, 1956) and *M. favosa* (Sommeijer, 1984). In these species, it was found that foraging is generally carried out by the older workers. From our study, we concluded that there is no age-dependent division of labor within the collection of the various items—the food products, plant resin, and water (Sommeijer et al., 1985).

M. favosa workers returning with nectar expose a drop between their mandibles and perform rapid movements provoking as many interactions as possible with surrounding workers. The returnee continuously offers food to her nestmates. It is only in this context that spontaneous food offering occurs. After this, she proceeds to an open nectar storage pot and discharges her load into this pot by a contraction of her abdomen. As in the honey bee, house bees dehydrate the nectar by taking it up from the pots. Sealed storage pots containing honey accumulated during periods of nectar flow serve as a stable basis for a continuous production of brood.

Returning pollen foragers also demonstrate excited behavior. After a few minutes of intense contact with surrounding bees, the pollen-loaded bee proceeds to a pollen storage pot and unloads. The pollen lumps are masticated with the mandibles and molded into the pollen mass by other bees. Within the pollen pot a fermentation process takes place that gives the pollen store its characteristic acid smell. Unfermented pollen is unsuitable for brood rearing. The uptake of fermented pollen from the storage pots is carried out by workers with a wide age range (Sommeijer, 1984).

9.5.2. *Social Interactions Between Adult Bees in Relation to the Distribution of Food in the Colony*

Trophallaxis, the exchange of alimentary liquid among members of a colony, has been little studied in stingless bees. Most of these studies pertain to the trophalactic transfer of food between the queen and the workers (Sakagami et al., 1977; Sommeijer et al., 1982; Sommeijer, 1985). Having noticed that a worker regurgitates food more often than once in the provisioning process of a single brood cell, we studied trophallaxactic behavior among workers of age-marked colonies of *M. favosa* (Sommeijer et al., 1984). House bees of *M. favosa* never offer food spontaneously to nestmates. All transfers between these nonforagers are initiated by the soliciting behavior of the receiver. Cell-provisioning bees of *M. favosa* obtain most of the larval food that is to be released into the brood cell from other bees. The bees that discharge on a particular day are not involved in the pollen uptake from food pots on that day. The dischargers trophallactically obtain a liquid food, probably with a high pollen content. Among house bees a group of food-preparing bees, next to the dischargers, has been distinguished (Fig. 9.5). During the mass-provisioning process of a cell, discharging *M. favosa* workers perform active food solicitations immediately after they have discharged. They often depart quickly from the cell that is being filled and solicit from bees at some distance from the cell. Because they are so successful in their soliciting, they are often able to perform more than one regurgitation in a single cell. Under favorable conditions, when more individuals participate, the number of discharges per cell is lower. Under these conditions the actual amount of larval food in the cell may be higher. Kerr et al. (1966) found a higher frequency of queens if the number of regurgitations is lower.

9.5.3. *Cell Provisioning in* M. favosa

Once the construction of a cell is completed, there are always a few workers that are already performing brief body insertions. After the queen has arrived, she may drum the inserting worker on her dorsal body parts, leading to a distinct contraction of the abdomen. This indicates the first discharge of larval food. Successive discharges by other bees follow immediately, resulting in the filling of the cell. In between discharges, there are also body insertions without contractions. These mere insertions may become especially frequent at the end of the provisioning. We suppose that food is regurgitated from the stomach during the contractions; whether a qualitatively different food supply is deposited during insertions without contractions is not known. In a study in which 152 provisioning processes were analyzed, an average number of 11.4 discharges by an average of 8.5 regurgitating workers was found (Sommeijer et al., 1982).

In *M. favosa,* there is a strong link at the individual level between building activity at a brood cell and subsequent provisioning of this cell. All bees that participate in provisioning of the cell have previously worked on construction of that cell. There are also bees that only build. Construction and provisioning of brood cells and operculation of provisioned cells are mostly performed by bees at an age of 8–12 days. Workers oviposit when they are between 9 and 27 days of age, but most eggs are laid by 15- to 16-day-old workers (Sommeijer, 1984). Apparently, the highly social nature of cell construction and provisioning is the mere result of a tight overlap of behavioral programs in the workers reminiscent of those of solitary bees.

9.5.4. *Interspecific Differences and Mechanisms in the Pattern of Cell Provisioning*

In a series of publications by Sakagami and associates from 1963 onward (reviewed by Sakagami and Zucchi, 1974), the brood cell provisioning and oviposition process has been described in detail for more than 20 species of stingless bees. These authors divided cell construction into three types: (1) *successive*—various cell stages are found at one time; (2)

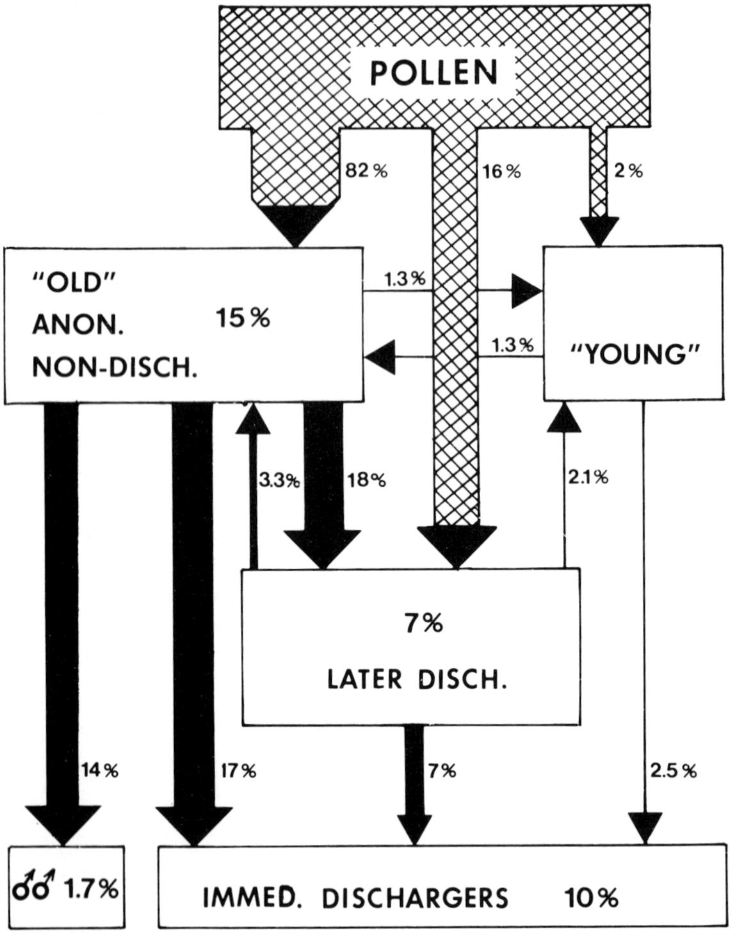

FIGURE 9.5. The social food flow in a colony of *M. favosa*. The direction of the food flow is indicated by the arrows. Their width indicates the number of transfers from bees of one category to bees of another in percentages of the total number of 253 transfers. (Food flows of less than 1% are omitted for reason of clarity. The percentages shown within the group blocks represent the number of transfers between individuals of the same groups.) (From Sommeijer et al., 1984.) The following categories are taken separately: (A) *immed. dischargers,* discharged in the provisioning and oviposition phase (POP) which immediately preceded or immediately succeeded the recorded transfer; (B) *later disch.,* bees observed to discharge later but who at the moment of transfer were older than the youngest observed dischargers (4 days, in this series); (C) *"young,"* bees who were younger than the youngest observed dischargers; (D) *"old," anon., non-disch.,* bees older than the

synchronous—cells are started synchronously, and those found at any time are all at the same stage; (3) *semisynchronous*—cells tend to be started successively, but differences gradually become smaller. The species could also be classified on the basis of the typical succession pattern of their ovipositions, e.g., exclusively batched, facultatively batched, or predominantly singular.

The number of food discharges per cell varies according to the general condition of the colony. Colonies with optimum food condition and a continuous supply of young bees have fewer regurgitations per cell than colonies under adverse conditions (Sommeijer et al., 1982).

There are very distinct interspecific differences concerning the number of regurgitations per cell. In *T. nigra* an average number of only 3.5 discharges fill the cell. As in other species with extended abdomens and few releases per cell, *T. nigra* nurses do not lay eggs (Sommeijer et al., 1985). In most species of *Melipona,* especially in those where worker ovipositions are frequent, about four or five times this number of food discharges are found. It is clear that such differences have their impact on the variability of the food inside the cell, and consequently on the extent in which trophic conditions act as a randomly fluctuating factor upon the development of female larvae.

9.6. Concluding Remarks

With regard to the hormonal regulation of caste determination, the Meliponini have been studied less profoundly than the other social insects. In a recent comparative treatise (Watson et al., 1985), they are hardly mentioned at all. From the present survey we can conclude that there are no indications that caste formation in stingless bees is of a unique nature. However, a genetic background rendering, under natural conditions, 75% of the females of *Melipona* unable to become a queen can still not be rejected.

Further research is needed to clarify a number of issues. Some questions concerning regulatory mechanisms at the individual level relate to the following matters:

---

oldest dischargers, together with unidentified bees, and bees who no longer participated in discharging but who were between the ages of youngest and oldest observed dischargers; (E) ♂♂, males. The participation of bees of these categories in the uptake of pollen from the storage pots is indicated by hatched arrows, according to the same convention (n = 186 visits to the pollen pots).

(1) The effect of stratification of the food in the cell should be investigated. As mentioned several times in this paper, *in vitro* experiments in which the stratification had been disturbed often yielded lower percentages of queens than expected. Hartfelder (1986), although offering a fourfold amount of food to *S. depilis* larvae, obtained queens which, upon emergence, on the average were only two-thirds of the weight of queens that were obtained from natural colonies.

(2) Variation in the composition of the food and its consequences for growth and development have been insufficiently studied. Data on the rate of growth in the different larval stages, published by da Cruz Landim (1983, on *M. quadrifasciata*), da Cruz Landim and Mello (1981, on *S. postica* workers), and Hartfelder (1986, on workers and *in vitro* queens of *S. depilis*) give no uniform impression of the gain in weight; the highest rate of increase is reported for the second, first, fourth, and second instar, respectively. Velthuis (1976) analyzed the effects on caste differentiation of a treatment with low concentrations of JH and concluded that the rate of development was related to sensitivity to JH. Additionally, the role of hypopharyngeal gland secretion needs further elucidation. Is it the royal jelly of stingless bees (Darchen and Delage-Darchen, 1977) enabling the happy few larvae that have received that factor in large quantity to become queens? Is it an enzyme facilitating the digestion of pollen or of carbohydrates? Or is it the protein that boosts all larvae through the first instars?

(3) The relation between colony conditions and the average quantity and composition of food in the cell should be analyzed. What is the significance of the number of workers provisioning a brood cell for the quality and composition of the food?

(4) Since in *Melipona* the rate at which queens emerge varies with the season, one might well expect seasonal variation in the food. What is the nature of this variation, and is it related to the number of provisioning bees? Through such a variable factor several extrinsic factors may have a combined influence: ecological conditions, the size of the colony determining the number of nurse bees, and the reproductive capacity of the queen (Sommeijer, 1984). In other cases, such as in *Scaptotrigona,* the same combination of factors could determine the occurrence of worker ovipositions leading to male production (Beig, 1972), instead of the production of trophic eggs only.

9.7. Summary

In this chapter our present knowledge on caste development in the stingless bees has been reviewed. Development of the queen caste requires that food present in the queen cells be more abundant than that in the cells from which workers or males arise. Additionally, in the genus *Melipona,* a genetic predisposition to become a queen in a part of the larvae should not be excluded.

First, in the Trigonini, there are the specific queen cells; these are several times larger than worker cells and contain at least three times as much food as is encountered in worker cells. Second, in bees that construct brood cells in a loose cluster arrangement rather than in a regular comb, such as *Leurotrigona* and *Frieseomelitta,* two cells may be constructed contiguously, allowing one larva to pierce the common wall and to obtain twice the normal ration of food, thereby becoming a queen rather than a worker. Third, in *Melipona,* where no special queen cells occur, queen development of predisposed larvae depends solely on variation in the amount of food deposited in the cell. Especially in this case, the regulatory mechanisms involved in cell provisioning are crucial in our understanding of why, during the year, queens are produced more frequently at certain times than at others.

Mass provisioning of a brood cell, characteristic of the Meliponinae, has several components of highly interactive sociality. Therefore, at the colony level, these social mechanisms regulate the food quantity to be deposited in a cell. Probably, also the fluctuations around an average composition of the food are determined in this way. We suggested that these fluctuations could play a more decisive role in caste differentiation. Related to this is the function of the hypopharyngeal gland secretion. In species where this constitutes only a few percent of the total amount of food, this secretion could function as an enzyme, facilitating digestion of the food, rather than as an alimentary food component.

Queen caste differentiation becomes apparent by the increased size and activity of the CA. In a mass-provisioning system, where amounts of food are most important, such differentiation in CA becomes detectable only at the end of larval life. In Trigonini, such as *Scaptotrigona,* having large queen cells, this may be somewhat earlier than in *Melipona.* Related to this is the sensitivity of worker larvae, at the end of larval life, to topical applications of JH or analogues. The experiments, mainly on *Melipona* species, on these caste inductions were fully discussed.

For *Melipona,* Kerr (1948, 1950) proposed a genetic model explaining the frequent occurrence of 25% queens in periods with favorable food conditions. *In vitro* experiments support this high frequency of queens. The model has been modified since then to include new information. In this review another version of the model was described that differs slightly from the one proposed by Kerr et al. (1975). In this model the genetic separation into individuals that can and those that cannot respond to food conditions favoring queen development is interpreted as a genetic predisposition of workers as well, in that they can cope in this way with unfavorable food conditions. Such conditions could lead to mortality in larvae predisposed to become queens. The new model is supported by evidence from the existing literature.

Acknowledgments

We are greatly indebted to Dr. Klaus Hartfelder and Prof. Wolf Engels for providing us with their unpublished results. They, Prof. Warwick E. Kerr, and Dr. Joop Beetsma commented on the original manuscript and were very instrumental in improving that draft. Anita van Vliet did much of the typing; Drs. Jan van der Blom prepared the graphs and Mrs. S. M. McNab, M.A., advised us in linguistic matters. To all of them we are very grateful.

References

Beig, D. 1971. Desenvolvimento embrionario de abelhas operárias de *Trigona* (*Scaptotrigona*) *postica* Latreille (Hymenoptera, Meliponinae). Arq. Zool. (São Paulo) 21: 179–234.

Beig, D. 1972. The production of males in queenright colonies of *Trigona* (*Scaptotrigona*) *postica*. J. Apic. Res. 11: 33–39.

Bonetti, A. M. 1982. Ação do hormônio juvenil sobre a expressão gênica em *Melipona* (Hymenoptera, Apidae, Meliponinae). Masters thesis, University of Ribeirão Preto, Brazil.

Brian, M. V. 1985. Comparative aspects of caste differentiation in social insects. Pp. 385–398 *in* J. A. L. Watson, B. M. Okot-Kotber, and C. Noirot (eds.), *Caste Differentiation in Social Insects*. Pergamon Press, Oxford and Elmsford, New York.

Bueno, O. C. and D. Beig. 1980. Biometric study of the corpora allata in *Scaptotrigona postica* during post-embryonic development. J. Apic. Res. 19: 219–223.

Buschinger, A. 1978. Genetisch bedingte Entstehung geflügelter Weibchen bei der sklavenhaltende Ameise *Harpagoxenus sublaevis* (Nyl.) (Hym., Form.). Insectes Soc. 25: 163–172.

Campos, L. A. [de Oliveira]. 1975. Determinacão de casta no gênero *Melipona* (Hymenoptera, Apidae): papel do hormônio juvenil. Masters thesis, University of São Paulo, Ribeirão Preto Campus, Brazil.

Campos, L. A. [de Oliveira], W. E. Kerr, and D. L. N. da Silva. 1979. Sex determination in bees. VIII. Relative action of genes X_a and X_b on sex determination in *Melipona* bees. Rev. Bras. Genet. 2(4): 267–280.

Campos, L. A. [de Oliveira], F. M. Velthuis-Kluppell, and H. H. W. Velthuis. 1975. Juvenile hormone and caste determination in a stingless bee: sex determination in bees. VII. Naturwissenschaften 62: 98–99.

da Cruz Landim, C. 1983. O corpo gorduroso da larva de *Melipona quadrifasciata anthidioides* Lep. (Apidae, Meliponinae). Naturalia (São Paulo) 8: 7–23.

da Cruz Landim, C. and R. de A. Mello. 1981. Desenvolvimento e envelhecimento de larvas e adultos de *Scaptotrigona postica* Latreille (Hymenoptera: Apidae): aspectos histológicos e histoquímicos. Publ. ACIESP (Acad. Ciências Estado São Paulo), 31: 1–118.

da Cruz Landim, C., S. M. F. dos Santos, and M. C. A. Höfling. 1980. Sex determination in bees. XV. Identification of queens of *Melipona quadrifasciata anthidioides* (Apidae) with the worker phenotype by a study of the tergal glands. Rev. Bras. Genet. 3(3): 295–302.

Darchen, R. and B. Darchen-Delage. 1971. Le déterminisme des castes chez les Trigones (Hyménoptères Apidés). Insectes Soc. 18: 121–134.

Darchen, R. and B. Darchen-Delage. 1974. Nouvelles expériences concernant le déterminisme des castes chez les Mélipones (Hyménoptères Apidés). C. R. Acad. Sci. Paris 278D: 907–910.

Darchen, R. and B. Darchen-Delage. 1975. Contribution à l'étude d'une abeille du Mexique *Melipona beecheii* B. (Hymenoptère: Apide). Apidologie 6: 295–339.

Darchen, R. and B. Darchen-Delage. 1977. Sur le déterminisme des castes chez les Mélipones (Hyménoptères Apidés). Bull. Biol. Fr. Belg. 111: 91–109.

da Silva, D. L. N. 1973. Estudos bionômicos em colônias mistas de Meliponinae (Hymenoptera, Apoidea). Ph.D. thesis, University São Paulo, Ribeirão Preto Campus, Brazil. [Published in Bol. Zool. Univ. São Paulo 2: 7–106 (1977).]

de Camargo, C. A. 1972. Aspectos da reprodução dos Apideos sociais. Masters thesis, University São Paulo, Ribeirão Preto Campus, Brazil.

de Camargo, C. A. 1976. Determinação do sexo e contrôle de reprodução em *Melipona quadrifasciata* Lep. (Hymenoptera, Apidae). Ph.D. thesis, University São Paulo, Ribeirão Preto Campus, Brazil.

de Camargo, C. A., M. G. de Almeida, M. G. N. Parra, and W. E. Kerr. 1976. Genetics of sex determination in bees. IX. Frequencies of queens and workers from larvae under controlled conditions (Hymenoptera: Apoidea). J. Kans. Entomol. Soc. 49: 120–125.

de Camargo, C. A., M. G. de Almeida, M. G. N. Parra, and W. E. Kerr. 1976. Genetics of sex determination in bees. IX. Frequencies of queens and workers from larvae under controlled conditions (Hymenoptera: Apoidea). J. Kans. Entomol. Soc. 49: 120–125.

Engel, M. S. 1979. Is caste determination in *Melipona quadrifasciata*, a stingless bee, influenced by 9-oxo-decenoic acid? Insectes Soc. 26: 273–278.

Hartfelder, K. 1986. Trophogene Basis und endokrine Reaktion in der Kastenentwicklung bei Stachellosen Bienen. Ph.D. thesis, University of Tübingen, Germany.

Hartfelder, K. 1987. Rates of juvenile hormone synthesis control caste differentiation in the stingless bee *Scaptotrigona postica depilis*. Wilhelm Roux's Arch. Dev. Biol. 196: 522–526.

Hartfelder, K. and W. Engels. 1989. The composition of larval food in stingless bees: evaluating nutritional balance by chemosystematic methods. Insectes Soc. 36: 1–14.

Imperatriz Fonseca, V. L. 1978. Studies on *Paratrigona subnuda* (Moure) Hymenoptera, Apidae, Meliponinae. III. Queen supersedure. Bol. Zool. Univ. São Paulo 3: 153–162.

Kerr, W. E. 1946. Formação de castas no gênero *Melipona* (Illiger 1806). An. Esc. Super. Agric. Luiz de Queiroz Univ. São Paulo 3: 299–312.

Kerr, W. E. 1948. Estudos sôbre o gênero *Melipona*. An. Esc. Super. Agric. Luiz de Queiroz Univ. São Paulo 5: 181–276.

Kerr, W. E. 1950. Genetic determination of castes in the genus *Melipona*. Genetics 35: 143–152.

Kerr, W. E., Y. Akahira, and C. A. de Camargo. 1975. Sex determination in bees. IV. Genetic control of juvenile hormone production in *Melipona quadrifasciata* (Apidae). Genetics 81: 749–756.

Kerr, W. E. and G. R. dos Santos Neto. 1956. Contribuição para o conhecimento da bionomia dos Meliponini. V. Divisão de trabalho entre operárias de *M. quadrifasciata* Lep. Insectes Soc. 3: 423–430.

Kerr, W. E. and R. A. Nielsen. 1966. Evidence that genetically determined *Melipona* queens can become workers. Genetics 54: 859–865.

Kerr, W. E., S. F. Sakagami, R. Zucchi, V. de Portugal Araújo, and J. M. F. de Camargo. 1967. Observações sobre a arquitetura dos ninhos e comportamento de algumas espécies de abelhas sem ferrão das vizinhanças de Manaus, Amazonas (Hymenoptera, Apoidea). Atas Simp. Biot. Amazônica 5: 255–309.

Kerr, W. E., A. C. Stort, and M. J. Montenegro. 1966. Importância de alguns fatôres ambientais na determinação das castas do gênero *Melipona*. An. Acad. Bras. Ciênc. 38: 149–168.

Lindauer, M. and W. E. Kerr. 1958. Die gegenseitige Verständigung bei den stachellosen Bienen. Z. Vgl. Physiol. 41: 405–434.

Martinho, M. R. 1975. Contribuição ao estudo da digestão do grão de pôlen em *Melipona quadrifasciata anthidioides* Lepeletier (Hymenoptera, Apidae, Meliponinae). Masters thesis, University São Paulo, Ribeirão Preto Campus, Brazil.

Michener, C. D. 1974. *The Social Behavior of the Bees*. Harvard University Press, Cambridge, Massachusetts.

Moure, J. S. 1951. Notas sôbre Meliponinae (Hymenopt. Apoidea). Dusenia 2: 25–70.

Moure, J. S. 1961. A preliminary supra-specific classification of the Old World meliponine bees (Hym., Apoidea). Stud. Entomol. 4: 181–242.

Perez, J. 1895. Sur le production des femelles et des mâles chez les Meliponides. C. R. Acad. Sci. Paris 120: 273–275.

Roubik, D. W. 1983. Nest and colony characteristics of stingless bees from Panama (Hym., Apidae). J. Kans. Entomol. Soc. 56: 327–355.

Ruttner, F. 1986. Geographical variability and classification. Pp. 23–56 *in* T. E. Rinderer (ed.), *Bee Genetics and Breeding*. Academic Press, Orlando, Florida.

Sakagami, S. F. 1982. Stingless bees. Pp. 361–423 *in* H. R. Hermann (ed.), *Social Insects*, Vol. 3. Academic Press, Orlando, Florida.

Sakagami, S. F., M. J. Montenegro, and W. E. Kerr. 1965. Behavior studies of the stingless bee, with special reference to the oviposition process. V. *Melipona quadrifasciata anthidioides* Lepeletier. J. Fac. Sci. Hokkaido Univ. Ser. VI Zool. 15: 578–607.

Sakagami, S. F. and R. Zucchi. 1974. Oviposition behavior of two dwarf stingless bees, *Hypotrigona* (*Leurotrigona*) *muelleri* and *H.* (*Trigonisca*) *duckei*, with notes on the temporal articulation of oviposition process in stingless bees. J. Fac. Sci. Hokkaido Univ. Ser. VI Zool. 19: 364–421.

Sakagami, S. F., R. Zucchi, and V. de Portugal-Araujo. 1977. Oviposition behavior of an aberrant African stingless bee, *Meliponula bocandei*, with notes on the mechanism and evolution of oviposition in stingless bees. J. Fac. Sci. Hokkaido Univ. Ser. VI Zool. 20: 647–690.

Salmah, S., T. Inoue, P. Mardius, and S. F. Sakagami. 1987. Incubation period and post-emergence pigmentation in the Sumatran stingless bee, *Trigona* (*Trigonella*) *moorei*. Kontyû 55: 383–390.

Sasaki, M., T. Tsuruta, and S. Asada. 1987. Role of the physical property of royal jelly in queen differentiation of honeybees. Pp. 306–307 *in* J. Eder and H. Rembold (eds.), *Chemistry and Biology of Social Insects*. Verlag Peperny, Munich.

Sommeijer, M. J. 1984. Distribution of labour among workers of *Melipona favosa* F.: age-polyethism and worker oviposition. Insectes Soc. 31: 171–184.

Sommeijer, M. J. 1985. The social behavior of *Melipona favosa* F.: some aspects of the activity of the queen in the nest. J. Kans. Entomol. Soc. 58: 386–396.

Sommeijer, M. J., F. T. Beuvens, and H. J. Verbeek. 1982. Distribution of labour among workers of *Melipona favosa* F.: construction and provisioning of brood cells. Insectes Soc. 29: 222–237.

Sommeijer, M. J., L. L. M. de Bruijn, and C. van de Guchte. 1985. The social food flow within the colony of the stingless bee, *Melipona favosa* (F.). Behaviour 92: 39–58.

Sommeijer, M. J., J. L. Houtekamer, and W. Bos. 1984. Cell construction and egg-laying in *Trigona nigra* var. *paupera* Provancher, with a note on the adaptive significance of the typical oviposition behaviour of stingless bees. Insectes Soc. 31: 199–217.

Terada, Y. 1973. Contribuição ao estudo de regulação social em *Leurotrigona muelleri* e *Frieseomelitta varia* (Hymenoptera, Apidae). Masters thesis, University São Paulo, Ribeirão Preto Campus, Brazil.

Velthuis, H. H. W. 1976. Environmental, genetic and endocrine infuences in stingless bee caste determination. Pp. 35–53 *in* M. Lüscher (ed.), *Phase and Caste Determination in Insects.* Pergamon Press, Oxford and Elmsford, New York.

Velthuis, H. H. W. and F. M. Velthuis-Kluppell. 1975. Caste differentiation in a stingless bee, *Melipona quadrifasciata* Lep., influenced by juvenile hormone application. Proc. K. Ned. Akad. Wet. Ser. C Biol. Med. Sci. 78: 81–94.

von Ihering, H. 1903. Biologie der stachellosen Honigbienen Brasiliens. Zool. Jahrb. 19: 179–287.

Watson, J. A. L., B. M. Okot-Kotber, and C. Noirot (eds.) 1985. *Caste Differentiation in Social Insects.* Pergamon Press, Oxford and Elmsford, New York.

Wille, A. 1979. Phylogeny and relationships among the genera and subgenera of the stingless bees (Meliponinae) of the world. Rev. Biol. Trop. 27: 241–277.

Wille, A. 1983. Biology of the stingless bees. Annu. Rev. Entomol. 28: 41–64.

Wille, A. and C. D. Michener, 1973. The nest architecture of stingless bees with special reference to those of Costa Rica. Rev. Biol. Trop. 21(Suppl. 1): 1–278.

Winston, M. and C. D. Michener, 1977. Dual origin of highly social behavior among bees. Proc. Natl. Acad. Sci. USA 74: 1135–1137.

10

Roles of Morphogenetic Hormones in Caste Polymorphism in Bumble Bees

PETER-FRANK RÖSELER

10.1. Introduction

Bumble bees together with honey bees and stingless bees belong to the family Apidae. The tribe Bombini includes two genera: the primitively social *Bombus* species and the socially parasitic *Psithyrus* species. Both genera are subdivided into several subgenera. More than 200 species are known. Most bumble bees occur in temperate regions of the Holarctic; only a few species are found in the tropics of South Asia and South America.

In temperate regions, bumble bees form annual colonies with only one generation a year (Free and Butler, 1959; Alford, 1975; Heinrich, 1979). Nests are started in the spring by single overwintered queens. During the preemergence period before the appearance of the first workers, the queen must perform all duties. The first brood consists of only workers that help the queen in brood rearing. They feed the larvae, stabilize nest temperature, defend the nest, and perform the energetically expensive and dangerous foraging tasks. When workers are present, the queen gradually gives up these tasks and then never leaves the nest. Since her activities are mainly limited to constructing egg cells, oviposition, and feeding young larvae, egg laying is accelerated to the queen's maximum capacity. The growth of the colony becomes exponential. After a phase of worker production, new reproductives—queens and drones—are reared. Shortly after copulation, young queens enter hibernation. The old queen dies, and the colony perishes. Only young queens hibernate and start a new colony in the next spring.

The colony cycle is adapted to diverse climates. In the short summer in the Arctic region, only the first eggs develop into workers; immediately thereafter, reproductives are reared (Richards, 1973). In the tropics, by contrast, perennial colonies exist (Sakagami, 1976). Some young mated queens rejoin the parental nest and start egg laying. They form a polygynous nest that gradually becomes monogynic through elimination of odd queens.

Psithyrus species are all social parasites; they have no worker caste. The females invade *Bombus* nests and supersede or kill the queen, so that the *Bombus* workers exclusively rear the offspring of the parasite.

Caste differentiation in bumble bees includes the determination into queens or workers during preimaginal development as well as regulation of caste-specific functions and social behavior during adult life. The annual colony cycle requires the early rearing of reproductives, i.e., before the end of the season. The adults must be able to regulate the development of the larvae, which must sensitively respond to the signals at the right time. Social life of adults must be regulated according to the needs of the colony. Competition for reproduction must be suppressed to make the division of labor possible. In the last 15 years, several reviews have appeared on various aspects of caste differentiation, including that in bumble bees (P.-F. Röseler, 1975, 1981; Brian, 1980; Nijhout and Wheeler, 1982; de Wilde and Beetsma, 1982; Plowright and Laverty, 1984; Fletcher and Ross, 1985; de Wilde, 1985; Wheeler, 1986). The present review concentrates on the extent to which hormones are involved in the control of caste differentiation during the whole life span. Up to date, investigations on hormonal regulation were only performed in the two species *Bombus hypnorum* and *B. terrestris.*

10.2 Characteristics of Castes

Queens and workers show a diverse but sometimes overlapping pattern of characteristics. Morphologically, the castes differ only in size. As a rule, queens are bigger than workers (Fig. 10.1). There are species in which size is a pronounced distinguishing feature, with queens being significantly larger than workers and with no intermediates. In other species, the sizes of castes broadly overlap: a large worker can be larger than a small queen. In a few species, queens and workers also differ in coloration.

Although castes are not readily separable externally, they are distinctly defined physiologically and behaviorally. Only young queens are able to hibernate and to start new colonies the next spring. During the first days after eclosion, they accumulate fat and glycogen in the fat body, which becomes very voluminous. The formation of reserves is finished between days 5 and 10 (see Fig. 10.3 in Section 10.4, below). In the hemolymph, the lipid content also increases during that period. Moreover, queens are characterized by a specific protein fraction in the hemolymph.

A

B

FIGURE 10.1. Size polymorphism in bumble bees. (A) In *Bombus hypnorum,* there is a gradual increase in body size from workers (on the left) to queens (on the right). (B) In *Bombus terrestris,* there is a marked dimorphism in body size, the three queens (on the right) are distinctly larger than workers.

Both castes also differ in behavior during the first days of adult life. From day 2 onward, workers are involved in feeding larvae and constructing the nest. At that age, they can even start collecting food. Young queens, in contrast, do not participate in the feeding and nest-building duties. They do not even defend the nest when it is disturbed, but rather try to escape beneath the combs or into the surrounding material. After day 3, the queens leave the nest. They are attracted by the pheromone-marked places along the flight routes of males and are ready for copulation. When workers are occasionally mounted by drones, they defend themselves by trying to cast off the drones or to sting them. Shortly after insemination, queens seek a suitable place for hibernation. The longevity of queens is thus greater than that of workers, which in the field can only live up to 3 weeks and in the laboratory only some 2–3 months. The queen maintains the social and reproductive dominance of the colony. She prevents workers from reproduction, so that she has the parentage of all offsprings as long as she is dominant.

Some of the attributes listed above are not exclusive to one caste. For instance, queenless egg-laying workers are also able to inhibit egg formation in other workers (P.-F. Röseler et al., 1981), and occasionally workers of *B. atratus* have been observed to copulate (Sakagami, 1976).

10.3. Differentiation of Castes During Development

Each female egg is bipotential: the larvae can develop into a queen or into a worker. There is no evidence that caste differentiation in bumble bees is under genetic control. The regime of nutrition enables the adults to regulate which caste is reared. Basically, when food is abundant, a larva develops into a queen; a suboptimally fed larva develops into a worker.

In simpler species, castes are determined only by the quantity of food (*B. pratorum:* Free, 1955; *B. rufocinctus* and *B. ternarius:* Plowright and Jay, 1968, 1976; *B. hypnorum*: P.-F. Röseler and I. Röseler, 1974). But it is not known whether the larvae perceive the amount of food or, more likely, the level of a certain component. Since bumble bees progressively feed larvae, the amount of food for larvae depends on the worker/larva ratio and, of course, on the amount of pollen stored in the nest. Larval feeding seems, moreover, to be regulated by workers, which record the food demand of larvae by inspection or by the movement of starving larvae (Pendrel and Plowright, 1981).

In *B. hypnorum,* a larva will develop into a queen at a worker/larva ratio of 1:2, whether a queen is present or not. Queen larvae of the last instar are fed at a faster rate (7 ×/h) than are worker larvae (4–5 ×/h). There are no specialized workers for feeding one caste, all the different age groups contributing to feeding to an extent corresponding to their frequency in the colony. The same worker may well change from feeding worker larvae to feeding queen larvae during one feeding sequence (Katayama, 1975). Since castes are determined by the quantity of food each individual is fed, queens of *B. hypnorum* are reared during colony development as soon as the worker/larva ratio of 1:2 is reached. Experimentally, queen production can be induced by a reduction in the ratio of larvae to workers. The sensitive period for caste determination occurs in the last instar; when larvae of prospective workers are optimally fed in that instar, they develop into queens.

In the more advanced species *B. terrestris*, caste development is not only regulated by the amount of food consumed by larvae but is additionally controlled by the queen (P.-F. Röseler, 1970). As long as she is dominant, she inhibits the development of queens, presumably via a pheromone. In the presence of a queen, larvae are irreversibly determined into workers just in the first instar, irrespective of the subsequent food regime. Even if the queen is absent during later stages, the larvae develop into workers. But when the queen is removed before hatching of the larvae, they develop into either of the castes according to the quantity of food supplied. Whereas the development of worker larvae is irreversibly determined, queen larvae remain sensitive until the last instar. When queen larvae are put in a nest with a dominant queen or treated with queen's food, they develop into adults of queen size but with the physiology of workers (P.-F. Röseler, 1976, 1977b). They do not prepare for hibernation but stay in the nest, and they are also involved in colony duties. That queen's caste control leads to large colonies in which reproductives are reared in late summer, after the queen has lost her dominance.

The different food regimes affect the activity of endocrine glands resulting in a modulation of hormone level in prepupae (Fig. 10.2) (P.-F. Röseler, 1977b; Strambi et al., 1984). In workers of *B. hypnorum,* the juvenile hormone (JH) titer rapidly increases, the peak being reached after 30–40 h. In queens, however, the rise of JH titer is delayed, the peak being after 50–60 h. Moreover, the JH level in queens is nearly twice as high as in workers. The difference between both castes can be determined by means of the *Galleria* bioassay as well as by the radioimmunoassay (RIA) method. The ecdysteroid peak in queens is later than in workers, resulting in a delayed pupal molt.

It has been experimentally confirmed that the elevated JH titer in prepupae of *B. hypnorum* induces the differentiation into a queen (P.-F. Röseler and I. Röseler, 1974; P.-F. Röseler, 1976). The first hours of prepupae are sensitive to JH treatment. By application of JH I, queen development was induced in prepupae of prospective workers when they were not older than 24 h (Fig. 10.2). The adults, of course, were of

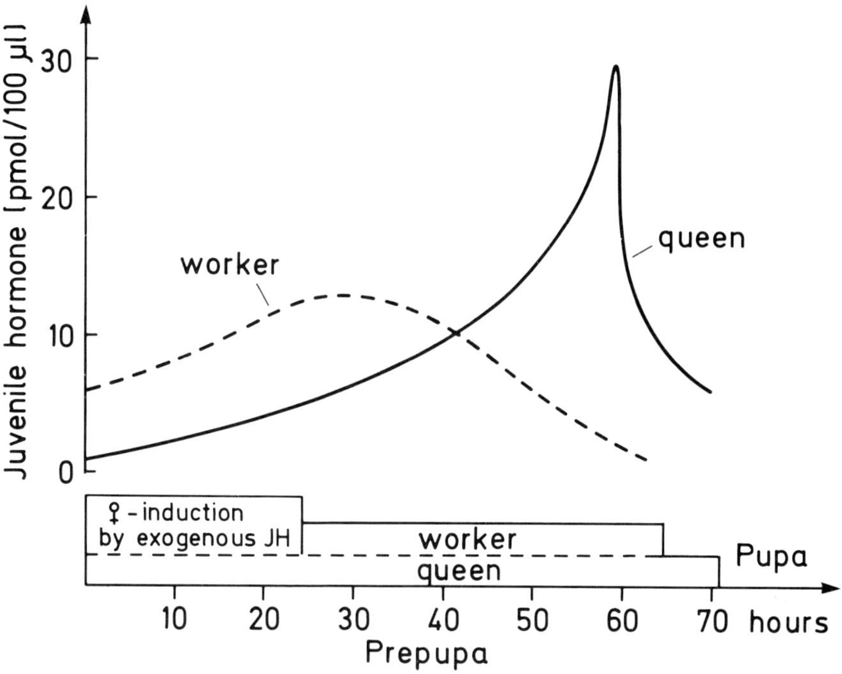

FIGURE 10.2 JH titers in hemolymph during prepupal development of queens and workers of *Bombus hypnorum*. JH was determined by radioimmunoassay. The length of the JH-sensitive period in workers is indicated. (Modified from Strambi et al., 1984.)

worker size, but they had the physiology and the behavior of a queen: they accumulated reserves in the fat body, and they did not participate in colony tasks but left the nest after some 3–5 days.

In *B. terrestris*, the temporal pattern of both hormones in prepupae is similar to that in *B. hypnorum*. In queen prepupae, the increase of hormones is delayed and with it the pupal molt. But in contrast to *B. hypnorum*, the level is nearly the same in both castes or even somewhat higher in workers. Application of JH, therefore, does not induce queen development in worker prepupae. In both species, JH III was found to be the endogenous JH.

Caste differentiation in bumble bees is regulated by a different endocrine program during the last instar. Larvae respond to changes in nutrition by modulating the endocrine activity. The exogenous signal food is transformed into an endocrine signal controlling development. Primarily, the period sensitive to nutrition occurs in the last instar. An increase in nutrition enhances the activity of the corpora allata (CA) resulting in a higher JH titer in prepupae, which induces development into queens. Starvation, on the other hand, leads to development into workers. By that late switch, small queens can develop, and these regularly appear during the colony cycle at the beginning of queen production. Such larvae had reached the last instar when the worker/larva ratio increased to the critical 1:2 level. But the early larval stages also seem to be sensitive to the amount of food consumed, for larval development of queens reared at a high worker/larva ratio from the first instar onward is shortened (P.-F. Röseler and I. Röseler, 1974; Plowright and Pendrel, 1977).

In *B. terrestris,* larvae are determined to become workers by the queen. These larvae are no longer sensitive to changes in nutrition and to JH treatment. It has not been tested whether larvae not influenced by the queen and suboptimally fed remain sensitive like conspecific queen larvae or worker larvae of *B. hypnorum*. The determination into the physiological castes also occurs during prepupal development. Both species show the same temporal pattern in endocrine activity. But so far, the influence of hormones on induction of caste characteristics remains completely unknown.

The different body sizes of queens and workers depends on the linkage between the quantity of food and caste differentiation. Only larvae that have obtained rich food will develop into queens, and thereby they also attain larger size. Whether there are phagostimulatory or phagoinhibitory factors involved regulating the intake of food by larvae is yet to be investigated.

10.4. Hormonal Regulation of the Functions of Adult Castes

Young queens are not involved in nest duties, but they do prepare for hibernation. During the first days after eclosion, their metabolism is adjusted to accumulate reserves in the fat body (Alford, 1969; Pouvreau, 1976; P.-F. Röseler and I. Röseler, 1986). The difference in metabolism between both castes can be seen in the lipid and glycogen content of the fat body. On the second day after emergence, the content of the reserves in queens starts increasing rapidly (Fig. 10.3). On day 10, the fat content of the fat body is 10-fold higher, the glycogen content 100-fold higher than in workers. The accumulation of glycogen in queens is brought about by a three- to four-fold higher activity of the enzyme UDP-glucose : glycogen 4-α-D-glycosyltransferase (EC 2.4.1.11) than in workers. Moreover, the affinity of the enzyme to UDP-glucose is twice as high as in workers. The activity of glycogen phosphorylase (EC 2.4.1.1), in contrast, is of the same magnitude in both castes. In the queen's hemolymph, the lipid content

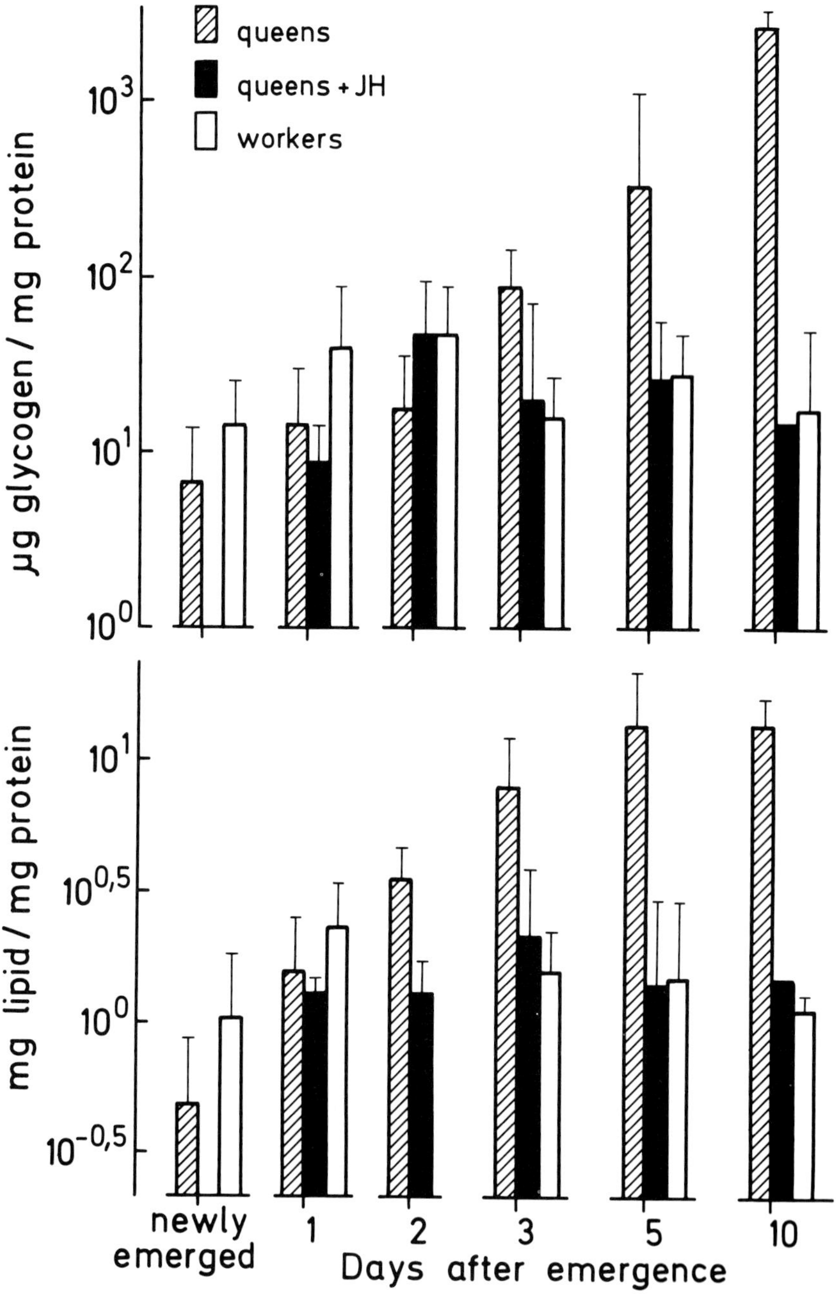

FIGURE 10.3 Glycogen content (above) and fat content (below) of fat body in untreated queens, JH-treated queens, and workers of *Bombus terrestris* during the first days after eclosion. Vertical bars indicate standard deviation (SD). (Data on glycogen from P.-F. Röseler and I. Röseler, 1986.)

also increases and a specific protein fraction occurs (I. Röseler and P.-F. Röseler, 1973; P.-F. Röseler and I. Röseler, 1974).

The physiology of young queens is based upon a specific endocrine program switched on during prepupal development. The metabolism and behavior leading to diapause are therefore not influenced by such environmental cues as daylength or temperature. The specific physiology of young queens depends on a low JH titer in hemolymph (P.-F. Röseler, 1976, 1977b; P.-F. Röseler and I. Röseler, 1986). The hormonal level in newly emerged queens was determined by *Galleria* bioassay to be <250 GU(*Galleria* units)/ml and thus 10-fold lower than in newly emerged workers (about 2700 GU/ml). In queens the titer increases to 600 GU/ml on day 3, but in workers up to 6000 GU/ml. When newly emerged queens are treated with JH I by topical application or by injection, they do not store reserves in the fat body (Fig. 10.3). Treated queens form eggs, and they can start egg laying after 1 week. Simultaneously, nesting behavior is induced. The queens stay in the nest, feed larvae, build on the comb, and occasionally collect food. But they remain ready to copulate and are attractive to drones. JH-treated queens, therefore, are not transformed into workers, as has been suggested previously, but into egg-laying queens. At present, it is not possible to induce and to maintain a worker-like level of JH in queens, all the treatments resulting in a strikingly higher JH titer in the hemolymph, triggering oogenesis.

Glycogen reserves are not stored because of the low level of glycosyltransferase. After JH treatment, the activity of that enzyme does not increase but remains on a low worker-like level. It seems unlikely however, that JH directly influences the activity of key enzymes regulating energy metabolism in the fat body. Studies on ovariolectomized queens show that JH induces synthesis of vitellogenins in the fat body, whereby the formation of reserves is no longer possible (P.-F. Röseler and I. Röseler, 1988). Vitellogenesis and accumulation of fat and glycogen seem to be alternative pathways that are mutually exclusive.

Occasionally, young queens collecting pollen are captured during late summer (Alford, 1975; I. Röseler and P.-F. Röseler, 1984). Dissection of such *B. terrestris* queens revealed that they had a small worker-like fat body and developed ovaries. The factors that inhibited preparation for diapause are not known. A possible explanation might be an increased CO_2 tension in large nests, as that gas is known to induce CA activity and egg formation (I. Röseler and P.-F. Röseler, 1984). But this explanation remains hypothetical since to date no measurements on CO_2 tension in nests have been done.

Workers and queens always show the appropriate behavior corresponding to their physiological status. Whether caste-specific behavior is also hormonally influenced or triggered by the ongoing metabolism is not yet clear. Foraging workers of unknown age captured in the field were found to have a low JH titer (3000 GU/ml). On the other hand, egg-laying workers with a high JH level do not collect food, but rather they stay in the nest in order to defend their egg batches. In colonies in which half of the newly emerged workers were treated with JH, the treated workers became foragers in the same percentage as that of untreated controls (van Doorn, 1987). There was only a slight effect on the age at which workers foraged for the first time: JH-treated workers started foraging somewhat later. Since exogenous JH is rapidly excreted (P.-F. Röseler and I. Röseler, 1978), it is not possible to maintain an elevated JH level over a long period. Probably the decreasing level of exogenous JH only delayed the start of foraging. Moreover, a possible influence of JH on foraging might be overcome by needs of the colony. It is, therefore, likely that the low hormone level in foragers is a consequence of that energy-expensive task. Another effect of JH was an increased activity. Treated workers more frequently interacted with their nestmates than did untreated workers. A similar situation was found in queenless groups. Workers injected with JH I were more likely to achieve the dominant position than were untreated workers (van Doorn, 1988). Probably, this effect is also brought about by a general increase in activity.

10.5. Control of Reproduction

In bumble bee colonies, the queen is the sole egg layer as long as she is dominant, preventing workers from reproduction. She inhibits egg formation in workers or eats the eggs laid by them at the end of colony development. In the presence of a queen, oocytes of workers slightly enlarge up to 0.15 mm on day 5 after emergence (Fig. 10.4). In queenless workers, a rapid oogenesis takes place. After day 5, the ovaries contain ripe eggs of 2.5–3.5 mm length depending on the body size (P.-F. Röseler, 1974, 1977a).

Oogenesis is controlled by JH titer in hemolymph (P.-F. Röseler, 1977a; P.-F. Röseler and I. Röseler, 1978). In newly emerged workers, the JH titer is about 2700 GU/ml. In the presence of a queen (*queenright*), the titer increases to 5000 GU/ml on day 5. In worker groups without a queen (*queenless*), the JH level rapidly rises. Twenty-four hours after eclosion the level was determined to be 9000 GU/ml, and on day 3 it was up to about 20,000 GU/ml. Queenless workers have larger CA than queenright workers, the difference can be threefold on day 5. In bumble bees, the volume of glands reflects the synthetic activity during the first 5 days after eclosion. This was tested by the JH synthesis *in vitro* assay. CA of queenright workers produce about 5–8 pmol JH/pair CA per hour, whereas synthesis in queenless workers increases to about 18 pmol JH/pair CA per hour (Fig. 10.4). That JH was determined to be JH III.

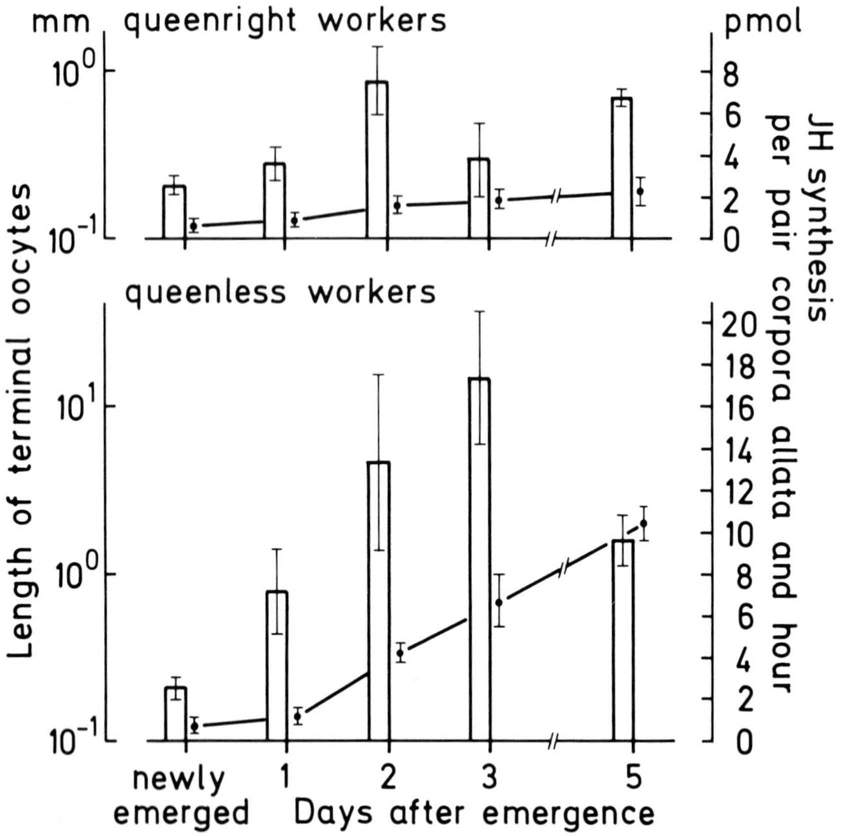

FIGURE 10.4 Synthesis of JH by CA *in vitro* (columns) and growth of terminal oocytes (lines) in queenright and queenless workers of *Bombus terrestris* during the first days after eclosion. Vertical bars indicate SD. (Data on JH synthesis from P.-F. Röseler and I. Röseler, 1978.)

Breakdown of JH in hemolymph does not occur, but the hormone is completely degraded in the hindgut and the metabolites (JH acid and JH diol) are then excreted. The rate of excretion is not affected by either the presence or the absence of a queen, so clearly the JH titer in adult bumble bees is mainly regulated by the synthetic activity of CA.

Since inhibition by the queen results in a low JH titer in workers, oogenesis can be induced by JH treatment even in the presence of a queen. The treated workers form eggs at the same rate as queenless workers do. Oogenesis takes place only as long as exogenous JH is present. Thereafter, oogenesis ceases and oocytes are resorbed. Endogenous JH production remains inhibited by the queen. Because of the high rate of excretion, about 50 μg JH I must be injected into newly emerged workers to ensure an elevated JH titer during the following 5 days needed for complete oogenesis. Lesser amounts of JH only induce vitellogenesis in the fat body and a small oocyte growth.

The queen controls the fertility of workers by a pheromone that is presumably produced in the mandibular glands and distributed over the body surface by grooming (P.-F. Röseler et al., 1981). This conclusion is derived from the following experiments: the mandibular glands of queens were excised and homogenized in chloroform : methanol, and then the extracts were compared with extracts obtained by washing the surface of a freshly killed queen. Both extracts presented on a dead queen to four newly emerged workers inhibited the activity of CA in those workers during the first day after emergence to the same extent as did a living or freshly killed queen. The activity was measured by means of the JH synthesis *in vitro* assay as well as by measuring gland volume.

The influence of the pheromone on the activity of CA is not yet properly understood. Since bumble bees do not lick each other, it is very likely that the volatile pheromone is not ingested but rather perceived by receptors. That external signal is transduced into an internal one that influences the neural centers controlling CA activity. The pheromone is only effective within a short distance. The queen, therefore, has to walk frequently over the comb, spreading the pheromone among workers in that way. But the larger a colony becomes, the more workers escape the pheromonal influence. Finally, egg-laying workers commonly occur late in colony development. At first, the oldest workers start egg laying, so it seems that workers might also become less sensitive to dominance signals with increasing age (van Honk et al., 1981; van Doorn and Heringa, 1986).

When workers with egg formation occur, the queen responds by a dominance behavior attacking the workers (P.-F. Röseler and I. Röseler, 1977; van der Blom, 1986). This behavior is also shown by queenless workers during establishment of a dominance hierarchy. But it is not known to what extent dominance behavior inhibits CA activity and egg formation. In no case can a role of pheromone be excluded. In general, physical attacks are used by bumble bees, when pheromones are not sufficient to suppress egg formation and to subordinate nestmates. Pheromones, therefore, also act as tranquilizers and prevent physical attacks.

10.6. Conclusions

Studies on the two bumble bee species *B. hypnorum* and *B. terrestris* have shown that the determination of larvae into queens or into workers as well as the functions of the adult castes are regulated by the endocrine system. Mainly JH contributes to caste differentiation, the role of ecdysteroids being unclear. Larvae are primarily determined into castes during the last instar by a different food regime that modulates the level and the temporal pattern of hormones in prepupae. Caste differentiation, therefore, must be seen in connection with larval–pupal transformation. During a certain sensitive period in the prepupal stage, the different hormone pattern induces the differentiation of castes. Since queens and workers are only physiologically defined, determination must take place at neural centers governing the activity of CA. An alternative program is switched on which controls the specific functions of adult castes. The main difference between castes of both species exists in the temporal pattern of developmental hormones. In prepupae of queens, pupal transformation is delayed. It is tempting to speculate that the neural centers are determined by the prolonged JH level. But the first response of those centers to hormone patterns in prepupae is as yet unknown. In the last few years, only some first results on hormonal regulation of the expression of cuticular proteins during larval–pupal transformation in butterflies have been published and various hypotheses on the role of hormones offered (Kiely and Riddiford, 1985; Wolbert, 1985; Riddiford et al., 1986).

Queens and workers emerge equipped with a different endocrine program. During the first period after emergence, this program is characterized by three JH levels:

(1) Low JH level in queens: preparation for diapause
(2) Intermediate JH level in workers: participation in colony tasks
(3) High JH level in workers: egg formation

Whereas the JH level in queens before overwintering is not influenced by the social environment, the worker-specific intermediate titer is attained by a pheromone of the queen. The presence of a queen causes the JH titer in workers to be lowered. In this way, the queen not only inhibits reproduction of workers, but the workers also remain sensitive to the needs of the colony. They perform the different tasks so that the queen can invest all her energy in reproduction. It has been shown that old nest workers with a high dominant position—presumably most of them with induced oogenesis and elevated JH titer—show least sensitivity to food shortage in the nest and do not change to foraging (van Doorn, 1987). To date, there is no clear evidence that JH directly influences behavior; all the specific behaviors could be induced by the physiological state.

Since the development and the functions of castes are hormonally regulated, all the controls affect the activity of the endocrine system. The various cues (nutrition, pheromones, behavior) are transformed into a signal influencing endocrine activity. The sensitivity of the endocrine system to exogenous factors was a prerequisite for caste differentiation in the evolution of social bees. Interestingly, the same hormone, JH, is used by bumble bees for caste differentiation in larvae as for regulation of caste functions in adults.

10.7. Summary

The castes of bumble bees are mainly defined by their physiology, though queens tend to be larger than workers. The differentiation of larvae into castes and the specific functions of adults are controlled by JH III. In larvae, a different food regime triggers a switch in endocrine activity, resulting in a caste-specific pattern of JH and ecdysteroids in prepupae. It is assumed that the prolonged and elevated JH level in prepupae of queens induces the development into that caste. In some species, caste differentiation is additionally regulated by the queen.

Queens and workers emerge with a different endocrine program. The low JH titer in queens controls the preparation for diapause after emergence. The functions of workers depend on an intermediate JH level that is controlled by the queen via a pheromone. In the absence of a queen, the activity of CA is increased and eggs are formed. The JH titer in adults is mainly regulated by CA activity.

References

Alford, D. V. 1969. Studies on the fat-body of adult bumblebees. J. Apic. Res. 8: 37–48.

Alford, D. V. 1975. *Bumblebees*. Davies-Poynter, London.

Brian, M. V. 1980. Social control over sex and caste in bees, wasps and ants. Biol. Rev. 55: 379–415.

de Wilde, J. 1985. Extrinsic control of caste differentiation in the honey bee (*Apis mellifera* L.) and in other Apidae. Pp. 361–369 *in* J. A. L. Watson, B. M. Okot-Kotber, and C. Noirot (eds.), *Caste Differentiation in Social Insects*. Pergamon Press, Oxford and Elmsford, New York.

de Wilde, J. and J. Beetsma. 1982. The physiology of caste development in social insects. Adv. Insect Physiol. 16: 167–246.

Fletcher, D. J. C. and K. G. Ross. 1985. Regulation of reproduction in eusocial Hymenoptera. Annu. Rev. Entomol. 30: 319–343.

Free, J. B. 1955. Queen production in colonies of bumblebees. Proc. R. Entomol. Soc. Lond. Ser. A Gen. Entomol. 30: 19–25.

Free, J. B. and C. G. Butler. 1959. *Bumblebees*. Collins, London.

Heinrich, B. 1979. *Bumblebee Economics*. Harvard University Press, Cambridge, Massachusetts.

Katayama, E. 1975. Egg-laying habits and brood development in *Bombus hypocrita* (Hymenoptera, Apidae). II. Brood development and feeding habits. Kontyû 43: 478–496.

Kiely, M. L. and L. M. Riddiford. 1985. Temporal patterns of protein synthesis in *Manduca* epidermis during the change to pupal commitment *in vitro:* their modulation by 20-hydroxyecdysone and juvenile hormone. Wilhelm Roux's Arch. Dev. Biol. 194: 336–343.

Nijhout, H. F. and D. F. Wheeler. 1982. Juvenile hormone and the physiological basis of insect polymorphism. Q. Rev. Biol. 57: 109–133.

Pendrel, B. A. and R. C. Plowright. 1981. Larval feeding by adult bumblebee workers (Hymenoptera: Apidae). Behav. Ecol. Sociobiol. 8: 71–76.

Plowright, R. C. and S. C. Jay. 1968. Caste differentiation in bumblebees (*Bombus* Latr.: Hym.). I. The determination of female size. Insectes Soc. 15: 171–192.

Plowright, R. C. and S. C. Jay. 1976. On the size determination of bumble bee castes (Hymenoptera: Apidae). Can. J. Zool. 55: 1133–1138.

Plowright, R. C. and T. M. Laverty. 1984. The ecology and sociobiology of bumble bees. Annu. Rev. Entomol. 29: 175–199.

Plowright, R. C. and B. A. Pendrel. 1977. Larval growth in bumble bees (Hymenoptera: Apidae). Can. Entomol. 109: 967–973.

Pouvreau, A. 1976. Contribution à la biologie des bourdons: étude de quelques paramètres écologiques et physiologiques en relation avec l'hibernation des reines. Thesis, University of Paris.

Richards, K. W. 1973. Biology of *Bombus polaris* Curtis and *B. hyperboreus* Schönherr at Lake Hazen, Northwest Territories (Hymenoptera: Bombini). Quaest. Entomol. 9: 115–157.

Riddiford, L. M., A. Baeckmann, R. H. Hice, and I. Rebers. 1986. Developmental expression of three genes for larval cuticular proteins of the tobacco hornworm, *Manduca sexta*. Dev. Biol. 118: 82–94.

Röseler, I. and P.-F. Röseler. 1973. Änderungen im Muster der Hämolymphproteine von adulten Königinnen der Hummelart *Bombus terrestris*. J. Insect Physiol. 19: 1741–1752.

Röseler, P.-F. 1970. Unterschiede in der Kastendetermination zwischen den Hummelarten *Bombus hypnorum* und *Bombus terrestris*. Z. Naturforsch. 25B: 543–548.

Röseler, P.-F. 1974. Vergleichende Untersuchungen zur Oogenese bei weiselrichtigen und weisellosen Arbeiterinnen der Hummelart *Bombus terrestris* (L.). Insectes Soc. 21: 249–274.

Röseler, P.-F. 1975. *Die Kasten der sozialen Bienen*. Steiner-Verlag, Wiesbaden, Germany.

Röseler, P.-F. 1976. Juvenile hormone and queen rearing in bumblebees. Pp. 55–61 *in* M. Lüscher (ed.), *Phase and Caste Determination in Insects*. Pergamon Press, Oxford and Elmsford, New York.

Röseler, P.-F. 1977a. Juvenile hormone control of oogenesis in bumblebee workers, *Bombus terrestris*. J. Insect Physiol. 23: 985–992.

Röseler, P.-F. 1977b. Endocrine control of polymorphism in bumblebees. Proc. 8th Int. Congr. IUSSI (Int. Union Study Soc. Insects), Wageningen, Netherlands pp. 22–23.

Röseler, P.-F. 1981. Caste differentiation in bees: the influence of behaviour, pheromones and food on the endocrine activity. Bull. Intérieur Sec. Fr. UIEIS (Union Int. Etude Insectes Soc.), Toulouse pp. 3–17.

Röseler, P.-F. and I. Röseler. 1974. Morphologische und physiologische Differenzierung der Kasten bei den Hummelarten *Bombus hypnorum* (L.) und *Bombus terrestris* (L.). Zool. Jahrb. Physiol. 78: 175–198.

Röseler, P.-F. and I. Röseler. 1977. Dominance in bumblebees. Proc. 8th Int. Congr. IUSSI (Int. Union Study Soc. Insects), Wageningen, Netherlands pp. 232–235.

Röseler, P.-F. and I. Röseler. 1978. Studies on the regulation of the juvenile hormone titer in bumblebee workers, *Bombus terrestris*. J. Insect Physiol. 24: 707–713.

Röseler, P.-F. and I. Röseler. 1984. Der Einfluss von CO_2 und der Kauterisation der Pars intercerebralis auf die Aktivität der Corpora allata und die Eibildung bei Hummeln (*Bombus hypnorum* und *Bombus terrestris*). Zool. Jahrb. Physiol. 88: 237–246.

Röseler, P.-F. and I. Röseler. 1986. Caste specific differences in fat body glycogen metabolism of the bumblebee, *Bombus terrestris*. Insect Biochem. 16: 501–508.

Röseler, P.-F. and I. Röseler. 1988. Influence of juvenile hormone on fat body metabolism in ovariolectomized queens of the bumblebee, *Bombus terrestris*. Insect Biochem. 18: 557–563.

Röseler, P.-F., I. Röseler, and C. G. J. van Honk. 1981. Evidence for inhibition of corpora allata activity in workers of *Bombus terrestris* by a pheromone from the queen's mandibular glands. Experientia (Basel) 37: 348–351.

Sakagami, S. F. 1976. Specific differences in the bionomic characters of bumblebees: a comparative review. J. Fac. Sci. Hokkaido Univ. Ser. VI Zool. 20: 390–447.

Strambi, A., C. Strambi, P.-F. Röseler, and I. Röseler. 1984. Simultaneous determination of juvenile hormone and ecdysteroid titers in the hemolymph of bumblebee prepupae (*Bombus hypnorum* and *B. terrestris*). Gen. Comp. Endocrinol. 55: 83–88.

van der Blom, J. 1986. Reproductive dominance within colonies of *Bombus terrestris* (L.). Behaviour 97: 37–49.

van Doorn, A. 1987. Investigations into the regulation of dominance behaviour and of the division of labour in bumblebee colonies (*Bombus terrestris*). Neth. J. Zool. 37: 255–276.

van Doorn, A. 1988. Factors influencing dominance behaviour in queenless bumblebee workers (*Bombus terrestris*). Physiol. Entomol. 14: 211–221.

van Doorn, A. and J. Heringa. 1986. The ontogeny of a dominance hierarchy in colonies of the bumblebee *Bombus terrestris* (Hymenoptera, Apidae). Insectes Soc. 33: 3–25.

van Honk, C. G. J., Röseler, P.-F., Velthuis, H. H. W., and J. C. Hoogeveen. 1981. Factors influencing the egg laying of workers in a captive *Bombus terrestris* colony. Behav. Ecol. Sociobiol. 9: 9–14.

Wheeler, D. E. 1986. Developmental and physiological determinants of caste in social hymenoptera: evolutionary implications. Am. Nat. 128: 13–34.

Wolbert, P. 1985. mRNA changes during imaginal determination of the epidermal cells in the pupa of *Galleria mellonella* L. Wilhelm Roux's Arch. Dev. Biol. 194: 385–389.

Roles of Morphogenetic Hormones in Caste Polymorphism in Ants

11

L. PASSERA AND
J. P. SUZZONI

This paper is dedicated to the late Dr. M. V. Brian who was a pioneer in the field of caste determination in ants.

11.1. Introduction

The major feature in the evolution of social insects has been the occurrence of reproductive division of labor among females. This trait has received considerable attention by all authors who have studied the fundamental characteristics of eusociality, and it is considered a salient point in the biology of termites, ants, and social bees and wasps (Wilson, 1971; Hermann, 1979; Brian, 1983; West-Eberhard, 1983; Passera, 1984; Fletcher and Ross, 1985). The most obvious manifestation of this division of labor is the occurrence of *castes,* a term used for the first time by Latreille (1797). According to Wilson (1979, 1985a), a caste is a group of individuals specialized in particular tasks for an extended period of their life. Usually each caste is distinguished by a set of morphological traits, but sometimes it is based on behavioral changes as a function of aging. In fact, the castes are either physical or temporal, or both. Physical castes involve the existence of *polymorphism,* which, according to Wilson (1971), is the coexistence of two or more castes within the same sex. In this chapter we will consider morphological variation of polymorphism.

In ants, polymorphism of males is unusual but has been known for a long time. The occurrence of two forms of males in *Ponera, Cardiocondyla,* and *Technomyrmex* was described by Le Masne (1956) and Terron (1972). These authors found normal winged males together with wingless males resembling the workers in the same colonies. Polymorphism occurs also among winged males, e.g., in the polygyne form of *Solenopsis invicta* (Ross and Fletcher, 1985) and *Formica exsecta* (Fortelius et al., 1987). In these species, small males (*micraners*) and large males (*macraners*) occur within the same nest.

In contrast to the apparently rare case of polymorphism in males, female polymorphism is well known. Generally speaking, there are only two castes within the female sex: one or several female reproductive individuals and the other females, which are more or less sterile. The former are termed *queens;* the latter, *workers.* In fact, queens can be defined as individuals of distinctive morphology (e.g., larger ovaries) conferring greater reproductive power (Crozier and Pamilo, 1986).

The situation can become more complicated when the worker caste is not monomorphic but differentiated into polymorphic subcastes. In this case, there is usually continuous size variation, being designated minors, medias, and majors, e.g., in *Camponotus aethiops* (Dartigues and Passera, 1979). The size differences are often accompanied by disproportionate growth in various body parts, especially the head, involving allometry; then, the majors often possess heads that are proportionately larger than in the case in the minors. This is obvious in some genera such as *Atta* (Wilson, 1985b). Polymorphism is still more pronounced in some species, such as the army ant *Eciton burchelli,* where there are four distinct physical castes of workers, namely, minors, medias, submajors, and majors, making this species one of the most polymorphic of known ant species (Franks, 1985). In a number of species the media workers have disappeared, resulting in a decrease in the number of subcastes. In this case there is complete worker dimorphism, without intermediates between the minor workers and the well-defined major workers usually named soldiers. Complete dimorphism has evolved in a few genera, the most widespread being *Pheidole* (Wilson, 1953; Passera, 1974).

A considerable number of investigations have been devoted to the study of the social and physiological factors determining caste in ants, mostly concerning queen/worker divergence, much less regarding the divergence of the worker subcastes. A first classic question is whether caste determination is genetic or epigenetic. Contrary to the opinion that was commonly accepted at the beginning of the century (Forel, 1921; see also Wheeler, 1986), the genetic hypothesis has rarely been verified; only one case is known in ants. In the European slavemaking ant, *Harpagoxenus sublaevis*, only larvae that are homozygous for a recessive allele appear able to develop into alate queens, provided that environmental conditions are favorable (Buschinger, 1978). In all other species studied, caste determination has proved to be nongenetic. Individuals become queens, minor workers, or major workers according to social factors (composition of the society) and environmental restrictions (climatic factors). According to the time when the divergence occurs, we can consider two cases:

(1) *Autogenic determination* (a term coined by Michener, 1961)—In this case, the bias has already occurred when eggs are laid and it is linked to a maternal influence during oogenesis or to a trophic effect of egg yolk, e.g., in *Formica of the group rufa* (Gösswald and Bier, 1953a,b), *Myrmica rubra* (Brian and Hibble, 1964; Brian and Kelly, 1967), *Monomorium pharaonis* (Petersen-Braun, 1975, 1977), and *Pheidole pallidula* (Passera, 1980a; Passera and Suzzoni, 1984).

(2) *Trophogenic determination*—In this case, the determination occurs later, depending on environmental stimuli received during larval life. The best studied species in this respect are *Myrmica rubra* (Brian, 1979, 1980), *Plagiolepis pygmaea* (Passera, 1969, 1980b), *Leptothorax nylanderi* (Plateaux, 1971), and *Solenopsis invicta* (Vargo and Fletcher, 1986, 1987). From these investigations and those of some 20 other species, the following general rules can be concluded:

- Because the divergence between castes occurs during the larval stage, a reasonable explanation invokes the importance of the food received by the larvae (trophogenic determinism).

- The workers determine the fate of the larvae by the quantity and/or quality of the food supplied.
- The mated queens are known to prevent the production of new queens. It is generally assumed that this inhibitory capacity of the queens is connected to the production of pheromones.
- When the colonies possess an overwintering brood, it is almost always from this brood that the queen larvae are reared. Moreover, the nurse workers best able to rear gynes are the ones just coming out of hibernation.

These investigations concern the social factors that act at the level of the colony. They tell us *when* and *how* a larva becomes queen or a worker, but they are not concerned with *why* the queen phenotype or the worker phenotype is expressed. The answer to this latter question requires physiological investigations at the level of the individual. The objective of this chapter is to present data concerning the role of hormones in female caste determination in ants.

11.2. Cytoanatomic Data

11.2.1. Adults

Several investigators have shown that the size of the corpus allatum (CA) is proportional to the size of the caste, e.g., in *Tapinoma erraticum* (Hultin, 1947) or in *Formica polyctena* (Schmidt, 1961). Gawande (1968) has shown in several species that the highest number of cells is found in the CA of the queen, followed by the male, then by the worker. In contrast, the number of cells in the CA of *Aphaenogaster senilis* is the same in the queen, the worker, and the male, although the proportional size differs according to the caste and sex (Bressac and Bitsch, 1969).

The most significant results were obtained in *Pheidole pallidula*, a species characterized by autogenic determination (Suzzoni, 1983). During the overwintering period, the total volume of the CA and the cellular volume (total volume/cell number) in the queens had means of 88.4 and 3.6 μm^3, respectively. After overwintering, the queens lay at first queen-biased eggs, then later switch to worker-biased eggs. Histological studies have shown that during the first stage after hibernation the total volume and the cellular volume of the CA increase to 98.1 and 4.1 μm^3, respectively. These values then decrease during the second stage to 68.7 and 2.8 μm^3, respectively. So the laying of queen-biased eggs is connected to the occurrence of large CA, whereas the laying of worker-biased eggs is connected to the occurrence of small CA.

11.2.2. *Larvae*

Because the development of polymorphism involves important changes during the larval stage, this phase has drawn the special interest of investigators.

In the ant *Eciton burchelli*, a species with strong polymorphism among workers, the size of the CA is directly proportional to the length of the larva (Lappano, 1958).

In *Myrmica rubra*, the volume of CA is higher in queens than in workers, but the relative volume of CA to the volume of the larva is higher in workers (Brian, 1959). In this species, there are also physiological differences between queens and workers, particularly concerning the brain, in the two lines of development: histological studies showed that in queen-determined larvae the pars intercerebralis releases a staining material later than in worker-determined larvae (Weir, 1959; Brian, 1959). The authors hypothesize that this difference is connected with the growth of the imaginal disks.

In the fire ant *Solenopsis invicta*, the CA are larger in sexual larvae than in worker larvae (fourth instar) but the number of the cells is the same (Petralia and Vinson, 1980). Nevertheless, the nuclei are distinctly larger in the CA of the former than in the CA of the latter. Unfortunately, the sexual larvae were not sexed.

Nevertheless, we should note that changes are seen when the determination of larvae is in progress. It will be interesting to examine the larvae just at the time when divergence occurs; this requires discrimination between castes and subcastes at an early stage. Unfortunately, such cases appear to be rather rare. Larval and pupal development, including the critical period of caste differentiation, were investigated in *Plagiolepis pygmaea* (Suzzoni and Grimal, 1981). During the undifferentiated period from the first to the third instar, the total volume of the CA increases from 910 to 2930 μm^3 but the weighted volume (reduced to the standard weight of 10 μg) decreases from 910 to 590 μm^3. During the same period, the cellular volume increases from 132 to 457 μm^3. The increase of the volume of the CA appears to be due solely to the increase in cellular volume, since the number of cells is unchanged, ranging from 7.1 to 7.3. During the differentiation period that extends from the end of the third instar to the prepupal stage, the data are as follows:

Worker-biased larvae. The total volume remains rather stable, ranging between 3000 and 4000 μm^3, and then decreases moderately from the prepupal stage to emergence (Fig. 11.1). The weighted volume also decreases, but somewhat less. In the same period, the cell number varies slightly between 8.1 at the end of overwintering to 6.2 during the adult stage. The cellular volume remains constant (about 500 μm^3) until the

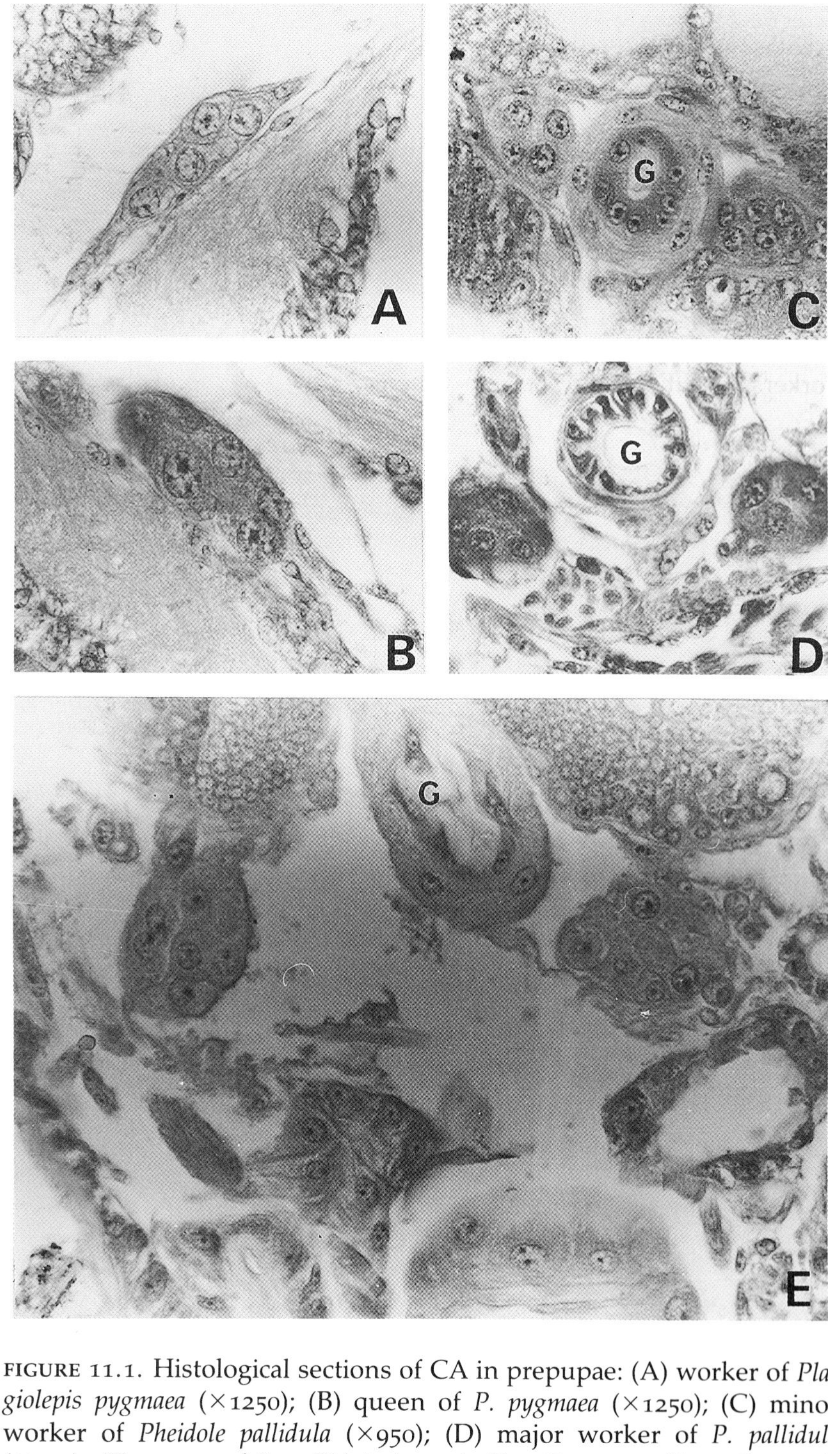

FIGURE 11.1. Histological sections of CA in prepupae: (A) worker of *Plagiolepis pygmaea* (×1250); (B) queen of *P. pygmaea* (×1250); (C) minor worker of *Pheidole pallidula* (×950); (D) major worker of *P. pallidula* (×950); (E) queen of *P. pallidula* (×950). *Key:* G = gut. (From Suzzoni, 1983.)

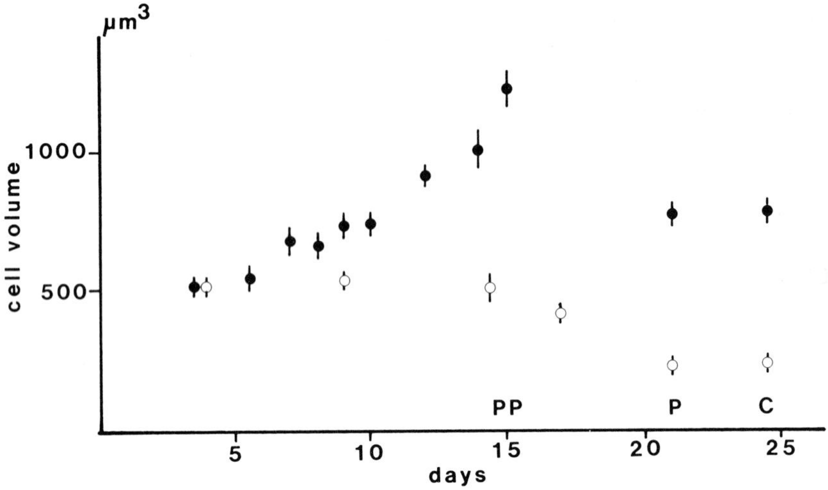

FIGURE 11.2. Changes in the cell volume of the CA in *Plagiolepis pygmaea* during the differentiation period from the last-instar larva (day 0) to callow (day 25). *Key:* C = callow; P = pupal stage; PP = prepupal stage; ● = queen-biased larvae; ○ = worker-biased larvae; bars = ±1 standard error (SE) about mean. (From Suzzoni and Grimal, 1981.)

prepupal stage, when it decreases moderately until the time of emergence (Fig. 11.2).

Queen-biased larvae. The total volume of the CA increases more than threefold, going from 3000 μm³ at the end of overwintering to almost 10,000 μm³ at the pharate pupal stage, and then decreases to 6500 μm³ at the imaginal stage (Fig. 11.1). As the weight of queen-biased larvae grows rapidly, the weighted volume decreases earlier and faster than in worker-biased larvae. The number of cells increases slightly, from 7.5 to 9.8 just before the prepupal stage, then decreases to 8.1 at the time of emergence. The most striking phenomenon is the increase in cellular volume, from 500 to 1240 μm³ (Fig. 11.2), perhaps resulting in higher JH production just at the time of queen differentiation.

Species exhibiting complete dimorphism of workers, leading to the occurrence of minor workers and soldiers, e.g., *Pheidole* spp. (Ono, 1982), also constitute promising material for such studies. In *Pheidole pallidula*, Suzzoni (1983) compared the CA of major and minor workers, from the third larval instar to the callows (Fig. 11.1). During the differentiation of these subcastes, at the end of the third instar, the total volume increases in both subcastes as a consequence of cell growth. At the time of the pupal stage, the total volume decreases in minor workers, whereas it continues to increase in major workers. This difference is due to the death of some cells and to the decrease of the cell volume in the CA of the minor worker.

By comparison with other insects, these changes in the retrocerebral system of the larvae at the time of determination suggest that the process is controlled by the interplay of two hormones, JH (juvenile hormone) and ecdysone.

11.3. Surgical Methods of Investigation

Because of their small size, it is very difficult to perform experiments by removing glands or other surgical procedures in larval ants. Furthermore, because of the social life of these insects, the individuals who have been operated on are usually neglected and discarded into the refuse pile or cannibalized. Therefore, the only such study is that of Brian (1974, 1976). In these experiments, large overwintered larvae of *Myrmica rubra* were used. These are queen-potential larvae and so possess imaginal wing disks. The degree of sexualization of the larvae was judged by measuring the surface area of the wing buds and comparing them with that of the leg buds, after 6 days' culture. As sexualization proceeded, the rate of growth of the imaginal wing disks increased whereas that of the legs slowed down. In controls (with a wound similar to that caused by extracting the CA), the surface area of wings buds was equal to that of the leg buds or larger. In contrast, in larvae treated either by cephalic ligature short-circuiting the CA or by surgical removal of the CA, the surface area of the wing buds was either equal to that of the leg buds or smaller and was never larger. This provides good evidence that the exclusion of the CA stops the inhibitory action normally controlling growth and differentiation of the legs, in favor of growth of the wing buds. The CA, and therefore JH, have a positive effect on development of the queen larvae.

11.4. Effects of Exogenous Hormones

11.4.1. *Queen/Worker Determination*

11.4.1.1. DETERMINATION DURING THE EGG STAGE (AUTOGENIC DETERMINATION)

It should be recalled that in this case determination has already occurred when the eggs are laid. For example, in *Pheidole pallidula* at the end of hibernation, the queens first lay queen-biased eggs and later only worker-biased eggs.

JHAs Supplied in the Food

Whole societies of *Pheidole pallidula* were used more than a month after the end of hibernation, that is, when they were producing only worker brood. They were fed *Tenebrio* larvae that had received an injection of 10 μg of a JH analogue (JHA). The treated societies produced sexual brood, whereas the controls only produced worker brood (Passera and Suzzoni, 1978a).

Topical Applications of JHAs

JHA dissolved in acetone or olive oil was used in topical applications. The queen of colonies rearing brood was treated weekly, 1 month after the end of hibernation, the period in which the sexual brood appeared (Passera and Suzzoni, 1978b, 1979). The colonies again gave rise to sexual larvae, whereas the controls continued to produce worker larvae. More often than not, sexual larvae appeared during the third week following treatment. This rather long interval shows that JHA acts on queens during oogenesis; indeed, if it were a case of action on the larvae, the effect would be immediate. In addition, treated larvae of the first, second, and third instar gave rise only to workers.

11.4.1.2. DETERMINATION DURING THE LARVAL STAGE (TROPHOGENIC DETERMINATION)

This type of determination is the most usual among ants (Brian, 1980; Passera, 1984), so the examples are numerous. When a fire ant colony (*Solenopsis invicta*) is exposed to insect growth regulators (IGRs), these compounds act just as they do in other insects, i.e., they disrupt metamorphosis. Cupp and O'Neal (1973), Vinson et al. (1974) and Banks et al. (1978), treated brood of *S. invicta* with two JHAs, ZR-512 (hydroprene) and ZR-515 (methoprene): numerous larvae died, resulting in a decrease in the worker population. Nevertheless, some larvae survived and pupated, exhibiting some characteristics that were indicative of hormonal play during caste divergence; as a matter of fact, in addition to pupae without legs or antennae or having other morphological abnormalities, Robeau and Vinson (1976) obtained several intercaste pupae intermediate between the major worker and queen.

Cockroaches injected with 100-μg JHA (ZR-512) were provided as food to colonies producing only workers. Ten days after the beginning of this treatment, 75% of the produced larvae were of the sexual type, most of them developing into males (Troisi and Riddiford, 1974).

Interesting results were obtained by Vinson and Robeau (1974), who fed ants JHAs. Sexual larvae were produced some 20–30 days after the beginning of the treatment. In the controls, sexual larvae were never produced. In treated colonies, those of larvae that survived and reached the pupal stage were all queens. This result is obviously different from that of Troisi and Riddiford (1974), who obtained many more males. In another series of tests, dequeened colonies possessing brood of all stages were fed by mixing JH-25 and oil (Robeau and Vinson, 1976). This treatment resulted in the production of winged queens. In control colonies, sexuals were never produced. These results provide evidence of an effect of JHA on sexualization of the brood; since queenless societies produce new queens, JHA does not act during oogenesis, but later during larval development. Experiments using units with discrete larval instars showed that only the first larval instar and perhaps the second are sensitive to the IGRs: it seems likely that caste determination occurs at this time (Banks et al., 1978).

In *Monomorium pharaonis*, laboratory societies exposed to food containing a JHA (methoprene) resulted in the appearance of sexual brood as in the fire ant (Edwards, 1987). But in the pharaoh's ant, the sexualization of the brood does not directly involve the JHA. In this case, JHA has another effect: it sterilizes the queen. Moreover, Edwards (1987) demonstrated that it is the lack of eggs which is responsible for the rearing of sexual larvae, since the production of sexuals can be inhibited by the introduction of eggs in a dequeened colony. Nevertheless, these results are quite different from those obtained on *S. invicta* in which the JHA acts directly on the brood, as was shown by Banks et al. (1978). In summary, JHAs supplied to the fire ant colony have a double effect:

- *Action on metamorphosis*—They disturb metamorphosis and induce morphological abnormalities.
- *Action on the caste*—If the dose delivered allows the queen to lay, the larvae that develop give rise to workers but more so to males and alate females.

The second species studied in this way was *Myrmica rubra*. Brian (1974) used large postwintering queen-potential larvae. Farnesic acid and farnesyl methyl ether were injected or applied topically. All the pupae resulting from both treated and control larvae were queens as expected, but the former weighed more than the latter. In a second experiment, Brian used medium-sized hibernated larvae, known to have less chance of developing into queens than their larger counterparts. He obtained more queen pupae from the treated larvae than from the control. So it appears that in *M. rubra*, JHAs delay metamorphosis, lengthening the growth period and hence the size of the queens. They also appear to increase the percentage of larvae differentiating into queens.

Ecdysone Supply

A few experiments have been performed with the other insect hormone, ecdysone. As in the JH tests, Brian (1974) injected α-ecdysone into post-wintering queen-potential larvae of *M. rubra*. The results yield three conclusions:

(1) As a classic effect of ecdysone, metamorphosis had been advanced.

(2) The weights of pharate pupae or pupae resulting from larvae treated with ecdysone were lower, so the queens obtained were small in size and sometimes would even be classified as microgynes.

(3) A caste effect was especially noteworthy. In control tests, Brian obtained only about 9% workers. From treated larvae he obtained 44% workers. In addition, he obtained numerous intercastes from the treated trials. So the ecdysone supply prevents the sexualization of the brood or leads to the occurrence of smaller reproductive females.

In contrast to these results on *M. rubra*, ecdysone supplied via feeding in fire ant colonies had no effect on caste determination (Vinson and Robeau, 1974).

11.4.2. *Minor Worker/Major Worker Determination*

11.4.2.1. JHAs SUPPLIED IN THE FOOD

Numerous species of the genus *Camponotus* are known to exhibit strong polymorphism within the worker caste. In a study conducted on field colonies, Fowler (1984) fed colonies of *C. pennsylvanicus* a suspension of RO-13-5223 in dilute corn syrup. Compared to controls, treated colonies had an altered worker balance. Unfortunately, it is not known if this change in the ratio of the two worker subcastes is owing to increased mortality of the minors or increased production of the majors.

Like some species of *Camponotus*, the workers of *Solenopsis invicta* show pronounced polymorphism with major workers (Porter and Tschinkel, 1985). To test JH effects on worker size, dequeened colonies containing mixed-brood stages were fed JH-25 (Robeau and Vinson, 1976). At the time of pupation, treated colonies contained more major workers than did control colonies. JH thus seems to induce the development of majors in *S. invicta*.

Other experiments to test the effect of the JH were performed with *Pheidole pallidula*. This species exhibits discrete worker dimorphism: the workers are divided into small-headed minor workers and large-headed major workers (Passera, 1974). Soldier determination occurs at the end of the third (last) instar, and it seems trophic. So in this species, queen/worker determination occurs very early, during the egg stage, even though minor/major determination occurs at the end of larval life. In a set of trials, Suzzoni et al. (1982) studied the effect of JHA. Experimental units containing 1000 minors and 1 queen without soldiers were set up. Larvae of *Tenebrio* injected with JHA were provided as food. Three or four weeks later the first pupae appeared. Treated colonies gave rise to a mean of 21 minors and 6 majors (22%) per week. Control colonies reared 17 minors and only 2 majors (10%) per week. Thus, the number of soldiers increased twofold when the ants were fed JHA. The studies of Edwards et al. (1981) showed a similar phenomenon in another species of *Pheidole, P. megacephala:* the use of bait containing JHA results in a large increase in the number of soldiers. However, as for the study by Fowler (1984) on *C. pennsylvanicus,* we can only speculate as to the reason for this increase; it may simply reflect a difference in longevity between soldiers and minors.

11.4.2.2. TOPICAL APPLICATION OF JHAs

Like most other species of the genus *Pheidole, P. bicarinata* is characterized by complete dimorphism within the worker caste. As in *P. pallidula,* soldier/minor worker determination takes place during the last larval instar (Wheeler and Nijhout, 1981). To examine the role of JH in soldier/minor divergence, Wheeler and Nijhout (1981, 1983) and Nijhout and Wheeler (1982) treated the last instar topically. They used a JHA (methoprene) diluted in acetone. The larvae were classified into six size classes from 0.6 to 1.3 mm in length and received one of three doses of JHA (10, 50, and 250 μg/g). As can be seen in Fig. 11.3, these results show that the treatment with the highest dose induced soldier differentiation in 75% of the larvae. With the medium dose, the larvae responded only when they were between 0.9 and 1.2 mm long. The lowest dose was ineffective. The JHA also had an effect on the size on the pupae. When the treated larvae pupated into minors, the pupae that resulted were larger than in the control (Fig. 11.4). The third effect produced by the supply of exogenous JHA is related to the timing of metamorphosis. The appearance of the first minor or soldier pupae following treatment was altered.

From these data, Wheeler and Nijhout (1983) developed the following hypothesis (Fig. 11.5): The future of a larva depends on the amount of circulating JH, during a sensitive period when the third larval instar grows from 0.9 to 1.2 mm. If the amount of JH is low, the morphogenetic program for minor development is initiated and the metamorphosis occurs as soon as the titer of JH allows prothoracicotropic hormone (PTTH) secretion. It is the standard way by which the minors appear. If the amount of JH is higher than the soldier-inducing threshold, the major morphogenetic program is triggered and the larva is allowed to

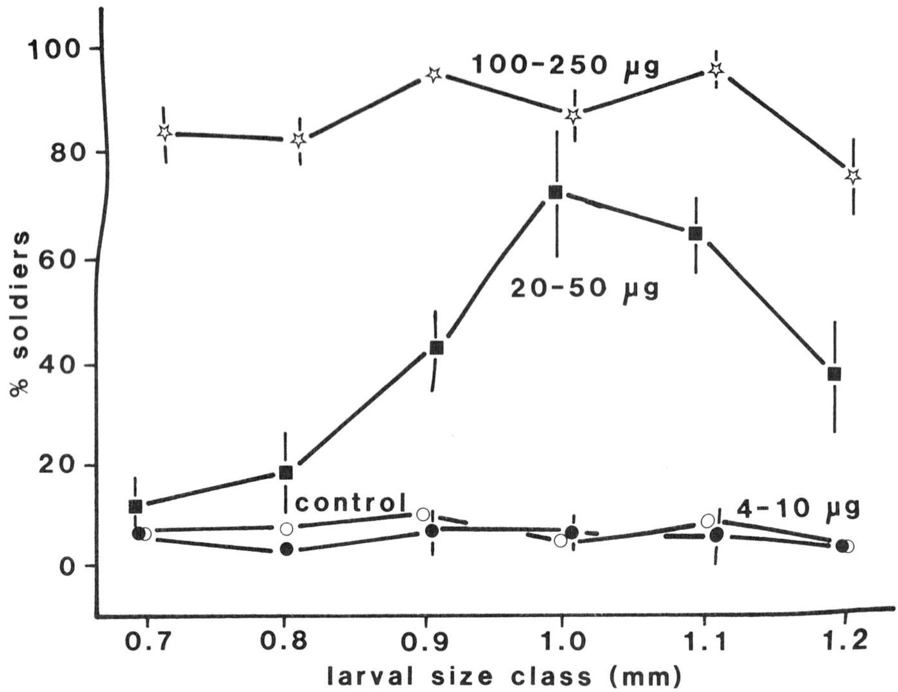

FIGURE 11.3. Developmental alternative of larvae after topical application of JHA (methoprene) on last-instar larvae in *Pheidole bicarinata*, a species exhibiting complete worker dimorphism (minor/major). Three doses were used: ● = 4–10 µg/g; ■ = 20–50 µg/g; ☆ = 100–250 µg/g; ○ = control (acetone); bars = ±1 SE about mean; *abscissa*—the treated larvae are classified into six size classes; *ordinate*—percentage of soldier pupae (major workers) produced. (From Wheeler and Nijhout, 1981.)

grow; when it reaches a critical size of 1.7 mm, the titer of JH decreases and soon allows PTTH secretion resulting in a soldier pupa. When JHA is supplied by topical application two cases are possible:

(1) If the larva receives a high dose of exogenous hormone (above the normal soldier threshold), the major worker program is initiated and in addition the growth period is delayed until the time when the degradation of JHA allows PTTH secretion; because of the extension of the growth period during which the larva is fed, the resulting pupa is larger than in normal development.

(2) If the larva receives a low dose of exogenous hormone inadequate to trigger the soldier developmental program, the growth period is nev-

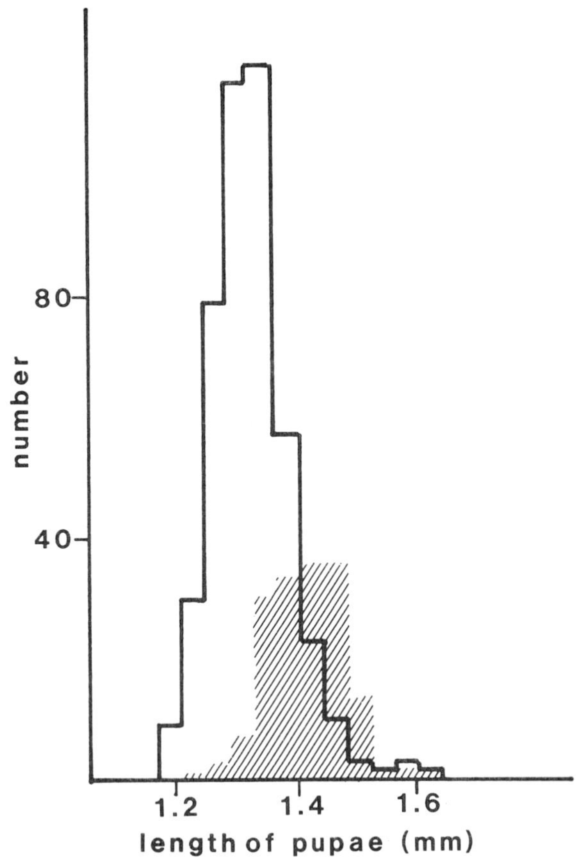

FIGURE 11.4. Effect of JHA on the size of worker pupae in *Pheidole bicarinata* after larvae have been treated with 1 mg/ml methoprene. Hatched histogram represents treated workers; open histogram represents control workers that received only acetone. (From Wheeler and Nijhout, 1983.)

ertheless delayed until the supplemental JHA is degraded, resulting in a minor worker larger than the normal minor.

Ono (1982) performed a comparable study in the Japanese species, *Pheidole fervida*. All the members of small units were treated with various solutions of a JHA (ZR-515). The lower doses failed to cause development of soldiers from the brood. The higher doses killed many larvae, but more soldiers developed from the survivors than in the control. In a

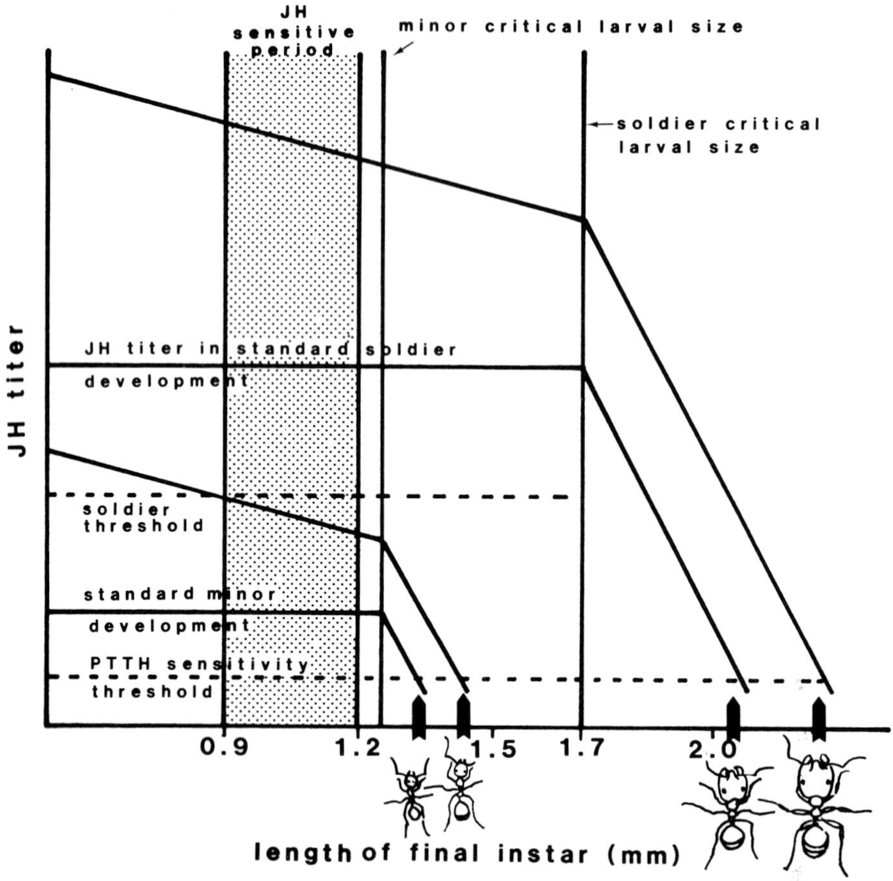

FIGURE 11.5. Theoretical model of minor/major determination in *Pheidole bicarinata*, a species exhibiting complete dimorphism within the worker caste. The future of a larva (minor or major worker) and the size of the resulting imago depend on the amount of circulating JH during a sensitive period (stippled area). (From Wheeler and Nijhout, 1983.)

more detailed experiment, only the last instar larvae were treated. Many soldier larvae appeared after this treatment. Thus, it seems that, as in *P. bicarinata*, the last instar is sensitive to the JHA and still has the competence to develop into a soldier.

Suzzoni et al. (1982) studied the same phenomenon in the European species, *P. pallidula*. Second- or third-instar larvae of various sizes were given to groups of queenless workers. Each larva received a topical application of JHA at various doses, but the JHA failed to induce soldier development. It may be in this species that the JH-sensitive period is very short. However, JHA affected the weight of the minor pupae as in *P. bicarinata:* the earlier the application, the heavier the resulting pupae; treatment of second-instar larvae produced pupae of 470 μg (mean weight), whereas treatment of the oldest third-instar larvae produced pupae of 284 μg.

11.5. Hormone Dosages

Data exist only for *Pheidole pallidula* and *Plagiolepis pygmaea* (Suzzoni et al., 1980, 1981, 1983).

11.5.1. *Pheidole pallidula*

Queens collected in early spring (those laying queen-biased eggs) or later in summer (those laying worker-biased eggs) were compared by radioimmunoassay (RIA) in total extracts. The rate of JH was 8.5 pmol/100 mg for queens collected in spring and 24.5 pmol/100 mg for those collected in summer. Dosages measured in total extracts of queen-biased or worker-biased young eggs less than 24 h old showed that they contained 4.7 and 10.3 pmol/100 mg, respectively. The JH measured in these eggs has a maternal origin and is incorporated during oogenesis. It appears from these data that the quantity of JH incorporated into each egg increases during the season: at the beginning of the spring, queens laying queen-biased eggs—and the queen-biased eggs themselves—have low levels of JH; later in the summer, queens laying worker-biased eggs—and the worker-biased eggs themselves—have a higher level of JH.

Concerning the ecdysteroids, comparisons between the two types of queens show great differences in the quantities of ecdysteroids and other compounds comigrating with ecdysone and ecydsterone (Fig. 11.6). However, when the percentage of RIA-active ecdysone and ecdysterone were compared, they were found to be roughly equal in both types of queens: 28% ecdysone and 14% ecdysterone in queens laying worker-biased eggs, and 27% and 11% for those laying queen-biased eggs. Eggs contained the same ecdysteroids as queens, but in different proportions (ecdysone, 41%; ecdysterone, 8.5%). Study of the variations in ecdysteroid titer in the different eggs as a function of time after laying showed (Fig. 11.7) the following:

(1) The amount of hormone is low during the first 3 days after the beginning of egg laying; then, in the second period of embryonic development (days 4–7), it rises considerably. It cannot be said whether this increase is due to synthesis or to unmasking of hormones in the embryo.

(2) Worker-biased eggs contain more ecdysteroids. Furthermore, we note that the longer the time after hibernation, the higher the ecdysteroid titer and the more the egg is likely to be worker biased.

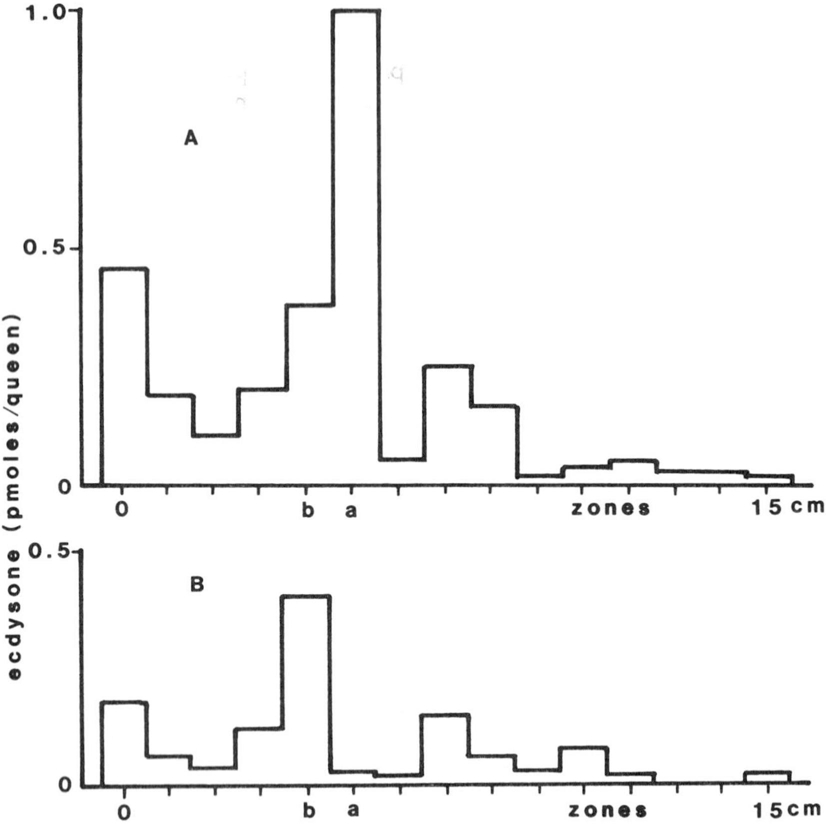

FIGURE 11.6. Distribution of RIA-active material (expressed as pico-moles of ecdysone equivalent per queen) in queens of *Pheidole pallidula* after thin-layer chromatography (TLC). The abscissa represents 1-cm zones scraped and eluted separately (0 cm, origin; 15 cm, front): (A) queens laying worker-biased eggs; (B) queens laying queen-biased eggs. *Key:* a = compounds comigrating with ecdysone; b = compounds comigrating with ecdysterone. (From Suzzoni et al., 1980.)

Additional experiments provide evidence for an interaction between the two developmental hormones (Suzzoni, 1983). Following topical application of queens with JH, the ecdysteroid titer decreases for 3 days and then increases again. These variations are connected with the future of the eggs laid by the queen (Fig. 11.8). Thus, JH appears to be an earlier signal acting on ecdysteroid levels, and this interaction probably plays a role during caste differentiation.

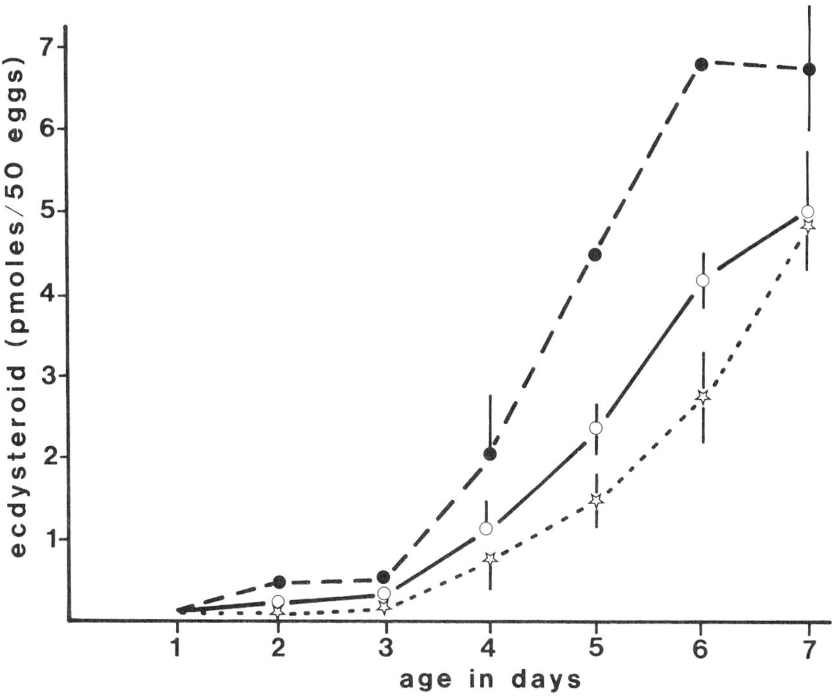

FIGURE 11.7. Increase in ecdysteroid levels during the embryonic development in the ant *Pheidole pallidula*. *Key:* ● = queen-biased eggs laid at the end of hibernation; ○ = worker-biased eggs laid at the end of hibernation; ☆ = worker-biased eggs laid 1 month after hibernation ended. (From Suzzoni et al., 1980.)

11.5.2. *Plagiolepis pygmaea*

In this species, divergence of gyne and worker developmental pathways depends on the presence of the queens: in dequeened societies, overwintered third-instar larvae develop into queens; in contrast, they develop into workers in queenright societies (Passera, 1969). Ecdysteroids were assayed in both queenless and queenright cultures (Suzzoni, 1983). The results are consistent with those obtained in *Pheidole pallidula*. In queen larvae, low values occur consistently throughout development of third-instar larvae, reaching 2.5 pmol/mg in larvae weighing 100–150 μg and peaking at 5.0 pmol/mg in larvae weighing 300–400 μg. The titer is much higher in worker larvae: the values reach 40 pmol/mg about midway through development, then decrease at the time of the prepupal stage. If the total immunoreactive ecdysteroids at the maximum titer in

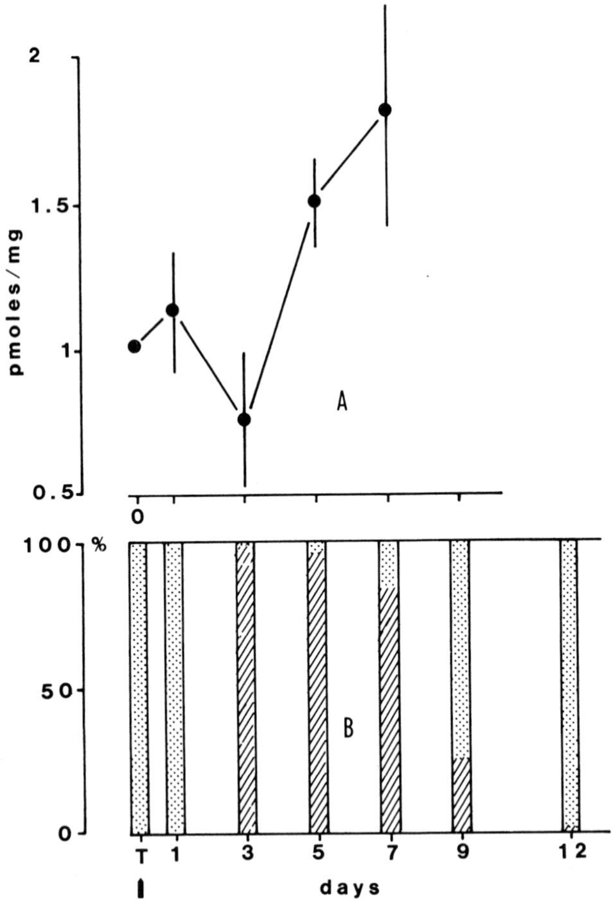

FIGURE 11.8. Change in the ecdysteroid level in the queen of *Pheidole pallidula* (A) and the ratio of queen larva to worker larva produced after topical application of JHA on the queens (B). *Ordinate*—(A) the ecdysteroid level is expressed in pmoles per milligram of queen fresh weight of ecdysterone equivalent; (B) the queen larvae number (hatched area) and the worker larvae number (stippled area) are expressed in percent of the total number of larvae. *Abscissa*—time in days since the topical application (T) at day 0. Bars = ±1 SE about mean. (From Suzzoni, 1983.)

the worker larvae is expressed as ecdysterone equivalents, the conjugates account for 30% of the RIA activity, ecdysone for 13%, and ecdysterone for 16%.

These experiments conducted on *Pheidole pallidula* and *Plagiolepis pygmaea* show that high ecdysteroid titers are connected with the development of the worker line.

11.6. Discussion

At the end of this review, it is striking to observe that there is rather little information concerning hormonal control of caste in ants. Of the 12,000–15,000 species (Snelling, 1981), hardly five or six have been investigated in regards to hormonal involvement. For instance, almost no studies have been devoted to species exhibiting continuous polymorphism among workers (Robeau and Vinson, 1976; Suzzoni et al., 1986b). Only queen/worker polymorphism and minor/major dimorphism in a few species possessing discrete polymorphism have been fairly well documented.

In spite of the small number of studies available, the role played by JH in caste determination under experimental conditions appears well established in several ants. It is generally consistent with that found in other social insects and especially in other social Hymenoptera. In lower termites and particularly in *Kalotermes flavicollis,* the involvement of hormones in the formation of castes is well documented (Lüscher, 1972, 1974; Lebrun, 1985). In the same way, hormones are involved in higher termites (Lüscher, 1976; Okot-Kotber, 1985; Lanzrein et al., 1985). In *Apis mellifera,* the best known social insect, control of queen determination occurs during a short period in the late third instar and JH plays an important part (Wirtz and Beetsma, 1972; de Wilde, 1976; Goewie, 1978; Beetsma, 1979; Copijn et al., 1979; Dietz et al., 1979; de Wilde and Beetsma, 1982; Engels, 1986). Hormonal implications have also been documented in bumble bees (Röseler, 1977; Röseler and Röseler, 1978) and in stingless bees (Velthuis and Velthuis-Kluppell, 1975; Campos, 1978, 1979). In polistine wasps, ecdysteroids and JH regulate the establishment of the social hierarchy and the division of reproduction (Röseler et al., 1984, 1985, 1986; Strambi, 1986).

JH acts effectively during sensitive periods when a morphogenetic program can be triggered. If the titer of JH is too low at that time, the endogenous factors responsible for a definite developmental program remain silent. The point in the cycle where the biological material becomes sensitive to JH is variable according to the species. Ants of the genus *Pheidole* provide an excellent example: queen/worker determination is the first to appear; it occurs very early, since during embryogenesis the eggs are already queen biased or worker biased according to the titer of hormone in the mother organism. A second sensitive period is responsible for the morphogenetic option within the worker caste; it occurs much later, when the worker-biased larva comes to the end of its larval life. At this time a supplement of JH triggers the soldier program.

When the worker caste is monomorphic, the determination is simpler, involving only one sensitive period, e.g., in *Myrmica rubra*. Furthermore, the sensitive period must be very short and difficult to localize; this probably explains why we have failed to obtain soldiers after topical application of *Pheidole pallidula*. In the same way, the failure of experiments aimed at modifying the brood orientation by massive doses of JH in several species, e.g., *Plagiolepis pygmaea, Temnothorax recedens, Diplorhoptrum fugax, Iridomyrmex humilis*, and *Colobopsis truncatus*, may result from the difficulty of finding the critical period (Passera and Suzzoni, unpublished data).

The results obtained with *M. rubra, P. pallidula*, and *P. pygmaea* suggest that ecdysteroids act in the opposite direction of JH by favoring the production of workers.

A similar hypothesis based on the relative variation of JH and ecdysteroids has already been proposed for termites by Lüscher (1976) and Noirot (1977).

It should be recalled at this point that trophic factors play an important role in caste differentiation. Most probably there is a link between the two factors: nutritional and hormonal. In the honey bee, the high sugar content of the royal jelly and the diet supplemented with sugars may stimulate CA secretion directly via abdominal stretch receptors (Asencot and Lensky, 1976; Goewie, 1978) or indirectly via brain neurosecretory cells (Dogra et al., 1977). As Wheeler (1986) has pointed out, the nutritional switch in ants should trigger the neuroendocrine system in the same way. Nevertheless, the exact mechanism of the linkage between food and endocrine glands is not yet known.

On the other hand, the amount of food given to larvae depends on the feeding activity of nurse workers. Generally, larvae are fed by trophallaxis in ants and the feeding behavior of workers is tightly controlled by the queen, e.g., in *P. pygmaea*: using honey containing radiolabeled gold, Bonavita-Cougourdan and Passera (1978) showed that postwintering larvae receive up to six times more food in queenless societies than in queenright societies. Of course, the amount of food delivered to larvae is also controlled by external factors, such as temperature, number of foragers, abundance of prey, etc. All in all, the development of castes in ants should be due to the sequence of events shown in Figure 11.9.

In a recent review, Hardie and Lees (1985) pointed out an important fact concerning the role played by JH among social insects. It is well documented that JH acts as a gonadotropic factor in many nonsocial

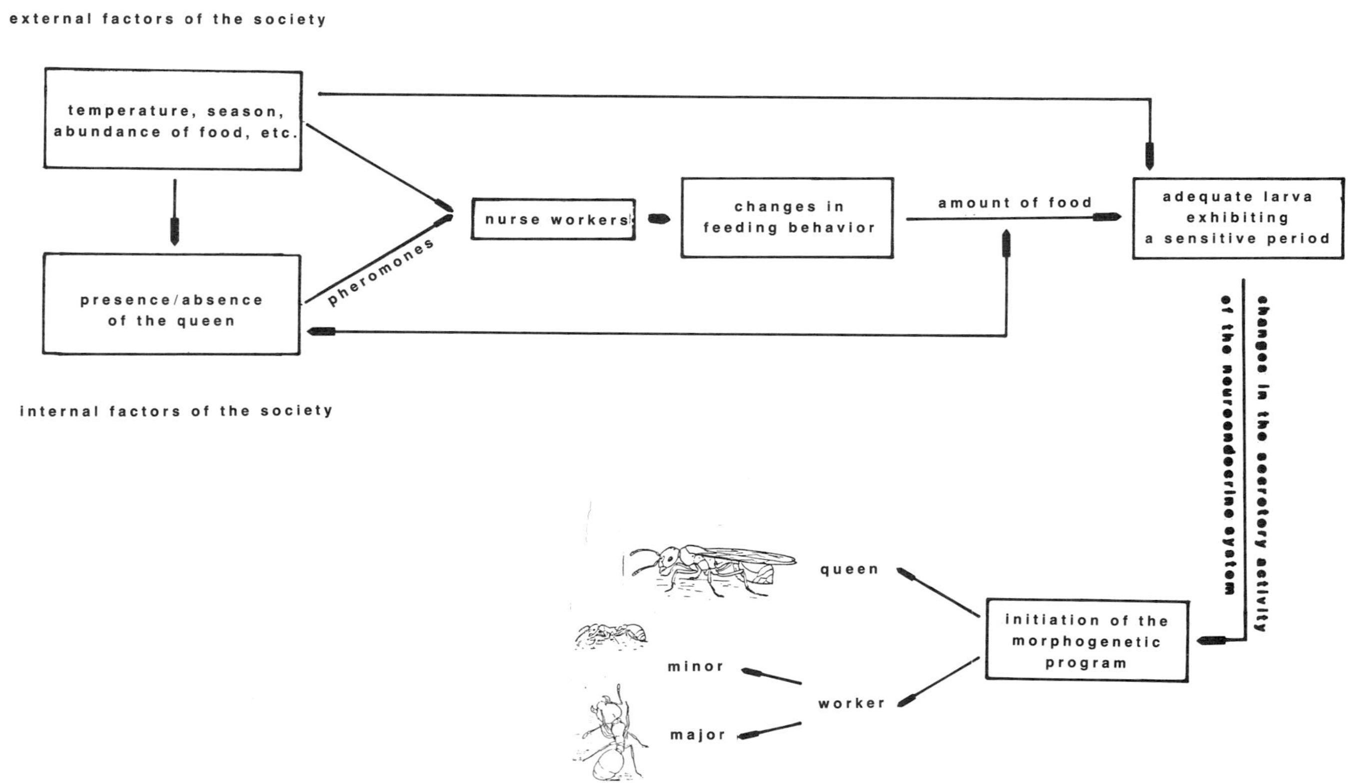

FIGURE 11.9. Hypothetical model for the development of castes in ants involving a connection between trophogenetic and neuroendocrine factors.

insects (reviewed in Gilbert, 1976). JHs also control ovary maturation in primitively social bees, e.g., *Lasioglossum zephyrum* (Bell, 1973), in bumble bees (Röseler, 1974), and in wasps (Barth et al., 1975). But in more complex societies, e.g., honey bees, the role of JH is more ambiguous: allatectomy does not prevent egg maturation (van Laere, 1974; Engels and Ramamurty, 1976), and treatment with JHA does not affect fecundity (Žďárek et al., 1976). In the same way, treatment of queens by precocene II, an anti-JH substance, is without effect (Fluri, 1983). Lastly, JH titers are not very different between egg-laying and nonlaying queens (Fluri et al., 1981). These data suggest that in honey bees JH is not the gonadotropic factor. According to Hardie and Lees (1985), JH could have been "captured" for an additional role as a determiner of polymorphism. This hypothesis is attractive, for several studies on ants also suggest that JH is not always involved in egg production as in other insects. For example, in *P. pygmaea*, after topical application of a JHA, the treated queens lay slightly more eggs than the control queens but their eggs are strongly deformed, indicating a side effect (Passera and Suzzoni, 1974).

In *Solenopsis invicta*, JH supplied in the food results in fewer larvae, because egg production stopped in 28–45 days (Vinson and Robeau, 1974; Banks and Harlan, 1982; Banks, 1986). The results obtained by Robeau and Vinson (1976) are even more definitive: after topical application of JHA, the treated queen stops laying for 2 days. Nevertheless, results obtained by Barker (1978) do not support the findings of these researchers. As a matter of fact, using extirpation of the CA of virgin queens, Barker showed a decrease in egg production, which can be restored by topical application of JHA. Whether or not JH has a gonadotropic action is still open to question in the fire ant.

Palma-Valli and Délye (1981) noted a relationship between the activity of the CA and the rate of egg-laying in queens of *Camponotus lateralis*. In addition, the amount of neurosecretory products was related to fecundity. These results are not in agreement with those obtained by Suzzoni et al. (1986b) in another species of *Camponotus*, *C. aethiops*: queens in incipient colonies as well as in mature colonies stop laying after receiving topical applications of JHA, but the effect is reversed after the application is ended. In the same species, workers begin laying after the society was dequeened. These workers stop laying after feeding on JHA mixed with honey.

Exposure of laboratory colonies of *Monomorium pharaonis* to food containing JHA (methoprene) made the queens sterile (Edwards, 1977, 1987; Rupěs et al., 1978). We have seen (above) the result of this event upon the production of sexual larvae.

So among several species of ants it is likely that JH has lost its gonadotropic action, but because there is as yet little information available concerning the vast majority of species, no general conclusion regarding the role of JH in ovary development can be drawn at the present time.

If the loss of ovary control by JH in ants is confirmed, it will be tempting to adopt the hypothesis of Hardie and Lees (1985) that JH has changed roles in eusocial Hymenoptera: instead of regulating fecundity, JH regulates caste differentiation and polymorphism of workers.

More studies carried out on many more species of ants are needed before the exact role of hormones in caste polymorphism will become clear.

11.7. Summary

The occurrence of castes among ants, a fundamental characteristic of eusociality, involves the interplay of two hormones, JH and ecdysone. Evidence in support of this assertion has been presented in this review, involving several observations. In species exhibiting caste differentiation during larval life, several parameters—e.g., total volume of the CA, cellular volume, and staining material in the pars intercerebralis—show a greater involvement of the retrocerebral system in queen-biased larvae than in worker-biased larvae. Removing larval CA favors development of worker larvae over development of queen larvae. Numerous experiments have been conducted by supplying JH analogues. When queens of species exhibiting queen/worker determination during oogenesis, e.g., *Pheidole pallidula*, are treated with JH by topical applications, they lay queen-biased eggs instead of worker-biased eggs. Similarly, when societies of species exhibiting queen/worker determination during larval life, e.g., *Solenopsis invicta*, *Monomorium pharaonis*, or *Myrmica rubra*, are fed a diet mixed with JH analogues, they rear numerous queen larvae. In contrast, ecdysone supply prevents sexualization of the brood.

JHs are also involved in the occurrence of subcastes, as in minor/major worker determination. In *Pheidole bicarinata*, a species exhibiting discrete worker dimorphism, when the last-instar larvae are treated by topical applications of JHA, a large number develop into major workers. In a similar way, societies fed a diet containing JHA rear numerous major workers.

Assays of JH and ecdysone titers at different phases of the queen/worker determination show that the amount of both hormones is connected with development of the worker and queen lines. JH appears to be an earlier signal acting on ecdysteroid levels; the interaction between the two hormones plays a role during caste differentiation.

These hormones act during sensitive periods during which a particular morphogenetic program can be triggered, and a hypothesis has been postulated concerning a link between nutritional factors and hormonal secretion at the time of these sensitive periods. So trophogenetic and neuroendocrine factors are tightly connected.

Finally, evidence has been presented regarding the loss of gonadotropic action of JH in some ants. In these species, like in the honeybee, JH may function exclusively as a determiner of caste polymorphism.

Acknowledgments

Sincere thanks are due to Laurent Keller for enlightening discussions, to Edward Vargo for invaluable help with the English language and critical reading, to Anne Grimal for technical assistance, and to Christiane Mont for typing and preparing this manuscript.

References

Asencot, M. and Y. Lensky. 1976. The effect of sugars and juvenile hormone on the differentiation of the female honeybee larvae (*Apis mellifera*) to queens. Life Sci. 18: 693–700.

Banks, W. A. 1986. Insect growth regulators for control of the imported fire ant. Pp. 387–398 *in* C. S. Lofgren and R. K. Vander Meer (eds.), *Fire Ant and Leaf-cutting Ants: Biology and Management*. Westview Press, Boulder, Colorado.

Banks, W. A. and D. P. Harlan. 1982. Tests with the insect growth regulator, CIBA–GEIGY CGA-38531, against laboratory and field colonies of red imported fire ants. J. Ga. Entomol. Soc. 17: 460–466.

Banks, W. A., C. S. Lofgren, and J. K. Plumley. 1978. Red imported fire ants: effects of insect growth regulators on caste formation and colony growth and survival. J. Econ. Entomol. 71: 75–78.

Barker, J. F. 1978. Neuroendocrine regulation of oocyte maturation in the imported fire ant *Solenopsis invicta*. Gen. Comp. Endocrinol. 35: 234–237.

Barth, R. H., L. Lester, P. Sroka, and R. Hearn. 1975. Juvenile hormone promotes dominance behaviour and ovarian development in social wasps (*Polistes annularis*). Experientia (Basel) 31: 691–692.

Beetsma, J. 1979. The process of queen/worker differentiation in the honeybee. Bee World 60: 24–39.

Bell, W. J. 1973. Factors controlling initiation of vitellogenesis in a primitive social bee, *Lasioglossum zephyrum* (Hymenoptera: Halictidae). Insectes Soc. 20: 253–260.

Bonavita-Cougourdan, A. and L. Passera. 1978. Etude comparative au moyen d'or radioactif de l'alimentation des larves d'ouvrières et des larves de reines chez la fourmi *Plagiolepis pygmaea* Latr. Insectes Soc. 25: 275–287.

Bressac, C. and J. Bitsch. 1969. Observations sur la structure du système nerveux céphalique (cerveau, masse sous-oesophagienne et complexe rétro-cérébral) de la fourmi *Aphaenogaster senilis* (Mayr, 1853) (Hymenoptera, Myrmicinae). Insectes Soc. 16: 135–148.

Brian, M. V. 1959. The neuro-secretory cells of the brain, the *corpora cardiaca* and the *corpora allata* during caste differentiation in an ant. Symp. on the Ontogeny of Insects (Acta Symp. Evol. Insect.), Prague, 1959 pp. 167–171.

Brian, M. V. 1974. Caste differentiation in *Myrmica rubra;* the role of hormones. J. Insect Physiol. 20: 1351–1365.

Brian, M. V. 1976. Endocrine control over caste differentiation in a myrmicine ant. Pp. 63–70 *in* M. Lüscher (ed.), *Phase and Caste Determination in Insects: Endocrine Aspects*. Pergamon Press, Oxford and Elmsford, New York.

Brian, M. V. 1979. Caste differentiation and division of labor. Pp. 121–222 *in* H. R. Hermann (ed.), *Social Insects*, Vol. 1. Academic Press, Orlando, Florida.

Brian, M. V. 1980. Social control over sex and caste in bees, wasps and ants. Biol. Rev. 55: 379–415.

Brian, M. V. (ed.) 1983. *Social Insects: Ecology and Behavioural Biology*. Chapman & Hall, London and New York.

Brian, M. V. and J. Hibble. 1964. Studies of caste differentiation in *Myrmica rubra* L. 7. Caste bias, queen age and influence. Insectes Soc. 11: 223–238.

Brian, M. V. and A. F. Kelly. 1967. Studies of caste differentiation in *Myrmica rubra* L. 9. Maternal environment and the caste bias of larvae. Insectes Soc. 14: 13–24.

Buschinger, A. 1978. Genetisch bedingte Entstehung geflügelter Weibchen bei der sklavenhaltenden Ameise *Harpagoxenus sublaevis* (Nyl.) (Hym., Form.). Insectes Soc. 25: 163–172.

Campos, L. A. 1978. Sex determination in bees. VI. Effect of a juvenile hormone analogue in males and females of *Melipona quadrifasciata* (Apidae). J. Kans. Entomol. Soc. 51: 228–234.

Campos, L. A. 1979. Sex determination in bees. XIII. The role of juvenile hormone in caste determination in *Partamona cupira* (Hym. Apidae). Ciênc. Cult. (São Paulo) 31: 65–69.

Copijn, G. M., J. Beetsma, and P. Wirtz. 1979. Queen differentiation and mortality after application of different juvenile hormone analogues to worker larvae of the honeybee (*Apis mellifera* L.). Proc. K. Ned. Akad. Wet., Ser. C Biol. Med. Sci. 82: 29–42.

Crozier, R. H. and P. Pamilo. 1986. Relatedness within and between colonies of a queenless ant species of the genus *Rhytidoponera* (Hymenoptera: Formicidae). Entomol. Gen. 11: 113–117.

Cupp, E. W. and J. O'Neal. 1973. The morphogenetic effects of two juvenile hormone analogues on larvae of imported fire ants. Environ. Entomol. 2: 191–194.

Dartigues, D. and L. Passera. 1979. Polymorphisme larvaire et chronologie de l'apparition des castes femelles chez *Camponotus aethiops* Latreille (Hymenoptera, Formicidae). Bull. Soc. Zool. Fr. 104: 197–207.

de Wilde, J. 1976. Juvenile hormone and caste differentiation in the honey bee (*Apis mellifera*). Pp. 5–20 *in* M. Lüscher (ed.), *Phase and Caste Determination in Insects*. Pergamon Press, Oxford and Elmsford, New York.

de Wilde, J. and J. Beetsma. 1982. The physiology of caste development in social insects. Adv. Insect Physiol. 16: 167–246.

Dietz, A., H. R. Hermann, and M. S. Blum. 1979. The role of exogenous JH I, JH III and anti-JH (precocene II) on queen induction of 4.5-day-old worker honeybee larvae. J. Insect Physiol. 25: 503–512.

Dogra, G. S., G. M. Ulrich, and H. Rembold. 1977. A comparative study of the endocrine system of the honeybee larvae under normal and experimental conditions. Z. Naturforsch. Sect. C Biosci. 32: 637–642.

Edwards, J. P. 1977. Control of *Monomorium pharaonis* with an insect juvenile hormone analogue. Proc. 8th Int. Congr. IUSSI (Int. Union Study Soc. Insects), Wageningen, Netherlands pp. 81–82.

Edwards, J. P. 1987. Caste regulation in the pharaoh's ant *Monomorium pharaonis:* the influence of queens on the production of new sexual forms. Physiol. Entomol. 12: 31–39.

Edwards, J. P., G. W. Pemberton, and P. J. Curran. 1981. The use of juvenile hormone analogues for control of *Pheidole megacephala* and other house-infesting ants. Pp. 769–779 *in* M. Kloza (ed.), *Regulation of Insect Development and Reproduction*. Wrocław Technical University Press, Wrocław, Poland.

Engels, W. 1986. Reproduction and caste development in social bees. Proc. 10th Int. Congr. IUSSI (Int. Union Study Soc. Insects), Munich pp. 275–281.

Engels, W. and P. S. Ramamurty. 1976. Initiation of oogenesis in allatectomized virgin honey bee queens by carbon dioxide treatment. J. Insect Physiol. 22: 1427–1432.
Fletcher, D. J. C. and K. G. Ross. 1985. Regulation of reproduction in eusocial Hymenoptera. Annu. Rev. Entomol. 30: 319–343.
Fluri, P. 1983. Precocene II has no anti–juvenile hormone effects in adult honey bees. Experientia (Basel) 39: 919–920.
Fluri, P., A. G. Sabatini, and H. Wille. 1981. Blood juvenile hormone, protein and vitellogenin titers in laying and non-laying queen honeybees. J. Apic. Res. 20: 221–225.
Forel, A. (ed.) 1921. *Le Monde Social des Fourmis*, Vol. 1. Kundig, Geneva.
Fortelius, W., P. Pamilo, R. Rosengren, and L. Sundström. 1987. Male size dimorphism and alternative reproductive tactics in *Formica exsecta* ants (Hymenoptera, Formicidae). Ann. Zool. Fenn. 24: 45–54.
Fowler, H. G. 1984. Colony-level regulation of forager caste ratios in response to caste perturbations in the carpenter ant, *Camponotus pennsylvanicus* (De Geer) (Hymenoptera: Formicidae). Insectes Soc. 31: 461–472.
Franks, N. R. 1985. Reproduction, foraging efficiency and worker polymorphism in army ants. Pp. 91–107 *in* B. Hölldobler and M. Lindauer (eds.), *Experimental Behavioural Ecology* (Fortsch. Zool. Vol. 31). Fisher Verlag, Stuttgart.
Gawande, R. B. 1968. A histological study of neurosecretion in ants (Formicoidea). Acta Entomol. Bohemoslov. 65: 349–363.
Gilbert, L. I. (ed.) 1976. *The Juvenile Hormones*. Plenum Press, New York.
Goewie, E. A. 1978. *Regulation of Caste Differentiation in the Honeybee* (*Apis mellifera* L.). Ph.D. thesis, Agricultural University, Wageningen (published in Meded. Landbouwhogesch. Wageningen), Netherlands.
Gösswald, K. and K. Bier. 1953a. Untersuchungen zur Kastendetermination in der Gattung *Formica*. Naturwissenschaften 40: 38–39.
Gösswald, K. and K. Bier. 1953b. Untersuchungen zur Kastendetermination in der Gattung *Formica*. 2. Die Aufzucht von Geschlechtstieren bein *Formica rufa pratensis* (Retz.). Zool. Anz. 151: 126–134.
Hardie, J. and A. D. Lees. 1985. Endocrine control of polymorphism and polyphenism. Pp. 441–490 *in* G. A. Kerkut and L. I. Gilbert (eds.), *Comprehensive Insect Physiology, Biochemistry and Pharmacology*, Vol. 8. Pergamon Press, Oxford and Elmsford, New York.
Hermann, H. R. 1979. Insect sociality: an introduction. Pp. 1–33 *in* H. R. Hermann (ed.), *Social Insects*, Vol. 1. Academic Press, Orlando, Florida.
Hultin, T. 1947. The corpora allata in various castes of ants. K. Fysiogr. Sällsk. Lund. Förh. 17: 1–7.
Lanzrein, B., V. Gentinetta, and R. Fehr. 1985. Titers of juvenile hormone and ecdysteroids in reproductives and eggs of *Macrotermes michaelseni*: relation to caste determination. Pp. 307–327 *in* J. A. L. Watson, B. M. Okot-Kotber, and Ch. Noirot (eds.), *Caste Differentiation in Social Insects*. Pergamon Press, Oxford and Elmsford, New York.
Lappano, E. R. 1958. A morphological study of larval development in polymorphic all-worker brood of the army ant *Eciton burchelli*. Insectes Soc. 5: 31–66.
Latreille, P. A. (ed.) 1797. *Essai sur l'Histoire des Fourmis de la France*. Bourdeaux Brive.
Lebrun, D. 1985. The role of hormones in social polymorphism and reproduction in *Kalotermes flavicollis* Fabr. Pp. 227–238 *in* J. A. L. Watson, B. M. Okot-Kotber, and Ch. Noirot (eds.), *Caste Differentiation in Social Insects*. Pergamon Press, Oxford and Elmsford, New York.
Le Masne, G. 1956. La signification des reproducteurs aptères chez la fourmi *Ponera eduardi* Forel. Insectes Soc. 3: 239–259.
Lüscher, M. 1972. Environmental control of juvenile hormone (JH) secretion and caste differentiation in termites. Comp. Endocrinol. Suppl. 3: 509–514.

Lüscher, M. 1974. Kasten und Kastendifferenzierung bei niederen Termiten. Pp. 695–739 *in* G. H. Schmidt (ed.), *Sozialpolymorphismus bei Insekten*. Wissenschaftliche Verlagsgesellschaft, Stuttgart.

Lüscher, M. 1976. Evidence for an endocrine control of caste determination in higher termites. Pp. 91–103 *in* M. Lüscher (ed.), *Phase and Caste Determination in Insects*, Pergamon Press, Oxford and Elmsford, New York.

Michener, C. D. 1961. Social polymorphism in Hymenoptera. Pp. 43–56 *in* J. S. Kennedy (ed.), *Insect Polymorphism* (Symp. No. 1, R. Entomol. Soc. Lond.). Bartholemew Press, Dorking, Surrey.

Nijhout, H. F. and D. E. Wheeler. 1982. Juvenile hormone and the physiological basis of insect polymorphisms. Q. Rev. Biol. 57: 109–133.

Noirot, Ch. 1977. Various aspects of hormone action in social insects. Proc. 8th Int. Congr. IUSSI (Int. Union Study Soc. Insects), Wageningen, Netherlands pp. 12–16.

Okot-Kotber, B. M. 1985. Mechanisms of caste determination in a higher termite, *Macrotermes michaelseni* (Isoptera, Macrotermitinae). Pp. 267–306 *in* J. A. L. Watson, B. M. Okot-Kotber, and Ch. Noirot (eds.), *Caste Differentiation in Social Insects*. Pergamon Press, Oxford and Elmsford, New York.

Ono, S. 1982. Effect of juvenile hormone on the caste determination in the ant *Pheidole fervida* Smith (Hymenoptera, Formicidae). Appl. Entomol. Zool. 17: 1–7.

Palma-Valli, G. and G. Délye. 1981. Contrôle neuro-endocrine de la ponte chez les reines de *Camponotus lateralis* Olivier (Hymenoptera, Formicidae). Insectes Soc. 28: 167–181.

Passera, L. 1969. Biologie de la reproduction chez *Plagiolepis pygmaea* Latreille et ses deux parasites sociaux *Plagiolepis grassei* Le Masne et *Plagiolepis xene* Stärke (Hymenoptera, Formicidae). Ann. Sci. Nat. Zool. Biol. Anim. [12] 11: 327–482.

Passera, L. 1974. Différenciation des soldats chez la fourmi *Pheidole pallidula* (Nyl.) (Formicidae, Myrmicinae). Insectes Soc. 21: 71–86.

Passera, L. 1980a. La ponte d'oeufs préorientés chez la fourmi *Pheidole pallidula* (Nyl.) (Hymenoptera–Formicidae). Insectes Soc. 27: 79–95.

Passera, L. 1980b. La fonction inhibitrice des reines de la fourmi *Plagiolepis pygmaea* Latr.: rôle des phéromones. Insectes Soc. 27: 212–225.

Passera, L. (ed.) 1984. *L'Organisation Sociale des Fourmis*. Privat, Toulouse.

Passera, L. 1985. Soldier determination in ants of the genus *Pheidole*. Pp. 331–346 *in* J. A. L. Watson, B. M. Okot-Kotber, and Ch. Noirot (eds.), *Caste Differentiation in Social Insects*. Pergamon Press, Oxford and Elmsford, New York.

Passera, L. and J. P. Suzzoni. 1974. Action d'un analogue de l'hormone juvénile sur la fécondité d'une fourmi (Hymenoptera, Formicidae). C. R. Acad. Sci. Paris 279D: 2079–2082.

Passera, L. and J. P. Suzzoni. 1978a. Sexualisation du couvain de la fourmi *Pheidole pallidula* (Hymenoptera, Formicidae) après traitement par l'hormone juvénile. C. R. Acad. Sci. Paris 286D: 615–618.

Passera, L. and J. P. Suzzoni. 1978b. Traitement des reines par l'hormone juvénile et sexualisation du couvain chez *Pheidole pallidula* (Nyl.) (Hymenoptera, Formicidae). C. R. Acad. Sci. Paris 287D: 1231–1233.

Passera, L. and J. P. Suzzoni. 1979. Le rôle de la reine de *Pheidole pallidula* (Nyl.) (Hymenoptera, Formicidae), dans la sexualisation du couvain après traitement par l'hormone juvénile. Insectes Soc. 26: 343–353.

Passera, L. and J. P. Suzzoni. 1984. La ponte d'oeufs préorientés par la fourmi *Pheidole pallidula* (Nyl.): caractéristiques biologiques des reines à la fin de l'hibernation. Insectes Soc. 31: 155–170.

Petersen-Braun, M. 1975. Untersuchungen zur sozialen Organisation der Pharaoameise *Monomorium pharaonis* (L.) (Hymenoptera, Formicidae). 1. Der Brutzyklus und seine Steuerung durch populationseigene Faktoren. Insectes Soc. 22: 269–292.

Petersen-Braun, M. 1977. Untersuchungen zur sozialen Organisation der Pharaoameise *Monomorium pharaonis* (L.) (Hymenoptera, Formicidae). 2. Die Kastendeterminierung. Insectes Soc. 24: 303–318.

Petralia, R. S. and Vinson, S. B. 1980. Internal anatomy of the fourth instar larva of the imported fire ant, *Solenopsis invicta* Buren (Hymenoptera, Formicidae). Int. J. Insect Morphol. Embryol. 9: 89–106.

Plateaux, L. 1971. Sur le polymorphisme social de la fourmi *Leptothorax nylanderi* (Förster). II. Activité des ouvrières et déterminisme des castes. Ann. Sci. Nat. Zool. Biol. Anim. [12] 13: 1–90.

Porter, S. D. and W. R. Tschinkel. 1985. Fire ant polymorphism (Hymenoptera: Formicidae): factors affecting worker size. Ann. Entomol. Soc. Am. 78: 381–386.

Robeau, R. M. and S. B. Vinson. 1976. Effects of juvenile hormone analogues on caste differentiation in the imported fire ant, *Solenopsis invicta*. J. Ga. Entomol. Soc. 11: 198–203.

Röseler, P. F. 1974. Vergleichende Untersuchungen zur oogenese bei weiselrichtigen und weisellosen Arbeiterinnen der Hummelart *Bombus terrestris* L. Insectes Soc. 21: 249–274.

Röseler, P. F. 1977. Endocrine control of polymorphism in bumblebees. Proc. 8th Int. Congr. IUSSI (Int. Union Study Soc. Insects), Wageningen, Netherlands pp. 22–23.

Röseler, P. F. and I. Röseler. 1978. Studies on the regulation of the juvenile hormone titer in bumblebee workers, *Bombus terrestris*. J. Insect Physiol. 24: 707–713.

Röseler, P. F., I. Röseler, and C. G. J. van Honk. 1981. Evidence for inhibition of *corpora allata* activity in workers of *Bombus terrestris* by a pheromone from the queen's mandibular glands. Experientia (Basel) 37: 348–351.

Röseler, P. F., I. Röseler, and A. Strambi. 1985. Role of ovaries and ecdysteroids in dominance hierarchy establishment among foundresses of the primitively social wasp, *Polistes gallicus*. Behav. Ecol. Sociobiol. 18: 9–13.

Röseler, P. F., I. Röseler, and A. Strambi. 1986. Studies of the dominance hierarchy in the paper wasp *Polistes gallicus* (L.) (Hymenoptera, Vespidae). Monit. Zool. Ital. [NS] 20: 283–290.

Röseler, P. F., I. Röseler, A. Strambi, and R. Augier. 1984. Influence of insect hormones on the establishment of dominance hierarchies among foundresses of the paper wasp *Polistes gallicus*. Behav. Ecol. Sociobiol. 15: 133–142.

Ross, K. G. and D. J. C. Fletcher. 1985. Genetic origin of male diploidy in the fire ant, *Solenopsis invicta* (Hymenoptera: Formicidae) and its evolutionary significance. Evolution 39: 888–903.

Rupěs, V., I. Hrdý, J. Pinterová, J. Žďárek, and J. Křeček. 1978. The influence of methoprene on pharaoh's ant, *Monomorium pharaonis* colonies. Acta Entomol. Bohemoslov. 75: 155–163.

Schmidt, G. H. 1961. Sekretionsphasen und cytologische Beobachtungen zur Funktion der Oenocyten während der Puppenphase verschiedener Kasten und Geschlechter von *Formica polyctena* Först. (Ins. Hym. Form.). Z. Zellforsch. 55: 707–723.

Snelling, R. R. 1981. Systematics of social Hymenoptera. Pp. 369–453 *in* H. R. Hermann (ed.), *Social Insects*, Vol. 2. Academic Press, Orlando, Florida.

Strambi, A. 1986. Endocrine control of caste determination in social Hymenoptera. Proc. 10th Int. Congr. IUSSI (Int. Union Study Soc. Insects), Munich pp. 271–274.

Suzzoni, J. P. 1983. Le polymorphisme et son déterminisme chez deux espèces de fourmis: *Plagiolepis pygmaea* Latr. (Formicinae) et *Pheidole pallidula* (Nyl.) (Myrmicinae): rôle des hormones du développement. Thesis, University of Toulouse.

Suzzoni, J. P. and A. Grimal. 1981. Variations biométriques des corps allates pendant la différenciation des castes reine et ouvrière chez *Plagiolepis pygmaea* Latr. (Hymenoptera Formicidae). Insectes Soc. 27: 399–414.

Suzzoni, J. P., A. Grimal, and L. Passera, 1986a. Modalités et cinétique du développement larvaire chez la fourmi *Camponotus aethiops* Latr. Bull. Soc. Zool. Fr. 111: 113–121.

Suzzoni, J. P., A. Grimal, and L. Passera, 1986b. La ponte de *Camponotus aethiops* (Hymenoptera Formicidae): interactions sociales et rôle de l'hormone juvénile. Actes Colloq. Insectes Soc. 3: 187–195.

Suzzoni, J. P., L. Passera, and A. Strambi. 1980. Ecdysteroid titer and caste determination in the ant *Pheidole pallidula* (Nyl.) (Hymenoptera, Formicidae). Experientia (Basel) 36: 1228–1229.

Suzzoni, J. P., L. Passera, and A. Strambi. 1981. Corps allates et ecdystéroïdes au cours de la différenciation de la caste chez la fourmi *Plagiolepis pygmaea*. Bull. Interne Sect. Fr. UIEIS (Union Int. Etude Insectes Soc.), Toulouse pp. 117–120.

Suzzoni, J. P., L. Passera, and A. Strambi. 1982. Etude morpho-anatomique et physiologique de la "soldatisation" chez la fourmi *Pheidole pallidula* (Nyl.). Bull. Interne Sect. Fr. UIEIS (Union Int. Etude Insectes Soc.), Barcelona pp. 147–156.

Suzzoni, J. P., L. Passera, and A. Strambi. 1983. Ecdysteroid production during caste differentiation in larvae of the ant, *Plagiolepis pygmaea*. Physiol. Entomol. 8: 93–96.

Terron, G. 1972. Observations sur les mâles ergatoïdes et les mâles ailés chez une fourmi du genre *Technomyrmex* Mayr (Hym., Formicidae, Dolichoderinae). Ann. Fac. Sci. Cameroun 10: 107–120.

Troisi, S. J. and L. M. Riddiford. 1974. Juvenile hormone effects on metamorphosis and reproduction of the fire ant, *Solenopsis invicta*. Environ. Entomol. 3: 112–116.

van Laere, O. 1974. Physiology of the honeybee corpora allata. 3. A new method of allatectomy of queens. J. Apic. Res. 13: 15–18.

Vargo, E. L. and D. J. C. Fletcher. 1986. Evidence of pheromonal queen control over the production of male and female sexuals in the fire ant, *Solenopsis invicta*. J. Comp. Physiol. A159: 741–749.

Vargo, E. L. and D. J. C. Fletcher. 1987. Effect of queen number on the production of sexuals in natural populations of the fire ant *Solenopsis invicta*. Physiol. Entomol. 12: 109–116.

Velthuis, H. H. and F. M. Velthuis-Kluppell. 1975. Caste differentiation in a stingless bee, *Melipona quadrifasciata* Lep. influenced by juvenile hormone application. Proc. K. Ned. Akad. Wet. Ser. C Biol. Med. Sci. 78: 81–94.

Vinson, S. B. and R. Robeau. 1974. Insect growth regulator: effects on colonies of the imported fire ant. J. Econ. Entomol. 67: 584–587.

Vinson, S. B., R. Robeau, and L. Dzuik. 1974. Bioassay and activity of several insect growth regulators on the imported fire ant. J. Econ. Entomol. 67: 325–328.

Weir, J. S. 1959. Changes in the retrocerebral endocrine system of larvae of *Myrmica,* and their relation to larval growth and development. Insectes Soc. 6: 375–386.

West-Eberhard, M. J. 1983. Sexual selection, social competition, and speciation. Q. Rev. Biol. 58: 155–183.

Wheeler, D. E. 1986. Developmental and physiological determinants of caste in social Hymenoptera: evolutionary implications. Am. Nat. 128: 13–34.

Wheeler, D. E. and H. F. Nijhout. 1981. Soldier determination in ants: new role for juvenile hormone. Science (Wash., DC) 213: 361–363.

Wheeler, D. E. and H. F. Nijhout. 1983. Soldier determination in *Pheidole bicarinata:* effect of methoprene on caste and size within castes. J. Insect Physiol. 29: 847–854.

Wheeler, W. M. (ed.) 1926. *Les Sociétés d'Insectes: Leur Origine, leur Evolution*. Douin, Paris.

Wilson, E. O. 1953. The origin and evolution of polymorphism in ants. Q. Rev. Biol. 28: 136–156.

Wilson, E. O. (ed.) 1971. *The Insect Societies*. Harvard University Press (Belknap), Cambridge, Massachusetts.

Wilson, E. O. 1979. The evolution of caste systems in social insects. Proc. Am. Philos. Soc. 123: 204–210.

Wilson, E. O. 1985a. The principles of caste evolution. Pp. 307–324 *in* B. Hölldobler and M. Lindauer (eds.), *Experimental Behavioural Ecology* (Fortschr. Zool., Vol. 31). Fisher Verlag, Stuttgart.

Wilson, E. O. 1985b. The sociogenesis of insect colonies. Science (Wash., DC) 228: 1489–1495.

Wirtz, P. and J. Beetsma. 1972. Induction of caste differentiation in the honeybee (*Apis mellifera* L.) by juvenile hormone. Entomol. Exp. Appl. 15: 517–520.

Ždárek, J., O. Haragsim, and V. Vesely. 1976. Action of juvenoids on the honey bee colony. Z. Angew. Entomol. 81: 392–401.

Roles of Androgenic Gland Hormone in Determining the Sexual Characters in Crustacea

12

GENEVIÈVE G. PAYEN

12.1. Introduction

The androgenic gland (AG) of crustaceans is the source of the masculinizing hormone. Several experiments carried out initially by Charniaux-Cotton (1954a,b, 1955, 1962) demonstrated that the testis is not involved in male differentiation. Thus, in the amphipod *Orchestia gammarella*, after castration, the males do not show any modification. However, an ovary implanted into an intact or castrated (= gonadectomized) male was transformed into a testis. In contrast, an ovary remains normal when transplanted into a male deprived of the AG.

Masculinization of females by implantation of the AG has been obtained in *Orchestia*, several isopods, and the decapod *Macrobrachium rosenbergii*. Such results have been reported more or less extensively in recent reviews by Katakura (1984), Charniaux-Cotton and Payen (1985), and Legrand et al. (1987), and also in a few papers dealing with the control of male genital activity (Payen, 1980, 1983, 1986).

In this chapter, special attention is devoted to the roles and modalities of action of the androgenic gland hormone (AGH) on the genital apparatus and external sex characteristics. Indeed, male differentiation of the genital apparatus and secondary sex characteristics in Malacostraca constitutes an original pattern in the animal kingdom since it is controlled only by a hormone whose source is external to the gonads.

Only recently, an active protein substance showing all the different functions of the AGH was extracted from the AG of the isopod *Armadillidium vulgare* (Katakura et al., 1975; Juchault et al., 1978; Hasegawa et al., 1987). These research efforts should permit a new approach to the study of sexual endocrinology of Malacostraca, a field that remained fruitless for several decades. Isolation of the gene coding for the AGH, i.e., the male-determining gene, can be also viewed as belonging to this line of work.

12.2. Characterization of the Androgenic Gland Hormone

The chemical nature of AGH remained unknown for a long time. Various investigations carried out on decapods since the early 1960s and involving injection of steroids in females (Carlisle, 1960; Sarojini, 1963, 1964; Tcholakian and Eik-Nes, 1971) or males (Nagabhushanam and Kulkarni, 1981) claimed that some similarities existed between the AGH and the sexual hormones of vertebrates.

However, no masculinizing effect was obtained with such compounds injected into female peracarids (Juchault, 1967; Charniaux-Cotton, 1972). Indeed, biochemical studies performed in decapods have revealed the presence of steroids or enzymes of steroidogenesis in the region that includes both the AG and the adjacent part of the sperm duct (Gilgan and Idler, 1967; Tcholakian and Eik-Nes, 1969, 1971; Godbillon and Balesdent, 1974; Veith and Malecha, 1983). According to Blanchet et al. (1972), the data on steroid metabolism would seem to concern the sperm duct rather than the AG. Moreover, the existence of steroid metabolism in the gland does not prove the steroid nature of the male hormone. It should be recalled that on the basis of ultrastructural and histological studies in normal and experimental males of various species, other authors have proposed that the AGH is proteinaceous (see Chapter 8, Sections 8.2 and 8.3, in Part 2).

Until now, attempts to isolate the AGH have been made only with the oniscoid isopod *Armadillidium vulgare* (Katakura et al., 1975; Juchault et al., 1978, 1984; Hasegawa et al., 1987) and the decapod *Carcinus maenas* (Berreur-Bonnenfant et al., 1973; Férézou et al., 1977b). With substances extracted from AG and tested for their masculinizing effects, different chemical natures, lipidic in the decapod and proteinaceous in the isopod have been reported.

I shall deal first with the results obtained in the isopod because (1) the proteinaceous nature of AGH agrees with the cytological features of the AG (abundant rough-surfaced endoplasmic reticulum and Golgi complexes that elaborate dense-cored vesicles and (2) the isolated substance exhibits an androgenic function, i.e., induces masculinization of both the external sexual characteristics and the internal reproductive organs of young females (see Section 12.4, below).

In 1975, Katakura et al. partially purified an active protein substance of some 15–17 kDa, isolated from 3000 pairs of male genital apparatuses, including the AG. This substance masculinizes the first pair of pleopods in young female *Armadillidium vulgare.* The water-soluble active substance was extracted and purified by a procedure that included ammonium sulfate precipitation, gel filtration through Sephadex, and column chromatography on DEAE(diethylaminoethyl)–cellulose. The proteinic nature was judged by proteolytic digestions. One *Armadillidium* unit was defined as the minimum amount necessary to masculinize abdominal appendages of one test animal. Masculinization of the female genital apparatus was obtained by several injections (Katakura and Hasegawa, 1983). When an extract of male reproductive organs from which the AG had been removed was injected into a young female, masculinization of the recipient was not induced (Katakura, 1984).

Juchault et al. (1978, 1984) extracted a water-soluble, dialyzable, thermostable factor from AG of intersexed males of the same species, *A. vulgare,* which display a hypertrophied AG. This factor, purified with the techniques of gel filtration, thin-layer chromatography, or high-performance liquid chromatography (HPLC), induced the appearance of external male characteristics in injected females. Its activity is destroyed by proteolytic enzymes, which confirms its protein nature as reported by Katakura et al. (1975). The latter authors used dialysis studies with cellulose membranes of known porosities and suggested a possible molecular weight range between 1.2 and 8.0 kDa.

Isolation and properties of the hormone have recently been reported by Hasegawa et al. (1987). This purification procedure includes pH adjustment, ammonium sulfate precipitation, gel filtration, ion-exchange chromatography, and three kinds of HPLC. AGH consists of two molecular forms, AGH_{I} and AGH_{II}, judged homogeneous by HPLC, sodium dodecyl sulfate–polyacrylamide gel electrophoresis (SDS–PAGE), and isoelectrofocusing. Their molecular weight was estimated to be, respectively, 17.0 ± 0.8 and 18.3 ± 1.0 kDa. Their chemical properties are alike, and their respective specific activities are 560 and 430 times that of the crude extract.

It appears that the two kinds of data concerning the various attempts at purification, in terms of the molecular size, cannot be attributed only to the incomplete purification of the AGH in intersexes. Indeed, they may also result from the difference of the molecular weights of AGH that could exist between normal males and intersexes, being due to the differences in their major molecular forms.

12.3. Purification and Role of a Lipoidal Factor from Androgenic Glands

Berreur-Bonnenfant et al. (1973) have isolated from the AG of *Carcinus maenas* a lipidic fraction that has provided a terpenic molecule inhibiting vitellogenesis in the ovaries of *Orchestia gammarella*. This substance also induces a male-type accumulation of carotenoid pigments in the antennae of females of *Talitrus saltator*. Two C_{18} isoprenoid ketones, hexahydrofarnesylacetone and farnesylacetone, have been identified (Férézou et al., 1977a,b) and shown to be exclusively synthesized by the AG (Férézou et al., 1978). Neither induction of spermatogenesis nor differentiation of male external sexual characteristics were obtained following injection of these substances into females.

Berreur-Bonnefant and Lawrence (1984) have described the effect of farnesylacetone on macromolecular synthesis in cultured ovaries. Leucine incorporation in ovaries is lowered in the presence of 10^{-7} *M* farnesylacetone. This molecule is more effective on immature ovaries than on ovaries in vitellogenesis. Farnesylacetone inhibits ovarian protein synthesis through the transcription step. In addition, uridine incorporation in ovaries is inhibited. At last, the effect of farnesylacetone was observable in the presence of actinomycin. The action of farnesylacetone is gonad specific and not species specific. The activity of the molecule on the gonads varies with the sexual cycle of the animals.

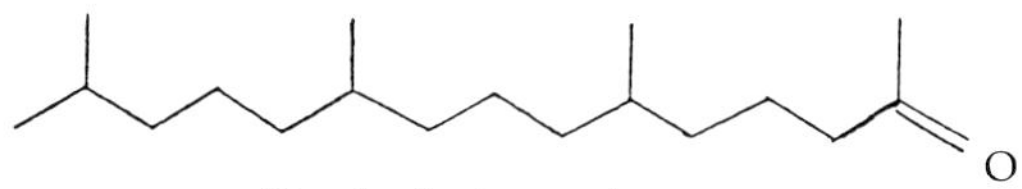

Hexahydrofarnesylacetone

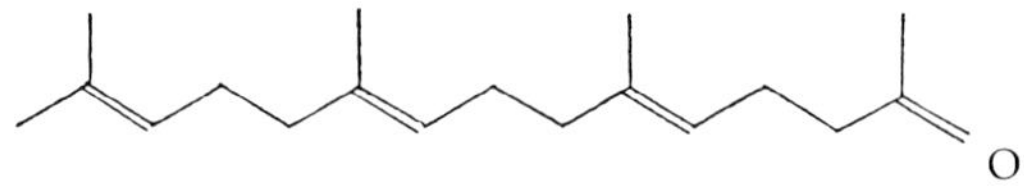

Farnesylacetone

Farnesylacetone = trimethyl-6,10,14-pentadecatriene-5*E*,9*E*,13one-2

In natural conditions, farnesylacetone does not act on ovaries but probably acts on the oocytes that appear in testes of males whose AG are small or have been removed. The AG have a double inhibitory action on this oogenesis. Farnesylacetone inhibits endogenous protein synthesis, chiefly in oocytes in previtellogenesis. AGH inhibits vitellogenin synthesis, which is why the oocytes are arrested in previtellogenesis and have a degenerative appearance.

During winter, farnesylacetone does not act on oocytes. That is probably why the oocytes are often more numerous and of good appearance in the testes of many species during genital rest. This result seems to be explained by a modification of oocyte receptors to farnesylacetone.

To summarize, nonfunctional hermaphroditism in males occurs not infrequently. The actions of AGH and farnesylacetone provoke the disappearance of most oocytes during the season of reproduction.

12.4. Role of the AGH in the Control of Sexual Differentiation

As already pointed out (see Section 12.1), the existence of a male hormone involved in the control of all the male differentiation and exclusively secreted by the AG was demonstrated in *Orchestia gammarella*, following inversion of ovaries implanted into normal and castrated males (Charniaux-Cotton, 1954a). The AGs are the only organs or tissues in the male that, when implanted into a pubescent or juvenile female, masculinize the ovaries and deferent ducts; simultaneously, secondary sexual characteristics become masculinized (Charniaux-Cotton, 1954b).

In the Isopoda Oniscoidea, the testes were first considered as the source of a male hormone inducing the secondary sexual characteristics (De Lattin and Gross, 1953; Legrand, 1954). However, AG were then described in *Asellus aquaticus* (Balesdent-Marquet, 1958). A short time later, AG in oniscoids were found attached to the apex of the three testicular lobes (Legrand, 1958; Katakura, 1959). The transformation of ovaries into testes were then obtained in *Armadillidium vulgare* by grafting AG into young females (Katakura, 1960), as well as in pubertal females of *Porcellio dilatatus petiti* and *Sphaeroma serratum* (Legrand et al., 1974; Raimond and Juchault, 1983).

In decapods, in spite of many implantations of AG into females, the masculinization of external sexual characteristics was obtained only in hermaphroditic prawns *Lysmata seticaudata* (Charniaux-Cotton, 1959) and *Pandalus borealis* (Berreur-Bonnenfant and Charniaux-Cotton, 1965), and occasionally in crabs *Carcinus maenas* (Charniaux-Cotton, 1958) and *Rhithropanopeus harrisii* (Payen, 1969, 1975). External masculinization was also obtained in the shrimp *Palaemonetes varians* and the crayfish *Procambarus clarki* (Charniaux-Cotton and Cazès, 1979; Nagamine and Knight, 1987). The complete role of AG in decapods was demonstrated by implantations and ablations in young females of *Macrobrachium rosenbergii* (Nagamine et al. 1980a,b). Nevertheless, in most malacostracans the action of circulating AGH is limited to the inhibition of vitellogenesis and oocyte degeneration. Indeed, it has been shown that this inhibition results from the disappearance of vitellogenin in the hemolymph of the grafted females (Besse et al., 1970; Meusy et al., 1971; Souty-Grosset and Juchault, 1987; see also the review by Meusy and Payen, 1988).

12.4.1. *Development of the Male Genital Tract and Onset of Spermatogenesis*

Circumstances regarding masculinization of females of *O. gammarella* by grafts of AG have been the object of several reviews by Charniaux-Cotton (1960, 1965, 1972). About 2 or 3 weeks after implantation of one or two androgenic glands, the germinative zone of the gonads releases germ cells that effect spermatogenesis in accordance with the same modalities as in normal testis. The follicular cells are transformed into "nurse cells," and their nuclei become polyploid. The first spermiogenesis is effected only when this transformation has taken place. The spermatozoids are fertile. In females engaged in reproduction, the first action of implanted AG is to inhibit the onset of the next vitellogenesis (Fig. 12.1). The implanted AG also provokes the development of rudiments of the sperm ducts through cellular multiplication. They grow to the last sternite, and genital papillae appear (Zerbib, 1964). The younger the female at the time of implantation, the more normal is the structure of the sperm duct, but the duct is not functional, because the genital opening is not in continuity with the duct. The existence of perfect thelygenic males (XX) makes possible the grafting of AG into females before the external differentiation of sex (Ginsburger-Vogel, 1972, 1975). If the implantation is carried out before the fourth

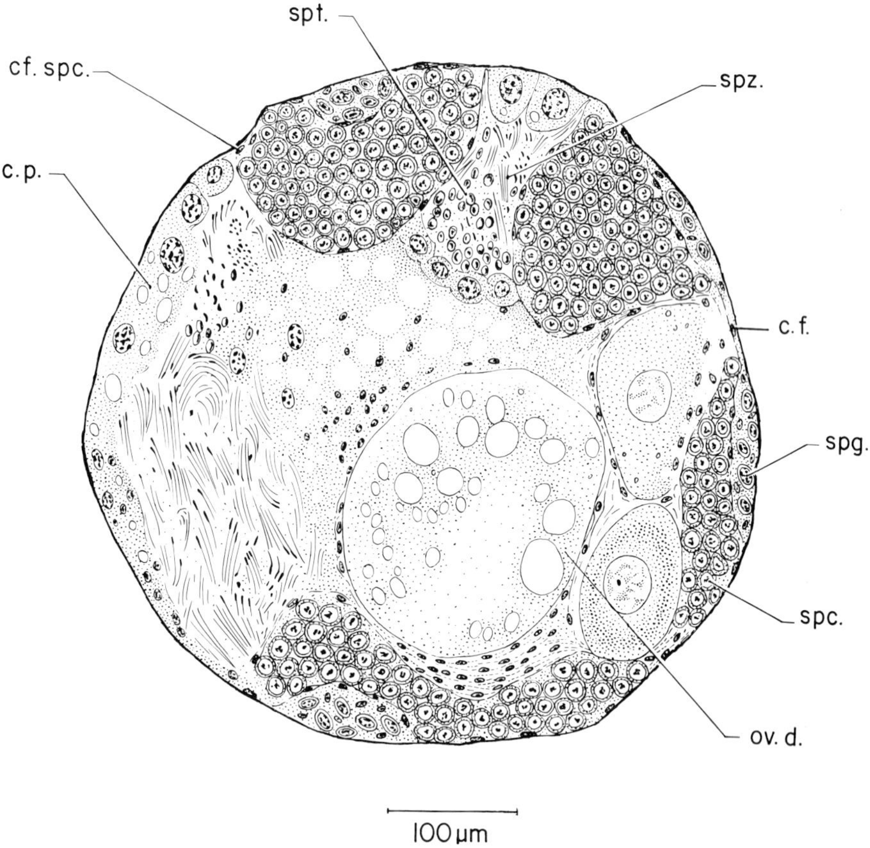

FIGURE 12.1. Transverse section of masculinized ovary in *Orchestia gammarella*. Gonad has been perserved after fifth molt, following implantation of androgenic gland. *Key:* c. f. = follicular cell; c. p. = polyploidic cell secreting mucopolysaccharids; cf. spc. = follicular cell surrounding spermatocytes; ov. d. = degenerating oocyte; spc. = spermatocyte; spg. = spermatogonium; spt. = spermatids; spz. = spermatozoa. (From Charniaux-Cotton, 1957, 1960.)

postembryonic molt, the females are transformed into functional neomales. It is important to note that these neomales have functional AG. After the complete removal of AG spermatogenesis decreases and then stops, some oocytes appear (Charniaux-Cotton, 1964).

Implantations of AG into young females of oniscoid isopods have shown that the implants induce the development of new, normally located AG (Katakura, 1960, 1961, 1967; Juchault and Legrand, 1964; see Legrand and Juchault, 1972, for a review). The transformation from ovaries into testes is due to the action of AGH diffusing from the newly formed glands and progressing along the genital tract. Actually, in species in which an AG is located at the anterior end of each of the three pairs of testes, each testicular anlage becomes organized as a testis only if a new gland has developed at its extremity. Comparable results have been obtained following injection of AGH into young females of *Armadillidium vulgare* (Katakura and Hasegawa, 1983).

Implantation of undifferentiated genital apparatus of *Helleria brevicornis* into pubescent males demonstrates equally well that the newly formed AG are responsible for male differentiation. Indeed, all undifferentiated genital apparatuses of *H. brevicornis* implanted into males differentiate AG and testicular tubes. However, when implants are deprived of presumptive androgenic area, they acquire an ovarian structure (Juchault, 1967).

The study of normal male sexual differentiation in *O. gammarella* and in oniscoids (see Chapter 8 in Part 2) suggests that this differentiation begins at the level of the AG primordia and progresses along the genital tract; this corroborates the concept of diffusion of AGH along this tract. It is probably the same in the reptantian decapods studied, even though no androgenic rudiment can be distinguished at the end of the sperm duct. Possibly in certain Reptantia, such as the crayfish *Pontastacus leptodactylus leptodactylus*, all mesodermal tissue of the genital tract is androgenic. In fact, sexual dimorphism begins at the level of this tissue, and it develops further in the future testis (Payen, 1973).

In most species, the transformation of pubescent ovaries into testes through circulating AGH, as occurs in mature female *Orchestia* after implantation of AG, appears to be an exceptional phenomenon that differs from what occurs under natural conditions.

In conclusion, AGH determines the male differentiation of genital tract by diffusion along this tract from the primordia of AG. AGH is also capable of stimulating the development of AG primordia in genetic females. These data allow us to conclude (in accord with Katakura, 1984) that in genetic males the hormone is first produced by the male-determining gene and is responsible for the development of the AG primordia.

12.4.2. *Development of the Secondary Male Characteristics*

The first proof of the control of the male external characteristics by AG was obtained in *Orchestia gammarella* (Charniaux-Cotton, 1954b). Juvenile or pubertal females implanted with one or two AG are completely masculinized. The appendages of the female progressively acquire the male form, their development being similar to that in a normal male. The masculinized females show male sexual behavior. The first experiments also demonstrated that the action of the AG is not mediated by the testes. Thus, an ovariectomized female implanted with AG acquires the male phenotype. The testes release no masculinizing hormone; a female implanted with a testis is not masculinized. Moreover, AG are the exclusive source of the male hormone; bilateral removal of these glands stops differentiation and growth of male appendages. An amputated gnathopod regenerates in an undifferentiated form (Charniaux-Cotton, 1955) (Fig. 12.2). The results obtained in *Orchestia* were the subject of detailed publications (Charniaux-Cotton, 1957, 1961).

Biometric study of the relative growth of the second gnathopod (Fig. 12.3) in normal and experimental *Orchestia* have shown that the AGH induces the male growth of this variant. Moreover, the changes in level of allometric growth appear to be due to changes in hormonal level. Thus, implantation of a supplementary AG in very young males at the beginning of the immature phase immediately triggers rapid growth characteristic of the intermediate phase (Charniaux-Cotton, 1961). The chela acquires a pubertal form but is shorter than in normal males; its length conforms to the allometry rule for the pubertal phase (Fig. 12.3). From these results one can interpret the effects of ablation of various areas of the protocerebrum in the isopods *Porcellio dilatatus* and *Idotea balthica* and in the crab *Carcinus maenas* (Juchault et al., 1965; Reidenbach, 1966; Demeusy, 1967). The acceleration of external sexual differentiation is probably due to an increase in the level of male hormone following suppression of a moderating control of the AG activity (see Section 12.6, below).

After the discovery of AG in *Asellus aquaticus* (Balesdent-Marquet, 1958), external masculinization of juvenile and pubertal females by transplantation of these glands was obtained in various isopods. As in amphipods, the order of appearance of the male characteristics corresponds to that occurring during normal development of a male (Legrand and Juchault, 1972).

Thus, in peracarids, development of secondary male characteristics is induced by circulating AGH. During the entire life span of the females, the target organs retain the capacity to respond to the male hormone. In contrast, the genital rudiments of isopods respond only to the diffusing hormone from AG during early development. In genetic males, testicular organogenesis precedes the development of external sexual characteristics. That is easily understandable because the secretion of the androgenic cells propagates as an embryonic inductor along the genital tract; for AGH to act on the target organs, its level must reach a certain threshold in the hemolymph. This threshold varies with each organ because, in a given species, the secondary sexual characteristics appear in a well-determined order.

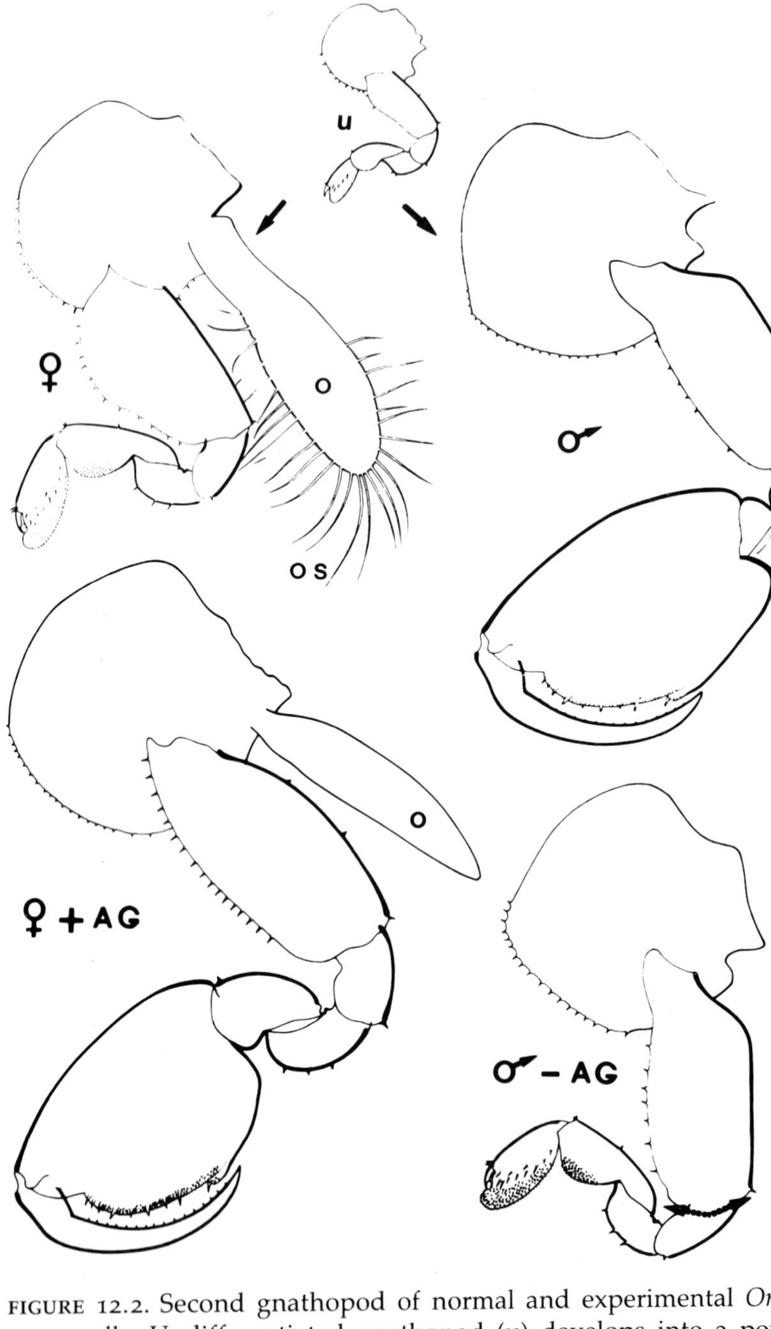

FIGURE 12.2. Second gnathopod of normal and experimental *Orchestia gammarella*. Undifferentiated gnathopod (u) develops into a powerful claw in the male. In the female, oostegite (o) is a permanent female characteristic and ovigerous setae (os) are temporary sex characteristics. After implantation of an androgenic gland into a female (♀ + AG), the second gnathopod is transformed into a male claw; the oostegite per-

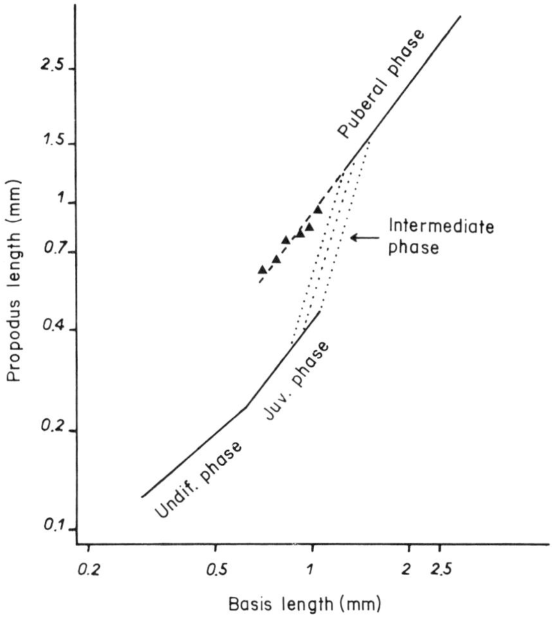

FIGURE 12.3. *Orchestia gammarella* males. Log length of propodus of second gnathopod plotted against log length of basis of third pereopod. Triangles indicate propodus of juvenile males that have received an additional androgenic gland. They acquire the pubertal form; their length is situated on the extension of the regressive line representative of the pubertal phase. *Key:* Juv. phase = juvenile phase; Undif. phase = undifferentiated phase. (From Charniaux-Cotton, 1957, 1961.)

In decapods, numerous experiments have shown that implantation of AG into prepubertal or pubertal females does not masculinize the external sexual characteristics. Masculinization of pleopods with development of the appendix masculina has been obtained in the pubertal shrimps *Palaemonetes varians* (Charniaux-Cotton and Cazès, 1979) and *Macrobrachium rosenbergii* (Nagamine et al., 1980b), as well as in certain hermaphroditic shrimps in the female phase, namely, *Lysmata seticaudata* and *Pandalus borealis* (Charniaux-Cotton, 1959; Berreur-Bonnenfant and

---

sists. After removal of androgenic gland (♂ − AG), the second gnathopod of the male regenerates into the undifferentiated form. Amputation was made between basipodite and ischiopodite. If an ovary (or a testis) from *Talitrus saltator* is implanted, an oostegite develops. Gill omitted. (Redrawn from Charniaux-Cotton, 1957, 1960.)

Charniaux-Cotton, 1965). In crabs, after implantation of AG, one female of *Carcinus maenas* and two females of *Rhithropanopeus harrisii* showed a partial masculinization of their pleon (differentiation of Pl_1, regression of the exopodite of Pl_2, loss of the articulation between tergites 3–4 and 4–5) (Payen, 1969, 1975). In the female crayfish *Procambarus clarki*, the partial transformation of the first pair of pleopods into male-type gonopods has also recently been observed after a similar operation (Nagamine and Knight, 1987).

Ablation of the AG in pubertal males of the shrimp *Palaemon serratus* (Touir, 1977a) leads to regeneration of pleopods 1 and 2 in an undifferentiated form. Andrectomized males of *Macrobrachium rosenbergii* initially lacking an appendix masculina and a male cheliped do not develop them. Those initially possessing external male characters do not lose them, but the appendages regenerate in an immature form (Nagamine et al., 1980b). The appendix masculina of hermaphroditic shrimps, in contrast to that of gonochoristic shrimps, disappears at the molt following the degeneration of the AG. In crayfishes, males of the subfamily Cambarinae exhibit a cyclic external sexual dimorphism, because they molt alternately between a sexually active form (called form I) and a sexually inactive one (called form II) in which the copulating organs (Pl_1 and Pl_2) regress. Extirpation of the AG from form II does not prevent the realization of form I (Kracht, 1980). In these decapods, the periodic change of the male pleopods does not appear to be controlled by the level of AGH. However, a correlation between the size of the AG and seasonal changes in male morphology of external characteristics has been noted by other authors (Carpenter and De Roos, 1970; Dudley and Jegla, 1978). A similar correlation seems to exist in the prawn *Macrobrachium idae* (Thampy and John, 1973).

To summarize, in most decapods it appears that the target territories lose their ability to respond to AGH more or less precociously. In crabs, the AG implantation must be performed in very young females. In gonochoristic Malacostraca, circulating AGH induces irreversible development of secondary male characteristics. In contrast, in proterandric hermaphroditic species these characteristics are reversed when AG degenerate. The maintenance of male sexual behavior also requires the presence of AGH. Moreover, AGH may regulate pheromone-mediated courtship displays in male crabs (Gleeson et al., 1987).

Analysis of the modalities of regeneration of the second gnathopods in *O. gammarella* and the second pleopods in *Lysmata seticaudata* and *Pandalus borealis* show that these appendages regenerate in a sexually undifferentiated form and then advance toward the male morphology (Charniaux-Cotton, 1957, 1967). Therefore, regenerating sexual appendages have no receptor to the male hormone until they acquire the undifferentiated form.

12.5. Role of the AGH in the Control of Spermatogenic Activity

As mentioned previously (see Section 12.4.1), the AGH is responsible for the differentiation of gonia into spermatogonia and for spermatogenesis. In the young male of *O. gammarella,* the initiation of spermatogenesis takes place when the AG are not yet separated from the sperm ducts and when external sexual dimorphism is not yet visible. The hormone reaches the gonia by diffusion along the genital tract.

In the decapod Reptantia, on the other hand, the young testis shows no spermatogenic activity for several intermolts. The initiation of spermatogenesis is probably due to circulating AGH. Indeed, the individualization of AG in crabs and European crayfishes is accompanied by synchronous modifications leading toward male puberty, that is, induction of the first spermatogenesis, beginning of glandular activity of the sperm ducts, and development of the gonopods (Payen, 1973, 1974a). Implantation of supplemental AG into young males of *Carcinus maenas* demonstrates very well the role of circulating AGH on the genital apparatus, since the latter, following the operation, shows accelerated development (Payen, 1968). In the shrimp *Macrobrachium rosenbergii,* the initiation of spermatogenesis likewise takes place after differentiation of the deferent ducts and individualization of the AG (C. Fauvel, unpublished data).

Completion of spermatogenic activity requires the continued presence of circulating AGH, probably in many species. Spermatogenesis stops in andrectomized males of *O. gammarella* (Charniaux-Cotton, 1964), of the marine isopod *Idotea balthica* (Reidenbach, 1971), of the American crayfish *Cambarus bartonii bartonii* (Puckett, 1964), and of the shrimp *Macrobrachium rosenbergii* (Nagamine et al., 1980a). In the crabs *Rhithropanopeus harrisii* and *Callinectes sapidus* and in the crayfish *Astacus,* spermatogenesis starts in testes only when the AG are individualized (Payen, 1973, 1974a). In a testis of *O. gammarella* cultured alone, spermatogenic activity ceases. When the testis is associated with an AG, the germinative zone once more liberates spermatogonia that undergo spermatogenesis (Berreur-Bonnenfant, 1968). As reported above (see Section 12.4.1), not only male differentiation of the germinative zone, but also spermatogenic activity require, in certain species, the continued presence of circulating AGH. In the absence of the male hormone or even in the presence of a low level of this hormone, the germinative zone discharges oogonia that carry on oogenesis. In certain species, nonetheless, spermatogenic activity, once initiated, seems capable of maintaining itself in the absence of AGH. Thus, testes of *Carcinus maenas,* when transplanted into females, continue to show normal spermatogenesis, at the very least, within the limits of the experiment (45 days) (Payen, 1974b).

In the sexually quiescent European crayfish *Pontastacus leptodactylus leptodactylus*, the same operation does not stop the release of a new cycle of spermatogenesis (Amato and Payen, 1976; Payen and Amato, 1978).

The intensity of spermatogenic activity is generally regulated by the level of AGH. During the reproductive season, the AG are well developed and spermatogenesis is intense. During genital repose, the AG are small and, correspondingly, the testes are reduced. The existence of such a correlation has been observed in *Orchestia gammarella* (Meusy, 1963), *Lysmata seticaudata* (Touir, 1973), and the crayfishes *Orconectes nais* and *P. l. leptodactylus* (Carpenter and De Roos, 1970; Payen, 1973).

12.6. Regulation of AG Activity

In eyestalked malacostracans, the growth and activity of the AG are themselves controlled by a moderating hormone whose source is situated in the eyestalks. Indeed, ablation of these organs in crabs during

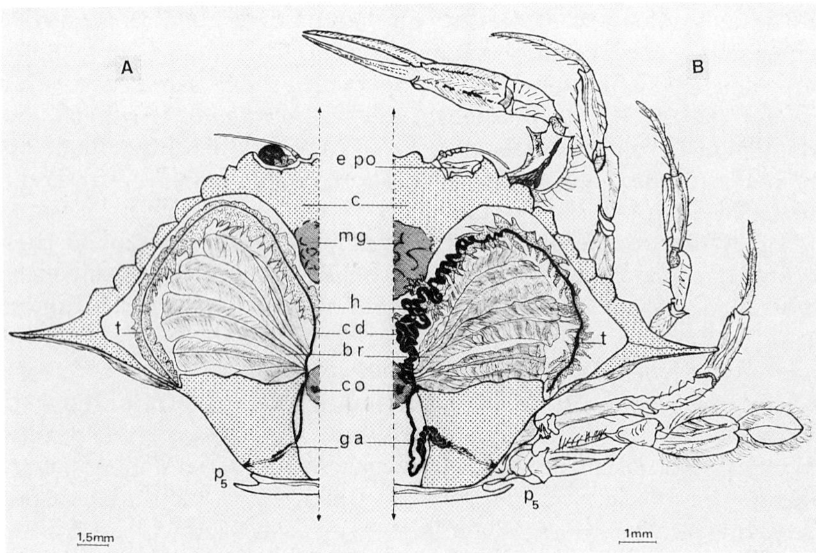

FIGURE 12.4. Male genital apparatus and androgenic gland of control (A) and destalked crab (B), *Callinectes sapidus*, sacrificed at the sixth postlarval intermolt. Eyestalk ablation was performed at the megalops stage; it leads to precocious development of the whole genital apparatus with hypertrophy of the androgenic gland and onset of spermatogenesis. Cephalothoracic length: 7 mm in the control crab; 9.5 mm in the operated one. *Key:* br = gills; c = carapace; c d = deferent duct; c o = heart; e po = location of the eyestalk; g a = androgenic gland; h = hepatopancreas; m g = gastric mill; p5 = fifth pereopod; t = testis. (From Payen et al., 1971.)

the zoeal, megalops, or juvenile stages elicits precocious spermatogenesis (Fig. 12.4) and an acceleration of sexual differentiation (Demeusy, 1960, 1967; Payen et al., 1971). The AG of young destalked males are hypertrophied and present ultrastructural characteristics related to hyperactivity. Comparable results have been obtained in immature isopods of the species *Idotea balthica* and *Porcellio dilatatus* after ablation of the optic lobe or of the protocerebrum, or both (Reidenbach, 1966; Legrand et al., 1968). Moreover, in the Flabellifera *Sphaeroma serratum*, removal of the median region of the protocerebrum in prepubertal males induces an anticipated differentiation of the appendix masculina that confirms the existence of a neurosecretory center controlling the functioning of the AG (Juchault, 1977). Note that three factors can intervene in the control of AG functioning in oniscoid isopods: one, inhibiting growth of the gland, is synthesized by the protocerebrum; a second, inhibiting synthesis of the hormone, is elaborated by neurosecretory cells dispersed in the entire nervous system; the third, which controls release of the hormone, is also synthesized in the entire nervous system (Juchault and Legrand, 1978; Legrand et al., 1982).

A moderating role of the neurosecretion from the crab ventral ganglionic mass on the development of AG seems also to exist. Such a role has been put forward following numerous observations on *Carcinus* parasitized by *Sacculina* (Rubiliani-Durozoi et al., 1980). A transitory hypertrophy of the androgenic organs always occurs simultaneously with the impairment of the ventral ganglionic mass (Payen et al., 1979), which is the first target of the parasite on entering into the host tissues (Payen et al., 1981).

12.7. Control of Disappearance of the AG in Hermaphroditic Malacostracans

In gonochoristic species, we can recapitulate that two neurohormones regulate the development and activity of the AG and genital apparatus (see Sections 12.4.1 and 12.5): a peduncular hormone has a moderating effect on AG activity, and a cerebral hormone ensures the maintenance of the male tract and AG. Bilateral removal of eyestalks from male *Pandalus platyceros* has been performed when the AG were in the process of atrophy and the gonads were changing into ovaries. Six to eight weeks afterward, the AG were hypertrophied and the male stage had been maintained. The gonads showed evidence of spermatogenesis (Hoffman, 1968). Similar experiments were performed in *L. seticaudata* (Touir, 1973); afterward, the animals were kept under observation for a period of 8 months. The external male characteristics, which had already regressed at the time of the operation, redeveloped. The AG had hypertrophied, and spermatogenesis was very active. Thus, after bilateral ablation of the eyestalks, sex reversal is suppressed, at least within the time limits of the experiments.

The peduncular hormone appears to be responsible for sex reversal. It probably acts by stopping the secretion of an androstimulating hormone. The existence of the latter hormone has been demonstrated by cauterization of the anteromedial region of the protocerebrum in *Lysmata* in the male phase (Touir, 1977a,b). Such a cauterization is followed by disappearance of the AG. Implantation of Bellonci's organ and nerve cord into the isopod *Anilocra physodes* in the female phase restores the male phase (Juchault and Legrand, 1965; Juchault, 1967). However, it must be pointed out that AG grafted into female shrimps *Lysmata* and *Pandalus* do not degenerate, at least within the time limits of the experiments, and do masculinize the host, i.e., external male characteristics develop and vitellogenesis is inhibited (Charniaux-Cotton, 1959, 1965). Note also that, after being destalked, these females of *Lysmata* are masculinized more rapidly than are normal females (Charniaux-Cotton, 1972). This may be explained by the absence of the moderating effect of the neurosecretory complex on mitosis and on secretion of the AG.

12.8. Summary

The endocrine control of male differentiation of both the genital apparatus and secondary sex characteristics by AGH (androgenic gland hormone) has been demonstrated experimentally in the amphipod *Orchestia gammarella*, several isopods, and the decapod *Macrobrachium rosenbergii*. The organs responsible for the synthesis of this hormone, the androgenic glands (AG), are external to the gonads. The testes are not endocrine organs. In the absence of AGH, female differentiation occurs spontaneously. In young females, the hormone stimulates the development of AG primordia, which start to synthesize AGH. Different results indicate that in genetic males the AGH promotes the development of AG rudiments.

In most species, AGH appears to act by diffusion along the genital tract; it induces male morphogenesis of the tract and the differentiation of gonia into spermatogonia. In some malacostracans, male differentiation of the germ cells is not definitive and requires the continued presence of AGH. The initiation of spermatogenic activity seems in some decapods to be due to the appearance of circulating AGH. Completion of spermatogenesis and its intensity are generally due to the gonadotropic action of AGH. In some Reptantia, however, spermatogenesis can continue in the absence of the hormone. Finally, the activity of the AG is regulated by a moderating neurohormone from protocerebrum.

The AGH recently isolated in the isopod *Armadillidium vulgare* is a polypeptide that consists of two molecular forms, AGH_I and AGH_{II}, whose molecular weight have been estimated to be 17.0 ± 0.8 and 18.3 ± 1.0 kDa, respectively. Another factor, farnesylacetone, which is without masculinizing effect but inhibits ovarian protein synthesis, has also been isolated from AG.

References

Amato, G. D. and G. G. Payen. 1976. Transplantations homoplastiques de testicules d'Ecrevisse *Pontastacus leptodactylus leptodactylus* (Escholtz, 1823) en repos sexuel dans des mâles et des femelles normaux et épédonculés: résultats préliminaires. C. R. Acad. Sci. 283D: 1783–1786.

Balesdent-Marquet, M. L. 1958. Présence d'une glande androgène chez le Crustacé Isopode *Asellus aquaticus* L. C. R. Acad. Sci. Paris 247: 534–536.

Berreur-Bonnenfant, J. 1968. Action de la glande androgène et du cerveau sur la gamétogenèse de Crustacés Péracarides. Arch. Zool. Exp. Gen. 108: 521–558.

Berreur-Bonnenfant, J. and F. Lawrence. 1984. Comparative effect of farnesylacetone on macromolecular synthesis on gonads of crustaceans. Gen. Comp. Endocrinol. 54: 462–468.

Berreur-Bonnenfant, J. and H. Charniaux-Cotton. 1965. Hermaphrodisme protérandrique et fonctionnement de la zone germinative chez la Crevette *Pandalus borealis* Kröyer. Bull. Soc. Zool. Fr. 40: 243–259.

Berreur-Bonnenfant, J., J.-J. Meusy, J.-P. Férézou, M. Devys, A. Quesneau-Thierry, and M. Barbier. 1973. Recherche sur la secretion sur la glande androgène des Crustacés Malacostracés: purification et propriétés d'une subtance à activité antrogène. C. R. Acad. Sci. Paris 277D: 971–974.

Besse, G., P. Juchault, J.-J. Legrand, and J.-P. Mocquard. 1970. Modifications de l'électrophorégramme de l'hémolymphe des Oniscoïdes (Crustacés Isopodes) par action de l'hormone androgène. C. R. Acad. Sci. 270D: 3276–3279.

Blanchet, M. F., R. Ozon, and J. J. Meusy. 1972. Metabolism of steroids, *in vitro*, in the male crab *Carcinus maenas* Linné. Comp. Biochem. Physiol. 41B: 251–261.

Carlisle, D. B. 1960. Sexual differentiation in Crustacea, Malacostraca. Mem. Soc. Endocrinol. 7: 9–16.

Carpenter, M. B. and R. De Roos. 1970. Seasonal morphology and histology of the androgenic gland of the crayfish, *Orconectes nais*. Gen. Comp. Endocrinol. 15: 143–157.

Charniaux-Cotton, H. 1954a. Implantation de gonades de sexe opposé à des mâles et des femelles chez un crustacé Amphipode (*Orchestia gammarella*). C. R. Acad. Sci. Paris 238: 953–955.

Charniaux-Cotton, H. 1954b. Découverte chez un Crustacé Amphipode (*Orchestia gammarella*) d'une glande endocrine responsable de la différenciation des caractères sexuels primaires et secondaires mâles. C. R. Acad. Sci. Paris 239: 780–782.

Charniaux-Cotton, H. 1955. Le déterminisme hormonal des caractères sexuels d'*Orchestia gammarella* (Crustacé Amphipode). C. R. Acad. Sci. Paris 240: 1487–1489.

Charniaux-Cotton, H. 1957. Croissance, régénération et déterminisme endocrinien des caractères sexuels d'*Orchestia gammarella* (Pallas) (Crustacé Amphipode). Ann. Sci. Nat. 19: 411–559.

Charniaux-Cotton, H. 1958. Contrôle hormonal de la différenciation du sexe et de la reproduction chez les Crustacés supérieurs. Bull. Soc. Zool. Fr. 83: 314–336.

Charniaux-Cotton, H. 1959. Masculinisation des femelles de la Crevette à hermaphrodisme protérandrique *Lysmata seticaudata*, par greffe glandes androgènes: interprétation de l'hermaphrodisme chez les Décapodes. Note préliminaire. C. R. Acad. Sci. Paris 249: 1580–1582.

Charniaux-Cotton, H. 1960. Sex determination. Pp. 411–447 *in* T. H. Waterman (ed.), *The Physiology of Crustacea*, Vol. 1. Academic Press, Orlando, Florida.

Charniaux-Cotton, H. 1961. La croissance et la morphogenèse des caractères sexuels des Crustacés supérieurs et l'hormone androgène. Bull. Soc. Zool. Fr. 86: 484–499.

Charniaux-Cotton, H. 1962. Androgenic gland of crustaceans. Gen. Comp. Endocrinol. 1: 241–247.

Charniaux-Cotton, H. 1964. Endocrinologie et génétique du sexe chez les Crustacés supérieurs. Ann. Endocrinol. 25: 36–42.

Charniaux-Cotton, H. 1965. Hormonal control of sex differentiation in invertebrates. Pp. 701–740 *in* R. De Haan and H. Ursprung (eds.), *Organogenesis*. Holt, New York.

Charniaux-Cotton, H. 1967. Régénération des appendices présentant un dimorphisme sexuel chez les Crustacés supérieurs. Bull Soc. Zool. Fr. 92: 361–372.

Charniaux-Cotton, H. 1972. Recherches récentes sur la différenciation sexuelle et l'activité génitale chez divers crustacés supérieurs. Pp. 128–178 *in* E. Wolff (ed.), *Hormones et Différenciation Sexuelle chez les Invertébrés*. Gordon & Breach, New York.

Charniaux-Cotton, H. and M. Cazès. 1979. Première masculinisation des femelles d'un Crustacé Décapode gonochorique, *Palaemonetes varians* (Leach), par greffes de glandes androgènes. C. R. Acad. Sci. Paris 288D: 1707–1709.

Charniaux-Cotton, H. and G. Payen. 1985. Sexual differentiation. Pp. 217–299 *in* D. E. Bliss and L. H. Mantel (eds.), *The Biology of Crustacea*. Academic Press, Orlando, Florida.

De Lattin, G. and F. J. Gross. 1953. Die Beeinflussbarkeit sekundärer Geschlechtsmerkmale von *Oniscus asellus* durch die Gonaden. Experientia (Basel) 9: 338–339.

Demeusy, N. 1960. Différenciation des voies génitales mâles du crabe *Carcinus maenas* Linné: rôle des pédoncules oculaires. Cah. Biol. Mar. 1: 259–278.

Demeusy, N. 1967. Modalités d'action du contrôle inhibiteur pédonculaire exercé sur les caractères sexuels externes mâles du Décapode Brachyoure *Carcinus maenus* L. C. R. Acad. Sci. Paris 265D: 628–630.

Dudley, H. G. and T. C. Jegla. 1978. Control of reproductive form in male crayfish. Gen. Comp. Endocrinol. 34: 108.

Férézou, J. P., M. Barbier, and J. Berreur-Bonnenfant. 1978. Biosynthèse de la farnésylacetone-(E-E) par les glandes androgènes du crabe *Carcinus maenas*. Helv. Chim. Acta 61: 669–674.

Férézou, J. P., J. Berreur-Bonnenfant, J. J. Meusy, M. Barbier, M. Suchy, and H. K. Wipf. 1977a. 6,10,14-Trimethylpentadecan-2-one and 6,10,14-trimethyl-5-*trans*,9-*trans*,13-pentadecatrien-2-one from the androgenic glands of the male crab *Carcinus maenas*. Experientia (Basel) 33: 290.

Férézou, J. P., J. Berreur-Bonnenfant, A. Tekitek, M. Rojas, M. Barbier, M. Suchy, W. Socar, and J. J. Meusy. 1977b. Biologically active lipids from the androgenic glands of the crab *Carcinus maenas*. Pp. 361–366 *in* D. J. Faulkner (ed.), *Marine Natural Products Chemistry*. Plenum Press, New York.

Gilgan, M. W. and D. R. Idler. 1967. The conversion of androstenedione to testosterone by some lobster (*Homarus americanus* Milne Edwards) tissues. Gen. Comp. Endocrinol. 9: 319–324.

Ginsburger-Vogel, T. 1972. Inversion des femelles d'*Orchestia gammarella* Pallas (Crustacé Amphipode Talitridae) en néo-mâles fonctionnels par greffe de glandes androgènes avant la mue de première différenciation externe du sexe. C. R. Acad. Sci. Paris 274D: 3606–3609.

Ginsburger-Vogel, T. 1975. Détermination génétique du sexe, monogénie et intersexualité chez *Orchestia gammarella* Pallas: étude des phénomènes de monogénie indépendants de l'intersexualité. Arch. Zool. Exp. Gen. 116: 615–647.

Gleeson, R. A., M. A. Adams, and A. B. Smith. 1987. Hormonal modulation of pheromone-mediated behavior in a crustacean. Biol. Bull. (Woods Hole) 172: 1–9.

Godbillon, G. and M. L. Balesdent. 1974. Mise en évidence d'une substance spécifique, de nature stéroïque, localisée au niveau du tractus génital du Crustacé Décapode *Carcinus maenas* Linné. C. R. Acad. Sci. Paris 278D: 2823–2826.

Hasegawa, Y. and Y. Katakura. 1981. Androgenic gland hormone and development of oviducts in the isopod crustacean, *Armadillidium vulgare*. Dev. Growth & Differ. 23(1): 59–62.

Hasegawa, Y., K. Haino-Fukushima, and Y. Katakura. 1987. Isolation and properties of androgenic gland hormone from the terrestrial isopod, *Armadillidium vulgare*. Gen. Comp. Endocrinol. 67: 101–110.

Hoffman, D. L. 1968. Seasonal eyestalk inhibition on the androgenic glands of a protandric shrimp. Nature (Lond.) 218: 170–172.

Juchault, P. 1967. Contribution à l'étude de la différenciation sexuelle mâle chez les Crustacés Isopodes. Ann. Biol. 6: 191–212.

Juchault, P. 1977. Corrélation entre différenciation sexuelle mâle externe et fonction andro-inhibitrice de la région médiane du protocérébron chez le Crustacé *Sphaeroma serratum* Fabr. (Isopoda, Flabellifera). C. R. Acad. Sci. Paris 285D: 179–182.

Juchault, P. and J.-J. Legrand. 1964. Mise en évidence d'un inducteur sexuel mâle distinct de l'hormone adulte et contribution à l'étude de l'autodifférenciation ovarienne chez l'oniscoïde *Helleria brevicornis*. C. R. Acad. Sci. Paris 258: 2416–2419.

Juchault, P. and J.-J. Legrand. 1965. Contribution à l'étude expérimentale de l'intervention des neurohormones dans le changement de sexe d'*Anilocra physodes* (Crustacé, Isopode, Cymothoidae). C. R. Acad. Sci. Paris 260: 1783–1786.

Juchault, P. and J.-J. Legrand. 1978. Etude du fonctionnement de la glande androgène dans le cas d'implantations croisées entre deux espèces de crustacés isopodes terrestres, *Porcellio dilatatus* Brandt et *Armadillidium vulgare* Latreille: notion de spécificité de l'hormone androgène et des neurohormones impliquées dans le contrôle de la fonction androgène. Gen. Comp. Endocrinol. 36: 175–186.

Juchault, P., J.-J. Legrand, and J.-P. Mocquard. 1965. Mise en évidence d'une inhibition protocérébrale de la glande androgène et de la croissance des variants sexuels mâles chez l'Oniscoïde *Porcellio dilatatus* Brandt. C. R. Acad. Sci. Paris 261: 116–118.

Juchault, P., J. Maissiat, and J.-J. Legrand. 1978. Caractérisation chimique d'une substance ayant les effets biologiques de l'hormone androgène chez le Crustacé Isopode terrestre *Armadillidium vulgare* Latreille. C. R. Acad. Sci. 268D: 73–76.

Juchault, P., J.-J. Legrand, and J. Maissiat. 1984. Present state of knowledge on the chemical nature of the androgenic hormone in higher crustaceans. Pp. 155–160 *in* J. Hoffmann and M. Porchet (eds.), *Biosynthesis, Metabolism and Mode of Action of Invertebrate Hormones*. Springer-Verlag, Berlin and New York.

Katakura, Y. 1959. Masculinization through implanting testes into the female *Armadillidium vulgare*, an isopod crustacean. Proc. Jpn. Acad. 35: 95–98.

Katakura, Y. 1960. Transformation of ovary into testis following implantation of androgenous glands in *Armadillidium vulgare*, an isopod crustacean. Annot. Zool. Jpn. 33: 241–244.

Katakura, Y. 1961. Hormonal control of development of sexual characters in the isopod crustacean, *Armadillidium vulgare*. Annot. Zool. Jpn. 34: 60–71.

Katakura, Y. 1967. Hormonal control of sex differentiation in the terrestrial isopod, *Armadillidium vulgare*. Gunma Symp. Endocrinol. 4: 49–64.

Katakura, Y. 1984. Sex differentiation and androgenic gland hormone in the terrestrial isopod *Armadillidium vulgare*. Symp. Zool. Soc. Lond. 53: 127–142.

Katakura, Y. and Y. Hasegawa. 1983. Masculinization of females of the isopod crustacean

Armadillidium vulgare following injections of an active extract of the androgenic gland. Gen. Comp. Endocrinol. 48: 57–62.

Katakura, Y., Y. Fujimaki, and K. Unno. 1975. Partial purification and characterization of androgenic gland hormone from the isopod crustacean *Armadillidium vulgare*. Annot. Zool. Jpn. 48: 203–209.

Kracht, D. 1980. Démonstration, par l'extirpation des glandes androgènes, de l'absence de corrélation entre le dimorphisme sexuel cyclique du mâle adulte et les variations annuelles de l'activité testiculaire de l'Ecrevisse *Orconectes limosus* (Rafinesque). C. R. Acad. Sci. Paris 290D: 465–468.

Legrand, J.-J. 1954. Induction des caractères sexuels secondaires mâles chez les femelles des Crustacés Isopodes terrestres par implantation testiculaire: premiers résultats. C. R. Acad. Sci. Paris 238: 2030–2032.

Legrand, J.-J. 1958. Mise en évidence histologique et expérimentale d'un tissu androgène chez les Oniscoïdes. C. R. Acad. Sci. Paris 247: 1238–1241.

Legrand, J.-J. and P. Juchault. 1972. Le contrôle humoral de la sexualité chez les Crustacé Isopodes gonochoriques. Pp. 179–218 *in* E. Wolff (ed.), *Hormones et Différenciation Sexuelle chez les Invertébrés*. Gordon & Breach, New York.

Legrand, J.-J., E. Legrand-Hamelin and P. Juchault. 1987. Sex determination in Crustacea. Biol. Rev. 62: 439–470.

Legrand, J.-J., P. Juchault, J.-P. Mocquard, and G. Noulin. 1968. Contribution à l'étude du contrôle neurohumoral de la physiologie sexuelle mâle chez les Crustacés Isopodes terrestres. Ann. Embryol. Morphog. 1: 97–105.

Legrand, J.-J., G. Martin, P. Juchault, and G. Besse. 1982. Contrôle neuroendocrine de la reproduction chez les Crustacés. J. Physiol. (Paris) 78: 543–552.

Legrand, J.-J., P. Juchault, J.-C. Artault, J.-P. Mocquard, and J.-L. Picaud. 1974. Le statut systématique de la "forme" *petiti* Vandel de *Porcellio dilatatus* Brandt récoltés à l'île Saint Honorat (Alpes-Maritimes): critères morphologiques, génétiques et physiologiques. Bull. Soc. Zool. Fr. 99: 461–471.

Meusy, J.-J. 1963. Description de la glande androgène chez un Crustacés Amphipode: *Paramysis nouveli* Labat (Mysidacé) et *Eocuma dollfusi* Calman (Cumacé). C. R. Acad. Sci. Paris 256: 5425–5428.

Meusy, J.-J. and G. G. Payen. 1988. Female reproduction in malacostracan Crustacea. Zool. Sci. 5(2): 217–265.

Meusy, J.-J., H. Junéra, and Y. Croisille. 1971. Recherche de la "fraction protéique femelle" chez les Crustacés Amphipodes *Orchestia gammarella* ayant subi une inversion expérimentale du sexe. C. R. Acad. Sci. Paris 273D: 592–594.

Nagamine, C. and A. W. Knight. 1987. Masculinization of female crayfish, *Procambarus clarki* (Girard). Int. J. Invert. Reprod. Develop. 11: 77–87.

Nagamine, C., A. W. Knight, A. Maggenti, and G. Paxman. 1980a. Effects of androgenic gland ablation on male primary and secondary sexual characteristics in the Malaysian prawn *Macrobrachium rosenbergii* (de Man) with first evidence of induced feminization in a non-hermaphroditic decapod. Gen. Comp. Endocrinol. 41: 423–441.

Nagamine, C., A. W. Knight, A. Maggenti, and G. Paxman. 1980b. Masculinization of female *Macrobrachium rosenbergii* (de Man) (Decapoda Palaemonidae) by androgenic gland implantation. Gen. Comp. Endocrinol. 41: 442–457.

Nagabhushanam, R. and G. K. Kulkarni. 1981. Effect of exogenous testosterone on the androgenic gland and testis of a marine penaeid prawn, *Parapenaeopsis hardwickii* (Miers) (Crustacea, Decapoda, Penaeidae). Aquaculture 23: 19–27.

Payen, G. 1968. Expériences de greffes de glandes androgènes sur le Crabe *Carcinus maenas* L. a. Premiers résultats. C. R. Acad. Sci Paris 266D: 1056–1058.

Payen, G. 1969. Expériences de greffes de glandes androgènes sur la femelle pubère du

Crabe *Rhithropanopeus harrisii* (Gould) (Crustacé, Décapode). C. R. Acad. Sci. Paris 268D: 393–396.

Payen, G. G. 1973. Etude descriptive des principales étapes de la morphogenèse sexuelle chez un Crustacé Décapode à développement condensé, l'Ecrevisse *Pontastacus leptodactylus leptodactylus* (Eschscholtz, 1823). Ann. Embryol. Morphog. 6: 179–206.

Payen, G. 1974a. Morphogenèse sexuelle de quelques Brachyoures (Cyclométopes) au cours du développement embryonnaire, larvaire et postlarvaire. Bull. Mus. Natl. Hist. Nat. Zool. 209(139): 201–262.

Payen, G. 1974b. Recherches sur la réalisation et le contrôle de la différenciation sexuelle chez les Crustacés Décapodes Reptantia. Doctoral thesis, CNRS No. 9930, University of Paris.

Payen, G. 1975. Effets masculinisants des glandes androgènes implantées chez la femelle pubère pédonculectomisée de *Rhithropanopeus harrisii* (Gould) (Crustacé, Décapode, Brachyoure). C. R. Acad. Sci. Paris 280D: 1111–1114.

Payen, G. G. 1980. Experimental studies of reproduction in Malacostra crustaceans: endocrine control of spermatogenic activity. Pp. 187–196 *in* W. M. Clark and T. S. Adams (eds.), *Advances in Invertebrate Reproduction,* Vol. 2. Elsevier/North-Holland Publ., Amsterdam and New York.

Payen, G. G. 1983. Endocrine regulation of male genital development in Malacostraca. Am. Zool. 23(4): 951.

Payen, G. G. 1986. Endocrine regulation of male and female genital activity in crustaceans: a retrospect and perspectives. Pp. 125–134 *in* M. Porchet, J. C. Andries, and A. Dhainaut (eds), *Advances in Invertebrate Reproduction*. Elsevier/North-Holland Publ., Amsterdam and New York.

Payen, G. G. and G. D. Amato. 1978. Données actuelles sur le contrôle de la spermatogenèse chez les Crustacés Décapodes Reptantia. Arch. Zool. Exp. Gen. 119: 447–464.

Payen, G. G., J. D. Costlow, and H. Charniaux-Cotton. 1971. Etude comparative de l'ultrastructure des glandes androgènes de Crabes normaux et pédonculectomisés pendant la vie larvaire ou après la puberté chez les espèces: *Rhithropanopeus harrisii* (Gould) et *Callinectes sapidus* Rathbun. Gen. Comp. Endocrinol. 17: 526–542.

Payen, G. G., C. Rubiliani, M. Hubert, and C. Chassard-Bouchaud. 1979. Données préliminaires relatives aux modifications induites par les racines des Rhizocéphales sur le système nerveux central des crabes hôtes: aspects structuraux et ultrastructuraux. C. R. Acad. Sci. Paris 288D: 705–708.

Payen, G. G., M. Hubert, Y. Turquier, C. Rubiliani, and C. Chassard-Bouchaud. 1981. Infestations expérimentales de crabes juvéniles par la sacculine: ultrastructure des racines parasitaires en croissance et relations avec la masse ganglionnaire ventrale de l'hôte. Can. J. Zool. 39: 1818–1826.

Puckett, D. H. 1964. Experimental studies on the crayfish androgenic gland in relation to testicular function. Diss. Abstr. 25: 3765.

Raimond, R., and P. Juchault. 1983. Masculinisation des femelles prépubères et pubères de *Sphaeroma serratum* Fabr. (Crustacé, Isopode, Flabellifère) par implantation d'une glande androgène de mâle pubère. Gen. Comp. Endocrinol. 50: 146–155.

Reidenbach, J.-M. 1966. Mise en évidence d'une intervention du complexe neurosécréteur céphalique dans la physiologie sexuelle mâle chez le Crustacé Isopode marin *Idotea balthica basteri* Audouin. C. R. Acad. Sci. Paris 262D: 682–684.

Reidenbach, J.-M. 1971. Les mécanismes endocriniens dans le contrôle de la différenciation du sexe, la physiologie sexuelle et la mue chez le Crustacé isopode marin: *Idotea balthica* (Pallas). Doctoral thesis, CNRS No. 4874, University of Nancy, France.

Rubiliani-Durozoi, M., C. Rubiliani, and G. G. Payen. 1980. Déroulement des gamétogenèses chez les crabes *Carcinus maenas* (L.) et *C. mediterraneus* Czerniavsky parasités par la Sacculine. Int. J. Invertebr. Reprod. 2: 107–120.

Sarojini, S. 1963. Comparision of the effects of androgenic hormone and testosterone propionate on the female ocypod crab. Curr. Sci. (Bangalore) 32: 411–412.

Sarojini, S. 1964. A note on the chemical nature of the crustacean androgenic hormone. Curr. Sci. (Bangalore) 33: 55–56.

Souty-Grosset, C. and P. Juchault. 1987. Etude de la synthèse de la vitellogénine chez les mâles intersexués d'*Armadillidium vulgare* Latreille (Crustacé Isopode Oniscoïde): comparaison avec les mâles et les femelles intactes ou ovariectomisées. Gen. Comp. Endocrinol. 66: 163–170.

Tcholakian, R. K. and K. B. Eik-Nes. 1969. Conversion of progesterone to 11deoxycorticosterone by the androgenic gland of the blud crab (*Callinectes sapidus* Rathbun). Gen. Comp. Endocrinol. 12: 171–173.

Tcholakian, R. K. and K. B. Eik-Nes. 1971. Steroidogenesis in the blue crab *Callinectes sapidus* Rathbun. Gen. Comp. Endocrinol. 17: 115–124.

Thampy, D. M. and P. A. John. 1973. Observations on variations in the male sex characters and their relation to the androgenic gland in the shrimp *Macrobrachium idae* (Heller). Acta Zool. (Stockh.) 54: 193–200.

Touir, A. 1973. Influence de l'ablation des pédoncules oculaires sur les glandes androgènes, les gonades, les caractères sexuels externes mâles et l'inversion sexuelle chez la Crevette hermaphrodite *Lysmata seticaudata* Risso. C. R. Acad. Sci. Paris 277: 2541–2544.

Touir, A. 1977a. Données nouvelles concernant l'endocrinologie sexuelle des Crustacés Décapodes Natantia hermaphrodites et gonochoriques. 1. Maintien des glandes androgènes et rôle de ces glandes dans le contrôle des gamétogenèses et des caractères sexuels externes mâles. Bull. Soc. Zool. Fr. 102: 375–400.

Touir, A. 1977b. Données nouvelles concernant l'endocrinologie sexuelle des Crustacés Décapodes Natantia hermaphrodites et gonochoriques. III. Mise en évidence d'un contrôle neurohormonal du maintien de l'appareil génital mâle et des glandes androgènes exercé par le protocérébron médian. C. R. Acad. Sci. Paris 285D: 539–542.

Veith, W. J. and S. R. Malecha. 1983. Histochemical study of the distribution of lipids, 3a- and 3b-hydroxysteroid dehydrogenase in the androgenic gland of the cultured prawn, *Macrobrachium rosenbergii* (de Man) (Crustacea: Decapoda). S. Afr. J. Sci. 79: 84–85.

Zerbib, C. 1964. Evolution post-embryonnaire de la voie déférente chez le mâle et chez la femelle normale et masculinisée *d'Orchestia gammarella* Pallas (Crustacé Amphipode). Bull. Biol. Fr. Belg. 98: 391–408.

Juvenile Hormone and Aphid Polymorphism

13

T. E. MITTLER

13.1. Introduction

The ancestors of aphids undoubtly were winged insects that reproduced sexually. As pointed out by Dixon (1987), parthenogenesis is thought to have arisen by macromutation early in the evolution of the group. Parthenogenesis now occurs in the seasonal life cycle of all aphids (Adelgoidea as well as Aphidoidea), and typically each sexual generation is separated by a few successive parthenogenetic generations. Such cyclic parthenogenesis is designated as *holocycly*. Several species, in addition to occurring as holocyclic strains, have deviated even further from this sexual/asexual life cycle and have evolved *anholocyclic* and *androcyclic* strains (those unable to produce sexuals or merely maintain the ability to produce males, respectively). Moreover, some species appear to be completely anholocyclic. Another derived state, in which sexual as well as asexual aphids may exist, is the wingless condition. (See also Shull, 1938).

Not all aphids share the capacity for viviparous reproduction. The adelgids and phylloxerids (among the Adelgoidea) have generations that reproduce parthenogenetically, but this is accomplished by the parthenogenetic development of yolky eggs after they have been deposited. In the Aphidoidea, on the other hand, parthenogenesis is characteristically linked to viviparity, because parthenogenetic oocytes initiate apomictic division soon after they are ovulated and the resulting embryos complete their development within their mothers. When parthenogenetic oocytes are ovulated and start to develop within these embryos, a further telescoping of generations results. This clearly provides an ideal situation for maternal influences on the development of subsequent generations.

Because of the sparsity of physiological information on the Adelgoidea, this chapter deals only with the Aphidoidea, specifically with a few species of one family of the Aphididae that have been studied in any detail. Hence, further reference to aphids herein will pertain only to members of the latter group.

We know a lot about the environmental and biotic conditions under which aphids propagate themselves by parthenogenetic viviparity instead of by sexual reproduction and under which they develop into apterae rather than alatae. However, we have only a very limited knowledge of the physiological mechanisms controlling the cytogenetic, developmental, and reproductive processes leading to the production of the various morphs in which aphids can exist.

It has long been recognized that aphid polymorphism may be regulated hormonally, and an increasing body of information indicates that juvenile hormone–type compounds (as yet undefined, but referred to here as JH) mediate various aspects of this phenomenon. Such a control mechanism could have arisen quite parsimoniously from the basic endocrine system with which insects are endowed for regulating their stepwise larval-to-adult development and for the maturation of their eggs.

In presenting the relevant pieces of information on the roles that JH may play in aphid polymorphism, I emphasize the diversion of these insects from their basic sexual and alate conditions into the derived states of parthenogenetic viviparity and apterousness. This evolutionary approach is the reciprocal of the one traditionally taken, but I feel that it will lead to a better understanding of the control mechanisms involved in the developmental biology of aphids.

For a comprehensive overview of aphid polymorphism and of the factors that affect this phenomenon, the reader is referred to the classical papers by Hille Ris Lambers (1966) and Lees (1966). Endocrinological aspects of the topic have been reviewed more recently by Hales (1976), Lees (1978), Nijhout and Wheeler (1982), Hardie (1984), and Hardie and Lees (1985).

13.2. Eggs and Viviparous Spring Generations

In temperate climates, aphids generally overwinter as large yolky eggs laid by sexual females (oviparae) in the autumn. Parthenogenetic development of unfertilized eggs has not been recorded, and fertilized eggs develop exclusively into females with a $2N + 2X$ chromosomal complement, because males have a $2N + 1X$ chromosomal makeup and spermatozoa without an X chromosome are inviable (see Blackman, 1987). Embryogenesis is generally completed in the spring after an obligatory diapause in the winter months. The first-generation aphids that hatch from the eggs develop into viviparae (parthenogenetic viviparous females), known as stem mothers or fundatrices, which are apterous in species with alary dimorphism. The next few generations also develop into viviparous females (which may be apterous or alate), despite the fact that the daily scotophases during their development in the spring are still relatively long (i.e., the scotophases would be of sufficient length to induce subsequent generations to produce males and oviparous females).

The maintenance by the spring generations of the parthenogenetic mode of reproduction has been considered to be subject to a "*facteur fundatrice*" (Bonnemaison, 1951) or "interval timer" (Lees, 1966). These terms imply that a factor has to dissipate or that a period of time must elapse before the immediate post-fundatrix generations can reproduce sexually. While there is no doubt that there is a time-related change in the responsiveness of these generations to natural photoperiodic conditions, the following examples clearly show that aphids do not necessarily have to go through a specific period of time before they can, or must, revert to sexual reproduction:

(a) Fundatrices of *Drepanosiphum platanoides* give birth to oviparae when they are exposed to scotophases much longer than those they would have encountered in spring (Dixon, 1971).

(b) Spring and later generations of some aphid species may be partially or completely inhibited from reverting to sexual reproduction even in a long-night (LN) environment if they are reared on young host plants and not crowded (Bonnemaison, 1972; Dixon and Glen, 1971). Under these biotic conditions, a holocyclic clone of *Aphis fabae* could be maintained for numerous generations under LN conditions (Tsitsipis and Mittler, 1977a).

(c) On the other hand, some species, e.g., *Dysaphis devecta*, revert to sexual reproduction within a few post-fundatrix generations even under short-night (SN) conditions unless they are prevented from doing so by being reared at low densities on young host plants (Forrest, 1970).

(d) Aphids do not revert to sexual reproduction under LN conditions if they are exposed to elevated temperatures (De Fluiter, 1950; Lees, 1966; Tsitsipis and Mittler, 1977b; Hardie, 1981a).

(e) Spring generations of *Schizaphis graminum* were found to engender males under SN conditions when chemically allatectomized with precocene III (Mittler, unpublished observation), although they did not produce males in response to LN exposures.

Various studies reviewed in this chapter indicate that high JH levels prevent aphids from engendering males and cause female aphids to reproduce parthenogenetically, to be more fecund, and to undergo an apterous development. The absence of sexuals among the spring generations and the greater fecundity of these aphids (Dixon, 1985a,b) therefore indicate higher JH levels in these aphids than in the later generations. The pronounced apterousness and fecundity of the fundatrices in some species would have to be ascribed to exceptionally high JH levels in the embryos (during their development in the eggs; see Dorn, 1983) and in larvae of this morph.

13.3. Apterous and Alate Viviparous Summer Generations

Johnson and Birks (1960) developed the concept that the basic course of development of wing-dimorphic aphids is an alatiform one and that a number of factors acting prenatally and/or postnatally may divert aphids from this developmental pathway. An increasing body of information suggests that prenatal diversion to apterousness may occur when aphids are exposed to elevated exogenous (maternal) levels of JH during the final stages of their embryonic development and that postnatal diversion may result from an increase in endogenous JH production by aphids early in their larval development.

In general, prenatal apterizing influences result in complete diversion to the apterous state, presumably because of an early activation of the genes for apterousness. On the other hand, postnatal influences on presumptively alate aphids may result in partial apterization or in juvenilization (i.e., the retention of features that characterize the early, apteriform, larval stages of alate as well as apterous aphids).

13.3.1. *Prenatal Diversion*

13.3.1.1. ALATE MATERNAL MORPH

The offspring of alatae (whether they are virginoparous and produce a further generation of viviparae or are gynoparous and produce oviparous females) typically undergo apterous development in aphids with wing dimorphism. That this may result from an increase in JH levels in the alatae when they attain adulthood, following a decline during their larval development to the winged state, is suggested by the finding that the volume of the corpora allata (CA) of virginoparous alatae under SN conditions doubled within 24 h of their molt to adulthood whereas during the previous larval stages of these aphids the volume of these glands did not increase appreciably (see Hales, 1976). However, for aphids as for other insects, a correlation between the size and the activity of the CA has been put in question (Leckstein, 1976), and these attributes may even bear an inverse relationship (Hardie, 1984, 1987b).

An increased JH level in young adult alatae is also suggested by the fact that their wing muscles degenerate a few days after the aphids attain adulthood and settle down to feed (Johnson, 1959a). Moreover, in alatiform aphids raised in isolation (a condition that may stimulate the activity of their CA: Shaw, 1970) or treated with the JH analogue kinoprene late in their larval development, the resulting inability of the aphids to fly has been attributed to the autolysis or incomplete development of their wing muscles (Mittler et al., 1976; Mittler and Lauritzen, 1989). JH-induced wing-muscle autolysis has also been observed in other hemipterans (see Rankin, 1978).

Since alatae generally do not initiate larviposition until they are at least a day or two into adulthood, the putatively high maternal JH levels would be able to exert their apterizing effect on the embryos for at least a day before they are born. Virginoparous alatae may however "revert" to the production of alatae when they are crowded and/or transferred from an SN to an LN regime (Lees, 1966; Tsitsipis and Mittler, 1977a; Mittler and Matsuka, 1985), presumably as a result of a reduction in maternal JH levels.

13.3.1.2. APTEROUS MATERNAL MORPH

In contrast to alatae, virginoparous apterae may produce alate offspring predominently, even under SN conditions, unless they are diverted from doing so by a variety of factors such as being reared at low densities or at elevated temperatures, or being fed on young host plants (Johnson, 1966a; Lamb and White, 1966; Lees, 1967; Sutherland, 1967, 1969; Mittler and Sutherland, 1969). Each of these conditions is considered to cause an elevation of maternal JH levels. On the other hand, apterae that produce apterous offspring as a result of such diversions may "revert" to the production of alatae under converse circumstances, presumably as a consequence of a decline in JH.

Diversion to Aptera Production

Whereas a certain maternal JH threshold may normally have to be exceeded before aphids are diverted prenatally to develop into apterae, aptera production may also ensue if an aphid's embryos are exposed for a sufficiently long time to JH levels that are below this threshold. This is suggested by the finding that when parturition by isolated alata-producing apterae is delayed by food deprivation, decapitation, or amputation of the rostrum or legs, the aphids switch over to aptera production, despite a putative decline in JH in starved or decapitated aphids (see Johansson, 1958; Lees, 1984; Hardie, 1987a; Grüber and Dixon, 1988).

A switchover to aptera production may not occur, however, when aphids are permitted to interact with one another while they are being starved (Lees, 1984) or if they are raised in a LN regime prior to decapitation (Johnson, 1966b), presumably because of an even lower JH level in these aphids. The finding of lower JH III titers in *Megoura viciae* raised under LN conditions than in those raised under SN conditions (Hardie et al., 1985) supports such a conclusion, as may the higher egg production of coccinellids fed on *Myzus persicae* raised under SN conditions than of those raised under LN conditions (Wipperfürth et al., 1987).

However, even when aphids are allowed to interact with one another they may produce apterae when they are starved (Mittler and Kleinjan, 1970) or deprived of essential nutrients, either directly, by being reared on nutritionally impoverished diets (Dadd, 1968), or indirectly, as a result of antibiotic treatments that derange the aphids' symbiotes (Mittler, 1971). As pointed out by Schaefers (1972), Mittler (1973, 1988), and Harrewijn (1978), these findings contradict the previously accepted dogma that poor nutrition causes aphids to produce alatae. Whether nutrients influence the endocrine system of aphids metabolically or through gustatory stimuli is generally difficult to establish. However, the extensive studies by Harrewijn (1976, 1978) on the monoamine metabolism of *M. persicae* reared on artificial diets show that various dietary compounds with serotonergic and adrenergic action can exert a direct influence on the neuroendocrine system and wing-dimorphism of aphids.

To date, the diversion of alata-producing aphids under SN conditions to aptera production by treating them with JH-related compounds has met with relatively little success (see Hales, 1976; Lees, 1980). An apterizing effect recorded for dendrolasin suggests that the enhanced production of apterae by ant-attended aphids (Johnson, 1959b; El-Ziady, 1960) may result in part from a transfer of such an ant-produced substance to aphids in the mandibular secretions of the ants when they palpate the aphids to solicit their honeydew excretion (Kleinjan and Mittler, 1975).

Reversion to Alata Production

The conditions under which apterae produce alatae under SN conditions have been among the most extensively studied factors in aphid polymorphism. Foremost among these are the interactions that occur between aphids as their densities and restlessness increase (Johnson, 1965; Lees, 1966, 1967) and the insects feeding on mature host plants (Johnson, 1966a; Sutherland, 1967, 1969). In heteroecious aphids, the production of alatae under LN conditions is of course an essential ecological response. Interestingly, this response is retained in anholocyclic clones of some aphids (e.g., *Aphis craccivora:* Johnson, 1966b).

A switch from aptera to alata production was found to occur when apterous virginoparae of *Acyrthosiphon pisum* and *Macrosiphum euphorbiae* were treated with precocene I, II, or III (Mackauer et al., 1979; Delisle et al., 1983; Kambhampati et al., 1984; Hardie, 1986). This effect presumably results from a deactivation of the CA of the treated mothers and/or of the larvae, because treatment of several aphid species with precocene III was found to cause complete allatectomy and precocious metamorphosis of their larvae (Hales and Mittler, 1981; Kambhampati et al., 1984). Although aptera-producing virginoparae of *M. persicae* and *A. fabae* did not switch to the production of alatae as a result of precocene III treatment, the precociously metamorphosing individuals among their larvae had several pigmentation characteristics of adult alatae (Hales and Mittler, 1981; Hardie, 1986).

13.3.2. *Postnatal Diversion*

13.3.2.1. DIVERSION TO APTEROUSNESS

Larvae that have not been diverted prenatally from their basic alatiform developmental pathway may be diverted to develop into apterae by various postnatal conditions. The extent to which the aphids are apterized depends on how early in their larval development the aphids are exposed to the morph-active conditions.

Under SN conditions, postnatal apterization may occur when the following situations prevail:

(a) *Larvae are reared in isolation*. This may result in the complete diversion of significant proportions of the aphids to the apterous state (White, 1968a; Shaw, 1970; Johnson, 1965; Mittler and Kunkel, 1971) or cause only minor external changes (e.g., in abdominal wax markings in *A. fabae*), while influencing, for example, the aphids' ability or willingness to fly (Shaw, 1970). Although the volume and ultrastructure of the CA of crowded and isolated aphids do not appear to differ (Leckstein, 1976), neuroendocrine mechanisms for the release of JH from these glands may nevertheless result in higher JH levels in isolated aphids.

(b) *Larvae are allowed to feed on young host plants* (Johnson and Birks, 1960; Johnson, 1966a). For *M. persicae* complete apterization could occur when young larvae were given access to seedling plants for as little as 1 h. This was demonstrated for aphids reared on a morph-neutral artificial diet on which they would otherwise have developed into alatae (Mittler and Sutherland, 1969; Mittler et al., 1970; Kunkel and Mittler, 1971). Presumably, chemicals that betoken the suitability of young host plants for maximum exploitation by the more fecund apterous morph provide gustatory stimuli that cause an activation of the aphids' CA. While such an activation may also occur under LN conditions in some aphids (e.g., *A. fabae*) when reared at low densities, others (e.g., *M. persicae*) are not apterized by young host plants under LN conditions even if they are raised on them in isolation.

(c) *Larvae are parasitized by hymenopterous parasites* (Johnson, 1959c). The possibility that apterization results from a stimulation of the aphids' CA by materials injected into the aphids when they are stung by the parasites or released from the eggs requires investigation (see also Lees, 1966; Beckage, 1985).

(d) *Larvae are exposed to JH-related compounds*. As with prenatal exposures, such materials have only occasionally been shown to promote aptera production, although they frequently cause the retention of juvenile features (Von Dehn, 1963; White, 1968b; Hales, 1976; Kohno and Takaoka, 1977; Lees, 1977, 1980; Cloutier, 1984); they have even been found to promote alata production (Hardie, 1986), possibly by deactivating the CA allata (see Elliott and McDonald, 1976).

In contrast, larvae produced by apterae raised under LN conditions may readily be diverted from their presumptive alatiform (and gynoparous) development into an apterous (and virginoparous) developmental pathway when treated with JH compounds, as well as on being transferred to SN conditions or elevated temperatures, shortly after their birth (Hardie, 1981a). These effects were ascribed to increases in endogenous JH levels in the treated aphids.

Although such treatments later in the larval life of presumptive alatae may not cause complete diversion to apterousness, they may nevertheless result in a number of morphological features (e.g., less pigmentation, fewer antennal sensoria, reduced wing muscle development, or enhanced embryogenesis) and behavioral attributes (e.g., early settling, feeding, and larviposition on summer host plants) that are more characteristic of the apterous than the alate morph (Mittler et al., 1976; Hardie, 1981b).

That JH compounds (and SN) are highly effective in apterizing presumptively gynoparous alate larvae, whereas presumptively virginoparous alate larvae are not as readily diverted to apterousness by these compounds, suggests that these materials do not affect the insect's genes for apterousness directly but exert their apterizing effects via the aphids' neuroendocrine system. A lower activity of the CA of the gynoparous larvae than of the alatiform virginoparous larvae (inferred from considerations presented below in Section 13.5) may be responsible for the differential responses of the two morphs to these compounds.

That JH levels are higher in larval apterae than in larval alatae is indicated by the fact that more embryos occur in the former (Tsitsipis and Mittler, 1976b), as well as by the high correlation between size of the nuclei in the CA and ovarian development in nymphs of both morphs (Elliott, 1975).

13.3.2.2. REVERSION TO ALATENESS

Larvae that have been diverted prenatally to develop into apterae as a result of their mothers having been reared at low densities under SN conditions may revert completely to an alatiform development if they are crowded (Sutherland and Mittler, 1971) or raised on mature host plants (Johnson, 1966a); however, they may merely develop some minor alatiform features (e.g., abdominal wax markings in *A. fabae*) on being crowded (Shaw, 1970).

An unexpected outcome of prenatal precocene III treatment of presumptively alatiform and apteriform virginoparae of *M. persicae* was that a considerable proportion (20–30%) of both morphs passed through five larval instars, instead of the normal four before attaining adulthood, while an almost equal proportion of aphids born at the same time to the same mothers metamorphosed precociously into alatiform or apteriform adultoids after passing through only one or two larval instars (Mittler and Hales, 1984). A similar response was observed with precocene II–treated *M. viciae* (Hardie, 1986).

The fact that the fourth-instar larvae preceding the supernumerary fifth-instar larvae closely resemble normal fourth instar larvae suggests that an increase in JH in the fourth instar (rather than in the third) leads to the production of the supernumerary fifth-instar larvae. This increase presumably occurred after a temporary decline in JH, because a few males were deposited early in the larviposition sequence of some of the sixth-instar adults (Mittler, unpublished observation; see also Section 15.4, below).

The supernumerary molting of precocene-treated aphids and their reversion to female production could have come about from a resurgence in the activity of their CA or from a collapse of these glands (see Miall and Mordue, 1980). Such events may however have already occurred in the third larval instar but only expressed themselves in the fourth instar, because JH treatments of presumptively alate virginoparae in the third instar (but not in the fourth) also resulted in supernumerary molts (White, 1968b; Cloutier and Perron, 1975; Lees, 1977, 1980; Hardie 1981a). Since hemipterans in general are less responsive to JH-related compounds in their last larval instar than in the previous instar (Staal, 1975), it may be that the aphids' genes for apterousness have to be activated at least two instars before they can express themselves.

In neither the precocene nor the JH treatments were the supernumerary larval instars observed to deposit larvae (cf. the pedogenetic reproduction by *S. graminum* larvae at elevated temperatures: Wanjama and Holliday, 1987).

13.4. Males

In general, male aphids are winged and produced by apterous mothers termed *androparae* (or *sexuparae*, if they also produce oviparous females) under autumnal temperatures and photoperiods. However, in some aphid species or biotypes, males may also occur to a limited extent under SN conditions and cool temperatures (see Lees, 1966; Dixon, 1971; Takada, 1982). That aphids have a higher propensity for producing males than sexual females is also reflected by the fact that the production of males is induced by autumnal photoperiodic conditions whose scotophases may be an hour shorter than those under which the sexual females are produced. Moreover, some biotypes are androcyclic and produce males but not sexual females under LN conditions. Whereas female aphids are produced either parthenogenetically or sexually and have a $2N + 2X$ chromosomal complement, males arise exclusively by parthenogenesis and have a $2N + 1X$ chromosomal makeup. The chromosomal differences between male and female oocytes are established during the first maturation division of the oogonia of their viviparous mothers (see Blackman, 1987).

The following observations support the view that this sex determination process is under JH control and that the temporal sequences in which males and females are deposited reflect changes in maternal JH levels, with males being generated when these levels fall below a certain threshold (Hales and Mittler, 1987):

(a) In the temporary production of a few males by some species or biotypes at SN (e.g., by *M. viciae*, *M. persicae*, or *Hypermyzus lactucae:* Lees, 1966; Takada, 1982; Harrington, 1984), the males are engendered when the JH levels in their mothers may be expected to have declined in relation to their metamorphosis to adulthood.

(b) The onset of male production is related to the time at which the aphids are transferred to LN conditions. Thus, it can be late in the reproductive sequence of aphids exposed to LN only at adulthood (Matsuka and Mittler, 1979), or it can occur in the first few days of reproduction when the aphids are transferred to LN earlier in their development (Crema, 1979; Searle and Mittler, 1981). In general, a few females precede the deposition of the males, presumably because the first ovulations in their apterous androparous mothers occur before the latter are born. Occasionally, however, androparae of a biotype of *M. persicae* that were exposed to LN conditions for several days before their birth produced males immediately and exclusively (Mittler, unpublished observation). The conclusion that these "pure" androparae had already been subjected to low maternal JH levels prior to birth is supported by the finding that their apterous mothers switched over to the production of alatae (gynoparae) immediately after the birth of the androparae.

(c) A few males were found to be produced by *M. viciae* and *A. pisum* under SN conditions when the aphids were temporarily starved (Mari and Orlando, 1984). A starvation-induced decline in JH production (Johansson, 1958; Hardie, 1985, 1987a; Grüber and Dixon, 1988) is probably responsible for this effect (and for ovipara production, see Section 13.6, below).

(d) Male production by *M. persicae* under LN conditions could be inhibited by treating presumptive androparae with kinoprene (Mittler et al., 1979).

(e) Males were generated under SN conditions in virginoparous apterae of a holocyclic biotype of *M. persicae* treated prenatally with precocene III (Hales and Mittler, 1983). Precocene-induced male production could also be inhibited by kinoprene (Hales and Mittler, 1987, 1988). These rescue experiments provide unequivocal evidence for the role of JH in the sex determination process. Kinoprene, presumably, mimics the aphid's natural JH sufficiently to enable it to act directly on the gonads. Precocene III treatment of apterous virginoparae of an anholyclic biotype of *M. persicae* also resulted in the generation of males (Mittler, unpublished observation). This indicates that such aphids have the genetic ability to produce males but do not normally do so in response to LN conditions because they do not register them or respond to them by reducing the JH level below the threshold for female production. This threshold may also be lower than in holocyclic aphids.

(f) The fecundity of androparae is mostly considerably lower than that of exclusively virginoparous aphids. The decline in JH level considered to bring about male determination may also be responsible for the reduced embryogenesis in androparae. A decrease in ovulations (see Hardie, 1987a), the occurrence of inviable oocytes (see below), and the fact that male embryos take longer to develop to term than do female embryos (Searle and Mittler, 1981) probably all contribute to the drastic fall in the larviposition rate of aphids for some days before the onset of male production (Dixon and Glen, 1971; Searle and Mittler, 1981; Mittler and Matsuka, 1985).

(g) In the ovarioles of androparae, an abnormal follicle frequently precedes the first of several male embryos in the ovarioles (*M. persicae:* Searle and Mittler, 1981); alternatively, it may occur instead of the single male embryo in the ovarioles (*M. viciae:* Crema, 1979). An intermediate titer of JH during the switchover from female to male production may result in abnormal chromosome behavior and inviability of the oocytes ovulated at that time. Although the possession by males of only one X chromosome may directly determine their alate development in heteroecious species, maternal JH levels may be too low at the time of the males' birth to divert them to apterousness. JH treatment of adult androparae may clarify this situation.

In early larval life, however, the CA of males may be more active than those of the gynoparous females similarly produced under LN conditions, for in *M. persicae* and *A. fabae* it was found that males but not gynoparae developed into precocious adultoids as a result of prenatal precocene III treatment (Hales and Mittler, 1981; Hardie, 1986). This inference is based on the premise that precocene exerts its cytotoxic action on active CA (Unnithan and Nair, 1979) and is supported by the finding that metathetelic development was not as readily induced by JH analogues in males as it was in females of *Macrosiphum euphorbiae* (Cloutier and Perron, 1975). However, the occurrence of spermatogenesis in the early larval stages of male aphids (see Blackman, 1987) may not be consistent with a high JH titer, since JH is generally antagonistic to this process in other insects (see Dumser, 1980).

13.5. Gynoparae

As is the case for males, the sexual oviparous females typically appear in the autumn when temperatures are cool and scotophases exceed a certain length. They are produced by viviparous mothers, i.e., gynoparae, which in host-alternating (heteroecious) species are winged and produce oviparae predominently. In autoecious species, the ovipara producers typically are wingless and are referred to as sexuparae when they also produce a substantial number of males. Ovipara producers may differ only slightly from their respective alate or apterous virginoparous morphs in their external morphology (Hardie, 1980), but may differ markedly from the latter in their ovarian development and, in the case of heteroecious species, in their host-plant preferences.

Aphids may function as gynoparae as a result of having been exposed to low maternal JH levels as embryos and/or because they have a low endogenous production of JH after birth. The alate development of gynoparae in heteroecious species may be another consequence of low JH levels in their apterous mothers. The production of males by the latter immediately following that of the gynoparae supports this contention.

In heteroecious species, oviparae are not produced by apterae. While this is to be expected on ecological grounds, it is surprising that the LN-induced decline in JH considered to result in the ovulation of males does not also permit at least some of the female embryos within apterae to develop into oviparae. Such a situation pertains to species in which only a few males occur within a sequence of females (Tsitsipis and Mittler, 1977a) as well as to species that switch from gynopara production to exclusive male production (Lamb and Pointing, 1975; Matsuka and Mittler, 1979; Searle and Mittler, 1981). In the latter case, the production of male oocytes under low JH levels may preempt the production of females that might have developed into oviparae under these hormonal conditions.

On the other hand, in autoecious species or in species transitional between heteroeciousness and autoeciousness (MacGillivray and Anderson, 1964), oviparae are produced by apterae under LN conditions. Males, if produced at all, tend to be few in number, and their deposition is generally followed as well as preceded by that of oviparae (see Crema, 1973). Presumably, the JH level in such sexuparae falls below the threshold for female production for only a short time in their development, while remaining low enough for their female embryos to develop into oviparae. In apterae of an English biotype of *A. pisum* under LN conditions, the older embryos develop into oviparae while the younger embryos develop into viviparae, apparently without any males being engendered (Corbitt and Hardie, 1985). Presumably this sequence of morphs is associated with an even smaller decline in JH. It could also indicate a resurgence of JH production, since a switch from ovipara to vivipara production also occurs when gynoparae are transferred from LN to SN conditions or treated with JH compounds (see below).

JH III has been detected in small amounts in *A. fabae* and *M. viciae* (Hardie et al., 1985). While JH III may not be the major JH-active hormone of aphids, the fact that it occurred at significantly lower levels in ovipara-producing apterae than in virginoparous apterae of *M. viciae* is consistent with the proposed hormonal control mechanism of oviparous/viviparous development.

That the CA of gynoparae are relatively inactive during the aphids' early larval development under LN conditions is indicated by the finding that gynoparae of *M. persicae* and *A. fabae* did not metamorphose precociously when treated prenatally with precocene III (Hales and Mittler, 1981; Hardie, 1986). However, when presumptive gynoparae of *M. persicae* were transferred to SN at birth, the aphids did metamorphose precociously, presumably as a result of an activation of their CA (Mittler and Hales, 1984).

Recent studies by Hardie (1987b) showed that the volume of the CA of apterous *M. viciae* increased considerably in relation to ovipara production by these aphids under LN conditions. This would indicate an inverse relationship between the size of the CA and the putative decline in JH production under these circumstances (cf. Hales, 1976). However a feedback mechanism may cause the CA to decrease in size in the presence of high JH levels (Elliott and McDonald, 1976).

Because of the small size of the CA in aphids, no direct physiological or biochemical assessment of the glands' activity has yet been made. However, comparative ultrastructural studies on these glands and on the neuroendocrine systems in general are quite feasible (see Hardie, 1987b). In conjunction with autoradiographic techniques using labeled JH precursors, it should be possible to relate morphological features of neurosecretion with the development, reproduction, and morph production of aphids exposed to different degrees of interaction, photoperiods, temperatures, JH compounds, and precocenes.

In gynoparae and virginoparae of *A. fabae* and *M. persicae,* one or two oocytes may already have been ovulated in each of their ovarioles when the aphids are born (Tsitsipis and Mittler, 1976b; Searle and Mittler, 1982). In gynoparae of *M. persicae,* only these oocytes (in positions 1 and 2, in each ovariole) undergo normal embryogenesis, and this accounts for the relatively low fecundity of the gynoparae compared to that of virginoparous aphids in which several additional oocytes undergo embryogenesis to term. Although two or three additional oocytes may also be ovulated in gynoparae of *M. persicae* during their larval development, such oocytes either do not undergo cleavage or, if they do, the resulting embryos degenerate during early embryogenesis (see Lees, 1966; Searle and Mittler, 1982), presumably because the JH levels remain low during the larval life of these aphids.

An increase in the fecundity of gynoparae of *M. persicae* was brought about by rearing the aphids in isolation or transferring them to SN conditions during their larval development (Searle and Mittler, 1982), as well as by exposing them to kinoprene during their last larval instar (Mittler et al., 1976). Whereas in the former two treatments only a slight increase in fecundity occurred as a result of a greater number of ovarioles producing an embryo in the first follicular position (and occasionally in the second), the fecundity of kinoprene-treated aphids increased considerably, for oocytes in the third and fourth follicular positions (but generally not in the second) were stimulated to undergo embryogenesis (Mittler and Lauritzen, 1989).

A slight increase in ovulation of oocytes has been observed *in vitro* in the ovarioles of virginoparous embryos of *A. fabae* in response to JH I (Hardie, 1987a; but see also Elliott and McDonald, 1976). In a number of aphid species, reductions in ovulation and embryo growth have been observed to result from starvation (Grüber and Dixon, 1988). Decapitation or precocene III treatment of *A. fabae* had similar effects, and these could be countered to some extent by JH treatments (Hardie, 1987a).

In addition to a diversion from an alate into an apterous developmental pathway, a switchover from the production of oviparae to that of viviparae was observed when presumptive gynoparae were transferred to SN conditions, exposed to elevated temperatures, or treated with JH or kinoprene (De Fluiter, 1950; Crema, 1973; Mittler et al., 1976, 1979; Tsitsipis and Mittler, 1976a; Hardie, 1981a; Hardie and Lees, 1983; Corbitt and Hardie, 1985). Partially effective treatments resulted in incompletely apterized individuals or in juvenilized individuals, and in ambiphasic intermorphs with embryos and sexual eggs.

The apterous development of oviparae may result from an increase in the activity of the CA of their gynoparous mothers in early adulthood (as in alate virginoparae: Hales, 1976). Such an increase may also account for the natural switchover from the production of oviparae to viviparae by ovipara producers in a number of species (Tsitsipis and Mittler, 1977a; Corbitt and Hardie, 1985) and for the relatively low numbers of males produced by ovipara producers.

The finding that the larvae produced by kinoprene- or JH-treated gynoparae developed into alatae or had alatiform characters (Mittler et al., 1976; Lees, 1978; Hardie, 1981a) at first appeared to be at variance with expectations. A possible explanation for this apparent anomaly is that the treatments caused a deactivation of the CA of the gynoparae (so that their larvae were not diverted prenatally to apterousness) while stimulating the viviparous development of their larvae. The LN conditions to which the treated gynoparae were exposed may also have contributed to an early deactivation of their CA because when kinoprene-treated gynoparae of *M. persicae* were transferred to SN conditions at the time of treatment they switched over to the production of virginoparous apterae (Mittler and Lauritzen, 1989).

The following treatments caused presumptively virginoparous aphids to produce oviparae under SN conditions:

(a) *Starvation of alatiform larvae of* A. fabae (Hardie, 1985). As in the case of the induction of males by starvation, a reduction in JH is thought to be responsible.

(b) *Extirpation of the median neurosecretory cells in the protocerebrum of* M.

viciae (Steel and Lees, 1977). The ablation was considered to have eliminated the production of a neurohormone (termed virginoparin) thought to influence ovarial development directly. However, recent studies indicate that the neurosecretory cells produce a hemolymph-borne factor that stimulates the CA to produce JH (Hardie, 1987b,c).

(c) *Precocene treatment of presumptively virginoparous alatiform larvae of a holocyclic clone of* A. fabae (Hardie, 1984). It would be of interest to determine whether or not virginoparae of anholocyclic aphids might also produce oviparae as a result of a similar inhibition of JH production.

13.6. Sexual Females

The oviparae of most aphid species are apterous and are generally produced in the autumn by their gynoparous or sexuparous mothers. Shortly after the oviparae have mated, they deposit their yolky eggs. These are relatively few in number in most aphids, because at best only one oocyte is ovulated and matures in each ovariole of an ovipara under LN conditions (Tsitsipis and Mittler, 1976b; see also Blackman, 1987). However, an increase in egg production was noted in oviparae of *M. persicae* raised under SN conditions (Searle and Mittler, 1982).

Since vitellogenesis in other insects generally is JH dependent, it appears probable that endogenous production of JH by the oviparae attains a sufficiently high level to promote the yolk deposition process. However, JH production may not have to be maintained in the oviparae to complete this process; in viviparae chemically allatectomized by prenatal treatment with precocene III, embryos continue to develop in the apparent absence of JH (Hales and Mittler, 1981). Moreover, in oviparae, as in the gynoparae, the CA appear to be refractory to prenatally applied precocene (Hales and Mittler, 1981; Hardie, 1986). The inferred inactivity of the CA during the early larval life of the oviparae may be related to the following: the glands fuse into a single organ only after the birth of the aphids (Bonvicini-Pagliai and Crema, 1975); and oocytes only begin to be ovulated in the second instar of this morph (see Tsitsipis and Mittler, 1976b; Blackman, 1987). This contrasts with the situation in viviparous aphids in which such a fusion occurs, oocytes are ovulated, and embryogenesis begins, all some days before the insects are born (Tsitsipis and Mittler, 1976b; Hardie, 1987a).

13.7. Conclusions

With few exceptions, hormonal studies on aphid polymorphism have been made on the summer and autumn generations of only a few aphid species. With recent improvements in methods for incubating and hatching the diapausing winter eggs of aphids (Wipperfürth and Mittler,

1986; Newton and Dixon, 1987), spring generations will become more amenable to experimentation.

Although many studies show that JH-related compounds are able to affect a number of developmental processes in aphids, our conclusions regarding the endocrine control of polymorphism in aphids will remain conjectural until we can demonstrate that the naturally occurring hormones in aphids act in the speculated manner. We keenly await the development of analytical methods for the characterization and quantitation of these hemipteran hormones. Studies on the mode of action of these hormones at the organismal, cellular, and genetical/molecular level are needed to provide us with a clearer understanding of this fascinating field of developmental biology.

13.8. Summary

From evidence based largely on the responses of aphids to topical applications of JH analogues and to chemical allatectomy with precocenes, it appears that high levels of endogenous JH (as yet chemically undefined) produced by the corpora allata of aphids may cause these insects to be diverted from an oviparous and sexual condition (considered to be primitive) to that of the viviparous and parthenogenetic state in which aphids most commonly occur. Associated with this is the suppression of male production by aphids. High levels of JH (acting prenatally and/or postnatally) may also be responsible for diverting aphids from an alate into an apterous developmental pathway.

Acknowledgments

I thank the following for past collaborative research efforts in acquiring some of the background information that has helped me to develop some of the ideas presented in this paper: R. H. Dadd, N. Davidson, J. Eisenbach, D. F. Hales, J. E. Kleinjan, H. Kunkel, M. Lauritzen, M. Matsuka, S. G. Nassar, G. B. Staal, O. R. W. Sutherland, J. A. Tsitsipis, J. Wilson, and T. Wipperfürth. I am grateful to N. Gorder for her help with this manuscript which was completed in February 1988.

References

Beckage, N. E. 1985. Endocrine interactions between endoparasitic insects and their hosts. Annu. Rev. Entomol. 30: 371–413.

Blackman, R. L. 1987. Reproduction, cytogenetics and development. Pp. 163–195 *in* A. K. Minks and P. Harrewijn (eds.), *Aphids: Their Biology, Natural Enemies, and Control,* Vol. 2A. Elsevier, Amsterdam and New York.

Bonnemaison, L. 1951. Contribution à l'étude de facteurs provoquant l'apparition des formes ailées et sexuées chez les Aphidinae. Ann. Epiphyt. (Paris) 2: 1–380.

Bonnemaison, L. 1972. Action combinée de l'effet de groupe et de la maturité de la plante-hôte sur la production des virginopares ailées du puceron due pois (*Acyrthosiphon pisum*) Ann. Soc. Entomol. Fr. 8: 291–297.

Bonvicini-Pagliai, A. M. and R. Crema. 1975. On the morphogenesis of corpus allatum in *Megoura viciae* Buckt. (Homoptera Aphid.). Boll. Zool. 42: 197–200.

Cloutier, C. 1984. Induction by precocene of alata production by aphids. Proc. 17th Int. Congr. Entomol. p. 165 (Abstr.).

Cloutier, C. and J. M. Perron. 1975. Polymorphism and sensitivity to a juvenile hormone analogue in the potato aphid, *Macrosiphum euphorbiae* (Homoptera: Aphididae). Entomol. Exp. Appl. 18: 457–464.

Corbitt, T. S. and J. Hardie. 1985. Juvenile hormone effects on polymorphism in the pea aphid, *Acyrthosiphon pisum*. Entomol. Exp. Appl. 38: 131–135.

Crema, R. 1973. Structure and determination of the ambiphasic ovary of *Acyrthosiphon pisum* (Homoptera: Aphididae). Entomol. Exp. Appl. 16: 427–432.

Crema, R. 1979. Egg viability and sex determination in *Megoura viciae* (Homoptera: Aphididae). Entomol. Exp. Appl. 26: 152–156.

Dadd, R. H. 1968. Dietary amino acids and wing determination in the aphid *Myzus persicae*. Ann. Entomol. Soc. Am. 61: 1201–1210.

De Fluiter, H. J. 1950. De invloed van daglengte en temperatuur op het optreden van de geslachtsdieren bij *Aphis fabae* Scop., de zwarte bonenluis. Tijdschr. Plantenziekten 56: 265–285.

Delisle, J., C. Cloutier, and J. N. McNeil. 1983. Precocene II–induced alate production in isolated and crowded alate and apterous virginoparae of the aphid, *Macrosiphum euphorbiae*. J. Insect Physiol. 29: 477–484.

Dixon, A. F. G. 1971. The "interval timer" and photoperiod in the determination of parthenogenetic and sexual morphs in the aphid *Drepanosiphon platanoides*. J. Insect Physiol. 17: 251–260.

Dixon, A. F. G. 1985a. Structure of aphid populations. Annu. Rev. Entomol. 30: 155–174.

Dixon, A. F. G. 1985b. *Aphid Ecology*. Chapman & Hall, London and New York.

Dixon, A. F. G. 1987. Evolution and adaptive significance of cyclical parthenogenesis in aphids. Pp. 289–297 *in* A. K. Minks and P. Harrewijn (eds.), *Aphids: Their Biology, Natural Enemies, and Control*, Vol. 2A. Elsevier, Amsterdam and New York.

Dixon, A. F. G. and D. M. Glen. 1971. Morph determination in the bird cherryoat aphid *Rhopalosiphum padi* L. Ann. Appl. Biol. 68: 11–21.

Dorn, A. 1983. Hormones during embryogenesis of the milkweed bug, *Oncopeltus fasciatus* (Heteroptera: Lygaeidae). Entomol. Gen. 8: 193–214.

Dumser, J. B. 1980. The regulation of spermatogenesis in insects. Annu. Rev. Entomol. 25: 341–369.

Elliott, H. J. 1975. Corpus allatum and ovarian growth in a polymorphic paedogenetic insect. Nature (Lond.) 257: 390–391.

Elliott, H. J. and F. J. D. McDonald. 1976. Effect of juvenile hormone analogue on morphology, reproduction and endocrine activity of the cowpea aphid, *Aphis craccivora* Koch. J. Aust. Entomol. Soc. 15: 1–5.

El-Ziady, S. 1960. Further effects of *Lasius niger* L. on *Aphis fabae* Scopoli. Proc. R. Entomol. Soc. Lond. A35: 30–38.

Forrest, J. M. S. 1970. The effect of maternal and larval experience on morph determination in *Dysaphis devecta*. J. Insect Physiol. 16: 2281–2292.

Grüber, K. and A. F. G. Dixon. 1988. The effect of nutrient stress on development and reproduction in an aphid. Entomol. Exp. Appl. 47: 23–30.

Hales, D. F. 1976. Juvenile hormone and aphid polymorphism. Pp. 105–115 *in* M. Lüscher (ed.), *Phase and Caste Determination in Insects*. Pergamon Press, Oxford and Elmsford, New York.

Hales, D. F. and T. E. Mittler. 1981. Precocious metamorphosis of the aphid *Myzus persicae* induced by the precocene analogue 6-methoxy-7-ethoxy-2,2-dimethylchromene. J. Insect Physiol. 27: 333–337.

Hales, D. F. and T. E. Mittler. 1983. Precocene causes male determination in the aphid *Myzus persicae*. J. Insect Physiol. 29: 819–823.

Hales, D. F. and T. E. Mittler, 1987. Chromosomal sex determination in aphids controlled by juvenile hormone. Genome 29: 107–109.

Hales, D. F. and T. E. Mittler. 1988. Male production by aphids prenatally treated with precocene: prevention by short-term kinoprene treatment. Arch. Biochem. Physiol. 7: 29–36.

Hardie, J. 1980. Reproductive, morphological and behavioural affinities between the alate gynopara and virginopara of the aphid, *Aphis fabae*. Physiol. Entomol. 5: 385–396.

Hardie, J. 1981a. Juvenile hormone and photoperiodically controlled polymorphism in *Aphis fabae:* postnatal effects on presumptive gynoparae. J. Insect Physiol. 27: 347–355.

Hardie, J. 1981b. The effect of juvenile hormone on host-plant preference in the black bean aphid, *Aphis fabae*. Physiol. Entomol. 6: 369–374.

Hardie, J. 1984. A hormonal basis for the photoperiodic control of polymorphism in aphids. Pp. 240–258 *in Photoperiodic Regulation of Insect and Molluscan Hormones* (Ciba Found. Symp. No. 104). Pitman, London.

Hardie, J. 1985. Starvation-induced oviparae in the black bean aphid, *Aphis fabae*. Entomol. Exp. Appl. 38: 287–289.

Hardie, J. 1986. Morphogenetic effects of precocenes on three aphid species. J. Insect Physiol. 32: 813–818.

Hardie, J. 1987a. Juvenile hormone stimulation of oocyte development and embryogenesis in the parthenogenetic ovaries of an aphid, *Aphis fabae*. Int. J. Invertebr. Reprod. Dev. 11: 189–202.

Hardie, J. 1987b. The corpus allatum, neurosecretion and photoperiodically controlled polymorphism in an aphid. J. Insect Physiol. 33: 201–205.

Hardie, J. 1987c. Neurosecretory and endocrine systems. Pp. 139–152. in A. K. Minks and P. Harrewijn (eds.), *Aphids: Their Biology, Natural Enemies, and Control*, Vol. 2A. Elsevier, Amsterdam and New York.

Hardie, J. and A. D. Lees. 1983. Photoperiodic regulation of the development of winged gynoparae in the aphid, *Aphis fabae*. Physiol. Entomol. 8: 385–391.

Hardie, J. and A. D. Lees. 1985. Endocrine control of polymorphism and polyphenism. Pp. 441–490 *in* G. A. Kerkut and L. I. Gilbert (eds.), *Comprehensive Insect Physiology, Biochemistry and Pharmacology*, Vol. 8. Pergamon Press, Oxford and Elmsford, New York.

Hardie, J., F. C. Baker, G. C. Jamieson, A. D. Lees, and D. A. Schooley. 1985. The identification of an aphid juvenile hormone, and its titre in relation to photoperiod. Physiol. Entomol. 10: 297–302.

Harrewijn, P. 1976. Role of monoamine metabolism in wing dimorphism of the aphid *Myzus persicae*. Comp. Biochem. Physiol. 55C: 147–153.

Harrewijn, P. 1978. The role of plant substances in polymorphism in the aphid *Myzus persicae*. Entomol Exp. Appl. 24: 398–414.

Harrington, R. 1984. Photoperiodic control of sexual morph production by the currant-sowthistle aphid, *Hyperomyzus lactucae*. Entomol. Exp. Appl. 35: 169–175.

Hille Ris Lambers, D. 1966. Polymorphism in Aphididae. Annu. Rev. Entomol. 11: 47–78.

Johansson, A. S. 1958. Relation of nutrition to endocrine-reproductive functions in the milkweed bug *Oncopeltus fasciatus* (Dallas) (Heteroptera: Lygaeidae). Nytt Mag. Zool. (Oslo) 7: 1–132.

Johnson, B. 1959a. Studies on the degeneration of the flight muscles of alate aphids. II. Histology and control of muscle breakdown. J. Insect Physiol. 3: 367–377.

Johnson, B. 1959b. Ants and form reversal in aphids. Nature (Lond.) 184: 740.

Johnson, B. 1959c. Effect of parasitization by *Aphidius platensis* Brethes on the developmental physiology of its host, *Aphis craccivora* Koch. Entomol. Exp. Appl. 2: 82–99.

Johnson, B. 1965. Wing polymorphism in aphids. II. Interaction between aphids. Entomol. Exp. Appl. 8: 49–64.

Johnson, B. 1966a. Wing polymorphism in aphids. III. The influence of the host plant. Entomol. Exp. Appl. 9: 213–222.

Johnson, B. 1966b. Wing polymorphism in aphids. IV. The effect of temperature and photoperiod. Entomol. Exp. Appl. 9: 301–313.

Johnson, B. and P. R. Birks. 1960. Studies on wing polymorphism in aphids. I. The developmental process involved in the production of the different forms. Entomol. Exp. Appl. 3: 327–339.

Kambhampati, S., M. Mackauer, and K. K. Nair. 1984. Precocious metamorphosis and wing formation in the pea aphid, *Acyrthosiphon pisum,* induced by precocene analogue 7-ethoxy-6-methoxy-2,2-dimethylchromene. Arch. Insect Biochem. Physiol. 1: 147–154.

Kleinjan, J. E. and T. E. Mittler. 1975. A chemical influence of ants on wing development in aphids. Entomol. Exp. Appl. 18: 384–388.

Kohno, M. and I. Takaoka. 1977. Effects of synthetic juvenile hormones through oral administrations on *Myzus persicae* (Sulzer). Kontyû 45: 132–136.

Kunkel, H. and T. E. Mittler. 1971. Einfluss der Ernährung bei Junglarven von *Myzus persicae* (Sulz.) (Aphididae) auf ihre Entwicklung zu Geflügelten oder Ungeflügelten. Oecologia (Berl.) 8: 110–134.

Lamb. R. J. and P. J. Pointing. 1975. The reproductive sequence and sex determination in the aphid *Acyrthosiphon pisum*. J. Insect Physiol. 21: 1433–1446.

Lamb, K. P. and D. White. 1966. Effect of temperature, starvation and crowding on production of alate young by the cabbage aphid *Brevicoryne brassicae*. Entomol. Exp. Appl. 9: 179–184.

Leckstein, P. M. 1976. The role of the corpus allatum in prenatal wing determination in *Megoura viciae*. J. Insect Physiol. 22: 1117–1121.

Lees, A. D. 1966. The control of polymorphism in aphids. Adv. Insect Physiol. 3: 207–277.

Lees, A. D. 1967. The production of the apterous and alate forms in the aphid *Megoura viciae* Buckton, with special reference to the role of crowding. J. Insect Physiol. 13: 289–318.

Lees, A. D. 1977. Action of juvenile hormone mimics on the regulation of larval–adult and alary polymorphism in aphids. Nature (Lond.) 267: 46–48.

Lees, A. D. 1978. Endocrine aspects of photoperiodism in aphids. Pp. 165–168 *in* P. J. Gaillard and H. H. Boer (eds.), *Comparative Endocrinology*. Elsevier/North-Holland Publ., Amsterdam and New York.

Lees, A. D. 1980. The development of juvenile hormone sensitivity in alatae of the aphid *Megoura viciae*. J. Insect Physiol. 26: 143–151.

Lees, A. D. 1984. Parturition and alate morph determination in the aphid *Megoura viciae*. Entomol. Exp. Appl. 35: 93–100.

MacGillivray, M. E. and G. B. Anderson, 1964. The effect of photoperiod and temperature on the production of gamic and agamic forms in *Macrosiphum euphorbiae* (Thomas). Can. J. Zool. 24: 491–510.

Mackauer, M., K. K. Nair, and G. C. Unnithan. 1979. Effect of precocene II on alate production in the pea aphid, *Acyrthosiphon pisum*. Can. J. Zool. 57: 856–859.

Mari, M. and E. Orlando. 1984. Some factors inducing male sex determination in the aphids *Megoura viciae* Buckton and *Acyrthosiphon pisum* Harris. Proc. 17th Int. Congr. Entomol., 1984 p. 428 (Abstr.).

Matsuka, M. and T. E. Mittler. 1979. Production of males and gynoparae by apterous viviparae of *Myzus persicae* continuously exposed to different scotoperiods. J. Insect Physiol. 25: 587–593.

Miall, R. C. and W. Mordue. 1980. Precocene II has juvenile-hormone effects in 5th instar *Locusta migratoria*. J. Insect Physiol. 26: 361–364.

Mittler, T. E. 1971. Some effects on the aphid *Myzus persicae* of ingesting antibiotics incorporated into artificial diets. J. Insect Physiol. 17: 1333–1347.

Mittler, T. E. 1973. Aphid polymorphism as affected by diet. Pp. 65–75 *in* A. D. Lowe (ed.), *Perspectives in Aphid Biology*. Entomol. Soc. N.Z. (Auckland) Bull. No. 2.

Mittler, T. E. 1988. Application of artificial feeding techniques for aphids. Pp. 145–170 *in* A. K. Minks and P. Harrewijn (eds.), *Aphids: Their Biology, Natural Enemies, and Control*, Vol. 2B. Elsevier, Amsterdam and New York.

Mittler, T. E. and D. Hales. 1984. Supernumerary larval instars and precocious metamorphosis in aphids induced by precocene, in relation to photoperiod. Proc. 17th Int. Congr. Entomol., 1984 p. 166 (Abstr.).

Mittler, T. E. and J. E. Kleinjan. 1970. Effect of artificial diet composition on wing-production by the aphid *Myzus persicae* J. Insect Physiol. 16: 833–850.

Mittler, T. E. and H. Kunkel. 1971. Wing production by grouped and isolated apterae of the aphid *Myzus persicae* on artificial diet. Entomol. Exp. Appl. 14: 83–92.

Mittler, T. E. and M. Lauritzen. 1989. Effects of the juvenile hormone analogue kinoprene and photoperiod on embryogenesis and polymorphism in gynoparae of the aphid *Myzus persicae*. *In* M. Tonner, T. Soldan, and B. Bennettorei (eds.), Proc. 4th Conf. on Insect Reprod., Zinkovy, Czechoslovakia, 1987.

Mittler, T. E., and M. Matsuka. 1985. Male production by apterous viviparae of the aphid *Myzus persicae* temporarily exposed to different scotoperiods. J. Insect Physiol. 31: 171–175.

Mittler, T. E., and O. R. Sutherland. 1969. Dietary influences on aphid polymorphism. Entomol. Exp. Appl. 12: 703–713.

Mittler, T. E., J. E. Kleinjan, and H. Kunkel. 1970. Apteriform development induced by radish seedlings in larvae of the aphid *Myzus persicae* reared on artificial diet. J. Insect Physiol. 16: 2119–2125.

Mittler, T. E., S. G. Nassar, and G. B. Staal. 1976. Wing development and parthenogenesis induced in progenies of kinoprene-treated gynoparae of *Aphis fabae* and *Myzus persicae*. J. Insect Physiol. 22: 1717–1725.

Mittler, T. E., J. Eisenbach, J. B. Searle, M. Matsuka, and S. G. Nassar. 1979. Inhibition by kinoprene of photoperiod-induced male production by apterous and alate viviparae of the aphid *Myzus persicae*. J. Insect Physiol. 25: 219–226.

Newton, C. and A. F. G. Dixon. 1987. Methods of hatching the eggs and rearing the fundatrices of the English grain aphid *Sitobium avenae*. Entomol. Exp. Appl. 45: 277–281.

Nijhout, H. F. and D. E. Wheeler. 1982. Juvenile hormone and the physiological basis of insect polymorphisms. Q. Rev. Biol. 57: 109–133.

Rankin, M. A. 1978. Hormonal control of insect migratory behavior. Pp. 5–32 *in* H. Dingle (ed.), *Evolution of Insect Migration and Diapause*. Springer-Verlag, Berlin and New York.

Schaefers, G. A. 1972. The role of nutrition in alary polymorphism among the Aphididae: an overview. Search Agric. (Geneva N.Y.) 2(4): 1–8.

Searle, J. B. and T. E. Mittler. 1981. Embryogenesis and the production of males by apterous viviparae of the green peach aphid *Myzus persicae* in relation to photoperiod. J. Insect Physiol. 27: 145–153.

Searle, J. B. and T. E. Mittler. 1982. Embryogenesis and oogenesis in alate virginoparae, gynoparae, and oviparae of the aphid *Myzus persicae,* in relation to photoperiod. J. Insect Physiol. 28: 213–220.

Shaw, M. J. P. 1970. Effects of population density on alienicolae of *Aphids fabae* Scop. III. The effect of isolation on the development of form and behaviour of alatae in a laboratory clone. Ann. Appl. Biol. 65: 205–212.

Shull. A. F. 1938. Time of determination and time of differentiation of aphid wings. Am. Nat. 72: 170–179.

Staal, G. B. 1975. Insect growth regulators with juvenile hormone activity. Annu. Rev. Entomol. 20: 417–460.

Steel, C. G. H. and A. D. Lees. 1977. The role of neurosecretion in the photoperiodic control of polymorphism in the aphid *Megoura viciae*. J. Exp. Biol. 67: 117–135.

Sutherland, O. R. W. 1967. Role of host plant in production of winged forms by a green strain of pea aphid *Acyrthosiphon pisum* Harris. Nature (Lond.) 216: 387–388.

Sutherland, O. R. W. 1969. The role of the host plant in the production of winged forms by two strains of the pea aphid, *Acyrthosiphon pisum*. J. Insect Physiol. 15: 2179–2201.

Sutherland, O. R. W. and T. E. Mittler. 1971. Influence of diet composition and crowding on wing production by the aphid *Myzus persicae*. J. Insect Physiol. 17: 321–328.

Takada, H. 1982. Influence of photoperiod and temperature on the production of sexual morphs in a green and red form of *Myzus persicae* (Sulzer) (Homoptera, Aphididae). I. Experiments in the Laboratory. Kontyû 50: 233–245.

Tsitsipis, J. A. and T. E. Mittler. 1976a. Influence of temperature on the production of parthenogenetic and sexual females by *Aphis fabae* under short-day conditions. Entomol. Exp. Appl. 19: 179–188.

Tsitsipis, J. A. and T. E. Mittler. 1976b. Embryogenesis in parthenogenetic and sexual females of *Aphis fabae*. Entomol. Exp. Appl. 19: 263–270.

Tsitsipis, J. A. and T. E. Mittler. 1977a. Influence of daylength on the production of parthenogenetic and sexual females of *Aphis fabae* at 17.5°. Entomol. Exp. Appl. 21: 163–173.

Tsitsipis, J. A. and T. E. Mittler. 1977b. Influence of temperature and daylength on the production of males by *Aphis fabae*. Entomol. Exp. Appl. 21: 229–237.

Unnithan, G. C. and K. K. Nair. 1979. The influence of corpus allatum activity on the susceptibility of *Oncopeltus fasciatus* to precocene. Ann. Entomol. Soc. Am. 72: 38–40.

Von Dehn, M. 1963. Hemmung der Flügelbildung durch Farnesol bei der schwarzen Bohnenlaus, *Doralis fabae* Scop. Naturwissenschaften 17: 578–579.

Wanjama, J. K. and N. J. Holliday. 1987. Paedogenesis in the wheat aphid *Schizaphis graminum*. Entomol. Exp. Appl. 45: 297–298.

White, D. F. 1968a. Cabbage aphid: effect of isolation on form and on endocrine activity. Science (Wash., D.C.) 159: 218–219.

White, D. 1968b. Postnatal treatment of the cabbage aphid with a synthetic juvenile hormone. J. Insect Physiol. 14: 901–912.

Wipperfürth, T. and T. E. Mittler. 1986. A method for hatching the eggs of the aphid *Schizaphis graminum*. Entomol. Exp. Appl. 42: 57–61.

Wipperfürth, T., K. S. Hagen, and T. E. Mittler. 1987. Egg production by the coccinellid *Hippodamia convergens* fed on two morphs of the green peach aphid, *Myzus persicae*. Entomol. Exp. Appl. 44: 195–198.

Roles of Morphogenetic Hormones in the Morphogenesis of Eyes in Insects

14

MICHEL MOUZE

14.1. Introduction

The visual system in insects is formed by two superposed parts (Fig. 14.1): the *compound eye,* composed of many ommatidia, which constitute the sensory receptor (although of epidermal origin and character, and probably because of its complex structure, the eye shows a very peculiar type of postembryonic growth, different from the growth of the remaining portion of the integument); and the *optic lobe,* to which each of the photosensitive cells of that eye sends information, and which transmits it to the brain after integration. Developmental growth of the three ganglia constituting the optic lobe must parallel that of one of the eyes, and it also is somewhat different from that of other nervous parts.

This chapter is divided into two main sections. The first (Section 13.2) includes the various modifications exhibited by the visual system during growth and metamorphosis; in it we shall focus on the processes that are probably correlated with the morphogenetic hormones. Then, in Section 13.3, we shall consider the results of experimental studies on the way these hormones control growth of the visual system. Because of the relative paucity of studies of that kind in most insects, a large part of this review will concern the results obtained in one group, Odonata.

14.2. Description of Eye Development in Insects

Growth of the visual system in insects can take place in two opposite ways. Development can be very gradual, starting during embryogenesis, continuing during the entire larval life, and ending at adult emergence; this kind of growth characterizes the hemimetabolous and some holometabolous insects. On the other hand, development can be sudden, i.e., the eye appears during metamorphosis, only in the preimaginal instar; the functional visual system of the larva, formed by very rudimentary stemmata, regresses and then disappears at this time. The latter kind of development can be observed in most holometabolous insects.

14.2.1. *The Compound Eye*

14.2.1.1. GRADUAL EYE GROWTH

Structure

Growth of the compound eye results from continuous addition of new ommatidia at its margin (anteriorly in the locust, dorsally in the dragonfly, and posteriorly in other insects) and results also from progressive growth of the differentiated ommatidia (Fig. 14.2). The new ommatidia originate from a special "proliferation zone." During postembryonic development, a continuous sequence of events occurs in the proliferation

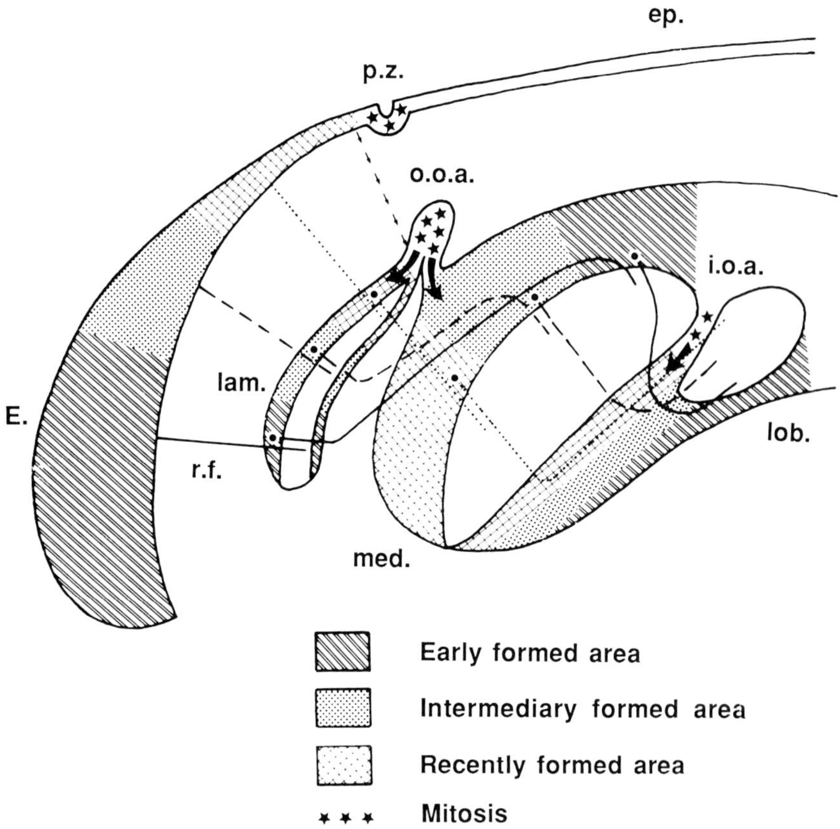

FIGURE 14.1. Schematic cross section of the visual system of a hemimetabolous insect. *Key:* E = eye; ep. = epidermis; i.o.a. = inner optic anlage; lam = lamina; lob. = lobula; med. = medulla; o.o.a. = outer optic anlage; p.z. = proliferation zone; r.f. = retinal fiber.

zone of the retina to transform undifferentiated cells into mature ommatidia. From the sagittal plane to the lateral region of the head, different zones can be distinguished: epithelial cells, the proliferation zone, the differentiation zone (ommatidia in various stages of development), and fully differentiated ommatidia.

The proliferation zone is a region of densely packed cells and of high mitotic activity. When a cell in this epithelium divides, it loses its attachment to the basement membrane and rounds out at the lumen surface; this is also observed in various other nervous structures in invertebrates or vertebrates. Because the spindle axis is always parallel to the surface, the cell divides in the plane of the epithelium and, after division, each

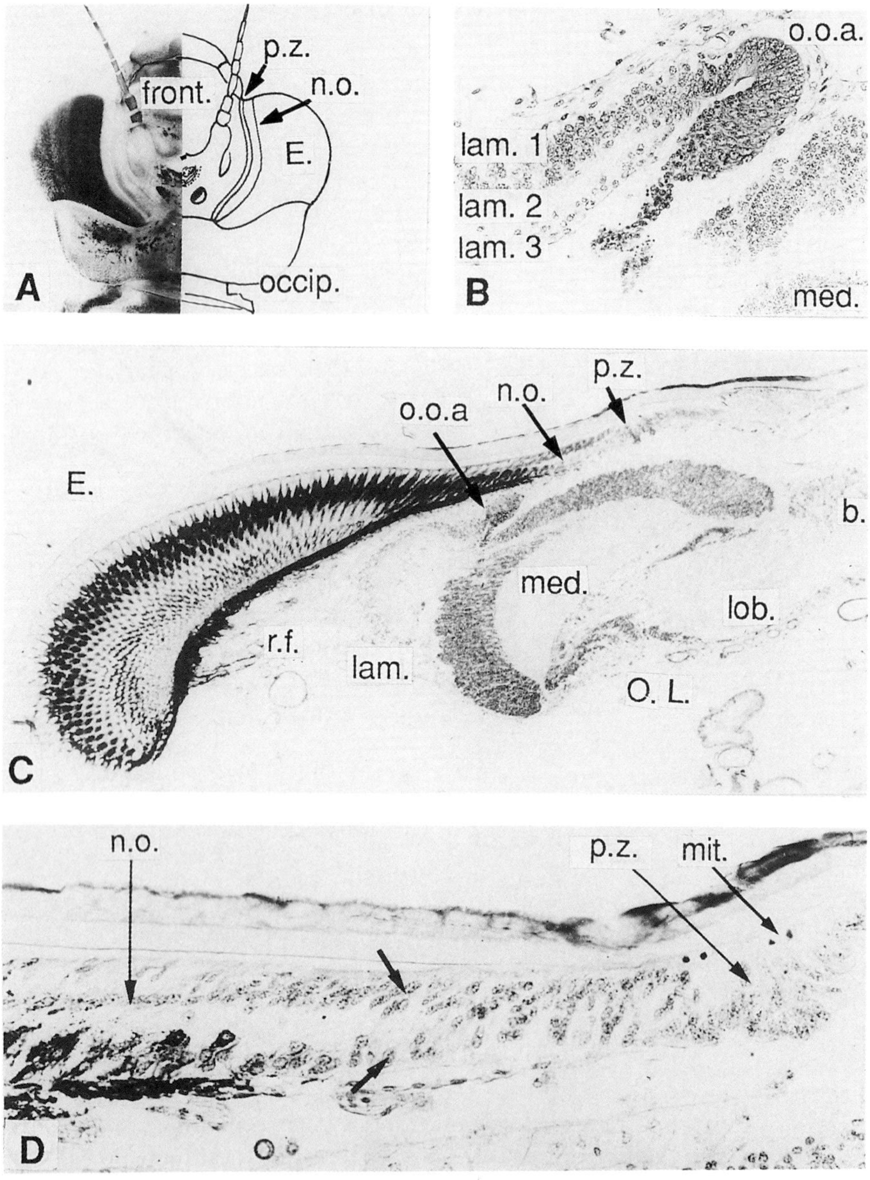

FIGURE 14.2. Morphology and structure of the visual system of the nymph of *Aeshna cynaea* (Odonata: Anisoptera). (A) Dorsal view of the head. ×6 (B) Outer optic anlage. ×300. (C) Cross section of the left side of the head. ×40. (D) Growth zone of the compound eye. ×220. *Key:* b. = brain; E = eye; lam. = lamina; lam. 1 = external layer of the lamina; lam. 2 = median layer; lam. 3 = internal layer; O.L. = optic lobe; lob. = lobula; med. = medulla; mit. = mitosis; n.o. = new ommatidia; occip. = occiput; o.o.a. = outer optic anlage; p.z. = proliferation zone; r.f. = retinal fiber.

daughter cell again extends a foot to the basement membrane. Thus, there is a progressive thickening of the retina as it extends from the margin of the eye, indicated by the increasing distance from the cornea to the basement membrane. In this differentiation zone, the cells gather in cartridges formed by more and more cells of different types and functions (crystalline, corneal, pigmental, and retinal cells), each cartridge prefiguring one ommatidia. These new ommatidia can be observed in various stages of development. The stage of differentiation of the ommatidia is also more and more advanced as the distance from the proliferation zone increases. In this zone the axons appear, growing from each retinal cell; neighboring axons group into bundles, which grow across the basement membrane of the eye to connect with the first optic ganglia (or lamina).

Cyclic Modifications During a Larval Intermolt Period

Mitotic activity. Several authors have studied changes in mitotic activity in the epidermis, proliferation zone, differentiation zone, and fully differentiated ommatidia throughout a larval instar (the locust *Schistocerca gregaria:* Anderson, 1978; the dragonfly *Aeshna cyanea:* Schaller, 1960, 1964; the cockroach *Periplaneta americana:* Nowel, 1981; Stark and Mote, 1981). Their results (Fig. 14.3) are comparable and can be summarized as follows: In the *proliferation zone,* dividing cells are observed throughout the molt cycle. From a low value just before ecdysis, the number of divisions rises steeply to a maximum, generally occurring between the first third and the middle of the intermolt period. Then, following a short plateau phase, the number of mitoses decreases slowly and generally maintains a low basal level until the end of the instar. In contrast, the cells of other regions resemble the pattern found in a wide variety of insect epidermal cells: cell divisions are confined to a discrete period within the molt cycle. In the *differentiation zone,* dividing cells appear only during a very restricted portion of the molt cycle: cell divisions generally coincide with the maximum number of mitoses in the proliferation zone, reach a peak in the middle course of the instar, then rapidly fall to zero, where they stay for the rest of the molt cycle. In the cockroach *Periplaneta,* Nowel (1981) observed that 97% of the total mitoses in this region occurred during the second quarter of the intermolt period. In *fully differentiated* ommatidia, virtually all mitotic activity occurs at the peak time in the maturation zone. Cells dividing among the mature ommatidia are probably secondary pigment cells, as the other components have a constant number at this stage.

In insects, cyclic mitotic activity in the epidermis is thought to be controlled by the hormonal milieu, which also varies cyclically in each instar, and it seems reasonable to assume that cell divisions in the three

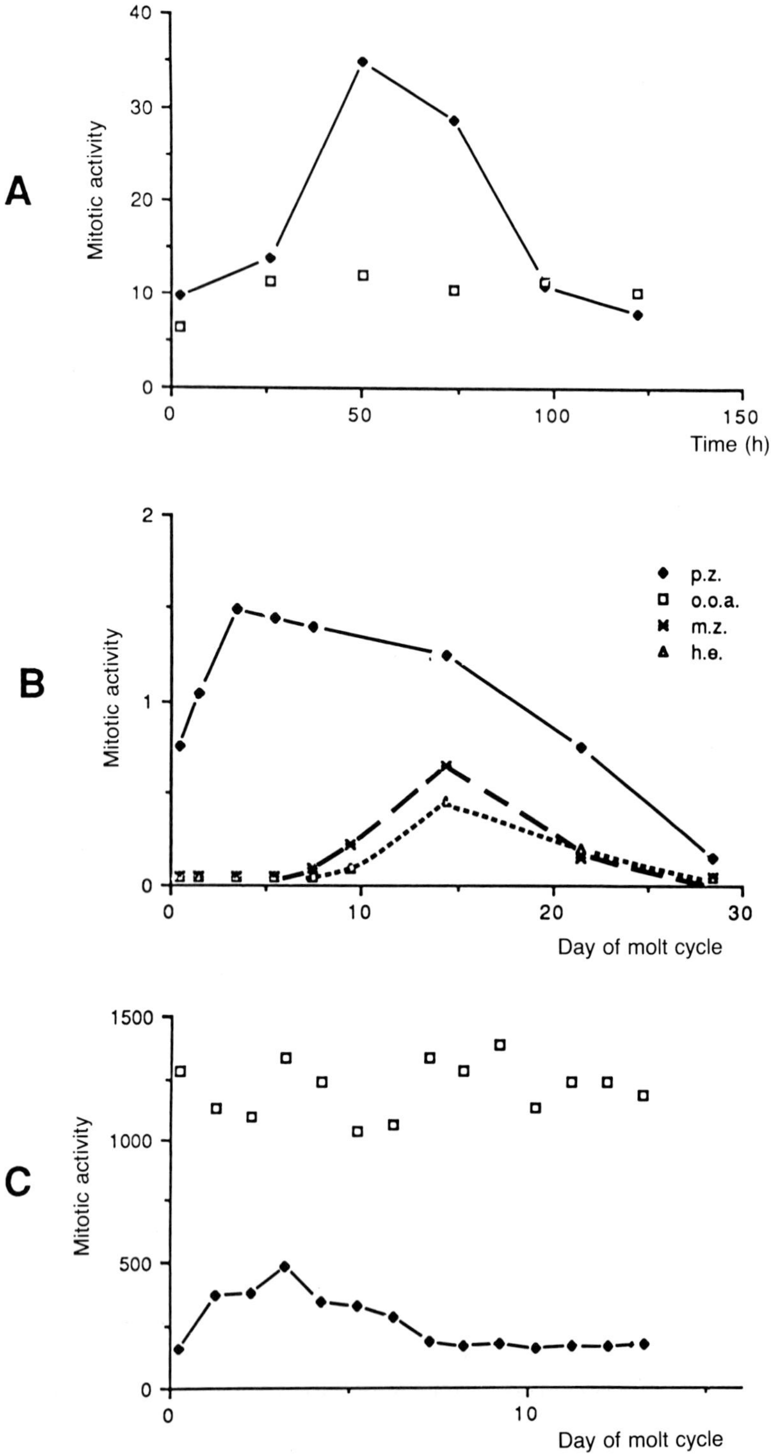
A
Mitotic activity
40
30
20
10
0
0
50
100
150
Time (h)
B
Mitotic activity
2
1
0
0
10
20
30
Day of molt cycle
p.z.
o.o.a.
m.z.
h.e.
C
Mitotic activity
1500
1000
500
0
0
10
Day of molt cycle

aforementioned regions are under the same hormonal control. Nevertheless, it remains unclear whether cell division in the proliferation zone depends entirely on hormonal influence, as suggested by the peak of mitotic activity, or whether it is partly independent of them, as Schaller (1964) suggested when considering the presence of a basal mitotic level in this proliferation zone on *all days* of the instar, or indeed whether it is extremely sensitive to a very low level of hormones, which would persist during the whole molt cycle. This point will be discussed later.

Cuticular cycle. Like all other insect epidermal regions, the proliferation zone secretes a cuticular layer in continuity with the head epidermal cuticle and with the eye cuticle (called the "corneal cuticle" owing to its transparency and its convex shape above each ommatidium).

Unlike what happens in other epidermal zones, however, the apolysis above the proliferation zone occurs around the beginning of each larval instar and progressively reaches the whole maturation zone. The continuous production of new cells by mitosis in this zone rapidly leads to a folding along the eye margin of the epidermal sheath, forming the proliferation zone; at the same time, the early apolysis above the maturation zone allows the continuation and completion of differentiation of the ommatidial cells (Mouze, 1972).

In some insects, during the course of each larval instar, external examination of the extension of eye pigmentation under the cuticle can therefore be used to precisely follow the progressive eye development during the instar.

FIGURE 14.3. Graphic showing the variation throughout one nymphal instar of mitotic activity within the head epidermis, the proliferation zone, and the maturation zone of the eye, and the outer optic anlage of the optic lobe. (A) In the fourth instar of the locust *Schistocerca gregaria: ordinate*—mean number of mitotic figures per 60 sections; *abscissa*—hours after the molt to the forth instar. (From Anderson, 1978.) (B) In intermediate stage (fifth-, sixth-, or seventh-instar nymphs) of the cockroach *Periplaneta americana: ordinate*—number of mitotic figures per examined section; *abscissa*—days after the previous molt. (From Nowel, 1981.) (C) In penultimate instar of the dragonfly *Aeshna cyanea: ordinate*—number of mitotic figures on a standard number of sections; *abscissa*—days after the previous moult. (From Schaller, 1964.) *Key:* h.e. = head epidermis; m.z. = maturation zone; o.o.a. = outer optic anlage; p.z. = proliferation zone.

Metamorphosis of the Eye

As in each larval instar, the mitotic activity of the *proliferation zone* increases steeply around the beginning of the instar and presents a peak in the first third of the intermolt period. Then the mitotic activity continues until the intermolt period is approximately half over but the dividing cells have become progressively rare; at this point the level of mitosis falls to zero, where it stays for the rest of this intermolt period. The proliferation zone appears at this time as a thickened portion of the epidermis at the margin of the eye.

In the differentiation zone and in mature ommatidia, mitosis can also be observed during a short period after the peak of activity in the proliferation zone and newly formed cartridges make up the cell number of the ommatidia. In Odonata (Anisoptera), all the ommatidia formed during the three last larval instars progressively differentiate and grow. This feature is particularly obvious if one observes the advance of the dark pigmented boundary of the eye under the corneal and dorsal cephalic cuticles.

Cuticle. In Odonata, during this last larval instar, the early apolysis is more extended than in the previous ones and covers a very large area on the dorsal side of the head. The cuticle, which covers the dorsal area between the eyes, is completely separated from the epidermis as early as the middle of the intermolt period, and this allows the eyes to extend under the cuticle (Fig. 14.4). By observing the immediate steps of this extension, Schaller (1960) established a developmental chart that enables one to figure out, by external examination, the precise physiological age and metamorphic progress in Odonata larvae (Fig. 14.5).

During the last days of this preimaginal stage, the entire eye undergoes its final modifications: completion of differentiation in the new ommatidia (growth of the crystalline cone; pigmentation of the pigment cells); completion of the extension of the microvillosity in the rhabdomeres; transformation of the larval functional ommatidia to adapt them to aerial vision (Lavoie et al., 1978a,b); growth of numerous tracheal tubules between the ommatidia of the whole eye. All these modifications result in an increase of the eye volume, and at the end of this instar both eyes are widely in contact on the sagittal plane of the head.

14.2.1.2. SUDDEN EYE GROWTH

In most holometabolous insects, typical growth of the eye involves "explosive" development during metamorphosis of a specialized imaginal disk. Differentiation of the cells intended to form ommatidia may begin late in larval life (in mosquitoes, sawflies, campodeiform Coleoptera, and other insects in which the larval visual system is functional) or may begin at the onset of metamorphosis (in honey bees, some Lepidoptera,

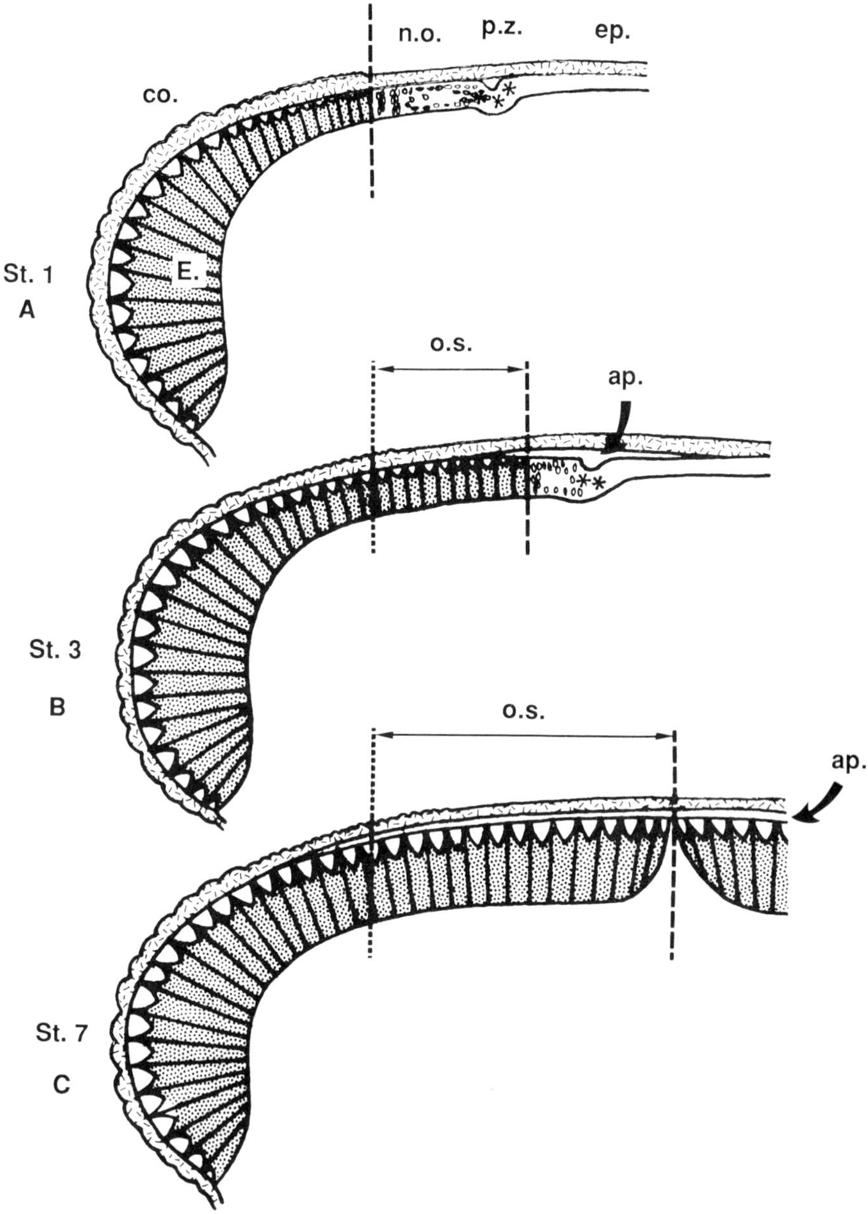

FIGURE 14.4. Schematic description of ocular metamorphosis of *Aeshna*. Successive steps during the last nymphal intermolt period (left side of the head). *Key:* ap. = apolysis; co. = cornea; E = eye; ep. = epidermis; n.o. = new ommatidia; o.s. = ocular spreading; p.z. = proliferation zone; St. 1, 2, 3 = steps 1, 2, and 3 (of ocular metamorphosis).

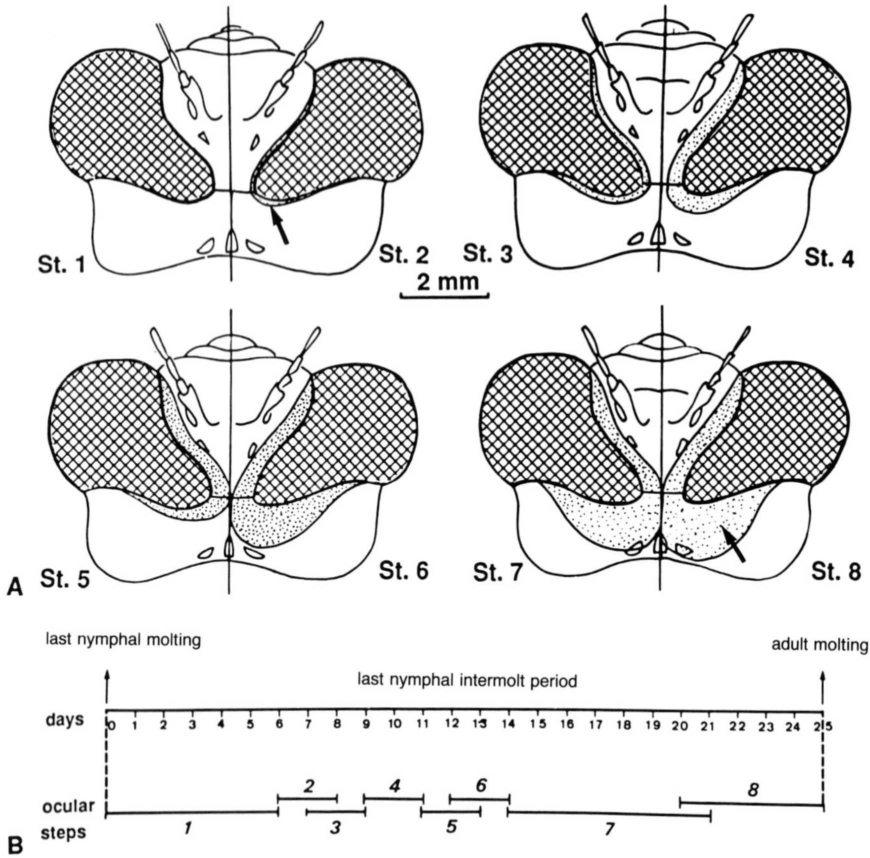

FIGURE 14.5. Last nymphal instar of *Aeshna cyanea*. (A) Successive steps (1–8) of ocular spreading (arrow); dorsal view of the head. (B) Standard chronology of the last nymphal instar and relation between ocular spreading and age of the nymph. (From Schaller, 1960.)

and other insects in which the visual system is rudimentary). However, regardless of the insect considered, the eye grows in a similar fashion.

Ommatidia are laid down as cell clusters after a wavelike mitotic invasion among the cells of the head epidermis. The sequence of patterned mitoses that generate ommatidial clusters has its origin at or close to the posterodorsal margin, in an autonomous region of the eye field that functions as a differentiation center. The formative eye spreads in a posteroanterior direction, the progressing border being limited by a dorsoventral furrow. Ahead of this furrow, in *Drosophila,* cell division occurs over the entire region, producing a supply of cells. Along the advancing front, unpatterned cells are first assembled into "preclusters," consisting of photoreceptor cells 2, 3, 4, 5, and 8, which undergo no further divisions. Behind the furrow other cells, surrounding the preclusters, divide in a mitotic wave, producing the remaining photoreceptor cells 1, 6, and 7, as well as others needed to complete the ommatidial set (i.e., cone cells, pigmentary cells).

In *Drosophila*, the peripodial sac will evert during metamorphosis, the surface of the disk becoming the outside of the eye. Throughout the larval life, the larval visual system (stemmata) send a slender nerve to the supraesophagal hemisphere, and at metamorphosis the centripetally fiber pathway between the retina and the first optic neuropile grows around this preexisting optic stalk. At that time, the stemmatal nerve, which has become thickened during late larval life, probably glides, carrying the remnants of the stemmatal pigment to the brain, where it remains permanently enclosed within the perineurial sheath.

14.2.2. *The Optic Lobe*

In insects the optic lobe is formed by the juxtaposition of three optic ganglia, i.e., from an external to an internal situation: the first optic ganglion, or *lamina* (into which penetrate the bundles of sensorial axons growing from ommatidia); the second optic ganglion, or *medulla* (receiving axons from the lamina); the third optic ganglion, or *lobula* (receiving axons from the medulla and connected with the brain).

14.2.2.1. GROWTH OF THE OPTIC LOBE IN HEMIMETABOLOUS INSECTS

Nymphal Growth

Ganglion cells of the optic lobe are essentially produced by two coiled rodlike aggregates of neuroblasts, the optic anlagen, already present in the newly hatched nymph, and which continue to divide throughout the postembryonic development (Fig. 14.2). The lamina and medulla grow by addition of new cells produced by mitotic activity in a common growth zone, the *outer optic anlage,* whereas the cells of the lobula arise from the *inner optic anlage.*

The outer optic anlage, the best defined and the easiest to study, is a characteristically folded structure lying beneath the proliferation zone of the retina and extending parallel to it along the optic lobe. It proliferates new cells from a stem cell population of neuroblasts, which divide asymmetrically to form another neuroblast and a ganglion mother cell. Subsequent ganglion mother cell divisions produce new ganglion cells that are continually displaced from the anlage by addition of yet newer cells. The progeny of ganglion mother cells become the ganglion cells of the lamina and medulla.

Differentiation of the ganglion cells can occur only if they establish synaptic contact with growing retinal axons. If not, these cells degenerate. This is a regulating mechanism by which the compound eye growth precisely controls the optic lobe growth (Mouze, 1979).

Neuroblasts and ganglion mother cells undergo mitosis every day of the instar without perceptible fluctuations corresponding to the molting cycle, and statistical analysis shows that there is no significant variation in mitotic activity between the days. Neuron production in the optic lobe therefore appears to be independent of cyclic aspects of the hormonal environment. This is the case in *Aeshna cyanea* (Schaller, 1964), *Schistocerca gregaria* (Anderson, 1978), and *Periplaneta americana* (Nowel, 1981).

Metamorphosis

In the first days of the last intermolt period, the volume of the anlage and the number of mitoses in neuroblasts begin to decline. For example, in *Odonata* the outer anlage begins to decrease on the fifth day of the last nymphal instar (whole duration of this instar: 25 days), regularly diminishes, and finally fully disappears on day 17.

The other important modification of the optic lobe is the histological transformation of the first optic ganglion, or lamina (Fig. 14.6). During the whole nymphal life, the lamina is structured in three layers: the external layer, formed by cellular bodies gathered in neuronal cartridges (each cartridge corresponding to one ommatidia); the median layer, exclusively composed of light striated fibers (the neuropile of the lamina); the internal layer, a thin sheath of nuclei and fibrous material.

During the first days of the last nymphal instar, mitoses appear in the internal layer: new cells produced by those divisions progressively invade the median layer, and this modification progresses during this intermolt period from the newly formed area (close to the outer anlage) to the older area of the lamina; at the end of this instar, the appearance of this first ganglion is very different from its larval structure.

In the same way, the ommatidial axons are progressively shortened, bringing the lamina closer to the eye, the lamina in the adult really being in contact with the basement membrane of the eye. Its structure at this stage consists only of two layers: the external one, formed by big vacuolar cells; and the internal one, originating from the fusion of the median and internal nymphal layers. The lamina therefore really undergoes a metamorphosis that probably also corresponds to alterations in the structure of the compound eye correlated with environmental changes.

14.2.2.2. GROWTH OF THE OPTIC LOBE IN HOLOMETABOLOUS INSECTS

Development of the adult optic lobe may proceed throughout the larval period in some insects (*Apis, Drosophila, Danaus*), whereas in others

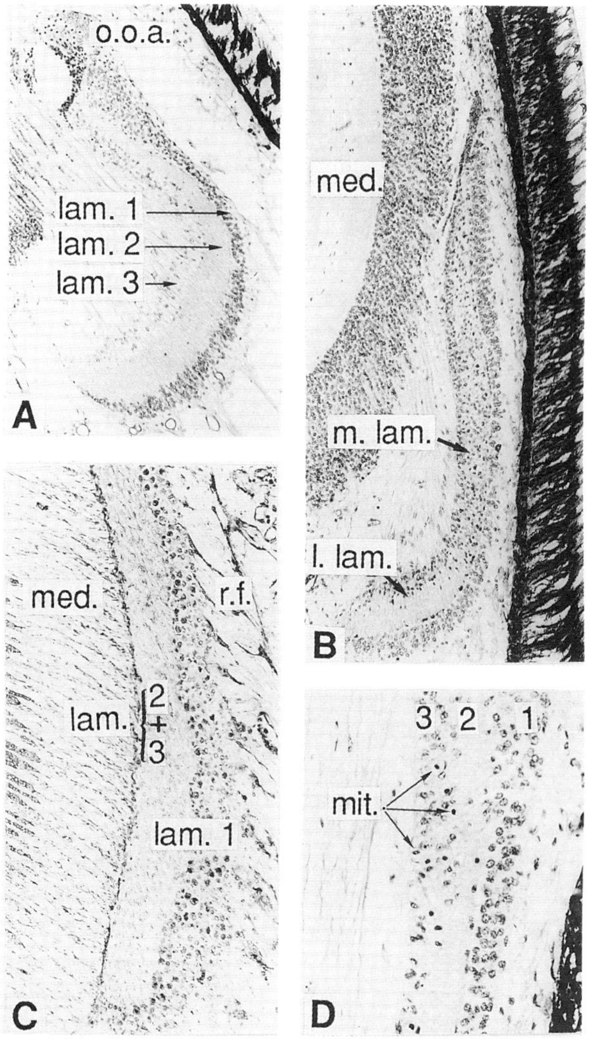

FIGURE 14.6. Metamorphosis of the lamina during the last nymphal intermolt of *Aeshna*. (A) Typical nymphal structure (step 1). ×90. (B) Advance of the transformation of the median layer of the lamina (step 5). ×90. (C) Structure of the lamina in the adult. ×165. (D) Mitoses in the internal layer (step 3). ×165. *Key:* lam. 1, 2, 3 = external, median, and internal layers of the lamina; l. lam = lamina of larval structure; m. lam = lamina of metamorphosed structure; med. = medulla; mit. = mitosis; o.o.a. = outer optic anlage; r.f. = retinal fiber.

(*Pieris, Ephestia*) it may be initiated in the pupal stage. When it develops continuously throughout postembryonic life, it grows independently of the molting cycle, as in hemimetabolous insects. In that case, optic lobe development is initiated early in larval life and accelerates as the animal enters metamorphosis. The divisions in the anlagen appear during the first days of larval life, and following pupation the mitotic activity decreases and neuroblasts degenerate (Nordlander and Edwards, 1969).

Cell production continues for only a few days after pupation, but cell enlargement and fiber growth continue for most of the pupal period. A very rapid laminar growth phase begins with the initiation of adult development and the introduction of ommatidial fibers.

14.3. Experimental Results

As soon as hormonal studies in insects began, some investigators were interested in the eyes: Bodenstein (1940, 1953) showed that, when eye disks of young *Drosophila* larvae were transplanted into the abdomen of adult male flies, they remained alive but were unable to grow. Conversely the disks continued to grow when a larval ring gland was transplanted simultaneously with the eye disk; the growth of the eye disk is therefore under control of a growth hormone released by the ring gland.

Bodenstein (1940, 1953) also shortened the "larval life" of eye disks by transplanting young disks into older larvae and found that they could still differentiate, the facet number however being reduced in proportion to the length of the developmental period. Nevertheless, when *Drosophila* eye disks are transplanted into hosts younger than the donor, the additional larval time thus provided allows for a prolonged growth period, resulting in an imaginal eye with more facets than normal.

Bounhiol (1938) and Fukuda (1944) showed that, when the corpora allata (CA) are removed from silkworm caterpillars, the animals pupate prematurely and give rise to small adults of apparently normal proportions, which suggests a reduction in eye size. The causal relationship between the time of larval eye growth and the onset of differentiation is thus brought into focus, showing that any disturbance of the normal time balance between these two processes has a decisive effect on the final size of the eye.

All these experiments therefore clearly show that hormones are special extrinsic factors on which eye growth depends and that the responsiveness of the eye disk changes during its period of growth. Since these first studies, a few other experiments have been performed to define the respective effects of the morphogenetic hormones on the visual system; we will now present some results obtained from various experimental studies performed on dragonfly larvae.

14.3.1. *Development of the Visual System in Insects Without Molting Hormone–Releasing Glands*

By ablation of the "molting hormone–releasing glands" at the beginning of the nymphal instar in *Aeshna cyanea* Müll. (Odonata: Anisoptera), Schaller (1960) obtained "permanent larvae." These animals were operated on either at the penultimate nymphal instar (typical nymphal instar) or at the last nymphal instar (metamorphosis) and generally lived without molting more than 6 months (standard duration of a nonoperated nymph in the natural condition: penultimate instar, 17 days; last instar, 25 days).

A morphological examination of the dorsal side of the head in such nymphs nevertheless revealed a distinct spreading of the eye boundaries under the cuticle. This observation was later completed by histological studies of the visual system in such experimental nymphs (Mouze and Schaller, 1971).

14.3.1.1. RESULTS

Compound Eye

In "permanent larvae" of the penultimate instar, the compound eye shows typical nymphal growth (Fig. 14.7). However, the eyes spreading under the cuticle, more extended than in control animals, result from continuous growth beyond the usual duration of the penultimate instar. In these nymphs, the proliferation zone keeps its typical structure and mitotic activity. In "permanent nymphs" of the last instar, the eye's growth (by external examination) does not seem to be much advanced; nevertheless the eye structure is much metamorphosed, much like that of a control animal at the end of the last nymphal instar (proliferation zone absent; fully differentiated ommatidia).

Optic Lobe

At the penultimate instar, the shape and structure of the optic lobes of permanent nymphs show a nymphal nature; more precisely, the optic anlage, which now appears reduced in size, still shows mitotic activity, resulting in growth of the optic lobes that is more developed than in the controls. In permanent nymphs of the last instar, the optic anlage progressively disappears, but its involution is slower than in a typical preimaginal instar (complete involution in 75 days, instead of 17 days). Yet the structure of the lamina always stays of nymphal nature, whether in the penultimate or in the last instar.

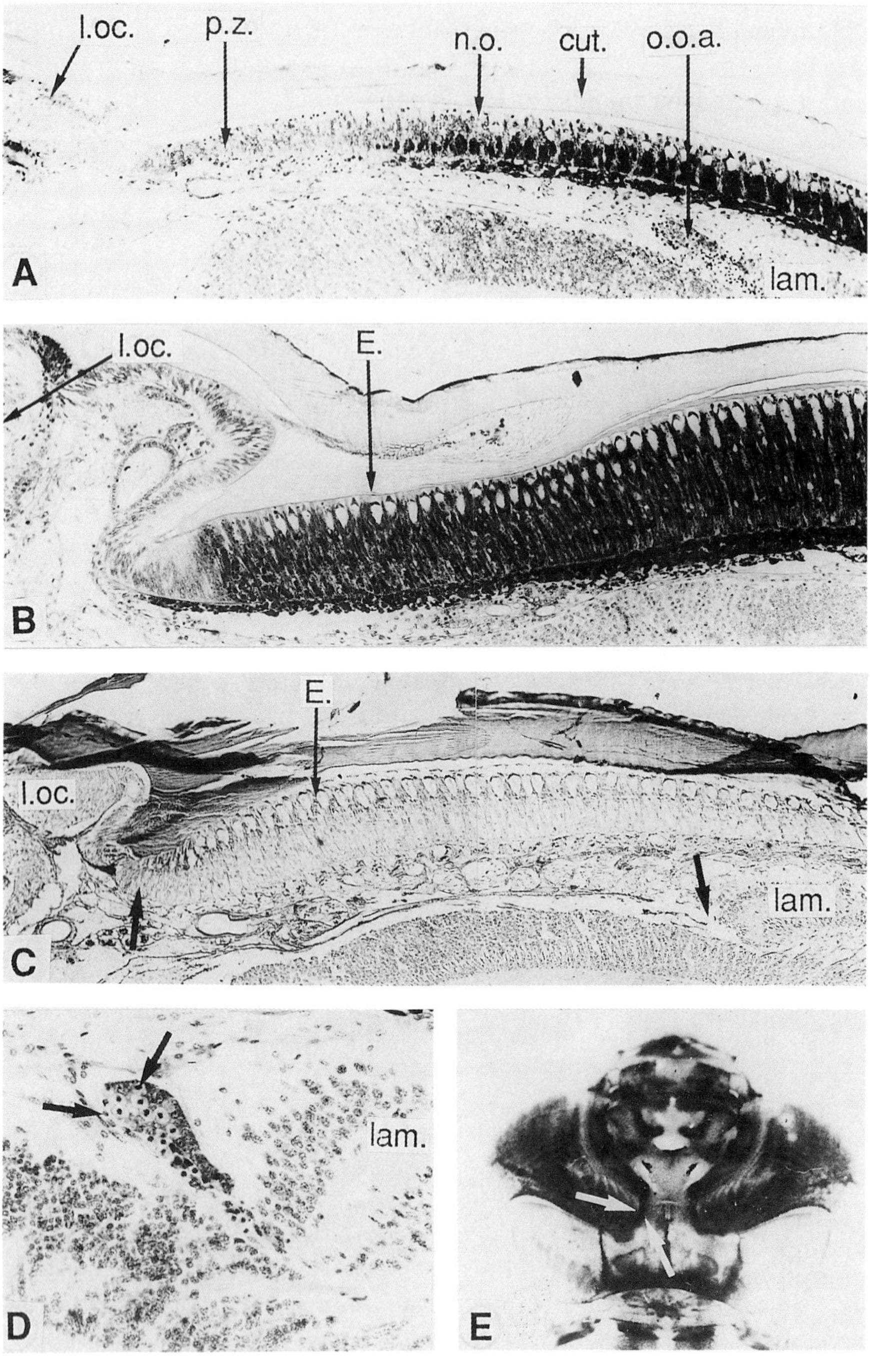

FIGURE 14.7. Results of ablation of the MH-releasing glands on development of the visual system in last-instar nymphs of *Aeshna.* (A,B,C) Evolution of the right eye in 27-, 72-, and 84-day-old permanent nymphs,

14.3.1.2. DISCUSSION

In permanent nymphs of *Aeshna cyanea*, growth of the visual system continues because of the persistence for some time of mitotic activity in all the proliferation zones.

Radioimmunoassays in such permanent nymphs, tested 80 days after ablation of the molting hormone(MH)–releasing glands, showed that the amounts of ecdysteroids are always excessively low, or indeed to the "background noise" limits of the technology used (Charlet, 1977).

The results, therefore, indicate that the developmental mechanisms in the visual system could be either completely independent of the MH or extremely sensitive to it, as reported by White (1961, 1963) and Schaller (1964); in the latter case, only traces of hormone would be enough for such development. This would explain the constant basic level of mitosis for each instar.

At every nymphal instar, the early peak of mitosis above this basic level might nevertheless accord with a peculiar sensitivity of cells in an ocular proliferation zone to ecdysterone. Indeed, it has been established in many insects that the main increase of the MH level is generally preceded by a first discrete peak. More precisely, similar observations made in *Aeshna cyanea* by radioimmunoassay (Charlet, 1977) indicate that the main peak of ecdysone (at day 10 of the penultimate instar and at day 20 of the last instar) is also preceded by a slight increase of the ecdysone level (days 4 and 10, respectively). These observations, therefore, might well agree with the previously stated assumption (White, 1961, 1963; Schaller, 1964).

The important extension of the eyes boundaries in permanent penultimate instar nymphs is easy to understand because of the very long time elapsed since the MH glands' ablation without molting. On the other hand, in permanent last instar nymphs, the externally visible early arrest of ocular spreading does not accord with the metamorphosed structure of that compound eye; indeed, even with fully differentiated adult ommatidia, the very limited apolysis at the ocular level related to the extremely low amount of ecdysone (or the very brief duration of its

---

respectively; note the increasing thickness of the cuticle and the disappearance of the outer optic anlage (the depigmentation of the eye in C results from the used histological technique). ×130. (D) Outer optic anlage of a 30-day-old permanent nymph; note the numerous mitoses (arrows) and the nymphal structure of the lamina. ×240. (E) Dorsal view of a permanent nymph showing the ocular spreading (arrows) blocked at step 3 of Schaller's (1960) chart (see Fig. 14.5). ×6. *Key:* cut. = cuticle; E = eye; lam. = lamina; l. oc. = lateral ocellus; n.o. = new ommatidia; o.o.a. = outer optic anlage; p.z. = proliferation zone.

action) does not allow the corresponding extension of the eyes under the cephalic cuticle.

In such nymphs, mitotic activity and ganglion cell differentiation also continued in the optic anlagen for a very long time in spite of the presumed absence of ecdysteroids. This observation agrees with two other findings: (1) constancy of the mitotic level in this anlage in each intermolt period; (2) the report of Schaller and Meunier (1967) of experimental results—obtained by *in vitro* cultures on medium without MH—of nymphal optic lobes associated with a brain of *Aeshna cyanea* (cellular divisions persist in the anlagen even after 15 days in culture). Nevertheless, Woolever and Pipa (1975) reported that the eye disks of very early prepupal *Galleria mellonella* readily differentiated *in vitro* in response to ecdysterone but did not do so at all without ecdysterone. In that instance, it could still be believed that the eye disk does not yet react like an ocular tissue but like any nondifferentiated insect epidermis requiring some MH for its development.

With regard to the lamina, it always keeps its nymphal structure in all permanent nymphs: its metamorphosis, i.e., the appearance and progress of mitotic activity in the different layers, seems to be inhibited by the absence—or by an insufficient amount—of ecdysteroids.

14.3.2. *Action of a Juvenile Hormone Mimic on Development of the Visual System*

By injection of different amounts of a juvenile hormone (JH) mimic, farnesyl methyl ether (FME), into nymphs on different days of the last nymphal instar, a continuous range of intermediate animals between nymph and adult can be obtained (Mouze, 1971). Use of Schaller's method (1960), based on external examination of the ocular spreading, enabled us to identify typical "supernumerary last-instar"—partially metamorphosed—adultoids or wholly metamorphosed nymphs (Fig. 14.8).

14.3.2.1. RESULTS

In supernumerary last-instar nymphs, the eye and the optic lobe are typically of a larval structure (Fig. 14.9). On the other hand, molting adultoids, the metamorphosis level of the visual apparatus (structure of the proliferation zone, ommatidia, optical anlagen, and lamina) is wholly similar to that observed in control larvae. Both are at the same step of ocular extension (according to Schaller's chart; see Fig. 14.5). In addition, the injected nymphs show symptoms of the next molt earlier than do control last-instar nymphs.

This level of adult transformations depends on the amount of FME injected and on the injection time; indeed, for identical amounts of FME

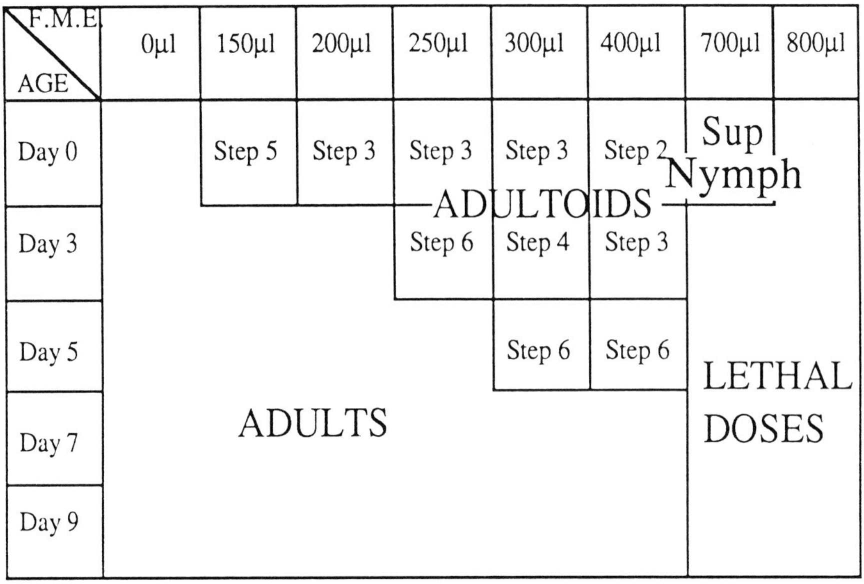

FIGURE 14.8. Result of injection of the JH mimic FME on visual system development of aged nymphs (at the last larval intermolt) of *Aeshna*. Kinds of insects obtained are correlated with the amount of FME injected (in microliters) and the ages of nymphs in the intermolt (in days) at the time of injection.

injected, the later the injection, the more imaginal structure is found in the visual system of the resulting adultoids.

A similar result is obtained after decreasing amounts of FME are injected into animals of the same age in the last larval instar. Adultoids cannot be obtained by injection after the fifth day of the last nymphal instar: a full metamorphosis is taking place, and only typical adults are achieved.

14.3.2.2. DISCUSSION

The precocity of the molt in supernumerary nymphs and adultoids (compared with the normal duration of the control last nymphal instar) probably corresponds to the "prothoracicotropic" action of FME, which results in ecdysone release and thereby in shortening of the intermolt period.

Only a substantial amount of JH analogue injected just after the molt (on the first day after the last nymphal instar) enables one to obtain typical (supernumerary) nymphs without any adult character. This result agrees with those obtained by Williams (1959), Emmerich (1969),

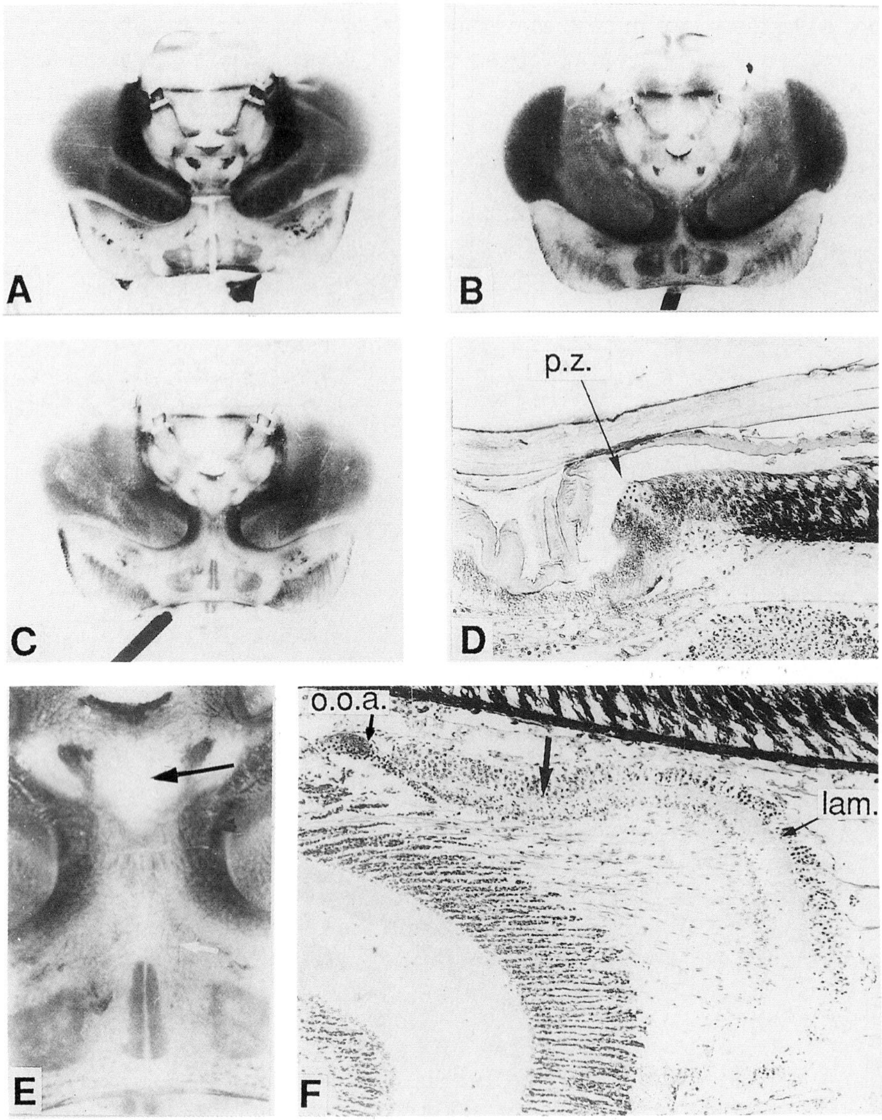

FIGURE 14.9. Results of injection of FME on development of the visual system in last-instar nymphs of *Aeshna*. (A) Head of a supernumerary nymph. ×6. (B,C) Heads of adultoids respectively at steps 3 and 5 of the ocular spreading chart (Fig. 14.5). ×6. (D) Proliferation zone in an adultoid (step 3). ×100. (E) Detail of C: a light adult pilosity appears (arrows). ×24. (F) Outer optic anlage and lamina in an adultoid (step 3). Note the progress of metamorphosis in the lamina (arrow). ×100. *Key:* lam. = lamina; o.o.a. = outer optic anlage; p.z. = proliferation zone.

and Willis (1969) in studies of the visual system of holometabolous insects, where they found it very difficult, if not impossible, to inhibit ocular metamorphosis. For instance, by injection of CA purified extracts (JH) to saturniid pupae, Willis (1969) also obtained a whole range of intermediate animals between pupa and adult. Nevertheless, whatever the amount of hormonal extract injected, the eyes always exhibited some adult characteristics, even in "deuteropupae" having a completely pupal appearance.

It seems, therefore, that the proliferation zone, like the optic anlage, only persists if a minimal amount of JH is present, which is the case only during the first 5 days of the last nymphal instar.

As soon as the start of the last nymphal intermolt period, the JH level in the hemolymph very rapidly decreases, virtually to zero (Hsiao and Hsiao, 1977).

The JH effect (or that of its mimics) on such organs, rendered indifferent to ecdysone and showing a continuous mitotic activity, could then be explained as follows (Fig. 14.10):

Injection of a large amount of FME just after the last nymphal molt, would, of course, correspond to an unusual level of JH at that time. Supposing that FME were eliminated by the insect's metabolism at the same rate as that of endogenous JH, the quantity of FME would not decrease under the minimal threshold, triggering the disappearance of the proliferation zone and optic anlage. The following molt, brought forward by the prothoracicotropic activity of FME, would take place even though those organs would still be of a typical nymphal nature. The injection of a small quantity (amount II) of FME, also performed just after the last nymphal molt, or a later injection of a larger quantity of this compound (amount III), would maintain, only for a period of time shorter than the duration of intermolt, a hormonal amount higher than the minimal level. The decrease of the proliferation zone and optic anlage would start only when that level is reached, i.e., just after the fifth day of the instar. The following molt would then stop the transformations of these organs, achieving in that way a more or less transformed adultoid (metamorphosis step II or III).

Injection of FME at this time, when the endogenous JH is below the necessary minimal level, could not maintain the proliferation zone and the optic anlage, because these tissues would already be programmed to disappear; that would explain why injections given after the fifth day of the last nymphal instar never result in adultoids, but only in typical adults.

In "permanent nymphs," obtained by ablation of MH-releasing glands (as previously described in Section 14.3.1.2), disappearance of the proliferation zone and optic anlage would also be provoked by decrease of the JH level. That would be a kind of ocular metamorphosis

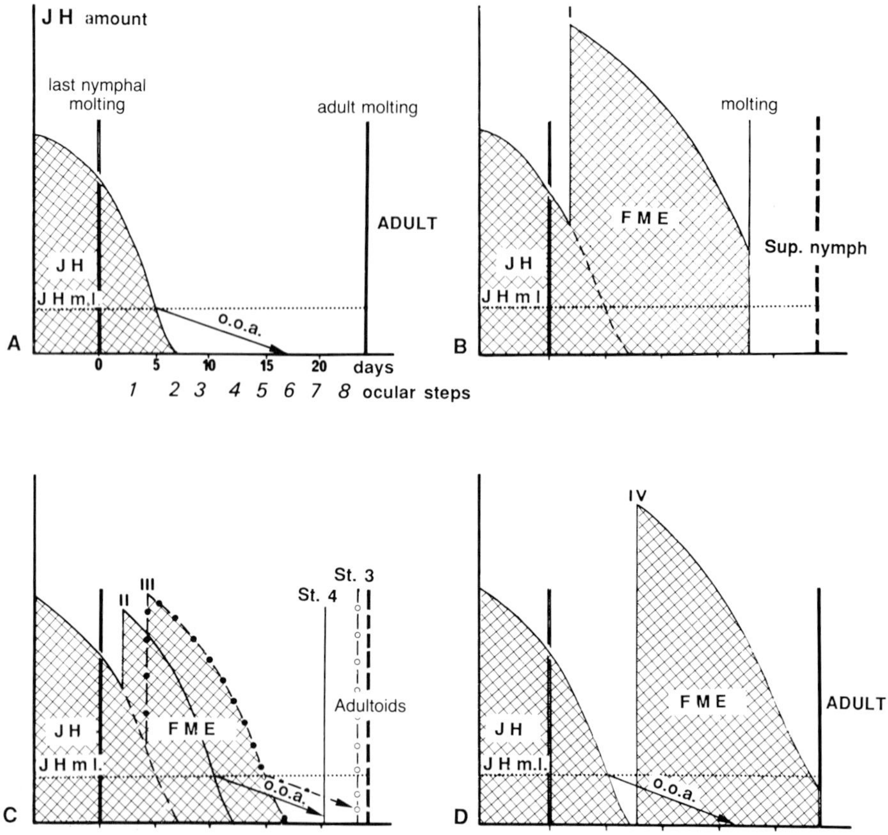

FIGURE 14.10. Graphic showing the hypothesis of JH action (and that of FME) on metamorphosis of the visual system in Odonata. (A) Typical last nymphal intermolt. The decrease of the outer optic anlage starts on day 5, as soon as the minimal level of the JH is reached. (B) Result of the injection of a large amount of FME just after the last nymphal molt: a typical ("supernumerary") nymph is obtained. (C) Result of the early injection of a small quantity (amount II) of FME, or of a later injection of a larger quantity of this chemical (amount III): adultoids are obtained. (D) Result of an injection performed too late (amount IV), when the endogenous JH is under the needed minimal level: adults are obtained. *Key:* FME = farnesyl methyl ether; JH = juvenile hormone; JH m.l. = juvenile hormone minimal level; o.o.a. = outer optic anlage; St. 3,4 = steps 3 and 4 of ocular spreading; sup. larv. = supernumerary nymph.

occurring in a very slow rhythm, probably because of the extremely low metabolism of such nymphs.

As for the lamina, which seems to be very sensitive to JH, the fall in level of this hormone would determine the progress of the mitoses in its different layers, provided that the MH is present; indeed, only the injection on the molting day of a large amount of JH mimic can totally inhibit its metamorphosis. In the preceding descriptive discussion (Section 14.2), the metamorphosis of this ganglion was depicted as progress of the mitosis area into the whole lamina, which was attributed to a diverse sensitivity to the JH, depending on the area considered. This assumption seems to be confirmed, for—as in the proliferation zone and the optic anlage—the lamina metamorphosis is blocked at different steps according to the FME amount injected or the time of injection, which would agree with a diverse sensitivity to the JH in each area of the ganglion.

14.4.2. *Effects of Ecdysteroid Injection on Metamorphosis of the Visual System*

Another way to approach the mechanism of growth and metamorphosis in the visual system was to study the results of injection of different amounts of MH into last-instar nymphs. Indeed, according to the injection time and the amount of ecdysteroid (ecdysone or ecdysterone) injected, different kinds of animals could be distinguished by external examination. As in the previous experiment (with FME injections), "supernumerary last-instar" nymphs, typical adults, and a range of adultoids were obtained (Andriès and Mouze, 1975).

14.4.2.1. RESULTS

The type of externally visible metamorphic perturbation essentially depends on the time of the hormonal injection; and only a very large amount of ecdysterone (80 μg/g) injected on the first day of the last instar enabled us to obtain "supernumerary last-instar" nymphs. All other types of injection resulted in animals with an apparently partly metamorphosed visual apparatus; more precisely, the later the hormonal injection was given, the more transformed were the eyes. In such adultoids, however, the features related to cuticular evolution must be distinguished from the actual histological development of the organs. If only the ocular spreading under the cuticle is considered, imaginal evolution does not seem to be very important. Nevertheless, if the histology of the visual system is studied (especially the step of differentiation of the ommatidia, as well as the structure of the lamina and the involution of the proliferation zone and optic anlage), it appears obvious that all these tissues are in a more advanced metamorphic phase than those in a control nymph showing a comparable ocular spreading.

In fact, in such adultoids, these different tissues seem to develop in a normal way, the course of metamorphosis apparently being undisturbed, but the ocular spreading is stopped at a less advanced step because of the early arrest of apolysis.

14.4.2.2. DISCUSSION

Still, how can we explain the possibility of obtaining supernumerary last-instar nymphs that do not show any sign of metamorphosis? Because this result is comparable to those obtained by injections on the first day of the last nymphal instar of large amounts of FME, it is logical to assume an identical mechanism of action. Indeed, some investigators (Siew and Gilbert, 1971; Willis, 1974) have showed that ecdysteroids could stimulate the release of JH by the CA. MH injection would involve a resumption of the secretory activity in those glands, resulting in an increase of JH in the hemolymph. Only injection of a large amount of ecdysteroid at the beginning of the last nymphal instar could induce JH secretion in a sufficient amount to prevent the triggering of the disappearance of the proliferation zone and optic anlage. Later or lower ecdysone injections would lead to the release of a small amount of JH that is still sufficient to slightly delay the start of metamorphosis.

However, in the adultoids obtained by FME injection, on the one hand, and by MH injection, on the other, some differences can be noted between cuticular evolution and histological evolution.

After ecdysteroid injection, the apolysis occurs very rapidly, followed by formation of the new cuticle. Instead of cuticular detachment near the eye, which in progressing allows ocular spreading, i.e., ommatidial extension, the mechanisms are speeded up, and the sliding of the chitinogen epithelium, as in the compound eye when it secretes the cornea, is stopped by its adhesion to the cuticle.

It is rather surprising to note that two opposite experiments, one involving ablation of MH-releasing glands and the other involving ecdysteroid injection, both result, on the basis of external examination, in animals with eyes that appear little advanced; nevertheless, as previously described, comparison of the histological structures of their respective visual apparatuses reveals very different results between these two experiments.

Ocular spreading, considered as a good indication of the step of metamorphosis in Odonata (and in some other insects), must be regarded as the result of two different mechanisms: completion of ommatidial differentiation and growth on the one hand, and advance of cuticular detachment near the eye, on the other. In the different kinds of experimentally obtained nymphs or adultoids, ocular spreading can be prevented either by lack of the MH (in permanent last-instar nymphs) or by a very early apolysis, immediately followed by the cuticular secretion resulting directly from ecdysteroid injection or indirectly from FME injection.

So, the externally observed step of ocular metamorphosis, even if it undoubtedly offers a good indication of developmental progress during metamorphosis of the control nymphs, cannot be used as an absolute pointer to the real physiological age of nymphs whose hormonal balance has been experimentally perturbed.

Other experiments disturbing the hormonal balance in insects (other than Odonata) also permitted dissection of the responses of different parts of the visual system and hence a better understanding of its mode of growth. Imaginal disks of mature *Drosophila* larvae are capable of responding to high levels of the MH by proceeding through cellular differentiation (Campos-Ortega and Gateff, 1976; Trujillo-Cenoz and Melamed, 1978). When eye imaginal disks are implanted into larvae ready to pupate, some parts of these metamorphosed explants will be of a typical adult structure whereas others will be of an abnormal structure, built by ommatidia with incomplete sets of cells. Some parts of the eye disk have been forced to interrupt their normal developmental course at a given time by transplantation into an hormonal milieu favoring morphological differentiation. These results indicate the existence of a sequence of differentiation abilities for the optic anlage, i.e., the whole disk does not reach differentiative ability simultaneously; in *Drosophila*, proximal parts of the disk become competent earlier than distal ones. The present author believes that the reason for the incomplete differentiation of adult structures, and the occurrence of ommatidia with incomplete receptor sets in certain parts of the optic anlage, is not the hormonal situation in the pupa, but rather may be, for example, the necessity of the cell's requirement of a definite number of mitoses in order to become definitely postmitotic. In these experiments, the hormones triggering and controlling metamorphosis will act on a population of cells with unfinished mitotic programs. This interpretation agrees with the descriptive results presented in Section 14.2, i.e., the occurrence of two consecutive mitotic waves across the eye field, these two waves being required to complete the cell number of each ommatidium.

14.5. Conclusion

It is not easy to gather all these descriptive and experimental results in a general synthesis elucidating the relationships with the hormonal system of a unit as complex as the visual apparatus of insects, the growth of its two constitutive organs (the eye and the optic lobe) having to be synchronized during the whole period of development. Schematically, however, two phases can be distinguished in the growth of the visual system, especially in hemimetabolous insects:

In the first phase, the nymphal period, the eye and the optic lobe show continuous growth (cellular divisions and differentiation) that appears to be almost independent of the molting cycle; only the cuticular processes shown by the eye and related to the molt seem to be under direct control of the MH. During this first nymphal phase, the eye imposes its differentiation rhythm on the optic lobe through the channel of its retinal nervous fibers, which regulate and adjust the differentiation of ganglionic cells in the first optic ganglion.

In the second phase, at the end of the nymphal period, the visual system undergoes its last transformations and achieves its full growth; during this period, the decreasing level of the JH triggers metamorphosis (disappearance or change) of the various organs. The synchronization of development in both parts of the visual system (eyes and optic lobes) is then provided not by nervous but by hormonal means.

14.6. Summary

The visual system of insects is a complex apparatus composed of two superposed organs—the compound eye (the photoreceptor part) and the optic lobe (the nervous integrating part). Both organs grow simultaneously, the rhythm of development in the optic lobe being under the control of the eye through the channel of its retinal nervous fibers. Growth of the visual system takes place throughout postembryonal life (in hemimetabolous insects) or particularly during the last phase of larval life (in holometabolous insects). Growth of the eye and optic lobe is essentially the result of mitotic activity in different growth zones, this activity remaining almost constant during the entire larval life, without any distinct cyclic change at each intermolt period. The insensitivity to the MH (or the extremely high sensitivity to low amounts of this hormone) observed in these growth zones is again found experimentally after ablation of the MH-releasing glands in larvae whose eyes continue to grow. The persistence of these growth zones during larval life would normally be controlled by the amount of JH present; the decline in this hormone titer during the last larval intermolt period simultaneously triggers the disappearance of all parts of the growth zone. Synchronization of metamorphosis in all the organs constituting the visual system is then achieved not by nervous control but by hormonal regulation.

References

Anderson, H. 1978. Postembryonic development of the visual system of the locust *Schistocerca gregaria*. I. Patterns of growth and developmental interactions in the retina and optic lobe. J. Embryol. Exp. Morphol. 45: 55–83.

Andriès, J. C. and M. Mouze. 1975. Action *in vivo* d'ecdysones sur la morphogenèse imaginale d'*Aeschna cyanea* (Odonata). J. Insect Physiol. 21: 111–135.

Bodenstein, D. 1940. Growth regulation of transplanted eye and leg discs in *Drosophila*. J. Exp. Zool. 84: 23–37.

Bodenstein, D. 1953. Postembryonic development. Pp. 822–865 *in* K. D. Roeder (ed.), *Insect Physiology*. Wiley, New York.

Bounhiol, J. J. 1938. Recherches expérimentales sur le déterminisme de la métamorphose chez les Lépidoptères. Bull. Biol. Fr. Belg. Suppl. 24: pp. 1–200.

Campos-Ortega, J. A. and E. A. Gateff. 1976. The development of ommatidial patterning in metamorphosed eye imaginal disc implants of *Drosophila melanogaster*. Wilhelm Roux' Arch. Entwicklungsmech. Org. 179: 373–392.

Charlet, M. 1977. Contribution à l'étude du contrôle endocrinien de la mue chez la larve de l'Insecte Odonate *Aeschna cyanea*. Doctoral thesis, University of Strasbourg, France.

Emmerich, H. 1969. Beeinflussung der Imaginalentwicklung von *Tenebrio molitor* durch farnesylmethyläther und actinomycin. Zool. Anz. 32: 519–526.

Fukuda, S. 1944. The hormonal mechanism of larval molting and metamorphosis in the silkworm. J. Fac. Sci. Tokyo Imp. Univ., Sect. IV, Zool. 6: 477–532.

Hsiao, T. H. and C. H. Hsiao. 1977. Simultaneous determination of molting and juvenile hormone titers of the greater wax moth. J. Insect Physiol. 23: 89–93.

Lavoie, J., J. G. Pilon, and M. A. Ali. 1978a. Etude histologique et morphométrique de la croissance de la partie optique de l'oeil composé d'*Enallagma boreale* Selys (Odonata: Coenagrionidae). Rev. Can. Biol. 37(3): 157–179.

Lavoie, J., J. G. Pilon, and M. A. Ali. 1978b. Etude histologique et morphométrique de la croissance de la partie photosensible de l'oeil composé d'*Enallagma boreale* Selys (Odonata Coenagrionidae). Biol. Vestn. 26(2): 141–151.

Mouze, M. 1971. Rôle de l'hormone juvénile dans la métamorphose oculaire de larves d'*Aeschna cyanea* Müll. (Insecte, Odonate). C. R. Acad. Sci. Paris 273: 2316–2319.

Mouze, M. 1972. Croissance et métamorphose de l'appareil visuel des Aeschnidae (Odonata). Int. J. Insect Morphol. Embryol. 1(2): 181–200.

Mouze, M. 1979. Rôle des fibres post-rétiniennes dans la croissance du lobe optique de la larve d'*Aeshna cyanea* Müll. (Insecte Odonate). Wilhelm Roux' Arch. Entwicklungsmech. Org. 184: 325–350.

Mouze, M. and F. Schaller. 1971. Métamorphose oculaire de larves d'*Aeschna cyanea* Müll. (Insecte, Odonate) privées d'ecdysone. C. R. Acad. Sci. Paris 273: 2122–2125.

Nordlander, R. H. and J. S. Edwards. 1969. Postembryonic brain development in the monarch butterfly, *Danaus plexippus plexippus* L. II. The optic lobes. Wilhelm Roux' Arch. Entwicklungsmech. Org. 163: 197–220.

Nowel, M. S. 1981. Postembryonic growth of the compound eye of the cockroach. J. Embryol. Exp. Morphol. 62: 259–275.

Schaller, F. 1960. Etude du développement post-embryonnaire d'*Aeschna cyanea* Müll. Ann. Sci. Nat. Zool. [12] 12: 755–868.

Schaller, F. 1964. Croissance oculaire au cours de développements normaux et perturbés de la larve d'*Aeschna cyanea* Müll. (Insecte, Odonate). Ann. Endocrinol. 25(5): 122–127.

Schaller, F. and J. Meunier. 1967. Résultats de cultures organotypiques du cerveau et du ganglion sous-oesophagien d'*Aeschna cyanea* Müll. (Insecte Odonate): survie des organes et évolution des éléments neurosecrèteurs. C. R. Acad. Sci. Paris 264: 1441–1444.

Siew, Y. C. and L. I. Gilbert. 1971. Effects of moulting hormone and juvenile hormone on insect endocrine gland activity. J. Insect. Physiol. 17: 2095–2104.

Stark, R. J. and M. I. Mote. 1981. Postembryonic development of the visual system of *Periplaneta americana*. I. Patterns of growth and differentiation. J. Embryol. Exp. Morphol. 66: 235–255.

Trujillo-Cenoz, O. and J. Melamed. 1978. Development of photoreceptor patterns in the compound eyes of muscoid flies. J. Ultrastruct. Res. 64: 46–62.

White, R. 1961. Analysis of the development of the compound eye in the mosquito *Aedes aegypti*. J. Exp. Zool. 148: 223–240.

White, R. 1963. Evidence for the existence of a differentiation center in the developing eye of the mosquito. J. Exp. Zool. 152: 139–148.

Williams, C. M. 1959. The juvenile hormone. I. Endocrine activity of the corpora allata of the adult *cecropia* silkworm. Biol. Bull. (Woods Hole) 116(2): 323–338.

Willis, J. H. 1969. The programming of differentiation and its control by juvenile hormone in saturniids. J. Embryol. Exp. Morphol. 22(1): 27–44.

Willis, J. H. 1974. Morphogenetic action of Insect hormones. Annu. Rev. Entomol. 19: 97–115.

Woolever, P. and R. L. Pipa. 1975. Eye disk differentiation in the wax moth: induction *in vitro*. J. Exp. Zool. 191(3): 359–364.

Morphogenetic Hormones and Regeneration in Arthropods

15

DÉSIRÉ BULLIÈRE AND
FRANÇOISE BULLIÈRE

15.1. Introduction

Regeneration, the ability to rebuild a lost part of the body is found in almost all animal groups, and nearly three centuries ago Réaumur (1710) put forward an explanation for this phenomenon. In arthropods, the initial shape is established during embryonic development and is stabilized by the cuticle. It was apparent very early that morphogenetic changes could occur only at the time of molts. Gradually it became evident, particularly in insects, whose endocrinology is best known, that the shape visible after the molt depends on the hormonal environment that controlled the preparative phases of the molt; from this arose the concept of morphogenetic hormones. Regeneration, which involves a change in form at the level of the stump, was consequently rapidly linked to the molt phenomenon. Pflugfelder (1939) and Bodenstein (1955) first clearly evoked the role of hormones in insect regeneration. Considering the strict relationship between regeneration and the larval state, Pflugfelder (1939) linked regeneration with the presence of juvenile hormone (JH). Subsequently, it was assumed by Bodenstein (1953, 1955), taking into account the fact that molt and regeneration go hand in hand, that molting hormone (MH) has a positive effect on regeneration. This idea was taken up by various authors. Studies done during the last 20 years, particularly on insects and crustaceans, have resulted in a much more precise perception of regeneration and of its relationship with hormones that regulate the molting processes. Moreover, it was established that regeneration sometimes provides a more suitable opportunity than embryonic development for study of morphogenesis, insofar as morphogenesis can be experimentally triggered and various parameters, particularly hormonal ones, can be controlled.

Can one speak of "morphogenetic hormones"? Not if one thinks solely in terms of a molecule which would realize a particular form, but certainly if one considers that realization of the form is closely correlated to the hormonal environment. A morphogenetic hormone must not be considered as an "architect" molecule that, so to speak, manages the assembly of any population of cells in order to construct a regenerate. Rather, the hormone *conditions* the development of the cells that will constitute the regenerate and also *controls* the triggering of regenerative events, such as the growth and differentiation of the regenerate (and thus the form) before it is released at the time of ecdysis.

15.2. Regenerative Morphogenesis in Arthropods

Before looking at the possible interactions between morphogenetic hormones and regeneration in the various arthropod groups, let us first examine the studies on regeneration performed in these groups. Morphogeneses were once obtained by regeneration in most of the arthropod groups, but insects and crustaceans were the more studied, and—to a lesser extent—arachnids and myriapods. In other groups, such as merostoms, various observations lead us to think that the ability to regenerate also exists in these animals.

15.2.1. *Insects*

Observations in various orders (Dictyoptera, Phasmidoptera, Orthoptera, Coleoptera, Lepidoptera) and several experiments have shown that regeneration is possible for locomotor appendages, wings, antennae, eyes, cerci, bucal appendages, and even imaginal disks in most insects (D. Bullière and F. Bullière, 1985). In most cases, the capability to regenerate was shown to exist both in larvae or nymphs and embryos (D. Bullière et al., 1969; Fournier, 1969). Adults, as they do not molt, are unable to regenerate.

Various insects are able to lose one or several appendages by autotomy when in danger and thereafter regenerate them. Regenerated elements are generally quite similar to those they replace. Thus, regenerative morphogeneses are usually not different from embryonic ones. Nevertheless, it must be emphasized that exceptional situations can occur. For example, the tarsus of cockroach nymphs, which is always pentamerous, regenerates, following an amputation through the first tarsal segment or above, according to a tetramerous pattern. Regenerated antennae after the first postoperative molt have a limited number of segments, this number increasing following later molts.

15.2.2. *Crustaceans*

Regeneration has principally been studied in three orders: Isopoda, Amphipoda, and especially in Decapoda. Mittenhal et al. (1985) showed that if the regenerative capability is high for pereiopods, it also affects other body parts such as pleopods, antennae, eyestalks, and to a lesser extent the tail fan. Nevertheless, note that in crustaceans, as in insects, regenerative ability disappears when animals have lost the ability to molt by entering terminal anecdysis (Carlisle and Dorn, 1953; see the review of Skinner, 1985). The need to follow animals during such cycles, which are sometimes very long, complicates studies and results in imprecise work. Nevertheless, some groups do not seem to exhibit terminal molt (Skinner, 1985).

15.2.3. *Arachnids*

Appendage regeneration exists in arachnids (Bonnet, 1930; Vachon, 1967; Randall, 1981) and has been principally studied in araneids. Missing appendages are restored, but as shown by Vachon (1967), who also studied scorpion regeneration (Vachon, 1957), the appendage may not right away show the shape it would have displayed without being amputated. In acarines, great variations seem to exist between the various suborders (Rockett and Woodring, 1972). In Ixodida the regenerative capacity is maximal, whereas in Uropodida it is less pronounced (the regenerate is reduced or incomplete). In some suborders, regenerative ability is very restricted. In others, the appendage amputation is not tolerated and leads to death of the amputated animal.

15.2.4. *Myriapods*

In this relatively little-studied class, the regeneration of appendages of various types is possible (Juberthie-Jupeau, 1964; Sahli, 1967). Regeneration seems to result in the formation of appendages with few differentiated characteristics, reminiscent of those of juveniles.

15.2.5. *Merostoms*

Various observations have revealed that these animals are also able to regenerate. Nevertheless, no systematic work has been performed in this class.

15.3. Regenerate Formation

Does appendage regeneration progress in the same manner in the various arthropod groups? In other words, can a general rule concerning the arthropods be enunciated? Regenerate formation in the two classes most highly studied from this point of view will be described next.

15.3.1. *Insects*

In insects, four steps were distinguished during regeneration (D. Bullière, 1972a) corresponding to (1) wound closure and cicatrization, (2) blastema formation, (3) regenerate growth, and (4) preparation for regenerate release.

Let us consider these steps following amputation at the base of the tarsus:

(1) *Wound closure and cicatrization.* This takes place in two stages. Within a few minutes, the wound is closed by hemocytes including coagulocytes and a clot is formed. Then, the integrity of the epithelium is restored so that the wound is healed.

(2) *Blastema formation.* The epithelium becomes detached from the cuticle, and the regeneration blastema is formed by reshaping of the stump extremity. During this period, intense DNA synthesis takes place in epithelial cells. Quite rapidly, the different segments of the tarsus are distinguishable. Their morphological individualization occurs at the

same time, and it is thus very difficult to speak of centrifugal or centripetal differentiation, as has been suggested by some reports.

(3) *Regenerate growth.* During this period, the regenerate grows intensively and a very active cell proliferation occurs. The stump epidermis continues to detach from the cuticle near the regenerate and retracts on itself, leaving more and more space for expansion of the regenerate, which has a smaller diameter than that of the place liberated by the stump retraction and coils up in an helicoidal manner. During this whole period, DNA synthesis and cell proliferation go hand in hand. Progressively, the regenerated structures gain not their definitive shape, which will be externalized only when the regenerate is released, but the possibility of realizing these forms. So, the components of the articulation condyle are put in position, as are the spines of the distal crown of the tibia.

(4) *Preparation for regenerate release.* When the regenerate is well formed, all the epidermal cells on the regenerate as well as those on the stump secrete the cuticle. This cuticle, wrinkled in appearance before ecdysis, is stretched out into a smooth surface, and the regenerate expresses the typical form of the appendage it replaces. Its size, a little smaller than that of a normal appendage, is a given percentage of the corresponding part of the contralateral, nonamputated, appendage (D. Bullière, 1967, 1968a).

To summarize: at first the wound is closed by a clot; then, under this protective sheath, epidermis continuity is restored, this cicatrization appearing to constitute a necessary prerequisite for regeneration. Development of the regenerate begins by the individualization of its own material: dedifferentiation precedes blastema formation. Subsequently, a growth phase takes place, and then the regenerate undergoes synchronous differentiation in preparation by the animal for molting.

Observation of the formation of tarsus regenerates in *Blaberus* embryos confirms these conclusions derived from histological study of regenerates. Figure 15.1 illustrates two steps in such regeneration.

15.3.2. *Crustaceans*

In crustaceans, the procedures of regenerate formation are less homogeneous. Two regeneration modalities exist: in the first, as in insects, the regenerate is built inside the distal part of the stump, in a cuticular muff liberated by retraction of the stump [this is the characteristic fashion in amphipods (Charniaux-Cotton, 1967), isopods (Needham, 1965), and some decapods (Skinner, 1985)]; in the second case, the regenerate develops at the end of the stump, covered only by a thin extensible membrane [this is the case in most decapods (Adiyodi, 1972; Laugier, 1984; Skinner, 1985)].

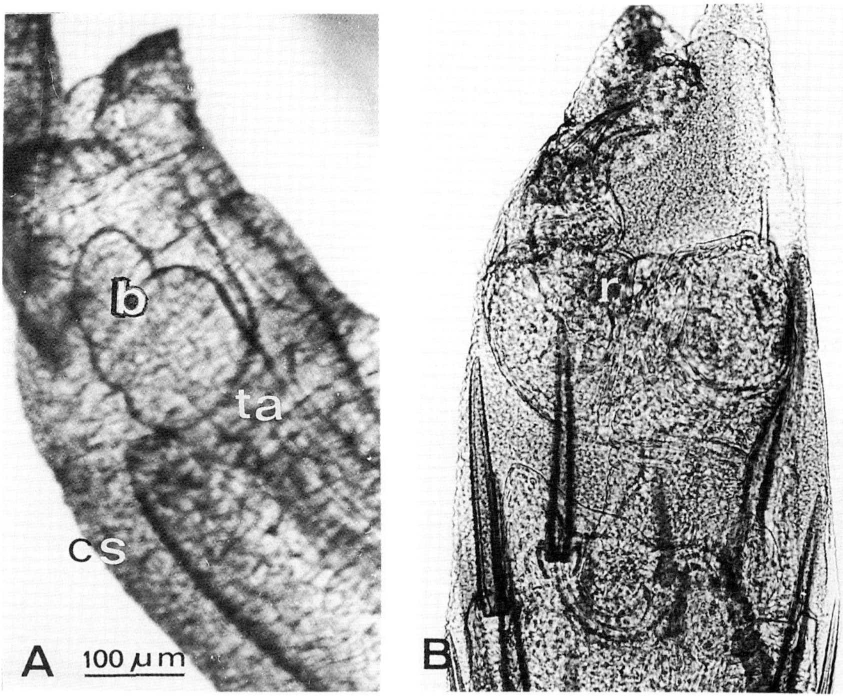

FIGURE 15.1. Embryonic regeneration in *Blaberus*. (A) After 7 days of *in vitro* culture, a blastema (b) is formed at the tibia apex (ta) inside the cuticle of the stump (cs). (B) Totally differentiated regenerate (r) after 28 days. ×60. (From D. Bullière and F. Bullière, 1985.)

In crustaceans in which the regenerate is interiorized, everything occurs as in insects. In the isopod *Helleria brevicornis*, for example (Hoarau, 1973), after wound closing by a clot, cicatricial epidermis takes place and afterward secretes a thin, fibrous protective cuticle, separating itself from the clot. Then, before the formation of the blastema, a latent period can occur, about 15 days in *Helleria* (Hoarau, 1973), but nonexistent in *Porcellio* (Maissiat, 1978). The blastema begins to be formed by the upthrust of the central zone of the cicatrization surface. Blastema growth can be more or less continuous and rapid until the phase of molting preparation, characterized by secretion of the new cuticle.

In crustaceans in which the regenerate is exteriorized, the processes are somewhat different (Adiyodi, 1972), in that the clot closing the wound breaks off and the chitinous layer constituting a hyaline membrane alone isolates the regenerate from the external environment. This layer secreted by the epidermal cells is very similar to that described by Hoarau (1973) in isopods. It becomes distended and thus allows the formation, at the end of the stump, of a kind of papilla which becomes larger and larger and inside which the regenerate develops.

The blastema is formed initially by the appearance of a papilla isolated by an invagination of the periphery of the cicatrization surface (Adiyodi, 1972) and then by the bulging up of the central part. During its growth, the regenerate remains essentially rectilinear in its envelope but is sometimes clearly folded on itself, as is found in brachyurans.

This development of the regenerate inside an external papilla enabled Bliss (1956, 1959) to follow the growth of what can be called the regeneration bud and to describe it by a value of *R*, the ratio of the length of the limb bud to the width of the cephalothorax. Thus, it was shown that in some decapods regenerate development exhibits a slowing or even a break, so that Bliss distinguished three steps in the regenerate evolution: (1) "basal limb growth" for $R < 6$; (2) an "advancing plateau" for $6 < R < 10$; and (3) "premolt growth."

In spite of these characteristics, essential differences do not seem to

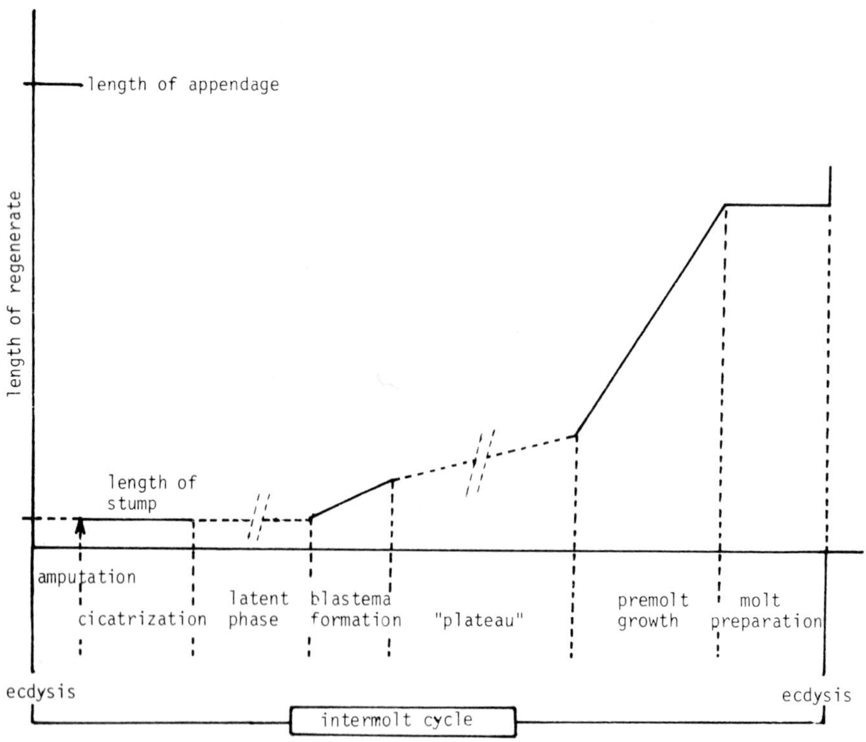

FIGURE 15.2. Representation of the regenerative events during an intermolt cycle in crustaceans.

exist between the basic course of development whether the bud is externalized or not. Such differences as there are affect essentially the chamber where the regenerate develops. As far as regenerate segmentation is concerned, it would seem that, as in insects, it is difficult to speak of basifugal or basipetal segmentation, or even of bidirectional segmentation from the middle of the blastema, as stated by various authors. Insofar as it is unlikely that fundamental mechanisms of regeneration differ from one group to another, probably the different segments are determined and their position defined very precociously in the blastema. The growth and differentiation that gives rise to the definitive form may be more or less readily observed along the blastema, depending on the groups of organisms studied, which led some authors to propose divergent interpretations. Figure 15.2 shows different stages of development of a regenerate in crustaceans.

15.3.3. *Other Arthropods*

In arachnids (Vachon, 1957, 1967) and myriapods (Sahli, 1967), the general processes of regenerate formation are quite similar to that found in insects. The regenerate develops at the end of the stump protected by the cuticle of the latter and is only released at the time of postoperative ecdysis.

15.3.4. *Conclusion*

Thus, regenerate formation in arthropods follows a general scheme that is valid for the various groups studied and is summarized in Fig. 15.3.

15.4. Regenerative Capabilities and Molting

Growth processes through molting are the same for all arthropods. These animals are thus subjected to the same constraints, and it is probable that the molting–regeneration interaction is the same in all groups.

The early investigators (Bordage, 1905; Emmel, 1907; Zuelzer, 1907) soon realized that in the various arthropod groups regeneration was possible only if the animal was still able to molt, as both phenomena seemed to be connected. Subsequently, they showed that animals able to molt regenerate before the postoperative molt only if the amputation has been performed early enough in the intermolt cycle. In order that a regenerate be elaborated, a certain minimal amount of time is required; moreover, the fact that at least one molt was found necessary implied that animals which no longer molt (e.g., adults in insects) do not regenerate. With the discovery of hormones, the hormonal environment was proposed as playing a major role in the process.

It may be deduced that the interactions between molting and regeneration are situated at two levels. On the one hand, to be able to

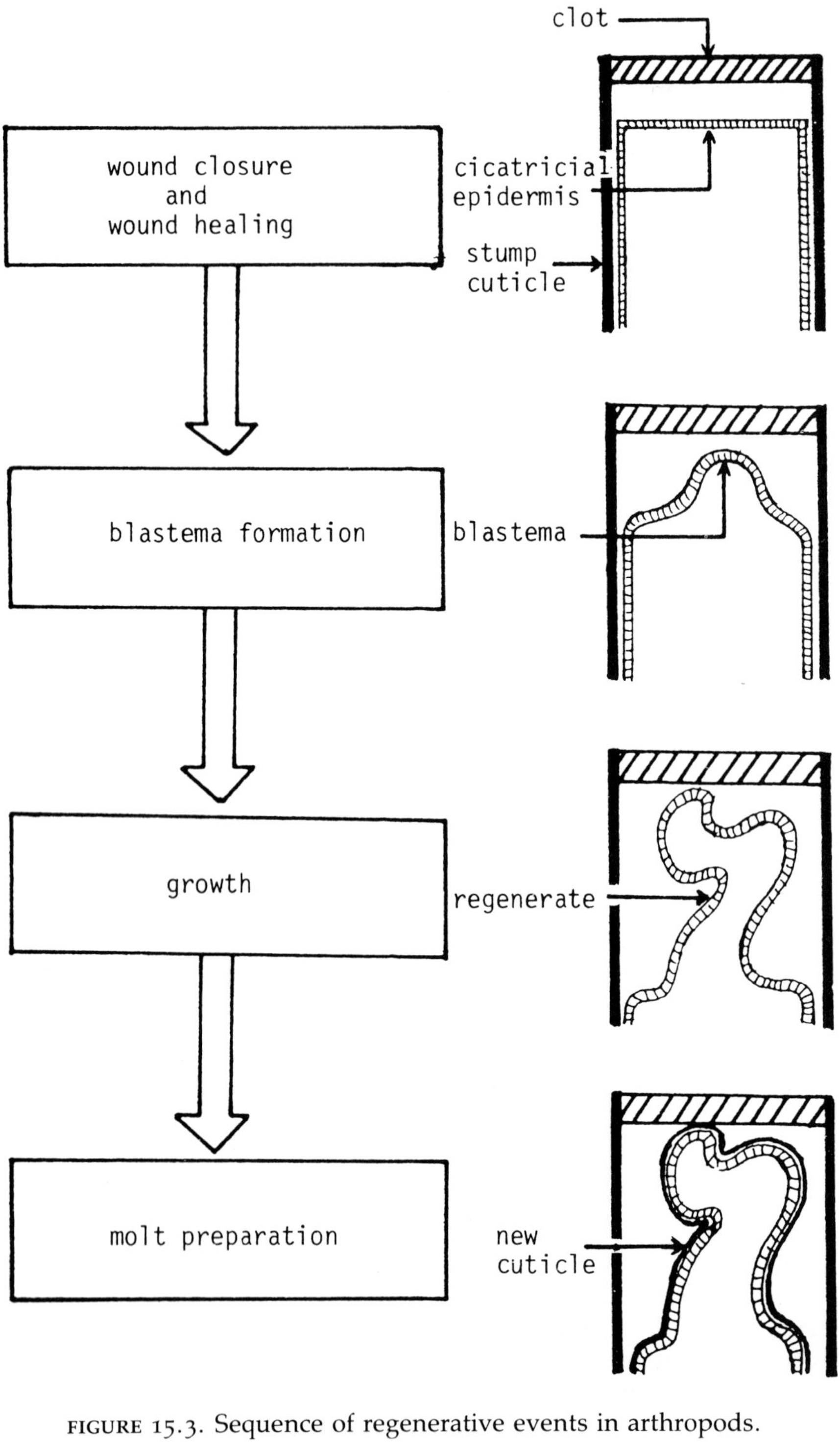

FIGURE 15.3. Sequence of regenerative events in arthropods.

regenerate, animals must be in a phase of their life where they can molt; so, they must be in a juvenile state (for example, the embryonic, larval, or nymphal state in insects); on the other hand, during a given molting cycle, the appendage loss must occur early enough in the cycle so that the regenerate is formed before the postoperative molt. Figure 15.4 shows the molting cycle in arthropods.

When regeneration and molting are taken into account together, it is apparent that two sets of quite different hormones act during morphogenesis: juvenile hormones (JHs), on the one hand, and molting hormones (MHs), on the other. However, morphogenesis does not escape the influence of a third group of hormones, i.e., those responsible for sexual dimorphism, which may occur during the morphogenetic determination of some regenerates. Note that hormones such as MIH

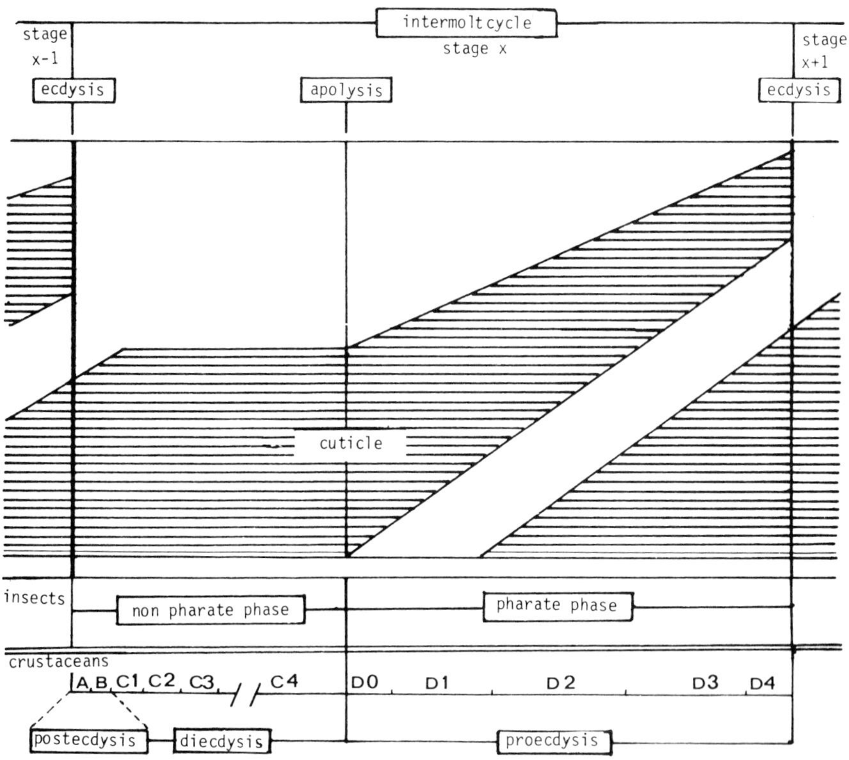

FIGURE 15.4. Schematic representation of the progression of an intermolt cycle in arthropods, showing the principal phases and their particular nomenclature for insects and crustaceans applicable to all the arthropod groups.

(molt-inhibiting hormone), thought to be specific to crustaceans but still hypothetical, will not be taken into account here.

15.4.1. *Regenerative Capabilities Throughout Life: Role of Juvenile Hormones*

In insects, JHs are found to be indispensable to the maintenance of molting. Probably their disappearance corresponds to the end of a state which can be termed juvenile and to the end of the ability to regenerate; or, in other cases, the fact that they do not disappear may explain why molting cycles do not stop and regenerative ability is maintained, as is the case in various crustaceans where mature animals continue to molt.

15.4.1.1. JUVENILE HORMONES AND MOLTING IN INSECTS

Pflugfelder (1939) was the first to envisage the role of JHs in regeneration. In fact, JHs do not seem linked to regenerative ability but rather to molting capacity. Other experiments on regeneration more directly implicate JHs in regeneration. By implanting corpora allata (CA), the source of JHs, in stick insects, Pflugfelder (1939) restored molting in adults, and these individuals consequently recovered regenerative capacity. O'Farrel and Stock (1964) obtained similar results by treating with farnesyl methyl ether the section surface of appendages that normally would have not regenerated.

In the cockroach *Blaberus craniifer,* the last molt, which produces the imaginal stage, results at the level of the appendage in a relatively characteristic change in shape: reduction of the diameter, lengthening of the tibia, and an alteration of the nature of the cuticle. When a regenerate is formed during the last intermolt cycle, i.e., the preimaginal intermolt, this regenerate appears with imaginal characteristics. If an appendage piece belonging to a last stage nymph is implanted into a nymph, which must molt two or three times before becoming an adult, a regenerate will be formed from the tissues of this piece, but will correspond to an appendage of nymphal type and not of imaginal type as would be the case if it would have been left in place or implanted in a last-stage nymph (D. Bullière, 1970).

Thus, the imaginal characteristics are gained only at the time of differentiation preceding the imaginal molt. Better yet, if such a piece is taken from an imago and implanted in a nymph still having to molt three times before becoming an imago, this piece will form a regenerate and that regenerate will have some nymphal characteristics. It displays adult characteristics only if the piece is implanted in a last-stage nymph that will consequently produce an imago at the time of molting (D. Bullière, 1970, and unpublished data).

The regenerate shape and its cuticle type depend on the nymphal stage during which the regenerate is built, and thus probably depend on the JH level. These results agree with those of Piepho (1939, 1963), Wigglesworth (1940, 1966), Lawrence (1966), Zlotkin and Levinson (1968), and Sehnal (1971, 1972) and correspond to what Slama (1975) termed "the old concept." We should emphasize that during regeneration epidermal cells proliferate and it is not exactly the same cells which secrete the new cuticle—so the reservations expressed by Schneiderman (1967) and Slama (1975) are not wholly justified.

15.4.1.2. JUVENILE HORMONES AND REGENERATION IN OTHER ARTHROPODS

JHs similar in action to those found in insects are most probably present in other arthropods. Laufer et al. (1987) have recently found them in crustaceans. In other arthropod groups such as araneids the presence of one or more hormones of the same type might well regulate the successive life stages—prelarval, larval, and nymphal—and explain the formation of regenerates of various types, each corresponding to an appendage shape characteristic of one stage in the life cycle (Vachon, 1967).

In diplopods, fluctuations in the level of comparable hormone might determine the reappearance of molts in adults, namely, the regressive molts that separate several adult stages, and in this manner restore regenerative ability (Sahli, 1967). To conclude, it can be said that the molting cycle is sustained or reestablished in order for regenerative ability to be sustained or recovered. Moreover, the presence or absence of JH would explain the type of regenerate obtained. Figure 15.5 summarizes this point of view.

15.4.2. *Regenerative Capabilities and Sexual Hormones*

In arthropods, relatively frequently, some appendages show sexual dimorphism. More often, these appendages, like the others, are able to regenerate. So one may ask whether during regeneration the regenerated appendage directly exhibits differentiated sexual characters or again goes through the steps of normal differentiation.

Studies relating to this subject are not very numerous in the literature. In crustaceans, Charniaux-Cotton (1957, 1967) worked on an amphipod, *Orchestia gamarella*, and a decapod, *Lysmata seticaudata*. In the decapod, she obtained the formation of pleopods II without appendix masculina, thus lacking the typical sexual characteristics, which appeared during the following molts. In *Orchestia*, the results were less clear-cut. A differentiated regenerate of male adult type may be formed directly when the amputation is performed very early, i.e., not long after the preoperative molt, but the regenerate exhibits a juvenile or intermediate aspect when the amputation is done later.

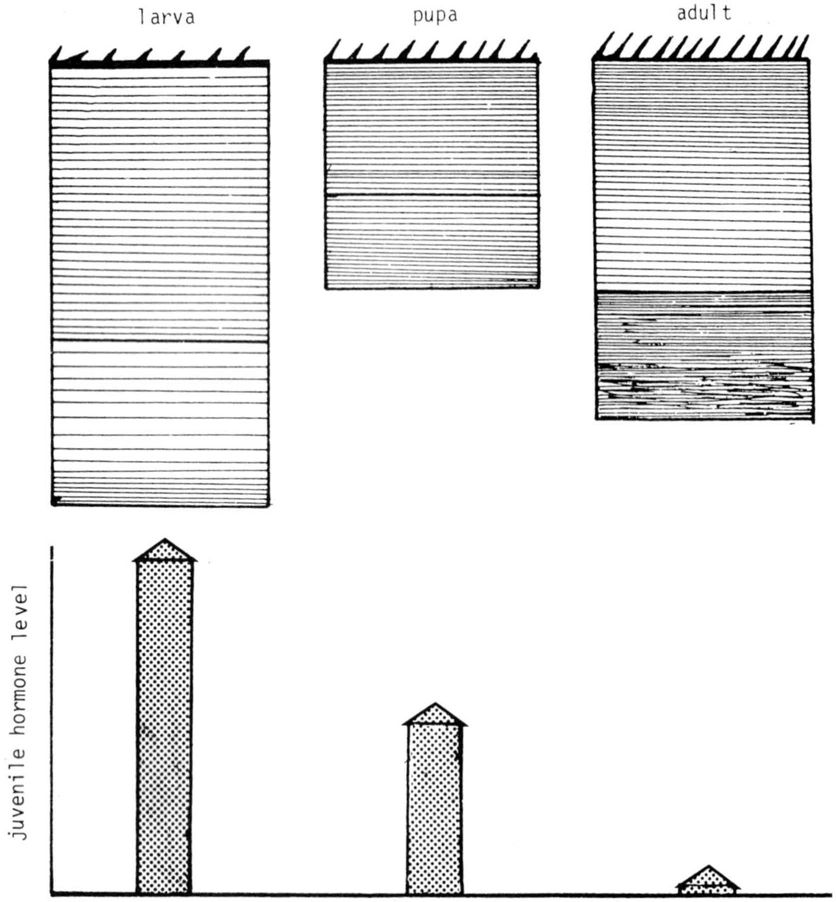

FIGURE 15.5. Juvenile hormone levels at the sensitive periods for the determination of the larval, pupal or adult cuticle.

Clearly sexual hormones have an influence over the shape of the regenerate, which is initially juvenile, then differentiates, but this differentiation may occur even before the postoperative molt.

In myriapods, Sahli (1967) showed that appendages with secondary sexual characters could regenerate during regressive molts and express relatively undifferentiated characters, sexual differentiation appearing again only during a new molt between the intermediate stage and adult.

In conclusion, it is obvious that sexual hormones act on growing regenerates to induce them to gain sexually differentiated forms.

15.4.3. *Regenerative Capabilities During an Intermolt Cycle: Role of Ecdysteroids*

For a regenerate to be released at the postoperative molt, loss of the appendage must not occur too late in the intermolt cycle. As fluctuations in ecdysteroid levels give rise to rhythms in the molting cycle, they consequently control the regenerative capability. However, does a strict relationship exist between the ecdysteroid titer and regenerate development? These relationships have been intensively investigated in insects and crustaceans. Investigators were able to distinguish two distinct aspects in these relationships—the first corresponding to the conditions necessary to initiate the regenerative processes, and the second corresponding to that necessary for development, growth, and then differentiation of the regenerate.

15.4.3.1. INITIATION OF REGENERATION AND MOLTING: TIMING OF THE CRITICAL PERIOD

Different considerations must be taken into account in a description of the relations existing between the start of regeneration and that of molting. First of all, it is clear that development of a regenerate, whatever it may be, takes some time. Thus, obviously an appendage lost or sectioned just before the molt cannot regenerate before that molt. Also, if the molt occurs when the epidermal cells of the regenerate have not begun to secrete the cuticle that might protect it, the regenerate will not be able to resist the increase of pressure during the ecdysis processes and there will exist a risk of bleeding leading to death of the animal. Therefore, in all groups of arthropods, a period is necessary before ecdysis during which an appendage loss is not followed by its immediate regeneration. Hence, molting cycles are divided into two periods—the first, at the beginning of the cycle, being favorable to immediate regeneration; the second, at the end of the cycle, a phase when immediate regeneration is no longer possible. These two periods are separated by what has been termed the *critical period.* How can this phase be understood? Is it a period of variable length, or is it simply a transitional phase from the period favorable to immediate regeneration to that which is unfavorable?

Taking into account the fact that the timing of the critical period during the intermolt depends on the level at which a section is made (D. Bullière, 1968a), it seems preferable to conceive of that period as one merely of changeover from conditions good for regeneration to those that are unsuitable. These conditions probably depend on the general hormonal environment and also on local conditions, as does the recovery of the continuity of the epidermis.

For immediate regeneration to occur, appendage loss must take place before the critical period. It seems that onset of the critical period is triggered by an increase in the ecdysteroid titer (see below). Thus, it may be assumed that regenerative ability is an intrinsic property of the cells and that it is always expressed, except when hormonal conditions act to block it.

Insects

The general rule pertaining to the critical period is entirely followed in insects. No single experiment has ever demonstrated that a regenerative factor has any positive effect. One must nevertheless bear in mind that the position of the critical period within the intermolt period depends on the level of section and that the more distal is the section, the more it is delayed (D. Bullière, 1968a,b).

Moreover, note that proximal amputations which break the coxa, resulting in a very wide surface of section, are very seldom followed by immediate regeneration. As it is unlikely that the hormone titer fluctuates along the appendage, it must be assumed that local processes interfere, such as restoration of the epidermal integrity before formation of the distal regenerate can be triggered. The surface of a section through the tibia or the coxa is larger than that of a section made through the tarsus, and consequently cicatrization takes longer. The hormonal environment that must be taken into account could be the one which exists locally when epidermal integrity is restored. Indeed, cicatrization always occurs whatever the time of amputation (except when the section is made just before ecdysis). Consequently, inhibition can occur only after this step. Depending on the surface area of the wound and its conformation, this obligatory step might last for a longer or a shorter time and advance to a greater or lesser extent the time of amputation in relation to the ecdysteroid titer of the critical period, so that immediate regeneration takes place. This point of view takes into consideration the local hormonal environment, whose role could be only permissive, an excessive hormone titer inhibiting regeneration (D. Bullière, 1972a; Maleville, 1978), and also takes into account cicatrization in which cell interactions, necessary to the qualitative determination of the regenerate, are implicated.

Nevertheless, this conclusion must be moderated, because for regeneration to occur a minimal level of tissue activity is necessary. A low titer of steroids seems to be necessary for minimal cell activity (see below).

Crustaceans

What about crustaceans? It is impossible to find a unifying pattern in a review of the literature. Indeed, it might be concluded that numerous particular cases exist. However, a detailed analysis of the various reports shows that more often than not different experimental conditions explain the divergences between the authors. It seems that, as in insects, a critical period exists after which an autotomy is not followed by regeneration before the postoperative molt; this phase divides the intermolt into two parts—one propitious to immediate regeneration, the other not. The critical period always occurs at the beginning of proecdysis: D_0 in *Orchestia gamarella* (Charniaux-Cotton, 1957); between D_0 and D_1 in *Leander serratus* (Tchernigovtzeff, 1965); at the end of $D_{1'}$ in *Carcinus maenas* (Démeusy, 1971); and during $D_{1'}$ in *Gecarcinus lateralis* (Skinner and Graham, 1972; Tchernigovtzeff, 1972, 1974; Holland and Skinner, 1976).

This loss in regenerative ability might be linked to an increase in the steroid titer, which may be similar to what occurs in insects. For example, Hoarau and Hirn (1978) place the critical period in the isopod *Helleria* in D_1, a little after the beginning of hemolymphatic ecdysteroid increase. That the place of the critical period in the intermolt cycle appears to fluctuate according to the group or even the species of crustaceans might be explained by the fact that the hormone level to be taken into account is that corresponding not to the time of amputation but to that of cicatrization, as in insects. Apparently, the critical period always occurs after an increase in circulating ecdysteroids but before the major peak, which triggers cuticle secretion and the next molt.

As in insects, it appears that a minimum hormone concentration is necessary for basal metabolism and for evolution of the cycle, and thus for regeneration (Noulin, 1984). The ability of the stump cells to reorganize themselves appears to be an intrinsic property of these cells, the hormone allowing the process and simply regulating its rate. Again we find a result already reported in insects, namely, that if the coxa is amputated or injured, regeneration is delayed until the following intermolt (Hodge, 1958; Gomez, 1964).

Other Arthropod Groups

There is little information concerning other groups of arthropods, such as diplopods and arachnids. Nevertheless, it seems that in araneids, for amputation to be followed by regeneration, it must be done before a given moment of the cycle. Moreover, the quality of the regenerate might well depend on the timing of the amputation with regard to that moment. As in insects, the position of the critical period can depend on the level of amputation. In acarines (Rockett and Woodring, 1972), ix-

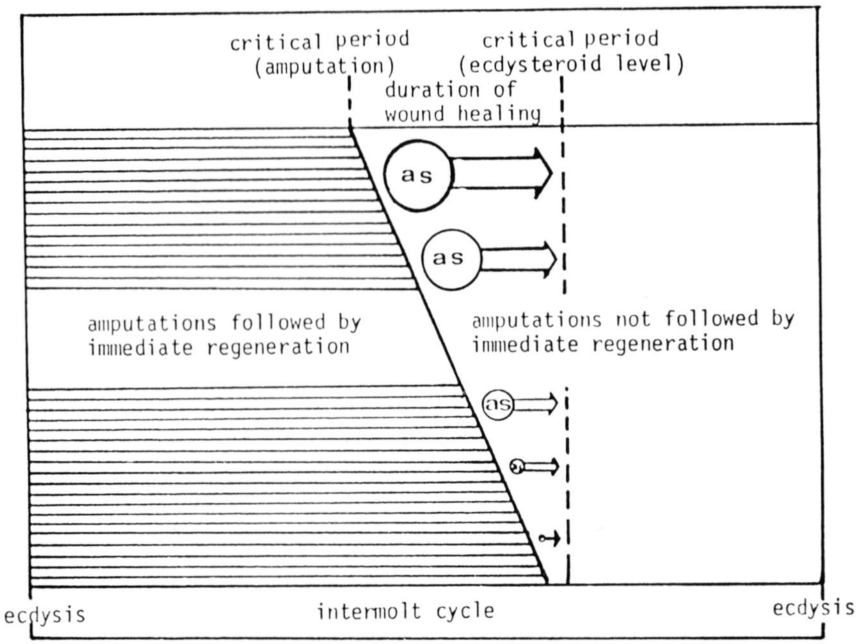

FIGURE 15.6. The place of the critical period in the intermolt cycle. If the MH intervenes and prevents regeneration from being triggered, one must envisage its action on epidermal cells after wound healing rather than just after amputation. *Key:* a s = amputation surface.

odids have the greatest regenerative capability, but tolerance to amputation varies during the cycle, and in *Amblysmura americanum,* for example, amputations performed during the active stage, which corresponds to about the first 10 days of the cycle, are followed by regeneration; amputations performed during the inactive, quiescent stage lead to death. Figure 15.6 presents the position of the critical period during the intermolt.

15.4.3.2. ORIGIN OF THE TRIGGERING OF REGENERATION

If it is clear that JH as well as ecdysteroids trigger regeneration, then their role is restricted to that of a control, the origin of such triggering being found at the level of cell contacts, as clearly demonstrated in insects by intercalary regeneration experiments of D. Bullière (1971), reworking an idea of Bart (1969). Indeed, when homologous levels are joined by graft (Fig. 15.7) regeneration is never triggered, whatever the time of amputation during the cycle and whatever the age of the animal. In contrast, when heterologous levels are joined (Fig. 15.7), this always

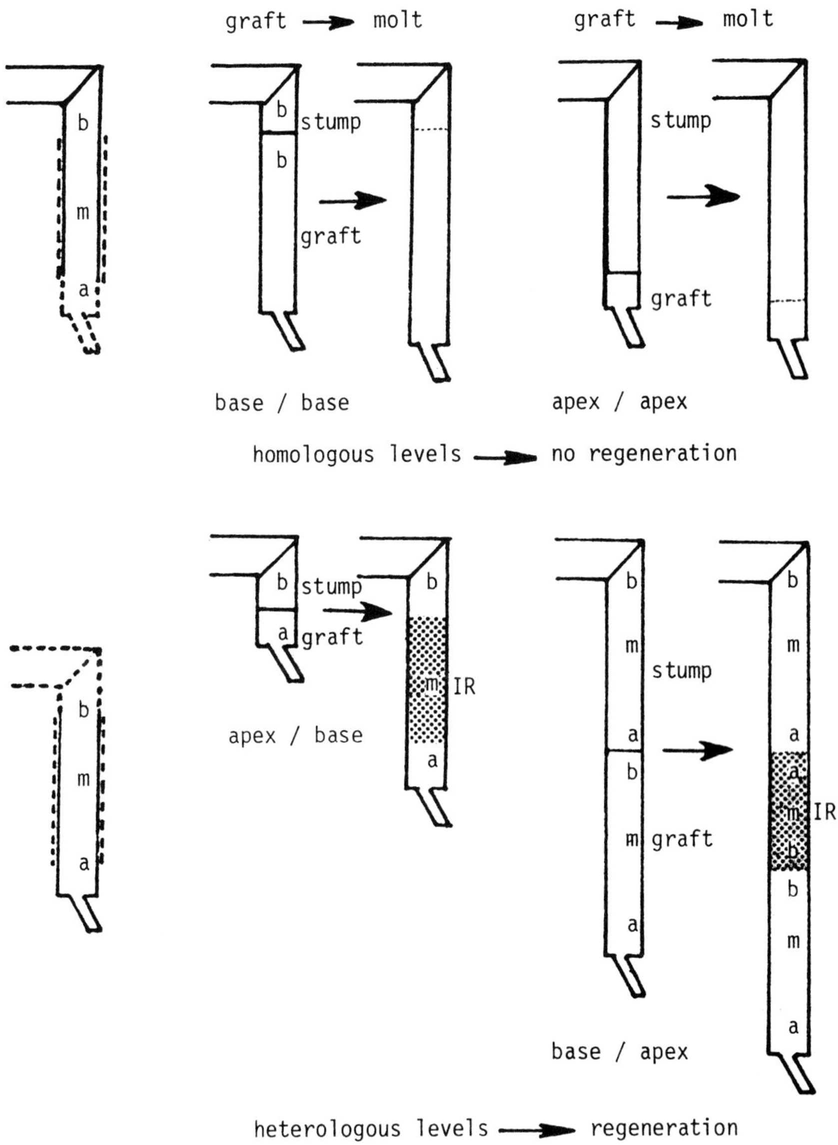

FIGURE 15.7. The joining by graft of two homologous levels triggers no morphogenesis other than wound healing, whereas the joining by graft of two nonhomologous levels leads to formation of a regenerate. (*Key:* a = apex; b = base; m = middle; IR = intercalary regenerate).

leads to formation of an intercalary regenerate, either before the postoperative molt, if the experiment is performed before the critical period, or following the intermolt, if performed after. Moreover, regeneration stops when "normal" contacts are restored by intercalary regeneration (Bohn, 1970; D. Bullière, 1971; French and D. Bullière, 1975; D. Bullière and F. Bullière, 1985). Thus, it is the local conditions and intrinsic capacities of the cells that determine the form, the hormones providing the necessary conditions for the cells to express themselves. Nevertheless, it is difficult to explain the incomplete, badly formed regenerates that are frequently found when the time of sectioning is near the critical period.

15.4.3.3. CONCLUSION

It seems true that in most arthropods, regenerate formation immediately follows amputation, autotomy, or loss of an appendage, providing these events occur before what is called the critical period, which separates each intermolt into two parts, the first being favorable to immediate regeneration and the other not (the timing of the latter period may depend on the level of amputation). Regeneration, which involves the formation of a blastema from the stump tissues, necessitates a change in the genomic activity of the cells that participate in this formation. If this change is inhibited, it is conceivable that such regeneration would no longer be possible. So, it is reasonable to conclude that the critical period might well correspond to such an inhibition state, probably due to the action of ecdysone (see below).

15.4.4. *Development of the Regenerate with Respect to the Molting Cycle*

A major question is whether each step of regeneration occurs at a precise time during the molting cycle. To answer this, it is necessary to follow the formation and development of regenerates in various groups of arthropods.

In insects, in some crustaceans, and especially in arachnids, it seems that regenerative processes are triggered as soon as an appendage loss occurs, providing it takes place before the critical period and continues during the whole intermolt cycle until molt preparation. During this period, differentiation of the epidermal cells of the regenerate shows no similarity with that of the stump cells, but during molt preparation, at the end of the cycle, a relatively strict synchronism between epidermal cell differentiation of the regenerate and that of the integument of the rest of the body seems to be obligatory.

15.4.4.1. ARACHNIDS

The best example of relative independence of regenerate formation with respect to the molting cycle is given by the experiments of Vachon (1967) on spiders. Indeed, depending on the time of amputation during the intermolt, he obtained a regenerate of larval type when the amputation was precocious or a regenerate of prelarva type II or even prelarva type I when amputation was performed very late.

Thus, clearly a regenerate that has at its disposal the whole duration of the intermolt in which to be constructed attains a more advanced stage of differentiation than a regenerate that has only the end of the intermolt in which to be formed. In the first case, the hormonal environment goes hand in hand with differentiation of a larval type appendage; in the second case, of a prelarval type appendage.

15.4.4.2. INSECTS

In insects, when a regenerate is released at the first postoperative molt, the cuticle which covers it is secreted synchronously with that which covers the rest of the body. The epidermal cells of the regenerate as well as those of the rest of the body have been induced to secrete the cuticle by a series of processes, apparently triggered by a high increase of the ecdysone titer. Consequently, during this last part of the cycle, all the regenerates undergo the same phase, which is preparation to molt. This phase also occurs when there is no regenerate formation but simply cicatrization. However, what happens before this? Must we admit that each part of the cycle corresponds to a given stage of regenerate development or, on the contrary, that regenerate development is largely independent of the intermolt cycle? Does a strict correspondence exist between regenerate development and the molting cycle? Rigorously speaking, regeneration should begin only after the critical period, as any loss of appendage before the critical period is followed by immediate regeneration. Moreover, the size and appearance of the regenerate should not depend on the precise time of the appendage loss, which is inconsistent with the fact that the time of amputation has an effect on the length of the regenerate (D. Bullière, 1968a,b). The shorter and sometimes the more badly formed a regenerate, the later was the amputation in relation to the critical period. Moreover, one might observe a latent period between the time of appendage loss and that of formation of the regenerate when this loss occurs at the beginning of the cycle, long before the critical period. From a regeneration point of view, the period of intermolt before this critical period might be *neutralized.* But such is not the case, inasmuch as it seems that regenerative processes begin from the time of amputation or appendage loss. In addition, it should be observed that formation of the regenerate is well advanced at the critical period. Indeed, the following experiments may be performed (Fig. 15.8): an appendage is cut at the base of the tarsus at the beginning of the cycle; then, the apical part of the tibia is removed at various times after the first ablation; and, finally, these animals are observed until the postoperative molt to determine whether a regenerate has been formed

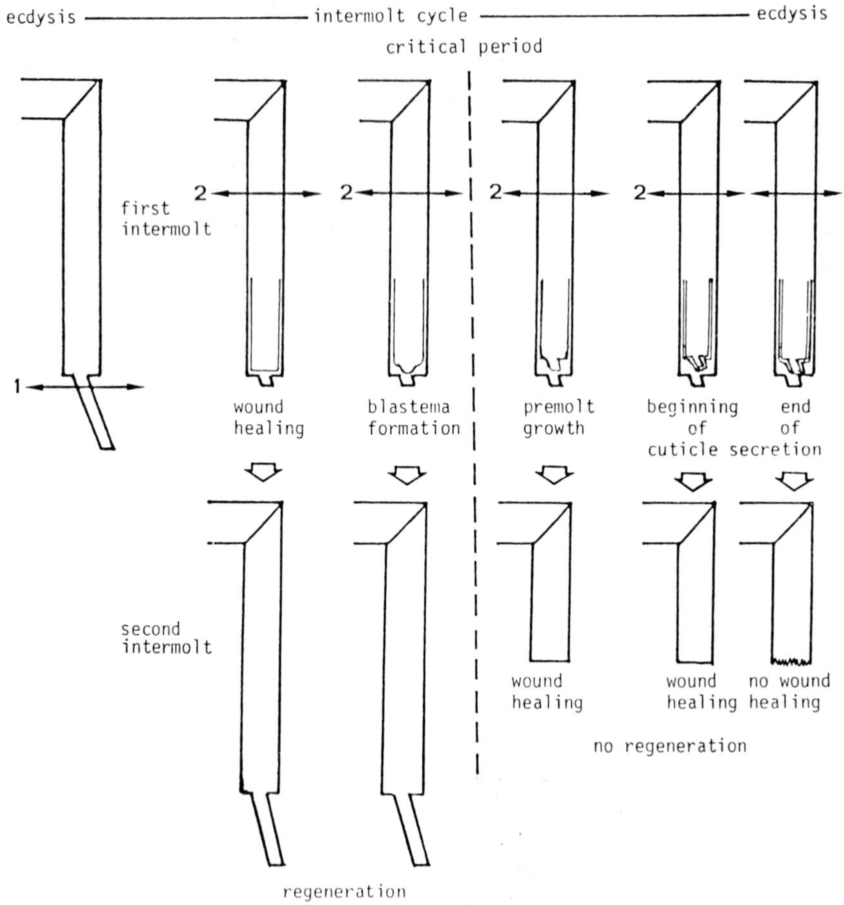

FIGURE 15.8. Correlation between the stages of regeneration and the molting cycle in arthropods. If the first section (1) is made at the beginning of an intermolt cycle, it provokes the formation of a regenerate at the extremity of the stump. If a second section (2) is then made at different moments during the intermolt, this enables us to determine the regenerative capacity for each moment and to observe the stage of regenerate development.

as a result of the second amputation at the tibia base. If this is the case, it demonstrates that the time of sectioning took place before the critical period. As in the apex of the tibia, one observes a more or less well-developed regenerate according to the time of the second section; it is thus evident that the different steps of regenerate formation do not correspond to a different moment of the intermolt cycle. Better yet, a regenerate can be experimentally induced and allowed to develop until shortly before the critical period; then, the distal part of this regenerate can be removed in order to obtain at the postoperative molt a regenerate whose base had the whole duration of intermolt in which to develop and the apex had less time. Under these conditions, the distal part of the regenerate appears relatively shorter and less developed than the proximal one (D. Bullière, 1972a, and unpublished data).

Thus, it is clear that regenerate development is relatively independent of the intermolt cycle. The hormonal environment seems to interfere at least twice to harmonize the intermolt cycle and regeneration: first, it suppresses any possibility of immediate regeneration, thus determining the critical period; later, it triggers the processes that result in epidermal cell differentiation, particularly cuticle secretion. This is confirmed by the fact that an appendage ablated while undergoing the growth phase of regeneration does not regenerate a second time. It does not even cicatrize when the operation is performed very late, 1 or 2 days before the postoperative molt, during the molting-preparation phase.

15.4.4.3. CRUSTACEANS

Within the class of Crustacea, diversity seems very high. Hoarau (1973) and Hoarau and Hirn (1981) described in an isopod, *Helleria brevicornis*, on days 4–5 after amputation, a latent phase of variable length before blastema formation: a 13-day mean for animals amputated at the beginning of the cycle (stages A and B of Drach, 1939); 15–17 days for animals amputated at stage C; 11 days for animals amputated at stage C but already regenerating another appendage; and 12 days for animals amputated at stage D_0. In decapods, other authors (Bliss, 1956; Skinner, 1962; Holland and Skinner, 1976) have described a latent phase called the *plateau* occurring after blastema formation, at the end of the anecdysis phase (stage C_4) but before the growth phase, which is shorter if the appendage was lost later (Charmantier-Daurès, 1980). All these observations do not favor the idea of a strict correspondence between regeneration and the molting cycle, as the regenerative processes, which take place before the plateau phase when the section is precocious, occur later in the case of a late section. Indeed, in *Gecarcinus lateralis*, some authors place the critical period between D_0 and D_1', that is, at the beginning of premolt (Skinner and Graham, 1972; Tchernigovtzeff, 1972, 1974; Holland and Skinner, 1976). Still more conclusive is the fact that late amputations (Tchernigovtzeff, 1974) or resections of regenerates (Tchernigovtzeff, 1974; McCarthy and Skinner, 1977a,b) lead to a halt in or a slowing down of the development of regenerates formed following precocious amputations and left in place. In other groups, the plateau phase may not exist at all, or it may be much less obvious (Pradeille-Rouquette, 1974, in *Pachygrapsus marmoratus*) or even optional. For example, in *Orconectes obscurus* (Stevenson and Henry, 1971), when the cycle is short (29 days) the regenerate shows a continuous growth, but when the cycle is long (71 days) growth may be interrupted for 5 days or more.

In fact, schematically, it seems that nearly everything occurs as it does in insects. Amputation is followed by the processes of wound closure and cicatrization, i.e., restoration of the integrity of the epidermis. Then, if these processes are achieved before the critical period, formation of the blastema starts, engaging the cell interactions necessary for formation of the regenerate. Development of the blastema can be more or less rapid and regular according to the group concerned, sometimes being very slow during the latent phases. As in insects, all these events take place before the critical period. Then, after the critical period, the growth phase occurs, during which the regenerate gains the size it will have after the postoperative molt. Evidently, any new appendage loss by autotomy or ablation during this period is not followed by immediate regeneration, as it occurs after the critical period. If the ablation is very late, cicatrization is no longer physically possible, as is the case in insects. It might be hypothesized that the development of regenerates occurs exactly in the same manner during this period and that the differences observed between them at the time of the postoperative molt (for example, minute regenerates observed when the appendage loss occurs very close to the critical period) are due to differences, at the critical period, in gains in size of the blastemas. Late loss of an appendage leads, due to lack of time, to formation of a minute regenerate; earlier loss leads to formation of a normal blastema and then to a normal regenerate. However, it might also be hypothesized that, when the triggering of blastema formation occurs just before the critical period, such blastema formation encroaches on the time required to regenerate growth, which will thus be incomplete. Few elements help us to choose between these two hypotheses. Nevertheless, if one admits that fluctuations of the ecdysteroid level control the intermolt cycle and regeneration, it is more logical to assume that evolution of the blastema is interrupted by an increase in the hormonal level that would modify the genomic functioning. Consequently, blastemas of different sizes, undergoing growth at the same rate, would evidently lead to the formation of regenerates of different sizes. A final argument against strict synchronism between the intermolt cycle and regeneration is given by the experiments of Charniaux-Cotton (1967) in *Orchestia gamarella*. Indeed, depending on the time of amputation, she obtained juvenile or male (corresponding to two steps of differentiation) gnathopods. This point of view, which brings together the various arthropod groups, is more persuasive, because it does not require different processes of regeneration to be imagined and allows regeneration to be thought of as deriving largely from the same origins as the embryonic processes. Figure 15.9 illustrates this interpretation.

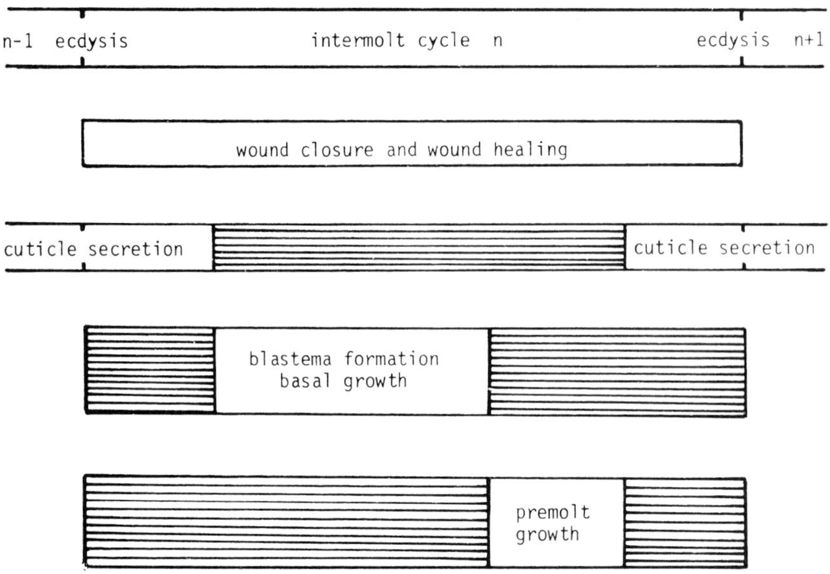

FIGURE 15.9. Periods of intermolt during which the different events of regeneration can take place (open areas) or not (striped areas).

15.5. Regeneration and Morphogenetic Hormones

Progress made in recent years has allowed investigators not only to isolate the hormones that regulate the molting cycle but also to determine their titers during the cycle and to make them act under various conditions.

15.5.1. *Molting Hormone Levels*

Hormonal levels have principally been studied in insects and crustaceans.

In insects, measurements taken throughout the whole duration of intermolt and performed in parallel with regeneration studies (Maleville, 1978; D. Bullière et al., 1979; Maleville and De Reggi, 1981; D. Bullière and F. Bullière, 1985) show primarily that the steroid level is never zero. Further, they demonstrate the existence of a first peak in the hormonal level toward the beginning of the second half of the cycle, corresponding to the critical period, and finally they show a rather considerable increase in the ecdysteroid level at the end of the cycle, which drops only before exuviation time. In some groups, or during some periods of development, a third peak, right at the beginning of the cycle, is observed.

In crustaceans, the situation is without doubt not very different, but it is quite difficult to formulate a coherent representation of it, as not only do results seem to fluctuate from one species to another, but even for the same species results often conflict among different investigators according to the techniques used. To summarize in a general way, the hormonal titer is maintained at a rather low level until the beginning of proecdysis (D_0–D_1) and then progressively increases to high values before falling again shortly before exuviation. The critical period would correspond to the increase in the hormonal titer. Before the end of the cycle, the high hormonal level triggers cuticle secretion. Figure 15.10 illustrates this interpretation in arthropods.

15.5.2. *Effect of Regeneration on the Intermolt Cycle and the Hormone Titer*

In insects, as well as in crustaceans, all authors reviewed by us have observed a feedback action of regeneration on the intermolt cycle and hormone titer.

In insects, all ablations lengthen the duration of the cycles (O'Farrel and Stock, 1954; D. Bullière, 1968b) and delay concomitantly the appearance of the hormonal peaks (Maleville, 1978; D. Bullière et al., 1979;

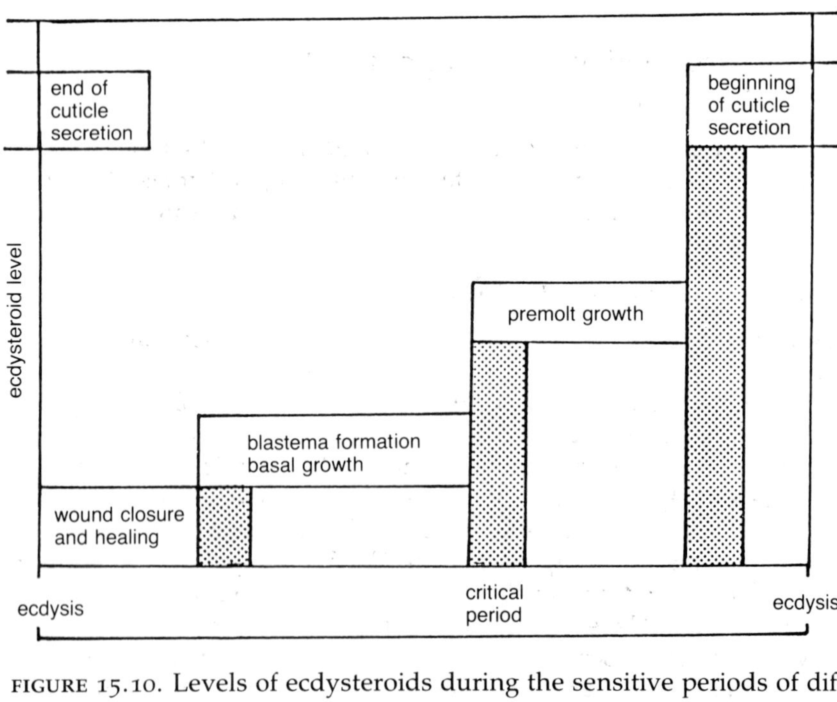

FIGURE 15.10. Levels of ecdysteroids during the sensitive periods of different stages of regeneration during an intermolt cycle in arthropods.

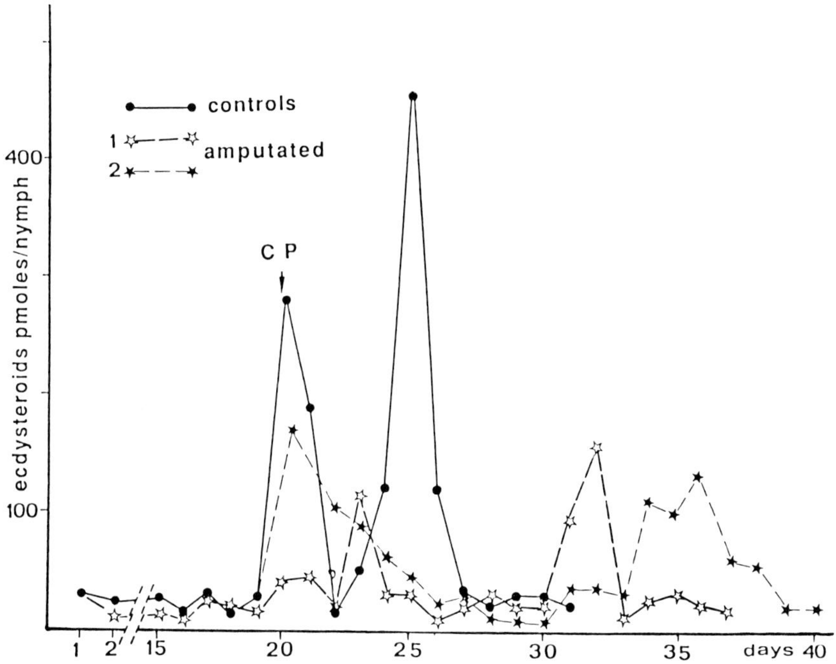

FIGURE 15.11. Evolution in *Blaberus* of the ecdysteroid titer in nonamputated nymphs (controls), nymphs amputated just after hatching (1), and those amputated just before the critical period (2). In control nymphs, the first peak occurs when the regenerative capacity disappears, i.e., at the critical period (CP). In experimental nymphs, early amputation leads to a decrease in the peaks that are delayed; late amputation followed by regeneration postpones the decrease of the first peak and delays the second and ecdysis.

Maleville and De Reggi, 1981; Roberts et al., 1983; D. Bullière and F. Bullière, 1985). The effect is all the more marked when the amputation is more severe and occurs nearer the critical period (Fig. 15.11).

In crustaceans, as already noted, there is a high diversity in experimental results and it is difficult to make a synthesis thereof—that is, to compare results of experiments involving single regenerations, intensive regenerations (namely, regeneration of several appendages) with either precocious or late amputations, performed on animals belonging to different groups, under variable experimental conditions (such as temperature). In some cases, like the isopod *Helleria* (Hoarau and Vernet, 1979; Hoarau and Hirn, 1981) and the decapods *Carcinus* (Demeusy, 1971) and *Gecarcinus* (Tchernigovtzeff, 1972, 1974), everything seems to occur as in insects: the more severe the amputation of one or several appendages and the nearer to the critical period, the greater the lengthening of the intermolt duration. In other cases, regeneration of only one appendage seems to have no effect, whereas intensive regeneration shortens the duration of the intermolt cycle if the amputation is made a relatively long time before the critical period and lengthens it if made just before the critical period (Skinner and Graham, 1972, in *Gecarcinus;* Skinner and Graham, 1972, and Fingerman and Fingerman, 1974, in *Uca;* Bennett, 1973, in *Cancer;* Bittner and Kopanda, 1973, in *Procambarus;* Stoffel and Hubschman, 1974, in *Palaemonetus;* Noulin, 1984 in *Porcellio*). Interestingly, precise data have been obtained by investigators who have followed the ecdysteroid levels of animals undergoing regeneration. They have consistently shown a strict correlation between changes in duration of the intermolt cycles and the positions of the MH peaks. In *Pachygrapsus* (Charmentier-Daurès, 1980), it has been clearly established that a relationship exists between the change in duration of the cycles and the position of the ecdysteroid peaks (Fig. 15.12). Hoarau and Hirn (1981) (Fig. 15.13) have shown that, as in insects, regeneration causes a lengthening of the intermolt cycle and a delay in the appearance of the ecdysteroid peaks. McCarthy and Skinner (1977a), in *Gecarcinus,* have demonstrated a parallelism between fluctuations of the steroid level and development of the regenerates. So, it can reasonably be proposed (de-

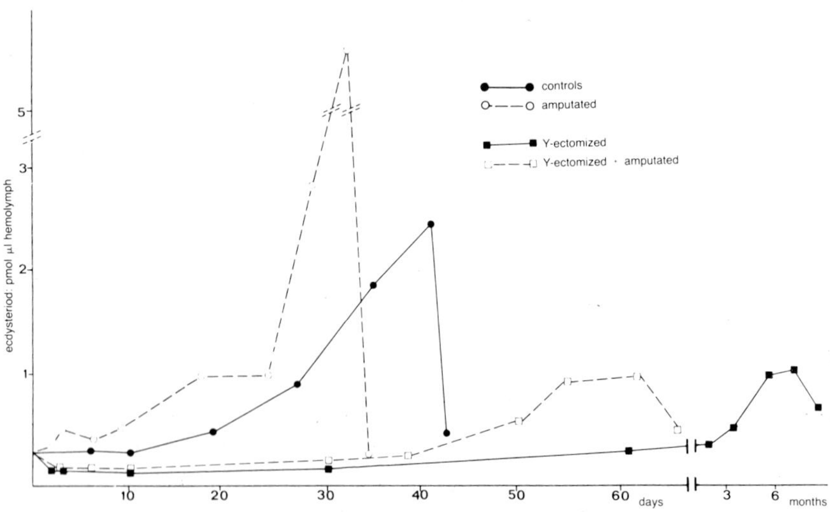

FIGURE 15.12. Variations in ecdysteroid level in control and operated *Pachygrapsus* (decapod). Depending on the operations, an increase or a decrease and a shifting of the hormone peak can be observed. (From Charmentier-Daurès, 1980.)

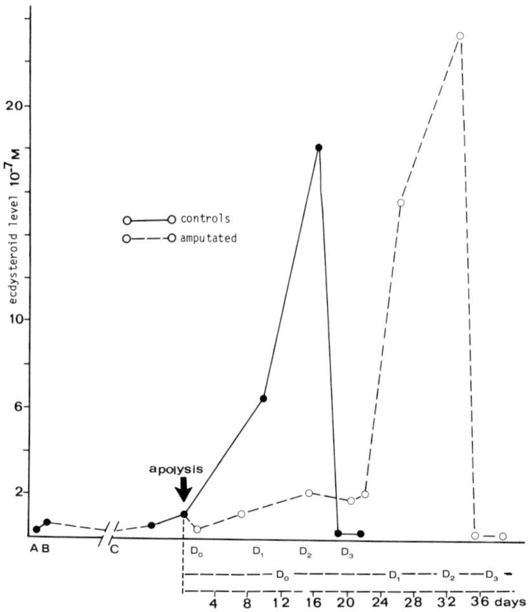

FIGURE 15.13. Variations in ecdysteroid level in control and amputated specimens of *Helleria* (isopod). See text.

spite the fact that some dosage data are not available) that a shortened cycle corresponds to a previous increase in the ecdysteroid titer and thus that fluctuations in the ecdysteroid level control the duration of intermolt cycles. As in insects, an ablation near the critical period lengthens the cycle by delaying the ecdysteroid peaks.

Shortening of the duration of the cycles induced by intensive regeneration, which is a characteristic of some crustacean groups, seems to be coupled with a change in time of ecdysteroid secretion and an increase of their level (Fig. 15.12) (Charmantier-Daurès, 1980).

In conclusion, we can reasonably assume that in all groups of arthropods intermolt cycles are controlled by fluctuations in ecdysteroid levels. Some steps of the regeneration are phased with the cycle by the ecdysteroid peaks. Figure 15.14 illustrates the correspondence between the hormonal pattern and stages of development.

15.5.3. *Changes in the Hormone Titer: Effects on Regeneration*

If the hormonal level controls the molting cycles, it is evident that anything disturbing these cycles must also affect regeneration. Depending

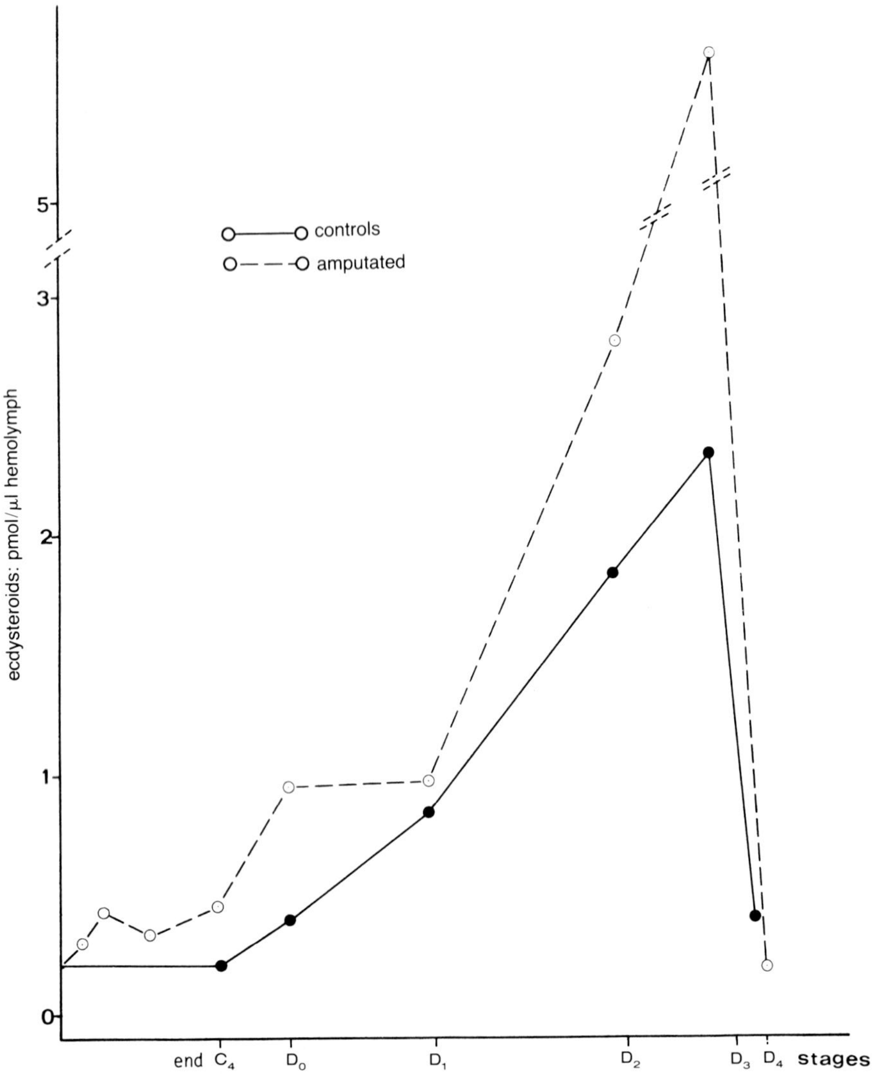

FIGURE 15.14. Variations in ecdysteroid level in control and amputated *Pachygrapsus* (decapod). See text. (From Charmantier-Daurès, 1980.)

on the groups concerned, various types of experiments have been performed.

Suppression of the Source of Hormone

Removal of molting glands. In crustaceans, Echalier (1956, 1959) first showed that removal of the Y-organs, considered as the major source of

MH in *Carcinus maenas,* abolishes regeneration by suppressing the capability to molt. Demeusy (1967) observed in the same species a beginning of regeneration when the animals were kept at 9°C, and Bazin (1973), in animals operated on in C_4, observed on histological slides the formation of a blastema, which was invisible externally and which did not continue to develop. In another species, *Sesarma reticulatum,* similar results were obtained (Passano and Jyssum, 1963): basal growth but no premolt growth took place in animals deprived of their Y-organ. Noulin and Maissiat (1974) in *Porcellio dilatatus,* following electrocoagulation of the Y-organs of animals in period C, did not observe the formation of a regeneration papilla. They allowed 10 days between the time of Y-organ destruction and that of ablation of a pereiopod. In *Pachygrapsus marmoratus* (Charmantier-Daurès, 1980), regenerate formation was obtained following ablation of a pereiopod 2 days after the Y-ectomy performed in C_3, C_4, D_0, or D_1. The end of the cycle, just before exuviation occurs, was reached after some 5–6 months, compared with 2 months normally (Fig. 15.12).

It seems that in crustaceans, apart from exceptions, suppressing the Y-organs stops the molting cycle and prevents the development of regenerates that do not evolve further than the stage of blastema formation.

Ligation experiments. In insects, owing to the technical difficulty involved in suppression of the molting glands, experimental results are few. However, it seems that when the hormonal source has been removed, the molting cycle as well as regeneration are interrupted (Bodenstein, 1955). Madhavan and Schneiderman (1969) obtained 65% regeneration of wing disks in permanent *Galleria melonella* larvae, resulting from a ligation between the pro- and mesothorax, which excludes any possibility of the hormone passing from the anterior to the posterior part. The regeneration rate dropped to 10% when ablation of the wing disks was made more than 6 days after the ligation. Regeneration of metathoracic appendages in *Blaberus craniifer* embryos cultivated *in vitro* during the second embryonic cycle (D. Bullière and F. Bullière, 1974) was not suppressed when the entire part anterior to the metathorax was removed. The cycle continued and an apparently normal cuticle was secreted. It must thus be admitted that prothoracic glands are not the unique source of MH.

15.5.4. *Action of Exogenous Hormone on Regeneration*

If the ecdysteroids really regulate the molt rhythm by controlling the cycle and in this case, regeneration, it is evident that exogenous ecdysteroids must also act on development of the intermolt cycle as well as on regeneration. Experiments have nearly all been done in insects and crustaceans.

In insects, the action of MH on regeneration was first demonstrated in embryos of *Blaberus* (F. Bullière and D. Bullière, 1971). They cultivated embryos and added exogenous hormone to the culture medium, which led to inhibition of regeneration. In nymphs, the results are less convincing—for, to obtain inhibition of regeneration, several injections are necessary shortly before the critical period and directly into the stump of the amputated appendage; otherwise, the hormones are ineffective, as they are without doubt metabolized (D. Bullière, 1972b). Madhavan and Schneiderman (1969) increased the regeneration rate of imaginal disks of permanent larvae of *Galleria melonella* (ligated between the pro- and meso-thorax) by injecting 0.75 μg ecdysone and inhibited it with 10 μg of that hormone. The regenerative ability of appendages of larvae of *Tenebrio* cultivated *in vitro* 72 h before being implanted in larvae is retained by adding 1 ng · ml^{-1} of ecdysone to the culture medium (Lenoir-Rousseaux, 1978). On the contrary, a concentration of 3–6 ng · ml^{-1} blocks regeneration. The same results were obtained with various other ecdysteroids, including 20-hydroxyecdysone and inokosterone. These, when given at rather high concentration, always trigger differentiation by creating a hormonal environment at the end of the intermolt cycle. Regeneration is thus inhibited in all cases. In embryos of *Blaberus* (F. Bullière and D. Bullière, 1973, 1977), if the regenerate is being formed, the precocious and accelerated differentiation prevents it from reaching normal size. A low concentration of hormone appears to be necessary in order to maintain the regenerative ability. Nevertheless, it must be emphasized that no experiment using exogenous hormone has allowed the capacity of regeneration to be restored in animals as imagoes that have already lost it. Still unclear is whether a low hormonal concentration provides suitable conditions for development of the cells or has a direct effect on regeneration.

In experiments using embryos, regeneration normally takes place even when appendages are separated from the presumed hormonal sources, namely the prothoracic glands. As the epidermal cells of these appendages secrete the cuticle at the end of the cycle and as variations in the ecdysteroid titer are shown to occur, the cultivated portions of the animal would secrete hormones.

In order to know if a minimal hormone level is required for regeneration, it would be necessary to have at one's disposal nymphs, larvae, or embryos unable to synthesize such steroid hormones. Taking into account that most cells are able to do this, this situation is difficult to achieve. Special conditions of rearing allow greatly lengthened intermolt cycles to be realized, up to 8 months for a cycle that normally lasts about 30 days. In such a case, amputated animals regenerate very slowly and their ecdysteroid titer is extremely low (D. Bullière, unpublished data).

In crustaceans, reports of experiments using exogenous hormones are numerous in the literature. As early as 1970, Krishnakumaran and Schneiderman showed that in *Procambarus* an injection of 20 $\mu g \cdot g^{-1}$ of 20-hydroxyecdysone induced a rapid molt (7–10 days later) without the appendages previously autotomized regenerating. Injecting a lower dose, 2–3 $\mu g \cdot g^{-1}$, allowed a small regenerate to develop.

In *Uca pugilator* (Rao et al., 1972; Rao, 1978), 1.5-g crabs were induced to autotomize their third pereiopod and 11 days later were injected twice with 50 μg of 20-hydroxyecdysone or inokosterone. Regeneration was inhibited in crabs that had not begun to regenerate and proecdysial growth was stimulated in the others, but this growth was rapidly interrupted by preparation for exuviation. In all cases, the regenerates were smaller than those of control animals.

Bazin (1977a,b) showed in *Carcinus maenas* that 10 $\mu g \cdot g^{-1}$ of 20-hydroxyecdysone triggers early exuviation and more or less inhibits regeneration. Similar results were obtained in other groups of crustaceans. Thus, regeneration was inhibited in *Porcellio* (Noulin and Maissiat, 1974; Maissiat, 1978; Noulin, 1984) by injecting 0.1 $\mu g \cdot mg^{-1}$ into animals in period C, autotomized before the injection. If the injection was performed late with regard to the amputation (more than 7 days), it had no effect. It was observed that injecting a low dose into Y-ectomized animals restored their ability to form a regenerate. Hoarau (1982a,b) obtained little effect on the regeneration of a pereiopd by injecting 100 μg of ecdysone to stage C *Helleria*.

Finally, it should be borne in mind that the exogenous hormones tested—20-hydroxyecdysone (or crustecdysone), ecdysone, and inokosterone—have similar effects *in vivo* as *in vitro* in both crustaceans and insects. These hormones at low concentration appear to be necessary for regeneration, probably by promoting and stimulating cell metabolism. In their absence, development of the intermolt cycles appears to be blocked. Thus, they have a positive effect. In contrast, at high concentrations, the hormones induce molt preparation and inhibit regeneration. In crustaceans, as in insects and doubtless in the other groups of arthropods (arachnids and merostoms) (Krishnakumaran and Schneiderman, 1970), exogenous hormones, by changing the hormonal environment, act on the molting cycle and consequently on regeneration.

15.6. Summary

What can be concluded as to the relation linking morphogenetic hormones and regeneration in arthropods? First of all, it is clear from all the work reviewed herein that regeneration is widely distributed in arthropods and cannot be dissociated from the molting cycle. From observations performed in various groups, it appears that the formation of an appendage during regeneration corresponds to the recovery of embryonic processes. The cells reorganize themselves, proliferate, and are distributed in such a manner that they form a quite normal appendage. Regeneration mechanisms thus imply complex cellular interactions necessary for the form to be realized, as during embryonic development. However, while all these processes occur, the animals must continue living. So all the regenerative processes must be compatible with their normal cycle and never constitute a danger to their survival. Consequently, some synchronism between regeneration and the molting cycle is fundamental so that harmful configurations cannot occur, and it is when this synchronism is established that what can be called morphogenetic hormones intervene.

If one considers that the capability to molt is *juvenile,* it is clear that the JHs, discovered and well studied in insects and recently found in crustaceans, by controlling the ability to molt, control regeneration and determine all morphogenesis linked to it. Nevertheless, how these hormones are implicated in the processes they control is still unclear and debated. MHs are found in most arthropod groups, and in numerous cases their fluctuations during the molting cycles are known. Moreover, their role in regeneration needs to be better established. Two effects must be considered: one positive and one negative.

First of all, regeneration implies differentiation of the amputated appendage, and this depends on a minimal ecdysone level. In the absence of this minimal level, the animal and its appendage stay as they are—i.e., unchanged. Regenerate growth appears to require, at least in crustaceans, a higher level of hormone than before. This higher level, while stimulating the growth of the regenerate blastema, might also *inhibit* any formation of a new blastema thereafter. Thus, any amputation performed after this time could only lead to cicatrization of the amputation surface. Finally, a still higher titer would trigger cuticle secretion in preparation for exuviation.

References

Adiyodi, R. G. 1972. Wound healing and regeneration in the crab *Paratelphusa hydrodromous.* Int. Rev. Cytol. 32: 257–289.

Bart, A. 1969. Recherches expérimentales sur le déclenchement et le développement de morphogenèses de type régénératrices chez un Insecte: *Carausius morosus.* Ph.D. thesis (Sci. Nat.) University of Lille, France.

Bazin, F. 1973. Formation de blastèmes de régénération des péréiopodes autotomisés chez des crabes *Carcinus mænas* privés de leurs glandes de mue. C. R. Acad. Sci. Paris 276D: 2585–2588.

Bazin, F. 1977a. Effets d'injections d'ecdystérone sur la mue et la régénération chez le Crabe *Carcinus mænas* (L.). C. R. Acad. Sci. Paris 284D: 765–768.

Bazin, F. 1977b. Action inhibitrice de l'ecdystérone sur la régénération chez le Crabe *Carcinus mænas* (L.). C. R. Acad. Sci. Paris 284D: 1211–1214.

Bennett, D. B. 1973. The effect of limb-loss and regeneration on the growth of the edible crab, *Cancer pagurus* L. J. Exp. Mar. Biol. Ecol. 13: 45–63.

Bittner, G. D. and R. Kopanda. 1973. Factors influencing molting in the crayfish *Procambarus clarkii*. J. Exp. Zool. 186: 7–16.

Bliss, D. E. 1956. Neurosecretion and the control of growth in a decapod crustacean. Pp. 56–75 *in* K. G. Wingstrand (ed.), *Bertil Hanström: Zoological Papers in Honour of His 65th Birthday*. Zoological Institute, Lund, Sweden.

Bliss, D. E. 1959. Factors controlling regeneration of legs and molting in land crabs. Pp. 131–144 *in* F. L. Campbell (ed.), *Physiology of Insect Development*. University of Chicago Press, Chicago.

Bodenstein, D. 1953. Studies on the humoral mechanisms in growth and metamorphosis of the Cockroach *Periplaneta americana*. I. Transplantations of integumental structures and experimental parabiosis. J. Exp. Zool. 87: 31–53.

Bodenstein, D. 1955. Contributions to the problem of regeneration in Insect. J. Exp. Zool. 129: 209–224.

Bohn, H. 1970. Interkalare regeneration und segmentale Gradienten bei den Extremitäten von *Leucophaea* larven (Blattaria). I. Femur und Tibia. Wilhelm Roux' Arch. Entwicklungsmech. Org. 165: 303–341.

Bonnet, P. 1930. La mue, l'autotomie et la régénération chez les Araignées. Bull. Soc. Hist. Nat. Toulouse 59: 613–939.

Bordage, E. 1905. Recherches anatomiques et biologiques sur l'autotomie et la régénération de divers Arthropodes. Bull. Biol. Fr. Belg. 39: 307–454.

Bullière, D. 1967. Etude de la régénération chez un Insecte Blattoptéroïde *Blabera craniifer* Burm. (Dictyoptère). I. Influence du niveau de la section sur la régénération de la patte métathoracique. Bull. Soc. Zool. Fr. 92: 523–536.

Bullière, D. 1968a. Etude de la régénération chez un Insecte Blattoptéroïde *Blabera craniifer* Burm. (Dictyoptère). II. Influence du moment de l'amputation dans l'intermue sur la régénération de la patte métathoracique. Bull. Soc. Zool. Fr. 93: 69–82.

Bullière, D. 1968b. Etude de la régénération chez un Insecte Blattoptéroïde, *Blabera craniifer* Burm. (Dictyoptère). III. Influence de la régénération d'une patte métathoracique sur la durée de l'intermue et le nombre de stades larvaires. Bull. Soc. Zool. Fr. 93: 251–257.

Bullière, D. 1970. Sur le déterminisme de la qualité régionale des régénérats d'appendices chez la Blatte, *Blabera craniifer*. J. Embryol. Exp. Morphol. 23: 323–335.

Bullière, D. 1971. Utilisation de la régénération intercalaire pour l'étude de la détermination cellulaire au cours de la morphogenèse chez les Insectes. Dev. Biol. 25: 672–709.

Bullière, D. 1972a. Etude de la régénération d'appendice chez un insecte: stades de la formation des régénérats et rapports avec le cycle de mue. Ann. Embryol. Morphog. 5: 61–74.

Bullière, D. 1972b. Action de l'ecdysone et de l'inokostérone sur la régénération d'appendices chez la larve de *Blabera craniifer* (Insecte Dictyoptère). C. R. Acad. Sci. Paris 274D: 1349–1352.

Bullière, D. and F. Bullière. 1974. Recherche des sites de synthèse de l'hormone de mue chez les embryons de *Blabera craniifer* Burm., Insecte Dictyoptère. C. R. Acad. Sci. Paris 278D: 377–380.

Bullière, D. and F. Bullière. 1985. Regeneration. Pp. 371–424 *in* G. A. Kerkut and L. I. Gilbert (eds.), *Comprehensive Insect Physiology, Biochemistry and Pharmacology*. Pergamon Press, Oxford and Elmsford, New York.

Bullière, D., F. Bullière, and M. De Reggi. 1979 Ecdysteroid titres during ovarian and embryonic development in *Blaberus craniifer*. Wilhelm Roux's Arch. Dev. Biol. 186: 103–114.

Bullière, D., F. Bullière, and P. Sengel. 1969. Régénération du tarse chez l'embryon de *Blabera craniifer* (Insecte Dictyoptère) en culture *in vitro*. C. R. Acad. Sci. Paris 269D: 355–357.

Bullière, F. and D. Bullière. 1971. Régénération, différenciation et ecdysones chez l'embryon de *Blabera craniifer* (Insecte Dictyoptère) en culture *in vitro*. C. R. Acad. Sci. Paris 273D: 955–958.

Bullière, F. and D. Bullière. 1973. Action de l'hormone de mue sur la différenciation des cellules épidermiques d'un appendice d'embryon de Blatte en culture *in vitro*. C. R. Acad. Sci. 276D: 2533–2536.

Bullière, F. and D. Bullière. 1977. Régénération, différenciation et hormones de mues chez l'embryon de Blatte en culture *in vitro*. Wilhelm Roux's Arch. Dev. Biol. 182: 255–275.

Carlisle, D. B. and P. F. R. Dorn. 1953. Studies on *Lysmata seticaudata* Risso (Crustacea Decapoda). II. Experimental evidence for a growth and moult accelerating factor obtainable from eyestalks. Publ. Stn. Zool. Napoli 24: 69–83.

Charmantier-Daurès, M. 1980. La mue et la régénération chez *Pachygrapsus marmoratus* (Fabricus, 1787) (Crustacé, Décapode, Brachyoure): interactions, contrôles endocrine et neuroendocrine. Ph.D. thesis (Sci.), University of Montpellier, France.

Charniaux-Cotton, H. 1957. Croissance, régénération et déterminisme endocrinien des caractères sexuels d'*Orchestia gamarella* (Pallas) Crustacé Amphipode. Ann. Sci. Nat. Zool. Biol. Anim. 19: 411–559.

Charniaux-Cotton, H. 1967. Régénération des appendices présentant un dimorphisme sexuel chez les crustacés supérieurs. Bull. Soc. Zool. Fr. 92: 361–372.

Demeusy, N. 1967. Glande de mue et blastème de régénération. Gen. Comp. Endocrinol. 9: 443–444.

Demeusy, N. 1971. Influence de la régénération sur la première exuviation postopératoire chez *Carcinus mænas* (Crustacé, Décapode, Brachyoure). C. R. Acad. Sci. Paris 273D: 1140–1143.

Drach, P. 1939. Mue et cycle d'intermue chez les Crustacés Décapodes. Ann. Inst. Océanogr. 19: 103–391.

Echalier, G. 1956. Influence de l'organe Y sur la régénération des pattes chez *Carcinus mænas* (L.). C. R. Acad. Sci. Paris 242: 2179–2180.

Echalier, G. 1959. L'organe Y et le déterminisme de la croissance et de la mue chez *Carcinus mænas* L., Crustacé Décapode. Ann. Sci. Nat. Zool. Biol. Anim. 1: 1–59.

Emmel, V. E. 1907. Relations between regeneration, the degree of injury and moulting in young lobster. Science (Wash., D.C.) 25: 785.

Fingerman, M. and S. W. Fingerman. 1974. The effects of limb removal on the rates of ecdysis of eyed and eyestalkless fiddler crabs, *Uca pugilator*. Zool. Jahrb. Abt. Allg. Zool. Physiol. Tiere 78: 301–309.

Fournier, B. 1969. Expériences d'amputations faites sur les embryons du Phasme *Carausius morosus* Br., en vue d'apprécier l'aptitude de la patte embryonnaire à reconstituer ses parties manquantes. C. R. Acad. Sci. Paris 269D: 2401–2404.

French, V. and D. Bullière. 1975. Nouvelles données sur le déterminisme de la position des cellules épidermiques sur un appendice de Blatte. C. R. Acad. Sci. Paris 280D: 53–56.

Gomez, R. 1964. Autotomy and regeneration in the crab *Parathelphusa hydrodromous*. J. Anim. Morphol. Physiol. 11: 97–104.

Hoarau, F. 1973. Comportement de l'hypoderme et progression de la différenciation au cours de la régénération d'un péréiopode chez l'Isopode terrestre *Helleria brevicornis* Ebner. Ann. Embryol. Morphol. 6: 125–135.

Hoarau, F. 1982a. Incidence des injections d'ecdystérone sur la durée du cycle de mue et la régénération du péréiopode chez *Helleria brevicornis* Ebner (Isopode terrestre). Biol. Ecol. Mediterr. 9: 29–40.

Hoarau, F. 1982b. Influence des injections d'ecdystérone sur la régénération du péréiopode chez *Helleria brevicornis* Ebner (Isopode terrestre). C. R. Acad. Sci. Paris Sér. III Sci. Vie 294: 403–406.

Hoarau, F. and M. Hirn. 1978. Evolution du taux des ecdystéroides au cours du cycle de mue chez *Helleria brevicornis* Ebner (Isopode terrestre). C. R. Acad. Sci. Paris 286D: 1443–1446.

Hoarau, F. and M. Hirn. 1981. Effects of amputation and subsequent regeneration of a leg on the duration of the intermoult period and the level of circulating ecdysteroids in *Helleria brevicornis* Ebner (ground isopod). Gen. Comp. Endocrinol. 43: 96–104.

Hoarau, F. and G. Vernet. 1979. Influence du moment de l'amputation sur la durée d'intermue et la régénération chez *Helleria brevicornis* (Isopode terrestre). C. R. Acad. Sci. Paris 289D: 347–350.

Hodge, M. H. 1958. Some aspects of the regeneration of walking legs in the land crab, *Gecarcinus lateralis.* Ph.D. dissertation, Radcliffe College, Cambridge, Massachusetts.

Holland, C. A. and D. M. Skinner. 1976. Interactions between molting and regeneration in the land crab. Biol. Bull. (Woods Hole) 150: 222–240.

Juberthie-Jupeau, L. 1964. Recherche du stimulus déclenchant la mue lors de l'ablation des antennes chez les Symphyles (Myriapodes). C. R. Acad. Sci. Paris 259: 658–659.

Krishnakumaran, A. and H. A. Schneiderman. 1970. Control of molting in mandibulate and chelicerate arthropods by ecdysones. Biol. Bull. (Woods Hole) 139: 520–538.

Laufer, H., D. Borst, F. C. Baker, C. Carrasco, M. Sinkus, C. C. Reuter, L. W. Tsai, and D. A. Schooley. 1987. Identification of a juvenile hormone-like compound in a crustacean. Science (Wash., D.C.) 235: 202–205.

Laugier, N. 1984. Régénération et cycle de mue chez le crabe *Acanthonyx lunulatus* (Risso) Crustacea decapoda oxyrhyncha: contribution à l'étude de leurs relations réciproques. Ph.D. thesis, University of Aix–Marseille, France.

Lawrence, P. 1966. The hormonal control of the development of hairs and bristles in the milkweed bug, *Oncopeltus fasciatus,* Dall. J. Exp. Biol. 44: 507–522.

Lenoir-Rousseaux, J. J. 1978. Perte de la capacité de restauration des ébauches imaginales de la patte de la prénymphe de *Tenebrio molitor:* action de l'ecdystérone. C. R. Acad. Sci. Paris 287D: 547–550.

McCarthy, J. F. and D. M. Skinner. 1977a. Proecdysial changes in serum ecdysone titers, gastrolith formation, and limb regeneration following molt induction by limb autotomy and/or eyestalk removal in the land crab *Gecarcinus lateralis.* Gen. Comp. Endocrinol. 33: 278–293.

McCarthy, J. F. and D. M. Skinner. 1977b. Interruption of proecdysis by autotomy of partially regenerated limbs in the land crab *Gecarcinus lateralis.* Dev. Biol. 61: 299–311.

Madhavan, K. and H. A. Schneiderman. 1969. Hormonal control of imaginal disc regeneration in *Galleria melonella* (Lepidoptera). Biol. Bull. (Woods Hole) 137: 321–331.

Maissiat, J. 1978. Contribution à l'étude de la mue et du rôle de l'hormone de mue dans divers processus physiologiques: vitellogenèse, régénération chez les Crustacés isopodes. Ph.D. thesis (Sci. Nat.), University of Poitiers, France.

Maleville, A. 1978. Recherches sur la glande prothoracique d'*Acheta domestica* L., en relation avec la mue dans les conditions normales et expérimentales. Ph.D. thesis (Sci. Nat.), University of Montpellier, France.

Maleville, A. and M. De Reggi. 1981. Influence of leg regeneration on ecdysteroid titres in *Acheta* larvae. J. Insect Physiol. 27: 35–40.

Mittenthal, J. E., J. Laforge, S. Hutto, and W. W. Trevarrow. 1985. Regeneration in the tail fan of crayfish. Wilhelm Roux's Arch. Dev. Biol. 194: 121–130.

Needham, A. E. 1965. Regeneration in arthropods and its endocrine control. Pp. 283–323 *in* V. Kiortsis and H. A. L. Trampusch (eds.), *Regeneration in Animals and Related Problems.* Elsevier/North-Holland Publ., Amsterdam and New York.

Noulin, G. 1984. Morphogénèse régénératrice chez le Crustacé Isopode *Porcellio dilatatus:*

histologie et cytologie, interactions avec la mue, formations surnuméraires. Ph.D. thesis (Sci. Nat.), University of Poitiers, France.

Noulin, G. and J. Maissiat. 1974. Etude du rôle de l'organe Y et de l'effet de l'ecdystérone dans la régénération d'un appendice chez l'Oniscoïde *Porcellio Dilatatus*. J. Insect Physiol. 20: 1963–1974.

O'Farrel, A. F. and A. Stock. 1954. Regeneration and the moulting cycle in *Blattella germanica* L. III. Successive regeneration of both metathoracic legs. Aust. J. Biol. Sci. 7: 525–536.

O'Farrel, A. F. and A. Stock. 1964. Some effects of the farnesyl methyl ether on regeneration and metamorphosis in the cockroach *Blatella germanica* (L). Life Sci. 3: 491–497.

Passano, L. M. and S. Jyssum. 1963. The role of the Y-organ in crab proecdysis and limb regeneration. Comp. Biochem. Physiol. 9: 195–213.

Piepho, H. 1939. Raupenhäutungen bereits verpuppter Hautstücke bei der Wachsmotte *Galleria mellonella*. Biol. Zentralbl. 58: 356–366.

Piepho, H. 1963. Wirkung des Juvenilhormons auf den Mitteldarm der Wachsmotte. J. Insect Physiol. 9: 713–719.

Pflugfelder, O. 1939. Beeinflussung von Regenerationsvorgängen bei *Dixippus morosus* Br. durch Exstirpation and Transplantation der *Corpora allata*. Z. Wiss. Zool. 152: 159–184.

Pradeille-Rouquette, M. 1974. Etude des différents facteurs intervenant dans la croissance ovarienne chez *Pachygrapsus marmoratus* (Fabricius). Ph.D. thesis (Sci. Nat.), University of Montpellier, France.

Randall, J. B. 1981. Regeneration and autotomy exhibited by the black widow spider, *Latrodectus variolus* Walckenaer: the legs. Wilhelm Roux's Arch. Dev. Biol. 190: 230–232.

Rao, K. R. 1978. Effects of ecdysterone, inokosterone and eyestalk ablation on limb regeneration in the fiddler crab, *Uca pugilator*. J. Exp. Zool. 203: 257–269.

Rao, K. R., M. Fingerman, and C. Hays. 1972. Comparison of the abilities of α-ecdysone and 20-hydroxyecdysone to induce precocious proecdysis and ecdysis in the fiddler crab, *Uca pugilator*. Z. Vgl. Physiol. 76: 270–284.

Réaumur, A. 1710. Observations sur les reproductions. Mém. Acad. Sci. Paris p. 466.

Roberts, B., S. L. Wentworth, and M. Kotzman. 1983. The levels of ecdysteroids in uninjured and leg-autotomized nymphs of *Blatella germanica* (L.). J. Insect Physiol. 29: 1849–1857.

Rockett, C. L. and J. P. Woodring. 1972. Comparative studies of acarine limb regeneration, apolysis and ecdysis. J. Insect Physiol. 18: 2319–2336.

Sahli, F. 1967. Régénération et péréiopodomorphose: expériences portant sur la première paire de pattes des mâles de *Tachypodoiulus albipes* (C. Koch), Diplopoda julida. Bull. Soc. Zool. Fr. 92: 385–401.

Schneiderman, H. A. 1967. Control systems in developing insects. Pp. 5–10 *in* R. T. Smith (ed.), *Ontogeny of Immunity*. University of Florida Press, Gainesville, Florida.

Sehnal, F. 1971. Rôle de l'hormone de mue et de l'hormone juvénile dans le contrôle de la métamorphose de *Galleria mellonella* (Lepidoptera). Arch. Zool. Exp. Gén. 112: 565–577.

Sehnal, F. 1972. Action of ecdysone on ligated larvae of *Galleria mellonella* L. (Lepidoptera): induction of development. Acta Entomol. Bohemoslov. 69: 143–155.

Skinner, D. M. 1962. The structure and metabolism of a crustacean integumentary tissue during a molt cycle. Biol. Bull. (Woods Hole) 123: 623–647.

Skinner, D. M. 1985. Molting and regeneration. Pp. 44–146 *in* D. E. Bliss and L. H. Mantel (eds.), *The Biology of Crustacea*, Vol. 9. Academic Press, Orlando, Florida.

Skinner, D. M. and D. E. Graham. 1972. Loss of limbs as a stimulus to ecdysis in *Brachyura* (true crabs). Biol. Bull. (Woods Hole) 143: 222–233.

Slama, K. 1975. Some old concepts and new findings on hormonal control of insect morphogenesis. J. Insect Physiol. 21: 921–955.

Stevenson, J. R. and Henry, B. A. 1971. Correlation between the molt cycle and limb regeneration in the crayfish *Orconectes obscurus* (Hagen) (Decapoda Astacidea). Crustaceana (Leiden) 20: 301–307.

Stoffel, L. A. and J. H. Hubschman. 1974. Limb loss and the molt cycle in the freshwater shrimp, *Palæmonetes kadiakensis*. Biol. Bull. (Woods Hole) 147: 203–212.

Tchernigovtzeff, C. 1965. Multiplication cellulaire et régénération au cours du cycle d'intermue des Crustacés Décapodes. Arch. Zool. Exp. Gén. 106: 377–497.

Tchernigovtzeff, C. 1972. Régénération et cycle d'intermue chez le crabe *Gecarcinus lateralis*. I. Etude de la relation entre la croissance préexuviale des bourgeons de pattes et les étapes de la morphogenèse des soies dans les épipodites branchiaux des maxillipèdes. Arch. Zool. Exp. Gén. 113: 197–213.

Tchernigovtzeff, C. 1974. Régénération et cycle d'intermue chez le crabe *Gecarcinus lateralis*. II. Situation du moment critique: incidence d'une régénération tardive sur le cours de la prémue. Arch. Zool. Exp. Gén. 115: 423–440.

Vachon, M. 1957. La régénération appendiculaire chez les Scorpions (Arachnides). C. R. Acad. Sci. Paris 244: 2556–2559.

Vachon, M. 1967. Nouvelles remarques sur la régénération des pattes chez l'araignée: *Cœlotes terrestris* Wid. (Agelenidae). Bull. Soc. Zool. Fr. 92: 417–428.

Wigglesworth, V. B. 1940. The determination of characters at metamorphosis in *Rhodnius prolixus* (Hemiptera). J. Exp. Zool. 17: 201–222.

Wigglesworth, V. B. 1966. Hormonal regulation and differentiation in insects. Pp. 180–209 *in* W. Beerman (ed.), *Cell Differentiation and Morphogenesis*. Elsevier/North-Holland Publ., Amsterdam and New York.

Zlotkin, E. and H. Z. Levinson. 1968. Influence of cecropia oil on the epidermis of *Tenebrio molitor*. J. Insect Physiol. 14: 1719–1723.

Zuelzer, M. 1907. Über dein Einfluss der Regeneration auf die Wachstumsgeschwindigkeit von *Asellus aquaticus* L. Wilhelm Roux' Arch. Entwicklungsmech. Org. 25: 361–397.

The Oenocytes of Insects: Differentiation, Changes During Molting, and Their Possible Involvement in the Secretion of Molting Hormone

16

FRANZ ROMER

16.1. Introduction

Oenocytes of insects are characterized by their volume and their regular eosinophil cytoplasm, which contains only a few inclusions. The increased volume of these cells is achieved by polyploidization. The position of oenocytes varies in different orders of insects. However, their position is similar within individual species of the same order. Oenocytes occur in all pterygote insects; they are thought to be absent in Thysanura, and no information about their presence or absence is available in Diplura, Protura, and Collembola.

Oenocytes are believed to have a number of functions. In earlier works (Weber, 1954), they were assigned regulative functions in maintenance of the physicochemical balance, based on their capability to quickly absorb and accumulate pigments injected into the blood. Several authors think that oenocytes are capable of detoxification, because in parasitized insects they respond quickly by hypertrophy (Boese, 1936); they are supposed to participate in detoxification in a way similar to that of the liver of vertebrates. In cockroaches, bugs, and beetles, activation of the oenocytes shortly before formation of the epicuticle has been observed (Wigglesworth, 1933, 1947, 1948). The oenocytes showed a staining behavior similar to that of the epicuticle. Another hypothesis by Koller (1929), Zavrel (1935), and Yokoyama (1936) assigns endocrine functions to these cells because they are the first to be activated in the molting cycle. Locke (1969) and Bonner-Weir (1970) thought that oenocytes were responsible for the production of steroid hormones, especially ecdysteroids, because they showed ultrastructural features rather like those of certain vertebrate cells.

16.2. Morphology

16.2.1. *Topographic Distribution of Oenocytes Within Various Insect Taxa*

In a large number of hemimetabolous insects, oenocytes are found in the epidermis: Ephemeroptera (Snodgrass, 1935), Libelluloidea, Plecoptera, Blattoidea (Kramer and Wigglesworth, 1950), and Hemipteroidea (Wigglesworth, 1933). In some others, they lie dissociated from the epidermis among the lobes of the fat body near the nervous system: Saltatoria (Philogène and McFarlane, 1967; Cassier and Fain-Maurel, 1969), Homoptera (Mackauer, 1965), and Thysanoptera (Risler and Kempter, 1961).

Among holometabolous insects, the oenocytes lie near the spiracles or alongside the tracheal trunks in Coleoptera (Romer, 1975). In the Hymenoptera, they are, with few exceptions, distributed over the fat body and show close contact with its cells (Schmidt, 1961; Boehm, 1965).

In the Neuropteroidea, they are predominantly grouped in clusters near the stigmata: Megaloptera (Huber, 1958), Trichoptera (Gee, 1911), Lepidoptera (Locke, 1969), and Diptera (Gnatzy, 1970). There are lots of observations on oenocytes in the older literature, most of them cited in the papers of Locke (1969), Gnatzy (1970), and Romer (1975).

A survey of the different positions of oenocytes in the abdomen is shown in Fig. 16.1A–D.

16.2.2. *Comparative Ultrastructure*

Up to now, the ultrastructure of oenocytes of about 20 insect species has been examined. The most characteristic feature common to all of them is the smooth, or agranular, *endoplasmic reticulum* (ER) (Figs. 16.2, 16.3A, and 16.4A), which is probably responsible for their homogeneous appearance under light microscopy. The ER is tubular in character. The individual tubules have a diameter of 20–40 nm. Depending on the packing density of the ER, the cells appear either dark or light, both types of cells occurring simultaneously. The smooth ER shows a tendency toward formation of vesicles (e.g., in *Tenebrio*), but so far it has not been established whether this has any functional significance or whether this is due to the hypoosmolar buffer solutions. Only Gnatzy (1970) has shown on microscopic evidence that in *Culex* dilatations of the agranular ER serve to store lipids. The smooth ER of *Pemphix tortrix* (Homoptera) shows in some areas parallelly oriented tubules, appearing hexagonally arrayed in cross sections. Rough ER in these cells is found only in small quantities (Fig. 16.3) and only at certain periods of the molting cycle. It is connected directly to the smooth ER. Usually only a few stacks of membrane are formed. The amount of rough ER seems too small to allow for participation in the formation of lipoproteins.

Oenocytes are among the largest somatic cells in insects. Closely connected to the growth of the oenocytes is the formation of their surface. The intensive development of the *plasma membrane system* can be regarded as compensation for the diminishing surface area. This has also been observed in a number of species with very large oenocytes, e.g., in *Calpodes* (Locke, 1969), *Culex* (Gnatzy, 1970), *Dacus* (Evans, 1967), *Pediculus* (Romer, 1973), *Tenebrio* (Romer, 1975). In Isoptera, Homoptera, and in some Diptera, only a few plasma membrane infoldings are present, whereas in other Hemimetabola none at all. *Mitochondria* are numerous in oenocytes. Usually they are of the cristate type, but tubular ones may also be found (e.g., in the embryo of *Oncopeltus:* Dorn and Romer, 1976). *Microbodies* could be detected in varying quantities in the oenocytes of all insects examined. In *Gryllus, Blaptica, Leucophaea, Heliothrips, Gerris,* and *Calpodes,* their formation from the smooth ER could be shown. Locke (1969) reported that he failed to prove the presence of peroxidase in *Calpodes.* In *Gryllus,* however, I noticed peroxidase near the clefts, where microbodies are very numerous (Fig. 16.3C).

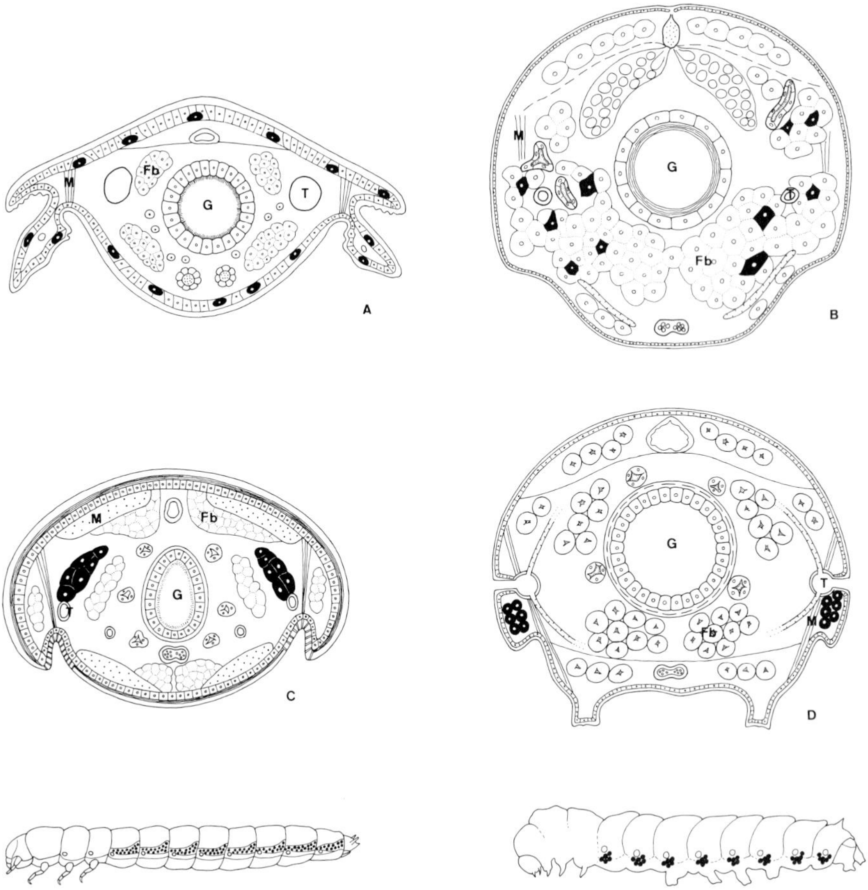

FIGURE 16.1. Position of oenocytes in various insect orders (dark cells). (A) *Carausius* (Phasmida): the oenocytes are wedged between the bases of the epidermis cells. (B) *Apis* (Hymenoptera): the oenocytes are dispersed between the cells of the fat body. (C) *Tenebrio* (Coleoptera): the cells are arranged along the upper side of the tracheal trunks. (D) *Bombyx* (Lepidoptera): the oenocytes are in clusters below the spiracles of each abdominal segment and are covered by dorsoventral muscle strains. *Key:* Fb = fat body; G = gut; M = muscles; T = tracheal trunks.

Golgi complexes can be found with varying frequency. In the larval oenocytes of *Tenebrio,* they could not be definitely proved. This may be accounted for by the dense packing of the smooth ER. In the pupa and imago of the same species, they were sometimes found. In other species, they occur with a certain regularity but without showing prominent stacks of cisternae. The scantiness of these structures clearly points to a

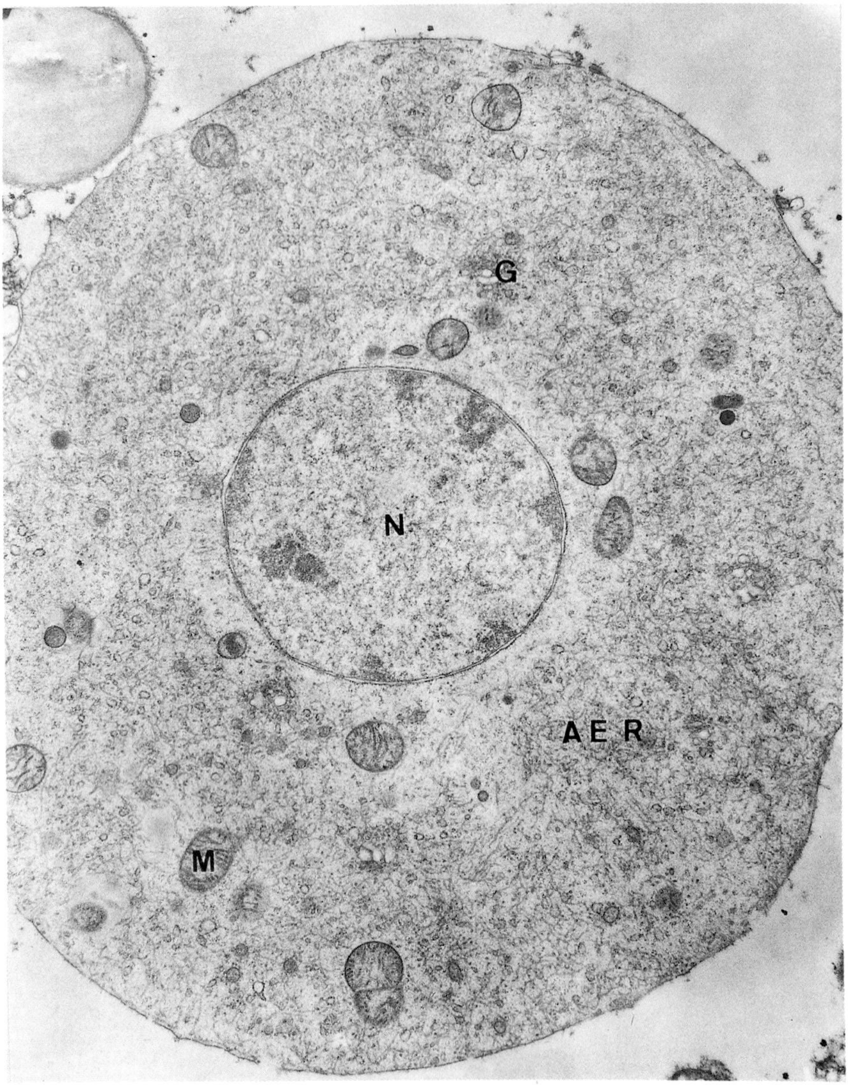

FIGURE 16.2. Survey of an oenocyte of the imaginal generation of *Oryzaephilus surinamensis* (pupa 61 h). The ER is developed only as smooth, or agranular (AER); Golgi complexes (G) are spread over the cell. *Key:* M = mitochondrium; N = nucleus. ×10,000.

FIGURE 16.3. (A) Part of an oenocyte of *Gryllus bimaculatus*, second instar, 25 h after the molt. The agranular reticulum is the predominant structure, interspersed with numerous polyribosomes (RIB). ×28,000. (B) Detail of an oenocyte with liposomes (Li) 12 h after ecdysis. ×9000. (C) Large cleft (Cl) with numerous microbodies (Mb) accumulated on both

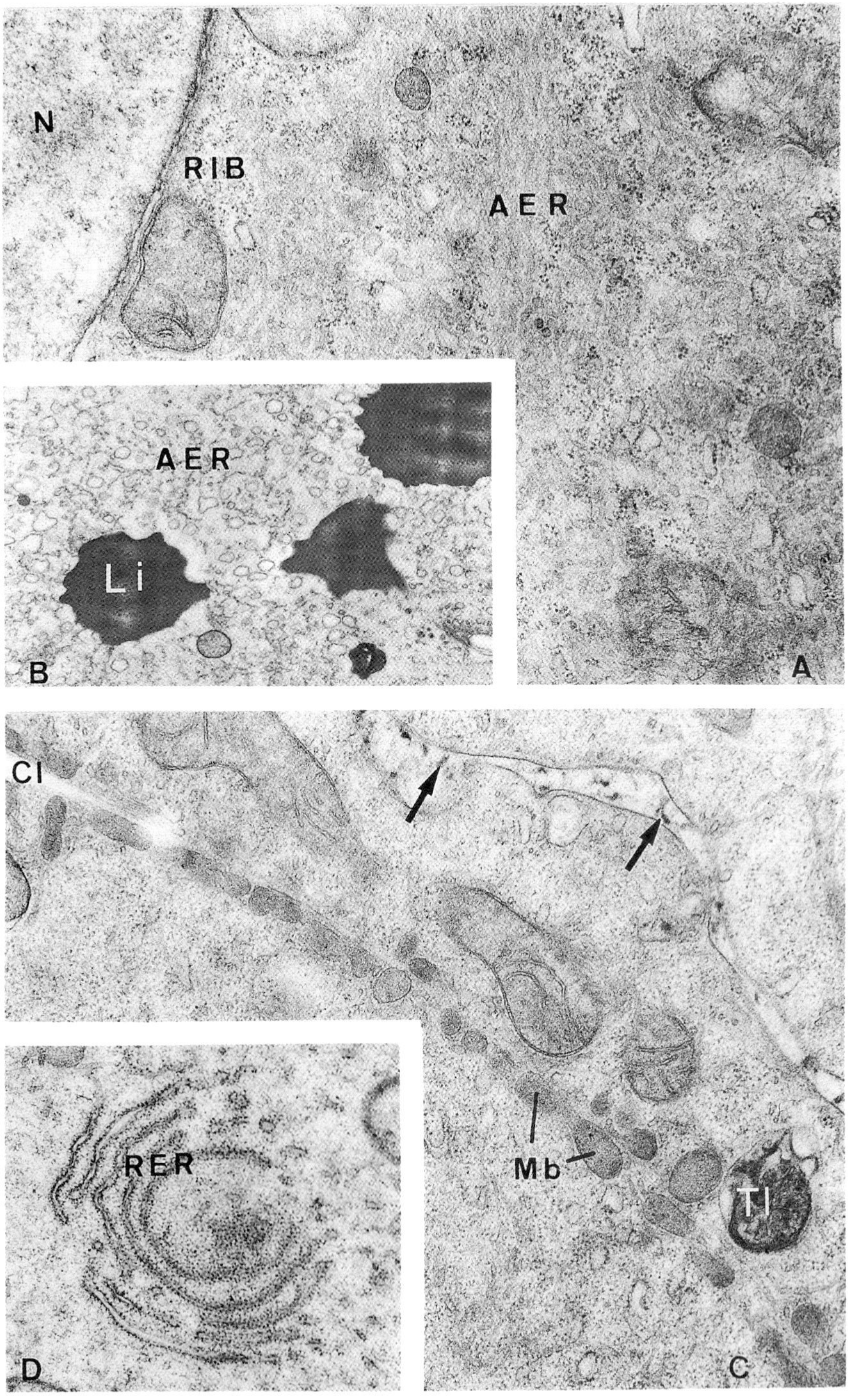

sides of the cleft. The extracellular space shows some particles originating from telolysosomes (Tl) (arrows). ×18,500. (D) Typical aspects of an oenocyte within the activation period, showing some traits of rough ER (RER). ×22,000.

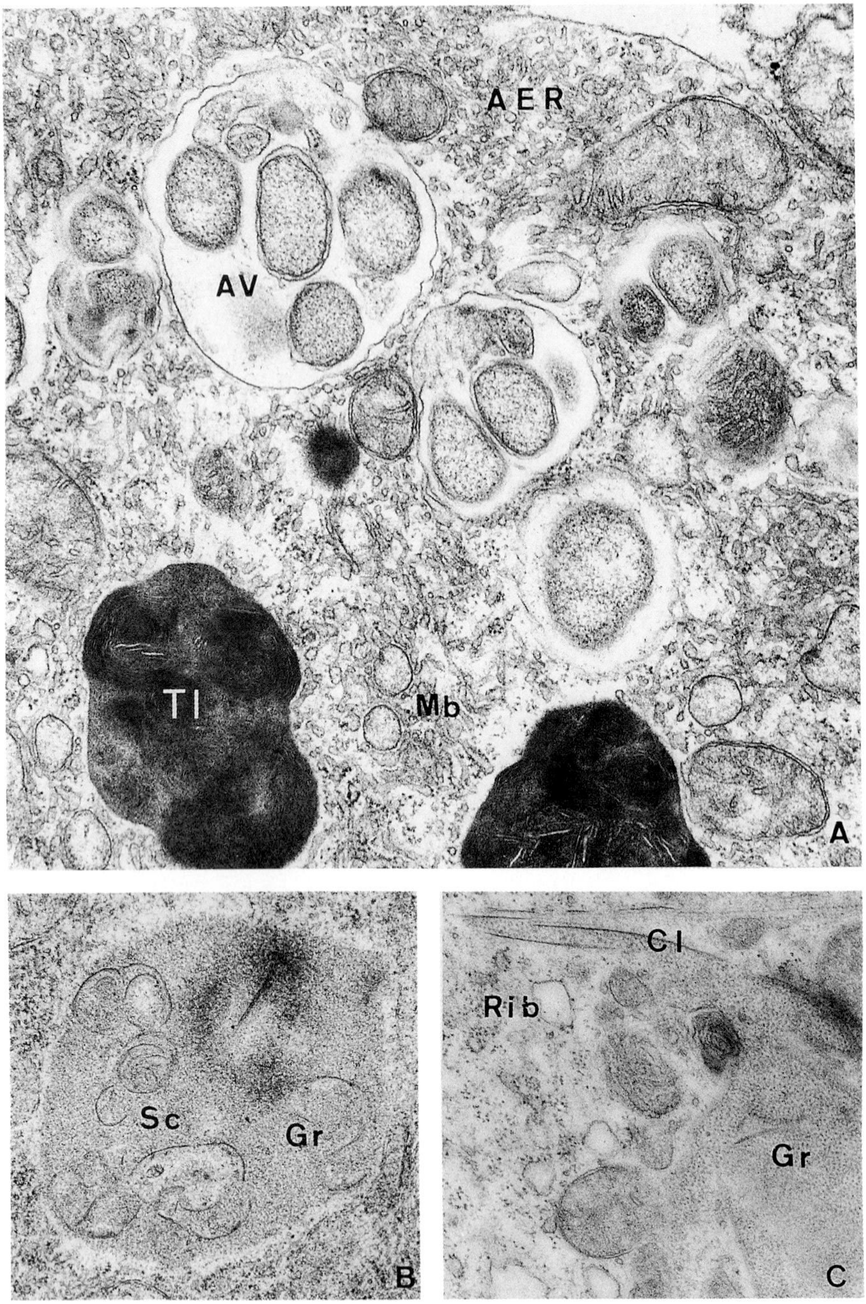

FIGURE 16.4. (A) Part of a larval oenocyte of *Oryzaephilus* 93 h after the molt. Beside agranular ER (AER), numerous microbodies (Mb) and different stages of autophagy are to be seen. The autophagic vacuoles (AV) have resorbed all kinds of organelles. *Key:* Tl = telolysosomes. ×27,000. (B) Formation of scrolls of an imaginal oenocyte of *Apis*. The scrolls

poor production and secretion of proteins. Locke (1969) succeeded in proving the formation of primary lysosomes in the Golgi complexes. Up to now, it is unknown as to where in *Tenebrio* the primary lysosomes are produced, because this species too shows a distinct formation of autophagosomes. Evidence of acid phosphatase at the ultrastructural level led to the conclusion that the lysosomal enzymes are packed directly in the ER, because vesicles similar to microbodies furnished a positive reaction to acid phosphatases (Romer, 1975). *Free ribosomes* and aggregates forming polyribosomes have been found in the oenocytes of all insect species examined. In *Tenebrio,* their frequency was relatively low; in *Gryllus* and *Schistocerca,* it was considerably higher, particularly in the region near the nucleus.

Clefts represent a special structure of their own. They have been observed in a number of insect species, but not in all [*Gryllus* (Figs. 16.3C and 16.4C), *Locusta, Schistocerca, Blaptica, Leucophaea, Apis,* and *Calpodes*]. Sometimes they permeate the whole cell (*Gryllus*) and are confined by a membrane. Clefts are formed in areas of granular inclusions, which Rinterknecht (1985) considered to be ribosomes. But when observed under higher magnification, the granules appear to be elongated (*Gryllus*) and thus can hardly be identified as ribosomes (Fig. 16.4B,C). Gnatzy (1970) found comparable structures in *Culex,* where they developed into ellipsoid and then vesicular formations, stainable with Sudan dyes. Locke (1969) considered them to be crystalline inclusions.

Glycogen deposits were observed during certain periods of the molting cycle in *Gryllus, Locusta, Schistocerca, Leucophaea, Heliothrips,* and *Apis.* Glycogen rosettes can occur individually, or they may be concentrated in whole glycogen fields. However, in other species (*Carausius, Tenebrio, Calpodes, Culex,* and *Drosophila*) glycogen was not found. A possible explanation for this would be that all species which show glycogen in the epidermis do so also in the oenocytes. *Liposomes* were found in some species [*Baetis, Gryllus* (Fig. 16.3B), *Dysdercus, Leucophaea,* and *Heliothrips*]. When the complete molting cycle was examined, it was possible to establish the occurrence of liposomes at a rather early stage of the cycle and their disappearance after the first visible activation. Whether these liposomes should be interpreted as cholesterol reservoirs has not yet been determined.

An obvious characteristic of oenocytes are the *autophagic vacuoles* or

---

emerge from areas composed of very small particles (Gr). ×22,000. (C) Formation of clefts in *Gryllus.* The same particles are engaged here as in *Apis.* They are obviously smaller than the ribosomes (Rib) in the adjacent cytoplasm. These particles are linearly arranged along the growing membranes. ×25,000.

their products, i.e., residual bodies or lipofuscin granules. These products can be found throughout the molting cycle. In *Gryllus bimaculatus*, their number during the first nymphal instar is rather low. After 45 h, their number increases considerably and reaches its maximum at the time of molt. Typical Golgi complexes with flat, prolonged cisternae are rare. Thus, the process of formation of the isolating membranes probably does not start from the disk-shaped Golgi cisternae but from the membranes of the smooth ER. Isolated organelles occur in the form of smooth ER, mitochondria, ribosomes, and—if present—microbodies. Apparently, the autophagic vacuoles eliminate most of the spent organelles at the end of each larval/nymphal instar, for at the beginning of the next stage only very few large residual bodies can be found despite the preceding massive autophagy. The temporal code for autolysis seems to vary throughout the insect species. Locke (1969) reported in *Calpodes* a massive autophagy during the last 36 h (overall duration: 192 h) of the fifth larval instar, apparently extending partly into the next larval stage. Early stages of autophagy in *Gryllus* can only be found near the end of the nymphal period; after ecdysis, most of the remains are eliminated. In *Tenebrio*, excretion of the telolysosomes occurs only 2 days after ecdysis.

But where do the telolysosomes remain? Certainly not within the cells, otherwise there should be an increase in their number with every new larval instar. Instead, after each molting, their number is reduced. Supposedly, they are drained by exocytosis into the hemolymph. Corresponding observations have been made in the pupa of *Tenebrio* and in the nymphs of *Gryllus* (Fig. 16.3C).

What happens to larval oenocytes during metamorphosis? In *Tenebrio* and *Oryzaephilus* (Fig. 16.4A), accumulation of telolysosomes in the pupal cytoplasm has been observed. In comparison with the larval period, the cells seem almost flooded with them. Until the early imago, their number is reduced considerably, so that after a while their appearance again resembles that of an oenocyte during the larval period. Apparently, many oenocytes survive the imaginal ecdysis and again become capable of operation in the imago. So far, no data about ultrastructural details during degeneration in other species are available.

16.2.3. *Ultrastructural Changes Within the Molting Cycle*

The most prominent feature of oenocytes is the smooth and predominantly tubular ER (Fig. 16.5). Comparable cells occur mainly as steroid-hormone-producing cells in vertebrates. Owing to this particular feature, the oenocytes were regarded as participating in the synthesis of steroid hormones, especially of ecdysteroids (Delachambre, 1966; Locke, 1969; Gnatzy, 1970; Romer, 1971, 1975). During the activation period, some parallels between oenocytes and cells of vertebrates occur. Nonactivated steroid-producing cells of mammals show large inclusions of glycogen and lipid vacuoles (Aoki, 1970). On the appearance of the activating hormone, these components disappear. Then, stacks of ribosome-covered ER appear, and these are finally replaced as more and more smooth ER is formed.

According to some authors (Baillie, 1964; Moses et al., 1969), the lipid droplets store large amounts of cholesterol. On the other hand, the absence of gonadotropins leads to the concentration of liposomes. Most probably, these lipids are formed by disintegration of membrane material and serve as depots. Only in the oenocytes of a few insect species are the same structures found as in the aforementioned cells of vertebrates. Rosettes of glycogen can be found alone or in clusters even 36 h after the molt. The membrane stacks with rough ER occur simultaneously. It is followed by a period of increased formation of smooth ER, and toward the end of the nymphal period, between 60 and 72 h, the autophagic vacuoles multiply. As liposomes and rosettes of glycogen occur only in a few species, it is seldom possible to see the full sequence of the signs of activation. Usually, the only hint before the increase of smooth ER is the occurrence of individual profiles of rough ER.

16.2.4. *The Different Generations of Oenocytes*

The groups of insects included under Hemimetabola (Ephemeroptera to Hemipteroidea) form, in the course of molting, a specific number of oenocytes in addition to the existing ones (Wigglesworth, 1933). So the existence is possible of oenocytes in the imago that had been differentiated during embryogenesis and had survived the molts of all nymphal stages. After differentiation, the oenocytes usually undergo several steps of polyploidization until they reach a limit. It is different though for the group of insects under Holometabola (Coleoptera to Diptera). So far as is known, the first generation of oenocytes is differentiated during embryonic development. During the individual larval instars, their number does not vary but their size is adapted to the needs of the larval dimensions by polyploidization. This so-called larval generation is maintained until the end of the pupal period and degenerates shortly after the imaginal molt (Coleoptera excepted). During early metamorphosis (pharate pupa), a new generation of oenocytes, smaller than the larval ones, is differentiated. With respect to the Holometabola, this is called the imaginal generation.

16.2.5. *Differentiation and Mode of Growth*

Development

Mackauer (1965) described the genesis of oenocytes in *Aphis pomae* (Homoptera) and noted that some 60–72 cells are formed in the embryo

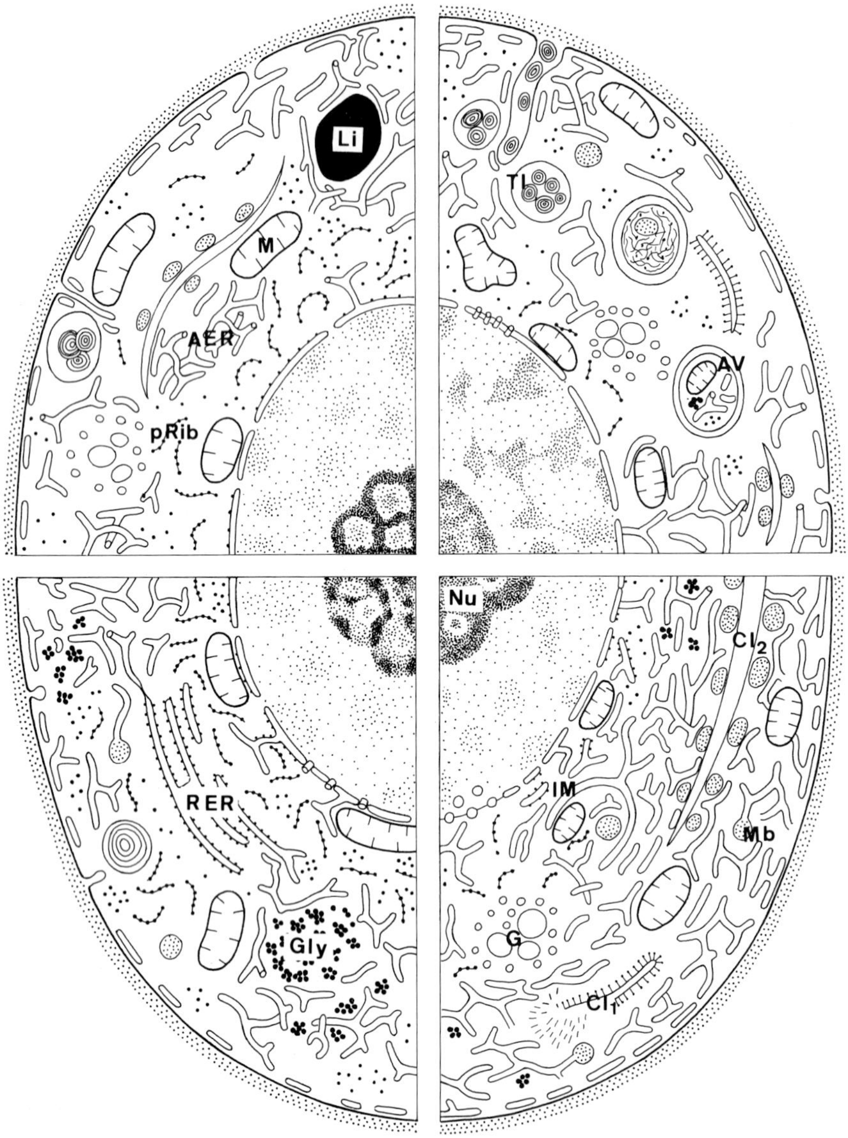

FIGURE 16.5. The four main phases of ultrastructural changes of an oenocyte within the molting cycle (arranged anticlockwise). The upper-left section shows the situation immediately after the last molt: agranular ER (AER) is poorly developed; liposomes (Li) occur at this time; numerous free ribosomes are to be found in the perinuclear space (pRib), followed by an activation of rough ER (RER) and an appearance of glycogen rosettes (Gly). In the third phase, a strong increase of the AER and an elevated formation and enlargement of microbodies is to be observed. The glycogen deposits have disappeared. The first signs of formation of autophagic vacuoles can be seen. The maximum molting

by separation from the epidermis of the abdomen immediately adjacent to the spiracles. Subsequently, they pass into the thoracic and abdominal fat body. No imaginal generation occurs. In *Oncopeltus fasciatus* (Dorn and Romer, 1976), the oenocytes differentiate gradually until hatching and then immediately become polyploid. They pass into the fat body and come to rest there individually or in small clusters. Usually, the oenocytes are a tissue restricted to the abdomen. However, it was reported (Sander, 1965) that oenocytes can also be formed in the thorax or even in the head. These reports refer to *Pyrilla perpusilla, Oncopeltus,* and *Tenebrio.*

Madhavan and Madhavan (1980) investigated the origin of imaginal oenocytes in *Drosophila.* Twenty-four hours after pupation, they found in whole mounts of insects in the so-called dorsal nest and less prominent in the ventral part a darker colored region in which oenocytes were formed. However, the authors were not able to determine whether the oenocytes were of truly ectodermal origin. Lawrence and Johnston (1982, 1986) came up with an impressive answer to this problem by producing mosaics of different mutants. The late syncytial blastoderm served as the donor, and early stages of cleavage as hosts. The cell autonomous mutant *sdh 8* (it shows a different activity for succinodehydrogenase) served as the marker gene. All cells possessing this gene would show a distinct coloration within a mosaic. Thus, the authors succeeded in proving the descent of the epidermis, oenocytes, and central nervous system (CNS), but not of muscular cells, from one original clone. Rinterknecht and Matz (1983) describe in detail the differentiation of oenocytes in *Leucophaea maderae* with respect to the ultrastructural aspects. In *Leucophaea,* oenocytes occur first around the tracheal invaginations during stage 15 after segmentation of the embryonic germ band. At that point of time, the oenocytes are located between the bases of the epidermal cells. Their ultrastructure shows a few traits of partly granular and partly smooth ER. Lipid droplets occur surrounded by tubules of the smooth ER. Golgi complexes too have been formed by this time. During stage 15 (10 days before the embryonic cuticle begins to form), a clear increase of smooth reticulum can be observed at the periphery of some cells. These cells must then be defined as oenocytes. In

---

hormone titer occurs at this time. The upper-right section shows that the last phase brings a pronounced formation of autophagic vacuoles, whose products are eliminated by infoldings of the plasma membrane system. This scheme is mainly based on the findings in *Gryllus. Key:* AV = autophagic vacuole; Cl_1 = cleft in formation; Cl_2 = cleft full-developed, with numerous microbodies; G = Golgi complex; IM = isolation membrane; M = mitochondrium; Nu = nucleolus; Tl = telolysosome.

stage 16, peroxisome-like structures appear. In stage 20, another generation of newly differentiating oenocytes occurs.

In *Leucophaea*, the oenocytes, which can be detected only after completed formation of the embryonic germ band, cannot possibly participate in formation of the embryonic membrane in stages 11–12. Before "dorsal closure" and formation of the epicuticle, a significant increase of agranular ER is observed. The function of oenocytes in this process, be it the synthesis of either epicuticle or hormones, has not yet been determined. But note that in *Schistocerca* (Lagueux et al., 1979), *Blaberus* (Bullière et al., 1979), and *Nauphoeta* (Imboden und Lanzrein, 1982), the formation of the first imaginal cuticle is accompanied by a maximum ecdysteroid titer. Dorn and Romer (1976) found that in *Oncopeltus* the oenocytes shortly after excorporation from the ectoderm (74 h after oviposition, with an overall duration of embryonic development of 123 h) bear all essential features of nymphal or imaginal cells and, unlike prothoracic glands, are already capable of operating. Thus, they are a possible source of the ecdysteroid peak occurring between 96 and 120 h. In *Gryllus*, new oenocytes are differentiated from the pleura near the stigmata in the course of each molt (Romer and Eisenbeis, 1983). Ultrastructural research (Romer, 1972) proved these cells to contain a high density of mitochondria and numerous lysosomes (as opposed to *Leucophaea*). Smooth ER can be identified initially along the infoldings of the basal cell membrane.

Mode of Growth

In Section 16.2.4, above, concerning the different generations of oenocytes, it has already been mentioned that, throughout the entire larval development in Holometabola, there is only one generation of oenocytes, which is formed embryonically and grows simply by gaining volume. In *Oryzaephilus surinamensis* (Romer, 1966), the oenocytes' DNA content (c) amounts to 4c during the first larval instar, to 16c during the second, to 32c during the third, and (in some) 64c during the fourth larval instar. Toward the end of the fourth larval instar, the imaginal generation differentiates from epidermal cells. In comparison with the larval generation it is considerably smaller. In our laboratory, data about the DNA content of *Tenebrio molitor* was obtained. Here, as is well known, the oenocytes are intimately bound up with the longitudinal tracheal trunks, on an average of 150 per segment. This number remains constant. With respect to diploid ganglion nuclei, the larval oenocytes of *Tenebrio* can reach up to 2048c, that is, 11 steps of replication. Population growth of the larval oenocytes is proportionate to the logarithm of the molting weight. Locke (1969) estimated the degree of polyploidization in the first larval instar to be 32–64n. The estimate of oenocytes for the fifth larval instar reaches some 4000n. About one or two replicative steps per larval instar would be sufficient in *Calpodes* to obtain this final result. In *Gryllus* (Romer and Eisenbeis, 1983), the population of oenocytes came to 4–32c during the first nymphal instar, a steady population of differentiated oenocytes of 16–64c was observed, undergoing just one more step of polyploidy (up to 128c) until the imago. To compensate for the growth of the insects from the third to the ninth instar to the imago, new oenocytes are excorporated from the pleural region, which then have to undergo several steps of polyploidization to finally reach the defined dimensions of functional oenocytes.

16.3. Possible Functions of Oenocytes

16.3.1. *Epicuticle Formation*

Ever since the detailed examination by Wigglesworth in *Rhodnius* (1933) and *Tenebrio* (1948), it has been maintained that oenocytes participate in formation of the cuticle. In the predatory bug *Rhodnius,* Wigglesworth succeeded in showing a clear correspondence between formation of the epicuticle (9 days after feeding) and activity of the oenocytes, as measured by their volume. Owing to their staining properties, Wigglesworth suggested that a lipoprotein constitutes the secretion of the oenocytes as well as being a component of the epicuticle. After more detailed research, he credited the oenocytes with the synthesis of cuticulin (1970). Cuticulin is described as a polyester composed of hydroxy-fatty acids containing partially unsaturated fatty acids. In a more recent publication, Wigglesworth (1985) described these substances as wax plus sclerotin. In the epicuticle, cuticulin is mixed with proteins and tanned by quinones. The same staining properties of oenocytes and epicuticle, i.e., after careful oxidative destruction by hypochlorite and subsequent staining with fat dyes (Sudan black B), led Wigglesworth to assume that cuticulin as well as the accompanying proteins (identified by ninhydrin) were produced by the oenocytes and were—during secretion of the oenocytes, visible in their reduction in volume—deposited as epicuticle and partly as impregnation of exo- and endocuticle.

In adults of *Rhodnius,* Wigglesworth detected heightened activity of the oenocytes. In females, chorionin is built into the chorion; in males, into the spermatheca. Kramer and Wigglesworth (1950) carried out a very impressive experiment with *Periplaneta americana.* They erased the epicuticle of one side with aluminum dust and succeeded in showing that the oenocytes under this area, where the aluminum dust had come into effect, were exhausted, i.e., had shrunk due to increased secretion. Such work of Wigglesworth and others is based on argument by analogy. The synchronism of certain events in oenocytes and epicuticle leads to the conclusion that both processes must be related to each other.

Only experiments employing tracer techniques achieved more definite results. In these experiments ^{14}C-labeled acetate offered to oenocytes, which in *Apis* are mixed with fat body cells, was incorporated into hydrocarbons and free wax acids. Wax esters or their components, acids and alcohols, were not labeled (Piek, 1964). Autoradiographic evidence was furnished that oenocytes incorporate acetates more rapidly than do the fat cells. Diehl (1973), by the incorporation of [^{14}C]acetate, furnished proof that in the fifth nymphal instar of *Schistocerca* fat body, which is rich in oenocytes, triglycerides and especially paraffins are labeled in addition to small amounts of phospholipids. The synthesis shows a constantly high level: no peak could be observed during the new formation of the cuticle. By gas chromatographic analysis, the paraffin fraction was identified as *n*-alcanes with a chain length of C_{21}–C_{33}. Epidermic cells, too, possess the ability to synthesize paraffins from acetate, but to a far lower degree. Parallel to the results in *Schistocerca*, Romer (1980) attempted incubation with [^{14}C]acetate in *Tenebrio*, because oenocytes of this species can be isolated and prepared with almost no fat tissue. If offered [^{14}C]acetate as precursor, isolated oenocytes of *Tenebrio* produced predominantly ^{14}C-labeled paraffins (Fig. 16.6). Bursell and Clements (1967) analyzed the cuticle lipids of *Tenebrio*, finding them to consist of 55% pentacosane-8,9-diol, 13% wax esters, 10% paraffins, 10% fatty acids, and 2% phospholipids. In addition to the paraffins in the incubates of oenocytes, small amounts of mono-, di-, and triglycerides were found. Thus, the other components of the cuticle must either be produced by other tissues or by oenocytes, but from other precursors. Unlike in *Musca* (Kon and Monroe, 1971), in *Tenebrio* amino acids do *not* serve as precursors for cuticular lipids. Apparently, the successive products of the [^{14}C]acetate injected, with respect to the cuticular lipids, were also only incorporated into paraffins. This was further supported by *in vivo* experiments.

Indirect evidence of the activity of oenocytes was furnished by Armold and Regnier (1975), who studied the incorporation of labeled acetates into paraffins in *Sarcophaga bullata*. After feeding, larvae were treated with [^{3}H]acetate for 10.5 h. The series treated with 2 μg of 20-hydroxyecdysone (20-OH-E) showed 1.3 times more incorporation of acetate into the hydrocarbon fraction than control series without any treatment. When the hormone was applied first by injection and the labeled acetate was offered 24 h later, the factor of incorporation rose by 3.3 in comparison with the control. Separation of integument and viscera in the hormone-treated insects increased the result even by a factor of 10. For these incubation experiments of Armold and Regnier, it was not possible to differentiate between the synthetic capacity of epidermal cells and that of oenocytes. With the aforementioned results with *Schistocerca* and *Tenebrio* in mind, it is reasonable to assume that in *Sar-*

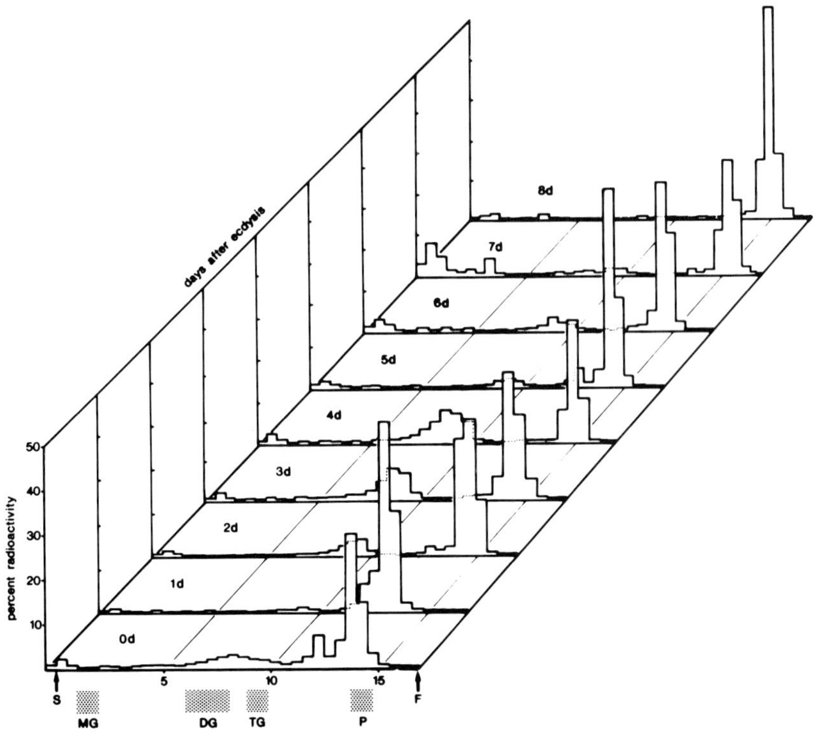

FIGURE 16.6. Incubation of isolated larval oenocytes of *Tenebrio* with [^{14}C]acetate within a molting cycle. The incubates are subjected to lipid extraction; then the extracts are separated by thin-layer chromatography (TLC), the silica gel is scraped off in 5-mm stripes, and the radioactivity determined in a liquid scintillation counter. The main synthesis products belong to the paraffin fraction (P). In addition, a small incubation rate is found in diglycerides. The results show that oenocytes synthesize paraffins within the whole molting cycle. *Ordinate*—percent radioactivity of the whole chromatogram. *Abscissa*—DG = di-, MG = mono-, TG = triglycerides; F = front; S = start region.

cophaga the paraffins are also produced by oenocytes. In *Tenebrio*, hormonal activation of the secretion of paraffins is not very probable, because the amount of paraffins produced is very steady.

16.3.2. *Secretion and Synthesis of Ecdysteroids in Various Insect Species*

Tenebrio molitor. For some authors, prothoracic glands represent the only source of molting hormones; others (e.g., Delbecque and Sláma,

1980) credit single parts of the insect with the ability to produce molting hormones. On the other hand, the oenocytes are also credited with this ability because of their ultrastructure. By incubating the glands of 15–20 larvae of *Tenebrio* with [^{14}C]cholesterol, Romer et al. (1974) succeeded in identifying 20-OH-E as predominant synthesis product of oenocytes, and ecdysone as that of prothoracic glands.

Parts of the active zones of ecdysone and 20-OH-E were recrystallized with authentic ecdysteroids and then acetylated with [^{3}H]acetic anhydride. The incubates of oenocytes produced tri- and tetraacetates of 20-OH-E; those of the prothoracic glands produced tri- and tetraacetates of ecdysone. Free ecdysone and 20-OH-E were synthesized only when unlabeled reference substances were present at the same time; otherwise conjugates of the two ecdysteroids were formed. This was demonstrated by digestion with *Helix* enzyme. So the capability of oenocytes of *Tenebrio* to form 20-OH-E under *in vitro* conditions from cholesterol was proved.

In recent work in our laboratory, J. Thierfelder (unpublished results; see Fig. 16.7) prepared secretion profiles and discovered that oenocytes are capable of passing ecdysteroids into the medium over a long period (about 7 h). To compare the rate of secretion of both sides of the body, the oenocytes of one side were fixed immediately after preparation and the usual rinsing in Ringer solution. The cells of the other side of the body were rinsed and then incubated to establish a secretion profile. The hormone level of the incubated cells (4607 pg 20-OH-E equivalents) turned out to be significantly higher than that of the cells fixed immediately (875 pg).

Gryllus bimaculatus. In *Gryllus,* the oenocytes are located mainly in the neural sinus and are interspersed with fat cells. Thus, it is impossible to obtain pure oenocyte preparates. Twenty hours of *in vitro* incubation of oenocytes of this region yielded between 2 and (at the most) 15 ng of ecdysteroid equivalents, depending on their age. Explanted prothoracic glands of this species do not exceed this level either. Only combination with parts of the thoracic fat body led to secretion of a larger amount of ecdysteroids into the medium. Low peaks of secretion occurred after 25 and 55 h, higher peaks after 95 and 120 h (Fig. 16.8). With respect to the compounds occurring in oenocytes, most of them must be classed as 20-OH-E. The largest peak of oenocytes (95 h) showed 220 ng ecdysteroid equivalents. At this time, in whole-body extracts approximately 1700 ng/g fresh weight can be traced. Since the oenocytes including fat body represent no more than 3–4% of the body weight, the peak could not have been caused simply by previously accumulated material (by which the level of ecdysteroid equivalents in the incubates could not have risen about 50 ng, presuming a uniform distribution of

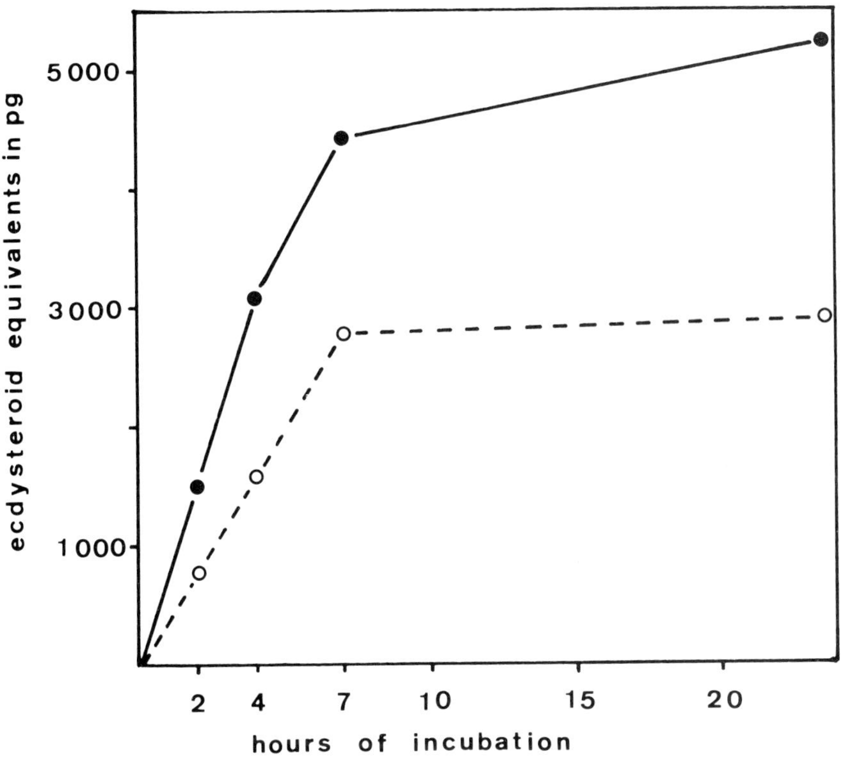

FIGURE 16.7. Secretion profile of isolated oenocytes of *Tenebrio* larvae: solid line—oenocytes of both sides; dashed line—oenocytes of one side. Immediately after preparation, the cells of one side contained 875 pg ecdysteroid equivalents. The secretion profiles were raised by replacement of one-third of the incubation fluid after 2, 4, 7, and 24 h. The values given in the graph refer to the content of the removed third.

the hormone). The synthesis of 20-OH-E was further evidenced by gas chromatographic (GC) analysis of the secretion products (Fig. 16.9).

Bombyx mori. By applying the *Calliphora* bioassay, homogenates of oenocytes of *Bombyx* were shown to contain considerable amounts of hormones (Ruhland und Romer, 1977).

In a new series of experiments, we incubated isolated oenocytes of *Bombyx in vitro* and established secretion profiles (Fig. 16.10). At the start of spinning activity it was still rather low, up to then not more than 6 ng were secreted into the medium. In correspondence with the development of the hemolymph titer, 1 day after the start of spinning the maximum of 23.5 ng was observed. Two days after the start of spinning, a

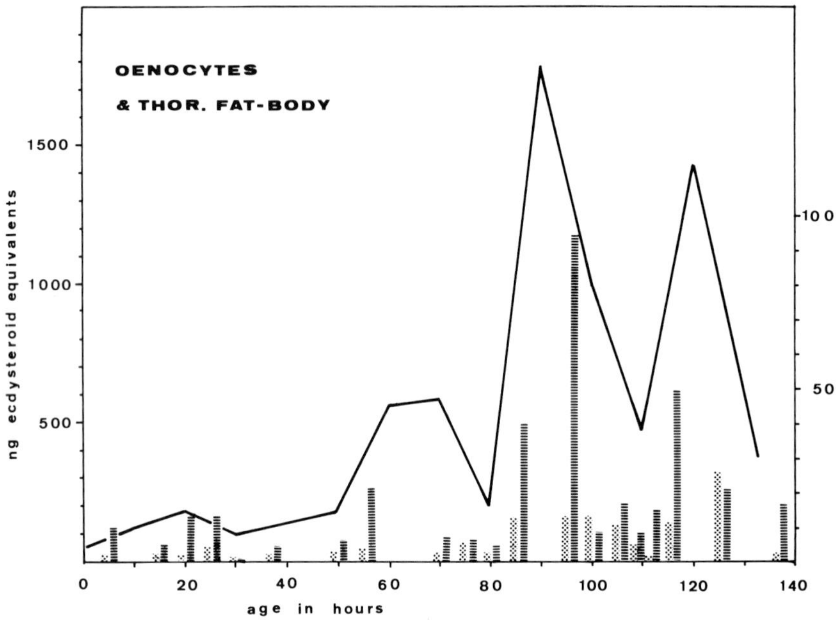

FIGURE 16.8. Ecdysteroid titer course (ecdysone + 20-OH-E) of *Gryllus bimaculatus* within the eighth nymphal instar: solid line—whole animal extracts according to Gnatzy and Romer (1984) (scale—left ordinate); bars—results of *in vitro* incubation of oenocytes together with thoracic fat body within the eighth nymphal instar (scale—right ordinate). The ecdysteroids were separated by TLC and quantified by radioimmunoassay (RIA): stippled bars—ecdysone; striped bars—20-OH-E. Most ecdysteroids were secreted between 85 and 95 h; a second maximum occurs between 115 and 125 h. Note the different scales of whole-animal extracts and *in vitro* incubations.

clearly detectable ecdysteroid concentration of 11.5 ng/7 h was still secreted into the medium. For those insects whose oenocytes were incubated 1 and 2 days after the start of spinning, an almost linear graph could be drawn in the secretion profiles within the first 7 h. This leads to the conclusion that the hormone had not been accumulated beforehand; otherwise a plateau-like course should have been noted in the graph. The attempt to incubate isolated oenocytes of *Bombyx* with cholesterol and thus to realize the conversion *in vitro* yielded no noteworthy amounts of labeled ecdysone and 20-OH-E. Therefore, if the oenocytes do synthesize the hormone secreted, their precursors must be other than cholesterol.

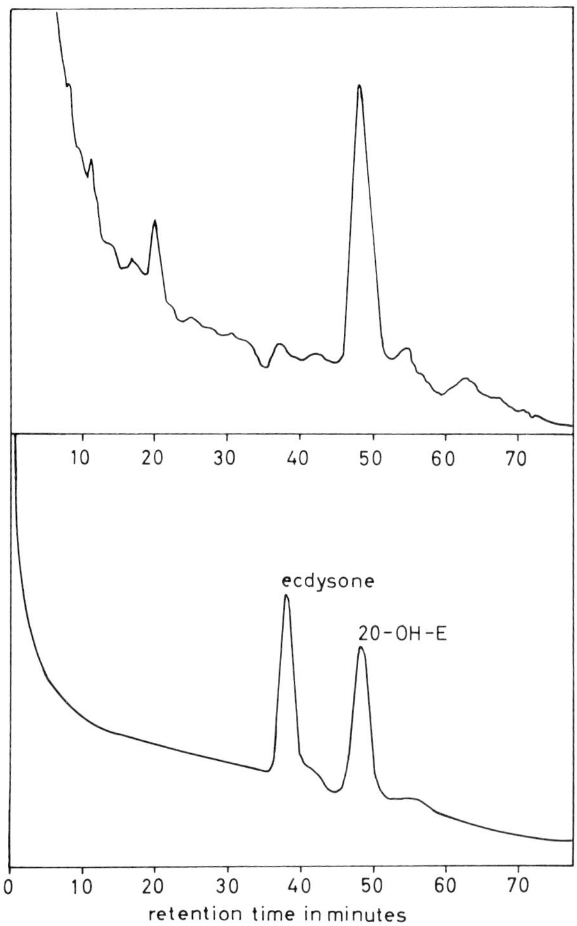

FIGURE 16.9. Gas chromatography of products secreted by oenocytes of *Gryllus* together with thoracic fat body (upper half). The main peak occurring at a retention time of 48 min represents 20-OH-E, as can be seen in the chromatogram with authentic ecdysteroids (lower half). Chromatographic conditions: [^{63}Ni]ECD (electron capture detector); column length, 200 cm; diameter, 4 mm; filling pretested 3% OV 101 by weight on 100/120 Gaschrom Q. Temperature: injector, 300°C; detector, 290°C.

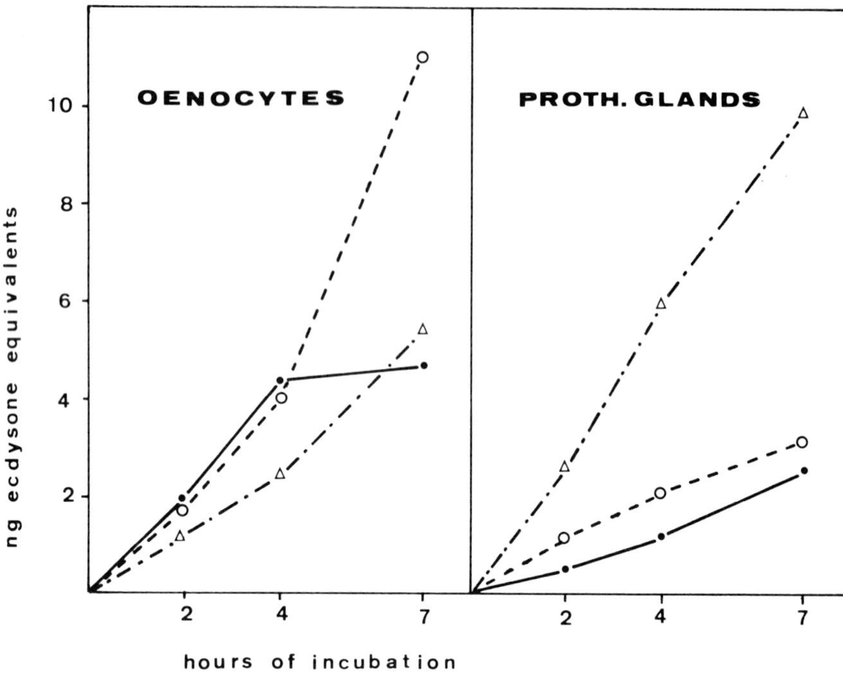

FIGURE 16.10. Secretion profile of oenocytes and prothoracic glands *Bombyx mori*. After 2, 4, and 7 h, 300 μl of incubation fluid was withdrawn from the whole quantity of 900 μl and replaced by fresh medium. The several points of the graphs represent the added amounts of ecdysone found in 300 μl by RIA: ●—● = start of spinning; ○—○ = 1 day after start of spinning; △—△ = 2 days after start of spinning.

16.3.3. *Detoxification*

The discussion of the role of oenocytes in detoxification was initiated by Clark and Dahm (1973), 35 years after Boese (1936) had pointed out that oenocytes of parasitized insects show a hypertrophic reaction. Clark and Dahm fed newly emerged house flies with 0.25 g of phenobarbital with and without addition of thioacetamide. The latter is supposed to inhibit the transport of RNA from the nucleus to the cytoplasm, thus preventing the formation of microsomal enzymes. A phenobarbital treatment of some 48- to 72-h duration led to increased formation of scrolls (cf. Sections 16.2.2, 16.2.3, and 16.2.5, above). The addition of thioacetamide prevented this. Like other xenobiotics, phenobarbital is capable of activating enzymes participating in detoxication, i.e., gluthathione-S-transferase, carboxyesterases, and mixed-function microsomal oxidases (Terriere, 1984). My own experiments on the application of ^{14}C-labeled phenobarbital did not produce the desired metabolization of the phenobarbital by induction of the appropriate enzymes. The conversion into other compounds of different structure and polarity was negligible (far below 1%).

16.4. Summary

Oenocytes are found in all stages of development in embryos, larvae, pupae, and adults. They never retain the diploid level. After differentiation, they no longer divide but grow according to their systematic position by endopolyploidy until they reach the volume specific to their species. In their ultrastructure, they are clearly defined, possessing predominantly smooth and only few traces of rough ER, numerous free ribosomes, polyribosomes, microbodies, and autophagic vacuoles. In the course of a molting cycle, the oenocytes show typical signs of activity, such as appearance of profiles in the rough ER, increase of smooth ER, and partial disintegration of cell structures at the end of a molting cycle. Oenocytes definitely originate from the ectoderm. A functional analysis of oenocytes can be accomplished only in those few species where they can be isolated from the surrounding tissue. With respect to their participation in the formation of the epicuticle, it has been demonstrated that they synthesize the hydrocarbon fraction from acetate. They are also credited with the formation of other components like cuticulin, but so far only histochemical evidence has been furnished for that. It will have to be determined whether precursors different from acetate are also included in the formation of cuticular lipids. Unlike what has been found in *Musca*, amino acids are not eligible as precursors of cuticular lipids in *Tenebrio*. The supposition that oenocytes synthesize lipoproteins seems less likely in regard to their ultrastructure and to the incubation experiments with amino acids. The synthesis of ecdysteroids from cholesterol as a precursor has been demonstrated in *Tenebrio*. In two other species, *Gryllus bimaculatus* and *Bombyx mori*, secretion of ecdysteroids by isolated oenocytes has been observed over a period of at least 7 h. The secretion products were quantified by radioimmunoassay (RIA) and identified by gas chromatography (GC) or high-pressure liquid chromatography (HPLC) as predominantly 20-OH-E, as in *Tenebrio*. At least in *Bombyx*, cholesterol as a precursor does not yield significant conversion rates.

There still remain a number of questions concerning the function of oenocytes. Which compounds serve as precursor molecules for the ecdysteroids secreted by the oenocytes? How is the activity of the oenocytes controlled? Is there a hormone similar to the prothoracicotropic hormone (PTTH), or is it even the same? Up to now, no influence of juvenile hormone (JH) on oenocytes has been detected. But hints are given by changes in the histology and state of activation of oenocytes in worker bees of *Apis* (Boehm, 1965): the oenocytes are highly active during the period of wax secretion and reduce their activity after secretion has finished. In flying bees forced to produce wax, the oenocytes could be reactivated. Unfortunately, the development of JH titer during this stage of development has not yet been determined. The JH values reported by Fluri et al. (1982) in *Apis* show an increase during the first 40 days of adult life. However, the JH values taken at 5-day intervals do not allow coordination with the wax secretion. The problem of detoxification also needs further research, because other toxic substances beside phenobarbital, which did not produce significant results, should be examined with respect to reactions involved in formation of certain key metabolites.

Acknowledgments

I thank Mrs. K. Rehbinder for drawing some of the illustrations. The manuscript was translated by Mrs. C. Delbasteh.

References

Aoki, A. 1970. Hormonal control of Leydig cell differentiation. Protoplasma 71: 209–225.

Armold, M. T. and F. E. Regnier. 1975. Stimulation of hydrocarbon biosynthesis by ecdysterone in the flesh fly *Sarcophaga bullata*. J. Insect Physiol. 21: 1581–1586.

Baillie, A. H. 1964. Further observations on the growth and histochemistry of the Leydig tissue in the postnatal mouse testis. J. Anat. 98: 403–419.

Boehm, B. 1965. Beziehungen zwischen Fettkörper, Oenocyten und Wachsdrüsenentwicklung bei *Apis mellifica* L. Z. Zellforsch. Mikrosk. Anat. 65: 74–115.

Boese, G. 1936. Der Einfluss tierischer Parasiten auf den Organismus der Insekten. Z. Parasitenk. 8: 243–284.

Bonner-Weir, S. 1970. Control of moulting in an insect. Nature (Lond.) 228: 580–581.

Bullière, D., F. Bullière, and M. De Reggi. 1979. Ecdysteroid titres during ovarian and embryonic development in *Blaberus cranifer*. Wilhelm Roux' Arch. Entwicklungsmech. Org. 186: 103–114.

Bursell, E. and A. N. Clements. 1967. The cuticular lipids of the larva of *Tenebrio molitor* L. (Coleoptera). J. Insect Physiol. 13: 1671–1678.

Cassier, P. and M. A. Fain-Maurel. 1969. Sur l'abondance d'un reticulum endoplasmique agranulaire et tubulaire dans les oenocytes de *Locusta migratoria migratorioides* (R. et F.) C. R. Acad. Sci. Paris 269: 1979–1981.

Clark, M. K. and P. A. Dahm. 1973. Phenobarbital-induced, membrane-like scrolls in the oenocytes of *Musca domestica* Linnaeus. J. Cell Biol. 56: 870–875.

Delachambre, M. J. 1966. Remarques sur l'histologie des oenocytes épidermiques de la nymphe de *Tenebrio molitor* L. C. R. Acad. Sci. Paris 263: 764–767.

Delbecque, J. P. and K. Sláma. 1980. Ecdysteroid titres during autonomous metamorphosis in a dermestid beetle. Z. Naturforsch. 35c: 1066–1080.

Diehl, P. A. 1973. Paraffin synthesis in the oenocytes of the desert locust. Nature (Lond.) 243: 468–470.

Dorn, A. and F. Romer. 1976. Structure and function of prothoracic glands and oenocytes in embryos and last larval instars of *Oncopeltus fasciatus* Dallas (Insecta, Heteroptera). Cell Tissue Res. 171: 331–350.

Evans, J. T. 1967. Development and ultrastructure of fat body cells and oenocytes of the Queensland fruit fly *Dacus tyroni* (Frogg.). Z. Zellforsch. Mikrosk. Anat. 81: 49–61.

Fluri, P., M. Lüscher, H. Wille, and L. Gerig. 1982. Changes in weight of the pharyngeal gland and haemolymph titres of juvenile hormone, protein and vitellogenin in worker honey bees. J. Insect Physiol. 28: 61–68.

Gee, W. P. 1911. The oenocytes of *Platyphylax designatus* Walker. Biol. Bull. (Woods Hole) 21: 222–234.

Gnatzy, W. 1970. Struktur und Entwicklung des Integuments und der Oenocyten von *Culex pipiens* L. (Dipt.). Z. Zellforsch. Mikrosk. Anat. 110: 401–443.

Gnatzy, W. and F. Romer. 1984. Cuticle: formation, moulting and control. Pp. 638–684 *in* J. Bereiter-Hahn, A. G. Matoltsy, and K. S. Richards (eds.), *Biology of the Integument*. Springer-Verlag, Berlin and New York.

Huber, M. 1958. Histologische und experimentelle Untersuchungen über die Oenocyten der Larve von *Sialis lutaria* L. Z. Zellforsch. Mikrosk. Anat. 49: 661–697.

Imboden, H. and B. Lanzrein. 1982. Investigations on ecdysteroids and juvenile hormones and on morphological aspects during early embryogenesis in the ovoviviparous cockroach *Nauphoeta cinerea*. J. Insect Physiol. 28: 37–46.

Koller, G. 1929. Die innere Sekretion von wirbellosen Tieren. Biol. Rev. 4: 269–306.

Kon, R. T. and R. E. Monroe. 1971. Utilization of dietary amino acids in lipid synthesis by aseptically reared *Musca domestica*. Ann. Entomol. Soc. Am. 64: 247–250.

Kramer, S. and V. B. Wigglesworth. 1950. The outer layers of the cuticle in the cockroach, *Periplaneta americana*, and the function of oenocytes. Q. J. Microsc. Sci. 91: 63–72.

Lagueux, M., C. Hetru, F. Goltzené, Ch. Kappler, and J. A. Hoffmann. 1979. Ecdysone titre and metabolism in relation to cuticulogenesis in embryos of *Locusta migratoria*. J. Insect Physiol. 25: 709–723.

Lawrence, P. A. and P. Johnston. 1982. Cell lineage of the *Drosophila* abdomen: the epidermis, oenocytes and ventral muscles. J. Embryol. Exp. Morphol. 72: 197–208.

Lawrence, P. A. and P. Johnston. 1986. Observations on cell lineage of internal organs of *Drosophila*. J. Embryol. Exp. Morphol. 91: 251–266.

Locke, M. 1969. The ultrastructure of oenocytes in molt/intermolt cycle of an insect. Tissue & Cell 1: 103–154.

Mackauer, N. 1965. Histologische und karyologische Untersuchungen der Oenocyten von *Aphis pomi* De Geer (Hom. Aphididae). Naturwissenschaften 52: 351–352.

Madhavan, M. M. and K. Madhavan. 1980. Morphogenesis of the epidermis of adult abdomen of *Drosophila*. J. Embryol. Exp. Morphol. 60: 1–31.

Moses, H., W. W. Davis, A. S. Rosenthal, and L. D. Garren. 1969. Adrenal cholesterol: localization by electronmicroscope autoradiography. Science (Wash., D.C.) 163: 1203–1205.

Philogène, B. J. R. and J. E. McFarlane. 1967. The formation of the cuticle in the house cricket, *Acheta domesticus* (L.), and the role of the oenocytes. Can. J. Zool. 45: 181–190.

Piek, T. 1964. Synthesis of wax in the honeybee (*Apis mellifera* L). J. Insect Physiol. 10: 563–572.

Rinterknecht, E. 1985. Cuticulogenesis correlated with ultrastructural changes in oenocytes and epidermal cells in the late cockroach embryo. Tissue & Cell 17: 723–743.

Rinterknecht, E. and G. Matz. 1983. Oenocyte differentiation correlated with the formation of ectodermal coating in the embryo of a cockroach. Tissue & Cell 15: 375–390.

Risler, H. and E. Kempter. 1961. Die Haploidie der Männchen und die Endopolyploidie in einigen Geweben von *Haplothrips* (Thysanoptera). Chromosoma (Berl.) 12: 351–361.

Romer, F. 1966. Zytophotometrische Untersuchungen des DNS-Gehalts in verschiedenen Geweben der Larve und Imago von *Oryzaephilus surinamensis* L. (Cucujidae, Coleoptera). Biol. Zentralbl. 85: 409–438.

Romer, F. 1971. Häutungshormone in den Oenocyten des Mehlkäfers. Naturwissenschaften 58: 324–325.

Romer, F. 1972. Histologie, Histochemie, Polyploidie und Feinstruktur der Oenocyten von *Gryllus bimaculatus* (Saltatoria). Cytobiologie 6: 195–213.

Romer, F. 1973. Feinstrukturelle Merkmale der Oenocyten pterygoter Insekten. Verh. Dtsch. Zool. Ges. 66: 65–70.

Romer, F. 1975. Morphologie und Ultrastruktur der larvalen Oenocyten von *Tenebrio molitor* L. (Insecta, Coleoptera) bei Larve, Puppe und Imago. Z. Morphol. Tiere 80: 1–40.

Romer, F. 1980. Histochemical and biochemical investigations concerning the function of larval oenocytes of *Tenebrio molitor* L. (Coleoptera, Insecta). Histochemistry 69: 69–84.

Romer, F. and I. Eisenbeis. 1983. DNA content and synthesis in several tissues and variation of moulting hormone-level in *Gryllus bimaculatus* DEG (Ensifera, Insecta). Z. Naturforsch. 38c: 112–125.

Romer, F., H. Emmerich, and J. Nowock. 1974. Biosynthesis of moulting hormones in isolated prothoracic glands and oenocytes of *Tenebrio molitor in vitro*. J. Insect Physiol. 20: 1975–1987.

Ruhland, U. and F. Romer. 1977. Nachweis von häutungsaktiven Stoffen in isolierten Prothorakaldrüsen und Oenocyten bei *Bombyx mori* während des 5. Larvenstadiums. Wilhelm Roux' Arch. Entwicklungsmech. Org. 181: 123–134.

Sander, K. 1965. The early embryology of *Pyrilla perpusilla* Walker (Homoptera) including some observations on the later development. Aligarh Muslim Univ. Publ. Zool. Ser. 4: 1–61.

Schmidt, G. H. 1961. Sekretionsphasen und cytologische Beobachtungen zur Funktion der Oenocyten während der Puppenphase verschiedener Kasten und Geschlechter von *Formica polyctena* Foerst. (Ins. Hym. Form.). Z. Zellforsch. Mikrosk. Anat. 55: 707–723.

Snodgrass, R. E. 1935. *Principles of Insect Morphology*. McGraw-Hill, New York.

Terriere, L. C. 1984. Induction of detoxication enzymes in insects. Annu. Rev. Entomol. 29: 71–88.

Weber, H. 1954. *Grundriss der Insektenkunde*, 3rd ed. Fischer Verlag, Stuttgart.

Wigglesworth, V. B. 1933. The physiology of the cuticle and of ecdysis in *Rhodnius prolixus* (Triatom., Hemipt.) with special reference to the function of the oenocytes and of dermal glands. Q. J. Microsc. Sci. 76: 269–318.

Wigglesworth, V. B. 1947. The epicuticle of an insect, *Rhodnius prolixus* (Hemipt.). Proc. R. Soc. Lond. B134: 163–181.

Wigglesworth, V. B. 1948. The structure and deposition of the cuticle in the adult mealworm *Tenebrio molitor* L. (Coleoptera). Q. J. Microsc. Sci. 89: 197–217.

Wigglesworth, V. B. 1970. Structural lipids in the insect cuticle and the function of the oenocytes. Tissue & Cell 2: 155–179.

Wigglesworth, V. B. 1985. Sclerotin and lipid in the waterproofing of the inset cuticle. Tissue & Cell 17: 227–248.

Yokoyama, T. 1936. Histological observations on a nonmoulting strain of silkworm. Proc. R. Entomol. Soc. Lond. A11: 35–44.

Zavrel, J. 1935. Endokrine Hautdrüsen von *Syndiamesa branicki*. Publ. Fac. Sci. Univ. Masaryk (Brno) 213: 1–18.

Roles of Morphogenetic Hormones in Spermatogenesis in Myriapoda

17

MICHEL DESCAMPS

17.1. Introduction

In Myriapoda, the endocrine control of spermatogenesis has been studied only in Symphyla and in Chilopoda. In Symphyla (Juberthie-Jupeau, 1960, 1963), the molt cycle and the spermatogenetic cycle are linked and some evidence exists for a bifactorial control of spermatogenesis. However, it is not possible to define more accurately the hormones involved in this control.

In Chilopoda, in which the spermatogenetic cycle and the molt cycle are not linked, results concerning the control of spermatogenesis were obtained in the Lithobiidae, particularly in *Lithobius forficatus*. Two aspects of spermatogenesis in these animals are of interest for experimental study: (1) during the growth phase the spermatocytes undergo a dramatic increase in diameter from about 10–15 to 70 (*Lithobius crassipes:* Beniouri et al., 1983) or 100 μm (*L. forficatus:* Descamps, 1971); (2) a physiological rest period occurs during late autumn and early winter. Thus, during this period, the changes in spermatocyte metabolism and/or ultrastructure are a good index for hormonal effects. In addition, a testis–blood barrier was found in *L. forficatus* (Beniouri, 1984).

In this chapter, the action of ecdysteroids, juvenile hormone or mimics, pars intercerebralis hormone(s), and cerebral gland hormone(s) will be described. Some of the aspects of the physiological activity of these hormonal factors may be metabolic in nature; nevertheless I have chosen to discuss them here.

17.2. Ecdysteroids

17.2.1. *Long-Term Influence*

Effects on the spermatogenetic cycle were studied (Descamps, 1977a) after single or repeated injections. In *L. forficatus,* single injection of ecdysteroids [0.45 or 0.9 μg of a mixture of ecdysone and 20-hydroxyecdysone (20-OH-E)] caused a delay in the production of spermatozoa (from 15 to 60 days compared with the controls (see Table 17.1).

In some experimental series (Fig. 17.1B,F), before the first complete spermatogenetic phase, an aborted phase was observed. Note that meiosis sometimes occurred (Fig. 17.1B,C) when spermatozoa of the previous phase had not yet moved to the seminal vesicles. This event (i.e., an overlapping spermatogenetic phase) was also observed in other experimental series and is not related to the supply of ecdysteroids.

Differences observed in the experimental series depend on two main factors: spermatocyte age and ecdysteroid dose. Indeed, injection of 0.45 μg was sufficient to induce abortion of a spermatogenetic phase in testes with spermatocytes about in the middle of their growth phase (Fig.

TABLE 17.1. Effects of Ecdysteroid Injection on the Spermatogenetic Cycle in *Lithobius forficatus*[a]

Experiments[b]		Events occurring before the first spermatogenetic phase	First spermatogenetic phase	Second spermatogenetic phase
A	Controls		75	135
A	1 × 0.45 μg	Abortion of meiosis and degeneration of spermatids between day 30 and day 45	90 (15)	165 (30)
B	Controls		30	90
B	1 × 0.9 μg	Abortion of meiosis between day 0 and day 15	45 (15)	120 (30)
C	Controls		60	105
C	1 × 0.45 μg		75 (15)	120 (15)
C	1 × 0.9 μg	Abortion of meiosis between day 45 and day 60	120 (60)	None
C	n × 0.9 μg		90 (30)	None

[a]Time corresponds to the number of days between the beginning of the experiments and the events observed. Numbers in parentheses: delay, compared with controls.

[b]Injections are performed: (A) during the spermatocyte growth phase; (B) during the meiotic phase; (C) at the beginning of the spermatocyte growth phase.

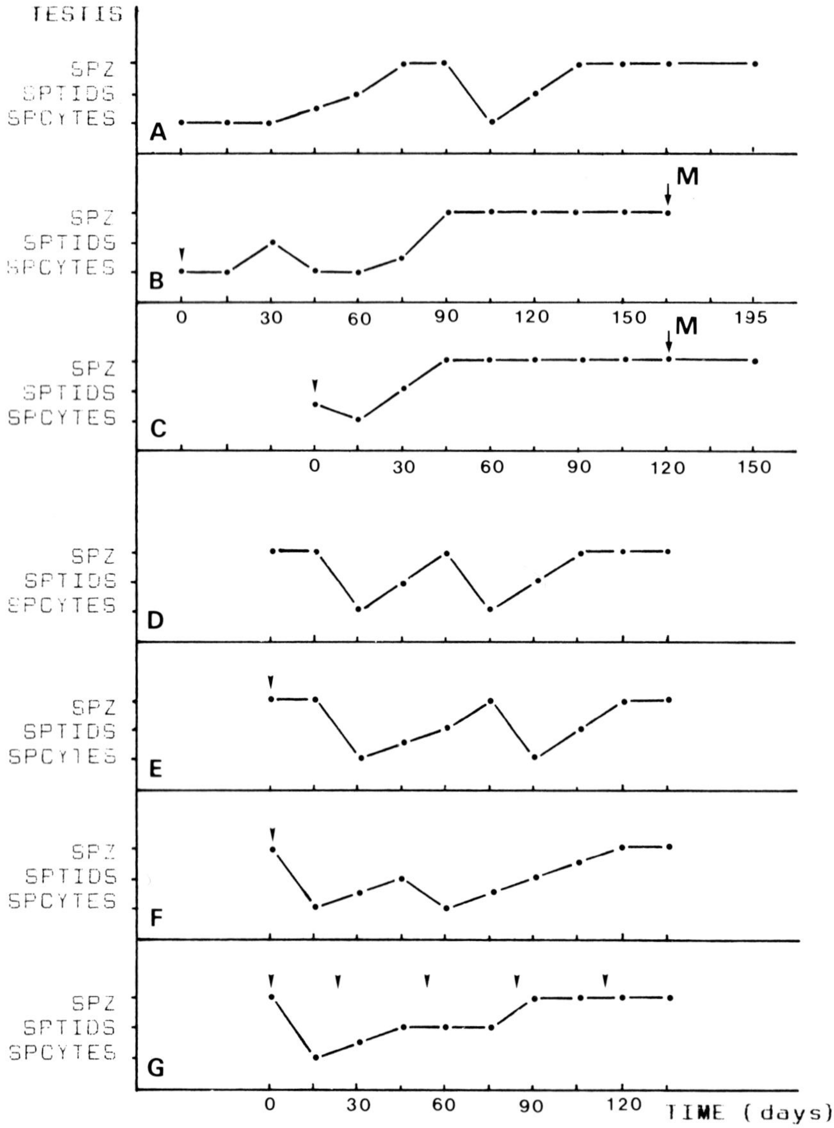

FIGURE 17.1. Influence of ecdysteroid injection on the spermatogenetic cycle in *Lithobius forficatus*. (A,D) Controls. (B,E) Single injection of 0.45-μg ecdysteroid. (C,F) Single injection of 0.9-μg ecdysteroid. (G) Repeated injections of 0.9-μg ecdysteroid. *Key:* arrowheads = time place of injections; M = beginning of meiosis for overlapping spermatogenetic phases; *y* axis = oldest stage observed in the testes. (Modified from Descamps, 1977a.)

17.1B); the same dose induced only a delay in triggering spermatogenetic events in testes with very young spermatocytes (Fig. 17.1E). In this latter case, 0.9 μg caused the abortion of the spermatogenetic phase (Fig. 17.1F). Last, when injected during meiosis, ecdysteroid induced abortion of the dividing spermatocytes about 2 weeks after injection (Fig. 17.1C). However, this result does not agree with those obtained in *L. crassipes* in which 20-OH-E supply during premeiotic phase did not lead to germinal cell degeneration (Beniouri et al., 1983).

A study of cell populations in the testes showed that ecdysteroid injection induced (1) numerous gonial mitoses during at least the first 15 days of the experimental series and (2) a great abundance of degenerating spermatocytes related to the abortion of spermatogenetic phases. When injections (0.9 μg each) were regularly performed all along the experimental series (arrowheads, Fig. 17.1G), meiosis was triggered about 15 days earlier than in controls (Fig. 17.1D). In addition, the second spermatogenetic phase was not observed during the period of the experiment. After a single injection, the study of cell populations showed numerous gonial mitoses during the first 15 days. Later injections of ecdysteroids were less efficient; for example, after the third, no effect was observed on the rate of gonial mitoses.

The triggering of gonial mitoses after ecdysteroid injection, as observed here, is comparable to that observed in other Arthropoda: Crustacea (Arvy et al., 1956; Charniaux-Cotton and Kleinholz, 1964) and Insecta (Schmidt and Williams, 1953; Yagi et al., 1969; Dumser, 1980; Richard-Mercier et al., 1986). Concerning the spermatogenetic cycle, effects observed several weeks or even months after the injection are the consequence of changes induced earlier, during the first hours or days of the experimental series; indeed, ecdysteroids are quickly metabolized by the animals.

17.2.2. *Short-Term Influence*

These experiments were conducted on *L. forficatus* (Descamps, 1981; Beniouri et al., 1985) and *L. crassipes* (Beniouri et al., 1983).

17.2.2.1. INFLUENCE ON UPTAKE OF PRECURSORS

Uptake of labeled precursors ([^{3}H]L-leucine or [^{3}H]uridine) was measured by a Leitz MPV cytophotometer after autoradiography. Because the experiments were done in late autumn, the labeling of the controls remained at about the same level during the whole period of the experiment (20 days).

In *L. forficatus,* in leucine experiments (Fig. 17.2A), an increase in uptake occurred from 24 to 72 h after the injection of 1 μg of 20-OH-E, with a maximum in 48-h animals. Then, a decrease was observed and control values were recorded in 20-day animals. In uridine experiments

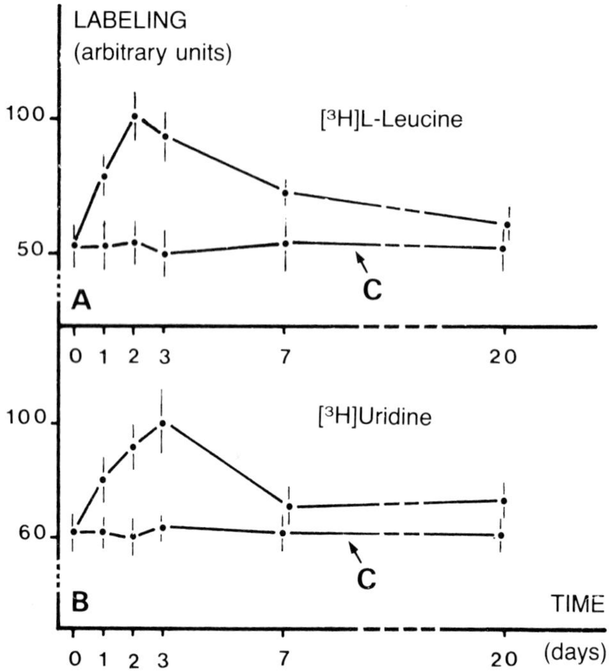

FIGURE 17.2. *L. forficatus:* cytophotometric measurements of spermatocyte labeling in controls (C) and in 20-OH-E–injected animals. (Adapted from Descamps, 1981.)

also (Fig. 17.2B) a marked increase in uptake was observed, particularly in 48- and 72-h animals. Compared with leucine experiments, control values were recorded earlier, in 7-day animals.

Comparable results were obtained in *L. crassipes*. However, only a slight increase in uptake was observed on day 2; the maximum was reached on day 7; and, on day 20, uptake was always significantly higher than in controls. Thus, in both species, injected ecdysteroids induce an increase in RNAs and protein syntheses.

17.2.2.2. INFLUENCE ON CELLULAR ULTRASTRUCTURE

Nucleus

Injection of 20-OH-E enabled us to obtain photomicrographs of activated cells. The nucleolus was surrounded by numerous extrusions, and sometimes strands and generous blebbing of nucleolar material were observed (Fig. 17.3B). Such aspects were not present in the controls (Fig.

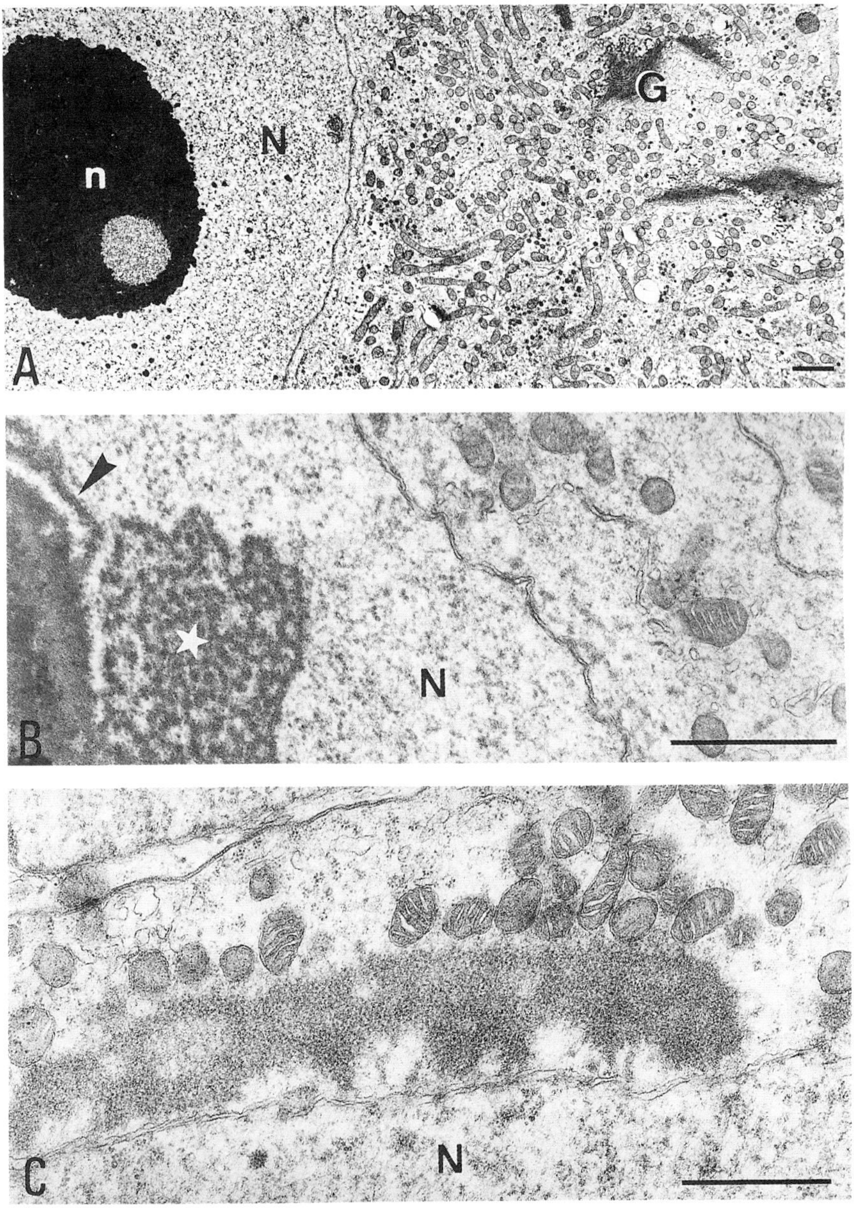

FIGURE 17.3. Spermatocyte aspects in *L. forficatus.* (A) Control: spermatocyte of 60–70 μm diameter. (B) Aspect of nucleolar and nuclear structures in a young spermatocyte, 24 h after supply of 20-OH-E. *Key:* arrowhead = strand of nucleolar material; Star = nucleolar blebbing. (C) Accumulation of nucleus-originated material in the cytoplasm of a young spermatocyte, 7 days after supply of ecdysteroid. *Key:* G = dictyosome; n = nucleolus; N = nucleus. Scale bar = 1 μm. (From Descamps, 1981.)

17.3A). Great amounts of nuclear material were passing through nuclear pores, and accumulations of fibrillo-granular material were observed in the cytoplasm (Fig. 17.3C).

Cytoplasm

Numerous secretion granules were observed as early as 24 h after injection. The dictyosomes were secreting abundantly, and the rough endoplasmic reticulum (RER) was well developed, particularly from 72 h on. In *L. crassipes,* dilated cisternae of RER were sometimes observed on day 1, 2, or 3.

Effects of 20-OH-E injection was observed over a longer period in *L. crassipes* than in *L. forficatus.* Syntheses were increased for at least 20 days in *L. crassipes,* in contrast to only 7 days (only 3 days if we consider uridine uptake!) in *L. forficatus.* These differences could be explained by a higher rate of metabolism in *L. crassipes* controls, which were more easily stimulated.

Plasma Membrane

These experiments, conducted in *L. forficatus,* were performed on freeze-fractured replicas, and only plasma membranes of spermatocytes between 50 and 80 μm in diameter were studied. The partition coefficient used here was the ratio of intramembranous particles (IMPs) density on the P face (PF) to that on the E face (EF). The total density of IMPs (PF + EF) was also used as one of the parameters for the characterization of the membrane.

In winter experiments (Table 17.2), after injection of 20-OH-E, the PF/EF ratio (4.38) was about the same as in the controls (4.47), but total IMP density was only 1010 ± 231 compared to 1738 ± 301 in controls. The IMP distribution on PF and EF was 822 ± 136 and 188 ± 95, respectively. In spring experiments, as in winter ones, the PF/EF ratio (2.55) was comparable to that of the controls (2.37), but the total IMP density was only 920 ± 238 and the IMP distribution on the PF and EF was 661 ± 48 and 259 ± 190, respectively. So, injection of ecdysteroids led, in winter as well as in spring, to total IMP densities and to PF and EF IMP densities of only about two-thirds those of controls, whereas PF/EF ratios stayed at comparable values.

Changes in the inner structure of the plasma membrane are related to the spermatogenetic cycle (differences between winter and spring controls) and to the injection of 20-OH-E. These changes, owing to shifts of IMPs in the lipid bilayer of the membrane, are probably related to transport capacity or/and to receptor linkage (for a complete discussion, see Beniouri et al., 1985). For this part, 20-OH-E injection induces a vertical shift of IMPs into the PF hemimembrane (for a comparison with neurohormone action, see Section 17.4.2.2, below).

TABLE 17.2. *Lithobius forficatus* Spermatocyte Plasma Membrane Characteristics (Winter Experiments)[a]

	Control		20-OH-E		PI electrostimulation	
Fracture faces:	*E*	*P*	*E*	*P*	*E*	*P*
IMP/μm^2	318 ± 48	1420 ± 253	188 ± 95	822 ± 136	290 ± 90	1492 ± 178
Student's *t* test	|--------	------- S -------	--------|			
		|--------	------- S -------	--------|		
	|--------	----------------	---- NS ----	----------------	--------|	
		|--------	----------------	---- NS ----	----------------	--------|
Total density	1738 ± 301		1010 ± 231		1782 ± 268	
PF/EF ratio	4.47		4.38		5.20	

SOURCE: Data from Beniouri et al. (1985).
[a]S = significant; NS = not significant.

17.2.3. *Correlations Between Ecdysteroid Levels and Spermatogenesis*

Synthetic activity is higher in *L. crassipes* controls than in those of *L. forficatus*, and related ultrastructural aspects are more pronounced. Total levels of ecdysteroids were investigated in these two species during the period of minimal physiological activity in order to avoid changes related to the molt cycle. Means and standard deviations (SD) obtained from 10 dosages were as follows: *L. crassipes* = 148.2 ± 69; *L. forficatus* = 287.5 ± 92.4. As the mean weight of *L. forficatus* is 10 times higher than in *L. crassipes*, there is, according to the weight, a total ecdysteroid level five times as high in *L. crassipes*. This difference is sufficient to explain the higher metabolism observed in the spermatocytes of *L. crassipes* and confirm the stimulating effect of ecdysteroids on spermatocytes.

17.2.4. *Periods Refractory to Ecdysteroid Influence*

The first evidence for a period refractory to 20-OH-E action was found in *L. crassipes* (Beniouri et al., 1983). During the premeiotic and meiotic phases, uptake of precursors by spermatocytes, even those too young to undergo meiosis, was lower than during the growth phase. Injection of 20-OH-E was without effect, except for a slight decrease of uptake on day 1. At the end of the meiotic phase, uptake resumed. A period refractory to hormonal action was confirmed in *L. forficatus* by dosage of cAMP (adenosine 3′:5′-cyclic monophosphate) in the testes (see Section 17.4, below).

17.2.5. *Role of Ecdysteroids in the Control of the Testis–Blood Barrier*

Such a permeability barrier and its control by, at least, 20-OH-E were demonstrated by Beniouri (1984). During the period of minimal testis permeability, electron-opaque tracers of <2-nm diameter stayed in internal intercellular spaces of the testis chamber. When the testicular physiology was activated (naturally or by injection of 20-OH-E), the electron-opaque tracers were also found inside the testicular chambers, which is evidence of a change in permeability. Note that spermatogonia, located just under the testis sheath, are not protected by the permeability barrier. The most likely structures providing this barrier are the septate junctions observed in the testicular chamber sheath.

17.3. Juvenile Hormone and JH Mimics

Presence of juvenile hormone (JH) in the hemolymph of *L. forficatus* has not yet been firmly established, but some preliminary work in our laboratory (unpublished) has demonstrated the presence of low amounts of putative JH.

17.3.1. *Long-Term Influence*

A JH mimic (farnesyl methyl ether, FME) was used in this series of experiments (Descamps, 1980). Adult males of *L. forficatus* were injected with sunflower oil solutions of FME. Single (40 nl/10 μl, 90 nl/10 μl, and 140 nl/10 μl) or repeated injections (40 nl/10 μl, every 30 days) were given; controls were injected with a single dose of 10 μl of sunflower oil.

After oil injection, the animals showed a significant lag in the spermatogenetic cycle. On the other hand, injections of FME accelerated spermatogenesis: single doses of 90 and 140 nl, but not 40 nl, induced overlapping spermatogenetic phases. Injection of a single dose of 40 nl had no apparent toxic effect, but higher doses or repeated doses of 40 nl induced more degenerating spermatocytes (1.5 times as high as in controls after 90-nl injection). Survival of the animals generally did not exceed 80 or 90 days after either a single 140-nl injection or four injections of 40 nl, respectively.

17.3.2. *Short-Term Influence*

The effects of JH III were studied (Descamps, 1985) at different doses (0.016, 0.16, and 0.8 μg in 5 μl of sunflower oil). Controls were injected with 5 μl of sunflower oil. The anti-JH substance precocene II was also tested (0.1 and 1.0 μg per animal).

17.3.2.1. INFLUENCE ON UPTAKE OF PRECURSORS

Injections of oil slowed down the uptake of precursors in the controls (Fig. 17.4). In contrast, injection of JH III increased this uptake (Fig. 17.5). At lower dose levels the increase reached a maximum on day 3, but at a dose level of 0.8 μg the uptake of both leucine and uridine continued to rise and reached a maximum on day 7.

In the precocene II group (Fig. 17.6), the uptake of precursors in the 0.1-μg-injected animals was comparable to that of the noninjected controls, but the 1.0-μg-injected animals showed an increase in the uptake of leucine on days 2 and 3. This was followed by a decrease. In uridine experiments, the maximum uptake occurred on day 7.

Thus, despite a moderating effect of oil on syntheses, JH III and precocene II can induce an increase of RNA and protein syntheses.

17.3.2.2. INFLUENCE ON CELLULAR ULTRASTRUCTURE

Micrographs confirmed the autoradiographic data: in JH-injected animals an increase in nucleolar blebbing, nuclear extrusions, Golgi secretion, and RER development was observed (Fig. 17.7A). Multivesicular and residual bodies were present in spermatocyte cytoplasm 20 days after injection (Fig. 17.7B).

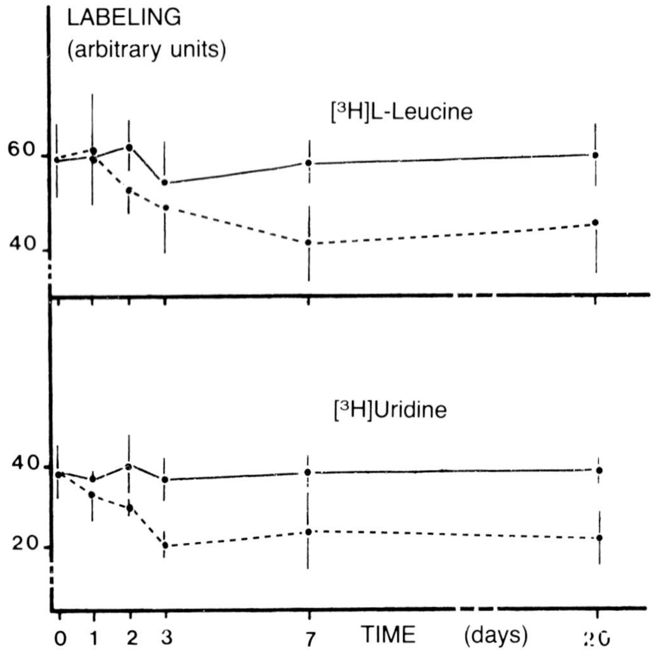

FIGURE 17.4. *L. forficatus:* cytophotometric measurements of spermatocyte labeling in noninjected animals (solid line) and sunflower oil–injected animals (dashed line). (Adapted from Descamps, 1985.)

In sunflower oil and the precocene II series, multivesicular and residual bodies were present within 2 days after injection; degenerating spermatocytes were more abundant than in JH III–injected animals.

The synthetic metabolism of spermatocyte growth after JH injection increases more slowly than after ecdysteroid injection and is related, at least partly, to an indirect action via the neurosecretory cells of the pars intercerebralis area of the protocerebrum; indeed, the latter are stimulated by JH III injection (Joly et al., 1986). However, it is also possible that a direct and rapid action may have been masked by the inhibitory effect of the sunflower oil.

Concerning the action of precocene II, it has been demonstrated in some insects that this compound acts by "chemical allatectomy" (Pener et al., 1978) and that there is a release of JH before the destruction of corpora allata (CA) (Friedman-Cohen and Pener, 1980; Miall and Mordue, 1980). This would provide an explanation for the data reported here: at a lower dose of precocene II, a slow destruction of the putative

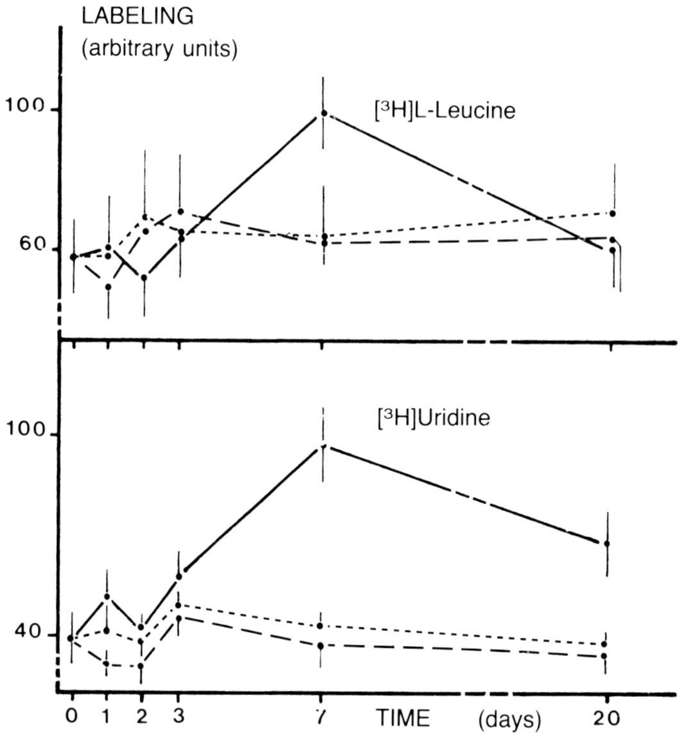

FIGURE 17.5. *L. forficatus:* spermatocyte labeling in JH III–injected animals. *Key:* dashed line (large) = 0.016 μg/ animal; dashed line (small) = 0.16 μg/animal; solid line = 0.8 μg/animal. (Adapted from Descamps, 1985.)

gland and a progressive release of JH might well occur, compensating for the effect of sunflower oil; at a higher dose, it has been shown that precocene II results in an increased uptake of uridine on day 7, the same day as it occurs in animals given an 0.8-μg injection of JH III, and is thus in good agreement with the foregoing hypothesis. But the "CA-like gland" of Chilopoda is still unknown.

17.4. Pars Intercerebralis Hormone(s)

Study of the effects of the neurohormones issued from the pars intercerebralis neurosecretory cells (PI NSCs) was carried out by microcauterization, followed or not by reimplantation of the same endocrine area (Descamps, 1974) and by electrical stimulation of these NSCs (Descamps, 1977b, 1978).

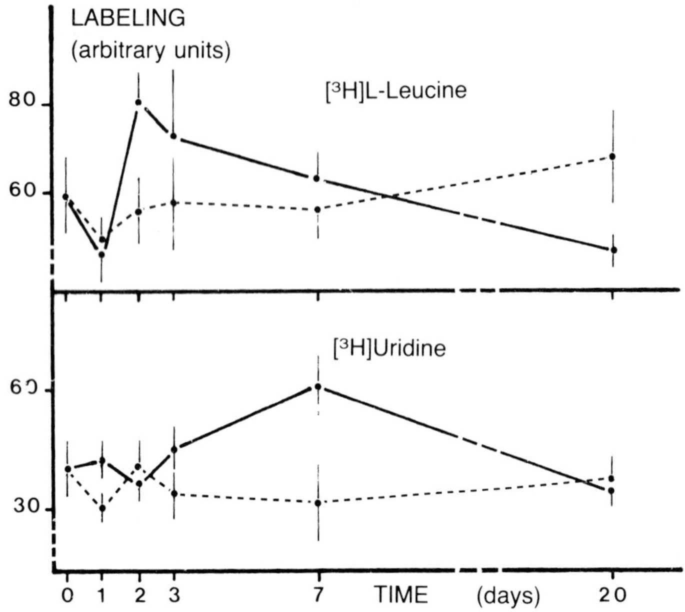

FIGURE 17.6. *L. forficatus:* spermatocyte labeling in precocene II–injected animals. *Key:* dashed line = 0.1 μg; solid line = 1.0 μg. (Adapted from Descamps, 1985.)

17.4.1. *Effects on the Spermatogenetic Cycle*

In mature animals, when the destruction of the PI NSCs was performed before meiosis, the spermatocyte growth phase was lengthened but neither meiosis nor spermiogenesis were blocked.

In addition, after 20–45 days, according to the experimental series, an increase in spermatogonial number was observed with a maximum (2.5 or 3.0 times as high as in controls) between day 75 and day 90. Concerning the spermatocytes, when the destruction of PI NSCs took place just at the beginning of the spermatocyte growth phase, the number of spermatocytes in the testis was increased but their mean diameter reached only 40–50 μm compared with 80–100 μm in the controls (Descamps and Joly, 1971).

If destruction of the PI NSCs was followed by reimplantation of two grafts of PI, the spermatogenetic cycle was comparable to that of the controls and the dramatic increase in spermatogonial number was never observed.

Destruction of PI NSCs during the meiotic phase did not lead to important disorders: divisions were not blocked, and the course of spermiogenesis was not changed significantly.

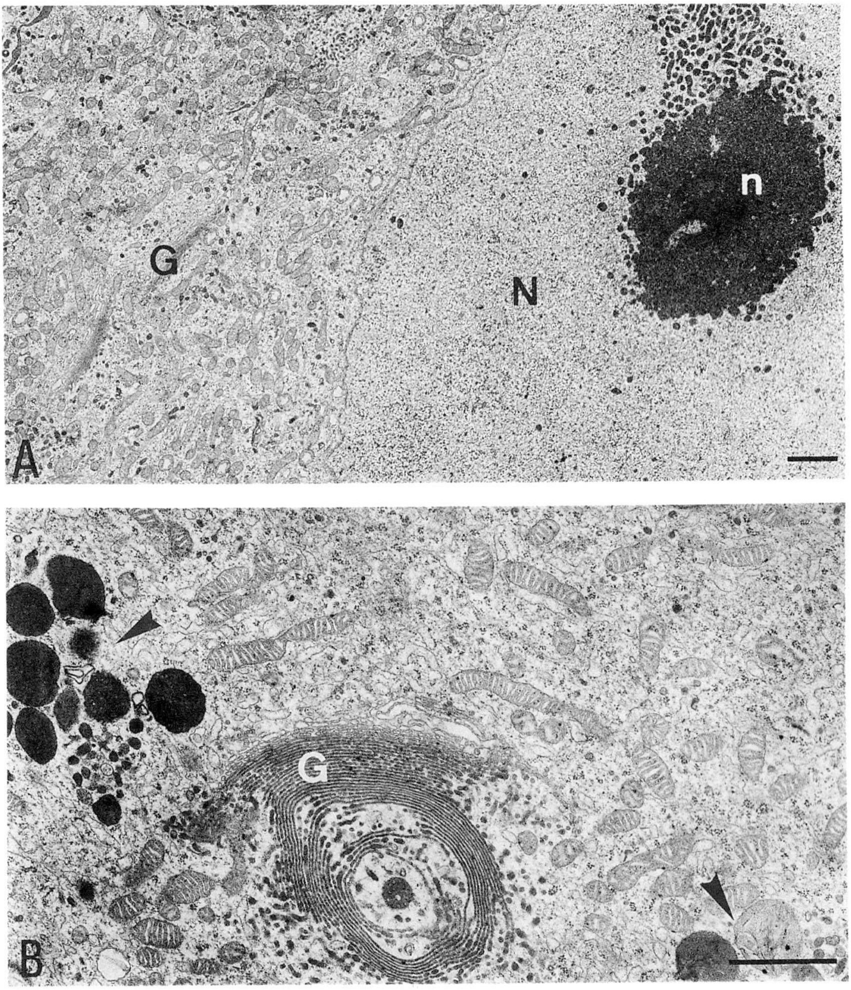

FIGURE 17.7. Spermatocyte aspects after injection of 0.8-μg JH III in *L. forficatus*. (A) Seven days after supply of hormone. Note the numerous nucleolar extrusions (for control aspect, see Fig. 17.3A). (B) Twenty days after hormone injection, residual bodies are observed (arrowheads). *Key:* G = dictyosomes; n = nucleolus; N = nucleus. Scale bar = 1 μm. (Descamps, unpublished observations.)

The stimulating effects of the neurohormones released by the PI NSCs were confirmed by electrical stimulation of this endocrine area. Indeed, it was possible to trigger meiosis during the winter rest, whereas in controls meiosis was never observed.

In immature animals, the first spermatogenetic cycle remains incomplete (Joly and Descamps, 1969): germinal cells degenerate during and after the meiotic phase. When the PI area of animals of different anamorphous stages (*immaturus, praematurus,* and *pseudomaturus* I and II) was electrically stimulated (Descamps, 1977b), it was possible to observe normal meiosis and spermiogenesis in *pseudomaturus* animals but not in the others (Table 17.3).

So, PI NSCs also play a role in the acquisition of puberty, as has been found in the insect *Locusta migratoria* (Girardie et al., 1975).

17.4.2. *Effects on Spermatocytes*

17.4.2.1. ROLE IN UPTAKE OF PRECURSORS

The synthetic metabolism of spermatocytes was studied both in long-term experiments (animals deprived of PI NSCs) and in short-term experiments (after electrical stimulation of the PI area). In the first case, the uptake of precursors was significantly lower than in controls, but after a period of decrease the curve became nearly parallel to that of the controls (Fig. 17.8). After electrical stimulation of the PI area, the curves of uptake showed a rapid increase in leucine experiments (Fig. 17.9A), with a maximum reached on day 3. Then a decrease occurred; nevertheless, in 20-day animals, the labeling remained higher than in controls. Comparable results were obtained in uridine experiments (Fig. 17.9B), but the maximum was reached on days 5–9 and control values were recorded in 20-day animals.

Note that there is a time lag of about 2 days between the maxima observed in leucine and uridine experiments. Conceivably, a two-step mechanism may be involved: First, a rapid and direct action on protein syntheses, using preformed RNAs. The rapidity of this response may be in close relation with the release of a neurohormone originated from the PI NSCs. Indeed, Joly and Descamps (1977) demonstrated that, at least for the PI axons ending in the cerebral gland, the release was increased 2 h after electrical stimulation, with a maximum after about 24 h. The second step may be due either to nucleocytoplasmic exchanges, the nucleus answering to the new cytoplasmic status, or to an indirect mechanism involving the ecdysial gland and a higher hemolymphatic titer of ecdysteroids (or both). It should be recalled that electrostimulation of PI NSCs leads to an increase of the molting rate (Joly and Descamps, 1977).

17.4.2.2. EFFECTS ON CELLULAR ULTRASTRUCTURE

Cytoplasm

In animals deprived of PI NSCs, the spermatocytes showed a cytoplasm poor in Golgi secretions and in RER. In contrast, after electrostimula-

TABLE 17.3. Influence of Electrical Stimulation of the PI Area on Course of the Spermatogenetic Cycle in *L. forficatus*

			No testicular evolution	Normal meiosis and spermiogenesis	Total No. of animals
Anamorphous stages	*immaturus* + *praematurus*	controls	meiosis never observed		
		stimulated	11	0	11
	pseudomaturus	controls	aborted meiosis		
		stimulated	13	9	22
Adults		controls	25	0	25
		stimulated	17	11	28

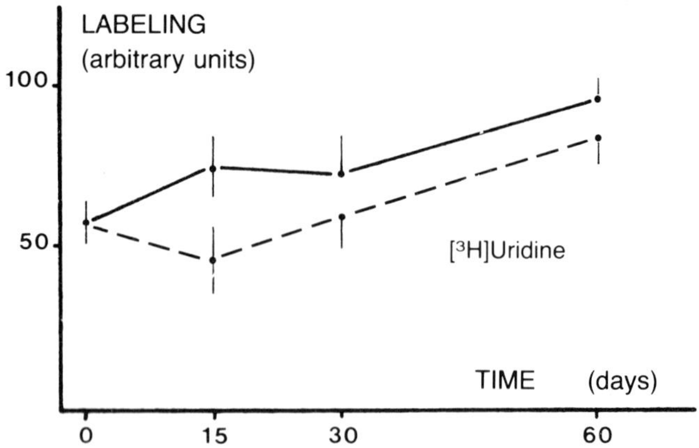

FIGURE 17.8. *L. forficatus:* spermatocyte labeling after [^{3}H]uridine uptake in controls (solid line) and in animals deprived of PI NSCs (dashed line). (Modified from Descamps, 1986.)

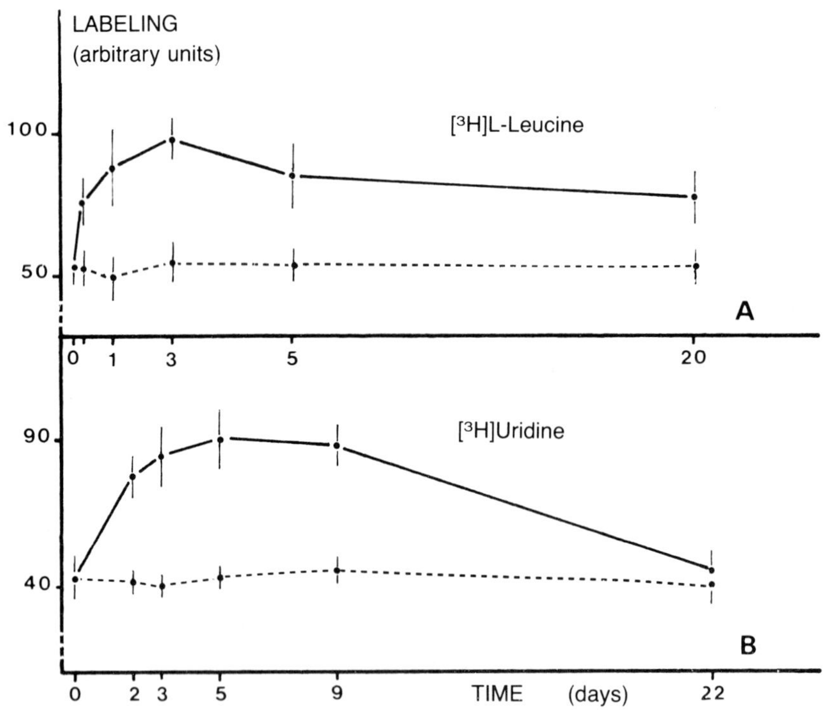

FIGURE 17.9. *L. forficatus:* Spermatocyte labeling after (A) [^{3}H]L-leucine or (B) [^{3}H]uridine uptake in controls (dashed line) and in PI-electrostimulated animals (solid line). (Descamps, unpublished observations.)

tion, the Golgi apparatus secreted profusely and the RER was abundant. Thus, the ultrastructural aspects confirm the stimulating role of PI neurohormones on spermatocyte growth.

Plasma Membrane

In winter experiments, the IMPs' total density (1782 ± 268) and the PF/EF ratio (5.20) were respectively comparable and higher than those of the controls (1738 ± 301; 4.47). This change in the ratio was related to a slight decrease in EF density and an increase in PF density. In spring experiments (Table 17.4), the total density (1684 ± 242) was higher than in controls (1577 ± 171) and the asymmetry between the two hemimembranes increased (PF/EF ratio of 3.57 in contrast to 2.37 in controls). It must be pointed out that the IMP distribution and partition in spermatocytes of electrostimulated animals are very different from those recorded in animals stimulated with 20-OH-E.

17.4.3. *Periods Refractory to Hormonal Influence*

It has previously been shown that the testis does not respond to ecdysteroids during the premeiotic or meiotic phases. A comparable lack of response to hormonal action was demonstrated for PI hormones by dosage of cAMP in the testes (Descamps et al., 1986) (Fig. 17.10).

17.4.4. *Interactions Between PI Hormones and Ecdysteroids*

PI hormone(s) and ecdysteroids both stimulate spermatocyte growth, but some differences exist, mainly concerning the duration of the effects, the characteristics of IMP distribution and partition in the plasma membrane, and the level of cAMP recorded in each kind of experimental stimulation. It has also been demonstrated (Descamps, 1986) that ecdysteroids do not operate independently of PI hormones. Indeed, in animals with an intact brain and injected with 1 μg of 20-OH-E, the labeling of spermatocytes was about twofold higher than in controls. In animals deprived of PI NSCs, 15 days after this destruction the labeling remained higher than in controls. Then, in day-30 and day-60 animals, the labeling was lower than in controls (Fig. 17.11). Thus, ecdysteroids have a stimulating influence on spermatocyte growth only when a sufficient amount of PI hormones is present; below a critical level of the latter, ecdysteroids have an inhibiting effect on spermatocyte growth, the opposite of the effect recorded in intact animals.

17.5. Cerebral Gland Hormone(s)

As for PI hormones, the effects of cerebral gland hormones (in fact, neurohormones and hormones released from the "protocerebral frontal

TABLE 17.4. *Lithobius forficatus* Spermatocyte Plasma Membrane Characteristics (Spring Experiments)[a]

	Control		20-OH-E		PI electrostimulation	
Fracture faces:	*E*	*P*	*E*	*P*	*E*	*P*
IMP/μm²	468 ± 18	1110 ± 153	259 ± 190	661 ± 48	368 ± 52	1316 ± 190
Student's *t* test	|------	------ S ------	------|			
		|------	------ S ------	------|		
	|------	------	------ S ------	------	------|	
		|------	------	------ NS ------	------	------|
Total density	1577 ± 171		920 ± 238		1684 ± 242	
PF/EF ratio	2.37		2.55		3.57	

SOURCE: Data from Beniouri et al. (1985).

[a] S = significant; NS = not significant.

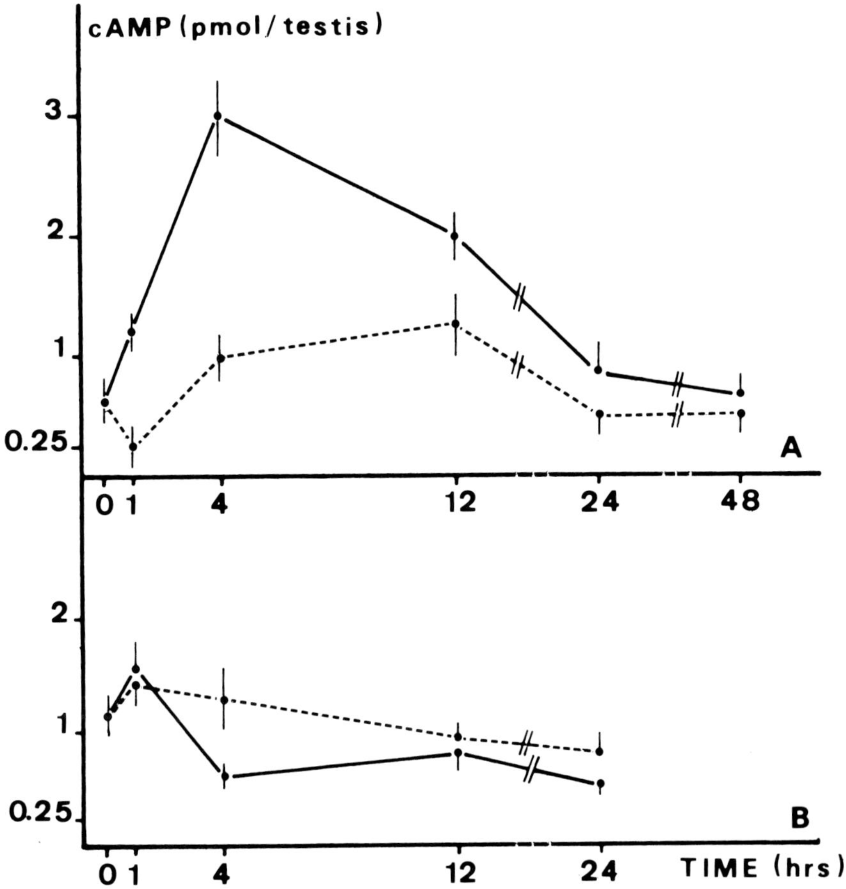

FIGURE 17.10. RIA dosage of cAMP in the testes of *L. forficatus* in the spermatocyte growth phase (A) or during the premeiotic phase (B). *Key:* solid line = PI-electrostimulated animals; dashed line = 20-OH-E–injected animals. (Data from Descamps et al., 1986.)

lobes NSCs–cerebral glands" complex) were studied after destruction of the secretory complex was carried out by surgical methods (Descamps, 1975, 1978).

17.5.1. *Effects on the Spermatogenetic Cycle*

The partial or complete destruction of the secretory complex led to a stimulation of the spermatogenetic cycle and to precocious spermiogenesis. For example, after complete destruction of the endocrine complex,

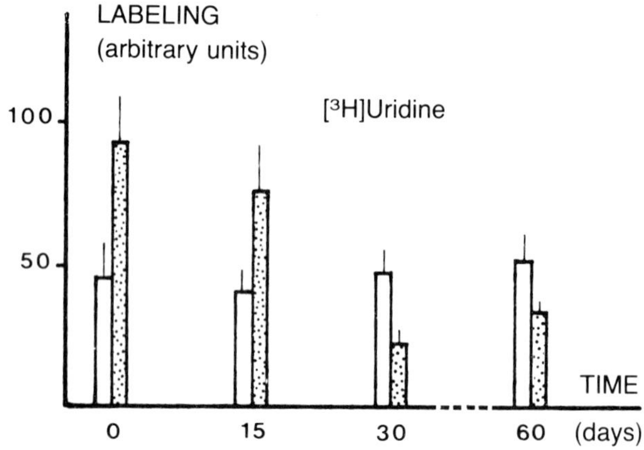

FIGURE 17.11. *L. forficatus:* spermatocyte labeling in unoperated controls (left columns) and in test animals (right column, stippled): day 0, animals injected with 1-μg 20-OH-E 2 days before fixation; days 15, 30, and 60, animals deprived of PI NSCs and injected with 1-μg 20-OH-E 2 days before fixation. (Adapted from Descamps, 1986.)

mature spermatozoa were observed in the testis at least 30 days before they were seen in controls; moreover, in operated animals, spermatogenetic phases occurred regularly, every 45 or 60 days. The rate of gonial mitoses largely rose before each spermatogenetic phase, related to a rapid renewal of testis cell populations. Results of experiments of massive implantation and those of unilateral removal of the cerebral glands or destruction of the NSCs of the protocerebral frontal lobes point to the role of hormone titers.

Experiments of electrical stimulation of the frontal lobe areas confirm previous results: when performed in early spring, a delay in the course of the spermatogenetic cycle was observed. So, all the results point in the same direction: the hormones and neurohormones secreted from the complex moderate both gonial mitoses and spermatocyte growth.

17.5.2. *Effects on Spermatocytes*

In animals deprived of cerebral glands, cellular activity is increased compared with controls. Dictyosomes of more than 15 saccules were observed earlier in the development of the germinal cells, and their secretion markedly raised. RER development was also enlarged.

17.5.3. *Interactions Between PI Hormones and Cerebral Gland Hormones*

When destruction of PI NSCs and removal of cerebral glands were performed in the same animals, the greater part of the spermatocytes degenerated. Then a large increase in the rate of spermatogonial mitoses occurred, such that a normal population of spermatocytes was recovered. Meiosis and spermiogenesis were triggered earlier in test animals than in controls. So, despite the events having to do with the massive degeneration, the effects observed are much more related to those obtained after removal of cerebral glands than to those observed after destruction of PI NSCs: moderating factors secreted by the cerebral glands are more effective than the stimulating factors secreted by PI NSCs!

17.6. Conclusions

Concerning germinal cells, it can be concluded that in *Lithobius*, as far as is known, hormones act only on spermatogonia and on spermatocytes. At the end of the spermatocyte growth phase, when entering in premeiotic phase, stimulating hormones cannot influence the course of further events. Neither meiosis nor spermiogenesis seem to be under hormonal control; nevertheless, repeated injections of 20-OH-E led to precocious meiosis.

The reason for the lack of hormonal sensibility remains unknown, but a couple of hypotheses can be proposed: (1) the testis–blood barrier might become impermeable by a mechanism that is not under the influence of ecdysteroids or PI hormones; or (2) hormonal receptors might be blocked (by antireceptors, by modification of the spacial configuration, or by some other means).

In short, spermatogenesis is, on the one hand, under the stimulating control of (at the least) PI hormones and ecdysteroids and, on the other hand, under the moderating control of cerebral gland hormones. In addition, JH, JH mimics, and the subesophageal ganglion (Descamps, 1979) have a stimulatory effect.

Testis–blood barrier modulation throughout the life cycle is also an aspect of spermatogenesis control, the modulation of this permeability barrier promoting or moderating the access of metabolites and hormones to the spermatocytes. Nevertheless, a direct action of hormones on spermatocytes is possible through gap junctions: in insects, molecules up to 1600–1900 Da pass through the gaps (Simpson et al., 1977). Hormonal actions are summarized in Fig. 17.12.

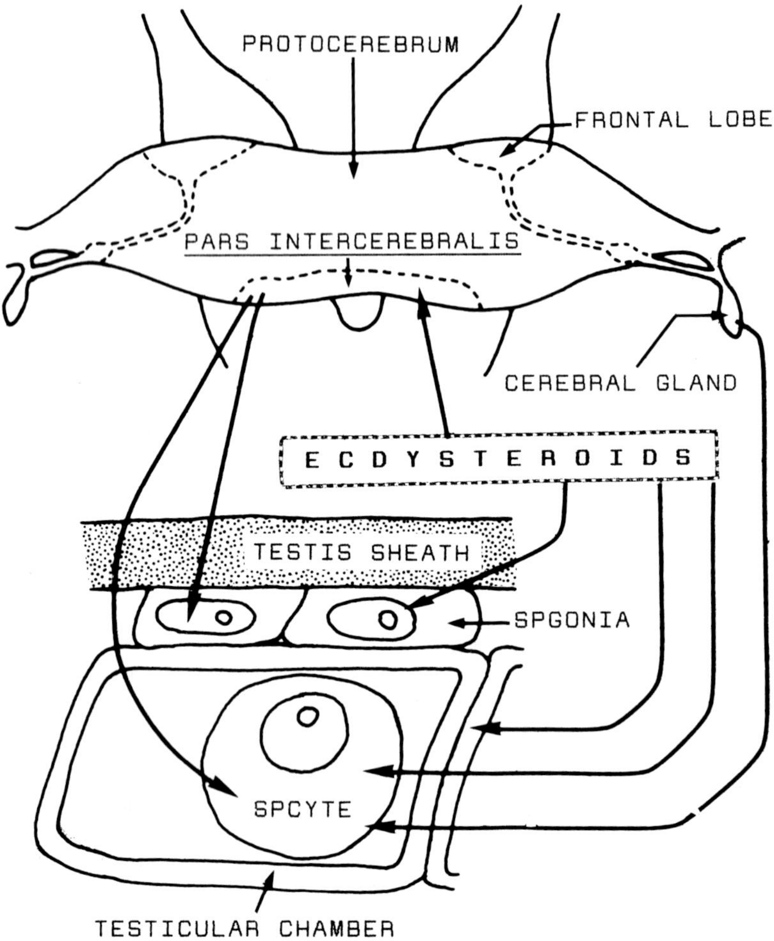

FIGURE 17.12. Hormonal influences on spermatogenesis in Lithobiidae (JH, JH mimics, and subesophageal ganglion influence not included). (Descamps, unpublished observations.)

17.7. Summary

In the Myriapoda, most of the results concerning the influence of hormones on spermatogenesis were obtained in the Lithobiidae (Chilopoda).

Spermatocyte growth is under the control of various hormones, either moderating (issued from the "protocerebral frontal lobes NSCs–

cerebral glands" complex) or stimulating [PI hormone(s) and ecdysteroids].

The double stimulating system is shown by differences observed in the metabolism, the plasma membrane structure of the spermatocytes, and in the level of cAMP in the testis in response to either electrical stimulation of the PI area or injection of ecdysteroids. However, (1) spermatocyte growth cannot be stimulated during premeiotic and meiotic phases of the spermatogenetic cycle; (2) ecdysteroids have a stimulating effect only when there is a sufficient amount of PI hormone(s). A testis–blood barrier, controlled at least partly by ecdysteroids, is also involved in the control of spermatogenesis.

References

Arvy, L., G. Echalier, and M. Gabe. 1956. Organe Y et gonade chez *Carcinides moenas* L. Ann. Sci. Nat. Zool. Biol. Anim. [11] 18: 263–267.

Beniouri, R. 1984. Testis–blood barrier control by 20-hydroxyecdysone in *Lithobius forficatus* L. (Myriapoda, Chilopoda). Cytobios 40: 159–170.

Beniouri, R., M. Descamps, P. Porcheron, and R. Joly. 1983. Corrélations naturelles et expérimentales entre croissance spermatocytaire et taux d'ecdystéroïdes chez les Lithobiidae (Chilopoda). Rev. Can. Biol. Exp. 42: 183–189.

Beniouri, R., M. Descamps, and G. Torpier. 1985. The spermatocyte membrane in *Lithobius forficatus* L. (Myriapoda Chilopoda): changes induced by hormonal actions. Preliminary results. Reprod. Nutr. Dév. 25: 83–91.

Charniaux-Cotton, H. and L. H. Kleinholz. 1964. Hormones in invertebrates other than insects. Pp. 135–198 *in* G. Pincus, K. V. Thimann, and E. B. Astwood (eds.), *Hormones*, Vol. IV. Academic Press, Orlando, Florida.

Descamps, M. 1971. Etude ultrastructurale des spermatogonies et de la croissance spermatocytaire chez *Lithobius forficatus* L. (Myriapode Chilopode). Z. Zellforsch. Mikrosk. Anat. 121: 14–26.

Descamps, M. 1974. Etude du contrôle endocrinien du cycle spermatogénétique chez *Lithobius forficatus* L. (Myriapode Chilopode): rôle de la *pars intercerebralis*. Gen. Comp. Endocrinol. 24: 191–202.

Descamps, M. 1975. Etude du contrôle endocrinien du cycle spermatogénétique chez *Lithobius forficatus* L. (Myriapode Chilopode): rôle du complexe "cellules neurosécrétrices des lobes frontaux du protocérébron–glandes cérébrales." Gen. Comp. Endocrinol. 25: 346–357.

Descamps, M. 1977a. Influence de la croissance somatique sur le cycle spermatogénétique de *Lithobius forficatus* L. (Myriapode Chilopode). Gen. Comp. Endocrinol. 33: 412–422.

Descamps, M. 1977b. Recherches expérimentales sur la régulation du cycle spermatogénétique au cours du développement post-embryonnaire chez *Lithobius forficatus* L. (Myriapode Chilopode). Arch. Biol. 88: 203–215.

Descamps, M. 1978. Rôle des centres endocrines céphaliques dans la régulation de la spermatogenèse chez *Lithobius forficatus* L. (Myriapode Chilopode). Bull. Soc. Zool. Fr. 103: 367–373.

Descamps, M. 1979. Influence of the ventral nerve cord on the spermatogenetic cycle in *Lithobius forficatus* L. (Myriapode Chilopode). Int. J. Invertebr. Reprod. 1: 205–208.

Descamps, M. 1980. Influence d'un mimétique de l'hormone juvénile sur le cycle spermatogénétique de *Lithobius forficatus* L. (Myriapode Chilopode). Bull. Soc. Zool. Fr. 105: 57–63.

Descamps, M. 1981. β-Ecdysone influence on the spermatocyte growth in *Lithobius forficatus* L. (Myriapoda Chilopoda): autoradiographic (optical level) and ultrastructural studies. Arch. Biol. 92: 53–65.

Descamps, M. 1985. Hormonal control of spermatogenesis in the myriapod *Lithobius forficatus*. Pp. 317–320 *in* B. Lofts and W. N. Holmes (eds.), *Current Trends in Comparative Endocrinology*. Hong Kong University Press, Hong Kong.

Descamps, M. 1986. Endocrine control of spermatogenesis in the Lithobiidae (Myriapoda Chilopoda). Pp. 145–149 *in* M. Porchet, J. C. Andries, and A. Dhainaut (eds.), *Advances in Invertebrate Reproduction*, Vol. 4. Elsevier, Amsterdam and New York.

Descamps, M. and R. Joly. 1971. Rôle de la *pars intercerebralis* dans le déroulement du cycle spermatogénétique de *Lithobius forficatus* L. (Myriapode Chilopode). C. R. Acad. Sci. Paris 273: 768–770.

Descamps, M., C. Cardon, and B. Leu. 1986. Influence of brain electrical stimulation or 20-hydroxyecdysone injection in the cAMP level in the testes of *Lithobius forficatus* L. (Myriapoda Chilopoda). P. 506 *in* M. Porchet, J. C. Andries, and A. Dhainaut (eds.), *Advances in Invertebrate Reproduction*, Vol. 4. Elsevier, Amsterdam and New York.

Dumser, J. B. 1980. *In vitro* effects of ecdysterone on the spermatogonial cell cycle in *Locusta*. Int. J. Invertebr. Reprod. 2: 165–174.

Friedman-Cohen, S. and M. P. Pener. 1980. Precocenes induce effect of juvenile hormone excess in *Locusta migratoria*. Nature (Lond.) 286: 711–713.

Girardie, A., J. Girardie, and M. Moulins. 1975. Preuves radiochimiques et physiologiques d'une activation par électrostimulation des cellules neurosécrétrices de la *pars intercerebralis* chez *Locusta migratoria* (Insecte Orthoptère). Gen. Comp. Endocrinol. 25: 416–424.

Joly, R. and M. Descamps. 1969. Evolution du testicule, des vésicules séminales et cycle spermatogénétique chez *Lithobius forficatus* L. (Myriapode Chilopode). Arch. Zool. Exp. Gén. 110: 341–348.

Joly, R. and M. Descamps. 1977. Influence de l'électrostimulation cérébrale sur l'histologie ultrastructurale et le rôle physiologique des glandes cérébrales chez *Lithobius forficatus* L. (Myriapode Chilopode). Arch. Biol. 88: 333–347.

Joly, R., M. Descamps, and C. Jamault-Navarro. 1986. Juvenile hormone effects on the brain in *Lithobius forficatus* L. (Myriapoda Chilopoda). Comp. Endocrinol. (Life Sci. Adv.) 5: 25–29.

Juberthie-Jupeau, L. 1960. Cycle d'émission des spermatophores et évolution des testicules et des vésicules séminales au cours de l'intermue chez *Scutigerella pagesi* Jupeau (Symphyles). C. R. Acad. Sci. Paris 250: 2285–2287.

Juberthie-Jupeau, L. 1963. Recherche sur la reproduction et la mue des Symphyles. Arch. Zool. Exp. Gén. 102: 1–172.

Miall, R. C. and W. Mordue. 1980. Precocene II has juvenile-hormone effects in 5th instar *Locusta migratoria*. J. Insect Physiol. 26: 361–364.

Pener, M. P., L. Orshan, and J. de Wilde. 1978. Precocene II causes atrophy of corpora allata in *Locusta migratoria*. Nature (Lond.) 272: 350–353.

Richard-Mercier, N., P. Porcheron, and R. Poulhe. 1986. Initiation des gamétogenèses et ecdystéroïdes chez les larves du Doryphore *Leptinotarsa decemlineata* Say (Coleoptère, Chrysomelidée). C. R. Acad. Sci. Paris 303: 223–228.

Schmidt, E. L. and C. M. Williams. 1953. Physiology of insect diapause. V. Assay for the growth and differentiation hormone of Lepidoptera by the method of tissue culture. Biol. Bull. (Woods Hole) 105: 174–187.

Simpson, E., B. Rose, and W. R. Loewenstein. 1977. Size limit of molecules permeating the junctional membrane channels. Science (Wash., D.C.) 195: 294–296.

Yagi, S., E. Kondo, and M. Fukaya. 1969. Hormone effect on cultivated insect tissues. I. Effect of ecdysone on cultivated testes of diapausing rice stem borer larvae (Lepidoptera: Pyralidae). Appl. Entomol. Zool. 4: 70–78.

Hormonal Control of Early Metamorphosis in Flies

18

J. ŽĎÁREK AND

P. SIVASUBRAMANIAN

18.1. Pupariation and Pupation in Flies

Metamorphosis in Diptera (Cyclorrhapha) has several distinct and unique features. A pupa develops inside the puparium, a barrel-like structure formed from the cuticle of the last (third) larval instar. Formation of the puparium (pupariation) and pupal molt are fundamentally different developmental events separated from each other by many coarctata hours.

Pupariation is a morphogenetic process during which the soft cuticle of a crawling larva turns into a hard tanned puparium. It involves stereotyped activities accurately timed with structural changes in the cuticle (Ždárek and Fraenkel, 1972). Five stages have been distinguished in a pupariating larva: (1) immobilization; (2) retraction of the anterior segments; (3) contraction to the ovoid puparium; (4) smoothening of the surface due to shrinkage of the cuticle; and (5) tanning of the puparium. Monitoring changes in hemocoelic pressure during pupariation revealed patterns characteristic for each stage (Fig. 18.1) (Ždárek et al., 1979). The pressure changes are basically of two types: pulsations of various kinds corresponding to different patterns of muscular contractions, and a steady increase of the pressure baseline caused by shrinkage and loss of elasticity of the cuticle due to its sclerotization. Pupariation is potentiated by the molting hormone, and its performance is orchestrated by the pupariation factors of neurosecretory origin.

The stage beginning with the puparium is recognized as a prepupa (Fraenkel and Bhaskaran, 1973). The prepupal stage is terminated by the larval–pupal apolysis, which takes place several hours after pupariation when the puparium becomes hard and rigid. At this stage, the pupa is called cryptocephalic (Wahl, 1914), because its head is invaginated and

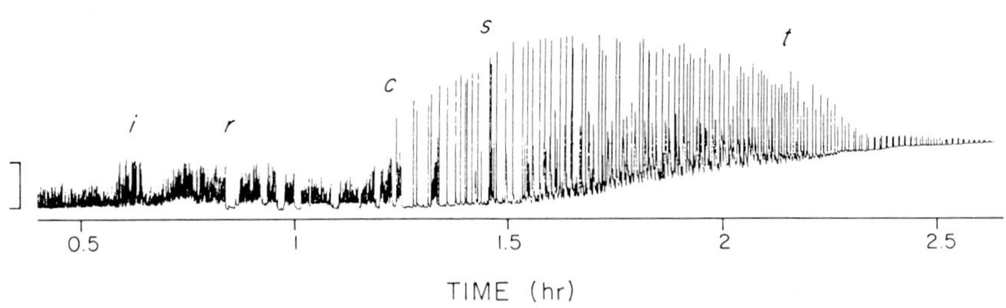

FIGURE 18.1. A record of hemocoelic pressure fluctuations (barogram) during pupariation in *Sarcophaga bullata*. *Key: i* = beginning of immobilization; *r* = beginning of retraction of the anterior segments; *c* = beginning of longitudinal contraction of the larval body; *s* = beginning of cuticular shrinkage; *t* = beginning of tanning. The vertical bar represents 10 kPa.

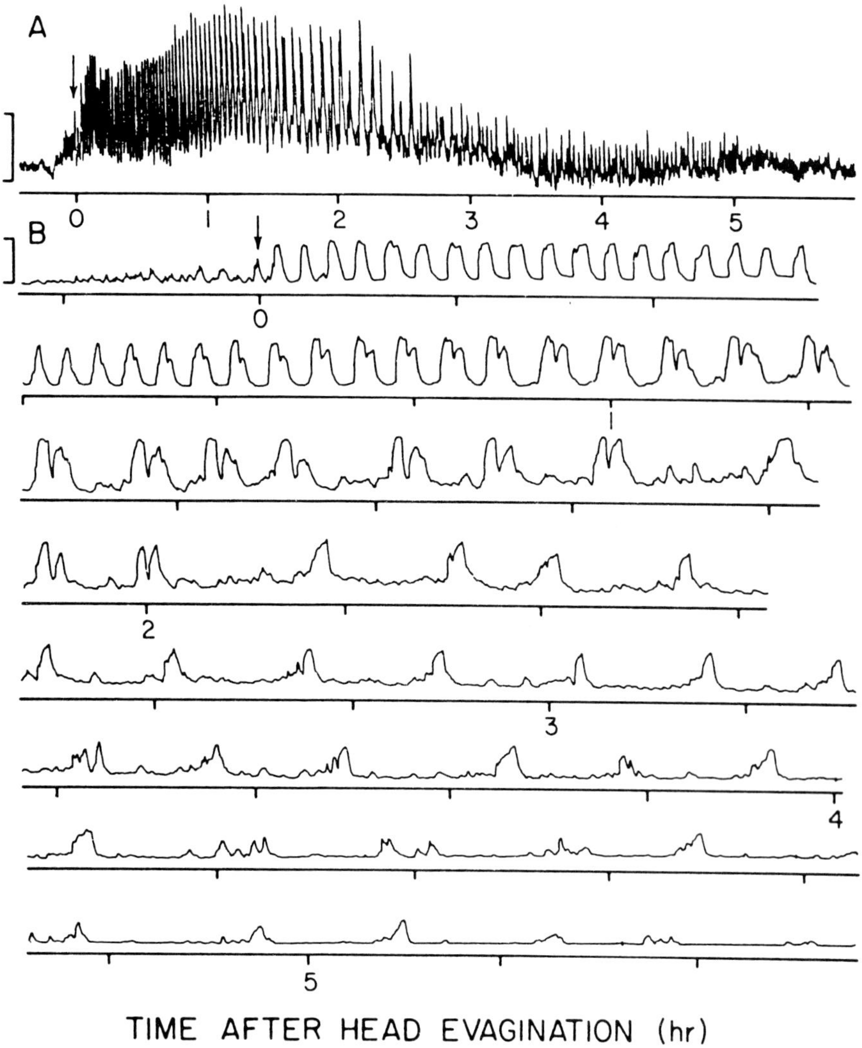

FIGURE 18.2. A barogram of hemocoelic pressure changes during pupal ecdysis in *S. bullata,* taken at a slow (A) and fast (B) chart speed. The vertical bars represent 1 kPa; the arrows, the time of completion of head evagination. (From Žďárek and Friedman, 1986.)

the thoracic appendages are undifferentiated and only partially everted. The cryptocephalic pupa transforms into a phanerocephalic (everted) pupa during a cataclysmic process consisting of head evagination and expansion of thoracic appendages. The event is analogous to pupal ecdysis of other Holometabola, even though the formation of the pupa inside the puparium is not followed by casting off of the pupal cuticle. The larval and pupal cuticles are shed simultaneously at the time of eclosion of the adult (Žďárek and Friedman, 1986).

Formation of the phanerocephalic pupa (pupal ecdysis) is associated with stereotypical behavior. Similarly as occurs in pupariation, the muscular movements during ecdysis are reflected in fluctuations of hemocoelic pressure (Fig. 18.2). The head evaginates before the most vigorous activity begins. Later, the pulses become regular and stronger, each reflecting a swing of the abdomen to one side or the other. The single pulses gradually change into double pulses, and their frequency decreases while their amplitude continues to increase. When the body attains its definitive pupal shape, the pulses produced by the abdominal musculature gradually diminish and finally cease entirely.

Muscular contractions of the abdomen are of two types. The hydrodynamic force developed by peristalsis of the abdominal musculature causes inflation of the pupal legs and wings, followed by evagination of the head. The second type of pattern, consisting mostly of asymmetrical abdominal contractions, is responsible for the completion of certain morphological rearrangements within the phanerocephalic pupa, such as cuticular expansion, retraction of larval tracheae, and movement of the internal organs within the newly expanded cuticle. The last of these patterns seems to be a special adaptation of flies to pupal ecdysis and morphogenesis of pupal structures within the limited space inside the puparium.

18.2. Role of Ecdysteroids in Regulation of Early Metamorphosis in Flies

It is generally known that during the molting cycle apolysis, cell division, differentiation of tissues, and deposition of the new cuticle are under the control of ecdysteroids. The existence of a puparium in higher Diptera complicates a direct application of the classical scheme of the endocrine control of metamorphosis to these insects. Several reports on ecdysteroid titers in various fly species have appeared during the past decade (Borst et al., 1974; De Reggi et al., 1975; Hodgetts et al., 1977; Briers and De Loof, 1980; Wentworth et al., 1981; Briers et al., 1983; Žďárek, 1985; Richards et al., 1987). The titer curves reported are similar in all of the species investigated (Fig. 18.3).

Each larval–larval molt is preceded by a surge in ecdysteroid titer,

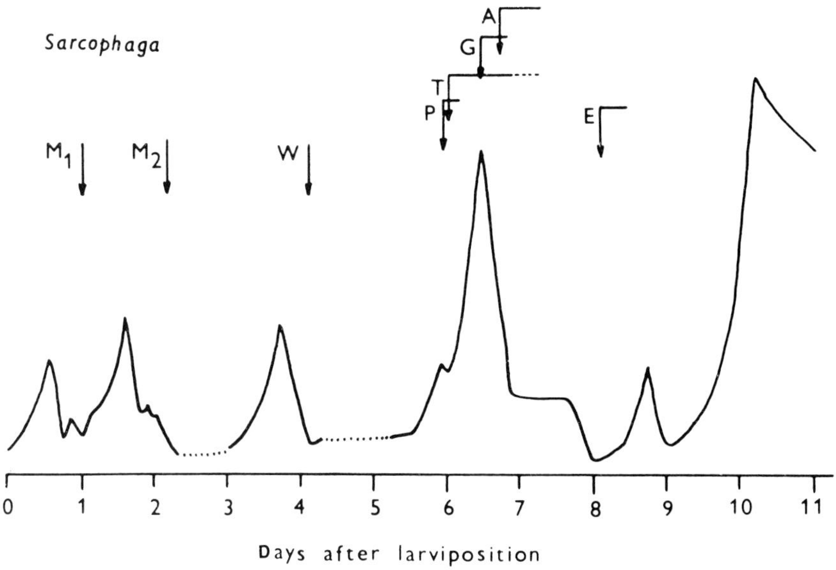

FIGURE 18.3. A compiled curve showing the level of ecdysteroids during postembryonic development of the flesh fly as related to various developmental events such as the larval–larval ecdyses (M_1, M_2), the onset of wandering stage (W), pupariation (P), tanning of the puparium (T), formation of the gas bubble in the intestine (G), apolysis of the pupal cuticle (A), and pupal ecdysis (E). (From Žďárek, 1988.)

whereas at the time of ecdysis the titer is low again. The exodus of the larva from food is associated with another ecdysteroid peak in the middle of the third instar. Toward the end of this instar, the hormone titer gradually rises, thus giving an irreversible hormonal stimulus for pupariation to begin. The increase of the hormone level is interrupted when the puparium is formed, and the increase continues only after the puparium is completed and its cuticle begins to sclerotize. The titer culminates in a narrow high peak, which is presumed to be connected with epidermal apolysis, degradation of larval tissues, deposition of the pupal cuticle, and pumping of gas into the intestine. Several hours later, the ecdysteroid titer is down again, reaching its lowest point shortly before evagination of the head and pupal ecdysis.

Ecdysteroids undoubtedly play a key role in regulation of the pace of metamorphosis and timing of larval molts. It has been shown first in *Tenebrio* (Sláma, 1980) and later in *Manduca* (Truman, 1981) that a successful molt is dependent not only on the presence of these hormones at the right time but also on the absence of the hormones at critical times prior to ecdysis, because exogenous ecdysteroids can delay or prevent ecdysis if administered at the wrong time. This conclusion is supported by the results of experiments on flies. Even small amounts of 20hydroxyecdysone injected during decline of endogenous ecdysteroid titer between pupariation and pupation in *Sarcophaga* were sufficient to block all later events, namely, withdrawal of the molting fluid, dissipation of the gas bubble, and all events of pupal ecdysis dependent on abdominal contractions (Žďárek and Denlinger, 1987).

In contrast to ecdysis, different hormonal conditions appear to be effective at pupariation. Administration of ecdysteroids at any time before pupariation cannot postpone or prevent this process. On the contrary, this hormonal treatment stimulates a precocious onset of pupariation when it is performed during the wandering or late feeding periods. However, the process does not start for several hours after the injection, no matter how large the dose. The existence of such a constant period of latency suggests that the hormone exerts modifying effects by potentiating the central nervous system (CNS) for a new behavioral program, which is then released by the action of the pupariation factors.

18.3. Pupariation Factors

18.3.1. Discovery

As already discussed, pupariation is a complex process involving several behavioral, morphological, and biochemical events. How can a single hormone released several hours earlier bring about these events that must be coordinated temporally and spatially? Apparently, this is accomplished by means of pupariation factors that are released into the system at appropriate times to bring forth their effects on specific targets.

Most of our knowledge on pupariation factors comes from the investigations of Fraenkel and his group at the University of Illinois, carried out between 1969 and 1984, and what follows is based on their studies using the flesh fly, *Sarcophaga bullata*. The *Calliphora* test of Fraenkel (1934), based on ligature of grown last-instar larvae, an assay that paved the way for subsequent isolation of molting hormone ecdysone from *Bombyx* pupae (Butenandt and Karlson, 1954), also formed the basis for the discovery of pupariation factors. However, there is a slight difference in the age of the larvae used for the assay. Whereas for the ecdysone bioassay the larvae are ligated behind the brain about 15 h before pupariation (i.e., before release of the hormone), for the investigation into pupariation factors the larvae are ligated around the sixth larval segment some 3–4 h before pupariation. This time can be recognized in all flesh fly larvae by the beginning of posterior spiracle tanning (red spiracle larvae).

A larva at the red spiracle stage, having already passed the critical time of ecdysone release, pupariates on both sides of the ligature. However, there is a difference in the timing of tanning. The posterior part lacking the brain–ring gland complex tans some 1–2 h after the anterior part, and therefore the quotient of time taken to pupariate the posterior part over time taken to pupariate the anterior part (*P*/*A*) is always more than 1. Žďárek and Fraenkel (1969) injected various fractions into the unligated as well as ligated hind parts of red spiracle larvae of *S. bullata* (Table 18.1) and concluded that a neurohormone from the pars intercerebralis of the brain is released into the hemolymph during pupariation and that this factor accelerates puparium formation and tanning. The injected hind parts tanned earlier than the anterior parts and hence the quotient *P*/*A* was well below 1. The injected nonligated red spiracle larvae also pupariated much earlier than the saline-injected control. That the effect was not due to ecdysone present in their fractions was obvious from the fact that injection of ecdysone alone not only failed to accelerate pupariation but delayed it sometimes when the dose was high.

The earlier work of Žďárek and Fraenkel (1969) assumed that the same

TABLE 18.1. Effect on the Acceleration of Pupariation of Various Tissue Preparations Injected into Ligated Hind Parts of Red Spiracle Larvae of *Sarcophaga bullata*

Materials injected	Pretanning period (min)		
	Anterior (A)	*Posterior (P)*	*P/A*
Distilled water	226	297	1.31
CNS homogenate	234	114	0.49
Brain homogenate	172	99	0.58
Ring gland homogenate	227	133	0.59
Corpora cardiaca homogenate	142	72	0.51
Pars intercerebralis	236	107	0.45
Brain minus pars intercerebralis	250	336	1.34
Muscle homogenate	236	307	1.28
Corpora allata homogenate	148	174	1.18
Corpora cardiaca of *Periplaneta*	157	86	0.55
Hemolymph of orange puparia	220	172	0.78
Hemolymph of wandering larvae	186	265	1.42
17-ng ecdysone	165	320	1.94

SOURCE: From Žďárek and Fraenkel (1969).

neurohormone accelerated all processes involved in pupariation, i.e., its behavioral components (immobilization, anterior retraction, contraction), as well as tanning of the cuticle. However, a subsequent study revealed the existence of two separate factors—one accelerating the retraction of the anterior segments, and the other accelerating tanning (Fraenkel et al., 1972). This was based on the source of active fractions, their relative concentrations, as well as the time of appearance and disappearance of their activity. The active fractions were obtained from CNS and hemolymph of the freshly formed (1-h-old orange-colored) prepupae of *S. bullata.* As shown in Table 18.2, although both fractions accelerate retraction of anterior segments as well as tanning, the CNS extract was stronger on the tanning effect whereas the hemolymph was more potent in accelerating retraction. The hemolymph exhibited strong retraction activity even after eightfold dilution, whereas tanning activity disappeared after fourfold dilution. Further, the time of disappearance of such activity from the hemolymph also suggested the existence of two different factors. The retraction activity disappeared sharply within 10 h after pupariation, whereas tanning activity disappeared more slowly and was still present at 16 h after pupariation. These preliminary experiments indicated that the acceleration of retraction and tanning was brought about by two different hormonal factors, which were later named the *anterior retraction factor* (ARF) and the *puparium-tanning factor* (PTF) by Sivasubramanian et al. (1974a).

Yet other hormonal factors involved in the mediation of pupariation events were reported later by Žďárek et al. (1981). One was called the *puparium immobilization factor* (PIF). It occurs in the hemolymph of prepupae of *S. bullata* 1–2 h after puparial contraction (orange puparia). The bioassay for PIF is based on its immobilizing properties on mature fly larvae. The flesh fly larvae are most sensitive 6–24 h before pupariation (Fig. 18.4). Its action seems to be unspecific in cyclorrhaphous flies. The question of true biological function of PIF has not yet been satisfactorily resolved. At first it seems logical to assume that the immobilization factor participates in the control of behavior associated with the beginning of pupariation, i.e., when the larva becomes immobile and retracted. However, the maximum PIF activity occurs in the hemolymph as late as 2 h after contraction, and no PIF activity can be detected before the white puparium is completed and begins to tan (Fig. 18.5). Hence, it has been suggested that this factor acts upon the larval muscles after formation of the puparium by inhibiting any further muscular contractions. This is quite important in view of the fact that the cuticle is now hardening and muscle contractions at this time would be mechanically harmful.

The other still less understood factor was found in the hemolymph of flesh fly larvae that are just beginning to contract (Žďárek et al., 1981;

TABLE 18.2. Effect of Injection into Red Spiracle Larvae of *Sarcophaga bullata* of Homogenates from the CNS of Red Spiracle Larvae and of Hemolymph from 1-h-old Puparia on the Manifestation of Various Events During Pupariation

	Become immobile	Mouth hooks withdrawn	Gradually contracting	White puparium	Tanning starts
Controls (Ringer injected)					
Average time[a]	159	185	190	200	234
Two CNS per larva					
Average time[a]	37	47	20	60	75
Percentage of controls	23.3	25.4	10.5	30.0	32.0
Hemolymph (4 μg/larva)					
Average time[a]	16.1	20.5	25	30	85
Percentage of controls	10.1	11.1	13.2	15.0	36.3

SOURCE: After Fraenkel et al. (1972).
[a] Time is in minutes after injection.

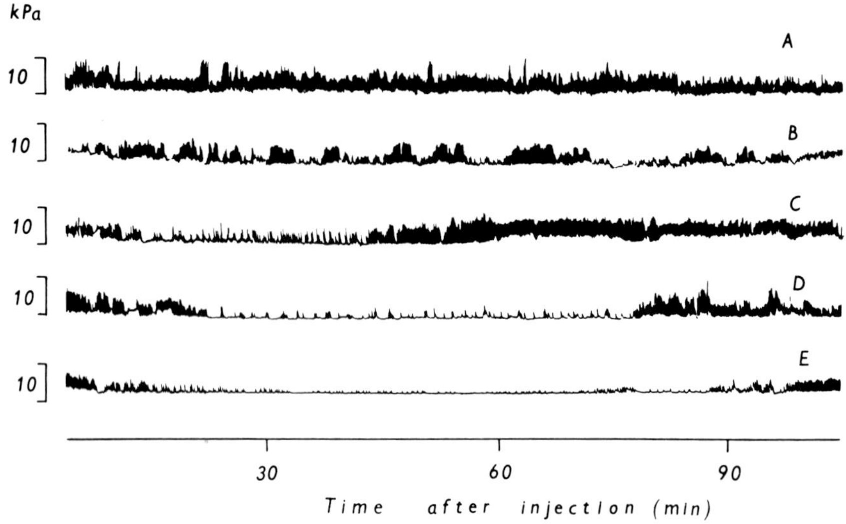

FIGURE 18.4. Barograms of hemocoelic pressure changes in wandering larvae of *S. bullata* at the age of approximately 2 days (A), 1 day (B), 18 h (C), 12 h (D), and 6 h (E) before pupariation, injected with hemolymph containing PIF (from orange puparia). (From Žďárek et al., 1981.)

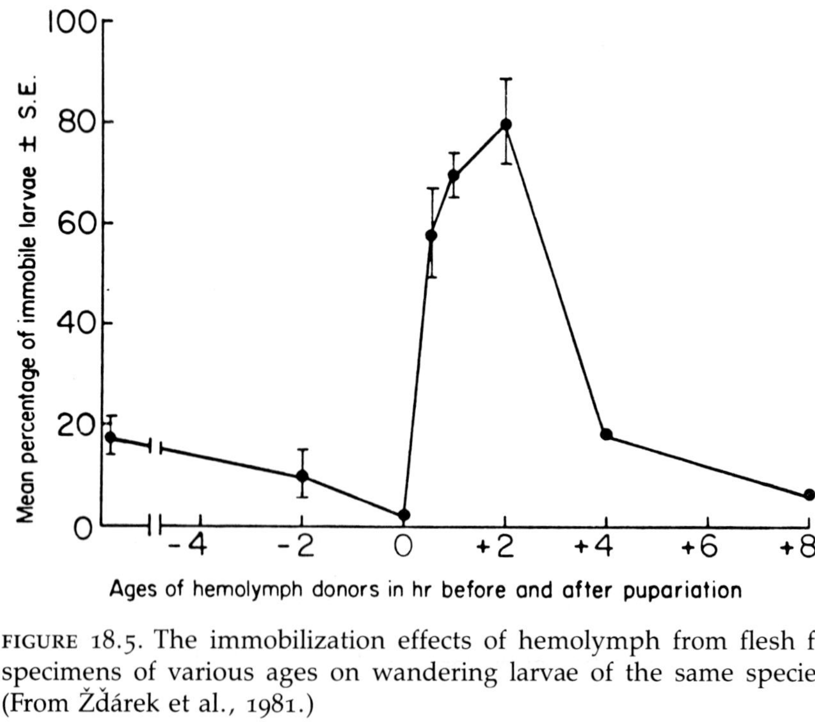

FIGURE 18.5. The immobilization effects of hemolymph from flesh fly specimens of various ages on wandering larvae of the same species. (From Žďárek et al., 1981.)

Žďárek, 1985). The type of response varies with the physiological age of the host (Fig. 18.6). The common feature of all the observed effects is a stimulation of the neuromuscular system. Hence, the active principle from the hemolymph of contracting larvae was called the *puparium-stimulating factor* (PSF). No further investigation of PSF has been done so far mainly owing to the lack of a convenient specific bioassay.

18.3.2. *Chemistry*

The schematic diagram presented in Fig. 18.7 illustrates the procedure used by Sivasubramanian et al. (1974a) to isolate and purify the hormonal factors from CNS and hemolymph, using ammonium sulfate precipitation, ultrafiltration through Amicon membranes, gel filtration

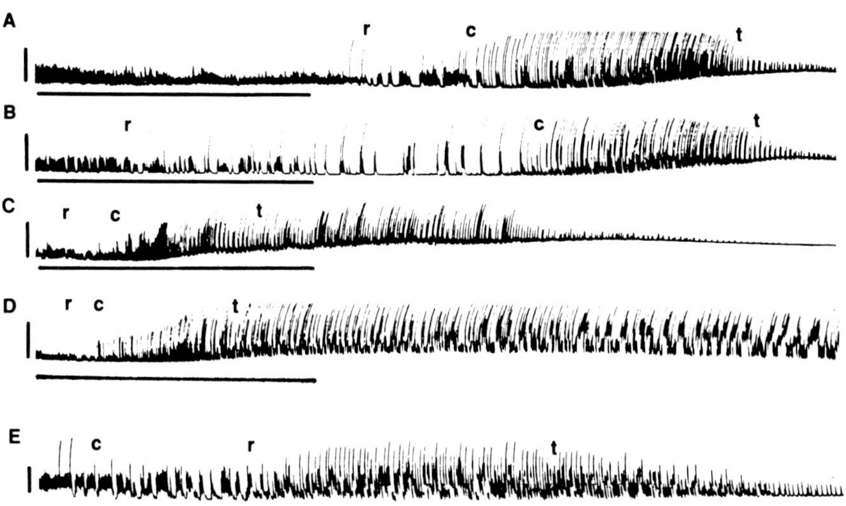

FIGURE 18.6. Barograms of hemocoelic pressure changes of pupariating larvae of *S. bullata* that were injected with various materials approximately 10 min before the record starts: (A) hemolymph from larvae of the same age (control); (B) hemolymph from contracting larvae supposedly containing PSF (note that it causes the stage of retraction to extend when injected more than 2 h before puparial contraction starts); (C) hemolymph from contracting larvae injected less than 2 h before the expected time of pupariation (note that it accelerates the whole process); (D) hemolymph from orange puparia (containing ARF and PIF); (E) a CNS homogenate (note that the two latter materials accelerate and extend pupariation behavior. *Key: r* = beginning of retraction of the anterior segments; *c* = beginning of longitudinal contraction of the larval body; *t* = beginning of tanning. Calibration: vertical, 10 kPa; horizontal, 1 h. (After Žďárek, 1985.)

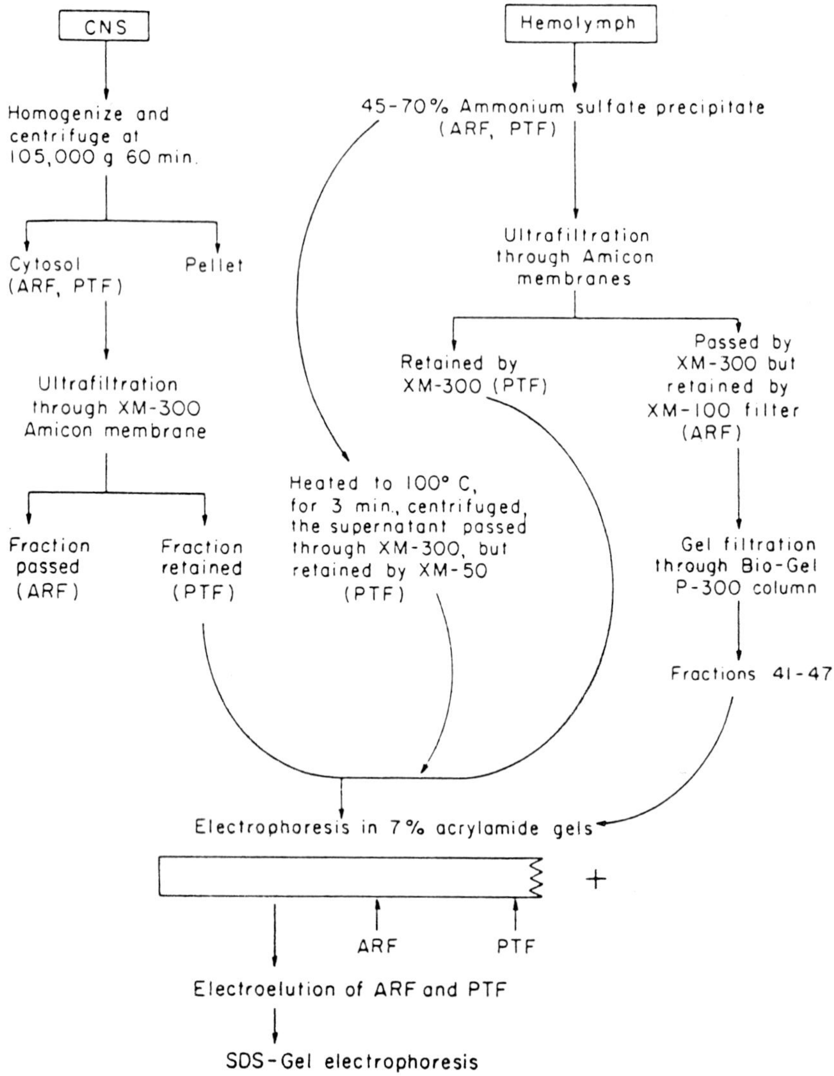

FIGURE 18.7. Scheme of the steps taken in the isolation of ARF and PIF from the hemolymph and CNS of *S. bullata*. (From Sivasubramanian et al., 1974a.)

through Bio-Gel P-300 columns, electrophoresis in 7% acrylamide gels, and SDS (sodium dodecyl sulfate) gel electrophoresis of electroeluted fractions. Two active fractions were obtained: one was specific for anterior retraction (ARF), and the other accelerated tanning (PTF).

A plot of molecular weight as a function of elution volume in a Bio-

Gel P-300 column using standard proteins with known molecular weights gave a value of 180–220 kDa for ARF (Fig. 18.8). The subunit molecular weight of the most active electroeluted ARF fraction as estimated by SDS gel electrophoresis was 90 kDa (Fig. 18.9). Thus, the ARF was estimated to have a molecular weight of 180 kDa, with two subunits of 90 kDa each. By the same procedure the subunit molecular weight of most active PTF fraction from CNS was found to be 80 kDa. Since the native fraction was retained by Amicon XM-300 filter, it was estimated that the PTF molecule must have at least four subunits with a molecular weight of 80 kDa each. Although both factors are proteins, as indicated by enzyme hydrolysis with trypsin and pepsin, PTF is relatively more resistant to heating to 100°C for 3 min, whereas ARF is inactivated by

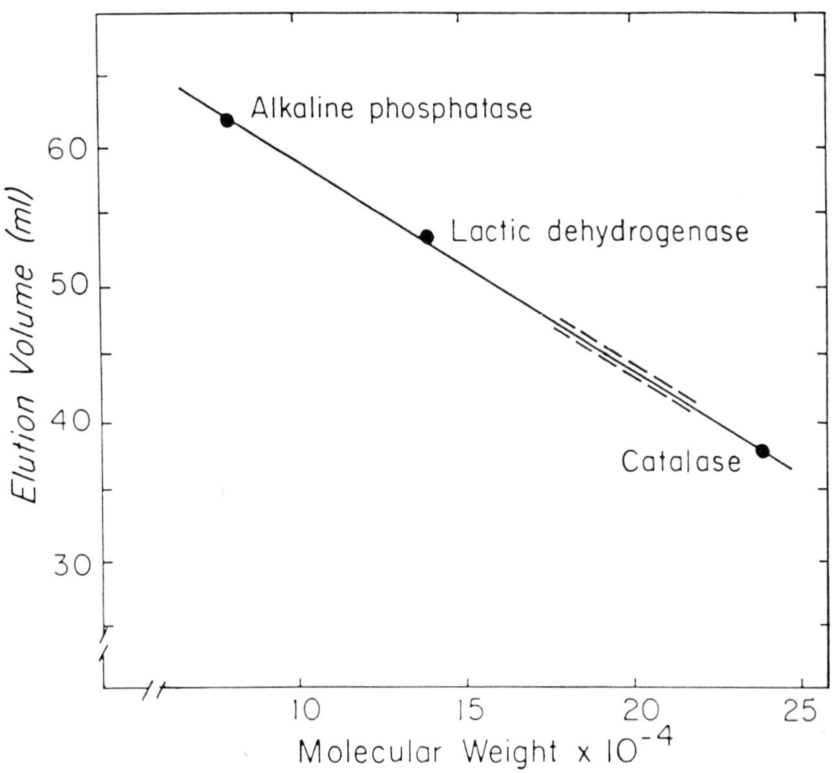

FIGURE 18.8. Molecular weight determination based on the flow rate through a Bio-Gel column. The area in dotted line represents the region where the hemolymph fractions were active for ARF. (From Sivasubramanian et al., 1974a.)

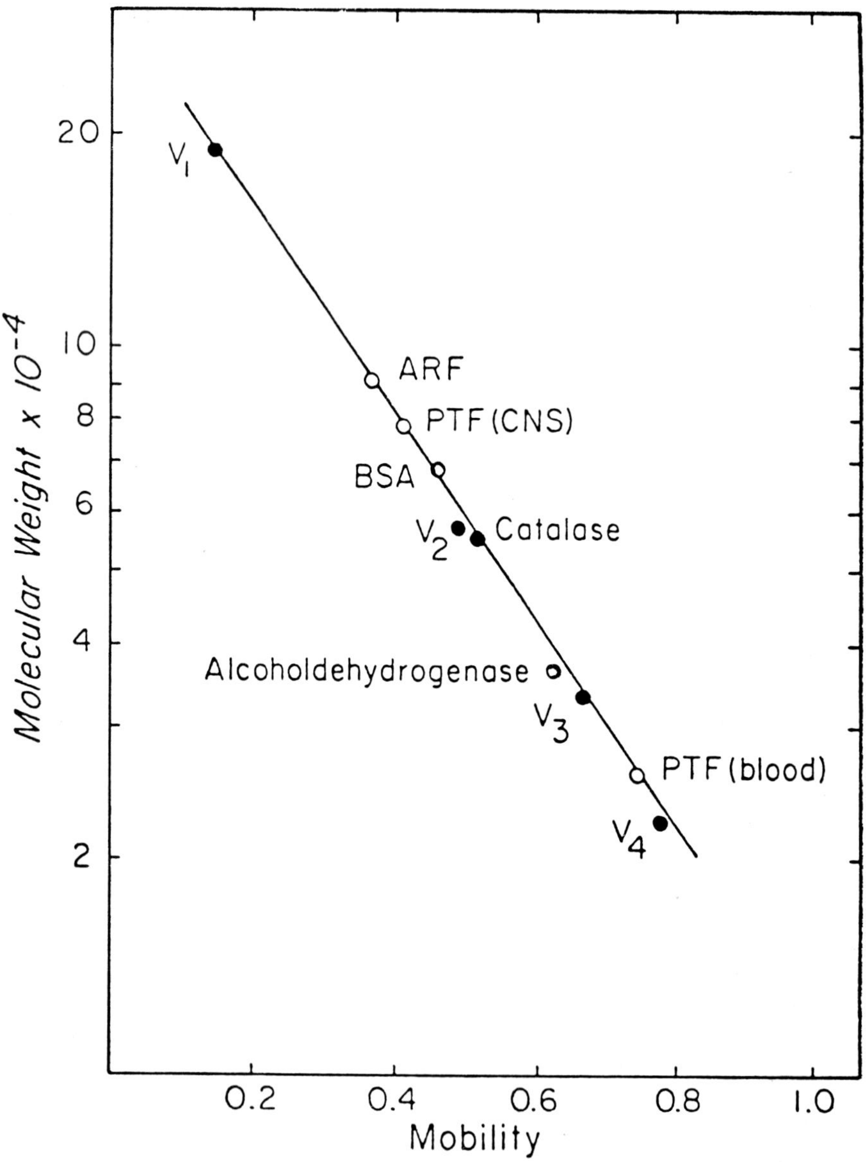

FIGURE 18.9. Semilog plot of molecular weight as a function of mobility of marker proteins using SDS-gel electrophoresis. *Key:* V_1–V_4 = proteins of potato yellow dwarf virus with known molecular weights; BSA = bovine serum albumin. (From Sivasubramanian et al., 1974a.)

TABLE 18.3. Comparison of Effects of Various Treatments on ARF and PTF from the Hemolymph of Orange Puparia of *Sarcophaga bullata*

Treatment	ARF	PTF
Heating at 100°C for 3 min	Activity lost	Activity remains
Ultrafiltration	Activity passed through XM-300 but retained by XM-100 filter	Activity retained by XM-300 filter
Acrylamide gel electroforesis	Activity in the middle third of the gel	Activity in lower third of the gel
Subunit molecular weight as determined by SDS-gel electrophoresis	90 kDa	26 kDa
Injection of the purest fraction obtained by electroelution	Accelerates the retraction of anterior segments only	Accelerates tanning only

SOURCE: After Sivasubramanian et al. (1974a).

such treatment. Other differences between the two factors are summarized in Table 18.3.

A similar analysis aimed at elucidation of properties of the active principle of PIF was done by Žďárek et al. (1981). It also has all the properties of a protein: a molecular weight of 100–300 kDa; it is heat labile, nondialyzable, inactivated by trypsin, and precipitated by ammonium sulfate. It may be identical with ARF, because in not a single instance did the properties of ARF and PIF differ when the conventional test for ARF was simultaneously applied.

18.3.3. *Function*

Sivasubramanian et al. (1974a) proposed a comprehensive scheme of pupariation in flies in order to put CNS activity, hormonal factors, and ecdysone activity into perspective (Fig. 18.10). This was based on the

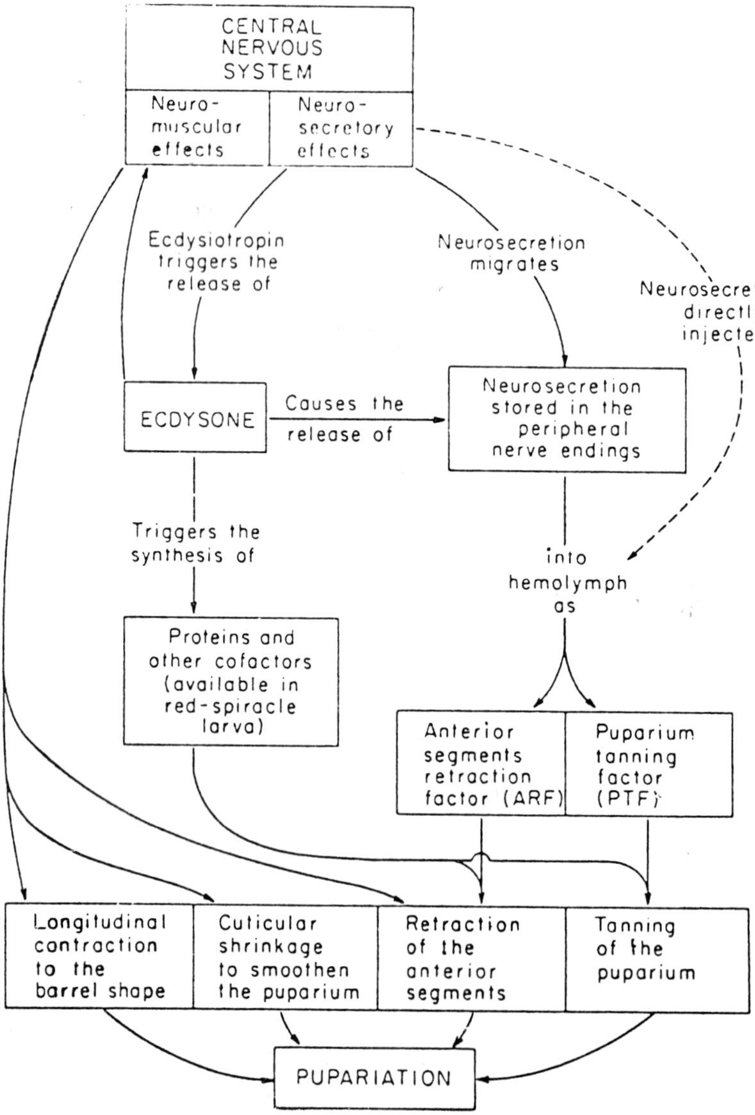

FIGURE 18.10. Relationships between neuromuscular and neuroendocrine events that occur during pupariation in *S. bullata*. (From Sivasubramanian et al., 1974a.)

following facts: (1) CNSs from younger larvae are active in pupariation bioassay. (2) The properties of ARF and PTF in the CNS are essentially identical to those that appear in hemolymph during pupariation (Table 18.4). (3) Pupariation factors appear in the hemolymph only at the time

TABLE 18.4. Comparison of the Properties of ARF and PTF from Hemolymph and CNS Extracts

Property	Hemolymph		CNS	
	ARF	*PTF*	*ARF*	*PTF*
Ultrafiltration through XM-300 filter	Passed through filter	Retained by filter	Passed through filter	Retained by filter
Heating at 100°C for 3 min	Destroyed	Not destroyed	Destroyed	Not destroyed
Acrylamide gel electrophoresis	Activity in the middle third of the gel	Activity in the lower third of the gel	—[a]	Activity in the lower third of the gel

SOURCE: From Sivasubramanian et al. (1974a).

[a]Since the concentration of ARF was very low in the CNS extracts, it was not isolated and hence the information is not available.

of pupariation. (4) ARF and PTF cannot induce pupariation in a precritical stage larva (i.e., before the release of ecdysone); ecdysone must be released first before the factors can act upon their targets (Sivasubramanian et al., 1974b). (5) The factors are released into the hemolymph during pupariation even if the larvae were ligated precritically; however, in such cases ecdysone must be injected into the ligated hind parts.

Thus, according to this model, ARF and PTF originate in the CNS as neurosecretory products and migrate via nerves to be stored in the peripheral nerve endings even before the critical period for ecdysone release. Storage of neurosecretion at nerve terminals for subsequent release is not uncommon in insects (Osborne et al., 1971). The role of ecdysone in this scheme is twofold: First, it triggers the synthesis of proteins, enzymes, and other cofactors essential for various aspects of pupariation in post-critical-stage larvae. Ecdysone-induced macromolecule synthesis during pupariation is well documented in the literature (Wyatt and Wyatt, 1971; Riddiford, 1980). It takes at least 8 h for ecdysone to do so (Sivasubramanian et al., 1974b). Then, just prior to pupariation, it releases the hormonal factors from the neurosecretory nerve terminals. Fig. 18.11 shows a temporal relationship between ecdysteroid titer and the release of ARF in *Sarcophaga bullata*. The timing of ecdysone secretion seems to be closely synchronized with the appearance of pupariation factors in the hemolymph and fluctuations of its titer critical for their action (Žďárek, 1985).

How pupariation factors bring about their effects is not well understood at this time. The very fact that they fail to accelerate pupariation in a pre-critical-stage larva speaks to the importance of ecdysone in their action. As suggested by Truman et al. (1983), perhaps the tissues must be sensitized by a low ecdysone titer prior to the action of the neurohormone. Or perhaps ecdysone-induced enzymes and proteins are essential to mediate the effects of pupariation factors. One such well-documented enzyme induction is that of dopa decarboxylase by ecdysone (Sekeris, 1965). This enzyme mediates the tanning process during pupariation. Or it is also possible that some of these macromolecules whose synthesis depends upon ecdysone may alter the target cell permeability and thus facilitate the entrance of hormonal factors into the target cells (Kambysellis and Williams, 1971).

18.3.4. Mechanism of Action

Attempts have been made to elucidate the mechanism of action of at least one of the pupariation factors—that of PTF on the cellular level. Fraenkel et al. (1977) duplicated the PTF effects by injecting cAMP (adenosine 3′:5′-cyclic monophosphate) into red spiracle larvae. The accelerated tanning was identical to normal tanning during pupariation. Further, theophylline, a phosphodiesterase inhibitor, enhanced cAMP

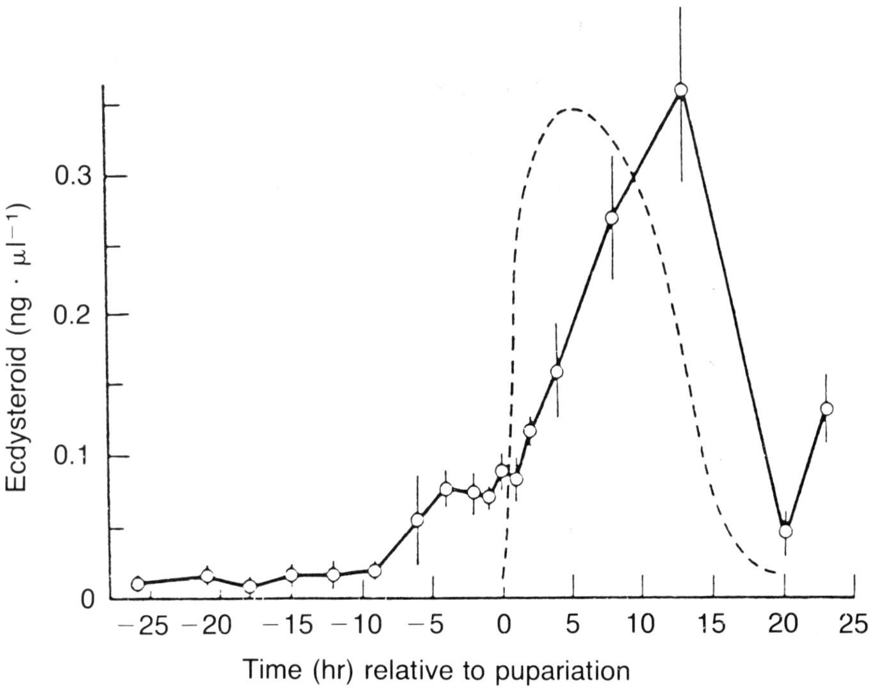

FIGURE 18.11. The titer of ecdysteroids (solid line) and ARF (dashed line) in the hemolymph of *S. bullata* over several hours before and after pupariation. ARF activity is expressed in relative units. (From Žďárek, 1985.)

effects in the tanning process, as would be expected if cAMP synthesis were an essential step in the action of PTF. Based on this, the aforementioned authors suggested that the cyclic nucleotide plays the role of a second messenger for PTF.

Seligman et al. (1977) proposed a complex scheme of hypothetical relationships based on their studies of the effects of transcriptional and translational inhibitors on pupariation behavior and tanning. They suggested that PTF is concerned with the regulation of two components of the tanning response: (1) acceleration of synthesis of a particular protein associated with the tyrosine hydroxylation complex, and (2) activation via cAMP of a component of the tyrosine-hydroxylating system. The mode of action of other factors (ARF, PIF, PSF) has not yet been elucidated. Presumably, they all have neurotropic activity on the level of the CNS acting as the releasers of specific behavioral patterns. The neuromuscular and integumental systems respond to the factors by executing species-specific responses, although the factors themselves are not species specific. For example, the active hemolymph of *Sarcophaga* stimulates phenolic tanning in the puparium of *Sarcophaga* or *Calliphora* but calcification of the puparial cuticle in *Musca autumnalis*. Similarly, the neurotropic factors from the hemolymph of *Sarcophaga* stimulate anterior retraction prior to longitudinal contraction in larvae of flesh flies, but in *Phormia* a reversed sequence of events (i.e., contraction prior to retraction) is induced by the active principle from *Sarcophaga*.

18.4. Summary

In cyclorrhaphous flies, metamorphosis takes place inside a puparium, which is a hard ovoid case formed from the cuticle of the last (third) larval instar. Formation of the puparium (pupariation) is a complex process involving a set of specific motor programs well timed with biochemical and biophysical changes in the cuticle. The muscular and cuticular systems work in concert, being under neurohormonal control. Inside the puparium, the prepupa undergoes morphological changes including larval–pupal apolysis, deposition of the pupal cuticle, inflation of the gut, and eversion of the imaginal disks. The prepupal stage is terminated by pupal ecdysis, during which the pupal head and thoracic appendages become evaginated and inflated and the tracheal lining is withdrawn. The process is brought about by stereotyped activity of the abdominal muscles.

Regulation of early metamorphosis is realized at two levels. Timing of pupariation with respect to environmental and internal conditions is determined by the CNS and executed through stimulation of molting hormone secretion. The actual process is ensured by the timely appearance in the hemolymph of the so-called pupariation factors, which have either neurotropic (ARF, PIF, PSF) or tanning-accelerating (PTF) activity. The factors are proteins of large molecular weight, presumably of neurosecretory origin. The prepupal morphological and biochemical transformations are stimulated by molting hormone, but actual ecdysis can only occur if the endogenous molting hormone titer drops. There are indications that ecdysial behavior inside the puparium is stimulated by a blood-borne factor.

References

Borst, D. W., W. E. Bollenbacher, J. D. O'Connor, D. S. King, and J. W. Fristrom. 1974. Ecdysone levels during metamorphosis of *Drosophila melanogaster*. Dev. Biol. 39: 308–316.

Briers, T. and A. De Loof. 1980. The molting hormone activity in *Sarcophaga bullata* in relation to metamorphosis and reproduction. J. Invertebr. Reprod. 2: 363–372.

Briers, T., E. van Beek, and A. De Loof. 1983. Ecdysteroid activity during metamorphosis and in male and female adults of four blowfly species. Comp. Biochem. Physiol. 74A: 521–524.

Butenandt, A. and P. Karlson. 1954. Über die Isolierung eines Metamorphose-Hormons der Insekten in kristallisierter Form. Z. Naturforsch. 9B: 389–391.

De Reggi, M. L., M. H. Hirn, and M. A. De Laage. 1975. Radioimmunoassay of ecdysone on application to *Drosophila* larvae and pupae. Biochem. Biophys. Res. Commun. 66: 1307–1315.

Fraenkel, G. 1934. Pupation of flies initiated by a hormone. Nature (Lond.) 133: 834.

Fraenkel, G. and G. Bhaskaran. 1973. Pupariation and pupation in cyclorrhaphous flies (Diptera): terminology and interpretation. Ann. Entomol. Soc. Am. 66: 418–422.

Fraenkel, G., A. Blechl, J. Blechl, P. Herman, and M. I. Seligman. 1977. 3′:5′-Cyclic AMP and hormonal control of puparium formation in the fleshfly *Sarcophaga bullata*. Proc. Natl. Acad. Sci. USA 74(5): 2182–2186.

Fraenkel, G., J. Žďárek, and P. Sivasubramanian. 1972. Hormonal factors in the CNS and hemolymph of pupariating fly larvae which accelerate puparium formation and tanning. Biol. Bull. (Woods Hole) 143: 127–139.

Hodgetts, R. B., B. Sage, and J. D. O'Connor. 1977. Ecdysone titers during postembryonic development of *Drosophila melanogaster*. Dev. Biol. 60: 310–317.

Kambysellis, M. P. and C. M. Williams. 1971. *In vitro* development of insect tissues. II. The role of ecdysone in the spermatogenesis of silkworms. Biol. Bull. (Woods Hole) 141: 541–552.

Osborne, M. P., I. H. Finlayson, and M. J. Rice. 1971. Neurosecretory endings associated with striated muscles in three insets (*Schistocerca, Carausius,* and *Phormia*) and a frog. Z. Zellforsch. Mikrosk. Anat. 166: 391–404.

Richards, D. S., J. T. Warren, D. S. Saunders, and L. I. Gilbert. 1987. Haemolymph ecdysteroid titres in diapause and non-diapause destined larvae and pupae of *Sarcophaga argyrostoma*. J. Insect. Physiol. 33: 115–122.

Riddiford, L. M. 1980. Insect endocrinology: action of hormones at the cellular level. Annu. Rev. Physiol. 42: 511–528.

Sekeris, C. E. 1965. Action of ecdysone on ARN and protein metabolism in the blowfly *Calliphora erythrocephala*. Pp. 149–167 *in* P. Karlson (ed.), *Mechanism of Hormone Action*. Thieme Verlag,, Stuttgart, and Academic Press, Orlando, Florida.

Seligman, M., A. Blechl, J. Blechl, P. Herman, and G. Fraenkel. 1977. Role of ecdysone, pupariation factors and cyclic AMP in formation and tanning of the puparium of the fleshfly *Sarcophaga bullata*. Proc. Natl. Acad. Sci. USA 74: 4697–4701.

Sivasubramanian, P., S. Friedman, and G. Fraenkel. 1974a. Nature and role of proteinaceous hormonal factors acting during puparium formation in flies. Biol. Bull. (Woods Hole) 146: 163–185.

Sivasubramanian, P., H. S. Ducoff, and G. Fraenkel. 1974b. Effects of X-irradiation on the formation of the puparium in the fleshfly, *Sarcophaga bullata*. J. Insect Physiol. 20: 1303–1317.

Sláma, K. 1980. Homeostatic function of ecdysteroids in ecdysis and oviposition. Acta Ent. Bohemoslov. 77: 145–168.

Truman, J. W. 1981. Interaction between ecdysteroid, eclosion hormone, and bursicon titres in *Manduca sexta*. Am. Zool. 21: 655–661.

Truman, J. W., D. B. Rountree, S. E. Reiss, and L. M. Schwartz. 1983. Ecdysteroids regulate the release and action of eclosion hormone in the tobacco hornworm, *Manduca sexta* (L.). J. Insect Physiol. 29: 895–900.

Wahl, B. 1914. Über die Kopfbildung cyclorrhapher Dipteren-larven und die postembryonale Entwicklung des Fliegen-kopfes. Arb. Zool. Inst. Univ. Wien 20: 661–664.

Wentworth, S. L., B. Roberts, and J. D. O'Connor. 1981. Ecdysteroid titres during postembryonic development of *Sarcophaga bullata* (Sarcophagidae: Diptera). J. Insect Physiol. 27: 435–440.

Wyatt, S. S. and G. R. Wyatt. 1971. Stimulation of RNA and protein synthesis in silkmoth pupal wing tissues by ecdysone *in vitro*. Gen. Comp. Endocrinol. 16: 369–374.

Žďárek, J. 1985. Regulation of pupariation in flies. Pp. 301–333 *in* G. A. Kerkut and L. I. Gilbert (eds.), *Comprehensive Insect Physiology, Biochemistry and Pharmacology*. Pergamon Press, Oxford and Elmsford, New York.

Žďárek, J. 1988. The mechanism of pupation in flies and its hormonal control. Pp. 615–624 *in* F. Sehnal, A. Zabza, and D. L. Denlinger (eds.), *Endocrinological Foundation in Physiological Insect Ecology*. Wrocław Technical University Press, Wrocaw, Poland.

Žďárek, J. and D. L. Denlinger. 1987. Pupal ecdysis in flies: the role of ecdysteroids in its regulation. J. Insect Physiol. 33: 123–128.

Žďárek, J. and G. Fraenkel. 1969. Correlated effects of ecdysone and neurosecretion in puparium formation (pupariation) in flies. Proc. Natl. Acad. Sci. USA 64: 565–572.

Žďárek, J. and G. Fraenkel. 1972. The mechanism of puparium formation in flies. J. Exp. Zool. 179: 315–324.

Žďárek, J. and S. Friedman. 1986. Pupal ecdysis in flies: mechanisms of evagination of the head and expansion of the thoracic appendages. J. Insect Physiol. 32: 917–923.

Žďárek, J., K. Sláma, and G. Fraenkel. 1979. Changes in internal pressure during puparium formation in flies. J. Exp. Zool. 207: 187–196.

Žďárek, J., R. Rohlf, J. Blechl, and G. Fraenkel. 1981. A hormone affecting immobilization in pupariating fly larvae. J. Exp. Biol. 93: 51–63.

Part III Taxonomic Index

Part III Subject Index